DISKS, PLANETESIMALS, AND PLANETS

COVER ILLUSTRATION:

NICMOS 1.1 μm image of HR 4796A after reprocessing to remove background due to PSF residuals. The outer white circle is the size of the coronagraphic hole, and the annulus is the width of the PSF. Thus the filled circle shows the region where the PSF is fully occulted. Most, if not all, of the lumpiness in the image is likely to be induced by incomplete and imperfect modeling of the PSF residuals and should not be interpreted as clumps in the ring (see p. 332).

A SERIES OF BOOKS ON RECENT DEVELOPMENTS IN ASTRONOMY AND ASTROPHYSICS

First Published 2000

ASTRONOMICAL SOCIETY OF THE PACIFIC
390 Ashton Avenue, San Francisco, California, USA 94112-1722
Phone: (415) 337-1100 **Fax: (415) 337-5205**
E-Mail: catalog@aspsky.org **Web Site: www.aspsky.org**

Printed by:
Sheridan Books, Inc., 613 East Industrial Drive, Chelsea, Michigan 48118

Library of Congress Catalog Card Number: 00-110481
ISBN: 1-58381-051-X

Madrid
2001

JUAN PEREZ-MERCADER.

ASTRONOMICAL SOCIETY OF THE PACIFIC
CONFERENCE SERIES

Volume 219

DISKS, PLANETESIMALS, AND PLANETS

Proceedings of a Euroconference held at
Puerto de la Cruz, Tenerife, Spain
24-28 January 2000

Edited by

F. Garzón
Instituto de Astrofísica de Canarias, E-38205 La Laguna, Tenerife, Spain

C. Eiroa
Universidad Autónoma de Madrid, Madrid, Spain

D. de Winter
Universidad Autónoma de Madrid, Madrid, Spain

and

T. J. Mahoney
Institute de Astrofísica de Canarias, E-38205 La Laguna, Tenerife, Spain

A listing of all other ASP Conference Series Volumes and IAU Volumes published by the ASP is cited at the back of this volume

Contents

Part 3. The Pre-Main Sequence and Main Sequence

Part 4. Exoplanets

Part 5. Future Projects

In Conclusion . . .

Preface

EXPORT (EXoPlanetary Observational Research Team) is made up of scientists from several, mainly European, countries interested in the detection and study of exoplanetary systems. In 1998 the group obtained the 5 per cent international time allocation of the Canarian Observatories' International Time Programme (ITP), which offers guaranteed (usually simultaneous) use of all nocturnal telescopes at the Roque de los Muchachos Observatory (La Palma) and the Teide Observatory (Tenerife) during 18 nights. This involves a considerable observational effort and allows more ambitious projects to be tackled than are normally within the reach of observational programs that are allotted time through the usual channels.

EXPORT's particular aim is to act as a precursor to *DARWIN* (a cornerstone future missions of the Horizon 2000+ program of the European Space Agency), an orbital infrared interferometer, whose objectives will include, the detection of traces of life on extrasolar planets. To achieve this goal, our group has concentrated on the observational study of the formation and evolution of planetary systems in an effort to understand the very different scenarios in which planets form around stars.

As has become the custom with ITP projects, we have decided to organize and conference for the presentation of our results with the logistic support of the Instituto de Astrofísica de Canarias (IAC). This proceedings gathers together the contributions presented by a distinguished group of astronomers with a shared interest in exoplanetary systems. We should mention that, although the conference has centered on the results obtained by EXPORT, this was only one aspect of the meeting, as may be seen by leafing through pages that follow.

Over five intensive days, more than a hundred participants, including many young researchers and a substantial number of experts, presented some of the latest findings in the field, both observationally , including the development of specific instrumentation, and theoretically.

The main part of the meeting was divided into three consecutive scientific sessions, each with specific theme, but with a certain degree of latitude. A concluding session was dedicated to an analysis of the immense future prospects that are now opening up in this field. The wide coverage of the conference by the local media clearly indicates public excitement for subject and its implications, which go beyond the bounds of the purely scientific.

This meeting was made possible by the generous support of a number of national and European institutions. The European Community, through its *High Level Scientific Conferences* under the Fifth Framework Programme, provided funding for a large number of young researchers in recognition of the high research training value of the conference. The IAC gave its full support to the event in the form of human resources and materials, in addition to direct financial assistance. Spain's Minister of Science and Education provided important additional funds, as did the Autonomous University of Madrid, the Spanish National Research Council (CSIC) and the Spanish Astronomical Society. The Banco Bilbao–Vizcaya and the petrol company DISA provided disinterested financial support. Without their help, this conference would not have been possible. Similarly, the organizers would like express their gratitude to the Local Government authorities of Tenerife and the Town Hall of Puerto de la Cruz for their generous collaboration at various phases of the organization. Last, but not least, we thank the efficient and constant hard work of Judith de Araoz and Tanja Karthaus, whose contribution as part of the Local Organizing Committee made the organization of this event an enjoyable experience.

The editors

Participants

ANTONIO ALBERDI, Instituto de Astrofísica de Andalucía, Apdo. Correos 3004, E-18080 Granada, Spain ⟨antxon@iaa.es⟩

JOÃO ALVES, European Southern Observatory, Karl-Schwarzschildstrasse 2, D-85748 Garching bei München, Germany ⟨jalves@eso.org⟩

JEAN-CHARLES AUGEREAU, Observatoire de Grenoble, Laboratoire d'Astrophysique, 414 Rue de la Piscine, BP 53, F-38041 Grenoble Cedex 9, France ⟨augereau@obs.ujf-grenoble.fr⟩

DAVID BARRADO Y NAVASCUÉS, Max Planck Institut für Astronomie, Königstuhl 17, D-69117 Heidelberg, Germany ⟨barrado@mpia-hd.mpg.de⟩

JEAN PHILLIPE BEAULIEU, Institut d'Astrophysique de Paris, 98 bis Boulevard Arago, F-75014 Paris, France ⟨beaulieu@iap.fr⟩

DAVID BENNETT, University of Notre Dame, Physics Department, 225 Nieuwland Science Hall, Notre Dame, IN 46556, USA ⟨bennett@bustard.phys.nd.edu⟩

JEAN-LOUP BERTAUX, Service d'Astronomie du CNRS, BP 3, F-91371 Verrihres-le Buisson, France ⟨bertaux@aerov.jussieu.fr⟩

WILLIAM BORUCKI, NASA Ames Research Center, MS 245 Moffett Field, CA 94035-1000, USA ⟨wboruck@mail.arc.nasa.gov⟩

ALAN BOSS, Carnegie Institution of Washington, 55241 Broad Branch Road NW, Washington DC, USA ⟨boss@dtm.ciw.edu⟩

STEPHANE BRILLANT, European Southern Observatory, Alonso de Córdoba 3107, Vitacura, Santiago de Chile, Chile ⟨sbrillan@eso.org⟩

TIMOTHY BROWN, HAO/NCAR, 3450 Mitchell Lane, Boulder, CO 80307, USA ⟨timbrown@hao.ucar.edu⟩

MARK CASALI, UKATC, Royal Observatory Edinburgh, Blackford Hill, Edinburgh, EH9 3HJ, Scotland, UK ⟨mmc@roe.ac.uk⟩

ANGELA COTERA, Steward Observatory, University of Arizona, Tucson, AZ 85721, USA ⟨cotera@as.arizona.edu⟩

VINCENT COUDÉ DU FORESTO, Observatoire de Paris, DESPA, 5 Place Jules Janssen, F-92195 Meudon, France ⟨vincent.foresto@obspm.fr⟩

DAVID CHARBONNEAU, Harvard-Smithsonian Center for Astrophysics, 60 Garden St., Cambridge, MA 02138, USA ⟨dcharbonneau@cfa.harvard.edu⟩

HIROKI CHIHARA, Osaka University, Dept. Earth and Space Science, 1-1, Machikaneyama, Toyonaka, Osaka, 560-0043, Japan ⟨chi@ess.sci.osaka-u.ac.jp⟩

JOHN KEITH DAVIES, Joint Astronomy Center, 660 N. A. Ohoku Pl., Hilo, Hawaii, USA ⟨j.davies@jach.hawaii.edu⟩

JAMES DE BUIZER, University of Florida, 211 SSRB, Gainesville, FL-32611-2055, USA ⟨debuizer@astro.ufl.edu⟩

DOLF DE WINTER, TNO-TPD/Space, Stieltjesweg 1, Postbus 155, NL-2600 AD Delft, The Netherlands ⟨dolf@xiada.ft.uam.es⟩

GREET DECIN, Instituut voor Sterrenkunde, Kuleuven, Celestijnenlaan, 200B, B-3001 Leuven, Belgium ⟨greet@ster.kuleuven.ac.be⟩

HANS J. DEEG, Instituto de Astrofísica de Canarias, Via Láctea, s/n, E-38205 La Laguna, Tenerife, Spain ⟨hdeeg@ll.iac.es⟩

MAGALI DELEUIL, Laboratoire d'Astronomie Spatiale, BP8, Les Trois Lucs, F-13376 Marseille Cedex 12, France ⟨magali.deleuil@astrsp-mrs.fr⟩

XAVIER DELFOSSE, Instituto de Astrofísica de Canarias, Via Láctea, s/n, E-38205 La Laguna, Tenerife, Spain ⟨delfosse@ll.iac.es⟩

MARTIN DOMINIK, Kapteyn Astronomical Institute, Postbus 800, NL-9700 AV Groningen, The Netherlands ⟨dominik@astro.rug.nl⟩

GASPARD DUCHÊNE, Observatoire de Grenoble, Laboratoire d'Astrophysique, Université Joseph Fourier, BP53, F-38041 Grenoble Cedex 9, France ⟨gaspard.duchened@obs.ujf-grenoble.fr⟩

ANNE DUTREY, Inst. Radio Astronomie Millimetrique (IRAM), 300 Rue de la Piscine, F-38406 Saint-Martin d'Heres, France ⟨dutrey@iram.fr⟩

CARLOS EIROA, Universidad Autónoma de Madrid, Departamento de Física Teórica, Módulo C-XI, Cantoblanco, E-28049 Madrid, Spain ⟨carlos@xiada.ft.uam.es⟩

ROGER FERLET, Institut d'Astrophysique de Paris, CNRS 98 bis Bd. Arago, F-75014 Paris, France ⟨ferlet@iap.fr⟩

ROBERT SCOTT FISHER, University of Florida, Infrared Astrophysics Group, 211 SSRB, Gainesville, FL 32611, USA ⟨fisher@astro.ufl.edu⟩

BERNARD FOING, ESA–ESTEC, Astrophysics Division, Space Science Department, Postbus 299, NL-2200 AG, Noordwijk, The Netherlands ⟨bfoing@estec.esa.nl⟩

JOSÉ GAITE CUESTA, LAEFF, ESA Satellite Tracking Station, PO Box 50727, E-28080 Madrid, Spain ⟨gaite@laeff.esa.es⟩

FRANCISCO GARZÓN, Instituto de Astrofísica de Canarias, Via Láctea, s/n, E-38205 La Laguna, Tenerife, Spain ⟨fgl@ll.iac.es⟩

PATRICK GODON, STScI, 3700 San Martin Drive, Baltimore, MD 21218, USA ⟨godon@stsci.edu⟩

YOLANDA GÓMEZ, UNAM, Campus Morelia, J J Tablada 1006, 58090 Morelia, Michoacan, México ⟨gocy@astrosmo.unam.mx⟩

GUILLERMO GONZÁLEZ, University of Wahsington, Box 351580, Seattle, WA 98195, USA ⟨gonzalez@astro.washington.edu⟩

Jean-François Gonzalez, Centre de Recherche astronomique de Lyon, École Normale Superieure 45, 46 Allée d'Italie, F-69364 Lyon Cedex 07, France ⟨ jean-francois.gonzalez ⟩

Carol Grady, NASA's Goddard Space Flight Cente, Code 681, GSFC, Greenbelt, MD 20771, USA ⟨ cgrady@echelle.gsfc.nasa.gov ⟩

Jane Greaves, Joint Astronomy Center, 660 N A'Ohoku Place, Hilo, Hawaii 96720, USA ⟨ jsg@jach.hawaii.edu ⟩

Vladimir Grinin, Astronomical Institute of St Petersburg, Bibliotechnaya Pl. 2, 198904 Petrodvorets, St Petersburg, Russia ⟨ grinin@vg1723.spb.edu ⟩

Tristan Guillot, Observatoire de la Cote d'Azur, BP 4429, F-06304 Nice, France ⟨ guillot@obs-nice.fr ⟩

Stephane Guilloteau, Institut Radio Astronomie Millimetrique (IRAM), 300 Rue de la Piscine, F-38406 Saint-Martin d'Heres, France ⟨ guillote@iram.fr ⟩

Alan Harris, DLR–German Aerospace Center, Rutherford Str. 2, D-12489 Berlin, Germany ⟨ alan.harris@dlr.de ⟩

Lee Hartmann, Center for Astrophysics, University of Harvard, 60 Garden St., Cambridge, MA 02138, USA ⟨ hartmann@cfa.harvard.edu ⟩

Thomas Henning, Universitäts Sternwarte Jena, Schillergaesschen 3, D-07745 Jena, Germany ⟨ henning@astro.uni-jena.de ⟩

William Herbst, MPI für Astronomie, Königstuhl 17, D-69117 Heidelberg, Germany ⟨ wherbst@wesleyan.edu ⟩

Paul T. P. Ho, Harvard–Smithsonian Center for Astrophysics, 60 Garden Street, Cambridge, MA 02138, USA ⟨ pho@cfa.harvard.edu ⟩

Keith Horne, St Andrews University, North Haugh, St Andrews, KY16 9SS, Scotland, UK ⟨ kdh1@st-and.ac.uk ⟩

Barrie Jones, The Open University, Department of Physics and Astronomy, Walton Hall, Milton Keynes, MK7 6AA, UK ⟨ b.w.jones@open.ac.uk ⟩

Pierre Kervella, European Southern Observatory, Karl-Schwarzschildstrasse 2, D-85748 Garching bei München, Germany ⟨ pkervell@eso.org ⟩

Hubertus Klahr, UCO/Lick Observatory, 122 Oxford Way, Santa Cruz, CA 95060, USA ⟨ klahr@ucolick.org ⟩

Wilhelm Kley, MPI für Astronomie, Königstuhl 17, D-69117 Heidelberg, Germany ⟨ kley@mpia-hd.mpg.de ⟩

Christian Koeberl, Institute of Geochemistry, University of Vienna, Althanstrasse 14, A-1090 Vienna, Austria ⟨ christian.koeberl@univie.ac.at ⟩

Valerij Kozhevnikov, Astronomical Observatory, Ural State University, Lenin Ave. 51, Ekaterinburg 620083, Russia ⟨ valerij.kozhevnikov@usu.ru ⟩

Alexander Krivov, MPI für Astronomie, Max-Planck-Strasse 2, D-37191 Katlenburg-Lindau, Germany ⟨ krivov@linmpi.mpg.de ⟩

Natalia Krivova, MPI für Astronomie, Max-Planck-Strasse 2, D-37191 Katlenburg-Lindau, Germany ⟨ natalie@linmpi.mpg.de ⟩

PIERRE-OLIVIER LAGAGE, CEA–Saclay, Service d'Astrophysique, F-91191 Gif-sur-Yvette, France ⟨ lagage@cea.fr ⟩

DAVID LATHAM, Harvard–Smithsonian Center for Astrophysics, 60 Garden Street, Cambridge, MA 02138, USA ⟨ dlatham@cfa.harvard.edu ⟩

ALAIN LECAVELIER, Institut d'Astrophysique de Paris, 98 bis Bd. Arago, F-75014 Paris, France ⟨ lecaveli@iap.fr ⟩

ROSARIO LÓPEZ HERMOSO, Universidad de Barcelona, Diagonal, 647, E-08028 Barcelona, Spain ⟨ rosario@mizar.am.ub.es ⟩

JANE LUU, Sterrewacht Leiden, Postbus 9513, NL-2300 RA Leiden, The Netherlands ⟨ luu@strw.leidenuniv.nl ⟩

TERENCE JOHN MAHONEY, Instituto de Astrofísica de Canarias, Via Láctea, s/n, E-38205 La Laguna, Tenerife, Spain ⟨ tjm@ll.iac.es ⟩

KOEN MALFAIT, Instituut voor Sterrenkunde Kuleuven, Celestijnenlaan, 200B, B-3001 Heverlee, Belgium ⟨ koen.malfait@ster.kuleuven.ac.be ⟩

EDUARDO L. MARTÍN, Caltech, Division of Planetary Sciences, 1200 E california Blvd. Caltech 150-21, Pasadena, CA 91125, USA ⟨ ege@gps.caltech.edu ⟩

FREDERIC MASSET, Queen Mary and Westfield College, Mathematics Department, 327 Mile End Road, London E1 4NS, UK ⟨ f.s.masset@qmw.ac.uk ⟩

MICHEL MAYOR, Observatoire de Genève, Ch. des Maillettes, CH-1290 Sauverny, Switzerland ⟨ michel.mayor@obs.unige.ch ⟩

TSEVI MAZEH, Tel Aviv University, Ramat Aviv, Tel Aviv 69978, Israel ⟨ mazeh@wise.tau.ac.il ⟩

FRANÇOIS MÉNARD, Canada–France–Hawaii Telescope Corp., PO Box 1597, Kamuela, HI 96743, Hawaii, USA ⟨ menard@cfht.hawaii.edu ⟩

BRUNO MERÍN MARTÍN, , ⟨ bruno@laeff.esa.es ⟩

LUIS F. MIRANDA, Instituto de Astrofísica de Andaluciía, Camino Bajo de Huetor, Apartado 3004, E-18080 Granada, Spain ⟨ lfm@iaa.es ⟩

DAVID MONTES, Universidad Complutense de Madrid, Departamento de Astrofísica, Facultad de Físicas, E-28040 Madrid, Spain ⟨ dmg@astrax.fis.ucm.es ⟩

BENJAMÍN MONTESINOS, LAEFF, ESA Satellite Tracking Station, PO Box 50727, E-28080 Madrid, Spain ⟨ bmm@laeff.esa.es ⟩

ALCIONE MORA FERNÁNDEZ, Universidad Autónoma de Madrid, Departamento de Física Teórica, Módulo C-XI, Cantoblanco, E-28049 Madrid, Spain ⟨ alcione@xiada.ft.uam.es ⟩

CLAIRE MOUTOU, European Southern Observatory, Alonso de Córdoba, 3107, Vitacura, Santiago de Chile, Chile ⟨ cmoutou@eso.org ⟩

ANTONELLA NATTA, Osservatorio di Arcetri, Largo Fermi 5, I-50125 Firenze, Italy ⟨ natta@arcetri.astro.it ⟩

ROBERT W. NOYES, Harvard–Smithsonian Center for Astrophysics, 60 Garden Street, Cambridge, MA 02138, USA ⟨ rnoyes@cfa.harvard.edu ⟩

Joana Oliveira, ESTEC/ESA, SCI/SO, Keplerlaan 1, Postbus 299, NL-2200 AG Noordwijk, The Netherlands ⟨joana@so.estec.esa.nl⟩

Javier Palacios Bermejo, Universidad Autónoma de Madrid, Campus de Cantoblanco, E-28049, Madrid, Spain ⟨javier@xiada.ft.uam.es⟩

Eric Pantin, CEA–Saclay, Service d'Astrophysique, Bat. 709, 91191 Gif-sur-Yvette Cedex, France ⟨epantin@cea.fr⟩

Tomasz Plewa, Nicolaus Copernicus astronomical Center, Barticka 18, PL-00716 Warsaw, Poland ⟨tomek@camk.edu.pl⟩

Elisha Polomski, University of Florida, Astronomy Department, Gainesville FL-32611-2055, USA ⟨elwood@astro.ufl.edu⟩

Andreas Quirrenbach, University of California at San Diego, Center for Astrophysics and Space Science, Mail Code 0424, 9500 Gilman Drive, USA ⟨qui@cassir.ucsd.edu⟩

Heike Rauer, DLR–German aerospace Center, Rutherfordstr. 2, D-12489 Berlin, Germany ⟨heike.rauer@dlr.de⟩

Rafael Rebolo, Instituto de Astrofísica de Canarias, Via Láctea, s/n, E-38205 La Laguna, Tenerife, Spain ⟨rrl@ll.iac.es⟩

Luis F. Rodríguez, Instituto de Astronomía, UNAM, J J Tablada 1006, Morelia, Michoacan 58090, México ⟨luisfr@astrosmo.unam.mx⟩

Neda Safizadeh, University of California at San Diego, 9393a Discovery Way, La Jolla, CA 92093, USA ⟨nsafi@ucsd.edu⟩

Victor J. Sánchez-Béjar, Instituto de Astrofísica de Canarias, Via Láctea, s/n, E-38205 La Laguna, Tenerife, Spain ⟨vbejar@ll.iac.es⟩

Glenn Schneider, Steward Observatory, University of Arizona, 933 N Cherry Avenue, Tucson, AZ 85721, USA ⟨gschneider@as.arizona.edu⟩

Sara Seager, Institute for Advanced Study, School of Natural Sciences, Olden Lane, Princeton, NJ 08540, USA ⟨seager@ias.edu⟩

Michael Shao, Jet Propulsion Laboratory, M/S 301-486, 4800 Oak Grove Drive, Pasadena, CA 91109-8099, USA ⟨mshao@huey.jpl.nasa.gov⟩

Peter Nick Sleep, The Open University, Department of Physics and Astronomy, Walton Hall, Milton Keynes, MK7 6AA, UK ⟨p.n.sleep@open.ac.uk⟩

Peter Sorensen, Isaac Newton Group of Telescopes, Apartado 321, E-38700 Santa Cruz de La Palma, Spain ⟨pms@ing.iac.es⟩

Rachel Street, St Andrews University, North Haugh, St Andrews, KY16 9SS, Scotland, UK ⟨ras11@st-andrews.ac.uk⟩

Paolo Tanga, Ossevatorio Astronomico di Torino, Strada Osservatorio 20, I-10025 Pino Torinese, Italy ⟨tanga@to.astro.it⟩

Charles Telesco, University of Florida, Department of Astronomy, SSRB-211, PO Box 112055, Gainesville, FL 32611, USA ⟨telesco@astro.ufl.edu⟩

Caroline Terquem, Institut d'Astrophysique de Paris, 98 bis Bd. Arago, F-75014 Paris, France ⟨terquem@iap.fr⟩

GELYS TRANCHO LEMES, Telescopio Nazionale Galileo, Apdo. Correos 565, E-38700 Santa Cruz de la Palma, Spain ⟨ trancho@tng.ll.iac.es ⟩

YIANNIS TSAPRAS, St Andrew University, North Haugh, St Andrews, KY16 9SS, Scotland, UK ⟨ yt2@st-andrews.ac.uk ⟩

MARIO VAN DEN ANCKER, Harvard–Smithsonian Center for Astrophysics, 60 Garden Street, MS 42, Cambridge, MA 02138, USA ⟨ mvandenancker@cfa.harvard.edu ⟩

LARRY WEBSTER, NASA Ames Research Center, MS 244-14 Moffett Field, CA 94035, USA ⟨ lwebster@mail.arc.nasa.gov ⟩

ALYCIA WEINBERGER, University of California, UCLA, Division of Astronomy, 405 Hilgard Ave., Los Angeles, CA 90095-1562, USA ⟨ alycia@astro.ucla.edu ⟩

PAUL SRON WESSELIUS, Kapeyn Astronomical Institute, Postbus 800, NL-9700 AV Groningen, The Netherlands ⟨ p.r.wesselius@sron.rug.nl ⟩

SEBASTIAN WOLF, Thueringer Landessternwarte Tautenburg, Sternwarte 5, D-07778 Tautenburg, Germany ⟨ wolf@tls-tautenburg.de ⟩

MARK WYATT, University of Florida, 211 Bryant Space Scence Center, Gainesville, FL 32611, USA ⟨ wyatt@astro.ufl.edu ⟩

RUSLAN YUDIN, Central Astronomical Observatory Pulkovo, Pulkovskoye Sh. 65, 196140 St Petersburg, Russia ⟨ ruslan@pulkovo.spb.su ⟩

MARÍA ROSA ZAPATERO OSORIO, Instituto de Astrofísica de Canarias, Via Láctea, s/n, E-38205 La Laguna, Tenerife, Spain ⟨ mosorio@ll.iac.es ⟩

Part 1

From Disks to Planets

Disks, Planetesimals and Planets
ASP Conference Series, Vol. 219, 2000
F. Garzón, C. Eiroa, D. de Winter and T. J. Mahoney, eds.

1998 La Palma International Time Programme: Formation and Properties of Planetary Systems

C. Eiroa, A. Mora, and J. Palacios
Dpto. Física Teórica, Universidad Autónoma de Madrid, 28049 Madrid, Spain

A. Alberdi and L. F. Miranda
Instituto de Astrofísica de Andalucía, Apartado de Correos 3004, 18080 Granada, Spain

A. Cameron, K. Horne, and Y. Tsapras
Physics & Astronomy, University of St. Andrews, North Haugh, St. Andrews KY16 9SS, Scotland, UK

J. K. Davies
Joint Astronomy Centre, 660 N. A'ohoku Place, Hilo, Hawaii 96720, USA

H. J. Deeg and F. Garzón
Instituto de Astrofísica de Canarias, 38205 La Laguna, Tenerife, Spain

D. de Winter
TNO/TPD-Space Instrumentation, Stieltjesweg 1, PO Box 155, 2600 AD Delft, The Netherlands

R. Ferlet
CNRS, Institute d'Astrophysique de Paris, 98bis Bd. Arago, 75014 Paris, France

C. A. Grady
NOAO/STIS, Goddard Space Flight Center, Code 681, NASA/GSFC, Greenbelt, MD 20771, USA

A. Harris and H. Rauer
DLR Institute of Planetary Exploration, Rudower Chaussee 5, 12489 Berlin, Germany

B. Merín, B. Montesinos, and E. Solano
LAEFF, VILSPA, Apartado de Correos 50727, 28080 Madrid, Spain

R. Oudmaijer
Department of Physics and Astronomy, E. C. Stoner Building, Leeds LS2 9JT, W Yorkshire, UK

A. Penny
Rutherford Appleton Laboratory, Didcot, Oxforshire OX11 0QX, UK

A. Quirrenbach
Department of Physics, Center for Astrophysics and Space Sciences, University of California San Diego, Mail Code 0424, La Jolla, CA 92093-0424, USA

J. Schneider
Observatoire de Paris, 92195 Meudon, France

P. R. Wesselius
SRON, Universiteitscomplex "Zernike", Landleven 12, PO Box 800, 9700 AV Groningen, The Netherlands

Abstract. The 1998 International Time of the Canary Islands Observatories was awarded to the project "Planetary Systems: their Formation and Properties", carried out by a joint European/American team—EXPORT (EXoPlanet Observational Research Team). A total of ≈ 75 observing nights, distributed over four observing runs, and five different telescopes were used to study the following topics: the formation and evolution of planetary systems, the spectral signatures of extrasolar planetary atmospheres and new planet searches. We were able to collect an enormous data set, including high- and intermediate-resolution optical spectroscopy, optical photopolarimetry, near-IR photometry, and optical CCD images. This contribution presents a short description of the EXPORT project and the observing runs. Some general and specific results are presented elsewhere in this volume.

1. Introduction

Studies of exoplanetary systems have received considerable impetus since the discovery of a Jupiter-like planet orbiting 51 Peg (Mayor & Queloz, 1995). To date, more than 30 Jupiter-like planets have been reported. The discoveries have been made using the radial-velocity method, which introduces a $\sin i$ uncertainty in the derived mass. The recently observed photometric transit across the star HD 209458 leaves no doubt that the object orbiting the star has a planetary mass. (Charbonnneau et al. 2000; Henry et al. 2000).

These planet discoveries have profound scientific significance and have opened up new field of research concerning the nature, formation, and evolution of planets and planetary systems. For example, the existence of Jupiter-mass planets orbiting very close to their parent stars indicates that the Solar System constitutes only one particular scenario of planetary systems; therefore, theoretical and observational studies dealing with alternative scenarios are stimulated. An intriguing point concerns the nature of those massive planets; i.e., are they gaseous objects more or less similar to the giant planets of the Solar System or are they large rocky planets? In addition, the success of the radial-velocity method provides a new impulse to alternative methods in the detection of planets, such as astrometry, transit photometry, or microlensing.

2. The EXPORT Project

Five per cent of the observing time at the Canary Islands Observatories is dedicated to international projects, which can benefit from the coordinated use of a group of telescopes and are of a kind unlikely to obtain time through the usual time-allocation process. A group of more than 20 astronomers from different countries (EXPORT: EXoPlanet Observational Research Team) submitted a proposal for the 1998 La Palma International Time slot in which the following main objectives were addressed:

Table 1. Log of EXPORT observations

Telescope	Instrument/ observing mode	May 98	Jul 98	Oct 98	Jan 99
WHT (ORM)	UES $R \sim 50000$ $\sim \lambda\lambda$ 3700–6100 Å	14–17	28–31	23–26	28–31
INT (ORM)	IDS $R \sim 10000$ $\sim \lambda\lambda$ 5800–6700 Å	14–17	28–31	24–28	29–31
NOT (ORM)	TurPOL Opt. photopolarimetry $UBVRI$	14–17	28–31	23–27	29–31
CST (OT)	IR phot. + CAIN IR photometry JHK	15–17	28–31	23–26	28–31
JKT (ORM)	CCD Camera Photometry R	–	25–31	24–1	–

- The formation and evolution of planetary systems through observational studies of protoplanetary disks around PMS stars and MS Vega-type objects.
- The search for spectral signatures of extrasolar planetary atmospheres, specifically in the spectra of τ Boo and 51 Peg.
- New planet searches by means of microlensing techniques and photometric transits in clusters.

Table 1 is a log of the EXPORT observations carried out at the Roque de los Muchachos Observatory (ORM) on La Palma. The telescopes, common-user equipment, and observing modes, as well as the observing dates, are detailed. In addition, the IAC80 Telescope at the Teide Observatory (OT) was used during several nights in summer 1998 to obtain direct CCD images of microlensing alerts. Sky conditions were in general very good and spectroscopic data were obtained during each single scheduled night; more than 50% of the total were good photometric nights.

3. EXPORT Results

We were able to collect an enormous data set, including high- and intermediate-resolution optical spectroscopy, optical photopolarimetry, near-IR photometry, and optical CCD images. Hence, the main observational objectives were fulfilled.

The 2.5 m William Herschel Telescope (WHT, ORM), the 2.5 m Isaac Newton Telescope (INT, ORM), the 2.56 m Nordic Optical Telescope (NOT, ORM), and the 1.55 m Carlos Sánchez Telescope (CST, OT) were used during the four observing runs to study protoplanetary systems, including Vega-type and PMS stars. Approximately 80 stars were observed simultaneously with all the telescopes and their corresponding observing modes. Simultaneity is necessary when studying PMS stars and their associated protoplanetary disks, for it is well known that these objects are variable. Also, the spectroscopic events observed in the direction of several Vega-type stars, which have been interpreted in terms of infalling solid bodies (e.g., in β Pic), are transient. Simultaneity allowed us to monitor the stars on timescales of hours, days, and months. The analysis of these data is still in progress though some preliminary results are presented in various contributions in this volume (Mora et al. p. 231; Oudmaijer et al. p. 238; Palacios et al. p. 255; Eiroa et al. p. 363; de Winter et al. p. 356; Mérin et al. p. 399; Harris et al. p. 454; Miranda et al. p. 405; Solano et al. p. 422).

The WHT was used on two consecutives nights in 1998 May and October to observe the predicted photometric transits of the planets around τ Boo and 51 Peg, respectively. The idea was to search for spectroscopic signatures due to the orbiting planet. The contribution of Harris et al. (this volume, p. 454) provides the details and results. In addition, some control spectra and photometry of both stars were taken with the INT and NOT.

Two observing runs on the 1.0 m Jacobus Kapteyn Telescope (JKT, ORM) were used to obtain R-band CCD images of stellar clusters to search for photometric transits. Several hundreds of images were taken. The status of this work is described in the contribution by Quirrenbach et al. (this volume, p. 566).

Finally, the 80 cm IAC80 Telescope was used to follow microlensing events during summer 1998. The results of these observations are also presented in this volume (p. 626) by Tsapras et al.

4. Conclusion

The observational objectives of the EXPORT project carried out within the framework of the 1998 Canary Islands International Time were fully achieved. In some respects, the observations are unique since they provide simultaneous spectroscopic and photopolarimetric data of a very large sample of PMS and Vega-type stars. Preliminary results are presented in several contributions in this volume. Detailed analyses are in progress and will be published elsewhere. We are considering various options for making these data available to the community since we consider their scientific content to be of interest to other investigations.

References

Charbonneau, D., Brown, T. M., Latham, D. W., & Mayor, M. 2000, ApJ, 529, L45

Henry, G. W., Marcy, G. W., Butler, R. P., & Vogt, S. S. 2000, ApJ, 529, L41

Mayor, M., & Queloz, D. 1995, Nat, 378, 355

Disks, Planetesimals, and Planets
ASP Conference Series, Vol. 219, 2000
F. Garzón, C. Eiroa, D. de Winter, and T. J. Mahoney, eds.

From Disks to Planets: An Overview

Alan P. Boss

Carnegie Institution of Washington, Department of Terrestrial Magnetism, 5241 Broad Branch Road, NW, Washington DC, 20015-1305, USA

Abstract. Astronomical observations of protoplanetary disks and extrasolar planetary systems have begun to transform the field of planetary origins from an isolated, theorist's playground into a thriving, world-wide effort. We now have examples of dozens of planets, and even planetary systems, outside the Solar System, with the promise of many more to come in the next two decades. Several of these extrasolar planets seem to be accompanied by circumstellar dust disks, indicative of the presence of an unseen distribution of smaller bodies that can be a source of the dust grains orbiting these main sequence stars. Debris disks have been imaged around stars with ages ranging from a few billion years to as young as about 10 million years, an age when terrestrial planets are expected to still be in the process of formation. Protoplanetary disks have also been imaged and probed around pre-main-sequence stars and young stellar objects, and their properties are crucial for confirming and constraining theories of planet formation. Planetary origins is thus beginning a period of rapid growth and development, similar to that experienced in the field of star formation in the last two decades, when examples of essentially all phases of protostellar and pre-main-sequence evolution were found and assembled into a coherent picture. We expect that current theories of planet formation will evolve rapidly as a result.

1. Introduction

For the last several decades, if not centuries, serious workers on the theory of planet formation have had only the example of our own Solar System to guide them in deciphering the origins of planetary systems. As a consequence, theoretical attention has been focused largely on the specific problem of the formation of our Solar System, to the detriment of the development of a more general theory of planet formation. In the absence of any reproducible evidence of extrasolar planetary systems, such theoretical single-mindedness was both understandable and perhaps even admirable—after all, there was more than enough of a challenge involved in explaining the origin of the Solar System's menagerie. As a result, we now have a reasonably mature theory of terrestrial planet formation, some fairly well developed ideas about how gas-giant planets form, and the hope that someday we will understand the formation of ice-giant planets. Meanwhile, this theoretical effort has been overtaken by events—the

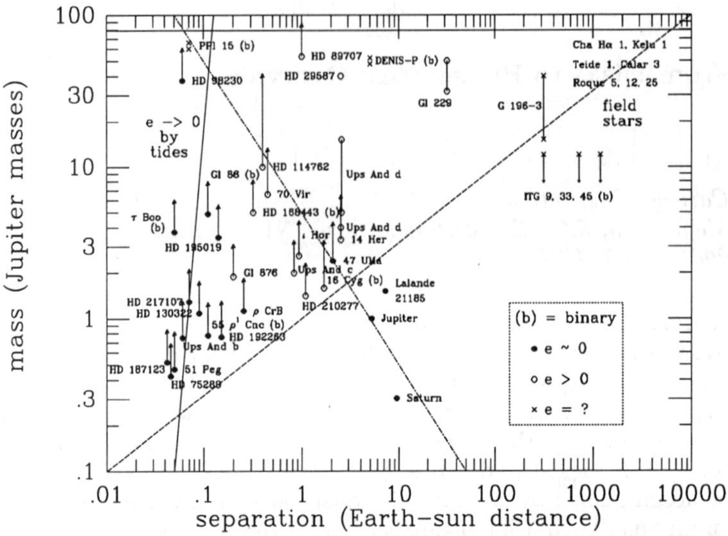

Figure 1. Discovery space for extrasolar planets and brown dwarf stars as of 2000 February. *Oblique dashed line:* Sensitivity limit for radial-velocity searches at a given accuracy. *Oblique dash-dotted line:* The same for astrometric searches. The former method preferentially finds short-period companions, while the latter prefers long-period companions. Nearly all of these objects were found by radial velocity (yielding only lower limits on the companion's mass); only Lalande 21185's putative planet, as yet unconfirmed, was found by astrometry. *Oblique solid line:* Systems expected to have circular orbits as a result of tidal dissipation over several billion years. Names in *top right-hand corner* are field brown dwarf stars, while the other objects are either companions to dwarf stars or are brown dwarf binaries. *Horizontal solid line:* Eighty Jupiter masses, the upper limit for the mass of a brown dwarf star.

discovery of extrasolar planets and debris disks orbiting nearby stars, and the beginnings of the characterization of protoplanetary disks around young stars. As a result of these observational advances, we can expect the more general theory of planet formation to evolve rapidly in the coming years. I will briefly review these developments in this overview of the wealth of forefront topics covered at the *Disks, Planetesimals, and Planets* Euroconference.

2. Extrasolar Planets

The modern era of extrasolar planet detection began in 1995 with the announcement of the discovery of a half-Jupiter-mass companion in orbit around the nearby star 51 Pegasi (Mayor & Queloz 1995). Five years into this era, the number of planetary-mass companions to solar-type stars has already risen to well over 30 (Figure 1; Marcy & Butler 1998).

Several major surprises are evident in Figure 1. First, 51 Pegasi's planet has an orbital period of 4.23 days, such that its semi-major axis is a factor of 100 times smaller than Jupiter's. By now, a number of such "hot jupiters" have been found, and in fact massive planets seem to be distributed throughout the entire range of semi-major axes from about 0.05 AU to 3 AU. Second, the planets with semi-major axes larger than about 0.3 AU have considerably more eccentric orbits than Jupiter ($e \approx 0.05$), a trait seemingly more akin to binary stars than to planets. Third, many of these objects are considerably more massive than Jupiter, by a factor of order 10, raising the question of whether theories developed to explain the formation of our Jupiter are likely to be adequate to account for these extrasolar behemoths. These results generally were not anticipated by theorists, leaving them largely in the unfortunate position of playing catch-up with the hard-charging observers.

However, there have been some very reassuring aspects to these discoveries in addition to the surprises. There seems to be a paucity of brown dwarf companions to these primarily solar-mass stars, as would be expected given the rarity of large mass ratios for binary stars. There also is an absolute dearth of companions with minimum masses roughly in the range of 10 $M_{\rm Jup}$ to 30 $M_{\rm Jup}$. This absence, however, strongly helps to make the case that the cluster of very low mass companions with masses in the range from about 0.5 $M_{\rm Jup}$ to 10 $M_{\rm Jup}$ should be termed "planets", as opposed to being merely a continuation of the brown dwarf star population to unexpectedly low masses. Estimates of the minimum mass of brown dwarf stars fall around 10 $M_{\rm Jup}$, and the most likely place to find a brown dwarf star is in orbit around another brown dwarf star—binary brown dwarfs are now beginning to be discovered (Basri & Martín 1999). Another strong argument in favor of the planetary interpretation of these objects was the discovery of the Upsilon Andromedae triple planet system (Butler et al. 1999). There are no stellar systems consisting of a massive primary orbited by three small stars—all stable multiple star systems are hierarchical, consisting of short-period binaries nested within longer-period binaries. Finally, the discovery of the first transiting planet, in a short period orbit around HD 209458 (Charbonneau et al. 2000; Henry et al. 2000) provides further confirmation of the planetary nature of these objects, and in particular their identification with gas-giant planets. The 1.7% dimming of the star's light during the transit leads to a direct estimate of the planet's radius, given the star's radius (Mazeh et al. 2000), which leads to a value in good accord with the expectations for a hot jupiter (Burrows et al. 2000).

3. Debris Disks

Several of the extrasolar planets appear to be accompanied by circumstellar dust disks. Trilling & Brown (1998) and Trilling et al. (2000) have made coronagraphic infrared images of 55 Rho^1 Cnc, Rho CrB, and HD 210277 that imply the presence of dust disks with diameters of order 100 AU around these stars. Given the ages of these main-sequence stars, and the much shorter lifetimes of circumstellar dust grains caused by rapid orbital evolution due to radiation pressure and Poynting–Robertson drag, these disks are thought to be debris disks rather than planet-forming disks. That is, these disks owe their existence to

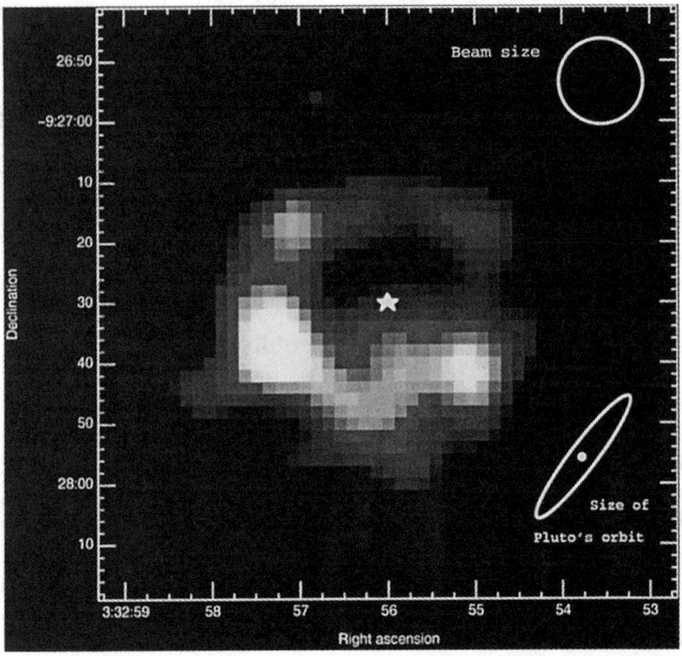

Figure 2. 850 μm image of a dust ring around the nearby star Epsilon Eridani, taken with the Submillimetre Common-User Bolometer Array (SCUBA) at the James Clerk Maxwell Telescope (Greaves et al. 1998). Image courtesy of Jane Graeves (Joint Astronomy Centre, Hawaii).

processes that create new dust grains from an unseen population of larger-mass bodies also in orbit, presumably a population of relatively small bodies such as asteroids or comets, whose mutual collisions and/or small periastra result in the production of fresh dust grains.

Debris disks were first detected in orbit around stars which were not known to have planets, and debris disks are now known to occur around stars with a wide range of ages, from 55 Rho^1 Cancri ($\sim 3 \times 10^9$ yr), to Epsilon Eridani ($\sim 1 \times 10^9$ yr), to Beta Pictoris ($\sim 2 \times 10^7$ yr), to HR 4796A ($\sim 8 \times 10^6$ yr; Jayawardhana et al. 1999). The fractional infrared luminosity increases strongly with younger ages in these stars (Barrado y Navascués et al. 1999), implying that debris disks fade away with time and as their sources of fresh dust grains are depleted. The Epsilon Eridani disk is clearly ring-like (Figure 2), with peak emission at a radius of 60 AU, and little emission inside 30 AU (Greaves et al. 1998). With a mass of at least 0.01 $M_\oplus$, this disk is comparable in mass to estimates of the Kuiper Belt of comets, and the non-axisymmetric nature of the emission hints at the presence of otherwise unseen sources of dust and possibly planets. HR 4796A has a dust ring with a radius of 70 AU and a width of only 17 AU (Figure 3), suggestive of orbital confinement by unseen low-mass

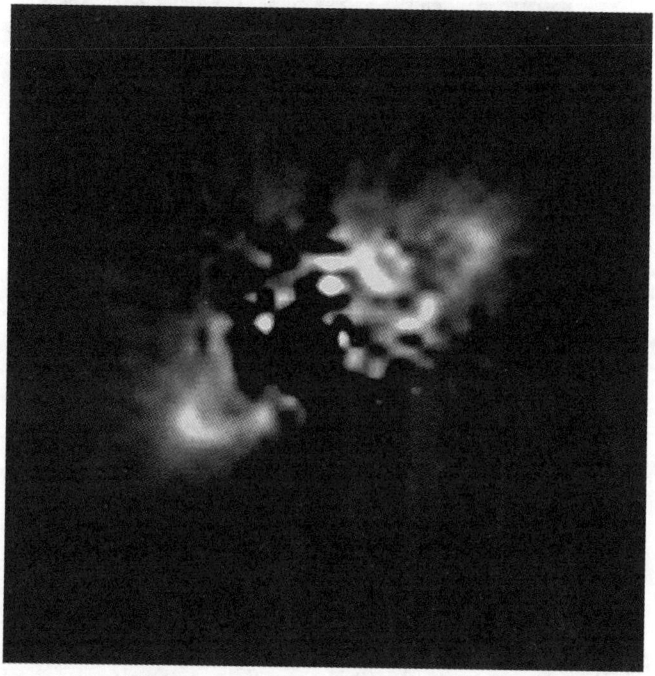

Figure 3. Near-infrared image of the disk orbiting HR 4796A, taken with the NICMOS camera of the *Hubble Space Telescope* (Schneider et al. 1999). Image courtesy of Brad Smith (University of Hawaii), Glenn Schneider (University of Arizona), and NASA.

companions (Schneider et al. 1999). The lack of gas emission from HR 4796A places an upper limit of less than about 0.01 $M_{\rm Jup}$ of gas in the disk, implying that if gas giant planets formed around HR 4796A, they must have taken no more than $\sim 8 \times 10^6$ yr to form (Greaves, Mannings, & Holland 2000).

4. Protoplanetary Disks

Disks around young stars provide even more direct constraints on the planet formation process. While millimeter-wave and infrared telescopes have pioneered the discovery and characterization of circumstellar disks, recently some spectacular optical images of disks in star-forming regions have emerged from the *Hubble Space Telescope* (*HST*). For example, the "proplyds" are flattened disks of dust that are seen silhouetted in front of the hot gases of the Orion Nebula. The absence of detectable molecular CO gas in these objects places an upper bound of about 0.01 $M_\odot$ on the amount of gas present, and the high rate of losses due to ultraviolet radiation places a limit on the lifetime of these objects of order 10^5 yr or less (Bally et al. 1998). Figure 4 shows an *HST* image of

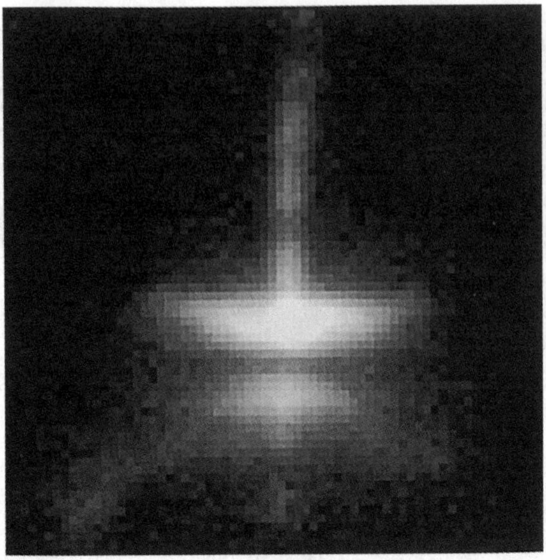

Figure 4. *HST* optical image of the HH 30 disk and jet in Taurus. The disk radius is about 250 AU (Burrows et al. 1996). Image courtesy of Chris Burrows (Space Telescope Science Institute), the WFPC2 Science Team, and NASA.

the disk and jet surrounding the HH 30 embedded young star that confirms the basic theoretical picture of a protoplanetary disk: a well-defined disk midplane, a flared surface, and a jet emanating in the direction perpendicular to the disk (Burrows et al. 1996). The disk mass is estimated to be less than 0.01 $M_\odot$ and to have a lifetime of $\sim 10^5$ yr. Whether or not planets can form as quickly as seems to be required in the Orion disks and in the HH 30 disk remains to be seen.

More massive disks, with masses in the range of 0.01 to 0.1 $M_\odot$, are commonly found in orbit around young stars with ages of several million years or less. By the time that young stars reach ages of $\sim 10^7$ yr, however, the evidence for planet-forming disks disappears, placing an upper bound on the time available for assembling planets from the gas and dust of the protoplanetary disk (Wolk & Walter 1996).

L1551 IRS 5 is a very young protobinary system where each binary protostar is orbited by a compact disk with a mass of ~ 0.05 $M_\odot$ and a radius of about 10 AU; the binary separation for comparison is about 45 AU (Rodríguez et al. 1998). Circumbinary disks have also been imaged (Guilloteau, Dutrey, & Simon 1999), such as the spectacular disk in orbit around one binary of the quadruple system GG Tauri (Figure 5). Evidently, binary stars should be able to form planets as well as single stars, and this implication is confirmed by the detection of several extrasolar planets in binary star systems (Figure 1).

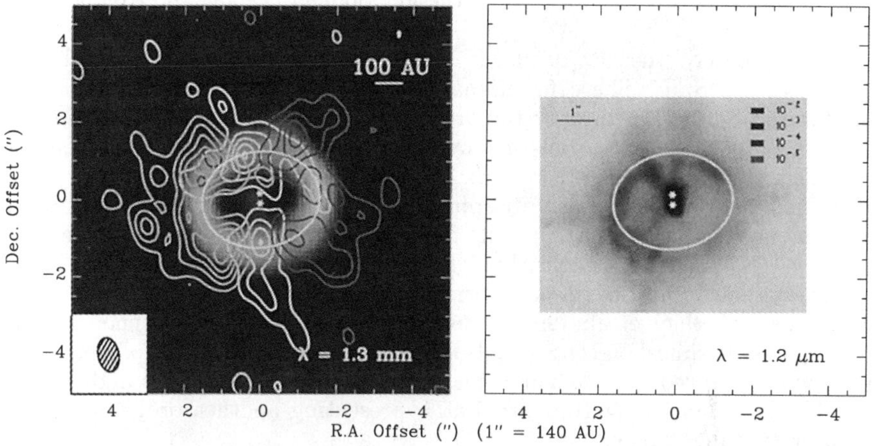

Figure 5. Image of the circumbinary disk orbiting one of the GG Tau binary systems (Guilloteau et al. 1999). The gray-scale images are dust continuum emission at the indicated wavelengths, while the contours in the left-hand image are ^{13}CO emission at three different velocities. Image courtesy of Michal Simon (SUNY, Stony Brook).

5. Planet Formation Theories

Given a suitable protoplanetary disk, that is one with at least 0.01 $M_\odot$ of gas and dust in Keplerian orbit around a solar-mass protostar, theoreticians have spent decades imagining how the planet formation process develops (Safronov 1969). Most work has concentrated on the "solar nebula" thought to have formed our Solar System (Wetherill 1990), but this is beginning to change as observations of other planetary systems have become available (Wetherill 1996).

The terrestrial planets are universally believed to have formed through the collisional accumulation of successively larger, solid bodies—micron-sized dust grains, kilometer-sized planetesimals, and Moon-sized planetary embryos, culminating in the formation of the Earth and other terrestrial planets. Growth from micron-size is thought to occur by sticking of dust grains, with growth to kilometer-size thought to occur in $\sim 10^4$ yr (Weidenschilling 1988). Thereafter self-gravity becomes important in holding objects together after they collide. These planetesimals are thought to reach lunar-size in about 10^5 yr through a runaway growth process, where the larger bodies grow at the expense of the smaller bodies because the former's gravity effectively increases the cross-sectional area for collisions. The lunar-sized planetary embryos accumulate all the mass in their orbital annuli, but further growth is limited by the circularity of the orbits of the embryos. Over a period of $\sim 10^7$ yr, the long-range gravitational interactions of the embryos results in pumping up of their orbital eccentricities, leading to further collisions (some large enough to form the Moon). By an age of 10^8 yr, the terrestrial planets have formed (Wetherill 1990), and thereafter the leftover debris pockmarks their lithospheres with craters. A typical outcome can be an earth and a venus along with a few smaller planets, but considerably different outcomes are also possible if the locations of the gas giant planets are altered (Wetherill 1996).

Unfortunately, theories of planet formation become less certain with distance from the Sun. The conventional wisdom is that the gas giant planets (Jupiter and Saturn) formed by the process of core accretion, where a roughly 10 Earth-mass solid core forms first by collisional accumulation, and then accretes disk gas (Mizuno 1980; Pollack 1984). Because this process must run to completion before the disk gas disappears, i.e., within a few million years, a phase of runaway accretion is usually invoked to speed the growth of the core (Lissauer 1987). With a sufficiently high surface density of solids, detailed models suggest that Jupiter could have formed by core accretion within about eight million years (Pollack et al. 1996), a time that is marginally short enough considering disk lifetimes. Recent models of the interiors of gas giant planets have significantly reduced the estimated masses of the cores of Jupiter and Saturn (e.g., Guillot, Gautier, & Hubbard 1997), presenting another problem for the core accretion mechanism.

Considering these concerns regarding core accretion, it would only be prudent to consider other mechanisms of gas giant planet formation as well. The only alternative appears to be disk instability, where gas giant protoplanets might form rapidly through a gravitational instability of the gaseous portion of the disk (Kuiper 1951; Cameron 1978). Disk instability was initially ruled out by the seemingly large core masses of Jupiter and Saturn, which are now much less of a concern. Also, a core can form in a giant gaseous protoplanet by sedimentation of dust grains to the center, prior to contraction of the protoplanet to planetary densities and temperatures (Boss 1997). The disk instability mechanism has also been neglected because of the danger that gravitational instability will only lead to rapid mass and angular momentum transport in the disk rather than to the formation of self-gravitating clumps capable of surviving to form gas giant planets (Cassen et al. 1981; Laughlin & Bodenheimer 1994). However, recent work has suggested that there might be circumstances in which disk instability can lead to the formation of gas giant protoplanets (Boss 1997, 1998b; Nelson et al. 1998), so the disk instability mechanism appears to be worthy of further investigation (Figure 6).

The surprises evident in Figure 1 regarding extrasolar gas giant planets have forced some revisions in theoretical ideas. The existence of hot jupiters appears to require inward orbital migration, considering the difficulty of forming them so close to their stars (Boss 1995, 1998a; Bodenheimer, Hubicky, & Lissauer 2000). The likely agent for this inward orbital migration is interactions between the planet and the gaseous portion of the disk, either before or after the planet becomes large enough to open a gap in the disk (Lin, Bodenheimer, & Richardson 1996; Ward 1997a,b). The eccentric orbits of the planets with orbits outside about 0.3 AU remains a major mystery, though gravitational interactions between planets that form too close together is one possibility (Weidenschilling & Marzari 1996). The large masses may be hard to explain in the context of the core accretion mechanism, where the opening of a disk gap eventually slows down the rate at which disk mass can be added to the planet, which must be considered a liability for a mechanism which is already slow at forming even a Jupiter-mass planet. Disk instability, on the other hand, works best in massive

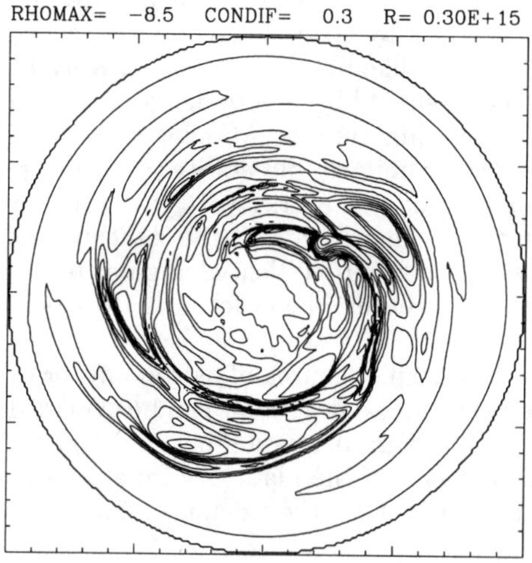

Figure 6. Density contours in the midplane of a 3D hydrodynamical model of a gravitationally unstable disk showing filaments and clumps that might form gas giant protoplanets. The disk is an initially marginally unstable disk ($Q_{\min} = 1.3$), with a mass of 0.09 $M_\odot$ within a radius of 20 AU (as shown). Each contour represents a factor of two change in density.

disks, and so should be able to form relatively massive planets, but may have trouble forming sub-Jupiter-mass planets.

The ice giant planets are believed to have formed by the collisional accumulation of icy solids, but the theory of their formation must still be considered to be embryonic (Levison, Lissauer, & Duncan 1998). In order to make any progress at all in modeling their formation in situ, the collisional cross sections of the growing planetary embryos must be artificially inflated, and other liberties taken as well. A more promising idea may be to form them closer in, between Jupiter and Saturn, and then to migrate them outward to their present locations (Thommes, Duncan, & Levison 1999). Observations of extrasolar ice giants should aid immeasurably in deciphering their origins.

6. Future Progress

While we have already made tremendous progress, it is clear that many more exciting discoveries are just a few years away. We can discover right now Jupiter-mass extrasolar planets by spectroscopic detection of radial velocity variations, but soon we should be able to find extrasolar planets by precise astrometric measurements of the positional wobbles induced by unseen planetary companions (Keck Interferometer and Very Large Telescope Interferometer). When taken to space, astrometry will find planets around huge samples of stars (e.g., the *Full-Sky Astrometric Explorer*, *GAIA*), and will begin the search for Neptune-mass and even Earth-mass planets (*Space Interferometry Mission*). Photometric detections of planetary transits by space telescopes are also planned and should be able to detect large Earth-like planets (*COROT*, *Kepler*). Ultimately, an infrared space interferometer should be able to image entire planetary systems and return crude spectra of their inhabitants (*Infrared Space Interferometer/Darwin*, *Terrestrial Planet Finder*). Similarly, our ability to study circumstellar disks will soon leapfrog over our current capabilities with the launch of the *Space Infrared Telescope Facility* and the construction of the Atacama Large Millimeter Array in Chile, the latter offering unprecedented angular resolution and sensitivity that will be invaluable for probing planet-forming regions and for searching for protoplanets.

It would be hard to imagine an area of astronomy, or of any science, where more rapid progress on such an exciting topic can be expected in the next decade. We can expect to develop a firm understanding of the planet formation process as a result, as well as a census of the planetary inhabitants of our corner of the galaxy, and perhaps even a hint for the existence of a nearby habitable planet.

References

Bally, J., Testi, L., Sargent, A. & Carlstrom, J. 1998, Astron. J., 116, 854

Barrado y Navascués, D., Stauffer, J. R., Song, I. & Caillault, J.-P. 1999, Astrophys. J., 520, L123

Basri, G. & Martin, E. L. 1999, Astrophys. J., 118, 2460

Bodenheimer, P., Hubickyj, O. & Lissauer, J. J. 2000, Icarus, 143, 2

Boss, A. P. 1995, Science, 267, 360

Boss, A. P. 1997, Science, 276, 1836

Boss, A. P. 1998a, Ann. Rev. Earth Planet. Sci., 26, 53

Boss, A. P. 1998b, Astrophys. J., 503, 923

Burrows, A., et al. 2000, Astrophys. J., in press

Burrows, C. J., et al. 1996, Astrophys. J., 473, 437

Butler, R. P., et al. 1999, Astrophys. J., 526, 916

Cameron, A. G. W. 1978, Moon Planets, 18, 5

Cassen, P. M., Smith, B. F., Miller, R. H. & Reynolds, R. T. 1981, Icarus, 48, 377

Charbonneau, D., Brown, T. M., Latham, D. W. & Mayor, M. 2000, Astrophys. J., 529, L45

Greaves, J. S., et al. 1998, Astrophys. J., 506, L133

Greaves, J. S., Mannings, V. & Holland, W. S. 2000, Icarus, 143, 155

Guillot, T., Gautier, D. & Hubbard, W. B. 1997, Icarus, 130, 534

Guilloteau, S., Dutrey, A. & Simon, M. 1999, Astron. Astrophys., 348, 570

Henry, G. W., Marcy, G. W., Butler, R. P. & Vogt, S. S. 2000, Astrophys. J., 529, L41

Jayawardhana, R., et al. 1999, Astrophys. J., 521, L129

Kuiper, G. P. 1951, Proc. Nat. Acad. Sci., 37, 1

Laughlin, G. & Bodenheimer, P., 1994, Astrophys. J., 436, 335
Levison, H. F., Lissauer, J. J. & Duncan, M. J. 1998, Astrophys. J., 116, 1998
Lin, D. N. C., Bodenheimer, P. & Richardson, D. C. 1996, Nature, 380, 606
Lissauer, J. J. 1987, Icarus, 69, 249
Marcy, G. W. & Butler, R. P. 1998, Ann. Rev. Astr. Astrophys., 36, 57
Mayor, M. & Queloz, D. 1995, Nature, 378, 355
Mazeh, T., et al. 2000, Astrophys. J., in press
Mizuno, H. 1980, Progr. Theor. Phys., 64, 544
Nelson, A. F., Benz, W., Adams, F. C. & Arnett, W. D. 1998, Astrophys. J., 502, 342
Pollack, J. B. 1984, Ann. Rev. Astron. Astrophys., 22, 389
Pollack, J. B., et al. 1996, Icarus, 124, 62
Rodríguez, L. F., et al. 1998, Nature, 395, 355
Safronov, V. S. 1969, Evolution of the Protoplanetary Cloud and Formation of the Earth and the Planets (Moscow: Nauka)
Schneider, G., et al. 1999, Astrophys. J., 513, L127
Thommes, E. W., Duncan, M. J. & Levison, H. F. 1999, Nature, 402, 635
Trilling, D. E. & Brown, R. H. 1998, Nature, 395, 775
Trilling, D. E., et al. 2000, in press
Ward, W. R. 1997a, Astrophys. J., 482, L211
Ward, W. R. 1997b, Icarus, 126, 261
Weidenschilling, S. J. 1988, in Meteorites and the Early Solar System, ed. J. F. Kerridge & M. S. Matthews (Tucson: Univ. of Arizona Press), 348
Weidenschilling, S. J. & Marzari, F. 1996, Nature, 384, 619
Wetherill, G. W. 1990, Ann. Rev. Earth Planet. Sci., 18, 205
Wetherill, G. W. 1996, Icarus, 119, 219
Wolk, S. J. & Walter, F. M. 1996, Astron. J., 111, 2066

Discussion

Lee Hartmann: It isn't clear that the Orion shadow disks can only be 10^5 yr old. They are evaporated only when they are within ~ 0.1 pc of θ^1C Ori. Thus, if the stars are not on elongated orbits, the disks won't get destroyed, and even if they are, they may only spend $\sim 10^5$ yr per 10^6 yr-long orbit near θ^1C. (Also, you probably only evaporate the outer disk, $R \gtrsim 10$ AU.)

Alan Boss: At any rate, they are certainly interesting systems with lifetimes at the short end of the range of expected disk lifetimes.

Tristan Guillot: How do the surface densities in your calculations compare to the work by Pollack et al.? I thought that you required quite a large disk mass for gravitational instabilities to occur.

Alan Boss: In my latest models, assuming a dust-to-gas ratio of 1:100, the surface density at 5 AU is initially within a factor of about 1.3 of that assumed in the standard core accretion model of Pollack et al. (1996) and falls within the range estimated for the solar nebula by Weidenschilling (1977).

Keith Horne: Has there been any work on the prospects of forming planets in the dusty disks around active galactic nuclei?

Alan Boss: Not that I know of. If the space density of gas and dust reaches about 10–10 g cm^{-3} or higher in these regions, then one might expect that dust grain coagulation would become important.

William Borucki: What do the high eccentricities found for the recently discovered planets tell us about planet formation?

Alan Boss: Currently, there are two explanations. First, it's possible that scattering is important. Second, for very massive disks, the interaction between the disk and massive (only) planets can excite high eccentricities.

Bernard Foing: The accretion rate is critically dependent on disk surface density, as illustrated by your reference to the Pollack models. As there is a large range of gas and dust densities observed (disk masses of 0.01–0.1 $M_\odot$, β Pic dust $\sim$ 10 000 zodiacal), how are models adapted to describe these widely different conditions?

Alan Boss: We don't yet have the spatial resolution to probe protoplanetary disks at 1–10 AU sizes to see what the true density distribution is. Theoretical models are often based only on the minimum-mass solar nebula.

Bernard Foing: Can you clarify the different surface densities in the Pollack (7.5–10 g cm^{-3} at 5 AU) and Boss (10^3 g cm^{-3} [?] at [?] AU)?

Alan Boss: Pollack: dust surface density only. Boss: gas and dust surface density with a dust-to-gas ratio of 1/100 assumed.

Willy Kley: You mentioned in your talk that the disadvantage of the core-instability model is the inability of the planet to accrete more mass after a gap has been opened in the disk by the planet. However, several calculations show that even after gap formation the planet may still accumulate mass and grow up to 5–10 $M_{\rm Jup}$.

Alan Boss: Yes, I have seen these simulations. I did not say that accretion through the gap is stopped, but that it is slowed down, and the question then is how long the disk survives as a reservoir for the growing exoplanet.

Disks, Planetesimals, and Planets
ASP Conference Series, Vol. 219, 2000
F. Garzón, C. Eiroa, D. de Winter, and T. J. Mahoney, eds.

Disk Evolution towards Planet Formation

Caroline Terquem

Institut d'Atrophysique de Paris, 98 bis Boulevard Arago, 75014 Paris, France

John C. B. Papaloizou and Richard P. Nelson

Astronomy Unit, School of Mathematical Sciences, Queen Mary & Westfield College, Mile End Road, London E1 4NS, UK

Abstract. We review our current knowledge of accretion-disk evolution and protoplanet migration, and present recent results about the effect of migration on giant-planet formation.

We first discuss different mechanisms leading to angular momentum removal from or redistribution within accretion disks and present α-disk models. We then review the current theories of protoplanet migration in the context of these models, focusing on the gaseous disk–protoplanet tidal interaction. According to these theories, the migration timescale may be shorter than the planetary formation timescale. Therefore we present the results of a recent study of planet formation in the context of a migrating core, in which both the growth of the core and the build-up of the envelope in the course of the migration are considered.

1. Introduction: Observations of Protosellar Disks

The presence of disks around low-mass young stellar objects has been inferred from their infrared excess (Adams, Lada, & Shu 1987). Recently they have also been imaged directly with the *Hubble Space Telescope* and ground-based adaptive optics (see Duchêne et al., this volume, p. 169). Surveys in the Orion Nebula (Stauffer et al. 1994) and the Taurus–Auriga dark clouds (Beckwith et al. 1990) indicate that these disks are common, apparently surrounding between 25 and 75% of the young stellar objects. Their infrared emission may be produced by the gravitational potential energy liberated by matter flowing inwards at a rate $\dot{M} \sim 10^{-8\pm1}\ M_\odot\ \mathrm{yr}^{-1}$ (Hartmann et al. 1998). The non-observation of disks around older T Tauri stars, together with these values of $\dot{M}$, suggest adisk lifetime of between 10^6 and 10^7 yr (Strom, Edwards & Skrutskie 1993). Masses between 10^{-3} and $10^{-1}\ M_\odot$ and dimensions in the range 10–100 AU have been estimated (Beckwith & Sargent 1996).

2. Disk Formation

It is believed that protostars and protoplanetary disks originate from interstellar matter contained in molecular clouds. Observations (Goodman et al. 1993) indicate that typical star–forming dense cores in dark molecular clouds have specific angular momentum $j > 6 \times 10^{20}$ cm^2 s^{-1}. When these clouds undergo gravitational collapse, because the collapse is fast, j is initially approximately conserved. Gas in the outer parts is thus unable to fall directly to the center ($r = 0$) if j is not zero.

To obtain an estimate of the initial size of the disk, we consider the idealized situation when the pre-collapse cloud is a cold rotating sphere of mass M and radius R, with a single axis of rotation. Matter located on the rotational axis has no angular momentum, so it can fall directly to the center. An estimate of the time required, $t_{\rm ff}$, is given by the time required to free-fall from rest through a distance R under the initial inward gravitational acceleration at the surface, $g = GM/R^2$. This gives $t_{\rm ff} = \sqrt{2R/g} = \sqrt{2R^3/(GM)}$. This may be expressed in terms of the initial mean density, $\overline{\rho}$,of the cloud as $t_{\rm ff} = \sqrt{3/(2\pi G\overline{\rho})}$. For molecular cloud cores with $M = 1\ M_\odot$ and $R = 0.1$ pc, this gives typically $t_{\rm ff} \sim 6 \times 10^5$ yr.

While matter located on the rotational axis can move directly to the center, matter in the equatorial plane at the surface of the sphere will be at the outermost radius $R_{\rm d}$ of a disk after collapse. Conservation of specific angular momentum determines $R_{\rm d}$. Assuming the total mass, M, to be concentrated at the center, the angular velocity at the outer edge of the disk will be given by Kepler's law such that:

$$\Omega^2 = \frac{j^2}{R_{\rm d}^4} = \frac{GM}{R_{\rm d}^3}. \tag{1}$$

Thus, $R_{\rm d}$ is given in terms of the conserved quantities j and M by $R_{\rm d} = j^2/(GM)$. Adopting $j = 6 \times 10^{20}$ cm^2 s^{-1} and $M = 1\ M_\odot$, we find that $R_{\rm d} \sim 180$ AU. The characteristic dimension of such a disk is about an order of magnitude larger than our present Solar System but is similar to those of protostellar disks now being observed by direct imaging.

3. Disk Evolution

In a Keplerian accretion disk, the specific angular momentum increases with radius. Therefore, a particle can be accreted only if its angular momentum is removed or transferred to particles located at larger radii. Whether angular momentum is removed from or redistributed within the disk depends on whether the disk is subject to external or internal torques. Possible external torques can be either magnetic (when an outflow is present) or tidal (in binary systems), whereas possible internal torques can be either gravitational (massive disks) or viscous (turbulent disks). Three of these mechanisms are briefly discussed below. These have been more extensively discussed by Papaloizou & Lin (1995).

3.1. Hydromagnetic Outflows

The formation of a protostellar disk through the collapse of a molecular cloud core takes 10^5–10^6 yr. During the early stages, when the disk is still embedded (class 0/I object) and has a significant mass compared to the central star, there may exist strong disk winds and bipolar outflows (e.g., Reipurth, Baly, & Devine 1997) with associated magnetic fields. During this stage, a hydromagnetic disk wind may be an important means of angular momentum removal for the system (Blandford & Payne 1982). Because of the action of magnetic torques, material ejected from the disk is able to carry away significantly more angular momentum than it originated with in the disk. Therefore, even a modest ejection rate can lead to a significant accretion rate through the disk. However, observations indicate that outflows may exist only in the early stages of disk evolution, so that this mechanism cannot account for angular momentum transport during the whole life of accretion disks. In addition, it may affect only the inner parts.

3.2. Gravitational Instabilities

When the mass of the disk is significant compared to that of the star, gravitational instabilities may develop, leading to outward angular momentum transport (Papaloizou & Savonije 1991; Heemskerk, Papaloizou, & Savonije 1992; Laughlin & Bodenheimer 1994; Pickett et al. 1998) that results in additional mass growth of the central star. This redistribution of mass may occur on the dynamical timescale (a few orbits) of the outer part of the disk and so may be quite rapid, of the order of 10^5 yr for $R = 500$ AU. The parameter governing the importance of disk self-gravity is the Toomre parameter, $Q \sim M_* H/(M_\mathrm{d} r)$, M_d being the disk mass contained within radius r and H being the disk semi-thickness. Typically $H/r \sim 0.1$ (Stapelfeldt et al. 1998) such that the condition for the importance of self-gravity, $Q \sim 1$, gives $M_\mathrm{d} \sim 0.1 M_*$.

The characteristic scale associated with growing density perturbations in a disk undergoing gravitational instability with $Q \sim 1$ is on the order of H, and the corresponding mass scale is $M_\mathrm{d}(H/r)^2 \sim M_*(H/r)^3$, which is of the order of one Jupiter mass for $H/r \sim 0.1$ and $M_* = 1\ M_\odot$. Gravitational instability has been proposed as a mechanism for directly forming giant planets by Cameron (1978) and more recently by Boss (1998). We note, however, that there are problems with this theory. Fragmentation, which follows the development of axisymmetric modes, requires $Q < 1$, whereas non-axisymmetric modes develop for values of Q slightly larger than unity. As noted by Laughlin & Bodenheimer (1994), it is unlikely that a disk would ever get massive enough for Q to reach unity. The disk builds up in mass as a result of the collapse of an envelope. The Q parameter therefore decreases with time from high values. When it reaches values slightly larger than unity, non-axisymmetric modes develop which redistribute the mass so as to increase Q, i.e., to restore gravitational stability. Therefore Q can never decrease to unity unless the disk can be cooled down or matter can be added on the dynamical timescale, which is unlikely to happen.

During the period of global gravitational instability, it is reasonable to suppose that the disk mass is quickly redistributed and reduced by accretion onto the central object such that the effects of self-gravity become negligible.

3.3. Viscous Evolution

When the disk mass is such that self-gravity can be ignored and the jet activity has significantly decreased, viscous torques may become the most important way of redistributing angular momentum in the disk.

At this stage, it is reasonable to regard the disk as an axisymmetric configuration in which, to a first approximation, material orbits in circles, conserving specific angular momentum, with centrifugal acceleration balanced by gravitational attraction towards the center. This gives the Keplerian angular frequency: $\Omega^2 = GM_*/r^3$. However, other weaker forces due to internal pressure, viscosity, or magnetic fields may also operate in the disk. These forces can redistribute angular momentum on a longer timescale. The angular momentum transport process determines the timescale on which mass accretion can occur and hence the evolutionary timescale of the disk.

Historically, the first angular momentum transport mechanism to be considered was through the action of viscosity (von Weizsäcker 1948). In order to result in evolution on astronomically interesting timescales, it is necessary to suppose that an anomalously large viscosity is produced through the action of some sort of turbulence. The magnitude of the viscosity is usually parameterized through the Shakura and Sunyaev (1973) α prescription. In this, we suppose the kinematic viscosity coefficient $\nu = \alpha c_s^2/\Omega \sim \alpha H^2 \Omega$, where α is a dimensionless constant which must be less than unity and c_s is the local sound speed. So far, only MHD instabilities (Balbus & Hawley 1991) have been shown to be able to produce and sustain turbulence in accretion disks and they do lead to an α-type disk (Balbus & Hawley 1998). However, because these instabilities develop only in an adequately ionized fluid, they may not operate everywhere in protostellar disks (Gammie 1996). Depending on the value of the magnetic Reynolds number and on whether there is a net magnetic flux, hydromagnetic instabilities produce α in the range 10^{-3}–10^{-1}.

In general, an accretion disk is not in a steady state but evolves diffusely according to the following equation (see Lynden-Bell & Pringle 1974):

$$\frac{\partial \Sigma}{\partial t} = \frac{3}{r}\frac{\partial}{\partial r}\left[r^{1/2}\frac{\partial}{\partial r}\left(\Sigma \langle \nu \rangle r^{1/2}\right)\right], \tag{2}$$

where $\langle \nu \rangle$ is the vertically averaged kinematic viscosity. From this we see that the characteristic diffusion timescale at radius r is $t_\nu \sim (r/H)^2 \Omega^{-1}/(3\alpha)$. For disks with approximately constant aspect ratio, H/r, as applies to the models considered here, t_ν scales as the local orbital period. One thus expects that the inner regions relax relatively quickly to a quasi-steady state which adjusts its accretion rate according to the more slowly evolving outer parts (see Lynden-Bell & Pringle 1974; Lin & Papaloizou 1985). For estimated sizes of protostellar disks of about 50 AU (Beckwith & Sargent 1996), the evolutionary timescale associated with the outer parts is about 30 times longer than that associated with the inner parts with $r < 5$ AU, which we consider here in the context of planetary formation. Thus, these inner regions are expected to be in a quasi-steady state through most of the disk lifetime.

4. Disk Models

It is very likely that each of the different processes discussed above operates at some point in the disk evolution, so that the α prescription probably cannot be used in the early stages of disk evolution. Later on, when the disk is at least partly turbulent, α is probably not constant through the disks, and it may be that only parts of these disks can be described using this α prescription for the viscosity. However, we may still learn about disks from these models in the same way as we learned about stars from simple polytropic models. Therefore, for the purpose of considering planet formation, as we are interested in here, we will use such models.

4.1. Vertical Structure

We consider thin disks which are in Keplerian rotation around a star of mass $M_* = 1\ M_\odot$. The opacity in the disk, taken from Bell & Lin (1994), has contributions from dust grains, molecules, atoms, and ions. We are free to choose two parameters to construct α-disk models. We take α and the local mass flow rate through the disk which is defined as $\dot{M}_{\rm st} \equiv 3\pi \langle \nu \rangle \Sigma$, where Σ is the surface mass density of gas. If the disk were in a steady state, $\dot{M}_{\rm st}$ would not vary with radius and would be the constant accretion rate through the disk. At a given radius r and for given values of $\dot{M}_{\rm st}$ and α, we solve the differential equations describing the disk vertical structure (equations of vertical hydrostatic equilibrium, energy conservation, and radiative transport) with appropriate boundary conditions to find the dependence of the temperature, pressure, mass density, and radiative flux on the vertical coordinate. An important point to note is that, as well as finding the disk structure, we also determine Σ for a given $\dot{M}_{\rm st}$, so that a relation between $\langle \nu \rangle$ and Σ is derived. In order to investigate the solutions of the diffusion equation which governs the disk evolution (see above) and for other purposes, we found it convenient to derive analytic piecewise power-law fits to this relation. Details of these fits, which can be used for a very wide range of disk parameters, are given in Papaloizou & Terquem (1999).

4.2. Steady-State Models

In the models we present here, we assume that the disk is immersed in a medium with background temperature $T_{\rm b} = 10$ K. In Figure 1 we plot H/r (where H is defined as the vertical height at approximately unit optical depth above which the atmosphere is isothermal), Σ, and the midplane temperature $T_{\rm m}$ versus r for $\dot{M}_{\rm st}$ between 10^{-9} and 10^{-6} $\rm M_\odot\ yr^{-1}$ (assuming this quantity to be the same at all radii, i.e., the disk is in a steady state) and for illustrative purposes we have adopted $\alpha = 10^{-2}$. The surface mass density of planetesimals can be derived from Σ by noting that in protostellar disks the gas-to-dust mass ratio is about 100. We note that, for a given gas accretion rate, $T_{\rm m}$ increases, and therefore the disk becomes thicker, when α decreases. However H/r does not change significantly (by less than a factor 1.5) when α is varied between 10^{-2} and 10^{-3} (e.g., Papaloizou & Terquem 1999).

In these models, reprocessing of the stellar radiation by the disk has not been included. The upper panel of Figure 1 indicates that the outer parts of

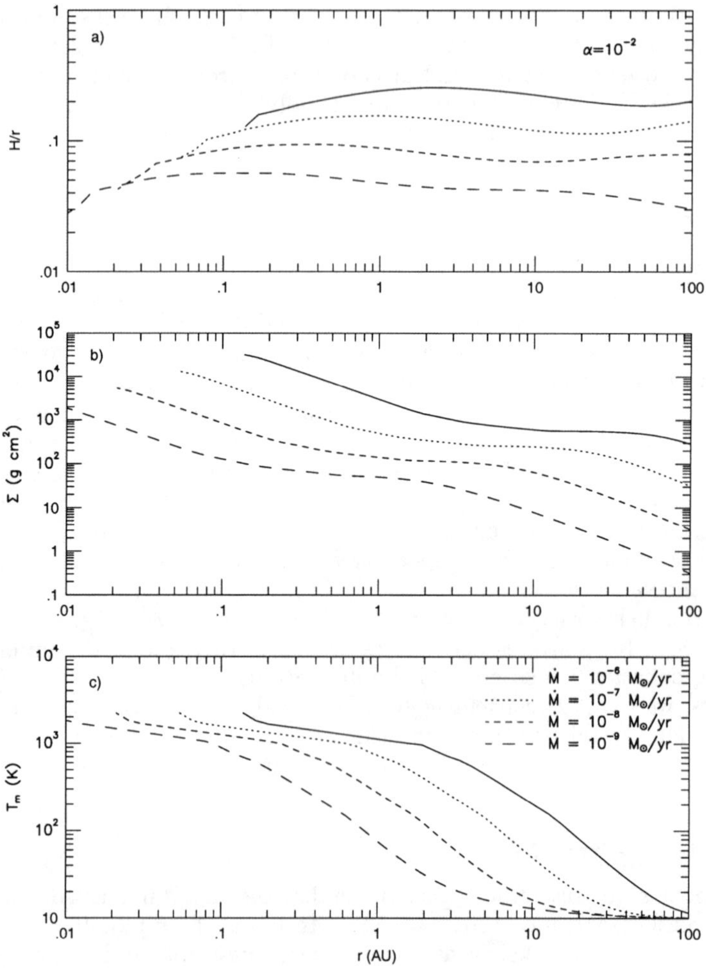

Figure 1. Shown is H/r (*upper panel*), Σ in g cm^{-2} (*middle panel*) and $T_{\rm m}$ in K (*bottom panel*) vs. r/AU using a logarithmic scale, for $\dot{M}_{\rm st}$ (in units $M_{\odot}$ yr^{-1}) $= 10^{-6}$ (*solid line*), 10^{-7} (*dotted line*), 10^{-8} (*short-dashed line*) and 10^{-9} (*long-dashed line*) and for $\alpha = 10^{-2}$.

the disk beyond some critical radius ($\sim$ 0.1 to a few AU, depending on α and $\dot{M}_{\rm st}$) are shielded from the radiation of the central star by the inner parts (the outermost parts may not be shielded but since they are optically thin they do not absorb the stellar radiation anyway). Since reprocessing of the stellar radiation by the disk is not an important heating factor below these radii, this indicates that this process can be self-consistently ignored. However, it is possible that a multiplicity of solutions exists when reprocessing is taken into account (Chiang & Goldreich 1997), with it being important for cases in which the disk is flared, as may be the case with HH 30 (Burrows et al. 1996), and unimportant when it is not, as may be the case with HK Tau (Stapelfeldt et al. 1998; Koresko 1998). In any case, reprocessing, when present, does not significantly affect the structure of the disk at $r \sim$ 1–5 AU, where planets are supposed to form.

5. Planet Migration and Formation

According to current models of planet formation (Safronov 1969; Wetherill & Stewart 1989), planetesimals form as the result of the coagulation of dust grains. Further accumulation through binary collisions then results in the formation of a core of several earth masses. An atmosphere in quasi-static and thermal equilibrium grows along with the core. By the time the core mass reaches the so-called critical core mass, the atmosphere has grown massive enough so that its energy losses can no longer be compensated for by the gravitational energy which the planetesimals entering the atmosphere release when they collide with the surface of the core. At that point the envelope has to contract gravitationally to supply more energy. This is a runaway process, leading to the very rapid accretion of gas onto the protoplanet and to the formation of giant planets such as Jupiter (Perri & Cameron 1974; Mizuno 1980; Bodenheimer & Pollack 1986; Pollack et al. 1996). The larger the planetesimal accretion rate onto the core, the larger the critical core mass.

5.1. Migration

Dynamical tidal interaction of a core or planet with the surrounding disk matter leads to phenomena such as inward orbital migration and gap formation (Lin & Papaloizou 1979; Goldreich & Tremaine 1980; Lin & Papaloizou 1993; Korycansky & Pollack 1993; Ward 1997). For conditions under which the tidal interaction with the disk is linear (cores with mass smaller than about 10 earth masses), the drift timescale of a protoplanet of mass $M_{\rm pl}$ undergoing migration (referred to as type I) in a uniform disk is (Ward 1986, 1997):

$$\tau_{\rm I}({\rm yr}) \sim 10^8 \left(\frac{M_{\rm pl}}{M_\oplus}\right)^{-1} \left(\frac{\Sigma}{\rm g\ cm^{-2}}\right)^{-1} \left(\frac{r}{\rm AU}\right)^{-1/2} \times 10^2 \left(\frac{H}{r}\right)^2. \qquad (3)$$

This timescale is insensitive to disk surface-density gradients. However $\tau_{\rm I}$ would be smaller if the disk midplane temperature increased inwards. Also, migration could reverse from inwards to outwards if the temperature decreased inwards more rapidly than linearly. In a disk with $\alpha = 10^{-2}$ and $\dot{M} = 10^{-7}\ M_\odot\ {\rm yr}^{-1}$, we have $H/r \simeq 10^{-1}$ and $\Sigma \simeq 600$ g cm^{-2} (see above), so that $\tau_{\rm I} \sim 10^5$ yr for a 1 $M_\oplus$ planet at $r = 1$ AU if we ignore the temperature gradient. In the same disk, we get $\tau_I \sim 10^4$ yr for a 10 $M_\oplus$ planet at $r = 5$ AU (where $\Sigma \simeq 300$ g cm^{-2} and $H/r \simeq 10^{-1}$). These timescales are much shorter than the disk lifetime or estimated planetary formation timescales.

When the mass of the protoplanet becomes large enough, the perturbation becomes nonlinear and it affects the structure of the disk around the orbit of the planet. If the density waves excited by the protoplanet are dissipated locally, the angular momentum they transport is deposited into the disk in the vicinity of the protoplanet, and a gap may be cleared out (Goldreich & Tremaine 1980; Lin & Papaloizou 1979, 1993, and references therein). The strongest Lindblad torques are exerted at a distance $\sim H$ from the protoplanet (at the location of the Lindblad resonances corresponding to $m \sim r/H$). Therefore, a gap of width $\sim H$ will open up if the waves launched at these locations dissipate before they can propagate significantly. If dissipation is due to nonlinearity of the waves, this requires (Lin & Papaloizou 1993; Korycansky & Papaloizou 1996; Ward 1997):

$$\frac{M_{\rm pl}}{M_*} \gtrsim 3 \left(\frac{H}{r} \right)^3 . \tag{4}$$

We note that the above discussion implicitly assumes that the Roche lobe of the planet is not significantly larger than H, i.e., $M_{\rm pl}/M_*$ is larger than but still comparable to $3(H/r)^3$. For larger $M_{\rm pl}/M_*$, the edges of the gap are set by the protoplanet's Roche lobe or the location of the 2:1 resonance rather than by the location of the Lindblad resonances corresponding to $m \sim r/H$ (Lin & Papaloizou 1979).

In order for the gap to be maintained, the tidal torque at its edges must exceed the intrinsic viscous torque, provided the tidally excited waves are completely damped. This requires (Lin & Papaloizou 1979):

$$\frac{M_{\rm pl}}{M_*} \gtrsim 40\alpha \left(\frac{H}{r} \right)^2 . \tag{5}$$

Once these two conditions are met, transfer of mass through the gap is reduced, or even switched off altogether (Bryden et al. 1999; Kley 1999). However, transfer of angular momentum from the disk inner parts to the protoplanet, and from the protoplanet to the disk outer parts is still taking place. Therefore, the protoplanet is locked into the angular momentum transport process of the disk. Unless the protoplanet is located in the disk outer parts, which may be diffusing outwards (Lynden-Bell & Pringle 1974), it is going to migrate inwards at a rate controlled by the disk viscous timescale:

$$\tau_{\rm II}({\rm yr}) = \frac{1}{3\alpha} \left(\frac{r}{H} \right)^2 \Omega^{-1} = 0.05 \frac{1}{\alpha} \left(\frac{r}{H} \right)^2 \left(\frac{r}{\rm AU} \right)^{3/2} . \tag{6}$$

This type of migration is referred to as "type–II". If we adopt $H/r = 10^{-1}$ and consider α in the range 10^{-2}–10^{-3}, we get $\tau_{\rm II} \sim 5 \times 10^2$–$5 \times 10^3$ or 6×10^3–6×10^4 yr at $r = 1$ or 5 AU, respectively. These timescales are again much shorter than the disk lifetime or estimated planetary formation timescales. The expression of $\tau_{\rm II}$ is independent of the protoplanet's mass and the surface mass density of the disk, but it is implicitly assumed here that the mass of gas within the characteristic orbital radius of the planet is at least comparable to the mass of the planet itself. If the disk is significantly less massive, then there is not enough gas in the vicinity of the planet to absorb its orbital angular momentum, and migration is slowed down (Ivanov, Papaloizou, & Polnarev 1999).

As pointed out above, a clean gap can be cleared out only if the perturbation is strongly non-linear. If the perturbation is not strong enough for the corotation region to be completely emptied out (a regime that we shall call "transitional"), we have only partial gap clearing and the evolution is controlled both by the disk viscosity and the action of the torques on the partially perturbed disk.

Migration rates, inward or outward, have not yet been consistently calculated in this transitional regime, which applies to protoplanets in the critical core-mass range. High-resolution numerical simulations are required and will be undertaken in the near future. The results are likely to be important for understanding planetary formation.

5.2. Planet Formation

As noted above, cores of several Earth masses form at about 5 AU and migrate inwards to small radii before they can become a giant planet. In doing so, they continue to grow. To know whether they can reach the critical mass on their way and thus begin to accrete significant amount of gas, we have to evaluate the mass of planetesimals they accrete as they migrate in.

Papaloizou & Terquem (1999) performed analytical estimates and numerical calculations to evaluate the fraction of planetesimals accreted by a core of mass $\sim$ 1–10 $M_\oplus$ migrating inward from $r \sim 1$ AU on a timescale $\sim 10^3$–10^4 yr. They found that the migrating protoplanet accretes about 25% of the planetesimals inside its orbit, as if it were in a homogeneous medium without a gap forming in the planetesimal distribution. Even more planetesimals are accreted when the migration is slower. Given that the disk models typically contain at least about 8 $M_\oplus$ interior to 5 AU (see above), this means that the accretion of planetesimals is high enough to maintain the core luminosity such that attainment of the critical core mass (and therefore gas accretion) does not occur as long as planetesimals are present, i.e., above $\sim$ 0.05–0.1 AU. Runaway gas accretion onto a small core can then occur at these radii. However, if the core is too small, the gas accretion phase may be longer than the disk lifetime. Even for core masses in the range 15–20 $M_\oplus$, the build-up of a massive atmosphere may take a time $\sim 10^6$ yr (Bodenheimer, Hubickyj, & Lissauer 2000). The reason for this is that once the core starts to accrete a significant atmosphere, energy production occurs through its gravitational contraction. The luminosity produced then slows down the evolution. An estimate of the evolutionary (Kelvin–Helmholtz) timescale $t_{\rm KH}$ at this stage has been calculated by Papaloizou & Terquem (1999). Typically, they found $t_{\rm KH} \sim 10^6 - 10^7$ yr for core masses in the range 10–20 $M_\oplus$ for radii larger than 0.075 AU. The core masses required to get such a characteristic timescale increase rapidly interior to 0.06 AU. However, we note that they decrease as the mass transfer rate through the disk does. The fact that fairly large core masses are required to give evolutionary timescales comparable to or less than the expected disk lifetime means that mergers of additional incoming cores may be required in order to produce a core of sufficient mass that real runaway gas accretion may begin.

6. Summary

In this paper we have briefly discussed different mechanisms which lead to angular momentum removal from or redistribution within accretion disks. Each of these processes probably operates at some location and at some point in the disk evolution. We have briefly described standard α-disk models and presented some numerical calculations for $\alpha = 10^{-2}$ and gas accretion rates through the disk varying from 10^{-9} to 10^{-6} $M_{\odot}$ yr^{-1}. Then we have discussed planet migration and formation in the context of these models.

We have reviewed the current theories of protoplanet migration, focusing on the gaseous disk–protoplanet tidal interaction. According to these theories, protoplanets migrate from the location where they begin to form, which is supposed to be $\sim$ 1–5 AU, down to the inner region of the disk, on a timescale significantly shorter than the disk lifetime or even the planetary formation timescale. Clearly the theory has to be extended to explain how at least some of the planets halt their migration before they plunge onto their parent star. It is likely that a proper examination of the transitional regime, as well as the incorporation of effects arising from the interaction of many protoplanetary cores may be important for our understanding here. However, present results suggest that planets may not form in situ but more probably form at the same time as they migrate.

Accordingly, we have considered both the growth of the core and the build-up of the envelope of a giant planet in the course of its migration. Our calculations indicate that accretion of planetesimals during migration is likely to be high enough to maintain the core luminosity such that attainment of the critical core mass (and therefore significant gas accretion) does not occur as long as planetesimals are present, i.e., above $\sim$ 0.05–0.1 AU. Although runaway gas accretion can then begin onto small mass cores at these small radii, the timescale for building up a massive envelope becomes longer than the disk lifetime if the core is too small. However, if it is supposed that some process such as behavior in the transitional regime, many-body effects, or a disk structural feature can halt the migration, cores massive enough can be built up through mergers of additional incoming cores on a timescale shorter than for in situ formation if the core were in isolation. The above considerations may lead to the preferential formation of short-period planets on a relatively short timescale in the early stages of the disk evolution.

References

Adams, F. C., Lada, C. J., & Shu, F. H. 1987, ApJ, 312, 788

Balbus, S. A., & Hawley, J. F. 1991, ApJ, 376, 214

Balbus, S. A., & Hawley, J. F. 1998, Rev. Mod. Phys., 70, 1

Beckwith, S. V. W., & Sargent, A. I. 1996, Nat, 383, 139

Beckwith, S. V. W., Sargent, A. I., Chini, R. S., & Guesten, R. 1990, AJ, 99, 924

Bell, K. R., & Lin, D. N. C. 1994, ApJ, 427, 987
Blandford, R. D., & Payne, D. G. 1982, MNRAS, 199, 883
Bodenheimer, P., Hubickyj, O., & Lissauer, J. J. 2000, Icarus, 143, 2
Bodenheimer, P., & Pollack, J. B. 1986, Icarus, 67, 391
Boss, A. P. 1998, ApJ, 503, 923
Bryden, G., Chen, X., Lin, D. N. C., Nelson, R. P., & Papaloizou, J. C. B. 1999, ApJ, 514, 344
Burrows, C. J. et al., 1996, ApJ, 473, 437
Cameron, A. G. W. 1978, Moon Planets, 18, 5
Chiang, E. I., & Goldreich, P. 1997, ApJ, 490, 368
Gammie, C. F. 1996, ApJ, 457, 355
Goldreich, P., & Tremaine, S. 1980, ApJ, 241, 425
Goodman, A. A., Benson, P. J., Fuller, G. A., & Myers, P. C. 1993, ApJ, 406, 528
Hartmann, L., Calvet, N., Gullbring, E., & d'Alessio, P. 1998, ApJ, 495, 385
Heemskerk, M. H. M., Papaloizou, J. C. B., & Savonije, G. J. 1992, A&A, 260, 161
Ivanov, P. B., Papaloizou, J. C. B., & Polnarev, A. G. 1999, MNRAS, 307, 791
Kley, W. 1999, MNRAS, 303, 696
Koresko, C. D. 1998, ApJ, 507, L145
Korycansky, D. G., & Papaloizou, J. C. B. 1996, ApJS, 105, 181
Korycansky, D. G. , & Pollack, J. B. 1993, Icarus, 102, 105
Laughlin, G., & Bodenheimer, P. 1994, ApJ, 436, 335
Lin, D. N. C., & Papaloizou, J. C. B. 1979, MNRAS, 188, 191
Lin, D. N. C., & Papaloizou, J. 1985, in Protostars and Planets II, ed. D. C. Black & M. S. Mathews (Tucson: University of Arizona Press), 981
Lin, D. N. C., & Papaloizou, J. C. B. 1993, in Protostars and Planets III, ed. E. H. Levy & J. I. Lunine (Tucson: University of Arizona Press), 749
Lynden-Bell, D., & Pringle, J. E. 1974, MNRAS, 168, 603
Mizuno, H. 1980, Prog. Theor. Phys., 64, 544
Papaloizou, J. C. B., & Lin, D. N. C. 1995, ARA&A, 33, 505
Papaloizou, J. C., & Savonije, G. J. 1991, MNRAS, 248, 35
Papaloizou, J. C., & Terquem, C. 1999, ApJ, 521, 823
Perri, F., & Cameron, A. G. W. 1974, Icarus, 22, 416
Pickett, B. K., Cassen, P., Durisen, R. H., & Link, R. 1998, ApJ, 504, 468
Pollack, J. B., Hubickyj, O., Bodenheimer, P., Lissauer, J. J., Podolak, M., & Greenzweig, Y. 1996, Icarus, 124, 62
Reipurth, B., Baly, J. & Devine, D. 1997, AJ, 114, 2708
Safronov, V. S. 1969, in Evolution of the Protoplanetary Cloud and Formation of the Earth and Planets (Nauka: Moscow, in Russian). Transl. Israel Program for Scientific Translations (1972), NASA TT F-677
Shakura, N. I., & Sunyaev, R. A. 1973, A&A, 24, 337

Stapelfeldt, K. R., Krist, J. E., Menard, F., Bouvier, J., Padgett, D. L., & Burrows, C. J. 1998, ApJ, 502, L65

Stauffer, J. R., Prosser, C. F., Hartmann, L., & McCaughrean, M. J. 1994, AJ, 108, 1375

Strom, S. E., Edwards, S., & Skrutskie, M. F. 1993, in Protostars and Planets III, ed. E. H. Levy & J. Lunine (Tucson: Univ. Arizona Press), 837

von Weizsäcker, C. F. 1948, Z. Naturforsch, 3a, 524

Ward, W. R. 1986, Icarus, 67, 164

Ward, W. R. 1997b, Icarus, 126, 261

Wetherill, G. W., & Stewart, G. R. 1989, Icarus, 77, 330

Discussion

William J. Borucki: Are you predicting that we will not find Neptune-mass objects at 0.05 AU?

Caroline Terquem: Yes!

Alan Boss (in response to Borucki's question: Models of terrestrial planet formation by collisional accumulation of refractory solids (Wetherill 1996) show that multiple-Earth planets could easily form at and inside 1 AU, especially if the disk mass is somewhat greater than the minimum-mass nebula.

Caroline Terquem: But again we would have to find a way of stopping the migration of these cores/planets.

H. Klahr: Isn't it possible that the MHD turbulence initiates hydroturbulence which transports angular momentum from 1–5 AU?

Caroline Terquem: That's very speculative! There are three arguments against what you suggest:

1. Why would MHD turbulence initiate hydroturbulence?
2. Even if it did, it's not clear that hydroturbulence would transport angular momentum outward.
3. It's even less clear that hydroturbulence would spread from 1 AU to 5 AU.

Disks, Planetesimals, and Planets
ASP Conference Series, Vol. 219, 2000
F. Garzón, C. Eiroa, D. de Winter, and T. J. Mahoney, eds.

The Evolution of Giant Protoplanets in Protostellar Disks

Richard P. Nelson[1], John C. B. Papaloizou, and Frédéric S. Masset

Astronomy Unit, Queen Mary & Westfield College, Mile End Rd, London E1 4NS, UK

Willy Kley[2]

Theoretisch-Physikalisches Institut, Universität Jena, Max-Wien-Platz 1, D-07743 Jena, Germany

Abstract. We present the results of non-linear two-dimensional hydrodynamic simulations of the interaction between a Jovian-mass protoplanet and a gaseous disk with aspect ratio and kinematic viscosity expected for protoplanetary disks. Three independent numerical codes were used. We find that the orbital migration is always inwards and the protoplanet reaches the central star in a near-circular orbit after a characteristic viscous timescale of $\sim 10^4$ initial orbital periods, independently of whether the planet is allowed to accrete mass or not. Maximally accreting protoplanets reached about four Jovian masses on reaching the neighborhood of the central star. Assuming some process such as termination of the inner disk through a magnetospheric cavity stops the migration, the range of masses estimated for a number of close-orbiting giant planets, as well as their inward orbital migration, can be accounted for by consideration of disk–protoplanet interactions during the late stages of giant planet formation.

1. Introduction

The recent discovery of a number of extrasolar giant planets orbiting nearby solar-type stars has stimulated renewed interest in the theory of planet formation. It is generally believed that planets form out of the gas and dust contained in the disks that are observed around T Tauri stars (Beckwith & Sargent 1996). The T Tauri disk phase is thought to last on the order of 10^6–10^7 yr, after which the disks appear to dissipate.

In the "critical core mass" model of giant-planet formation, formation proceeds through the build–up of a critical $\sim 15\ M_\oplus$ solid core, beyond which mass rapid gas accretion occurs leading to the formation of a gas-giant planet. Because this process must occur prior to the loss of the gas from the disk it is expected

[1]Presented by F. S. Masset.

[2]Also at Max Plank Institut für Astronomie, Königstuhl 17, D-69117 Heidelberg, Germany.

that the cores of gas giants form beyond a radius of $r \sim 4$ AU, the so-called "ice condensation radius". If giant planets begin to form beyond this radius, then orbital migration must have occurred in order to explain the existence of the closely orbiting extrasolar giant planets. In this paper we explore the idea that the required orbital migration occurs through the gravitational interaction between a protoplanet and the protostellar disk out of which it forms.

When the response of the disk to the presence of an embedded planet is non-linear, and the internal viscous torques are overwhelmed by tidal torques, then an annular gap is expected to form in the disk (Papaloizou & Lin 1984), and inward migration is expected to occur at the viscous evolution rate (Lin & Papaloizou 1986). If the mass of the planet becomes larger than the local disk mass, then the inertia of the planet becomes important in determining the migration rate of the planet. This scenario has been investigated by Syer & Clarke (1995) and more recently by Ivanov, Papaloizou, & Polnarev (1999).

In this paper we employ three independent Eulerian hydrodynamic codes (NIRVANA, FARGO, and RH2D) to examine the non-linear evolution of a combined star, disk, and embedded-protoplanet system. These codes and further details of this work are described in Nelson et al. (2000, hereafter NPMK2000). The main questions that we address are: 1) what the timescale is for a giant protoplanet to migrate in to the vicinity of its host star; 2) how much mass the planet is able to accrete from the disk during this time, and how the mass evolution of the planet affects its orbital migration rate; and 3) whether the tidal interaction between the disk and planet causes the growth of orbital eccentricity.

The physical parameters that we focus on are appropriate to a 1 $M_{\rm Jup}$ protoplanet embedded in a minimum-mass solar nebula model containing $\sim$ 2 $M_{\rm Jup}$ within 5 AU, the initial orbital radius of the protoplanet.

2. The Physical Model

The full equations of motion used in this paper and the details of the numerical procedures are described in NPMK2000. We use a locally isothermal equation of state such that the disk aspect ratio, H/r, is constant. For the NIRVANA and FARGO runs $H/r = 0.04$, and 0.05 for the RH2D runs. We use a constant kinematic viscosity $\nu = 10^{-5}$, expressed in dimensionless units. The unit of mass is taken to be the sum of the mass of the central star, M_*, and the initial mass of the protoplanet, $m_{\rm p}$, and the unit of length is taken to be the initial orbital radius of the protoplanet, a_0. We set the gravitational constant $G = 1$. The unit of time then becomes $t_0 = \{a_0^3/[G(M_* + m_{\rm p})]\}^{1/2}$ and the initial orbital period of the planet is $P_0 = 2\pi t_0$. The surface density is constant initially in the runs using NIRVANA and FARGO, and is normalized such that there are 2 $M_{\rm Jup}$ within a_0. The RH2D runs use $\Sigma \propto r^{-1/2}$, normalized such that Σ at a_0 was about three times smaller than that used in the NIRVANA/FARGO runs. The initial mass ratio between the protoplanet and the central star was $q = m_{\rm p}/M_* = 10^{-3}$, corresponding to a Jupiter-mass planet orbiting a solar-mass star. The protoplanet was started on a circular orbit of radius $r_{\rm p} = 1$.

The accretion of gas by the protoplanet is modeled by removing a fraction of the material that resides within the inner half of the planet's Roche lobe at each time-step. The rate of gas removal corresponds to a maximal accretion rate

Figure 1. The evolution of a protoplanet embedded in a protostellar disk for calculation N3. The relative surface density of disk material is represented by the gray-scale. The white circle represents the position of the protoplanet. The disk is initially unperturbed at time $t = 0$.

(NPMK2000). Simulations were also performed in which the accretion rate was set to zero.

3. Numerical Calculations

The results of the simulations are presented in table 1. Particularly significant are the estimated migration times, τ_{mig}, and final masses m_{final} (in units of the initial planet mass m_{init}). τ_{mig} was obtained by measuring the planet's inward velocity at the end of the run and extrapolating forward in time assuming that it remains constant, thus estimating the time for the planet to reach the central star. The final mass was obtained by measuring the accretion rate at the end of the run and extrapolating forward for a time τ_{mig}, estimating the mass of the planet when it reaches the central star. Note that the assumption of there being a constant migration or accretion rate is not strictly correct since these will decrease as the system evolves.

We will now describe the results of an individual run to illustrate the main results of our simulations. The evolution of run N3 is shown in Figure 1. At time $t = 0$ the disk surface density was unperturbed. The tidal force of the planet leads to the formation of trailing spiral shock waves, and an $m = 2$ spiral wave pattern may be observed. The transfer of angular momentum between the disk and protoplanet leads to the formation of a gap in the vicinity of the planet's orbit, which requires about 200 orbits to clear.

Table 1. Column 1 gives run label (N*i* NIRVANA, F*i* FARGO, R*i* RH2D), column 2 gives number of grid cells used, and column 3 indicates whether a planet accretes from the disk. Column 4 indicates whether the run started with an initial gap in the disk. Column 5 gives estimated migration time (in units of 10^4 P_0) and column 6 gives estimate of the final mass of the planet. Note that estimated times are given at different evolutionary times of the runs

Run	Resolution $N_r \times N_\phi$	Accretion on?	Initial gap?	$\tau_{\rm mig}$ $\times 10^4 P_0$	$m_{\rm final}$ $(m_{\rm init})$
N3	80 × 235	yes	no	1.00	3.2
N4	80 × 235	no	no	1.20	1.0
N5	130 × 384	yes	no	0.84	2.7
N6	130 × 384	no	no	1.0	1.0
N7	80 × 235	yes	yes	0.94	3.03
N8	80 × 235	no	yes	1.07	1.0
F3	80 × 236	yes	no	1.08	3.85
F4	80 × 236	no	no	1.33	1.0
F5	130 × 384	yes	no	0.85	3.27
F6	70 × 180	yes	yes	2.03	4.87
R1	128 × 128	(no)	gap	1.45	1.0
R2	128 × 128	yes	gap	1.48	4.63
R3	128 × 288	yes	gap	1.36	4.10

Eventually the inner disk drains through the open inner boundary. Tidal interaction between the disk and protoplanet leads to the disk interior to $r_{\rm p}$ exerting positive torques on the protoplanet, and that exterior to $r_{\rm p}$ exerting negative torques on it. The loss of the inner disk causes the interaction with the outer disk to be strongly dominant, and the planet migrates inwards as it loses angular momentum.

The evolution of the protoplanet orbital radius is shown in the first panel of Figure 2, and that of the planet mass in the second panel. The evolution of the protoplanet is rapid early on during gap formation. This phase lasts for a few hundred orbits and results in rapid accretion that almost doubles the mass of the protoplanet within $t = 300$ P_0. When the gap has been cleared, the evolution slows down and the planet spirals in towards the star at an almost constant rate while accreting gas from the disk at an almost steady rate. The migration rate slows down slightly as the run proceeds since the increasing mass of the planet increases its inertia. The calculation was initiated with local disk mass $M(r_{\rm p}) \sim 2m_{\rm p}$, but as the planet migrates and accretes from the disk $m_{\rm p}$ becomes $\geq M(r_{\rm p})$. This situation was studied by Ivanov et al. (1999), and the evolution is controlled by the viscous evolution of the disk and the inertia of the planet, with a more massive planet migrating more slowly, as described in NPMK2000.

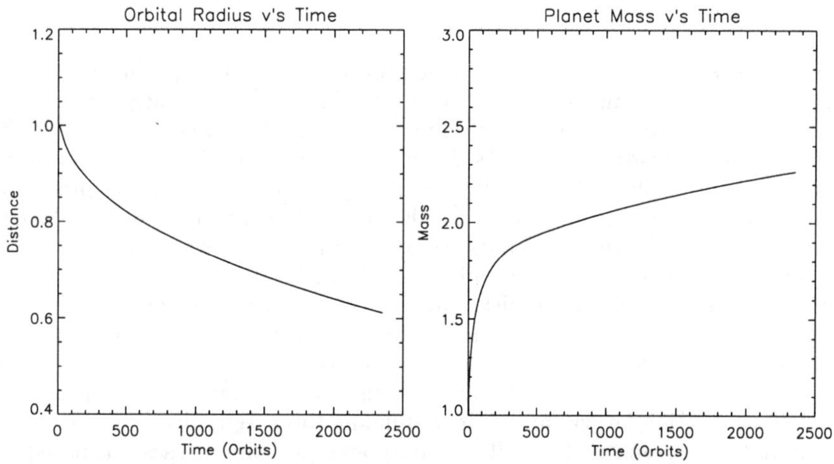

Figure 2. The *first panel* shows the evolution of the planet orbital radius for calculation N3 (*solid line*) and N7 (*dotted line*). The *second panel* shows the increase in mass as the protoplanets accrete gas from the disk for the same calculations, with the same line styles as described above

The accretion rate decreases as the protoplanet mass increases because the disk is tidally truncated more effectively by a more massive planet (e.g., Bryden et al. 1999). After a time of $t \sim 2400 \quad P_0$, the protoplanet has migrated to $r_\mathrm{p} \sim 0.6$ and has accreted $\sim 1.3\ M_\mathrm{Jup}$. By extrapolating the migration and accretion rates forward in time, we estimate that the planet will reach the central star after $t \sim 1.0 \times 10^4 \quad P_0$, by which time $m_\mathrm{final} \sim 3.2\ M_\mathrm{Jup}$. The viscous evolution time of the disk is given by $\tau_\nu = 2r^2/(3\nu)$. At $r = 1$, this corresponds to $\tau_\nu \sim 10^4\ P_0$, similar to the migration times of the protoplanets in our simulations.

We note that the planet remains in an almost circular orbit throughout its evolution and shows no sign of eccentricity growth since the gap region still contains sufficient corotating material that is able to damp the eccentricity growth caused by the outer Lindblad resonances (Goldreich & Tremaine 1980; Artymowicz 1993a,b).

Examining the results presented in Table 1, we see that the estimated migration times and the final planetary masses do not vary greatly in our simulations. In particular, the results are relatively insensitive to the code used, whether or not an initial gap was present, whether or not the planet was allowed to accrete, or what the assumed surface density profile was. Calculation F6 was run for about four times longer than the other runs (i.e., 10 000 P_0), and shows the effect of extrapolating the migration and accretion rates at later evolutionary times.

4. Summary

We have studied the interaction of a protoplanet of 1 $M_{\rm Jup}$ initially with a gaseous disk whose mass, aspect ratio, and kinematic viscosity are those expected for a minimum-mass solar nebula. The problem was studied with three independent hydrodynamic codes which were found to give consistent results. The direction of the orbital migration is always inwards and the protoplanet reaches the central star in a near circular orbit after a time of $\sim 10^4\ P_0$, which is characteristically the viscous timescale at the initial orbital radius. This is independent of whether the protoplanet is allowed to accrete mass or not, or the surface density profile in the disk. Inward migration was assisted by the accretion of the inner disk onto the central star. When the protoplanet was allowed to accrete at a near maximal rate the mass was found to increase to about 4 $M_{\rm Jup}$ as it reached the central star. Because of deep gap formation and lower accretion rates for larger masses it is difficult to exceed this mass in the kind of simulations presented here. It would appear that the masses estimated for a number of close-orbiting giant planets (Marcy, Cochran, & Mayor 1999; Marcy & Butler 1998) as well as their inward orbital migration, can be explained by the operation of the processes considered here during the late stages of giant-planet formation.

Acknowledgments. This work was supported in part (FM) by the European Commission under contract number ERBFMRX-CT98-0195 (TMR network "Accretion onto black holes, compact stars and protostars"), and by the Max-Planck-Gesellschaft, Grant No. 02160-361-TG74 (for WK). Computational resources of the Max-Planck-Institute for Astronomy in Heidelberg were available and are gratefully acknowledged. We thank Udo Ziegler for making a FORTRAN version of his code NIRVANA publicly available.

References

Artymowicz, P. 1993a, ApJ, 419, 155

Artymowicz, P. 1993b, ApJ, 419, 166

Beckwith, S. V. W., & Sargent, A. 1996, Nat, 383, 139

Bryden, G., Chen, X., Lin, D. N. C., Nelson, R. P., & Papaloizou, J. C. B. 1999, ApJ, 514, 344

Goldreich, P., & Tremaine, S. 1980, ApJ, 241, 425

Ivanov, P. B., Papaloizou, J. C. B., & Polnarev, A. G. 1999, MNRAS, 307, 79

Lin, D. N. C., & Papaloizou, J. C. B. 1986, ApJ, 309, 846

Marcy, G. W., & Butler, R. P. 1998, ARA&A, 36

Marcy, G. W., Cochran, W. D., & Mayor, M. 1999, in Protostars and Planets IV, ed. V. Mannings, A. P. Boss, & S. S. Russell (Tucson: University of Arizona Press), p. 1285

Nelson, R. P., Papaloizou, J. C. B., Masset, F., & Kley, W. 2000, MNRAS, in press (NPMK2000)

Papaloizou, J. C. B., & Lin, D. N. C. 1984, ApJ, 285, 818

Syer, D., & Clarke, C. J. 1995, MNRAS, 277, 758

Discussion

Alan Boss: These particular models would suggest that the shortest-period planets would have migrated the most and so gained the most mass, yet the shortest-period planets are not the most massive, and this cannot be a selection effect. What else must be going on to avoid this outcome?

Frédéric Masset: Obviously if you start systematically with a one-Jupiter-mass giant core at 5 AU embedded in a minimum-mass protoplanetary nebula, the viscosity and aspect ratio of which are always the same, and use the maximal accretion scheme that we have used, then what you are saying is true. However the "starting" conditions and disk parameter may differ widely from one system to the other, and it is likely that the actual accretion rate is smaller than the one we have used here. Furthermore, I think you have to make a distinction between the "light" hot jupiters and the other exoplanet population.

Tristan Guillot: How does the viscous accretion timescale of gas in the inner disk onto the planet translate into an accretion rate in your model? How does that compare to observations?

Frédéric Masset: The value of the viscosity we have employed in our runs is precisely derived from observational estimates of the accretion rate onto T Tauri stars. The value of alpha it corresponds to, at Jupiter's orbital radius, is about 6×10^{-3}. The accretion rate onto the primary it corresponds to, in our models, is roughly speaking $\sim 4 \times 10^{-8}\ M_{\odot}\ \mathrm{yr}^{-1}$.

Disks, Planetesimals, and Planets
ASP Conference Series, Vol. 219, 2000
F. Garzón, C. Eiroa, D. de Winter, and T. J. Mahoney, eds.

Modeling of Infrared/Millimeter Images of Circumstellar Disks with an Embedded Planet

Sebastian Wolf

Thüringer Landessternwarte Tautenburg, Sternwarte 5, D–07778 Tautenburg, Germany

Thomas Henning

Astrophysikalisches Institut und Universitäts-Sternwarte, Schillergässchen 2-3, D-07745 Jena, Germany

Wilhelm Kley

Max-Planck-Institut für Astronomie, Königstuhl 17, D-69117 Heidelberg, Germany

Abstract. We present radiative transfer (RT) simulations for a circumstellar disk around a solar-type star with an embedded planet of 1 Jupiter mass. The three-dimensional density structure in the disk results from hydrodynamical simulations. For the simulation of the RT, we applied a self-consistent three-dimensional continuum RT code. Based on images and visibility curves derived from the calculations, we discuss the possibilities of observing gaps in a disk—caused by a planet's motion—by direct imaging and interferometry.

1. Introduction

Studies of the evolution of protoplanets in a protostellar disk have shown that, depending on the hydrodynamical parameters of the planet and disk, those protoplanets may open a gap in the disk (e.g., Kley 1999, 2000). The radial extent of the gap may be a few AU and is therefore much larger than the planet. Thus, the question arises if it is possible to find such a gap as an indicator of the presence of a protoplanet. Based on RT simulations, we discuss this possibility. Spiral density waves, also caused by the planet's motion, are considered.

2. Model

The density structure of the protostellar disk results from hydrodynamical simulations for which the disk is assumed to be flat and non-self gravitating. The mutual gravitational interaction between the planet and the central star, and the gravitational torques of the disk acting on the planet and the central star are included. With the prospect of the subsequent RT simulations, we introduce a vertical density structure of the disk:

$$\rho(r,\phi,z) = \rho(r,\phi,z=0)\,\exp\left(-\frac{z^2}{2\,h(r,\phi)}\right), \tag{1}$$

where $\rho(r,\phi,z=0)$ is the midplane density and $h(r,\phi)$ is a disk scale height parameter. For $|z|/r > 1/20$ the density is zero.

The mass of the star is assumed to be 1 $M_\odot$, the mass of the planet is 1 Jupiter mass, and the mass of the disk is $0.05\,M_\odot$. For the dust-to-gas mass ratio we take a value of 1:100. The diameter of the disk is 104 AU. The planet is located at a distance of 5.2 AU from the star. We consider the disk after 150 orbits of the planet (initial midplane density distribution: $\Sigma \propto r^{-1/2}$).

3. Radiative Transfer Technique and Parameters

We simulate the RT with a 3D continuum RT code. Based on the Monte Carlo (MC) method, images, spectral energy distributions, and polarization maps can be computed. The spatial density and temperature structure of the models can be defined on a grid which can be chosen to be identical with the grid the hydrodynamical simulations are based on. Thus, any additional discretization error resulting from the "transfer" of the density/temperature structure to the RT code is avoided. For further information concerning the RT code, we refer to Wolf, Henning, & Stecklum (1999) and Wolf & Henning (2000b).

In addition to the results from the hydrodynamical simulations, we introduce the following RT parameters: spherical dust grains consisting of "astronomical" silicates (optical data from Draine & Lee 1984, radius $0.12\,\mu$m); star: $T = 5500\,$K, $L = 1\,L_\odot$; wavelength range for the simulation of the radiative transfer: 0.03–2000 μm.

4. Images

In Figures 1 and 2, the images of the disk seen under different inclination angles, i, (= 0°—face-on, 30°, 60°, and 90°–edge-on) in the MIR ($10\,\mu$m), in the submm ($700\,\mu$m) and mm wavelength range (1.2 mm) are shown. In Figure 1, the whole disk is shown, while in Figure 2 the inner region (diameter 28 AU) is magnified. Each frame consists of 101×101 pixels. Assuming a distance of 140 pc of the object (which is the distance to the nearby star-forming region in Taurus), the pixel scale is 7 mas/pixel (Figure 1) resp. 2 mas/pixel (Figure 2).

Due to the huge intensity peak caused by warm/hot dust located very close to the star (1–2 AU), the outer regions of the disk remain dark at $10\,\mu$m. The gap located at 5 AU can hardly be seen. In contrast to this, the brightness distribution is much smoother in the submillimeter/millimeter wavelength range. This can be explained by the behavior of the Planck function, $B_\lambda(T)$. The ratio $B_\lambda(T_1)/B_\lambda(T_2)$ ($T_1 > T_2$) strongly decreases with increasing wavelength. Thus, the flux ratio and the brightness contrast between regions characterized by different dust temperatures (T_1 and T_2) decreases with increasing wavelength λ. In the $700\,\mu$m and the 1.2 mm images the gap can be clearly seen—even in the case of large inclinations of the disk. However, direct submillimeter imaging

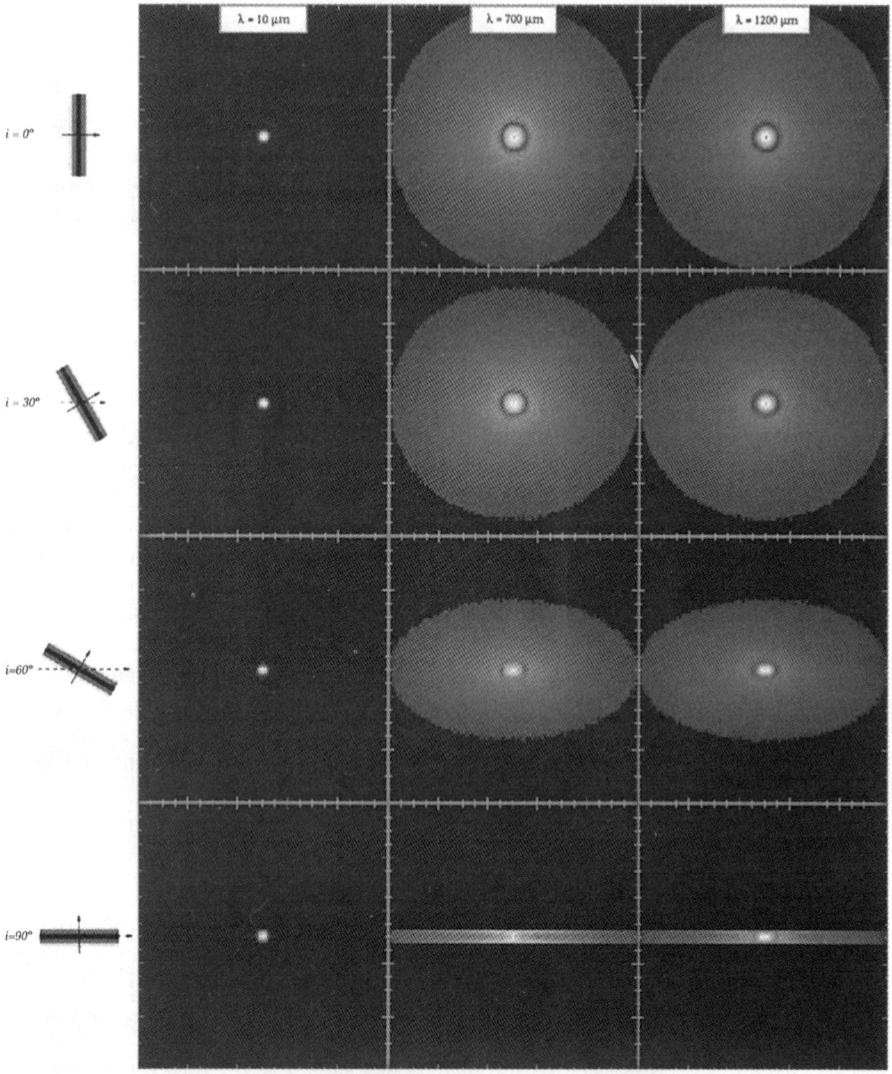

Figure 1. Images of the circumstellar disk with an embedded planet of 1 Jupiter mass at wavelengths $\lambda = 10, 700$, and $1200\,\mu$m. The different inclinations of the disk $i = 0°$ (face on), $30°, 60°$, and $90°$ (edge on) are symbolized in the left column. To increase the dynamic color range not the flux, F, but $F^{1/4}$ is shown. Due to the high brightness contrast at $\lambda = 10\,\mu$m, the gap can hardly be seen. In contrast to this, the gap is clearly visible at $\lambda = 700$ and $1200\,\mu$m (see Section 4.).

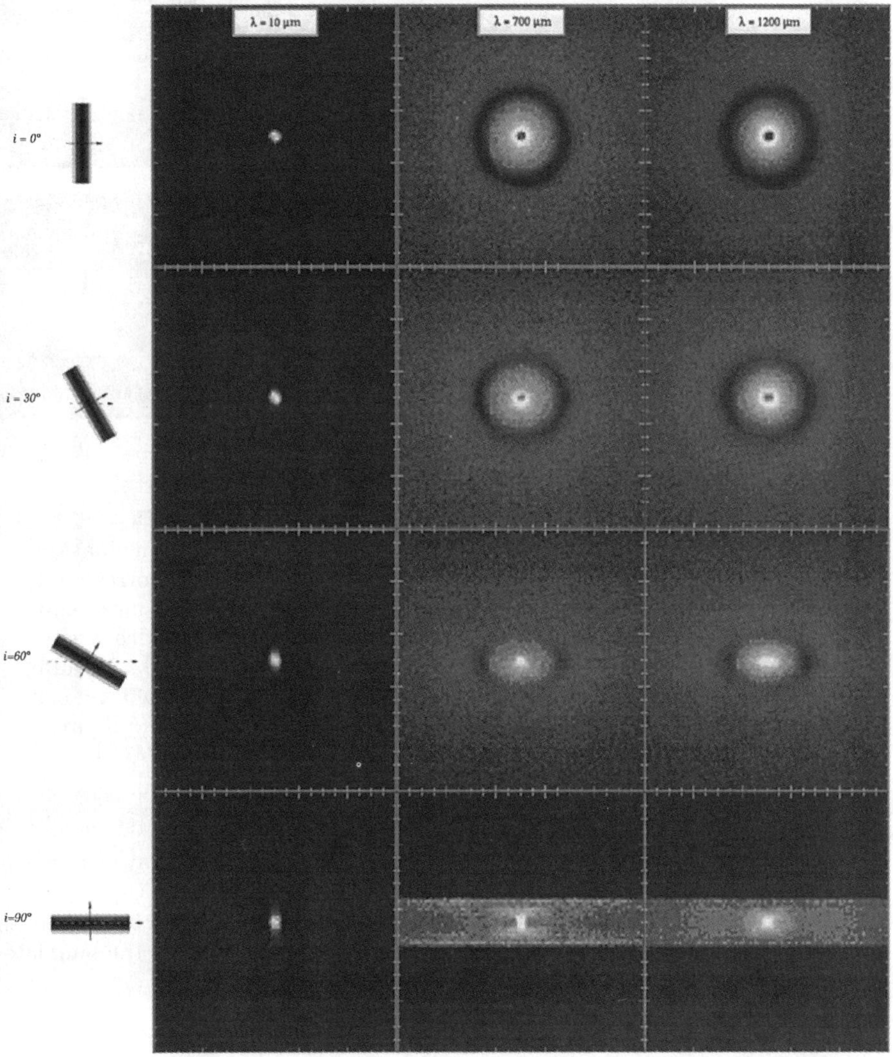

Figure 2. Images of the inner region (diameter 28 AU) of the disk. The spiral density waves in the disks are weakly visible. For comparison and further remarks see Figure 1 and Section 4.

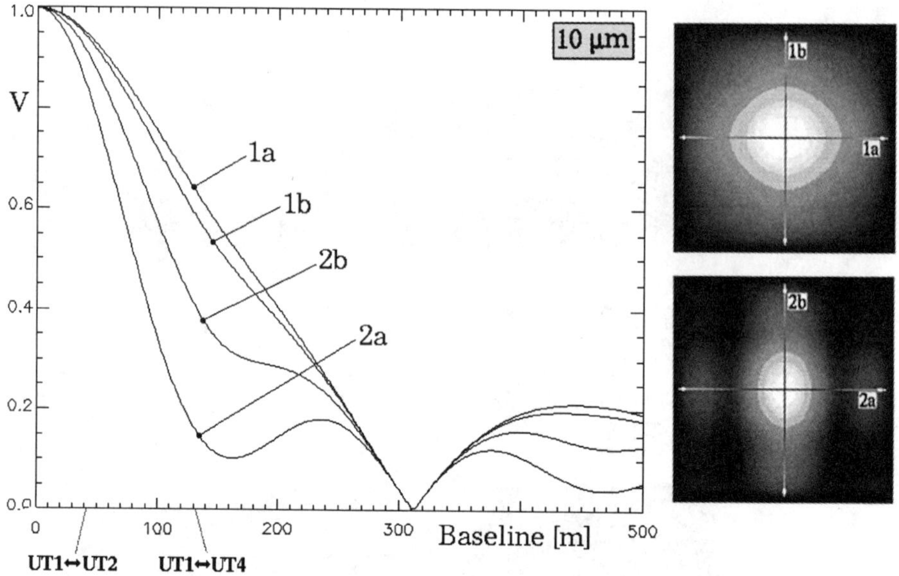

Figure 3. Visibility tracks at $\lambda = 10\ \mu$m for disks with an inclination of $i = 60°$ (assumed distance: 140 pc). Tracks $1a$ and $1b$ correspond to the image of a disk without a gap (but with the same mass and inclination), tracks $2a$ and $2b$ are derived from the disk with a gap shown in Figure 1 ($\lambda = 10\,\mu$m, $i = 60°$). The visibility tracks $1a$ and $1b$ resp. $2a$ and $2b$ are perpendicular to each other in the u–v plane. The inner parts of the u–v planes are shown in the right-hand figures. The distances between the telescopes UT1 and UT2 resp. UT1 and UT4 of the VLTI are marked.

(e.g., with SCUBA—Submillimeter Common-User Bolometer Array) will not be possible because of the required resolution.

Finally, we wish to mention that the spiral structures—caused by the motion of the planet—can be seen very weakly only (weak contrast) in the simulated images. They will be much harder to find than gaps.

5. Visibilities

For illustration of the power of interferometry, the visibility, V, for the disk seen under an inclination of $i = 60°$ was calculated. We consider the effect of a gap in the visibility track by comparing the visibilities for a disk both with and without a gap (same inclination and mass).

In Figure 3, the inner parts of the u–v plane resulting from the Fourier transformation of the disk images are shown. Furthermore, the lines along which the visibilities were determined ($1a$, $1b$, $2a$, $2b$) are marked. The tracks $1a$ and $1b$ result from a disk without gap. The small difference in the shape of the visibility tracks along two perpendicular directions in the u–v plane is caused by the inclination of the disk. In the case of a disk with a gap, the visibility track shows a steeper decrease (for comparison see the 10 μm image for $i = 60°$ in Figure 1). If the disk is seen nearly face on, an additional local minimum at lower baselines arises. For larger inclinations the additional local minimum can be found only along selected visibility tracks (e.g., visibility tracks $2a$ and $2b$).

In Figure 3 the distances between telescopes UT1 and UT2 resp. UT1 and UT4 of the VLTI (VLT Interferometer) are shown. Considering uncertainties of "real" measurements (e.g., with MIDI—MIR interferometric instrument for the VLTI), a distinction between different disk models (with/without gap) based on significant differences between the visibility tracks is possible only by a beam combination of the most distant telescopes of the VLTI.

In addition to this, simulations of ALMA (Atacama Large Millimeter Array) show that—based on the combination of 64 antennas—the gap can be clearly seen in the (sub)mm wavelength range on the reconstructed images. Taking into account real noise, the planet gap will be resolved and seen on the synthesized images at $\lambda = 1.2$ mm after several hours of observations. Further results of the calculations for ALMA will be discussed by Wolf, Gueth, & Henning (2000a).

6. Conclusions

Based on self-consistent 3D radiative transfer simulations, we generated MIR, submm and millimeter images of a circumstellar disk with an embedded planet surrounding a solar-type star. Because of the extreme brightness contrast at 10 μm in the innermost region of the disk, the gap is difficult to detect in the near- and MIR. However, even in the (sub-)millimeter wavelength range where the gap can be seen very clearly in the simulated images, an extremely high angular resolution is required (≈ 10 mas).

In contrast, MIR and (sub)mm interferometers which will become available in the near future (such as MIDI at the VLTI and ALMA) will achieve this goal. While MIDI can be used to distinguish between different disk models on the basis of the observed visibilities, ALMA will provide the basis for the reconstruction of an image of a gap. Thus the search for massive protoplanets in circumstellar disks can be based on the indication of a gap in cicrumstellar disks.

Acknowledgments. We wish to thank F. Gueth for performing simulations for ALMA. This research was supported by the DFG grant Ste 605/10 within the program "Physics of star formation".

References

Draine, B. T., Lee, H. M. 1984, ApJ, 285, 89

Kley, W. 1999, MNRAS, 303, 696

Kley, W. 2000, MNRAS, 313, L47

Wolf, S., Gueth, F., & Henning Th. 2000a, in preparation

Wolf, S., & Henning, Th. 2000b, Comp. Phys. Comm., in press

Wolf, S., Henning, Th., & Stecklum, B. 1999, A&A, 349, 839

Discussion

Bernard Foing: Could you quantify the availability of the features in your simulated intensity maps with pixel 2 mas (profiles and gaps) with the resolution of ALMA, VLT, *NGST*, and *DARWIN*?

Sebastian Wolf: That will be done in the published article for this proceedings.

Hubertus Klahr: Have you ever compared the calculated temperature of the radiation transport with that resulting from the hydro calculation?

Sebastian Wolf: Yes, but in the case of disks with a high optical depth (τ_ν [$\lambda =$ 550 nm] $\gg 10^3$ in the plane), the star—as an additional heating source—has a remarkable influence on the resulting dust temperature only in the inner region, which is small compared with the total size of the disk.

Lee Hartmann: How would your results change if you used a distribution of grain sizes?

Sebastian Wolf: The spectral temperature distribution would be a function of the particular grain size. But the "quantitative" results (images and polarization maps) would be very similar to the case of a single dust-grain size.

Disks, Planetesimals, and Planets
ASP Conference Series, Vol. 219, 2000
F. Garzón, C. Eiroa, D. de Winter, and T. J. Mahoney, eds.

High-Resolution Simulations and Visualization of Protoplanetary Disks

Paweł Ciecieląg, Tomasz Plewa, and Michał Różyczka

Nicolaus Copernicus Astronomical Center, Bartycka 18, 00-716 Warsaw, Poland

Abstract. The problem of mass flow in the immediate vicinity of a planet embedded in a protoplanetary disk is studied numerically in two dimensions. The large differences in the temporal and spatial scales involved suggest that a specialized discretization method for the solution of hydrodynamical equations may offer great savings in computational resources and could make extensive parameter studies feasible. Preliminary results obtained with the help the of adaptive mesh refinement technique and a high-order explicit Eulerian solver are presented. This combination of numerical techniques appears to be an excellent tool which allows for direct simulations of mass flow in vicinity of the accretor at moderate computational cost. In particular, it is possible to resolve the surface of the planet and to model the process of planet growth with a minimal set of assumptions. Some issues related to visualization of the results and future prospects are briefly discussed.

1. The Method

The extremely small temporal and spatial scales involved in the problem of accretion onto a protoplanet necessitate the use of non-uniform discretization in the vicinity of the accretor. In our study we used the adaptive mesh refinement (AMR) method combined with a high-resolution Godunov-type advection scheme (AMRA, Plewa & Müller 2000). The AMR discretization scheme follows the approach of Berger & Colella (1989). The computational domain is covered by a set of completely nested *patches* occupying *levels.* The levels create a refinement hierarchy. As one moves toward higher levels, the numerical resolution increases by a prescribed integer factor (separate for every direction). The net flow of material between patches at different levels is carefully accounted for in order to preserve the conservation properties of hydrodynamical equations. Boundary data for generated patches are obtained either by parabolic two-dimensional conservative interpolation of parental data or set according to prescribed boundary conditions.

Hydrodynamical equations are solved with the help of the direct Eulerian piecewise–parabolic method (PPMDE) of Colella & Woodward (1984), as implemented in the HERAKLES solver (Plewa & Müller 2000). Simulations have been done in spherical polar coordinates in a frame of reference corotating with the protoplanet. HERAKLES guarantees exact conservation of angular momentum,

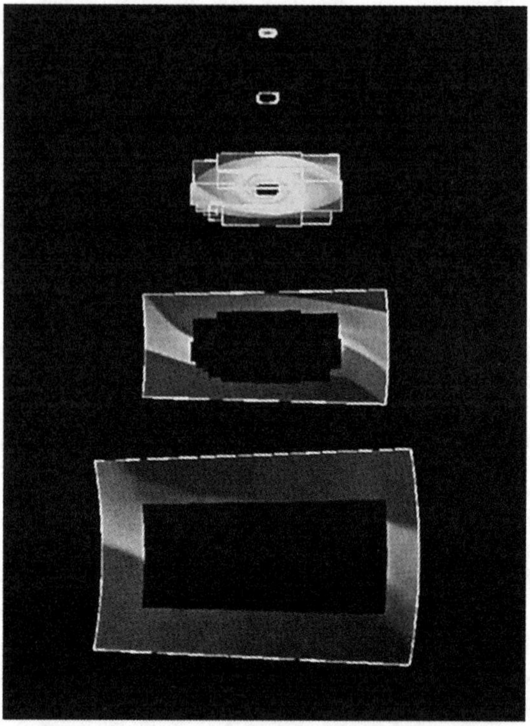

Figure 1. Breakdown of the finest five levels of the refinement hierarchy.

which is particularly important in the numerical modeling of disk accretion problems. The use of its multifluid option with tracer materials distributed within the disk (not presented here) allows us to identify the origin of the material accreted onto protoplanet. The AMRA code is written purely in FORTRAN 77 and has been successfully used on both vector supercomputers and superscalar cache-oriented workstations. Its parallelization on shared memory machines exploits microtasking (through the use of vendor-specific directives) or the OpenMP standard.

2. Simulation Set-Up

The computational domain extends from 0.25 to 2.5 radii of the planet's orbit. We employ seven levels with refinement ratios ranging from (2,4) to (4,4). The base level contains the protoplanetary (circumstellar) disk while the seventh level contains the planet and its immediate vicinity. The base grid consist of 128×128 cells uniformly distributed in r and θ. The effective resolution at the seventh level is 131072×524288 in r and θ, respectively. The topmost five levels are shown schematically in Figure 1. The white lines are boundaries of the patches. There are 1, 1, 1, 1, 12, 4, and 49 patches at levels 1–7, respectively.

The structure of the grid at level 7 is shown in Figure 2*f* with individual cell boundaries drawn with white lines (the dark disk shows size of the planet).

3. Physical Model

The simulation is initialized with a Keplerian disk. Originally the disk has a mass of 0.01 $M_\odot$, a constant h/r ratio of 0.05, and a surface density proportional to $r^{-1/2}$. The temperature is a fixed function of r throughout the simulation. There is no explicit viscosity in the disk. At the outer and inner boundary of the base grid the gas is allowed to flow freely from the computational domain. No inflow is allowed for. The accretion onto the planet is accounted for in a very simplified way. At every time-step the mean value of the density within two planetary radii is calculated, and whenever it is higher than a preset value, the excess gas is removed. At $t = 0$ a planet of one Jupiter mass in inserted into the disk in a circular orbit. The radius of the orbit and the mass of the planet remain constant throughout the simulation. The disk is allowed to evolve for 100 planetary orbits. A gap is cleared in it, and a secondary circumplanetary disk is formed.

The sequence of surface plots in Figure 2 shows the final structure of both disks (the surface-density distribution is displayed). The peak in Figure 2a is the unresolved image of the very dense circumplanetary disk. We have been able for the first time to see the details of the latter (Figures 2*c*–*d*). The streams of gas flowing across the gap from left and right edge of the frame (light gray) collide with the outer part of the circumplanetary disk. The collision regions (green wedges) bear a strong resemblance to hot-spots in cataclysmic binaries. In every region two strong shock waves are excited, one of them propagating into the stream, and the other into the disk. The shocked gas flows from the collision region along a loosely wound spiral towards the planet (Figure 2*e*). This picture is significantly more detailed than the one recently published by Lubow, Seibert, & Artymowicz (2000). Streamlines of the flow around the planet are shown in Figure 3, and they are in good accordance with those of Lubow et al.

Our simulation is of a preliminary nature, and its sole purpose is to demonstrate the capabilities of AMRA. Currently, we are improving the physics of the model. One of the problems we are going to attack is the calculation of the accurate value of the gravitational torque from the disk onto the planet in the phase preceding gap formation.

4. Visualization

To visualize the complicated AMRA output, we have chosen the AVS/Express environment for visual programming. It allows the user to build simple applications quickly employing standard library modules. Advanced users can develop their own, highly specialized modules and applications. Our AMRA-visualization application (VISA) is partly based on modules written by Favre, Walder, & Follini (1998), which have been substantially modified, and partly on our own modules. A screenshot of VISA is shown in Figure 4. The panel and the viewer are contained in the two topmost windows, while the bottom window contains the

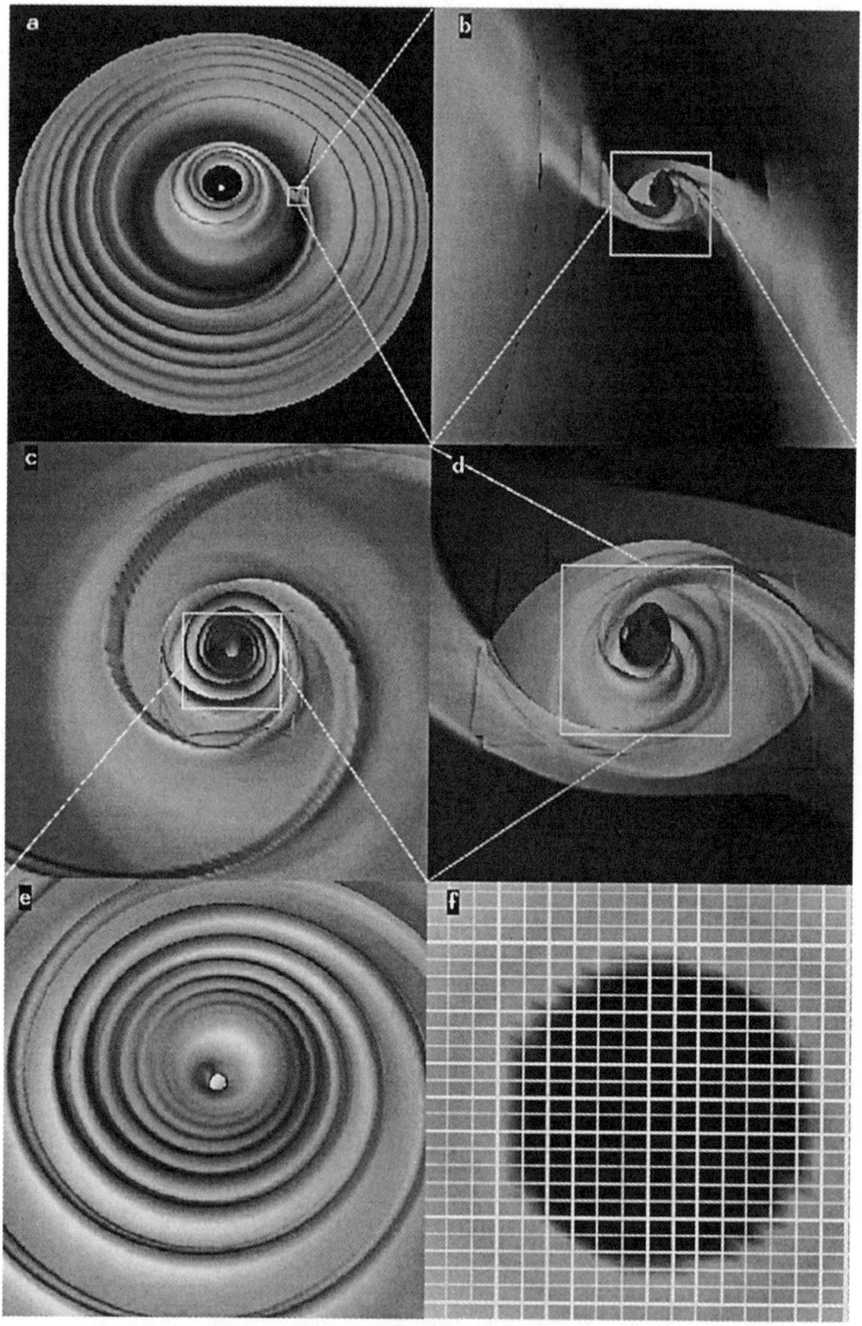

Figure 2. Surface-density distribution in the final model (frames *a*–*e*) and the distribution of grid cells at level 7 in the vicinity of the planet (frame *f*). The radius of the planet spans ≈ 8 grid cells.

Figure 3. Streamlines of the flow near the circumplanetary disk.

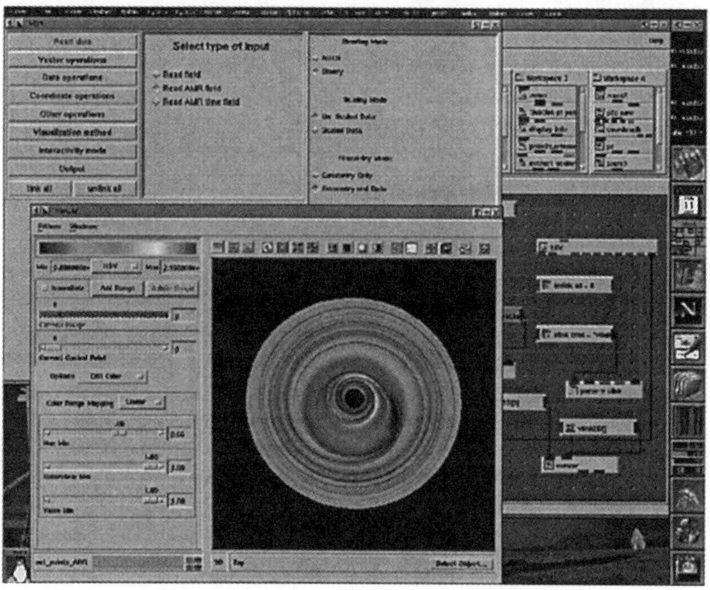

Figure 4. A screenshot of the VISA application.

AVS/Express programming platform. Currently we are able to read the AMR data, extract components, perform mathematical operations on data sets and coordinates, extract any subset of levels or patches, and apply to them various visualization technique (e.g., 2-D plot, surface plot, isolines, slice). Streamlines can also be calculated. The application is still under development, and new options are being added.

Acknowledgments. This research is supported by the Polish Committee for Scientific Research through the grant 2.P03D.004.13.

References

Berger, M. J., & Colella, P. 1989, J. Comp. Phys., 82, 64

Colella, P., & Woodward, P.R. 1984, J. Comp. Phys., 59, 264

Favre, J. M., Walder, R., & Follini, D. 1998, in Proc. 40th Cray User Group Conf,, Cray User Group Inc., Stuttgart

Lubow, S.H, Seibert, M., & Artymowicz, P. 2000, ApJ, 526, 1001

Plewa, T., & Müller, E. 2000, Comp. Phys. Commun., in preparation

Disks, Planetesimals, and Planets
ASP Conference Series, Vol. 219, 2000
F. Garzón, C. Eiroa, D. de Winter, and T. J. Mahoney, eds.

Orbital Motion of Resonant Clumps in Dusty Circumstellar Disks as a Signature of an Embedded Planet

Nick N. Gorkavyi

NRC/NAS, Code 685, NASA/GSFC, Greenbelt, MD 20771, USA

Leonid M. Ozernoy

5C3, Computational Sciences Institute and Department of Physics & Astronomy, George Mason U., Fairfax, VA 22030-4444;
also *Code 685, NASA/GSFC, Greenbelt, MD 20771*

John C. Mather

Code 685, NASA/GSFC, Greenbelt, MD 20771, USA

Sara Heap

Code 681, NASA/GSFC, Greenbelt, MD 20771, USA

Abstract. We have applied a powerful numerical approach to compute, with a high resolution, the structure of dusty circumstellar disks with embedded planets. We emphasize some testable implications of our simulations which would verify the presence of a planet via thermal emission of one or more dusty clumps which are in mean motion resonances with the planet. In particular, our simulations indicate that Vega may have a massive planet of $m \sim 2\ M_{\rm Jup}$ ($M_{\rm Jup}$ being Jupiter's mass) at a distance of 50–60 AU, and Epsilon Eri may have a less massive planet of $m \sim 0.2\ M_{\rm Jup}$ at a similar distance of 55–65 AU. This conclusion is testable: each resonant feature is stationary in the reference frame corotating with the planet, but it is not so for the observer at Earth. Therefore, if our interpretation of asymmetric clumps in circumstellar disks as dynamical resonant structures is correct, the above pattern revolves around the star with an angular velocity of $(1.2\text{–}1.6)^\circ$/yr (Vega) and $(0.6\text{–}0.8)^\circ$/yr (ϵ Eri)—a prediction that can be tested on a timescale of several years.

1. Introduction

Dusty disks around stars are a very common cosmic phenomenon. These disks, with scales of tens to hundreds (up to a thousand) of astronomical units, demonstrate a great variety in structure; sometimes they have a central "hole" void of gas and dust, and often are highly asymmetric. The dusty circumstellar disks are thought to accompany planetary systems and, at the same time, to hide them from the observer. Knowledge of dust characteristics is of prime importance for future NASA space missions, such as the *NGST* (*Next Generation Space Telescope*) and the *Terrestrial Planet Finder*. It is generally expected that a significant limitation to unambiguous planet detection and study will be contaminating thermal emission from dust in the target systems. We propose to turn this hazard into an advantage for detecting and characterizing planetary systems. Both the stellar radiation drag and stellar wind drag, as well as residual gas in the disk, tend to induce dust inflow toward the star. As dust particles pass by the planets in their infall, they interact with them by accumulating in planetary resonances, which are observable as clumps or belts in the dust disk.

2. Numerical Simulations

We focus on the conditions favorable to formation of resonant rings near the orbit of just one planet orbiting the parent star. Recently, we have elaborated a novel, very efficient approach to numerical modeling of distributions of test particles in an external gravitational field (Ozernoy, Gorkavyi, & Taidakova 2000; Gorkavyi et al. 2000b; Taidakova & Gorkavyi 1999). Our approach removes the major obstacle to reliable numerical simulations—-the particle-number limitation. In brief, our approach (which has a number of common elements with the "particle-in-cell" computational method) is as follows. Let us consider, for simplicity, a stationary particle distribution in the frame corotating with the planet. The locus of the given dust particle's positions (taken, say, as 10^4 positions every revolution about the star) are recorded and considered as the positions of *many other particles* produced by the same source of dust but *at a different time.* After this particle "dies" (as a result of collisions, infall, or ejection from the system by a planet-perturber), its recorded positions sampled over its lifetime form a stationary distribution as if it were produced by *many* particles. Typically, each run includes 10^3 revolutions, i.e. $\sim 10^7$ positions of a dust particle, which is equivalent, for a stationary distribution, to 10^7 particles. Allowing for 1000 sources of dust, we deal, after 1000 runs, with $\sim 10^{10}$ particle positions as if they were real particles. In the present paper, we immediately sort this information into $\sim 10^7$ spatial cells (each cell containing 10^3–10^4 particles), thereby forming a 3D grid that models the dust cloud around the star. We have computed about 300 model disks exploring the effects of various resonances given the location of dust sources and adopting as parameters the mass of the planet, initial orbital characteristics of dust particles, and particle radius, r (in μm), coupled with stellar luminosity, L_*, and mass, M_* (in solar units), as described by the parameter $\beta \approx 0.3\ L_*/(M_* r)$.

The modeling has shown that a planet, via resonances and gravitational scattering, produces an asymmetric resonant dust belt with one or more clumps plus one or more cavities, as well as a central region devoid of dust. These features can serve as indicators of a planet embedded in the circumstellar dust disk and, moreover, can be used to determine the mass of the planet and even some of its orbital parameters. Our simulations in more detail, as well as related work, are described by Ozernoy et al. (2000a).

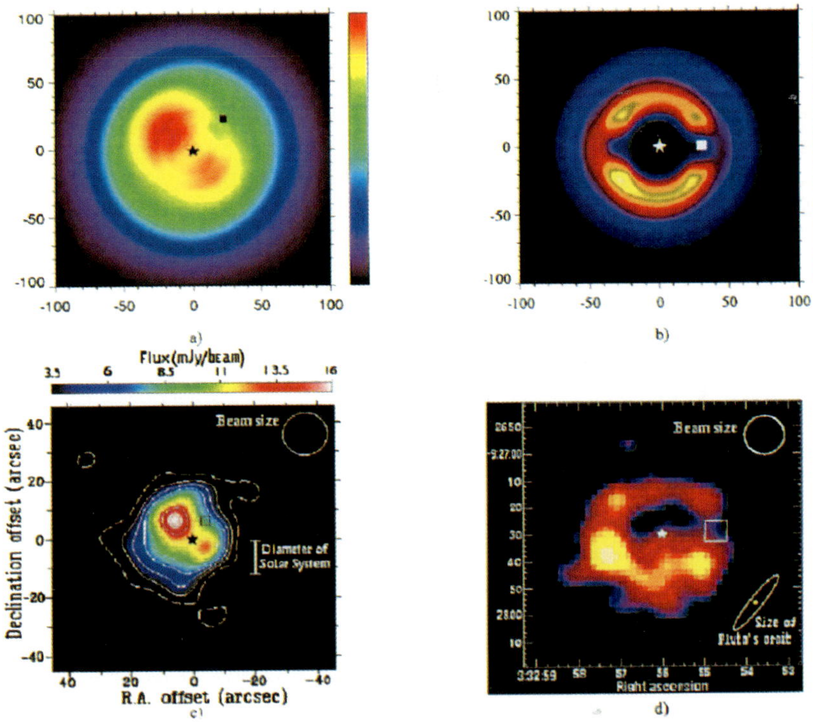

Figure 1. Thermal emission ($\lambda = 850$ μm) from simulated circumstellar disks with the resonant structures vs. observations. The star's location is (0,0). The unit of length = 1/30 of the planet's radius. The planet (shown in *a* and *b* as a *square*) revolves, like the disk, counterclockwise. In *c* and *d*, a *square* indicates the expected planet's location. To compare our simulations with observations, averaging with the beam size (50 units of length in *a* and 20 units of length in b is applied to the simulations. *a)* The simulation reminiscent of the disk of Vega: $M_\star = 2.5\ M_\odot$; $m_{\rm pl} = 2\ M_{\rm Jup}$; $\beta = 0.3$; the lifetime of dust particles $\tau = 2 \times 10^4$ planetary revolutions; the number of dust sources (comets) is 486, and the number of particle positions is 3×10^{10}. The adopted size of the central cavity is 10 units of length. For a version of this simulation with the beam size of 40 and the central cavity size of 12 units of length, see Ozernoy et al. (2000a). *b)* The simulation reminiscent of the disk of ϵ Eridani: $M_\star = 0.8\ M_\odot$; $m_{\rm pl} = 0.2\ M_{\rm Jup}$; the resonances 2:1 and 3:2 (populated in approximately equal proportions); $\beta = 0.002$; $\tau = 10^2$ planetary revolutions; the number of dust sources (kuiperoids) is 2910 (of these, 1250 sources form the upper branch, and 1660 sources form the lower branch of the pattern), and the number of particle positions is 2×10^9. *c)* Circumstellar disk of Vega (Holland et al. 1998). *d)* Circumstellar disk of ϵ Eridani (Greaves et al. 1998).

3. Comparison with Observations

The results of this study reveal a remarkable similarity to various types of highly asymmetric circumstellar disks observed at submillimeter wavelengths around ϵ Eri and Vega. Figures 1*a* and *b* show the thermal emission from simulated disks (seen face on) with an embedded planet, which are to be compared with the available observational data on Vega and ϵ Eri shown in Figures 1*c* and *d* (the parent stars are seen almost pole on). The sources of dust are assumed to be alike (gravitationally scattered) comets in the Vega's disk and alike (resonant) kuiperoids in the ϵ Eri disk. A planet as massive as 2 $M_{\rm Jup}$ (or somewhat less if the role of resonant comets is appreciable) produces, from the cometary dust, two clumps (Figure 1*a*), which is reminiscent of the observed asymmetric clumps near Vega shown in Figure 1*c*. A smaller mass ~ 0.2 $M_{\rm Jup}$ induces, from the kuiperoidal dust, two asymmetric arcs, with four clumps at its edges (Figure 1*b*). A similar asymmetric structure has been revealed in submm imaging of the ϵ Eri disk shown in Figure 1*d*.

Given the particular resonant pattern as the site for the captured dust, the appearance of the resonant structure in a dusty disk depends on the beam size and the presence/absence of a central cavity in the dust distribution. The latter can be produced as a result of gravitational scattering of the dust by a massive planet in the inner part of the planetary system, although the emission of dust from inner comets may mask its presence.

The above modeling—indicating that Vega may have a planet of mass ~ 2 $M_{\rm Jup}$ at a distance of 50–60 AU, and ϵ Eri may have a less massive planet of $m \approx 0.2$ $M_{\rm Jup}$ at a similar distance of 55–65 AU—is testable: each resonant feature is stationary in the reference frame corotating with the planet, but it is not so for the observer at Earth. Therefore, if the proposed interpretation of asymmetric clumps as dynamical resonant structures is correct, the above asymmetric feature revolves around the star with an angular velocity of $(1.2-1.6)^\circ$/yr (Vega) and $(0.6-0.8)^\circ$/yr (ϵ Eri)—a prediction that can be tested within several years. By the middle of 2000, one could expect the resonant features to be shifted, compared to their original positions in 1997–1998, by 4–5° in the Vega's disk and by $\sim 2^\circ$ in the ϵ Eri disk. In practical terms, an azimuthal brightness distribution at $R =$ 50–60 AU in the disk is to be measured and cross-correlated on a timescale of a few years. This could make it possible to reveal the revolution of the resonant pattern, to determine its direction, and evaluate the rate of motion.

If confirmed, the proposed interpretation of the structure in Vega- and ϵ Eri-like circumstellar disks seen face on would make it possible not just to reveal the embedded planet and determine its semi-major axis, but also to constrain its other basic parameters, such as the planet's mass and even to pinpoint the position of the planet. Our work suggests important observational tasks as follows:

A. to determine the direction of orbital revolution for dusty clumps;

B. to estimate the angular velocity of those clumps, and

C. to reveal a faint peripheral structure external to the clumps.

As our modeling indicates, an answer to task A would enable us to eliminate any uncertainty in the position of the predicted planet; its mass could be estimated more accurately as well. Task B would be important in determining the planet's orbital radius. Finally, task C would enable us to determine which particular resonances are the strongest and to pinpoint the position of the cavity in which the planet is located.

So far the most successful method of detecting extrasolar planets, such as the precision Doppler measurements, is biased toward comparatively small distances from the star. Our novel method offers a complementary approach for revealing invisible *outer* planets in circumstellar disks and determination of planetary parameters using the visible morphology of the *outer* part of the disk.

Acknowledgments. This work has been supported by NASA Grant NAG5-7065 to George Mason University. NG has been supported through NAS/NRC Associateship Research program. We thank to W.S. Holland and J.S. Greaves for permission to reproduce their SCUBA images.

References

Gorkavyi, N. N., Ozernoy, L. M., Mather, J. C., & Taidakova, T. 2000, in E. P. Smith & K. S. Long (eds.), ASP Conf. Ser., Vol. 207, NGST Science and Technology Exposition (San Francisco: ASP), p. 462

Greaves, J. S., Holland, W. S., Moriarty-Schieven, G., Jenness, T., Dent, W. R. F., Zuckerman, B., McCarthy, C., Webb, R. A., Butner, H. M., Gear, W. K., & Walker, H. J. 1998, ApJ, 506, L133

Holland, W. S., Greaves, J. S., Zuckerman, B., Webb, R. A., McCarthy, C., Coulson, I. M., Walther, D. M., Dent, W. R.F ., Gear, W. K., & Robson, I. 1998, Nat, 329, 788

Ozernoy, L. M., Gorkavyi, N. N., Mather, J. C. & Taidakova, T. A. 2000a, ApJ, submitted

Ozernoy, L. M., Gorkavyi, N. N., & Taidakova, T. 2000b, Planet. Sp. Sci., in press

Taidakova, T. A. & Gorkavyi, N. N. 1999, in B. A. Steves and A. E. Roy (eds.), The Dynamics of Small Bodies in the Solar Systems: A Major Key to Solar Systems Studies, (Dordrecht: Kluwer), p. 393

Disks, Planetesimals, and Planets
ASP Conference Series, Vol. 219, 2000
F. Garzón, C. Eiroa, D. de Winter, and T. J. Mahoney, eds.

Grain Diffusion and Chemical Evolution in Protoplanetary Disks

Th. Henning, M. Ilgner, and R. Schräpler

Astrophysical Institute and University Observatory (AIU), Friedrich Schiller University, Schillergäßchen 3, D-07745 Jena, Germany

Abstract. In this contribution, we treat the diffusion and vertical sedimentation of dust particles in a turbulent protoplanetary accretion disk. Based on the derivation of the diffusion coefficient, we calculate the vertical distribution of the grains. Furthermore, we develop a model which couples the disk dynamics with the chemical evolution of the gas-phase species. To begin with, we consider the time-dependent chemistry in a 1+1 D accretion disk model. The calculations demonstrate that vertical abundance gradients develop. We then include vertical diffusion processes. This is an important step towards a realistic modeling of multi-dimensional chemical processes in protoplanetary disks.

1. Introduction

A deeper understanding of the processes leading from micron-sized grains to kilometer-sized planetesimals requires the investigation of particle dynamics and growth (Beckwith, Henning, & Nakagawa 2000). An important process is the diffusion of grains in the turbulent gas and their vertical sedimentation, which determines the observational appearance of disks. In this paper, we will derive the diffusion coefficient from the dust continuity equation and the vertical scale height of the dust distribution.

Chemical processes in protoplanetary accretion disks depend on the grain distribution determining the ionization degree and the thermal structure of the disks. In addition, the grains provide "surfaces", which are important to the freeze-out process and various surface reactions. The radial distribution of the different gas-phase species determines the final composition of the planetary bodies forming from the disk. First radial models for the chemical evolution have been provided by Willacy et al. (1998) and Aikawa et al. (1999). A main result of these investigations is that, through reactions with ions, considerable amounts of CO and N_2, which are initially the dominant species in the disk, are transformed into CO_2, CH_4, NH_3, and HCN. In the temperature regions where freeze-out occurs, these components accumulate in grain mantles.

The next step has to be the study of the vertical distribution of molecular species. Strong vertical-density and temperature gradients will lead to a different chemical evolution at different radii. Aikawa & Herbst (1999) used a simplified three-layer model to treat the vertical structure of the disk and to determine the chemical evolution at these layers. Shortcomings of this model are both the neglect of diffusion processes and the simplified treatment of the vertical structure of the disks. In this contribution, we will calculate the vertical structure self-consistently and will also consider diffusion processes.

2. Diffusion and Vertical Sedimentation of Dust Grains

The development of a dust population in space and time can be described by the continuity equation of the dust distribution function f:

$$\frac{\partial f}{\partial t} + \nabla_{\vec{x}}(\underline{v}_{\rm d} f) = S, \qquad (1)$$

with S the coagulation source term and $\underline{v}_{\rm d}$ the velocity of the dust particles. The dust distribution function $f(\underline{x}, t, V)$ is a 3-parameter distribution function, where $\underline{x}$, t, V are the space vector, the time, and a parameter characterizing the size of the dust particles (compact volume V). The continuity equation describes the hydrodynamic aspects of the dust development and therefore, among other things, the dust diffusion. In the following, a short summary of the calculation of the diffusion coefficient is given.

Using the Reynolds method, $\underline{v}_{\rm d}$ and f are split into a mean (indicated by $^{-}$) and a fluctuating (indicated by $'$) part:

$$\underline{v}_{\rm d} = \overline{\underline{v}_{\rm d}} + \underline{v}'_{\rm d} \qquad (2)$$

$$f = \overline{f} + f' \qquad (3)$$

and inserted in Eq. (1). Reynolds averaging of the resulting equation gives the mean part of Eq. (1):

$$\frac{\partial \overline{f}}{\partial t} + \frac{\partial}{\partial x_i}(\overline{v_{\rm di}}\,\overline{f}) + \frac{\partial}{\partial x_i}(\overline{v'_{\rm di} f'}) = \overline{S}. \qquad (4)$$

Here, Einstein's summation convention is used. The dust diffusion tensor we are interested in is related to the correlation term $\overline{v'_{\rm di} f'}$. Assuming diffusion, caused by a spatial gradient, this term can be described by the ansatz

$$\overline{v'_{\rm di} f'} = -D_{ij}\frac{\partial \overline{f}}{\partial x_j}. \qquad (5)$$

In order to obtain the turbulent diffusion tensor and (assuming isotropy) the diffusion coefficient, the correlation $\overline{v'_{\rm di} f'}$ has to be calculated.

After extensive manipulations and some estimations (Schräpler 2000), the calculated diffusion coefficient follows as:

$$D = \nu_{\rm T}(1 + St)^{-1}, \qquad (6)$$

where $\nu_{\rm T}$ is the turbulent viscosity of the gas. The Stokes number is defined by the ratio of the friction time of the dust particles to the correlation time of the gas

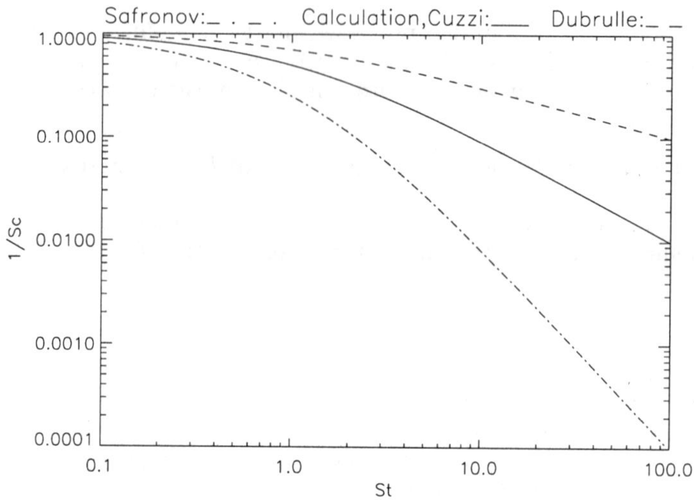

Figure 1. The Schmitt number $Sc = D/\nu_{\rm T}$ as a function of St.

turbulence $St = \tau_{\rm f}/\tau_{\rm c}$. This result confirms the estimate of Cuzzi, Dobrovolskis, & Champney (1993) and disproves the estimates of Dubrulle, Morfill, & Sterzik (1995), $D = \nu_{\rm T}(1+St)^{-1/2}$ and Safronov (1969) $D = \nu_{\rm T}(1+St)^{-2}$, which differ in the exponent of $(1+St)$. Figure 1 shows the Schmitt number $Sc = D/\nu_{\rm T}$ as a function of St.

The effect of D on the development of the protostellar disk is checked using the approximate stationary solution for the half-thickness of the dust sub-disk (Dubrulle et al. 1995):

$$h/H = \sqrt{\frac{D}{H^2\omega^2\tau_{\rm f}}}. \tag{7}$$

Here the quantities h and H are the dust and gas scale heights, respectively. The observationally relevant ratio h/H is shown in Figure 2.

3. Vertical Disk Structure and Chemical Evolution

In order to calculate chemical models for protoplanetary accretion disks, we have to specify their temperature and density distributions. Here we assumed that the disk will be in steady state and is optically thick. For the turbulent viscosity, we follow the α prescription. To derive the vertical temperature structure, one has to take into account that the gas, approximated by plane-parallel layers, is heated/cooled by different energy transport mechanisms: radiative, convective, and conductive transfer. Because of momentum conservation, pressure and density structures are based on the hydrostatic equilibrium. For a more detailed description on how the vertical structure has to be calculated, we refer to Meyer

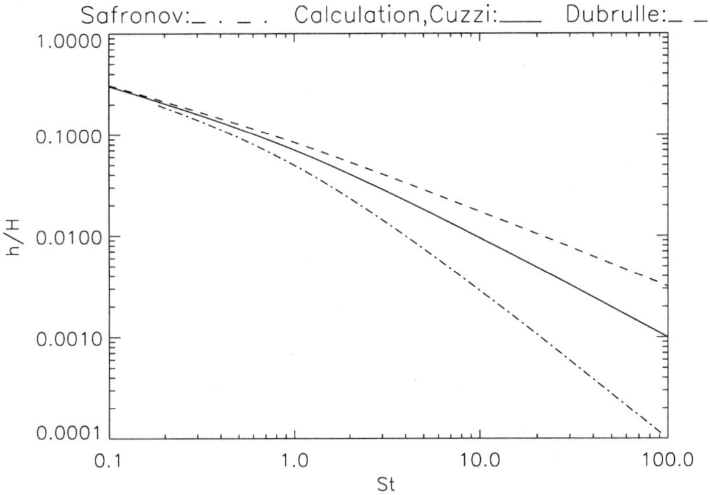

Figure 2. The ratio of dust and gas scale height over the Stokes number at 1AU. For the turbulent properties of the disc an α-model is used where the frequency of the biggest turbulent eddy, ω, equals the Keplerian frequency, Ω.

& Meyer-Hofmeister (1982). The disk is well characterized by a radial series of vertical-structure models because it is geometrically thin (1+1 D model).

The temperature regime at a radial position of 1 AU is characterized by temperature values below 700 K. The density ranges between 10^{-9} and 10^{-12} g cm^{-3} (based on an accretion rate of $\dot{M} = 10^{-7}\ M_{\odot}$ yr^{-1}, and an α value of 10^{-2}). In addition, there is a vertical ionization profile.

Considering the temperature and density profile at 1 AU, the reaction rates are too small to achieve a chemical equilibrium. Using the UMIST database, we included a coupled system of ordinary differential equations to calculate the time-dependent evolution of the chemical species in the gas phase:

$$\begin{aligned} \frac{d}{dt}N_i &= -N_i \sum_j k_{ij} N_j + \sum_{jl} k_{ijl} N_j N_l \\ &\quad -N_i \sum_{jl} k_{ijl} N_j N_l + \sum_{jlm} k_{jlm} N_j N_l N_m. \end{aligned} \tag{8}$$

Here, the quantities $k_{ij(k)}$ denote the rate coefficients and N_i is the number density of the ith molecule. This system of ordinary differential equations is a stiff system because of the very different rate coefficients (timescales) for the individual reactions. We applied the B(ackward)D(ifferentiation)F(ormulae) which is a flexible multi-step implicit method (Byrne & Hindmarsh 1975). At the maximum, the method allows 6th order in the steps.

In order to demonstrate that the different temperature and density layers affect the chemical state of the disk, we consider the chemical evolution of the

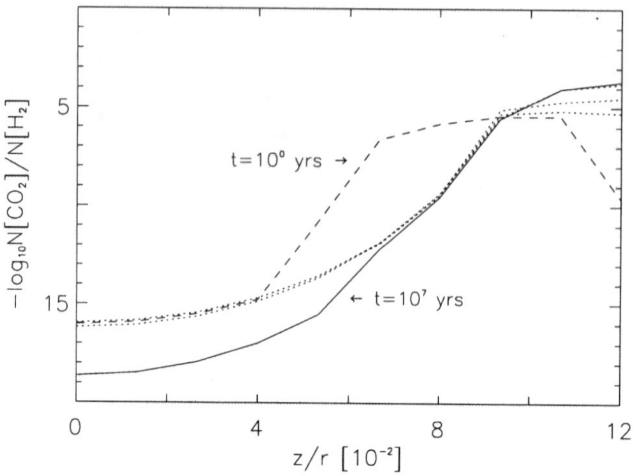

Figure 3. Time evolution of the vertical gradients of CO_2 without diffusion. The intermediate time steps are at 10^4, 10^5, and 10^6 years.

important species CO and CO_2 as an example. The results are summarized in Figures 3 and 4. Based on the results obtained with a chemical network containing 187 species and 2025 reactions, we conclude that the time-dependent evolution of CO_2 is strongly affected by the density and temperature profile. Due to the reaction pathways between CO_2 and CO ($CO + O_2 \rightarrow CO_2 + O$), the CO abundance also varies, especially near the disk surface.

Because of the existence of the vertical abundance gradients of most of the species, we also consider the possibility of diffusion processes among the gas-phase species. In this case, we have to add a diffusion equation for the different species:

$$\frac{\partial c_i}{\partial t} = \frac{1}{\varrho}\mathrm{div}\left(\varrho D \nabla c_i\right). \tag{9}$$

Here, the quantity c_i is the concentration of the ith species, D the diffusivity, and ϱ the total density. With respect to the temperature and pressure, the idealization of Eq. (9) does not restrict our model because diffusion caused by concentration gradients is the dominant part of the diffusion process. According to the vertical disk structure we approximate the diffusivity, D, by turbulent mixing; i.e., D is given by Equation (6).

For the numerical computation, we used the approach of the "fractional step method". This kind of modeling splits the whole process into a sequence of chemical and diffusion processes where special care has to be taken for the control of the time-step. Figure 5 shows a first result of these calculations for the CO molecule. This result demonstrates that the vertical density gradient is considerable reduced due to the diffusion process after about 10^3 years.

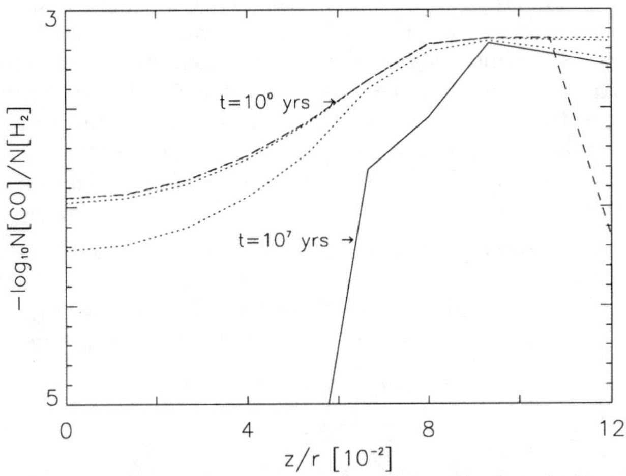

Figure 4. Time evolution of the vertical gradients of CO without diffusion. The intermediate time steps are at 10^4, 10^5, and 10^6 years.

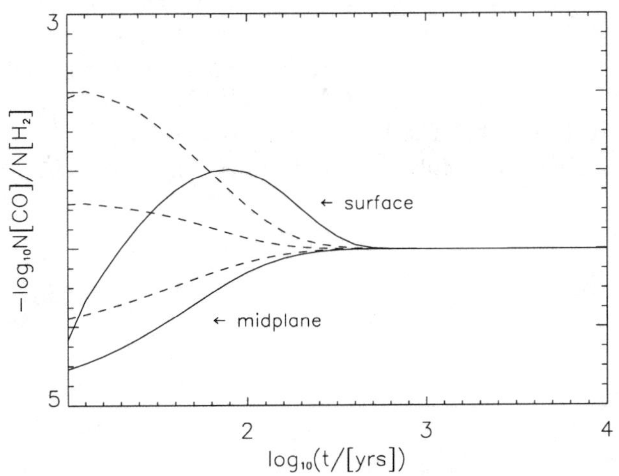

Figure 5. Time evolution of the vertical gradients of CO with diffusion. The time evolution is given for five different vertical grid points from the midplane to the surface.

4. Conclusions

In this paper, we have derived the diffusion coefficient from the continuity equation of the dust distribution. Furthermore, we treat the chemistry in a protoplanetary disk, which is determined by a complicated coupling between the disk structure and the chemical reaction network. The contribution especially demonstrates the importance of diffusion processes. Future calculations have to find out which are the most important reaction pathways and how they are influenced by diffusion. In addition, a more detailed ionization structure at the various radial grid points has to be included in the calculations.

Acknowledgments. The investigation of the chemical structure of protoplanetary accretion disks is supported by the grant He 1935/5-3 from the German Science Foundation (DFG) within the program "Physics of Star Formation." We acknowledge the on-going collaboration with the UMIST group.

References

Aikawa, Y., & Herbst, E. 1999, A&A, 351, 233

Aikawa, Y., Umebayashi, T., Nakano, T., & Miyama, S.M. 1999, ApJ, 486, L51

Beckwith, S., Henning, Th., & Nakagawa, Y. 2000, in Protostars and Planets IV, ed. V. Mannings, A. P. Boss, & S. S. Russell (Tucson: University of Arizona Press), p. 533

Byrne, G. D., & Hindmarsh, A.C. 1975, ACM Trans. on Math. Software, 1, 71

Cuzzi, J. N., Dobrovolskis, A. R., & Champney, J. M. 1993, Icarus, 106, 102

Dubrulle, B., Morfill, G., & Sterzik, M. 1995, Icarus, 114, 237

Meyer, F., & Meyer-Hofmeister, E. 1982, A&A, 106, 34

Safronov, V. S. 1969, Evolution of the Protoplanetary Cloud and the Formation of the Earth and Planets (NASA TTF-677)

Schräpler, R. 2000, PhD Thesis, University of Jena

Willacy, K., Klahr, H., Millar, T., & Henning, Th. 1998, A&A, 338, 995

Disks, Planetesimals, and Planets
ASP Conference Series, Vol. 219, 2000
F. Garzón, C. Eiroa, D. de Winter, and T. J. Mahoney, eds.

Angular Momentum Transport in Protoplanetary Accretion Disks by Hydrodynamical Turbulence

Hubertus Klahr and Peter Bodenheimer

UCO/Lick Observatory, University of California, Santa Cruz, CA 95064, USA

Abstract. In this paper we present local and global 3D radiation hydrodynamical simulations of protoplanetary accretion disks that led to the discovery of the source for hydrodynamical turbulence and vorticity generation in too-weakly ionized disks, which cannot be affected by magnetic fields. The careful treatment of the realistic global thermal structure of the disks naturally produces a radially outward-pointing entropy gradient, which makes the disk baroclinic. It is meanwhile known (Klahr & Bodenheimer 2000) that this global baroclinic instability leads to a non-axisymmetric flow, vorticity, and turbulence generation, including vortices, Rossby waves, and radially outward-bound angular-momentum transport. This allows for self-sustained turbulence and mass accretion onto the central object.

1. Introduction

The transport of angular momentum away from the central object is a sufficient condition for a protoplanetary disk to accrete matter onto the star and spin it down. Magnetic fields cannot be of importance for this process in a large part of the cold and dusty disk, where the planets supposedly form. Our new hypothesis on the angular-momentum transport based on radiation hydrodynamical simulations (TRAMP: Klahr, Henning, & Kley 1999) is as follows. Radiation transport and thermal convection produce a radially outward-pointing entropy gradient. This is a baroclinically unstable situation, which generates turbulence (Klahr & Bodenheimer 2000). The turbulence transports angular momentum outward and finally allows for an accretion process and dissipation of excess kinetic energy to ultimately sustain the thermal structure and entropy gradient.

In Klahr & Bodenheimer (2000) we presented the global baroclinic instability as a source for vigorous turbulence in Keplerian accretion disks. We showed there by analytical considerations and 3D radiation hydro simulations that protoplanetary disks in particular have a radially outward-directed entropy gradient, which makes them baroclinic. 2D numerical simulations showed that this baroclinic flow is unstable and produces turbulence. These findings were tested for numerical effects by performing barotropic simulations, which show that the initially imposed turbulence rapidly decays. The turbulence in baroclinic disks draws energy from the background shear, transports angular momentum outward, and creates a radially inward-bound accretion of matter, in this way forming a self-consistent process. The classical accretion paradigm of transforming gravitational energy via turbulent processes into heat is confirmed. We measure accretion rates of $\dot{M} = 10^{-9} - 10^{-7}\ M_{\odot}\ \mathrm{yr}^{-1}$ and viscosity parameters of $\alpha = 10^{-4}$–10^{-2}, which fit perfectly together and agree reasonably with observations. The turbulence creates pressure waves, Rossby waves, and vortices in the $r - \phi$ plane of the disk. We demonstrate in a global simulation that these vortices tend to form out of little background noise and hypothesize that these long-lasting features possibly lead straight to the formation of planets.

We present in Section 3 a local 3D radiation hydro calculation which is close to the Stone & Balbus (1996) simulations. The flow is baroclinic in our simulation (in contrast to that of Stone & Balbus, but the small azimuthal extent of the simulation (7°) limits the size of the wavenumber to $m > 52$, which prevents the non-axisymmetric global baroclinic instability to occur. A second simulation (Section 4), which is by far more global, allows for wave numbers down to $m \geq 4$, and indeed turbulence and positive Reynolds stresses can be observed.

2. Special Features of TRAMP

2.1. Shearing Disk

The shearing-box approximation used by Hawley, Gammie, & Balbus (1995) is possible only for pseudo-Cartesian coordinate systems, which themselves strongly limit the scale of the computational domain; i.e., the box size has to be small in comparison to the local radius. In order to overcome this problem, TRAMP calculates on a spherical (r, θ, ϕ) grid; thus large portions of the disk can be simulated. In the *shearing-disk* approximation one makes two assumptions: first, that the mean values of each quantity follow a power-law scaling with the radius r (e.g., $\rho \sim r^{-\beta_\rho}$), which is usually the case in steady-state accretion disks for a certain range of radii; and second, that the fluctuations are proportional to the mean value. The radial boundary condition then reads:

$$\rho(0) = \rho(NX - 1)(r(0)/r(NX - 1))^{-\beta_\rho}. \tag{1}$$

In addition we damp the innermost and outermost five grid cells in order to prevent a resonant amplification of waves in the box.

2.2. Reynolds Averaging

We measure the $r - \phi$ component of the Reynolds stress tensor (= turbulent angular-momentum transport) and scale it by the square of the sound speed. The result is α, or, in other words, a measure of how much viscosity would be needed in order to simulate the turbulent angular momentum transport by viscous diffusion of angular momentum in a laminar flow:

$$\alpha = \frac{\langle u'_\phi (\rho u_r)' \rangle}{\bar{\rho} c_s^2} \quad \text{with} \quad \langle u'_\phi (\rho u_r)' \rangle = \langle \rho u_\phi u_r \rangle - \bar{u}_\phi \langle \rho u_r \rangle, \tag{2}$$

with

$$\langle u'_\phi (\rho u_r)' \rangle = \langle \rho u_\phi u_r \rangle - \bar{u}_\phi \langle \rho u_r \rangle, \tag{3}$$

where $\langle\ \rangle$ represents a spatial and time average.

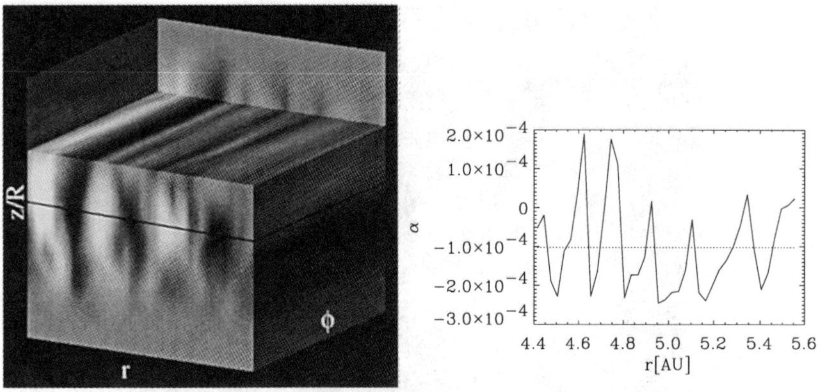

Figure 1. *Left:* Three-dimensional rendering of the vertical mass flux. *White* denotes positive velocities, and *black* denotes negative. *Right :* Vertically-, azimuthally- and time-averaged Reynolds stress over 17 orbits, measured in units of an effective alpha in a quasi-axisymmetric (small azimuthal extent) simulation.

3. Small Boxes

Starting from simulations similar to that published by Stone & Balbus (1996), we find that the *negative alpha* values produced by quasi-axisymmetric thermal convection lead to strong fluctuations in the rotational profile (without explicit viscosity); i.e., the azimuthal frequency $\Omega \sim r^{-q}$, varies as strongly as $0.0 < q < 2.5$. In addition it is a natural result that a negative entropy gradient is being created. This entropy gradient makes the disk baroclinic and hydrodynamically unstable. In the case of Stone & Balbus (1996) and Balbus, Hawley, & Stone (1996), however, there can be no global radial entropy gradient (barotropic disk) because of the shearing-box approximation, which prevents the instability from occurring. If the disk is baroclinically unstable one still needs to simulate a sufficiently large portion of azimuth (we find $\approx 15°$) in order to allow for the non-axisymmetric global instability effectively to grow.

4. Large Boxes

Extending the computational domain in radial (3.5 AU $< r <$ 6.5 AU) and azimuthal direction ($\Delta\varphi = 90^o$) shows the development of turbulent instabilities, wave phenomena, vortices, and positive Reynolds stresses. We used a $102 \times 40 \times 128$ grid.

Figure 2 shows the flow pattern and density distribution in the midplane of the disk. Figure 3 displays in addition the vorticity distribution in the midplane, and the two vortices and the Rossby waves become clearly recognizable. These vortices are possibly the seeds for planets (e.g., Tanga et al. 1996; Godon & Livio 1999; Adams & Watkins 1995). The positive Reynolds stresses can be seen in Figure 4 (*left*). Figure 4 (*right*) gives an impression of the global character of

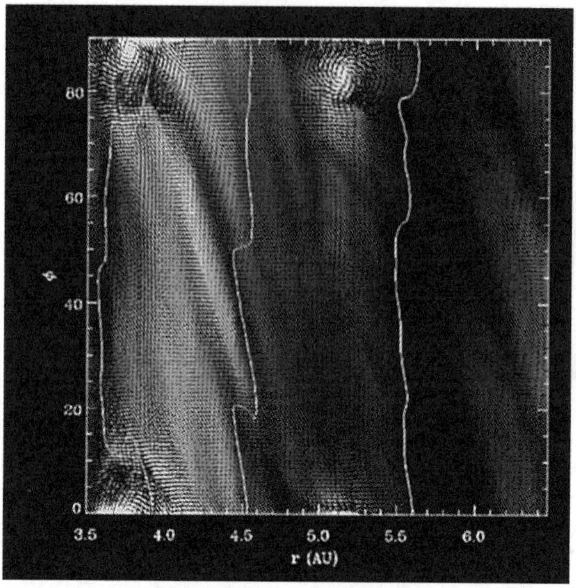

Figure 2. Flow-pattern in the midplane of the disk. Brightness indicates density. The contours are isothermal lines.

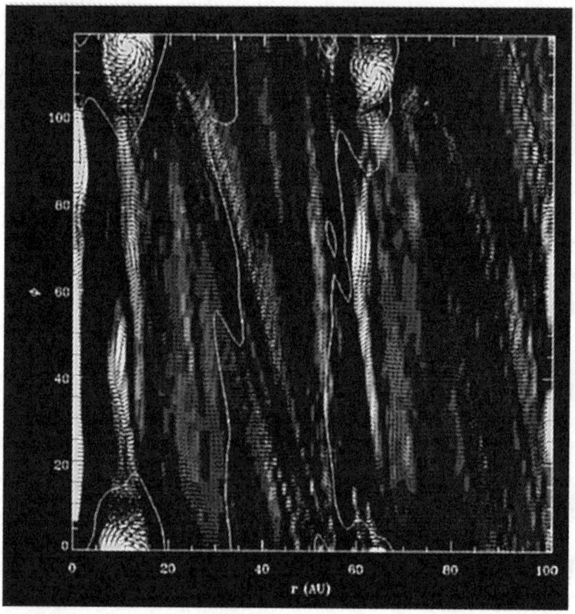

Figure 3. Vorticity (Rossby) waves. Brightness indicates local vorticity. Contours are lines of constant density.

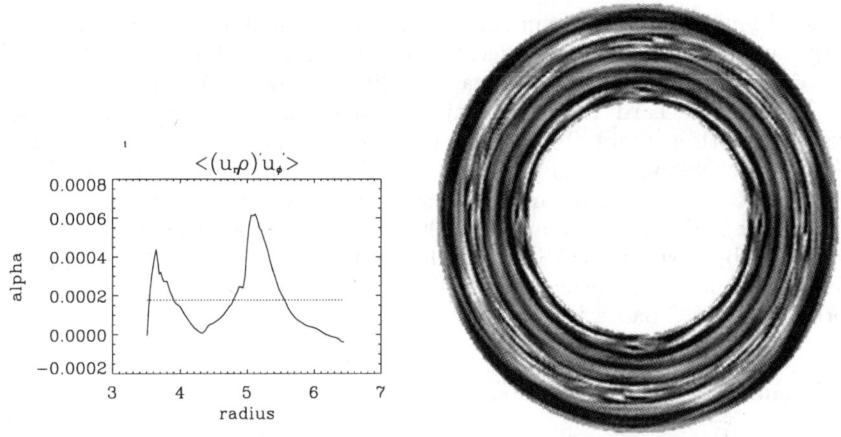

Figure 4. *Left:* Vertical-, azimuthal-, and time-averaged Reynolds stresses measured in units of an effective alpha over one orbit. *Right: White* (*black*) indicates positive (negative) convective velocities.

our simulation by plotting the shape of the convective cells.

5. Conclusions

We present a new theory of thermal convection, turbulence and angular-momentum transport especially, for the dust-rich zones of protoplanetary accretion disks. Even though neither convection nor barotropic turbulence can drive an accretion process, we show that the thermodynamics of the disk naturally produces a global entropy gradient. This leads to a global baroclinic instability resulting in geostrophic turbulence, which creates reasonably strong angular-momentum transport in the radially outward direction, allowing mass to be accreted. The Reynolds stresses we measure so far are of the order of $\langle\alpha\rangle \approx 10^{-4}$–$10^{-2}$, which is in agreement with the current interpretation of observations (Hartmann et al. 1998). The key points of the new theory are the following. Thermal convection is only indirectly related to the baroclinic instability, as convective and radiative transport are responsible for the radial temperature distribution and therefore the radial entropy gradient. An isothermal disk would always be barotropic, but we are considering only the situation where the optical depth is larger than one, so the disks are usually not isothermal. The optical depth greater than one is also necessary for thermal convection, but a vertically convectively stable purely radiative disk can also establish a radial entropy gradient. If convection is present, then it can be important in generating the initial non-axisymmetric perturbations induced in the Keplerian disk that are necessary to set off the baroclinic instability. On the other hand, convection produces negative Reynolds stresses; however, these turn out to be orders of magnitude weaker than that created by the baroclinic instability.

Acknowledgments. We want to thank Steve Balbus, Robbins Bell, Axel Brandenburg, Geoff Bryden, William Cabot, Pat Cassen, Jeff Cuzzi, Sandy

Davies, Steve Desch, Christian Fendt, Patrick Godon, John Hawley, Willy Kley, Greg Laughlin, Doug Lin, Mordecai-Mark Mac Low, Günther Rüdiger, Michal Różyczka, Jim Stone, Paolo Tanga, Caroline Terquem, David Trilling, Dotty Woolum, and Richard Young for fruitful criticism and inspirational discussions over the last three years. The results are based on calculations performed at the HLRZ (Hochleistungsrechenzentrum) in Jülich, Germany and on the NPACI grant CSC217 and CSC222 in San Diego. This research has been supported in part by the NSF through grant AST-9618548, by NASA through grant NAG5-4610, and by a special NASA astrophysics theory program which supports a joint Center for Star Formation Studies at NASA-Ames Research Center, UC Berkeley, and UC Santa Cruz.

References

Adams, F. C., & Watkins, R. 1995, ApJ, 451, 314

Balbus, S. A., Hawley, J. F., & Stone, J. M. 1996, ApJ, 467, 76

Godon, P., & Livio, M. 1999, ApJ, 523, 530

Hartmann, L., Calvet, N., Gullbring, E., & D'Alessio, P. 1998, ApJ, 495, 385

Hawley, J., Gammie, C. F., & Balbus, S. A. 1995, ApJ, 440, 742

Klahr, H. H., Henning, Th., & Kley, W. 1999, ApJ, 514, 325

Klahr, H. H., & Bodenheimer, P. 2000, ApJ, submitted

Stone, J. M., & Balbus, S. A. 1996, ApJ, 464, 364

Tanga, P., Babiano, A., Dubrulle, B., & Provenzale, A. 1996, Icarus, 121, 158

Disks, Planetesimals, and Planets
ASP Conference Series, Vol. 219, 2000
F. Garzón, C. Eiroa, D. de Winter, and T. J. Mahoney, eds.

The Orbital Evolution of Planets in Disks

Wilhelm Kley

Max-Planck-Institut für Astronomie, Königstuhl 17, D-69117 Heidelberg, Germany

Abstract. The orbital parameters of observed extrasolar planets differ strongly from those of our own Solar System. The differences include planets with high masses, small semi-major axes and large eccentricities. We performed numerical computations of embedded planets in disks and follow their mass growth and orbital evolution over several thousand periods.

We find that planets do migrate inwards on timescales of about 10^5 years on nearly circular orbits, during which they may grow up to about 5 Jupiter masses. The interaction of the disk with several planets may halt the migration process and lead to a system similar to the solar planetary system.

1. The Model

The evolution of protoplanets can only be studied simultaneously with that of the protostellar disk in which the planets are still embedded early in their lifetime. The disk is assumed to be flat and non-self-gravitating, and is modeled by the planar two-dimensional $(r - \varphi)$ Navier–Stokes equations. The mutual gravitational interaction of the planets and the star, and the gravitational torques of the disk acting on the planets and the central star are included. The time-dependent hydrodynamic equations are integrated by using a finite difference scheme (Kley 1998, 1999). The computations cover the radial range from 1.2 to 20 AU. The initial surface density distribution is axisymmetric and follows $\Sigma \propto r^{-1/2}$; to speed up the computations an annular gap is imposed (Kley 1999). The material orbits the star with Keplerian speed. The planets of one Jupiter mass are initially placed at typical distances of several AU from the star. The whole system (disk, star, and planets) is then evolved for several thousand orbital periods.

2. Results for a Single Planet

The presence of the planet perturbs the axisymmetric density by generating trailing spirals (Figure 1). In the situation where the tidal torques are greater than the internal viscous torques in the disk, and the disk response becomes non-linear, an annular gap, or surface density depression, will form at the radius of the planet. The radial width of the gap is determined by the equilibrium of gravitational (gap-opening) torques and viscous (gap-closing) torques. For a planet of one Jupiter mass ($M_{\rm Jup}$) and a typical disk viscosity of $\alpha \approx 10^{-3}$ the width of the gap is about two Roche-lobe sizes of the planet.

Although the density around the planet is greatly reduced in the gap region, material from upstream may enter the Roche lobe and eventually be accreted onto the planet (Figure 2). Typical values for the viscosity ($\alpha \approx 10^{-3}$) and mass of the disk ($M_{\rm disk} = 0.01\ M_\odot$) yield a mass accretion rate onto the planet of $4 \times 10^{-5}\ M_{\rm Jup}\ {\rm yr}^{-1}$ (Kley 1999). The rate increases with larger disk viscosity

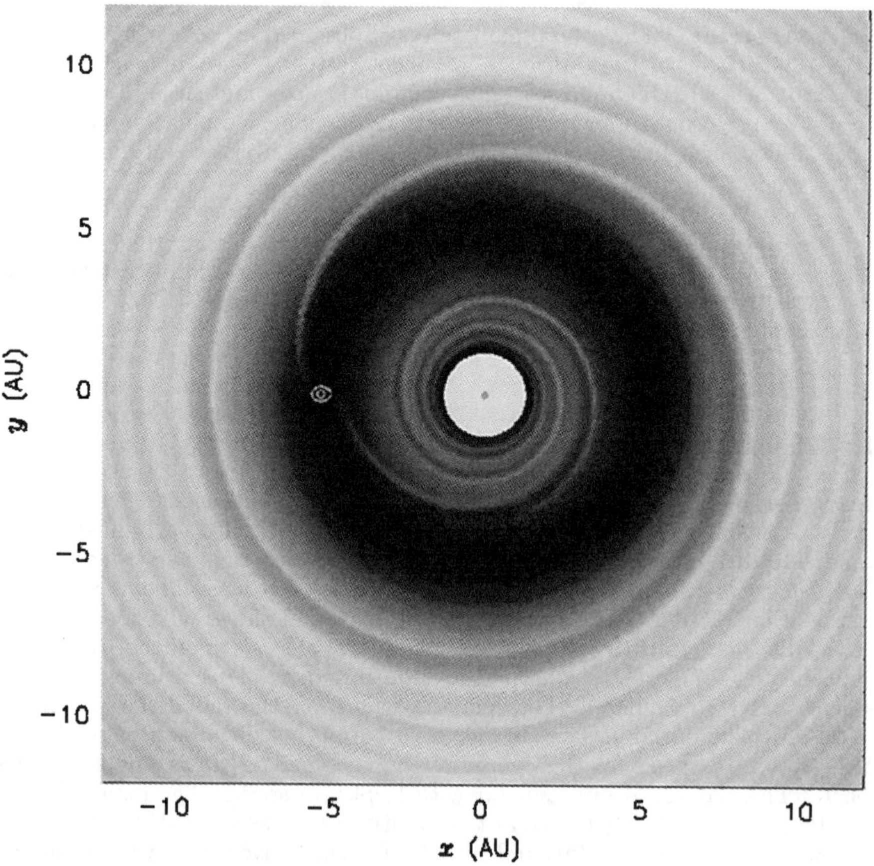

Figure 1. Gray-scale plot of the surface density after 200 orbits of the planet, which is located at $x = -5.2$, $y = 0.0$. The gravitational torques due to the planet acting on the disk lead to the excitation of spiral density waves.

and decreases when the mass of the planet grows. For very low disk viscosity and larger planetary masses the mass accumulation eventually terminates, and the maximum mass the planet may reach is about 5 $M_{\rm Jup}$; see also Bryden et al. (1999) and Nelson et al. (2000).

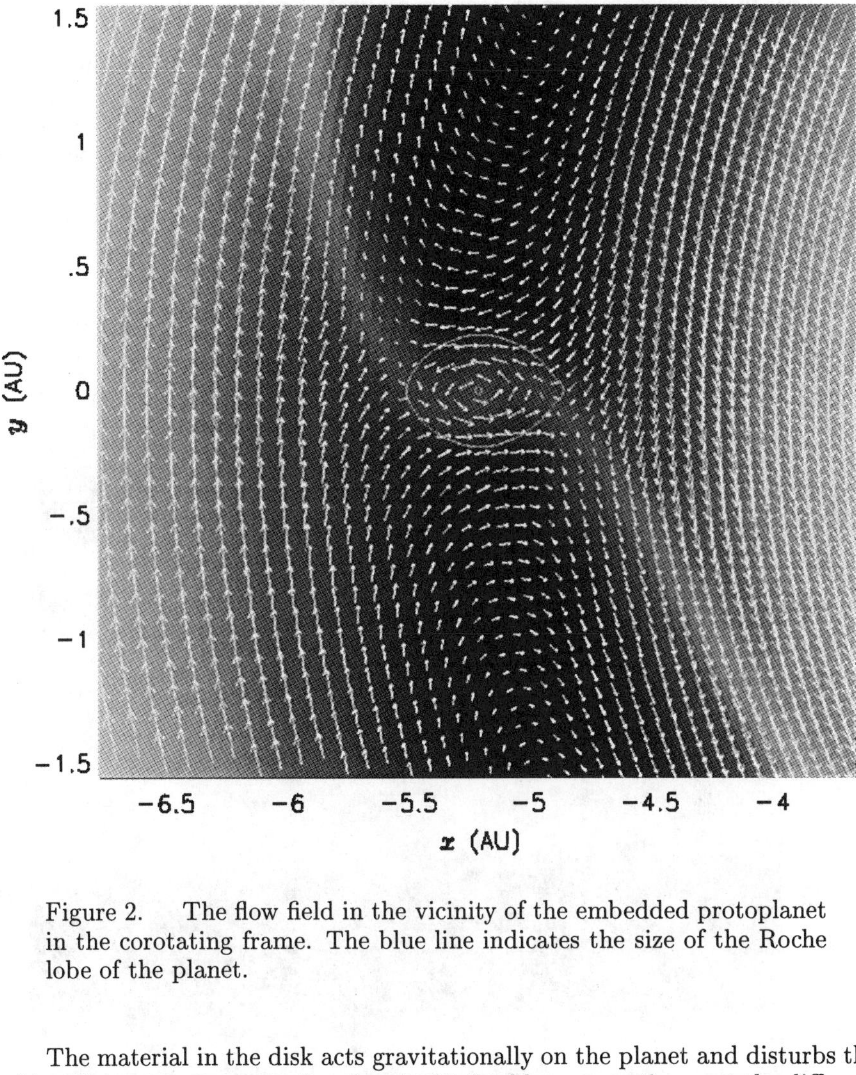

Figure 2. The flow field in the vicinity of the embedded protoplanet in the corotating frame. The blue line indicates the size of the Roche lobe of the planet.

The material in the disk acts gravitationally on the planet and disturbs the orbit. The net torque onto the planet, obtained by summation over the different annuli, is negative, which leads to an inward migration on typical timescales of 10^5 yr (Nelson et al. 2000).

3. Evolution of Two Planets

As it is believed that more eccentric orbits are caused by the interaction of several objects, we have performed a run with two embedded Jupiter-type planets. Initially they were located at $1a_{\rm Jup}$ and $2a_{\rm Jup}$ in opposition to each other, i.e., separated by $\Delta\varphi = 180°$. The two planets were assumed to be in circular orbits initially.

Two planets create a much more complicated pattern of wave-like disturbances in the density distribution of the disk than just one planet does. In a calculation of one planet in a fixed circular orbit, the wave pattern induced in the disk is stationary in the corotating frame. As seen in Figure 3, in the case of two planets (the Roche lobes are indicated) it changes strongly with time. As both planets accrete essentially all of the mass which enters their Roche lobes, their masses grow in time. The material initially located radially between the planets is consumed by the planets within the first few hundred orbital periods.

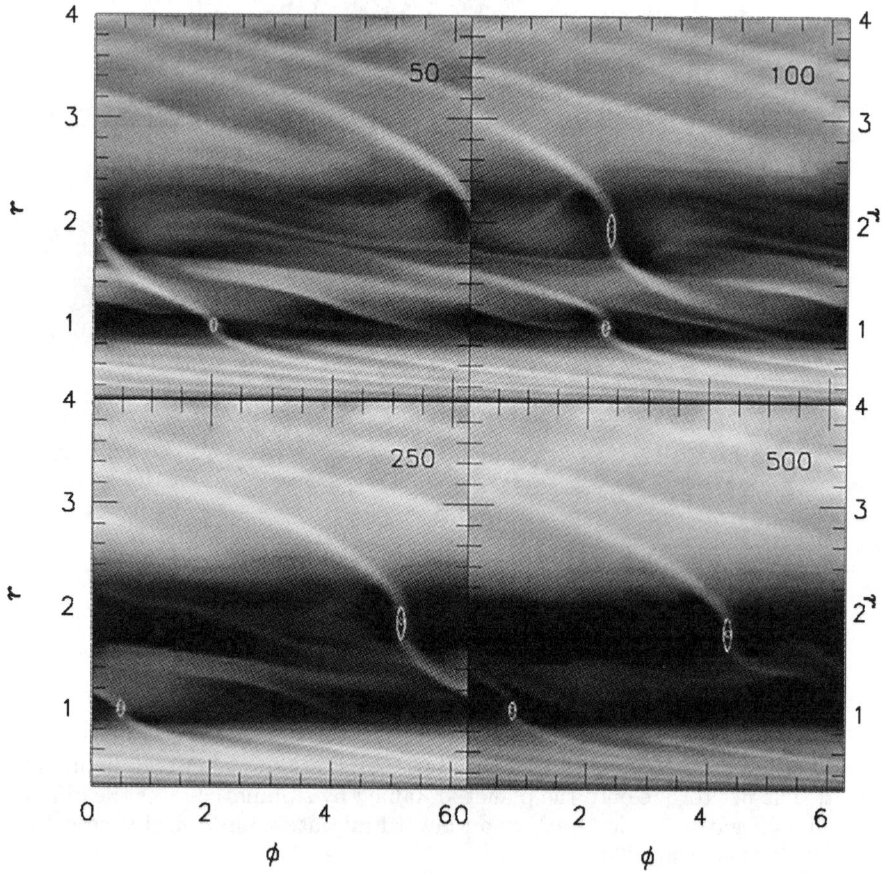

Figure 3. Surface density distribution in r–φ display at four evolutionary times. The panels are labeled consecutively by the number of orbits of the inner planet.

Finally, for the inner planet, the only gas available is that which flows through the gap created by the outer planet. Hence the mass of the inner planet grows at a smaller rate than the outer one, see Figure 4 (*bottom*). At the same time, the gravitational interaction of the two planets, the star, and the disk leads to changes in the orbital elements of the objects. The change in the semi-major axis of each planet is given in Figure 4 (top). During the evolution, the fluctuations of the radial distances to the star strongly increase, indicating a growth in eccentricity of the planet's orbit.

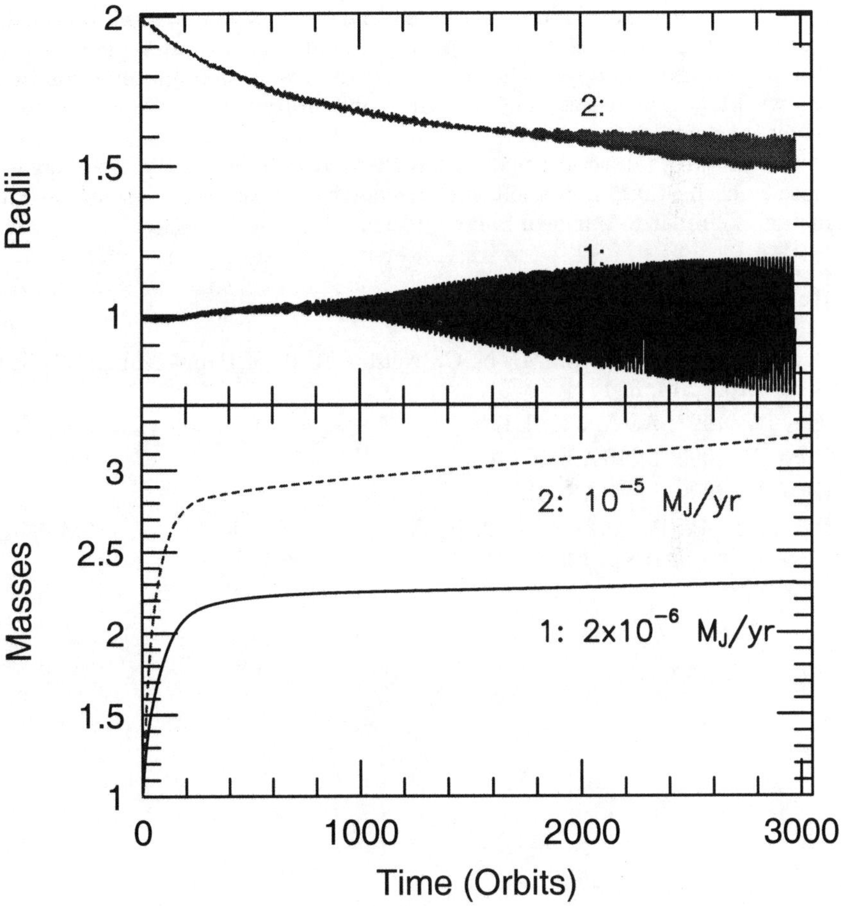

Figure 4. *Top:* Evolution of the semi-major axis of the two planets. *Bottom:* Evolution of the masses of the two planets (*1:* inner planet, *2:* outer planet). The inferred mass accretion rates during the long term evolution are indicated. Time is given in units of the initial period of the inner planet.

4. Conclusions

By initially considering one planet in a fixed circular orbit, it was shown that even though the planet opens up an annular gap in disk it is nevertheless able to accrete more mass from its surroundings. As more massive planets induce a wider and deeper gap the mass accumulation essentially terminates, which puts the upper limit to the mass of the planet mass at about 5 to 10 $M_{\rm Jup}$, in good agreement with the observations.

By considering the evolution of a system consisting of two Jupiter-type planets, we show (Kley 2000) that by mutual gravitational interaction the inward motion of the inner planet may come to a halt. As the outer planet continues to move inward and approaches the inner, the gravitational interaction between the two planets increases. The resulting configuration may be unstable, and lead to systems similar to υ And.

If the protoplanetary nebula has dissipated before the planets come very close to each other, one is left with a system of massive planets at several AU distance, similar to our own Solar System.

References

Bryden, G., Chen, X., Lin, D. N. C., Nelson, R. P., & Papaloizou, J. C. B. 1999, ApJ, 514, 344

Kley, W. 1998, A&A, 338, L37

Kley, W. 1999, MNRAS, 303, 696

Kley, W. 2000, MNRAS, 313, 47

Nelson, R. P., Papaloizou, J. C. B., Masset, F., & Kley, W. 2000, MNRAS, in press (astro-ph/9909486)

Disks, Planetesimals, and Planets
ASP Conference Series, Vol. 219, 2000
F. Garzón, C. Eiroa, D. de Winter, and T. J. Mahoney, eds.

FARGO: a Fast Eulerian Transport Algorithm for Differentially Rotating Disks

Frédéric S. Masset

Astronomy Unit, School of Mathematical Sciences, Queen Mary & Westfield College, 327 Mile End Road, London E1 4NS, UK

Abstract. I present an efficient and simple modification of the standard transport algorithm used in explicit Eulerian fixed polar grid codes, aimed at getting rid of the average azimuthal velocity when applying the CFL condition. This results in a much larger time-step than the usual procedure and is particularly well suited to describing a Keplerian disk, where one is traditionally limited by the very demanding CFL condition on the fast orbital motion at the inner boundary. In this modified algorithm, the time-step is limited by the perturbed velocity and the differential-rotational shear. The speed-up resulting from the use of the FARGO ("Fast Advection in Rotating Gaseous Objects") algorithm is problem-dependent. It is particularly efficient at describing the evolution of a low mass companion embedded in a gaseous disk. Indeed in the example presented here, which shows the evolution of a Saturn-sized protoplanet embedded in a minimum-mass protoplanetary disk, the FARGO algorithm is more than thirty times faster than a traditional transport scheme, with a much smaller numerical diffusivity. The accuracy of the algorithm for the embedded protoplanet case is also discussed.

1. Introduction

We want to model the hydrodynamical (HD) evolution of a disk on a fixed polar Eulerian grid. For the sake of simplicity we consider only a two-dimensional non-magnetized Keplerian disk, but the algorithm can be extended with little additional effort to any gaseous thin or thick disk in differential rotation. Usually the time-step is limited by the Courant–Friedrich–Levy (CFL) condition at the inner boundary, where the motion is fast. Indeed, the ratio of the distance swept by the material in one time-step to the zone width (called hereafter the CFL ratio) must be lower than unity over the whole grid. Since in most cases the region of interest is located much further than the grid inner boundary, the CFL ratio in the region of interest is much smaller than unity, which corresponds to a waste of computing time, and, as we are going to see below, to a higher numerical diffusivity along the orbital motion. I describe hereafter a simple method which enables one to work on a fixed polar grid and to get rid of the CFL condition on the average azimuthal velocity at each radius.

2. Notations and Standard Method

We consider a polar grid composed of N_s sectors and N_r rings The radial velocity is denoted v^r_{ij} and the azimuthal velocity is denoted v^θ_{ij}. Throughout this proceeding the algebra on the j coordinate is meant in $Z/N_s Z$ to account for the periodicity in azimuth). Usually the time-step is split in two parts (Stone & Norman 1992). The first part is composed of Eulerian substeps which update the HD quantities through the source terms in the evolution equations, and which include all the physical processes at work: pressure, gravity, viscosity, etc. The second part is the transport substep, in which the quantities are conservatively moved through the grid according to the flow $[(v^r_{ij}), (v^\theta_{ij})]$. In the example shown hereafter, the van Leer (1977) transport method is used, although other schemes (e.g., PPA) could be used as well. The strongest restriction on the time-step in the case of a thin Keplerian disk not too strongly perturbed is given by the azimuthal phase of the transport substep.

3. Modified Azimuthal Transport Algorithm

3.1. Overview

Let us take as an example the angular-momentum conservation equation:

$$\frac{\partial J}{\partial t} + \underbrace{\frac{1}{r}\frac{\partial(v^\theta J)}{\partial \theta}}_{\text{azim. transport}} + \underbrace{\frac{1}{r}\frac{\partial(r v^r J)}{\partial r}}_{\text{rad. transport}} = \text{source terms}, \tag{1}$$

where $J = \rho r v^\theta$. The transport equation of any HD quantity ξ will look the same as the LHS of eq. (1).

Now without loss of generality we can rewrite Eq. (1) as:

$$\frac{\partial J}{\partial t} + \frac{1}{r}\frac{\partial[(v^\theta - u)J]}{\partial \theta} + \frac{u}{r}\frac{\partial J}{\partial \theta} + \frac{1}{r}\frac{\partial(r v^r J)}{\partial r} = \text{source terms} \tag{2}$$

where u can be any quantity which does not depend on θ. No assumption has been made on the behavior of J up to this point, and Eqs. (1) and (2) are strictly equivalent. If we take u to be the average azimuthal velocity, $\overline{v}^\theta$, then Eq. (2) can be described as a composition of different steps, and each of them can be worked out independently with the well known operator-splitting technique: (i) a source step, (ii) a radial transport step, (iii) an azimuthal transport step with the velocity $v^\theta - \overline{v}^\theta$, which we are going to call the azimuthal *residual* velocity, and (iv) an additional step which corresponds to the following PDE:

$$\frac{\partial J}{\partial t} + \frac{\overline{v}^\theta}{r}\frac{\partial J}{\partial \theta} = 0. \tag{3}$$

It is an easy matter to check that the solution of this last equation can be written in a general way as:

$$J(\theta, t) = J\left(\theta - \frac{\int_0^t \overline{v}^\theta dt}{r}, 0\right), \tag{4}$$

which means that the solution of this equation at any time t looks like the initial profile ($t = 0$), except for a shift of $-\int_0^t \overline{v}^\theta dt/r$ in azimuth. It should be noted that this is true *whatever the profile of* J, which can even contain discontinuities (i.e., shocks). In particular, *no assumption has to be made* on the linearity of the flow (i.e., on the relative amplitude of the perturbed quantities). The idea on which FARGO is based is precisely to evolve the HD quantities through operators which mimic in a discrete way the different terms of Equation (2). The source step, the radial transport step and the residual azimuthal velocity transport step are performed in a standard way (e.g., Stone & Norman 1992). Now the last step in the operator-splitting described above, which corresponds to a simple shift amounting to $\overline{v}^\theta \Delta t/r$ in one time-step, can be implemented in such a way that the matter can sweep an arbitrary number of zone widths in one time-step. A detailed description of the scheme is given in Masset (2000).

3.2. Time-Step Limit

In the standard transport method, the time-step limitation arises from the combination of four different constraints (see, for example, Stone & Norman 1992); namely, the fact that a flow advected test particle in zone $[i, j]$ should not sweep a distance longer than the shortest zone size in one time-step and that the wavefront of any wave present in the system should not travel across more than one zone over one time-step (Richtmyer & Morton 1957). With the modified azimuthal transport algorithm, the constraint on the azimuthal motion has to be modified slightly. Following Stone & Norman's notation, instead of writing $\delta t_3^{ij} = \Delta y_i / v_{ij}^{\theta a}$, we write :

$$\delta t_3^{ij} = \frac{\Delta y_i}{v_{ij}^{\theta a} - \overline{v}_i^\theta} = \frac{\Delta y_i}{v_{ij}^{\theta \mathrm{res}}}, \tag{5}$$

which means that the time-step limitation comes now from the perturbed azimuthal velocity, which results in a much higher absolute value of δt_3. Another limitation arises from the shear. Indeed we do not want the shear to disconnect the two neighboring flow advected test particles in zones $[i, j]$ and $[i + 1, j]$ after one time-step. We write this condition as :

$$\delta t_{\mathrm{shear}}^{ij} = \frac{1}{2}\left(\frac{v_{ij}^{\theta a}}{\Delta y_i} - \frac{v_{i+1j}^{\theta a}}{\Delta y_{i+1}}\right)^{-1}. \tag{6}$$

Following Stone & Norman's notations, we finally adopt:

$$\Delta t = C_0/\{\max_{ij}[(\delta t_1^{ij})^{-2} + (\delta t_2^{ij})^{-2} + (\delta t_3^{ij})^{-2} + (\delta t_4^{ij})^{-2} + (\delta t_{\mathrm{shear}}^{ij})^{-2}]^{1/2}\}. \tag{7}$$

4. Two-Dimensional Example: the Embedded Protoplanet Case

4.1. Example Runs

We present a result obtained from a run performed with a one-solar-mass primary and a one-Saturn-mass protoplanet on a fixed circular orbit at $r_0 = 10$ AU embedded in a standard protoplanetary nebula (see, for example, Masset 2000).

The grid has an inner radius at 2 AU and an outer radius at 25 AU. The radial spacing is geometric (log-grid), with $N_r = 201$ and $N_s = 500$. The grid is fixed in a non-Galilean non-rotating frame centered on the primary. Its outer boundary is rigid and its inner boundary allows outflow but no inflow. The disk aspect ratio is set to $3 \cdot 10^{-2}$ everywhere. The planet potential is smoothed on a length scale which amounts to 40% of its Hill radius. In Eq. (7) we choose $C_0 = 0.5$.

We plot in Figure 1 the disk surface density after one Saturn orbit. We clearly see that the disk response in the FARGO and the standard cases are extremely similar, even in the innermost parts of the disk, and that the immediate vicinity of the planet exhibits much better behavior in the FARGO case.

4.2. Is There a "TOOFARGO" Problem in the Embedded Protoplanet Case?

One might believe that the time-step limitation is not restrictive enough in the embedded planet case, since the wave pattern (for a circular orbit) corotates with the planet and its associated tidal potential, the velocity of which, in the local frame, is $r|\Omega_K - \Omega_p|$, where "K" and "p" refer to "Keplerian" and "planet" rotation, respectively. Therefore, this velocity can be highly supersonic, and one could a priori imagine a situation in which a flow-advected test particle is located just before the shockfront at the beginning of a time-step and far beyond the shock region at the end of the time-step; this flow-advected test particle would "miss" the shock region and the disk response would probably be strongly affected in such a case. However, this kind of situation is strictly forbidden by the time-step limits provided by Eqs. (5)–(7). Indeed, the normal velocity after a shockfront is always subsonic, and the time-step limit of Eq. (5), along with the standard radial motion and sound-speed limits (respectively δt_2 and δt_1 in Stone & Norman 1992), does take that into account and prevents the flow-advected test particle mentioned above from being further than one zone size from the shockfront after one time-step. Now the shockfront is tightly wound far from the planet in a small-aspect-ratio disk. This tight winding of the shock spreads it geometrically (by a factor $1/\sin\alpha$, where α is the pitch angle) over a large number, N, of zones in azimuth. In any case, N/C_0 turns out to be larger than the speed-up ratio achieved by the use of FARGO; hence, the flow-advected test particle can sweep a large number of zones in azimuth (the transverse velocity is conserved across a shock and can be supersonic) and still will not "miss" the shock. This can be seen in a different way in the linear limit where the disk response is the sum over spiral density waves excited at all m. The group velocity of these waves can be inferred from their dispersion relation and shown to be of the order of the sound speed, so that the time-step limit provided by δt_1 is correct. Now the last concern has to do with the fast-moving external perturber; i.e., the tidal potential, which indeed moves at a high velocity w.r.t. the matter. No CFL criterion has to be applied to the external perturber to ensure stability. However, all the m-components of the perturber need to be resolved for the system response to be accurate. The condition for this reads $2\pi/m \gg \Delta t|\Omega_p - \Omega_K^{ex}|$, where Ω_K^{ex} is the orbital frequency at the radius at which the m-folded spiral density wave is launched. This condition simply means that the perturber must sweep out a distance much smaller than its characteristic size $2\pi/m$ in one time-step. If we replace Ω_θ^{ex} by the orbital frequency of the

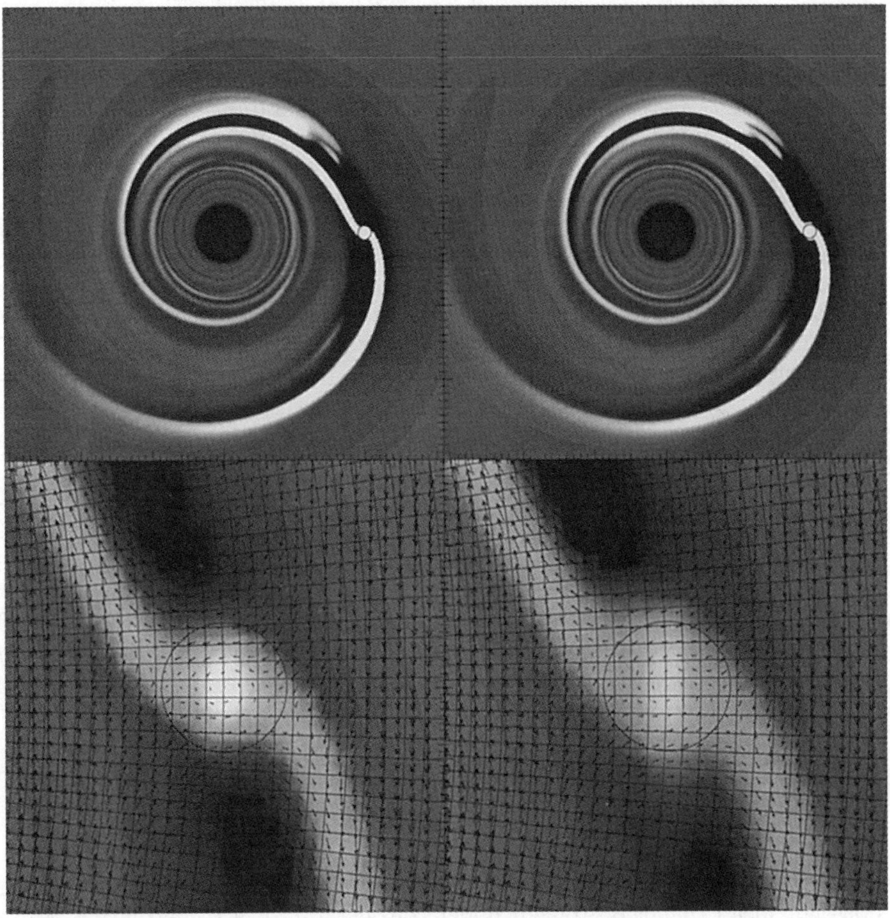

Figure 1. FARGO (*left column*) vs. standard algorithm (*right column*). The gray-scale intensity was scaled linearly between $0.85\Sigma_0$ (Σ_0 being the initial uniform surface density) and $1.15\Sigma_0$ in the *top row* plots in order to stress any difference between the two maps. Another gray-scale intensity coding is applied for the *bottom rows* which is the same for both close-ups. We clearly see the wake triggered by the planet in the disk, as well as transient features. One can easily spot the higher orbital diffusivity of the standard advection procedure in the *bottom-right* plot. A run performed in the planet corotating frame gives a result that looks very much like the FARGO results. The FARGO scheme for these results turned out to be 33 times faster than the standard method. This amounts to be $r/H = v_{\rm K}/c_{\rm s}$, and is therefore the maximum speed-up ratio that can be expected. It is actually reached here because the planet mass is small enough and the inner boundary has a much smaller radius than the perturbed region (and hence the limiting region is still close to the inner boundary).

associated mth inner or outer Lindblad resonance, we get: $\Omega_p \Delta t \ll 2\pi$ (i.e., the planet must describe less than one orbit in one time-step), a condition which is obviously always fulfilled. Even if the high-m Lindblad resonance shift due to pressure effects is taken into account, one can check that the sound-speed time-step limit, δt_1 (following Stone & Norman's notation) will allow any m-component to be correctly resolved, whatever the value of m.

From the above one can conclude that there is no "TOOFARGO" problem in the embedded planet case, and that FARGO does provide a reliable description of the disk response to the planet, which is in turn suitable for a torque evaluation. More than one year of hands-on daily experience in our group with widely varying disk parameters and various codes, all of which can be run in standard or FARGO mode, have confirmed the excellent agreement between the different methods, and have validated the suitability and accuracy of FARGO for the embedded-companion problem.

5. Conclusion

The FARGO algorithm for azimuthal transport turns out to be able to speed up by an order of magnitude or even more the numerical simulation of a differentially rotating disk, with a smaller numerical orbital diffusivity than the usual transport algorithm. It has been validated by many tests on the embedded protoplanet problem. It must be used with a good understanding of the physical processes at work in the system. In particular, the time-step given by Eq. (7) must be short compared to all the physical timescales relevant for the system. In the case we have presented this is automatically ensured by Eqs. (5)–(7), but if additional physics is to be added (magnetic field, radiative transfer, etc.), the time-step limit needs to be carefully worked out. Furthermore, no advantage is gained in using FARGO in problems where the perturbed velocity is comparable to the rest velocity. On the other hand, the FARGO algorithm appears to be very well suited to all the cases where the perturbed velocities in any differentially rotating disk are small compared to the unperturbed velocities (which does not mean that the problem under consideration has to be linear). More generally this scheme can be used to describe the HD evolution of any sheared fluid on a fixed orthogonal Eulerian grid.

Acknowledgments. I thank M. D. Snellgrove for a careful reading of this proceeding. This work was done in the research network "Accretion onto black holes, compact stars and protostars" funded by the European Commission under contract number ERBFMRX-CT98-0195. Computational resources of the Grand HPC consortium were available and are gratefully acknowledged.

References

Masset, F. S. 2000, A&AS, 141, 165

Richtmyer, R. D., & Morton, R. W. 1957, Difference Methods for Initial-Value Problems, 2nd. edn. (New York: Wiley Interscience)

Stone, J. M., & Norman, M. L. 1992, ApJ, 80, 753

van Leer, B. 1977, J. Comput. Phys., 23, 276

Part 2
Pre-Main-Sequence Stars

Disks, Planetesimals, and Planets
ASP Conference Series, Vol. 219, 2000
F. Garzón, C. Eiroa, D. de Winter, and T. J. Mahoney, eds.

Observational Aspects of Protostellar Disks

Paul T. P. Ho
Smithsonian Astrophysical Observatory, 60 Garden Street, Cambridge, MA 02138, USA

Abstract. Protostellar disks span the size scale from about 100 AU to maybe 10 000 AU. On the large scale, millimeter-wavelength interferometers have been successful in resolving the structures and the kinematics of the circumstellar cloud cores and disks. In particular, spectroscopy has revealed evidence for infall, outflow, and rotation. On the smaller scales, imaging becomes more difficult, although both *HST* and centimeter-wavelength interferometry have been able to see the inner dusty circumstellar disks. However, because of the low luminosity of such inner disks, kinematics and mass determinations remain very difficult. Recently, a new way to study these inner disks has become possible through proper-motion studies of water masers associated with the disks. In this paper, we will review the observational status of these various studies.

1. Introduction

The study of protostellar disks is important for a number of reasons. In the standard picture of star formation, cloud contraction proceeds to the formation of flattened structures because of the conservation of at least part of the angular momentum. Dissipation processes within circumstellar disks allow material to accumulate onto the protostar. The motions within the disks, the infall rate, the mass and size of the disks, the density and temperature within the disks, are all important constraints on theoretical models of star formation. Furthermore, with the rapid pace of the detection of exoplanets, we are reminded that circumstellar disks will eventually evolve into protoplanetary disks. The structures and motions within such protoplanetary disks may provide important hints on the formation and evolution of planetary bodies. The techniques which are being developed to examine the more massive circumstellar disks may become applicable for examining the protoplanetary disks as well.

2. Existence of Circumstellar Disks

The earliest suggestions on the presence of circumstellar dust structures came in the form of infrared and millimeter-wavelength excesses in the spectral energy distributions observed toward young stars. A cooler component was found to accompany the much hotter stellar radiation. However, without imaging, the spatial structure of this excess emission was open to debate. Finally, with the *HST*, circumstellar dust disks were directly seen in absorption against the Orion

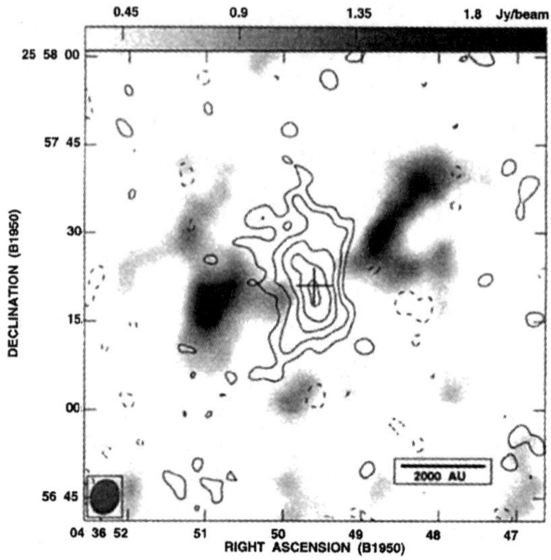

Figure 1. Integrated $^{12}C^{18}O$ emission in contour superposed on ^{12}CO molecular outflow in gray-scale (Ohashi et al. 1997a). The optically thin rarer isotope can pick out the flattened envelope of gas surrounding the central exciting star, while the higher-velocity wings in the main isotope are dominated by the outflow.

Nebula. Here, there can be no doubt (cf. McCaughrean et al. 1998). Spectacular structures, from elliptical to edge-on disks, some even with embedded stars, were detected. Of couse, while the dust emission (and extinction) were able to define the morphology of the disks, only the gas component holds the potential for studying the kinematics via spectroscopy. In this regard, great progress has been made via millimeter-wavelength interferometry. Observations of the continuum and line emission have been very successful toward the larger circumstellar disks, in defining the spatial structures and the accompanying motions. It is now clear that circumstellar disks are common, span a range of size scales and properties, and are flattened and rotating structures, just as predicted by theory.

3. Structures at 1000 AU

The nominal size scale of a circumstellar disk of 100 AU, located in a nearby low-mass star-forming region like Taurus at 100 pc, would subtend 1″. This small angular size scale, coupled with a typical temperature on the order of 100 K, implies a pretty faint emission signal. This is the reason why progress has been made mostly for the larger structures.

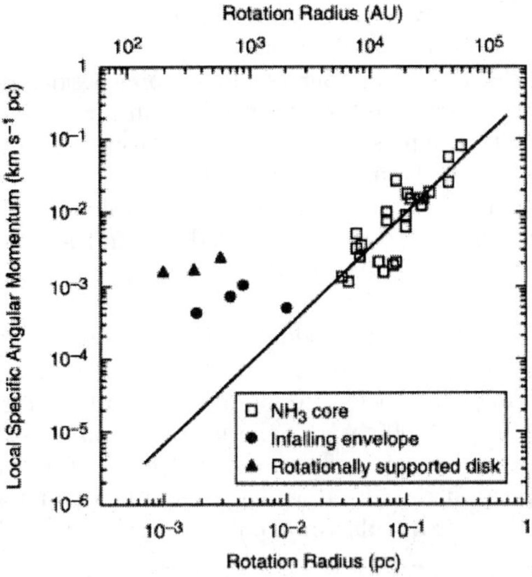

Figure 2. Specific angular momentum plot from Ohashi et al. (1997b). The local specific angular momentum seems to vary with radius until about 0.03 pc (6000 AU). At that scale, the specific angular momentum seems to flatten at the value of 10^{-3} km s^{-1} pc. This constant value seems to imply a size scale at which either dynamical collapse occurs or the disks become rotationally supported.

One of the earliest and best studied system is the disk associated with HL Tauri. Three groups independently imaged the region in ^{13}CO emission with different millimeter-wavelength interferometers (Sargent & Beckwith 1991; Hayashi, Ohashi, & Miyama 1993; Cabrit et al. 1996). There is agreement with the basic observational results. A 2000 AU disk is seen perpendicular to the optical jet. A velocity gradient is seen along the major axis of the flattened structure suggesting rotation. Motions are also seen along the minor axis parallel to the jet axis, with the same sense as the outflow seen in the optical jets. The geometry of the jet and the supposedly perpendicular disk, imposes constraints on the interpretation of the motions. Since the north side of the jet is blueshifted, it is tipped away from the plane of the sky toward us. So if the disk is perpendicular to the jet, then the south side of the disk must be in the foreground and nearer to us. Here, the interpretations diverge. If the motions along the minor axis represent motions within the disk, then their blueshifts toward the north and redshifts toward the south suggest that the disk is contracting. However, if the minor-axis motions are outside the disk, then they could be simply part of the outflow. This difficulty is at once the conundrum of this field while holding forth the promise of revealing the underlying physics. It should be noted that the use of ^{13}CO emission to study these regions was explicitly for the purpose of suppressing the contamination by outflows which would dominate the main ^{12}CO isotope emission. The larger motions of the outflow typically overwhelm the gravitationally bound motions. However, even in the rarer isotopes, we continue to worry about the contribution from the lower-velocity components of the outflows close to the central exciting sources.

Over the past decade, a number of cloud cores associated with low-mass stars have been imaged with millimeter-wavelength interferometers. While the early work was in ^{13}CO, the most recent work has been concentrating on the still rarer isotope $^{12}C^{18}O$ (e.g., Ohashi et al. 1997a). This line is almost certainly optically thin. The emission structures continue to appear to be flattened, while both infall and rotational motions are detected. Since the size scales are still at the 2000 AU level, and the structures do not appear thin, these may be more like flattened envelopes rather than disks. The real circumstellar disks are ten times smaller and are probably embedded within these envelopes.

Perhaps one of the most interesting results has been the observation that the specific angular momentum for cloud cores seems to decrease with the size scale until about 0.03 pc (6000 AU), where it remains constant at a level of about 10^{-3} km s^{-1} pc down to size scales of 200 AU (Ohashi et al. 1997b). This suggests a typical size scale at which dynamical collapse proceeds, with conservation of the specific angular momentum.

4. Circumstellar Disks at 100 AU

We have discussed already the direct imaging of dusty disks via their silhouettes against the bright nebular emission of Orion. It has now also become possible to directly image the disk emission itself. In the centimeter-wavelength range, we are in the Rayleigh–Jeans part of the blackbody curve for the dust emission. The spectral index of the dust emission is between 2 and 4, depending on the dust properties. Hence the contribution of the dust is rising very rapidly with frequency as compared to that from any stellar-wind component, which typically has a spectral index of 0.6, or from free–free emission, which is typically optically thin and has a spectral index of –0.1. Furthermore, at 7 mm, the Very Large Array operated by the National Radio Astronomy Observatory can achieve a maximum angular resolution of 0.04″. This corresponds to a spatial resolution of 5 AU at the distance of Taurus. A very successful series of experiments over the past few years has imaged the circumstellar disks around a number of nearby stars. These included the HL Tauri source, where a 100 AU disk was resolved and distinguished from the central 10 AU jet (Wilner, Ho, and Rodríguez 1996). Another interesting result was the imaging of the binary disks in L1551 (Rodríguez et al. 1998). The small sizes of these disks, 50 AU and their close projected separation of 100 AU suggest that gravitational interactions might have truncated the formation of the outer disks. A 100 AU disk was also imaged at 7 mm around TW Hya, which is only 50 pc away. The great advantage of the radio continuum interferometry is the high angular resolution. Good constraints can be made on the density distributions within these disks by these imaging studies. Luis Rodriguez will discuss these studies in much greater detail in his paper (this volume, p. 108).

The dust emission continues to get stronger as we go to shorter wavelengths. Successful experiments have now been made at 3 mm with the OVRO millimeter-wave interferometer, and at 1 mm with the IRAM interferometer. Disks can be seen toward a lot of the T Tauri systems. In particular, the spectacular circumbinary ring seen toward GG Tauri, with a radius of about 200 AU, can be seen via its dust emission at 1 mm (Guilloteau, Dutrey, and Simon 1999). Here, the evacuated central part of the disk may suggest the dynamical effects of the central stars. Anne Dutrey discusses more on this particular source (this volume, p. 128), while Gaspard Duchene has a poster paper on circumbinary disks (this volume, p. 169).

5. Disk Kinematics at 100 AU

Studying the kinematics at the scale of 100 AU and below, remains a most challenging problem. For angular resolutions under an arcsecond and velocity resolutions under 1 km s^{-1} brightness temperature sensitivities are typically on the order of a few K with current millimeter-wavelength interferometers. To resolve the kinematics at smaller spatial scales, we need a large increase in sensitivity, which will be available only when the international ALMA project is completed in the next decade, as will be discussed by Stephane Guilloteau (this volume, p. 645).

In the meantime, maser emission offers an unique tool to probe the kinematics of disks on the AU scale. The non-thermal nature of interstellar masers means that the brightness temperatures can be extraordinarily high, easily exceeding 10^9 K. With such high brightness temperatures, very long baseline interferometry (VLBI) can be used to achieve angular resolutions less than 1 mas. That is enough to see proper motions of 10 km s^{-1} at a distance of 100 pc on a timescale of one month. This will allow us to track motions of the gas within the disk in order to distinguish whether rotation, contraction, or outflow, is dominant within the disk.

5.1. CH_3OH Masers

Over the past few years, a series of experiments on the 6.7 GHz CH_3OH masers, using the Australian Telescope Compact Array, showed that the masers appear to lie in a straight line, often projected across the face of a compact H II region. Furthermore, the masers show a systematic velocity gradient across its linear spatial distribution. The typical size scale is about 1000 AU, and the typical range of motions is on the order of 10 km $^{-1}$. If such a morphology and kinematics can be interpreted as a rotating disk, the mass required to gravitationally bind the motions would be on the order of 100 solar masses. This could be reasonable if we are seeing the disks around fairly massive stars. VLBI measurements of the motions of these CH_3OH masers would be very useful, but they are not yet available.

5.2. H_2O Masers

It is well known that H_2O masers are closely associated with the early stages of star formation. The nature of this association has become clearer as higher angular resolution imaging has become available. By using the H_2O maser line at 1 cm as a phase reference, the associated continuum emission of the young star can be imaged with the VLA at a resolution of better than $0.1''$. The achievement of such high angular resolution depends on the ability to calibrate the atmospheric

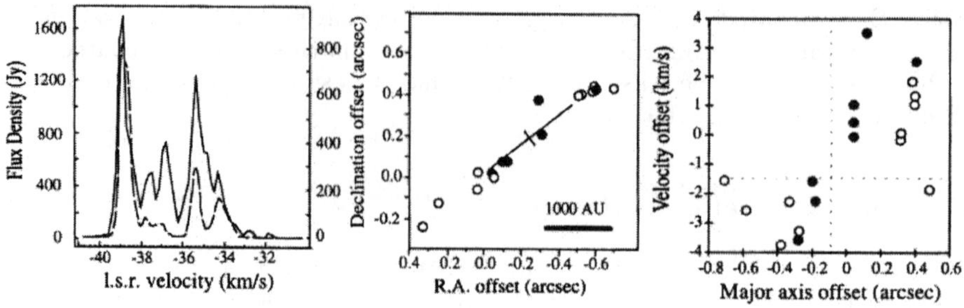

Figure 3. CH_3OH data towards G339.88-1.26 from Norris et al. (1998). *Left:* CH_3OH spectra. *Middle:* Relative positions of the maser spots. *Right:* Position-velocity diagram of the maser emission. This figure demonstrates the linear spatial distribution of the masers and a linear velocity gradient. The interpretation is a rotating disk around the central H II region.

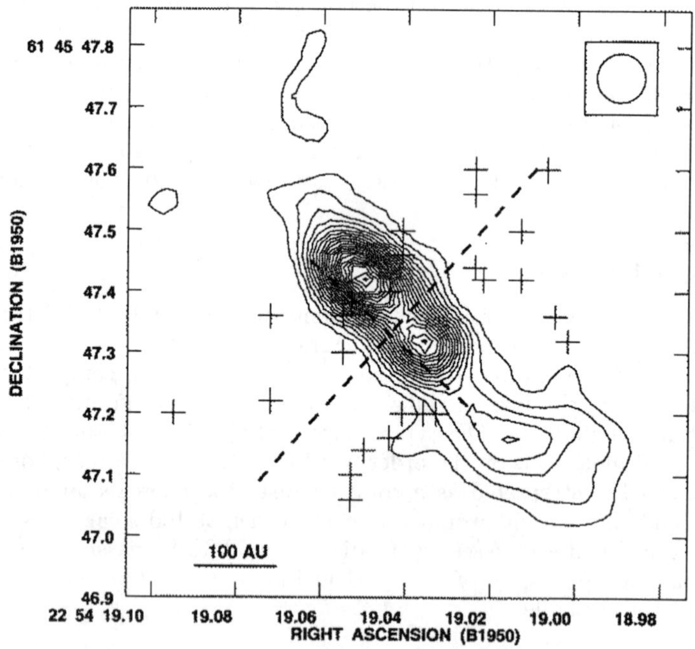

Figure 4. The spatial distribution of H_2O masers in Cepheus A (Torrelles et al. 1996). Note that the masers appear to be distributed perpendicularly to the radio continuum jet depicted in contours. Note that there is a certain thickness to the disk of H_2O masers. VLBI measurements are under way to track the proper motions of these masers.

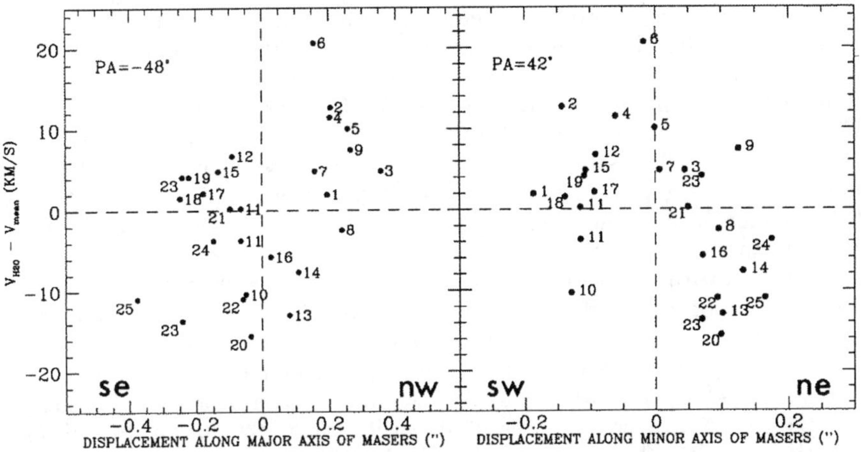

Figure 5. Position–velocity plots of H_2O masers in Cepheus A along the major and minor axes of the disk. There appear to be motions both along the continuum jet and perpendicular to it.

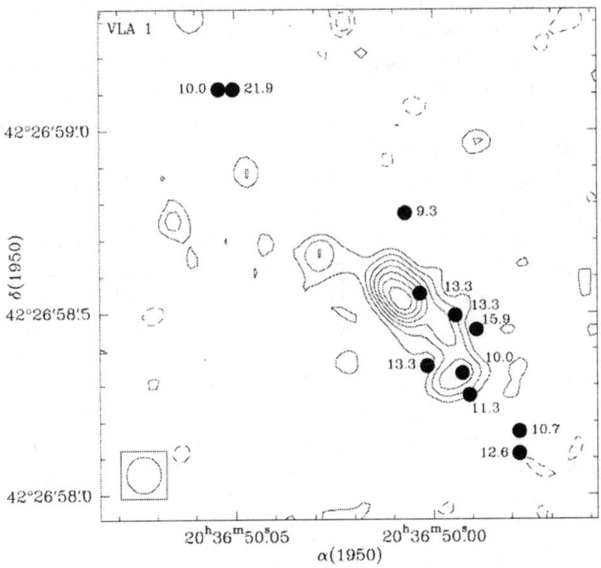

Figure 6. The spatial distribution of H_2O maser spots in W75N (Torrelles et al. 1997). Here the masers seem to lie along the axis of the jet shown in contours, rather than perpendicular to it. The differences in spatial morphology of the H_2O masers with respect to the underlying jet may indicate an evolutionary effect within the circumstellar disk. VLBI measurements are also under way to track the proper motions of the masers in this system.

fluctuations at 1 cm. The simultaneous observation of the strong maser line with the continuum guarantees essentially perfect phase compensation, thereby allowing the diffraction limit to be achieved on the 30 km baseline of the VLA.

The very high resolution continuum maps have resolved highly collimated jet structures as in the case of Cepheus A (Torrelles et al. 1996). The size scale of the jets, the observed fluxes, the morphology, are all consistent with a wind model. The distribution of the H_2O masers with respect to the jets is quite impressive, some of the masers seemingly perpendicular to the jet with a disk-like morphology, and some lying along the jet as in a sheath. Moreover, for some of the masers which are distributed in a disk, their kinematics appear to be consistent with rotational motions of the disk, with a linear velocity gradient along the major axis of the disk. There are even motions along the minor axis that might be consistent with contraction or expansion of the disk.

We note that the spatial distribution of the H_2O masers is much more compact than the case for CH_3OH masers by about a factor of 10. This suggests that the H_2O masers could represent an earlier stage of evolution. It is also possible that the CH_3OH masers are associated with more massive stars with much larger circumstellar disks. The fact that the H_2O masers are found sometimes perpendicular to the central jet and sometimes along the jet, also suggests different stages of evolution. While the statistics are still quite limited, there is an impression that when the spatial distribution of the masers are more compact, they tend to show a more disk-like morphology. When the maser distribution is more extended, then they seem to follow the outline of the continuum jet. If this picture holds up, then the spatial distribution of the H_2O masers may be useful as an indicator of the evolutionary status of the central star.

The cross calibration of the H_2O masers with the continuum emission results in relative positional accuracies on the order of a few milliarcseconds. VLBI with intercontinental baselines provides more than two orders of magnitude in improvement. With an angular resolution on the order of 0.3 mas, we can measure relative positions to an accuracy of 0.01 mas. A velocity of 10 km s^{-1} at a distance of 2 kpc will result in a proper motion of 0.1 mas in the time of one month. Hence, we can directly measure the relative motions of the masers within the disks or jets in order to understand their nature and evolution. Such measurements are under way, and we will know in the next year just what kind of motions we are seeing in the circumstellar disks. For example, in the case of IRAS 21391+5802, Patel el al. (1999) have found a ring of masers on the scale of 1 AU which might be expanding. Outflowing motions have also been suggested for a massive protostar in Orion by Greenhill et al. (1998).

VLBI measurement of the proper motions of H_2O masers is in fact a very well developed technique. Proper motions have been used to measure the distance to various regions including the Galactic center (e.g. Reid et al. 1988), and to limit the mass of black holes (cf. Miyoshi et al. 1995). In the last few years, more efforts have been devoted to studying the kinematics of circumstellar disks. However, heretofore, proper motions have mostly been found to be outflows. As we examine earlier and earlier cases, and perhaps in the lower-mass cases, motions within the circumstellar disks may become discernible. Setting limits to the mass inflow rate, and finding evolutionary effects within the disks, would be very exciting prospects.

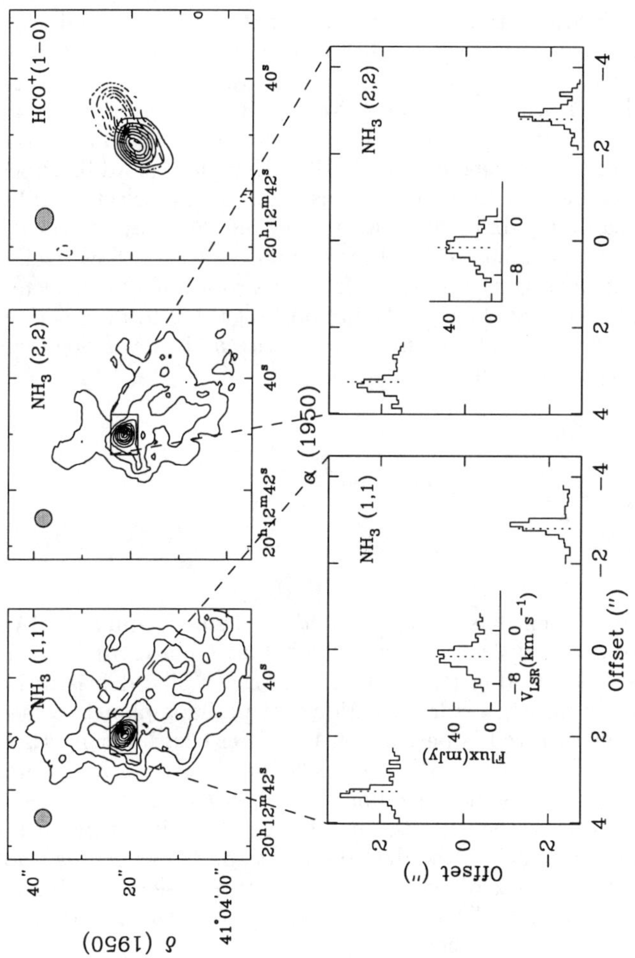

Figure 7. The circumstellar disk around IRAS 20126+4104 from Zhang et al. (1999). The central disk is seen to be embedded within a more diffuse envelope. The spectra show Keplerian motions with a rotational velocity gradient and enhanced line broadening toward the center. The disk lies perpendicular to a molecular outflow seen in HCO^+.

6. Disks Around Massive Stars

One of the drawbacks in studying disks around low-mass stars is that they have faint signatures. Since the stellar masses involved are small, the disk masses are also small, and the gravitational effects are also weak. The direct detection via dust emission is difficult as we have seen above, and tracing the kinematics is even more difficult. In contrast, the disks around massive stars can be much larger, and the larger gravitational forces mean that relevant bound motions, whether in contraction or rotation, can be substantially larger. The difficulty is that the more massive stars are on average further away, so that our intrinsic ability to resolve them spatially is poorer. Nevertheless, interferometry in the past few years has made great progress. A good example is the high-mass star IRAS 20126+4104 studied by Zhang et al. (1999). Here, as should be expected, the central disk is embedded in a more diffuse cloud. By selecting different transitions of NH_3, sampling different excitation regimes, one can see the diffuse envelope, the much more compact central disk, and even the outflow itself. The disk is quite large, on the order of 10 000 AU, and its rotational motions appear to be Keplerian and bound by 20 $M_{\odot}$. The important thing to note is that the signatures are quite strong, and the inner kinematics and morphology can be studied with higher spatial resolutions.

7. Debris Disks

A huge amount of excitement has been generated by the beautiful SCUBA submillimeter images of debris disks from the JCMT. Jane Greaves reviewed these observations during this meeting (this volume, p. 296). The debris disks are of course the leftover stuff after star formation and perhaps planet formation. The amount of matter involved is perhaps on the order of a lunar mass. That it can be detected at all is testimonial to the sensitivity of receivers as well as the rising spectrum of the dust in the submillimeter wavelengths. That it is of great interest may be due to the expectations that solid bodies in the disk may cause significant perturbations on the dust distribution. While we may not be able to see the planetary bodies directly, we may be able to see their dynamical effects on the surrounding dust disks, much like the gaps in the Saturnian rings which pointed to the presence of shepherding moons.

Our own excitement on the debris disks is driven by our efforts to build the Submillimeter Array on the top of Mauna Kea. This interferometer will have approximately the same collecting area as the JCMT, but with 10–100 times more angular resolution. We hope to be able to resolve such debris disks as are being detected with the JCMT. The resulting images, in a few years, should be most illuminating. Of course, a much more powerful instrument, the Atacama Large Millimeter Array will be built in Chile, with perhaps 100 times more collecting area. The images which we will be able to make in the next decades will be very important for understanding how planets are made.

References

Cabrit, S., Guilloteau, S., Andre, P., Bertout, C., Montmerle, T., & Schuster, K. 1996, A&A, 305, 527

Greenhill, L. J., Gwinn, C. R., Schwartz, C., Moran, J. M., & Diamond, P. J. 1998, Nat, 396, 650

Guilloteau, S., Dutrey, A., & Simon, M. 1999, A&A, 348, 570

Hayashi, M., Ohashi, N., and Miyama, S. 1993, ApJ, 418, L71.

McCaughrean, M. J., Chen, H., Bally, J., Erickson, E., Thompson, R., Rieke, M., Schneider, G., Stolovy, S., & Young, E. 1998, ApJ, 492, L157

Miyoshi, M., Moran, J. M., Herrnstein, J., Greenhill, L., Nakai, N., Diamond, P., & Inoue, M. 1995, Nat, 373, 127

Norris, R. P., Byleveld, S. E., Diamond, P. J., Ellingsen, S. P., Ferris, R. H., Gough, R. G., Kesteven, M. J., McCulloch, P. M., Phillips, C. J., Reynolds, J. E., Tzioumis, A. K., Takahashi, Y., Troup, E. R., & Wellington, K. J. 1998, ApJ, 508, 275

Ohashi, N., Hayashi, M., Ho, P. T. P., & Momose, M. 1997a, ApJ, 475, 211

Ohashi, N., Hayashi, M., Ho, P. T. P., Momose, M., Tamura, M., Hirano, N., & Sargent, A. I. 1997b, ApJ, 488, 317

Patel, N. A., Greenhill, L. J., Herrnstein, J. R., Zhang, Q., Moran, J. M., Ho, P. T. P., & Goldsmith, P. F. 1999, in Proc. Star Formation 1999, ed. T. Nakamoto (Nobeyama: Nobeyama Radio Observatory), 300

Reid, M. J., Schneps, M. H., Moran, J. M., Gwinn, C. R., Genzel, R., Downes, D., & Roennaeng, B. 1988, ApJ, 330, 809

Rodríguez, L. F., D'Alessio, P., Wilner, D. J., Ho, P. T. P., Torrelles, J. M., Curiel, S., Gómez, Y., Lizano, S., Pedlar, A., Cantó, J., & Raga, A. C. 1998, Nat, 395, 355

Sargent, A. I., & Beckwith, S. V. W. 1991, ApJ, 382, L31

Torrelles, J. M., Gómez, J. F., Rodríguez, L. F., Curiel, S., Ho, P. T. P., & Garay, G. 1996, ApJ, 457, L107

Torrelles, J. M., Gómez, J. F., Rodríguez, L. F., Ho, P. T. P., Curiel, S., & Vázquez, R. 1997, ApJ, 489, 744

Wilner, D. J., Ho, Paul T. P., & Rodríguez, L. F. 1996, ApJ, 470, L117

Zhang, Q., Hunter, T. R., & Sridharan, T. K. 1999, in Proc. Star Formation 1999, ed. T. Nakamoto, (Nobeyama: Nobeyama Radio Observatory), 225

Discussion

Lee Hartmann: The work by Jack Welch suggests that the large-scale ^{13}CO structure in the Tau region is actually part of a bubble centered near XZ Tau, demonstrating the importance of understanding the structure on many scales in complicated regions.

Paul Ho: The complicated kinematics from the presence of more than one type of phenomenon are indeed confusing. This is of course the reason why one must have a number of different probes sensitive to different cloud conditions in order to test any one model. Conditions in a disk, for example, must rely on tracers which are sensitive to the disk itself. In addition, a good model must be able to account for all available data.

Alan Boss: What portions of disks do H_2O masers trace—the disk midplane, or atmosphere, or envelopes, or what?

Paul Ho: That remains unclear at this point, awaiting detailed proper motions in specific cases. In a number of cases, the H_2O masers seem to trace the outflow, or the impact of winds on ambient matter. In other cases, the masers have both the spatial morphology and the apparent kinematics that trace a disk-like structure. However, even in the best cases for disks, it is still unclear as to which parts of the disk (e.g., surfaces) the masers are tracing.

Caroline Terquem: The temperature at 1 AU is a few hundred K rather than 10 K.

Paul Ho: You are right of course! The disk is 10 K only in the outer parts and will become much warmer towards the center. However, for structure at 1 AU resolution, with temperatures at 10–100 K or more, the power radiated, or flux, is still very faint for most instruments, especially in the spectral line mode. Hence, to measure kinematics and structures at 1 AU scales, non-thermal phenomena such as masers have some advantages.

Disks, Planetesimals, and Planets
ASP Conference Series, Vol. 219, 2000
F. Garzón, C. Eiroa, D. de Winter, and T. J. Mahoney, eds.

Accretion Disks

Lee Hartmann
Harvard-Smithsonian Center for Astrophysics, 60 Garden St., Cambridge, MA 02138, USA

Abstract. I review the observational constraints on disk accretion in low-mass, pre-main-sequence stars, and consider their implications for theories of planet formation. T Tauri stars accrete roughly $0.01M_{\odot}$ over their lifetimes of a few Myr, in approximate agreement with estimates from mm-wave dust emission. However, the best simple models which reconcile observed disk masses, accretion rates, and sizes suggest relatively fast viscous evolution, which could pose problems for planet formation. These difficulties could be eased if observations of dust emission substantially underestimate the amount of mass present in disks, especially at small radii, which might occur if we have underestimated the amount of grain growth that has occurred in these disks.

1. Introduction

From the time of the seminal work by Lynden-Bell & Pringle (1974) it was recognized that accretion disks are not static entities but are dynamically evolving; not only is accretion emptying out the mass of the disk onto the central star, but viscous disks must expand to conserve angular momentum as accretion proceeds. Despite this, for a long time most models of planet formation considered disks kto be relatively slowly evolving or essentially static systems, and that the planets of the Solar System formed more or less where they are today. However, the discovery of extrasolar giant planets orbiting near their central stars naturally suggested that planets may migrate considerable distances inward from the places where they are originally formed (see discussion in Marcy & Butler 1998). As discussed in Terquem's paper (this volume, p. 19; see also Ward 1997 and references therein), two types of generally inward migration can be distinguished. Type II migration, in which the planet opens up a gap in the disk, reflects the viscous evolution of the accretion disk surrounding it.

Both Type I and especially Type II migration can in principle help explain the close-in exoplanets by allowing them to form at large distances as expected by current theories (Guillot et al. 1996; Wuchterl, Guillot, & Lissauer 2000), and then migrate inward as the disk evolves (Lin & Papaloizou 1986; Lin, Bodenheimer, & Richardson 1996; Murray et al. 1998; Trilling et al. 1998, 2000). However, it is possible to have too much of a good thing and have most of the disk mass accrete before planets form, or have most of the planets be accreted into the central star. The "conveyor belt" model of planetary formation, in which planets keep being accreted until the "belt" (accretion) is turned off for some unknown reason, has become popular (e.g., Lin & Papaloizou 1986; Lin et al. 1996; Trilling et al. 1998, 2000) because it avoids the problem of stopping the exoplanet at a particular distance. The resulting variations in the distributions of giant planet orbits among differing systems represent statistical fluctuations in the accretion process, in the properties of disks, where and when and with what masses planets are formed, etc. However, this model also contains some potential difficulties. If viscous evolution is too rapid, formation of giant planets at large distances might be required for them to survive the accretion process with any significant probability; but surface densities are low at large distances, which may imply unacceptably long timescales for planetary formation.

Observations show that typical low-mass, pre-main-sequence T Tauri stars do accrete significant masses from their disks during their active lifetimes, whether in the form of planets or disk gas and dust. Constraints on the rate of viscous evolution are much less certain, but the current estimates suggest problems for forming planets, unless we are not detecting a substantial fraction of the mass in protoplanetary disks.

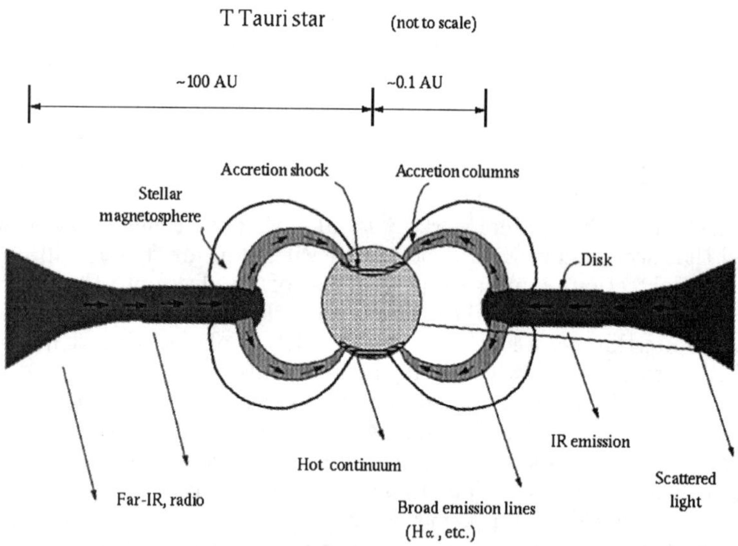

Figure 1. Schematic outline of the structure of an accreting T Tauri star. The accretion disk is truncated by the stellar magnetic field at a distance of a few stellar radii; material is channeled along magnetic field lines, acquiring near free-fall velocities as it moves inward toward the star, until it shocks at the stellar surface. The hot continuum emission produced at the accretion shock provides a good measure of the accretion luminosity and thus the mass transfer rate. From Hartmann (1998).

2. Disk Accretion Rates

In recent years the accretion processes in T Tauri stars have become much better understood. For most T Tauri stars, accretion rates cannot be inferred from disk emission because most of the disk heating is due to the absorption of radiation from the central star, not to local viscous dissipation. Instead, accretion rates are measured inside the inner edge of the disk, where the mass is added to the star. The stellar magnetic field truncates the disk at a few stellar radii (Figure 1), at which point the accreting material falls in along magnetospheric flux tubes.

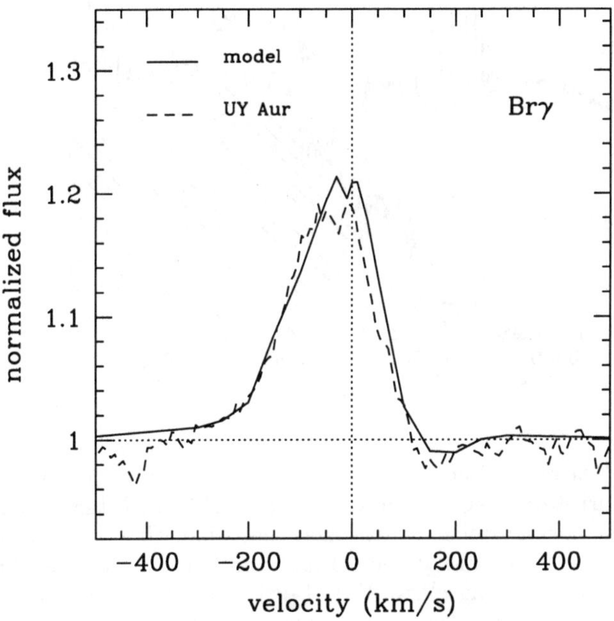

Figure 2. Profile of the Brackett γ emission line in the T Tauri star UY Aur, compared with a theoretical model of magnetospheric accretion. The inverse P Cygni profile characteristic of infall is evident, and the maximum velocities observed are comparable to the expected free-fall velocity at the stellar surface. From Muzerolle, Hartmann, & Calvet (1998c).

The accreting gas moves inward at nearly free-fall velocities, accelerating to $\sim$ 200–300 $\mathrm{km\,s^{-1}}$ before shocking at the stellar surface, radiating away the accretion energy in hydrogen continuum emission and a variety of emission lines. A variety of lines of evidence support this picture; among these are the observation that only stars with inner disks (as measured from infrared excess emission) exhibit the excess line and continuum emission; and careful modeling of improved observational data show that the line profiles of most emission lines can be understood as coming from the high-velocity infalling magnetospheric gas (cf. Muzerolle et al. 1998a,b,c; Figure 2).

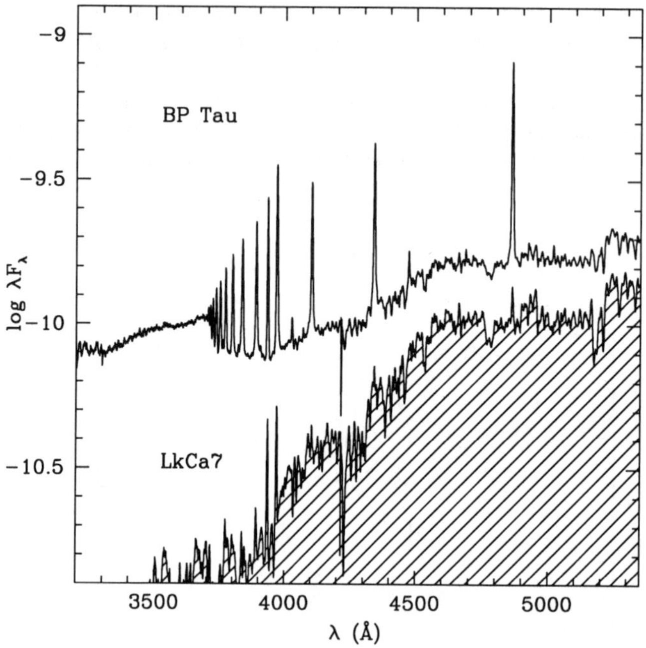

Figure 3. The excess line and continuum emission produced by accretion in T Tauri stars. LkCa 7 is a weak-emission T Tauri star (WTTS); it exhibits the normal stellar photospheric emission, plus weak chromospheric emission at H_β and at the Ca II H and K (3933, 3968 Å) lines. In contrast, the accreting classical T Tauri star (CTTS) BP Tau exhibits excess Balmer and Paschen continuum emission along with much stronger hydrogen line emission, due to accretion from its circumstellar disk. The relative level of the LkCa 7 spectrum has been scaled to the photospheric emission levels present in BP Tau, as determined from the weakened stellar photospheric absorption features.

Most of the energy of accretion is radiated away at the accretion shock, where the magnetospheric gas crashes into the star; thus measurements of the excess luminosity due to accretion can be used to infer the mass accretion rate through the relation

$$L_{\rm acc} \approx \frac{GM_*\dot{M}}{R_*}. \tag{1}$$

Although the concept is straightforward, it is not so easy to determine excess luminosities in this matter. For example, consider the spectra shown in Figure 3. The excess emission in BP Tau is quite evident, but extracting this excess involves the following (see Hartigan et al. 1991 for a discussion):

(a) separating the stellar photospheric emission from the excess emission as a function of wavelength; (b) comparing the true stellar photospheric spectral energy distribution (SED) to that of an unreddened standard star of the same spectral type to determine the extinction to the system (the excess cannot be dereddened because its intrinsic SED is unknown a priori); (c) correct the excess luminosity for extinction; and (d) extrapolate the spectrum by means of models or some other observations to determine the total excess luminosity.

It is apparent that undertaking the above procedure is not completely trivial. In some stars, the excess emission is weak, making it relatively easy to determine the extinction, but hard to measure the excess; in others, the accretion energy is large, making it hard to determine the stellar SED for purposes of extinction determinations. Ultraviolet observations of the excesses are relatively rare and generally not simultaneous, a problem since T Tauri stars vary significantly, adding further difficulty to the problem of determining the excess emission over a wide range of wavelengths. Finally, the complex geometry of accretion (Figure 1) complicates the relationship between the total accretion luminosity and the flux observed along an individual line of sight.

Despite these difficulties, significant progress has been made in measuring accretion excesses and estimating accretion rates in T Tauri stars over the past few years. Figure 4 summarizes the results (Calvet, Hartmann, & Strom 2000; Hartmann et al. 1998; Gullbring et al. 1998). There is a large scatter of accretion rates at a given "age". Some of this scatter is undoubtedly real, though a significant fraction (a factor of 2–3) can probably be attributed to difficulties of measurement, variability, etc. There is a weak trend toward decaying accretion rates with increasing age (Hartmann et al. 1998), though there is a great deal of scatter both in the accretion rates and in the ages of individual stars, which are probably not much more accurate (in a relative sense) than a factor of two or so.

3. Disk Evolution

The results of the accretion rate measurements discussed above lead to the following general and important conclusion. Since the accreting T Tauri stars (the "classical" or CTTSs) constitute roughly half or more of the total population of $\sim$ 1 Myr-old stars (in the Taurus and Chamaeleon regions used in Figure 4), the accretion phase must typically last for 1 Myr or more. If we adopt a median accretion rate of $10^{-8}\ M_\odot\,\mathrm{yr}^{-1}$ at 1 Myr from inspection of Figure 4, the implication is that the typical CTTS accretes $\sim 0.01\ M_\odot$ from its circumstellar disk. Accounting for the decay of accretion with time does not change this general conclusion (Hartmann et al. 1998).

At first sight, this result is reassuring because estimates of disk masses from mm-wave emission *also* suggest masses of $\sim 0.01\ M_\odot$ (e.g., Beckwith et al. 1990 = BSCG; see review in Wilner & Lay 2000). These masses ($\sim 10\ M_{\mathrm{Jup}}$) are comparable to a minimum-mass solar nebula and are not excessively large in view of the detection of exoplanets of several Jupiter masses. But most of this emitting mass resides at fairly large radial distances of order 100 AU. If the mass that is being accreted by CTTSs is really moving in from 100 AU in 1 Myr, the transit time in to the star from 10 AU must be very short indeed. Accretion

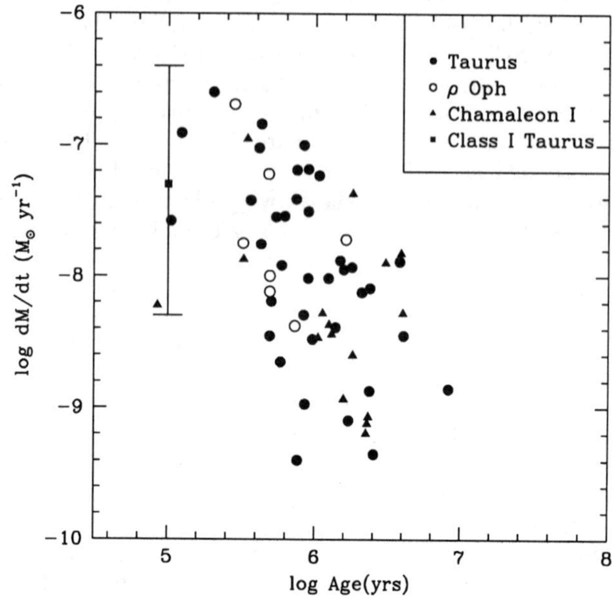

Figure 4. Accretion rates for T Tauri stars and protostars. The *vertical line* indicates the range of accretion rates for Class I (protostellar) objects. Accretion rates for the heavily extinguished Class I protostars are estimated from infrared emission lines rather than unobservable optical and ultraviolet emission excesses. From Calvet et al. (2000).

must be turned off very rapidly to stop a planet at 5 AU, let alone 0.5 AU. In addition, if giant planets must be formed at distances $\gg$ 10 AU, where most of the mass is, and which survives the necessary millions of years, other problems arise; for instance, the surface densities in the outer disk are quite low, and current models of giant planet formation suggest that there may be difficulties in forming objects fast enough (e.g., Wuchterl et al. 2000).

To make this a bit more specific, consider some simple solutions for disk evolution (Hartmann et al. 1998). We used the similarity solutions of Lynden-Bell & Pringle (1974) for the case where the viscosity $\nu \propto R$. While there is no reason to assume that the viscosity is truly a power law with radius, this form simplifies the problem considerably for purposes of developing physical insight (rather than detailed fits). Since the temperature distributions in the outer regions of CTTS disks vary as $T \approx R^{-1/2}$ (BSCG; Kenyon & Hartmann 1995) as a result of disk irradiation (D'Alessio et al. 1999), this choice for the power-law dependence of ν is consistent with a roughly constant value of the Shakura–Sunyaev viscosity parameter α.

The similarity parameter is

$$T = 1 + t/t_s, \tag{2}$$

where the viscous time t_s is given by

$$t_{\rm s} \;=\; \frac{1}{3}\,\frac{R_1^2}{\nu_1}. \tag{3}$$

In this equation ν_1 is the viscosity evaluated at a fiducial radius R_1, which is related to the initial size of the disk (see below). Numerically, in terms of typical T Tauri parameters for the stellar mass M_* and disk temperature structure (Hartmann et al. 1998),

$$t_{\rm s} \;=\; 8\times10^5\,{\rm yr}\left(\frac{\alpha}{10^{-2}}\right)^{-1}\left(\frac{R_1}{100\,{\rm AU}}\right)\left(\frac{M_*}{0.5\,M_\odot}\right)^{1/2}\left[\frac{T(100\,{\rm AU})}{10\,{\rm K}}\right]^{-1}. \tag{4}$$

The viscous disk expands to take up the angular momentum lost by accretion. The radius R_t in the disk marking the transition between net accretion and expansion is

$$R_t \;=\; 50\,{\rm AU}\;T\left(\frac{R_1}{100\,{\rm AU}}\right). \tag{5}$$

Thus at $t = 0$, the transition or characteristic radius is $R_1/2$. The disk mass is then

$$M_{\rm d}(R,T) = \frac{M_{\rm d}(0)}{T^{1/2}}\left[1-\exp\left(-\frac{R}{2\,R_t}\right)\right], \tag{6}$$

where $M_{\rm d}(0)$ is the total disk mass at $t = 0$. Note that at any given time 40% of the total disk mass lies interior to R_t in this model.

From Equation (5) it can be seen that if we try to match typical disk radii of order 100 AU at $t \sim 1$ Myr, the maximum viscosity is about $\alpha \sim 10^{-2}$; otherwise the disks would expand rapidly and become too large. If $R_t \sim$ 100 AU, the accreting material can be drawn from much of the disk detected by mm continuum observations. The accretion rate onto the star is given by the time derivative of Equation (6) with $R/R_1 \to 0$. Since the initial disk mass is unknown, it is convenient to express the result as

$$\frac{M_{\rm d}(T,0)}{|\dot{M}_{\rm d}(T,0)|} = 2\,t_{\rm s}\,T = 2(t_{\rm s}\;+\;t). \tag{7}$$

Consider first a disk with an initial characteristic radius $R_1 = 100$ AU. With $\alpha = 10^{-2}$, and $t = 1$ Myr, $M_{\rm d}(T,0) = 3.6\,{\rm Myr}\times|\dot{M}_{\rm d}(T,0)|$. For a typical accretion rate of $10^{-8}M_\odot\,{\rm yr}^{-1}$ (Figure 4), this implies a current disk mass of $\sim 0.04\;M_\odot$, in adequate agreement with observational estimates from mm emission (BSCG).

If on the other hand $\alpha = 10^{-3}$, for the same parameters, $M_{\rm d}(T,0) = 18\,{\rm Myr}|\dot{M}_{\rm d}(T,0)| \sim 0.18\;M_\odot$. In contrast to the previous case, this disk mass is much larger than estimated from mm-wave observations, and roughly at or beyond the limit for gravitational instability (Larson 1985). This occurs because with a small viscosity, the only regions of the disk which accrete rapidly are at radii where the viscous time is comparable to or less than the age, which for $\alpha = 10^{-3}$ and $t = 1$ Myr is roughly at $R \lesssim 10$ AU (Equation [4]). By drawing mostly on the innermost regions of the disk to supply the accreted material for the required time, the total disk mass must be significantly larger than in the high-viscosity case.

Alternatively, consider disks with initial radii $R_1 = 10$ AU. Now the disks expand significantly at $t = 1$ Myr for both $\alpha = 10^{-2}$ and $\alpha = 10^{-3}$. In the former case, $M_d(T,0) = 2.2\,\text{Myr} \times |\dot{M}_d(T,0)| \sim 0.02\ M_\odot$, still in reasonable agreement with the mm estimates. The $\alpha = 10^{-3}$ case also works, for $M_d(T,0) = 3.6\,\text{Myr} \times |\dot{M}_d(T,0)| \sim 0.04\ M_\odot$, because this disk is small, $R_t \sim 9$ AU, and so the disk mass is concentrated at small radii where it can be accreted effectively at this viscosity. However, this model conflicts with observations which indicate that the disk emission in most objects arises from radii much larger than 10 AU (Dutrey et al. 1996; Lay, Carlstrom, & Hills 1997; Wilner & Lay 2000).

Figure 5 displays some similarity solutions in graphical form as a function of age. Here the solutions are compared with the accretion rates found by Gullbring et al. (1998) and Hartmann et al. (1998), along with estimated disk masses from BSCG and Osterloh & Beckwith (1995)—the latter corrected for an error of 2.5 (Hartmann et al. 1998). The solid curves display results for $\alpha \sim 10^{-2}$; by varying the initial disk mass from 0.1 $M_\odot$ (heavy solid curve) to 0.003 $M_\odot$ (light solid curve), the observed range of disk masses can be explained, while the accretion rates are only slightly lower than observed. The comparison of $t \times \dot{M}$ vs. M_d shown in the lower left panel shows that these models span the observed range of values fairly well, as anticipated from the discussion of the previous paragraphs.

The analysis immediately above suggested that solutions with a lower viscosity, say $\alpha = 10^{-3}$, do not agree as well with the data. For example, consider the solution for $M_d(0) = 0.05\,M_\odot$; and $R_1 = 100$ AU (short dashes in Figure 5). Only the innermost regions of this disk evolve on timescales of a few Myr or less, and so the disk transition radius hardly expands. Similarly, the disk mass and accretion rate stay nearly constant throughout the relevant period. The disk mass is on the high end of the range of observed values while the accretion rate is low compared with average T Tauri stars. However, as shown in the lower left-hand panel of Figure 5, there is so much scatter in the $t\dot{M}$ vs. M_d relation that it is difficult to completely rule out these models.

Next, consider the low-α case but with $R_1 = 10$ AU (long dashed lines). The variation of disk mass and accretion rate as a function of age is now very similar to that of the $\alpha = 10^{-2}$ case; by making the disk ten times smaller, the viscous timescales of the two models are now similar. The potential problem with this solution is that the size of the disk does not begin approaching 100 AU until 10 Myr.

From these considerations one can see that while the simple similarity model with $\alpha = 10^{-2}$ gives results in better agreement with the observations than the $\alpha = 10^{-3}$ case, the scatter in the results does not absolutely preclude a lower viscosity in all cases. It is also clear that it is crucial to find and study older populations of stars with circumstellar disks. One of the main constraints on the viscosity (at least in the outer disk) is the expansion of the disk with age. The low-viscosity case seems to require disk sizes much smaller than consistent with observation, *assuming that the mass distribution in the disk is approximately that of the similarity model* (see Section 5). Conversely, the high-α models imply that disks around 10 Myr-old stars (if they last that long) should have outer radii of

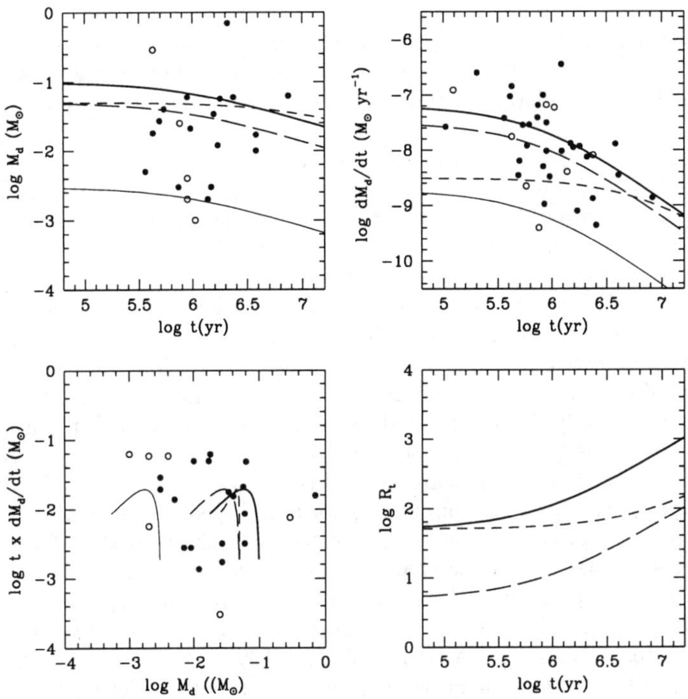

Figure 5. Similarity solutions for the masses (*upper left*), accretion rates (*upper right*), and transition radii (*lower right*) of viscously evolving disks, plus a comparison between the disk mass and the product of age and accretion rate, as described in the text. The *solid curves* denote solutions for $\alpha = 10^{-2}$; *heavy solid line*, initial disk mass $M_d(0) = 0.1\,M_\odot$; *light solid line*, $M_d(0) = 0.003\,M_\odot$. The *dashed curves* are solutions for $\alpha = 10^{-3}$ and $M_d(0) = 0.05\,M_\odot$; *short dashes*, initial disk radius $R_1 = 100$ AU; *long dashes*, $R_1 = 10$ AU.

order 1000 AU or more. The results of Weinberger et al. (1999) suggest that the disk of the nearby $\sim$ 10 Myr-old star TW Hya is at least 150 AU; but if its outer radius is not much larger than this, estimates of α would have to be revised substantially downward.

4. Disk Viscosity, Type II Drift, and Planet Formation

As discussed above, evolving disk models with $\alpha \sim 10^{-2}$ currently provide the best match to observations. However, in such models the viscous time is less than 0.1 Myr at $R \leq 10$ AU. If giant planets form and open gaps in their disks, subsequently migrating on the local viscous time of the disk (Lin & Papaloizou 1986; Ward 1977; Terquem, this volume, p. 19), it is difficult to see how any giant planets would avoid being accreted into the central star, or be left behind at 5–10 AU. In the conveyor belt model, it is possible in principle always to have enough giant planets around by simply continually making enough of them until the "belt" stops moving; but then one must consider whether giant planets are easily formed at large distances and low surface densities over the entire lifetime of the disk.

Trilling et al. (1998, 2000) claim to be able to avoid this problem in their studies of planetary migration with a statistical ensemble of disk properties, assuming viscosities near $\alpha \sim 0.005$. However, key assumptions of their models are that (1) giant planets are automatically and regularly produced at a fixed distance, irrespective of time, (2) planet masses are relatively independent of the mass of the disk producing them, and (3) the disk gas is removed by some unknown process at a specific time. The last assumption may be reasonable, although we don't really know at present how to do this; the old idea that stellar winds remove the disk has sunk under the evidence that the strong winds of T Tauri stars are highly collimated perpendicular to the disk plane (see, for example, reviews in Reipurth & Bertout 1997). However, the validity of the assumptions (1) and (2) is unclear. One of the ways that Trilling et al. (1998) avoid rapid migration into the central star for a high disk viscosity is by adopting a sufficiently large planetary mass relative to the disk, such that the disk gap opened by the planet is so wide that the torques of the disk on the planet are negligible. This however begs the question of whether such a planet would not have smaller gaps and thus very fast migration in the course of its growing to such a mass (e.g., Takeuchi, Miyama, & Lin 1996; Bryden et al. 1999), or whether the planet can even grow to a such mass by depleting such a large fraction of the disk.

Alternatively, it could be that the "observations" are misleading, especially estimates of disk masses which depend entirely upon the assumed dust opacity. Beckwith & Sargent (1991) show that the spectral indices of the dust emission near 1 mm suggest that grains have grown from their interstellar medium values, a process which is both expected and probably necessary to provide the seeds for planet formation. The calculations of Miyake & Nakagawa (1995) suggest that the dust opacities adopted in the disk mass estimates by BSCG and Osterloh & Beckwith (1995) could be overestimated, depending upon just how large the dust grains become. If disk masses can be revised upward, then viscosities can be revised downward (Section 4), slowing the rate of viscous transport.

An even better scenario for planet formation could result from the "dead zone" model of Gammie (1996). Gammie pointed out that because the ionization state in protoplanetary disks is so low in many regions, the magnetorotational instability invoked for producing a magnetic viscosity cannot operate. In the dead zone model, cosmic rays can ionize the upper layers of the inner disk (and perhaps all of the disk at large radii), but do not penetrate the disk interior at radii between about 0.2 AU and 10–30 AU. The inert material, which cannot accrete, then piles up into a very dense region or dead zone.

The high surface densities and zero radial transport of the dead zone make it an ideal region to form planets. One disadvantage of this model is that current mm-wave observations provide little evidence for such a high density inner region (Lay et al. 1997; Wilner & Lay 2000). Reconciling this model with mm and sub-mm interferometer images may require a very substantial amount of grain growth. Another potential problem is that this model might prevent *any* Type II migration if there is too much mass in the dead zone compared with the rest of the accretion disk.

One additional piece of evidence for lower viscosities comes from the multiple system GG Tau. This quadruple system has both a large circumbinary disk and circumstellar disk(s) around one or both of the close primaries (Guilloteau, Dutrey, & Simon 1999), which are separated by about 0.26 arcsec $\sim$ 35 AU in projection (Leinert et al. 1993). Since at least one or both of the stars in the close binary are accreting at an age of $\sim$ 1 Myr, and if the disk is truncated by the binary orbit at $\sim$ 15 AU or so, the viscosity would have to be less than about $\alpha \sim 3 \times 10^{-3}$ to avoid depleting the disk too fast. However, this argument depends upon the assumption that no material can leak in from the outer circumbinary ring, which may not be correct (Artymowicz & Lubow 1996).

In general, then, the possibility of substantial Type II migration which carries giant planets inward far from their natal sites as accretion proceeds remains a very strong possibility, simply because T Tauri stars accrete many Jupiter masses-worth of material over a few Myr. Whether the disk viscosity is sufficiently large to pose a problem for planet formation is currently unclear; we need to understand just how much dust particle growth has occurred in these disks. Better yet, molecular tracers might give us much better estimates of disk masses, if we can understand the chemical evolution involved. Observations which limit just how far disks expand at ages of 10 Myr will really constrain viscosities of outer disks. And finally, it would be useful to know when and how protoplanetary disks dissipate their mass.

Acknowledgments. I wish to acknowledge useful conversations with Doug Lin. This work was supported in part by NASA grant NAG5-4282.

References

Artymowicz, P., & Lubow, S. H. 1996, ApJ, 467, L77

Beckwith, S. V. W., & Sargent, A. I. 1991, ApJ, 381, 250

Beckwith, S. V. W., Sargent, A. I., Chini, R. S., & Güsten, R. 1990, AJ, 99, 924 (BSCG)

Bryden, G., Chen, X., Lin, D. N. C., Nelson, R. P., & Papaloizou, J. C. B. 1999, ApJ, 514, 344

Calvet, N., Hartmann, L., & Strom, S. E. 2000, in Protostars & Planets IV, eds. V. Mannings, A. Boss, & S. Russell, in press

D'Alessio, P., Calvet, N., Hartmann, L., Lizano, S. & Cantó, J. 1999, ApJ, 527, 893

Dutrey, A., Guilloteau, S., Duvert, G., Prato, L., Simon, M., Schuster, K. & Menard, F. 1996, A&A, 309, 493

Gammie, C. F. 1996, ApJ, 457, 355

Guillot, T., Burrows, A., Hubbard, W. B., Lunine, J. I., & Saumon, D. 1996, ApJ, 459, L35

Guilloteau, S., Dutrey, A. & Simon, M. 1999, A&A, 348, 570

Gullbring, E., Hartmann, L., Briceño, C. & Calvet, N. 1998, ApJ, 492, 323
Hartigan, P., Strom, S., Edwards, S., Kenyon, S., Hartmann, L., Stauffer, J., & Welty, A. 1991, ApJ, 382, 617
Hartmann, L. 1998, Accretion Processes in Star Formation (Cambridge: Cambridge University Press)
Hartmann, L., Calvet, N., Gullbring, E. & D'Alessio, P. 1998, ApJ, 495, 385
Kenyon, S. J., & Hartmann, L. 1995, ApJS, 101, 117
Larson, R. B. 1985, MNRAS, 214, 379
Lay, O. P., Carlstrom, J. E. & Hills, R. E. 1997, ApJ, 489, 917
Leinert, C., Zinnecker, H., Weitzel, N., Christou, J., Ridgway, S. T., Jameson, R., Haas, M. & Lenzen, R. 1993, A&A, 278, 129
Lin, D. N. C., & Papaloizou, J. C. B. 1986, ApJ, 309, 846
Lin, D. N. C., Bodenhimer, P., & Richardson, D. C. 1996, Nat, 380, 606
Lynden-Bell, D., & Pringle, J. E. 1974, MNRAS, 168, 603
Marcy, G. W. & Butler, R. P. 1998, ARA&A, 36, 57
Miyake, K. & Nakagawa, Y. 1995, ApJ, 441, 361
Murray, N., Hansen, B., Holman, M., & Tremaine, S. 1998, Science, 279, 69
Muzerolle, J., Calvet, N., & Hartmann, L. 1998a ApJ, 492, 743
Muzerolle, J., Hartmann, L., & Calvet, N. 1998b, AJ, 116, 455
Muzerolle, J., Hartmann, L., & Calvet, N. 1998c, AJ, 116, 2965
Osterloh, M., & Beckwith, S. V. W. 1995, ApJ, 439, 288
Reipurth, B., & Bertout, C. 1997, (eds.), Herbig–Haro Flows and the Birth of Low-Mass Stars, IAU Symposium 182 (Dordrecht: Kluwer)
Takeuchi, T., Miyama, S. M., & Lin, D. N. C. 1996, ApJ, 460, 832
Trilling, D. E., Benz, W., Guillot, T., Lunine, J. I., Hubbard, W. B., & Burrows, A. 1998, ApJ, 500, 428
Trilling, D. E., Lunine, J. I., & Benz, W. 2000, ApJ, in press
Ward, W. R. 1997, ApJ, 482, L211
Weinberger, A. J., Schneider, G., Becklin, E. E., Smith, B. A., & Hines, D. C. 1999, BAAS, 194, 6904
Wilner, D. J., & Lay, O. P. 2000, in Protostars & Planets IV, ed. V. Mannings, A. Boss, & S. Russell, in press
Wuchterl, G., Guillot, T., & Lissauer, J. 2000, in Protostars & Planets IV, ed. V. Mannings, A. Boss, & S. Russell, in press

Discussion

Caroline Terquem: You can stop the type II migration by assuming a very low viscosity, but not type I migration.

Lee Hartmann: Well, maybe you can avoid type I migration if you build the core up on a very short timescale.

Caroline Terquem: It would have to be a very, very short timescale...

Lee Hartmann: How do you stop type I migration, then?

Caroline Terquem: I don't know.

Lee Hartmann: Well, I have my own problems as well!

Willy Kley: When you presented a radially dependent α-type viscosity you indicated that α increases outward (from 0.0001 within $r = 1$ AU up to 0.01 for $r \gtrsim 100$ AU). From the typical temperature dependence in disks (T falls off with radius), wouldn't you expect the opposite behavior (i.e., α decreases with r)?

Lee Hartmann: In Gammie's model, the disk can be ionized by cosmic rays to a column density of about 100 g cm^{-2}. Thus, the outer disk regions can in principle sustain enough ionization to couple the magnetic field to disk material and thus have a significant magnetic viscosity.

Mark Casali: Given the redistribution of angular momentum that must occur in going from a rotating cloud to a TT disk in which the greater part of the angular momentum is in the low-mass disk, there should be a spectacular increase in the disk size to carry this angular momentum. How does this fit in with later presumed inward accretion and planet migration?

Lee Hartmann: This is a very good question. My guess is that most of the angular momentum of a typical cloud core goes into forming a binary system. The Solar System may have come from an exceptionally low angular momentum cloud, or perhaps it got ejected from a multiple system.

Tristan Guillot: It seems to me that another piece of the puzzle may be the recent discovery from *Galileo* probe data (Owen et al. 1999, Nat, 402, 269) that Ar, Xe, Kr, and C are enriched by a factor ~ 3 compared to the solar value. This should imply that very low T grains ($T < 30$ K were brought in, possibly from large radial distances.

Lee Hartmann: In principle, this would suggest an accretion from large distances, as you suggest.

Hubertus Klahr: Would the measured α of $10^{-3} > \alpha > 10^{-4}$ from Klahr & Bodenheimer (this volume, p. 63) be enough to agree with observation?

Lee Hartmann: Yes, and even better with respect to the migration of planets.

Disks, Planetesimals, and Planets
ASP Conference Series, Vol. 219, 2000
F. Garzón, C. Eiroa, D. de Winter, and T. J. Mahoney, eds.

Observations of Protoplanetary Disks at 7 mm

Luis F. Rodríguez

Instituto de Astronomía, UNAM, Campus Morelia, A.P. 3-72 (Xangari), Morelia, Michoacán 58090, Mexico

Abstract. I present results obtained at 7 mm with the Very Large Array toward young stars suspected of being surrounded with protoplanetary disks. These observations provide an unprecedented angular resolution of 50 milliarcsec. On the other hand, their interpretation requires a good knowledge of the morphology and continuum spectrum of the source at centimeter and millimeter wavelengths. This information enables to find out how much of the 7 mm emission is due to dust and how much to free–free from ionized outflows. While in HL Tau a "normal" protoplanetary disk with diameter of order 100 AU is found, in L1551 IRS5 a binary system of compact dust disks is observed, each with a radius of order 10 AU. The compactness of these disks is attributed to tidal truncation. Their small size may inhibit the formation of outer, icy planets but not of inner, terrestrial planets or of giant gas planets. We also present new results obtained for TW Hya, the closest analog to the early solar nebula.

1. Introduction

The last few years have witnessed remarkable advancements in observations of protoplanetary disks in the optical, infrared, and radio regimes. Many of these advancements have come about as a result of the consolidation of techniques that provide subarcsecond angular resolution.

In the radio regime, we have now extensive spectroscopic observations of the molecular content and kinematics of the disks (e.g., Dutrey et al. 1999), as well as continuum observations of the dust in them (Looney, Mundy, & Welch 2000). Searching for the highest angular resolution possible, the Very Large Array (VLA) of NRAO[1] has been progressively equipped with receivers for the wavelength of 7 mm as a result of a collaboration between the USA, Mexico, and Germany. For the largest VLA baselines of 35 km, this translates into an angular resolution of 50 milliarcseconds (mas). This remarkable angular resolution is the greatest advantage of the VLA at 7 mm. On the other hand, observing at mm wavelengths with large baselines requires excellent weather and the use of observational techniques such as fast switching (Carilli & Holdaway 1997), which exact a large overhead from the allocated time. Furthermore, while

[1]The National Radio Astronomy Observatory is a facility of the National Science Foundation operated under cooperative agreement by Associated Universities, Inc.

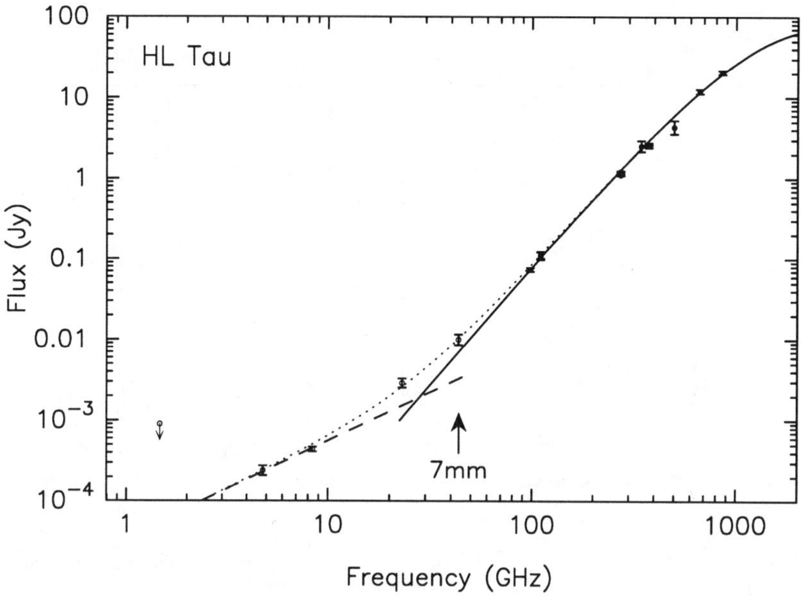

Figure 1. Spectrum of HL Tau from centimeter to submillimeter wavelengths, including measurements from various authors (see Wilner et al. 1996). The *solid line* shows a parametric disk model and the *dashed line* shows a power-law fit to the 6 and 3.6 cm fluxes. The *dotted line* shows the sum of these two components.

at wavelengths longer than 7 mm it is clear that the continuum emission from young stars is dominated by free–free emission from thermal jets (Rodríguez 1997), or gyrosynchrotron emission from coronal solar-like magnetic activity (Feigelson & Montmerle 1999), and at wavelengths shorter than 7 mm by thermal emission from dust (Looney et al. 2000), at 7 mm one can have any of these processes dominating and a detailed knowledge of the continuum spectrum plus information on the size and morphology of the source are needed to elucidate the nature of the emission. In this paper we present results on several sources where the 7 mm emission is believed to trace protoplanetary disks.

2. HL Tau

This young star in the L1551 dark cloud has been one of the preferred targets of protoplanetary disk researchers. Its millimeter emission has been imaged at 0.87 mm (Lay et al. 1994), 2.7 mm (Mundy et al. 1996), and 7 mm (Wilner, Ho, & Rodríguez 1996). All three studies agree on the dimension ($\sim$ 100 AU) and orientation (P.A. $\simeq$ 125°) of the major axis of the disk, as well as in a total mass of the order of 0.1 $M_{\odot}$.

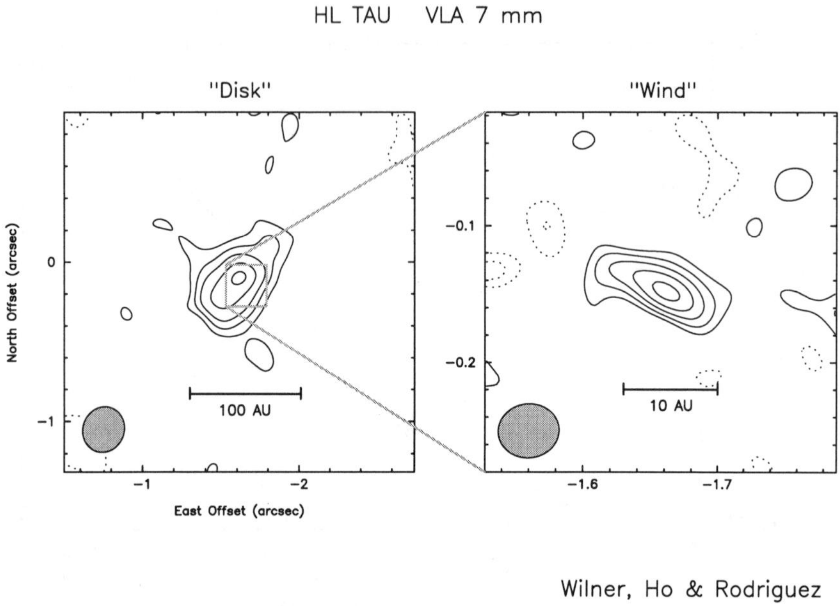

Figure 2. Images of HL Tau made at 7 mm with the VLA in the B configuration (*left*) and in the A configuration (*right*). The image on the left is interpreted as tracing the disk, while that on the right is believed to be tracing the jet (Wilner et al. 2000).

In Figure 1 we show the cm–mm spectrum of HL Tau, taken from Wilner et al. (1996). As can be seen from this spectrum, at 7 mm there are significant contributions both from the disk and from the ionized outflow. The recent study of Wilner, Ho, & Rodríguez (2000) indicates that the dust emission (tracing the disk) comes from an extended structure, while the free–free emission (tracing the ionized outflow) comes from a compact jet (see Figure 2). In this source, the combination of observations made with the VLA with different configurations allows the the extended structure of the disk to be "resolved out" and the more compact jet to be seen. The image of the jet in HL Tau at 7 mm has an angular resolution of $\sim$ 7 AU and shows that at this scale the collimation is clearly present, favoring models such as that of Shu et al. (1994), where the collimation and acceleration take place close to the star.

3. L1551 IRS5

A protostar embedded, as HL Tau, in the L1551 dark cloud, L1551 IRS5 has been studied extensively. For many years it was the prototype of a young, single star powering a remarkable bipolar outflow (Rodríguez et al. 1986). However, the 2.6 mm studies of Looney, Mundy, & Welch (1997) suggested that the source was actually double on the subarcsecond scale. Finally, the 7 mm observations

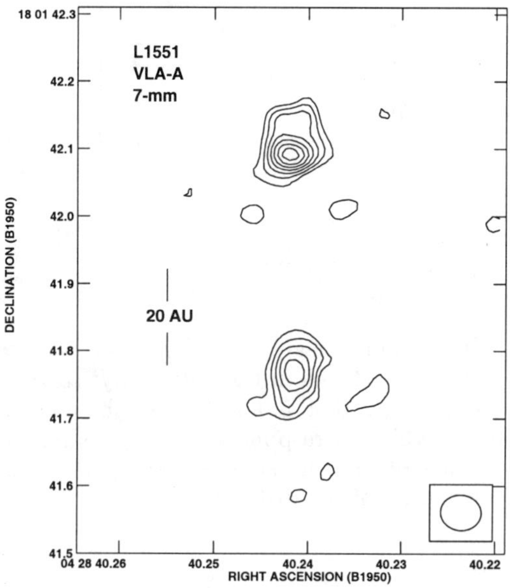

Figure 3. Cleaned, natural-weight VLA map of the L1551 IRS5 region at 7 mm. Two compact sources are evident in the map. Contours are –4, –3, 3, 4, 5, 6, 7, 8, and 9 × 0.24 mJy beam^{-1}, the rms noise of the map. The emission shown in this map is interpreted to arise from two compact protoplanetary structures forming a gravitationally-bound binary system.

of Rodríguez et al. (1998) clearly showed that L1551 IRS5 is a binary source (see Figure 3).

The sources have a projected separation of 45 AU. Both sources are spatially resolved and appear elongated perpendicular to centimeter emission that traces ionized jets from the stars. Both sources have semi-major axis dimensions of about 10 AU. The observed orientations, as well as the flux densities, imply that these structures are compact dust disks around the members of a binary system. The masses of the individual disks are estimated to be of order 0.05 solar masses, enough to form planetary systems like our own. However, these disks are an order of magnitude smaller in radial extent than those observed around single stars (see, for example, the image of HL Tau in Figure 2). This smaller size is attributed to tidal truncation. These results suggest that the binary nature of young stars such as L1551 may inhibit the formation of outer, giant planets but not of inner, terrestrial planets.

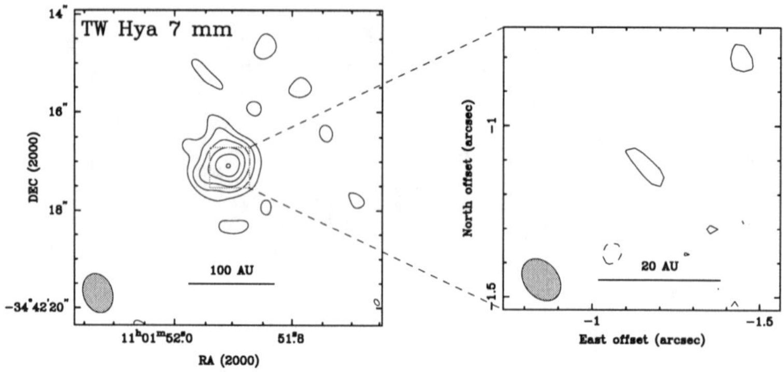

Figure 4. VLA 7 mm images of TW Hya at two resolutions. The synthesized beams are shown in the *bottom left corners of the panels*. Contour levels are −2, 2, 3, 4, 5, and 6 times the noise levels of 0.5 and 0.4 mJy beam^{-1}, which correspond to brightness temperatures of 0.66 and 20 K. The lack of compact structure suggests a shallow power law for the surface density of the disk.

How common are compact protoplanetary disks? Are they confined to binary systems? At present we are undertaking additional observations to address these issues.

4. TW Hya

Possibly the closest analog to our early solar nebula, TW Hya is part of the TW Hya association, at only 50 pc from the Sun. Because of its proximity and interesting age, this association has proved to be fertile ground for observations of phenomena related to the structure and evolution of circumstellar disks (see Wilner et al. 2000). We recently undertook VLA observations of this star that clearly show the nearly face-on disk of this source also seen in *Hubble Space Telescope* NICMOS and WFPC2 images (Weinberger et al. 1999; Krist et al. 1999)

The dust emission from TW Hya is best fitted with a power-law disk model having surface density falling as r^{-1}, as shallower or steeper power laws produce too little or too much signal at high resolution, especially at radii smaller than 10 AU. The observed structure conforms to that derived in self-consistent calculations of gaseous accretion disks with heating dominated by irradiation from the central star (D'Alessio et al. 1997, 1998).

Acknowledgments. LFR acknowledges the continued support from DGAPA, UNAM, and CONACyT, Mexico.

References

Carilli, C. L., & Holdaway, M. A. 1997, Application of Fast Switching Phase Calibration at mm Wavelengths on 33 km Baselines, VLA Scientific Memo No. 173 (NRAO: Socorro)

D'Alessio, P., Calvet, N., & Hartmann, L. 1997, ApJ, 474, 397

D'Alessio, P., Cantó, J., Calvet, N., & Lizano, S. 1998, ApJ, 500, 411

Dutrey, A., Guilloteau, S., Prato, L., Simon, M., Duvert, G., Schuster, K., & Menard, F. 1999, A&A, 338, L63

Feigelson, E. D., & Montmerle, T. 1999, ARA&A, 37, 363

Krist, J. E., Stapelfeldt, K. R., Burrows, C. J., Menard, F., & Padgett, D. L. 1999, AAS Meeting # 194, 69.05

Lay, O. P., Carlstrom, J. E., Hills, R. E., & Phillips, T. G. 1994, ApJ, 434, L75

Looney, L. W., Mundy, L. G., & Welch, W. J. 1997, ApJ, 484, L157

Looney, L. W., Mundy, L. G., & Welch, W. J. 2000, ApJ, 529, 477

Mundy, L. G., Looney, L. W., Erickson, W., Grossman, A., Welch, W. J., Forster, J. R., Wright, M. C. H., Plambeck, R. L., Lugten, J., & Thornton, D. D. 1996, ApJ, 464, L169

Rodríguez, L. F., Cantó, J., Torrelles, J. M., & Ho, P. T. P. 1986, ApJ, 301, L25

Rodríguez, L. F. 1997, in IAU Symp. 182, Herbig-Haro Flows and the Birth of Low Mass Stars, eds. B. Reipurth & C. Bertout (Dordrecht: Kluwer), 83

Rodríguez, L. F., D'Alessio, P., Wilner, D. J., Ho, P. T. P., Torrelles, J. M., Curiel, S., Gómez, Y., Lizano, S., Pedlar, A., Cantó, J. & Raga, A. C. 1998, Nat, 395, 355

Shu, F. H., Najita, J., Ostriker, E., Wilkin, F., Ruden, S., & Lizano, S. 1994, 1994, ApJ, 429, 781

Weinberger, A. J., Schneider, G., Becklin, E. E., Smith, B. A., & Hines, D. C. 1999, AAS Meeting # 194, 69.04

Wilner, D. J., Ho, P. T. P., & Rodríguez, L. F. 1996, ApJ, 470, L117

Wilner, D. J., Ho, P. T. P., Kastner, J. H., & Rodríguez, L. F. 2000, to appear in ApJ (Letters)

Wilner, D. J., Ho, P. T. P., & Rodríguez, L. F. 2000, in preparation

Discussion

Thomas Henning: Did you see any evidence for sources where the dust continuum radiation dominates even at cm wavelengths?

Luis Rodríguez: Remarkably, there are a handful of sources where a single power law fits the continuum spectrum from the millimeter down to 3.6 cm. Since it is not expected that dust emission will still be important at those long wavelengths, we need to study these few sources in great detail to determine the origin of the centimeter emission.

Paul Ho: Luis, with reference to the question earlier today about setting limits on the central density enhancements, what do the TW Hya data say?

Luis Rodríguez: We have made 7 mm maps of the source both with high and moderate angular resolution, and from these we can rule out that more than 20% of the total flux is coming from the inner 5 AU. The data are quite consistent with a surface density dropping as radius^{-1}.

Lee Hartmann: Can you tell where the outer edge of the TW Hya disk is?

Luis Rodríguez: We fit the 7 mm observations better by truncating the disk at a radius of 100 AU. Other indicators could well show a different size.

François Ménard: Did you derive a mass for TW Hya?

Luis Rodríguez: Yes, with the regular assumption we get a total mass (gas + dust) of $M_{\rm d} = 0.03\ M_{\odot}$.

Disks, Planetesimals, and Planets
ASP Conference Series, Vol. 219, 2000
F. Garzón, C. Eiroa, D. de Winter, and T. J. Mahoney, eds.

Rotation in Young Stars and the Evolution of Disks

Mark Casali
UK Astronomy Technology Centre, Royal Observatory, Blackford Hill, Edinburgh EH9 3HJ, UK

Thomas P. Greene
NASA Ames Research Center, MS 245-6, Moffett Field, CA 94035, USA

Carlos Eiroa
Dpto. Fisica Teorica C-XI, Facultad de Ciencias, Universidad Autónoma de Madrid Cantoblanco, E-28049 Madrid, Spain

Philippe Andre
CEA, Service d'astrophysique, Saclay, F91191, Gi sur Yvette Cedex, France

Abstract. A sample of obscured and unobscured YSOs has been observed spectroscopically in the infrared to determine their rotational velocities. By using the slope of the CO bandhead at 2.29 μm and the relative strength of the lower vibrational–rotational lines, rotational velocities were measured to better than 20 km s^{-1}. The resulting velocity distribution shows that the majority of objects are slow rotators, indistinguishable from classical T Tauri stars. Furthermore, rotational velocities decrease with decreasing CO band strength caused by veiling from a hot inner disk, consistent with other optical studies. The weight of the evidence is in favor of the existence of strong torques and rotational braking even in very young obscured stars, possibly through magnetic interaction with their surrounding disks.

1. Introduction

Studies of the rotational velocities and periods of young stars in clusters of differing ages have provided a great deal of information on the angular-momentum evolution of stars to the main sequence. Studies include clusters of different ages such as IC 2391/2602 (25 Myr, Stauffer et al. 1997a), Alpha Persei (50 Myr, Prosser 1992), the Pleiades (100 Myr, Queloz et al. 1998) and the Hyades (600 Myr, Radick et al. 1987).

A relatively new application for rotational studies has been in the area of T Tauri stars which are very young (<10 Myr) and not significantly affected by the wind braking which is believed to dominate as low-mass stars approach the main sequence. Furthermore, since stars undergo rapid changes in radius during the early pre-main-sequence phase, major changes are expected to occur in rotational velocities at this time. One of the most important results has been that young stars with indicators of a hot inner disk (mainly classical T Tauri stars—CTTS) are relatively slow rotators with periods of 5–12 days, while the supposedly more evolved Weak-lined T Tauri stars (WTTSs) rotate in 3–7 days (Ghosh 1995). This difference has been interpreted as evidence for an interaction between a strong stellar magnetic field and accretion disk in which magnetic torques act to slow the star down.

Attempts to study whether slow rotation is typical in stars which might be even younger than classical T Tauri stars run up against two problems. First, the youngest objects tend to be embedded in dark clouds and can therefore be obscured by many magnitudes of visual extinction. Second, young stars suffer significant veiling both in the UV/optical and the infrared, so that absorption lines can be difficult to detect. In order to overcome the former, observations need to be made in the infrared. The second problem is more difficult to overcome but can be eased if high S/N ratios can be achieved, allowing the detection of even weak lines. So in an attempt to extend rotational studies to younger, obscured objects a program of echelle IR spectroscopy was carried out. The sample for our study was taken from various well-studied young regions of star formation in Ophiuchus, Serpens, and Taurus.

2. Observations and Data Reduction

Observations of the YSOs were made in 1994 September and 1995 June, using the cooled grating spectrometer CGS4 on the United Kingdom Infrared Telescope (UKIRT) at Mauna Kea, Hawaii. All observations were with the echelle grating, together with a 62 × 58 InSb array in September and a 256 × 256 InSb array in 1995 June. The single-pixel wide slit gave instrumental resolutions of approximately 20 km s^{-1}.

In most cases the CO bandhead was clearly visible with a continuum signal-to-noise ratio > 10. Simulations with various degrees of broadening showed that there are two spectral features which are very useful in determining $v \sin(i)$. These are i) the slope of the normally sharp bandhead at 2.293 μm and ii) the peak-to-peak amplitude of the resolved vibrational–rotational lines longward of the bandhead relative to the depth of the bandhead itself. The latter indicator depends sensitively on broadening (i.e, the individual lines vanish altogether for broadening velocities greater than about 30 km s^{-1}) and is very useful for measuring the lowest rotational velocities, while for large velocities only the slope of the bandhead could be fitted.

The determination of $v \sin(i)$ proceeded using a direct fitting procedure for each spectrum. In all cases a 2.29 μm spectrum of the K7 giant star BS 341 was used as a template and was artificially broadened using a standard limb-darkened rotational broadening function (Gray 1992) until the best fit was achieved with the YSO spectrum. The depth of the CO feature was also adjusted continuously for the best fit since the feature is much weaker in T Tauri stars than in giants due to its inverse surface-gravity dependence. An example of the fitting procedure applied to two YSOs in the sample is shown in Figure 1.

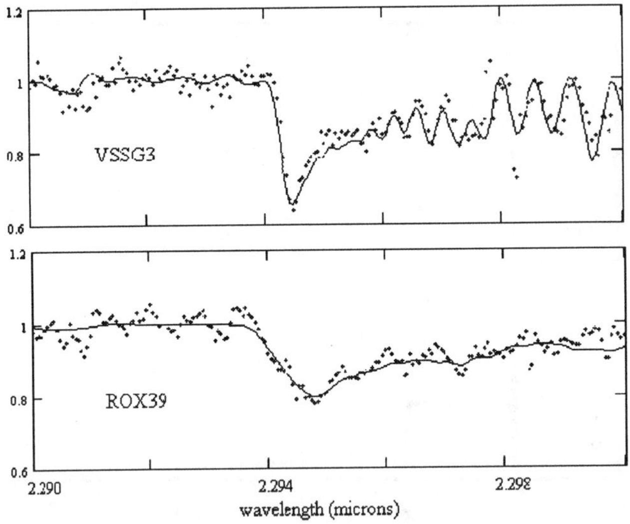

Figure 1. The fitting procedure for rotational velocity applied to VSSG3 and ROX39.

3. The Velocity Distribution

Figure 2 shows a histogram of the observed velocity distribution. There is a smooth decrease in number from the lowest- to highest-velocity sources (ROX 39 in ρ Oph). This is broadly consistent with existing (visible) observations of rotational velocities in young clusters as shown in Table 1. The table shows the velocities of the 50% (median) and 90% (9th decile) partitions. Note that the velocities are *observed* velocities only. Although a much larger amount of data is now available in the literature for rotational periods, these are not straightforward to convert to velocities since both stellar radii (and hence ages) and the projection sin(i) contribution need to be factored in.

The table shows that IR-determined $v \sin(i)$ are consistent with those observed by visible spectroscopy in other young clusters (e.g., Taurus–Auriga). In particular, there appears to be an absence of fast rotators compared to older clusters such as Alpha Persei in which substantial spin-up of cluster members has occurred. This confirms that nothing peculiar appears to happen while YSOs are still obscured. Our data support the idea that, on average, obscured young stars have similar rotational characteristics to clusters of visible T Tauri stars.

4. CO Depth

CO depth increases with decreasing spectral type or decreasing stellar mass at a given age. However, a stronger factor affecting the depth of this band as discussed by Casali & Eiroa (1996) is the IR veiling often present in YSOs and T Tauri stars. The veiling, probably due to dust emission, takes the form of

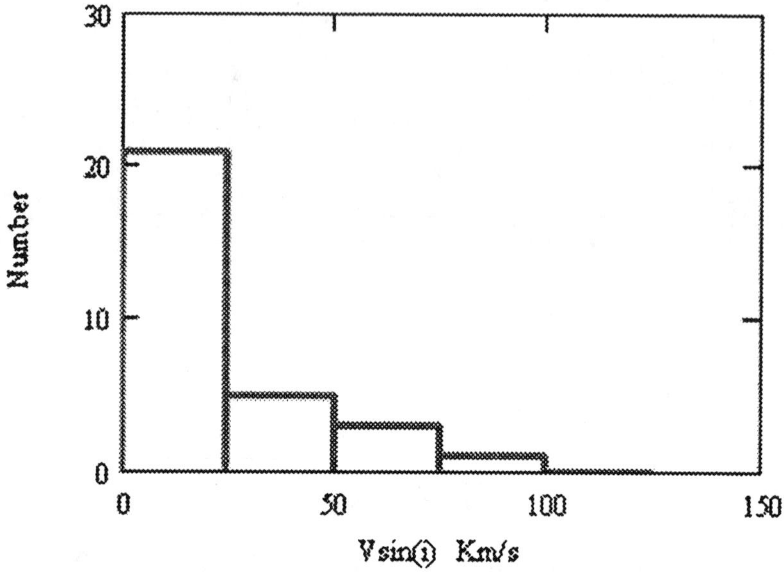

Figure 2. Histogram of $v \sin(i)$ for the observed YSOs.

Table 1. Rotational velocities in young clusters

Study	$v \sin(i)$ 50% (km s^{-1})	$v \sin(i)$ 90% (km s^{-1})	Age (Myr)	Notes/refs.
current	<20	50±7	few	
Taurus–Auriga	18	50	2	all masses (1)
IC2391/2602	20	90	30	(2)
Alpha Persei	15	80	50	(3)
Pleiades	6	65	100	0.6–0.9 $M_{\odot}$(4)
	24	65	100	1.1–1.5 $M_{\odot}$(4)
Hyades	<6	12	600	(5)

References: 1. Bouvier et al. (1986) 2. Stauffer et al. (1997a) 3. Stauffer et al. (1985) 4. Queloz et al. (1998) 5. Stauffer et al. (1997b) an added continuum and can completely dominate over the photosphere giving a typically featureless near-IR spectrum in Class I sources. For Class II objects, such as CTTSs, the veiling results in a weakening of observed spectral features, especially the CO bandhead, since it lies in the K band, where dust emission contributes significantly. The CO depth can therefore be taken as an approximate measure of the degree of veiling, and hence of the amount of hot circumstellar material. The measured $v \sin(i)$ is plotted against the CO band depth for each object in Figure 3. Despite significant scatter, there is a fairly clear trend, in that objects with strong CO tend to be faster rotators. The trend for higher rotational velocities to be found in YSOs with strong CO can be interpreted as a correlation between the rotational velocities and the absence of hot circumstellar material. This is in apparent agreement with other studies (Ghosh 1995), which find that the fastest rotators are those without evidence of infrared excesses and other CTTS features, an observation which has a plausible explanation in forced corotation of the stellar photosphere with an inner circumstellar disk (if present), via strong magnetic fields.

5. Fast Rotators in the ρ Ophiuchi Core

In a VLBI study of the core of ρ Oph, Andre et al. (1992) found a population of apparently diskless objects with strong gyrosynchrotron radio emission. They suggested the strong magnetic fields may be relic fields left over from the compression of the field in the ISM during protostellar collapse. A subset of these (ROX39, DoAr21, VSSG11, ROX31, and VSSG22) was among our sample of YSOs studied spectroscopically. They are all revealed as relatively fast rotators with a $v \sin(i)$ of 90, 75, 70, 55 and 20 km s^{-1}, respectively. Their CO bands have little veiling, indicating little if any circumstellar hot dust.

References

Andre, P., Deeney, B. D., Phillips, R. B., & Lestrade, J. F. 1992, ApJ, 401, 667

Bouvier, J., Bertout, C., Benze, W., & Mayor, M. 1986, A&A, 165, 110

Casali, M. M. & Eiroa, C. 1996, A&A, 306, 427

Gray, D. 1992, The Observation and Analysis of Stellar Photospheres (Cambridge: Cambridge University Press)

Ghosh, P. 1995, MNRAS, 272, 763

Prosser, C. F. 1992, AJ, 103, 488

Queloz, D., Allain, S., Mermilliod, J. C., Bouvier, J., & Mayor, M. 1998, A&A, 335, 183

Radick, R. R, Thompson, D. T., Lockwood, G. W., Duncan, D. K., & Bagget, W. E. 1987, ApJ, 321, 459

Stauffer, J. R. et al. 1997a, ApJ, 479, 776

Stauffer, J. R. et al. 1997b, ApJ, 479, 776

Stauffer, J. R., Hartmann, L. W., Burnham, J. N., & Jones, B. F. 1985, ApJ, 289, 247

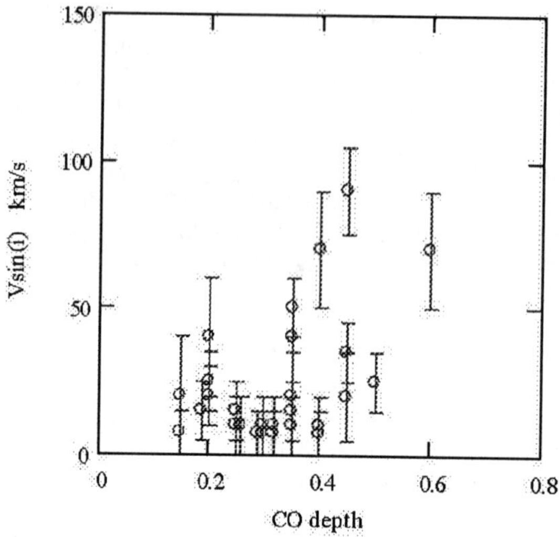

Figure 3. The CO depth plotted as a function of $v\,\sin(i)$.

Discussion

William Borucki: Isn't the depth of the CO band at 2.1 μm a function of metalicity as well as veiling?

Mark Casali: The measurements are all for nearby stars with metalicity similar to the Sun.

Bernard Foing: The veiling at 2.3 μm you determine would come from dust emission at $\sim$ 1000 K. How do you distinguish the effect of CO emission from the inner part of the disk?

Mark Casali: CO emission is rare and generally broadened by $\sim$ 100 km s^{-1} since it comes from the inner disk. This would make a very distinctive spectral signature (e.g., emission on the short-wavelength side of the bandhead), which we do not see.

Disks, Planetesimals, and Planets
ASP Conference Series, Vol. 219, 2000
F. Garzón, C. Eiroa, D. de Winter, and T. J. Mahoney, eds.

Photometric Studies of Young Stars in Clusters: Rotation and Disks

W. Herbst[1]

Astronomy Dept., Wesleyan University, Middletown, CT 06459, USA

K. L. Rhode

Dept. of Astronomy, Yale University, New Haven, CT 06520, USA

L. A. Hillenbrand

Astronomy Dept., California Institute of Technology, Pasadena, CA 91125, USA

J. A. Maley

Astronomy Dept., Wesleyan University, Middletown, CT 06459, USA

C. Bailer-Jones, R. Mundt, K. Meisenheimer and R. Wackermann

Max-Planck-Institut für Astronomie, Königstuhl 17, 69117 Heidelberg, Germany

Abstract. We report on photometric monitoring programs aimed at three extremely young clusters in the vicinity of the Sun: the Orion Nebula Cluster (ONC), IC 348, and NGC 2264. Rotation periods for more than 200 pre-main-sequence stars with masses between ~ 0.1 and 1 $M_\odot$ and ages between ~ 1 and 3 Myr have been discovered as a result of these programs, and this number will soon double. The distribution of rotation periods in IC 348 is indistinguishable from the ONC and the combined sample of stars with $M > 0.25$ is clearly bimodal with peaks near 2 days and 8 days. NGC 2264 is different, possibly because of a higher age. A correlation between infrared excess emission, indicative of an accretion disk, and corotation radius in units of the stellar radius is shown, which supports the canonical view that disk locking is an important phenomenon in the angular momentum evolution of stars. The half-life for disks based on our analysis is only ~ 1.3 Myr, which puts severe constraints on the time available for giant planet formation.

[1]Also at Max-Planck-Institut für Astronomie, Königstuhl 17, 69117 Heidelberg, Germany.

1. Introduction

The question of exactly when and how a star manages to rid itself of the excess angular momentum contained in a typical interstellar cloud, which it needs to do to end up on the main sequence rotating with less than critical velocity (v_c) is an important and long-standing one. Spectroscopic studies of the youngest optically visible stellar objects, T Tauri stars, showed in the 1980s that these pre-main-sequence (PMS) objects typically rotate at only $\sim 0.1\ v_c$ (Vogel & Kuhi 1981; Hartmann et al. 1986; Bouvier et al. 1986). It was proposed, by Königl (1991) and later by Shu et al. (1994), that a magnetic connection between the star and the disk could regulate the star's angular velocity to the Keplerian value at some "corotation" radius. This phenomenon, often called disk locking, has become an important ingredient in modeling the evolution of angular momentum during the first 1–10 Myr of the life of low-mass stars (cf. Bouvier, Forestini, & Allain 1997; Krishnamurthi et al. 1997), since it can act as an efficient drain of stellar angular momentum during the crucial early phase of pre-main sequence-contraction.

During the 1990s, an increasing number of rotation periods have been found for 1–10 Myr old PMS stars by detecting periodic variations in optical brightness caused by the rotation of their spotted surfaces. (see Herbst et al. 2000b, and references therein). Two results from this work bear directly on the possible relationship between stellar rotation and disks. First, the period distribution is bimodal, suggesting a population of stars which is either still locked to their disks or only recently released—the slow rotators—and a population which is not interacting with its disk and has not done so for some time—the rapid rotators. Bouvier et al. (1993) argued that the weak T Tauri stars (WTTSs) in their sample, drawn mostly from the Tau–Aur region, rotated faster than the classical T Tauri stars (CTTSs), in agreement with this suggestion. Second, Edwards et al. (1993) showed that near-infrared excess emission, attributable to disks, was found only around stars with longer rotation periods in the ONC, also supporting the idea of disk locking. Stassun et al. (1999) questioned the hypothesis based on their study of a large area in Orion which included the ONC, since their period distribution was not significantly bimodal. However, Herbst et al. (2000b) showed that there was no discrepancy between the datasets for ONC stars when allowance is made for the mass of the stars and its influence on rotational evolution. Here we summarize, update, and combine the results from published work (and work in progress) with emphasis on the evidence linking stellar rotation to accretion disks. Of particular interest in the context of this meeting is the lifetime of accretion disks since that bears directly on planetary-formation scenarios, as reviewed by Boss elsewhere in this volume (p. 7).

2. Observations

The observational data available for this study are summarized by Herbst et al. (2000b) and Herbst, Maley, & Williams (2000a). The distance to IC 348 is uncertain since traditional methods, as reviewed by Herbig (1998), point to a larger value than is obtained when *Hipparcos* distances to nine cluster members are averaged (Scholz et al. 1999). This makes the age uncertain, since it is determined by comparing the location of pre-main-sequence stars on the HR diagram with model calculations. Here we adopt the Herbig (1998) distance and age, which makes IC 348 only slightly older than the ONC, at a mean age of 1.3 Myr. Of course, in all of the clusters there is significant scatter in the derived luminosity of pre-main-sequence stars of a given spectral type which, if interpreted as an age spread, would mean the age spread is as large as or larger than the mean age of the cluster. We choose not to regard individual ages of stars as significant, however, since the random and systematic errors of determining them are large enough to explain all of the scatter, in our opinion. Unfortunately, there is no comprehensive study of NGC 2264 yet published which can provide masses, ages, and disk properties for stars with rotation periods in that cluster. We therefore have only a rough estimate of its age, ~ 3 Myr (Makidon et al. 1996), which suggests that it is about 3 times older than the ONC and IC 348.

A major monitoring program, which is not yet completed, will soon increase our database of rotation periods by a factor of 2 or 3. More than 50 nights of data were obtained on the ONC with the newly commissioned wide-field imager (WFI) built by the Max-Planck-Institute für Astronomie in Heidelberg. The instrument is attached to the 2.2 m Telescope at ESO in Chile and was used during the period 1998 December 25 to 1999 February 28 to monitor the ONC every clear night, with multiple observations on some nights. An intermediate-band filter centered near 790 nm was used, eliminating the brighter lines from the nebula and roughly matching the Cousins I band, in which all of the other optical monitoring has been done. The field of view of this instrument is ~ 30 $'$ on a side, so the entire cluster is included on an image. Typical seeing was less than 1 $''$ and the weather was generally good, so there are no large gaps in the time coverage. This is a uniquely valuable dataset on the ONC which promises to yield many new rotation periods and confirmation of existing ones. Data reduction was not sufficiently advanced yet at the time of writing to allow inclusion of any rotation periods directly in this study; however, we note that ~ 380 stars with significant periods have been found, more than doubling the number known.

3. Results

Figure 1 (*top panel*) shows a principal result of this work—the period distribution for stars with masses exceeding 0.25 $M_{\odot}$ in the ONC and IC 348. Combining results for these clusters is justified by their nearly identical distributions (Herbst et al. 2000a) and ages. We exclude lower-mass stars based on the discussion by Herbst et al. (2000b) who argue that deuterium burning is keeping them somewhat inflated and, therefore, spinning slower at a given age (near 1 Myr) than their more massive counterparts. Mixing the two mass ranges (i.e., stars with $M > 0.25M_{\odot}$ and those with less than that value) is largely responsible for producing the apparent differences in the period distributions in the ONC between Stassun et al. (1999) and earlier work (cf. Herbst et al. 2000b). The distribution in Figure 1 (*top*) represents our best estimate of the distribution of rotation periods of stars with masses between 0.25 and 1 $M_{\odot}$ at an age of ~ 1 Myr. It is clearly broad, with a range of a factor of 10, and bimodal, with peaks near 2 days and 8 days. What is the origin of this peculiar distribution, and can

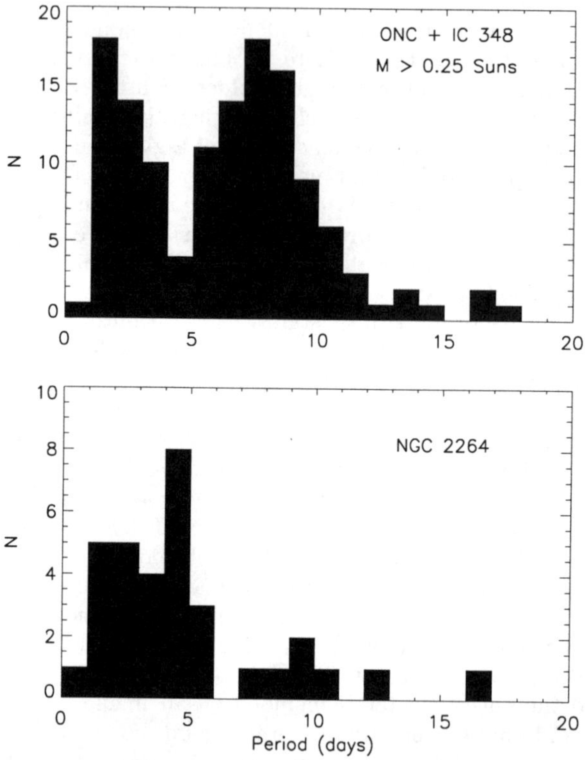

Figure 1. The distribution of rotation periods for stars in three extremely young clusters. Data are from Herbst & Kearns (1998), Stassun et al. (1999), Herbst et al. (2000b) and Herbst et al. (2000a).

its evolution with time be understood? We briefly address the latter question before spending most of the remainder of this contribution on the former.

The *bottom panel* of Figure 1 shows the period distribution for NGC 2264 from Kearns & Herbst (1998) and it is clearly different from the other clusters. Lacking masses and ages for stars in NGC 2264, we can speculate that the difference is basically an age effect, although mass could be involved as well. The bulk of stars rotating with periods of around 8 days at an age of 1 Myr should spin up to about 4 days at 3 Myr if they are decoupled from their disks. This phenomenon could account for the concentration of stars to periods near 4 days in NGC 2264. About 20% of the stars have much longer periods, indicating that they are still interacting with their disks, or were for most of the last 3 Myr. This is roughly consistent with the fraction of long-period stars in the ONC that still have accretion disks according to infrared-excess emission data (see below). It suggests that the typical half-life for a disk around 0.25–1 solar mass stars in clusters is $\sim$ 1.3 Myr. The lack of a very rapidly rotating set of stars—the spun-up versions of the stars in the short-period peak of the ONC and IC 348 distributions—may be an observational selection effect. They should have periods of about 1 day, which are hard to find when observing from a single longitude on Earth, since the power in the window function of the periodogram is large there. The further evolution of angular momentum as the stars age is beyond the scope of this paper and is unlikely to involve disks in any event.

Returning to the question of the origin of the breadth of the period distribution and its bimodality we note that it must, in some way, be connected with disks since it arises within the first 1 Myr of a star's existence, during which time an accretion disk must certainly have been present around most or all stars. Elsewhere in this volume, Casali (p. 115) discusses his discovery that embedded sources not yet visible at optical wavelengths have a $v \sin i$ distribution similar to the TTSs, generally supporting a role for disk locking in the earliest PMS or protostar phases. Here we follow Edwards et al. (1993) and Herbst et al. (2000b) to inquire whether a direct link between slow rotation and the presence of an accretion disk can be established for stars in the ONC. First, we note that such a correlation cannot be present in IC 348, since none of the stars with known rotation periods are CTTSs—they are all WTTSs. This illustrates one of the difficulties in establishing a direct link; namely, that the presence of an accretion disk makes it difficult to find the rotation period of a star. CTTSs usually show erratic photometric behavior, which is caused by accretion. They will be, therefore, underrepresented in any sample of periodic stars relative to the WTTS of the same mass and age. In IC 348, which has only been monitored for a single season, 17 of the 49 WTTSs in the monitored samples have rotation periods, but none of the 16 CTTS have them. In the ONC we do not yet have a division of our stars into WTTSs and CTTSs based on Hα emission equivalent widths owing to the complications of the nebular emission. However, by comparing the $v \sin i$ distributions for stars with known rotation periods and those without them (Rhode, Herbst, & Mathieu 2000) it may be possible to gain insight into this issue.

Figure 2 (from Herbst et al. 2000b) shows the relationship between corotation radius (in units of stellar radius) and excess infrared emission, $\Delta(I-K)$, as defined by Hillenbrand et al. (1998) for ONC stars with known rotation periods. Large values of this quantity are thought to indicate the presence of an accretion disk, while small values do not prove that a disk is absent, since inclination effects and holes at the center can reduce the contrast of the disk relative to the photosphere. There is considerable scatter in this diagram, as expected, but a clear correlation is also visible. A Spearman rank order correlation test confirms that the correlation is significant, arguing for a direct connection between disks and rotation.

What does this correlation imply for the disks around TTSs in the ONC? The somewhat surprising answer is that it links a larger central hole in the disk (scaled to the size of the star) to greater near-IR emission. The size of the hole in the more extreme examples is so large that the disk should be disrupted well beyond the point at which the disk temperature (even if heated only by the star) would be too small to contribute significantly to the K-band emission. The disk temperatures at the point of corotation, based on heating by the star alone, is less than 900 K for all of the stars with significant IR excesses. Therefore, to understand the correlation in Figure 2, we must presume that either the dust is heated significantly in the vicinity of the corotation radius or that there

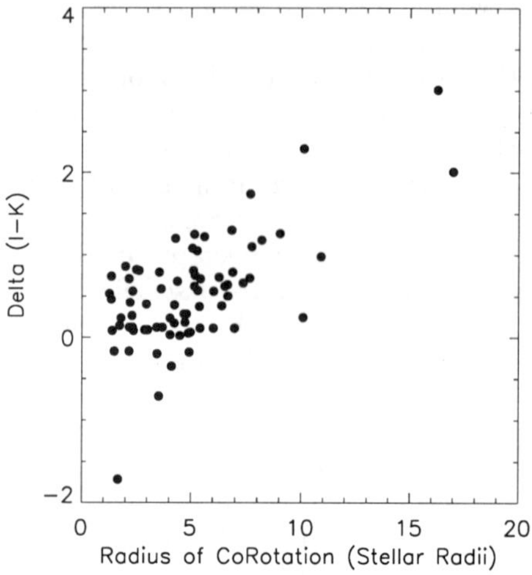

Figure 2. Infrared excess attributable to an accretion disk (see Hillenbrand et al. 1998) versus co-rotation radius as calculated from the rotation period and radius of the star (see Herbst et al. 2000).

is a considerable amount of dust present in the system beyond that which is contained in the truncated disk. This could be dust which is inside the corotation radius or, more likely, dust which is dragged by the gas into accretion columns. Observations at longer wavelengths and with time resolution in the infrared are needed to determine exactly what is happening.

4. Summary and Conclusions

We conclude that, at present, the disk-locking hypothesis seems to be the best explanation for a bimodal period distribution among the $\sim 1\ M_{\odot}$ stars at $\sim$ 1 Myr age in clusters. Further observations at longer infrared wavelengths are needed to ascertain the details of what is happening to the dust in disks which are truncated by a stellar magnetosphere. If rotation period can actually be used as a proxy for determining whether an accretion disk is present or absent, then the typical lifetime for a disk around a $\sim$ 0.75 solar mass star in a young cluster is only $\sim$ 1.3 Myr and is less for more massive stars. This puts a serious constraint on the amount of time that such stars have to form planets. Of course, this constraint may not exist for stars formed in loose associations such as Tau–Aur if their disks last longer. It will be important to compare, in the future, what the rotation studies of associations are telling us and whether it is consistent with the cluster studies.

References

Bouvier, J., Bertout, C., Benz, W., & Mayor, M. 1986, A&A, 165, 110
Bouvier, J., Cabrit, S., Fernandez, M., Martin, E. L., & Matthews, J. M. 1993, A&AS, 101, 485
Bouvier, J., Forestini, M., & Allain, S. 1997, A&A, 326, 1023
Edwards, S. et al. 1993, AJ, 106, 372
Hartmann, L., Hewett, R., Stahler, S., & Mathieu, R. D. 1986, ApJ, 309, 275
Herbig, G. H. 1998, AJ, 497, 736
Herbst, W., Maley, J. A., & Williams, E. C. 2000a, AJ, submitted
Herbst, W., Rhode, K. L., Hillenbrand, L. A., & Curran, G. 2000b, AJ, 119, 261
Hillenbrand, L. A., Strom, S. E., Calvet, N., Merrill, K. M., Gatley, I., Makidon, R. B., Meyer, M. R., & Skrutskie, M. F. 1998, AJ, 116, 1816
Kearns, K. E., & Herbst, W. 1998, AJ, 116, 261
Kögnil, A. 1991, ApJ, 370, L39
Krishnamurthi, A., Pinsonneault, M. H., Barnes, S., & Sofia, S. 1997, ApJ, 480, 303
Makidon, R. B., Adams, M. T., Strom, S. E., Kearns, K., Herbst, W., & Jones, B. F. 1996, BAAS, 189.4910
Rhode, K. L., Herbst, W., & Matheiu, R. D. 2000, in preparation
Scholz, R.-D., Brunzendorf, J., Ivanov, G., Kharchenko, N., Lasker, B., Meusinger, H., Preibisch, T., Schilbach, E., & Zinnecker, H. 1999, A&AS, 137, 305
Shu, F., Najita, J., Ostriker, E., Wilkin, F., Ruden, S., & Lizano, S. 1994, ApJ, 429, 781
Stassun, K. G., Mathieu, R. D., Mazeh, T., & Vrba, F. J. 1999, AJ, 117, 2941
Vogel, S. N., & Kuhi, L. V. 1981, ApJ, 245, 960

Discussion

Bernard Foing: How do you distinguish the effect of accretion hotspots or chromospheric faculae from cool spots? Do you use color photometric variability?

William Herbst: Our cluster monitoring is normally done in only one color—Cousins I. However, most periodic pre-main-sequence stars are weak T Tauri stars, whose variations are known (by the color behavior) to be caused by cool spots.

Lee Hartmann: K is a rather short wavelength to see disk emission, particularly with inner disk holes, so the poor correlation with $I - K$ might be better in terms of $I - L$, etc.

William Herbst: Yes.

Paul Ho: Are the statistics good enough to say whether the IR excess (longer period)–disk systems have more unusual behavior?

William Herbst: No, except for some weak correlation with position in the cluster. The disk systems and the long-period stars seem to concentrate towards the center of the Orion Nebula Cluster, which is also where the youngest stars concentrate.

Disks, Planetesimals, and Planets
ASP Conference Series, Vol. 219, 2000
F. Garzón, C. Eiroa, D. de Winter, and T. J. Mahoney, eds.

Images of Protoplanetary Disks from Millimeter to Near-Infrared Wavelengths

Anne Dutrey

Institut de Radio Astronomie Millimétrique 300 rue de la picine, F-38406, Saint-Martin-d'Hères, France

Abstract. Circumstellar disks around young stellar objects (YSOs) are thought to be the site of planetary formation. Over the past ten years they have been intensively observed at either millimeter (mm) or optical and near-infrared (NIR) wavelengths. I review here the physical properties inferred from observations at high angular resolution. I mainly focus on properties of disks surrounding pre-main-sequence (PMS) stars a few Myr old.

1. Introduction

Since 1990, the circumstellar gas and dust found around YSOs of a few ~ 0.1 to ~ 10 Myr have been intensively studied from the optical to mm wavelengths. Among young stellar objects, the so-called Class II phase and, to a lesser extent, the Class I phase (Adams et al. 1987), are likely to be very important for the formation of planetary systems. Class I and II low-mass stars (around 0.5–2 $M_\odot$) are indeed analogous to our Sun in its infancy, surrounded by a flattened structure of primary rotating gas and dust, the so-called protosolar nebula which provided the material to build the Solar System.

In the Class I phase, the protostar is still embedded inside the parent molecular cloud and is surrounded by an optically thick CO envelope while the central engine is emitting a powerful jet.

In the Class II phase, the pre-main-sequence low-mass star has become a classical line T Tauri star (CTTS). Contrary to Class I protostars, Class II objects are visible stars which have dissipated their molecular envelopes, and their main outflow activity is over.

Observations of circumstellar disks are not easy due to their small sizes (apparent size $\sim$ 1–6″) and to confusion with foreground and background gas and dust associated with parent molecular clouds. In such disks, except at a few AU from the star, the gas and dust, heated by the central star, remain at relatively low temperatures and radiate at long wavelengths, from the far infrared to millimeter. At optical and NIR wavelengths, dust also scatters the stellar light impinging the disk surface. In the last ten years, many disks have been observed with mm interferometers in the thermal continuum of the dust and CO lines. These observations reveal large disks in Keplerian rotation. In the meantime, the *Hubble Space Telescope* (*HST*) and adaptive optics systems on large ground-based optical telescopes have provided the first optical images of resolved dust disks in the scattering regime.

I will review here the observational properties of YSOs at mm wavelengths and compare them to the NIR and optical characteristics. For this purpose, I have chosen to focus on a few of the best known objects. More details on the observed properties of dust disks in the NIR and optical domains can be found in the contribution by Ménard et al. (this volume, p. 140).

2. Tracing Gas and Dust around YSOs

Most of the nearest star-forming regions, which form low-mass stars, such as the Taurus–Auriga clouds are located at 140 pc ($1'' = 140$ AU on a linear scale). At such a distance, the angular size of our Solar System (including the Kuiper Belt) would be about ~ 0.2–$0.3''$. The best angular resolutions currently available are ~ 0.3–$1''$ (~ 45–140 AU) and ~ 0.2–$0.5''$ (~ 30–70 AU) for mm and optical telescopes, respectively. Therefore studying in situ the formation of planets is beyond the scope of what we can do today.

Current images, either in the mm or optical domains, sample the properties of what we can call the "outer disk" as opposed to the standard protosolar nebula, which would correspond to the inner part of protoplanetary disks.

Figure 1 summarizes the different domains in circumstellar disks sampled today by mm arrays and large optical telescopes. The disk is associated with a Class II T Tauri star located at the distance of Taurus. For comparison with our Solar System, the approximate location of the Kuiper Belt is given.

2.1. The Gas Component

Carbon monoxide is the most abundant molecule in disks after H_2. CO, excited by collision with molecular hydrogen, has a low critical density (about 10^3 cm^{-3} at 20 K) which is easily reached in protoplanetary disks. Since CO remains thermalized, a simple 3D radiative transfer model assuming LTE conditions and an appropriate disk geometry following hydrostatic equilibrium allows us to analyze CO images produced by mm arrays.

As expected from chemical evolution (Aikawa et al. 1998), molecules appear to be depleted in disks with respect to the molecular abundance found in parent clouds (Dutrey, Guilloteau, & Guélin 1997). Hence, only CO interferometric images have the required sensitivity to allow quantitative studies of the disk properties.

The ^{12}CO $J = 1 \rightarrow 0$ and $J = 2 \rightarrow 1$ lines are optically thick and only sample the kinetic temperature $T_{\rm k}(r)$ of the disks. On the contrary, the ^{13}CO and C^{18}O $J = 1 \rightarrow 0$ and $J = 2 \rightarrow 1$ lines, which are partially optically thin, in the Rayleigh–Jeans approximation sample the product of the kinetic temperature by the surface density law $T_{\rm k}(r) \times \Sigma(r)$, hence the mass of the gas.

Finally, the kinematics of disks can be addressed only by spectroscopy. Resolved images of CO provide unique evidence that disks are in Keplerian rotation.

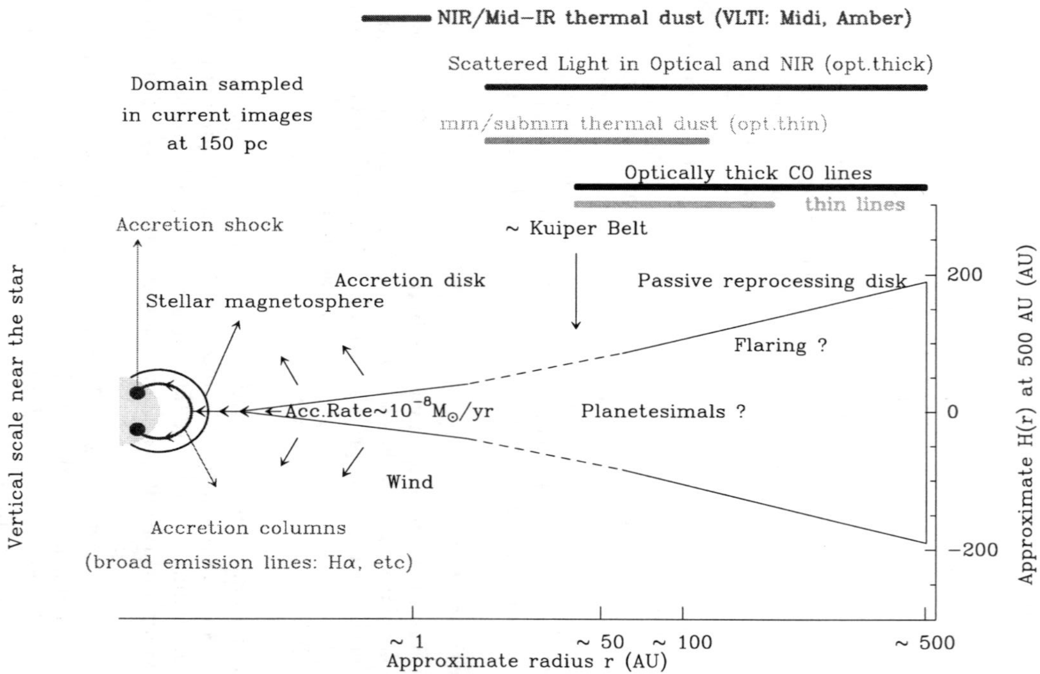

Figure 1. Global picture of a circumstellar disk surrounding a CTTS of $\sim 10^6$ years old (Vintage 2000). Partially adapted from Hartmann (1997). *Top:* domain sampled in images coming from current mm arrays and large optical telescopes, scaled for $D \simeq 150$ pc. *Bottom:* The star + disk are seen edge on. For comparison with our Solar System, the approximate location of the Kuiper Belt is given.

2.2. The Dust Component

The observed properties of dust in circumstellar disks are indeed different from those seen in the interstellar medium as expected for planetary formation. Dust particles (of radius a) are efficient absorbers of short-wavelength radiation (the UV–optical domain with $\lambda \leq a$). In equilibrium between heating and cooling, they re-emit a continuous spectrum at longer wavelengths that closely resembles a thermal spectrum. At shorter wavelengths, scattering of the light from external sources dominates, the limit between both regimes being around $\sim$ 3–5 μm. Very close to the star ($\sim$ 0.1–5 AU) the temperature is high ($\sim$ 500–1000 K) and the NIR/optical continuum is again dominated by the thermal regime (e.g., NIR interferometric observations of Fu Orionis by Malbet et al. 1998).

Thermal Regime In the mm/submm domain, the dust emission is mainly optically thin and a proper fit of the dust emissivity would in theory allow assessment of the total disk mass. Several groups (Beckwith et al. 1990; Mannings et al. 1994; Dutrey et al. 1996) now agree that the best fits of the absorption coefficient $\kappa_\nu = \kappa_o \times (\nu/10^{12}\ \mathrm{Hz})^\beta$ are obtained for $\kappa_o = 0.1$ with $\sim 0.4 \leq \beta \leq 1$ (including all material dust + gas, with a gas-to-dust ratio = 100). This value of β, smaller than the value found in molecular clouds ($\beta \simeq 2$), implies that grain growth is occurring. Indeed, current dust models fail to reproduce the high value of κ_ν and the low β observed in the mm/submm range (Pollack et al. 1994). Note also that inappropriate value of the gas-to-dust ratio, assumed to be similar to that in molecular clouds (100), may add important uncertainties in the mass determination. However, even takin these uncertainties into account, mm/submm observations of disks show that many of them have the critical mass to form planetary systems (Beckwith et al. 1990) with masses ranging from $\sim$ 0.001 to a few 0.1 $M_\odot$.

Scattering Regime In disks similar to those surrounding HH 30 or GM Aur, the column densities are high and optical depths of $\sim$ 1 are quickly reached in the optical and NIR, which thus mainly probe the disk surface. Contrary to mm/submm data, these observations do not allow an estimate of the disk mass without a large opacity correction and detailed modeling (single- or multiple-scattering models). Depending of the dust phase function, forward scattering is easier to produce than backward scattering as shown by many examples (e.g., GG Tau or UY Auriga, see Figure 4). Polarization, due to single or multiple scattering, also puts strong constraints on the dust disk properties and geometry. More details can be found in the contribution of Ménard et al. (this volume, p. 140).

Finally, to constrain the dust properties properly, multi-wavelength studies will give a clear improvement. Particularly, one would like to know if the dust can be fitted by a single size distribution explaining the properties from the mm to the optical range and how particles are located within the disk; e.g. do larger particles settle in the disk plane as expected?

3. Gas and Dust Morphology in YSOs

Observations of YSOs, either in the mm or the NIR and optical range, show confusion with remnant material along the line of sight. This confusion comes either from the parent cloud or from the molecular outflow associated to the young star. This can be a serious limitation for CO observations because CO lines are easily optically thick.

Furthermore, detailed studies require the selection of objects located in front of clouds or in regions devoid of dust and gas. Contrary to Class I protostars, this is the case for many Class II objects, even if some of them are still too embedded to allow quantitative analysis.

Figures 2 and 3 illustrate the problem of confusion by showing ^{12}CO $J = 2 \rightarrow 1$ channel maps of a few typical embedded objects (Figure 2) and Class II objects free of confusion (Figure 3).

3.1. Embedded Objects

Many embedded objects have now been observed in the mm domain and in NIR or optical. We present here the morphology of a few objects.

HH 30 Optical images from the *HST* (Burrows et al. 1996, their Figure 1) reveal a nearly edge-on flared disk of radius ~ 250 AU from which a collimated jet is emanating. *HST* monitoring of the disk reveals variable asymmetries within the reflection nebulosity of a timescale of around $\sim$ 3 years (Stapelfeldt et al. 1999). Such variability might originate in orbital motions of inner clumps. However, as shown in Figure 2, the high-angular-resolution observations of the CO line around HH 30 do not resemble a Keplerian CO disk. This object is still embedded in its parent molecular cloud and the CO emission is dominated by the CO flow originating from the young star (e.g., the channels on the right in Figure 2).

As in many cases for embedded objects, comparing the optical and mm properties (kinematics but also temperature, density, etc.) would require interferometric observations in optically thinner CO isotopes such as ^{13}CO or even $C^{18}O$.

YLW 15 is a Class I protostar strongly embedded in the Rho Oph molecular cloud. As in the case of HH 30, *HST* observations reveal a reflection nebula with a sharp absorption lane of $\sim 10''$ seen east–west, suggesting an optically thick disk nearly edge on. In this case, the $C^{18}O$ $J = 2 \rightarrow 1$ observations performed with the IRAM array show an east–west feature at the location of the optical dark lane. The kinematics of the $C^{18}O$ data are consistent with a rotating disk around a central object of at least $\sim 2\ M_\odot$ (Grosso 1999, Figure 15).

3.2. Class II Objects

Since Class II objects are less embedded (even though some confusion can remain, e.g., DL Tau, Figure 3) the more abundant isotope of CO, ^{12}CO, allows in many cases detailed studies of the properties and kinematics. Figure 3 shows that the CO patterns present strong similarities from one source to another. Indeed, all the CO structures presented here are consistent with disks in Keplerian

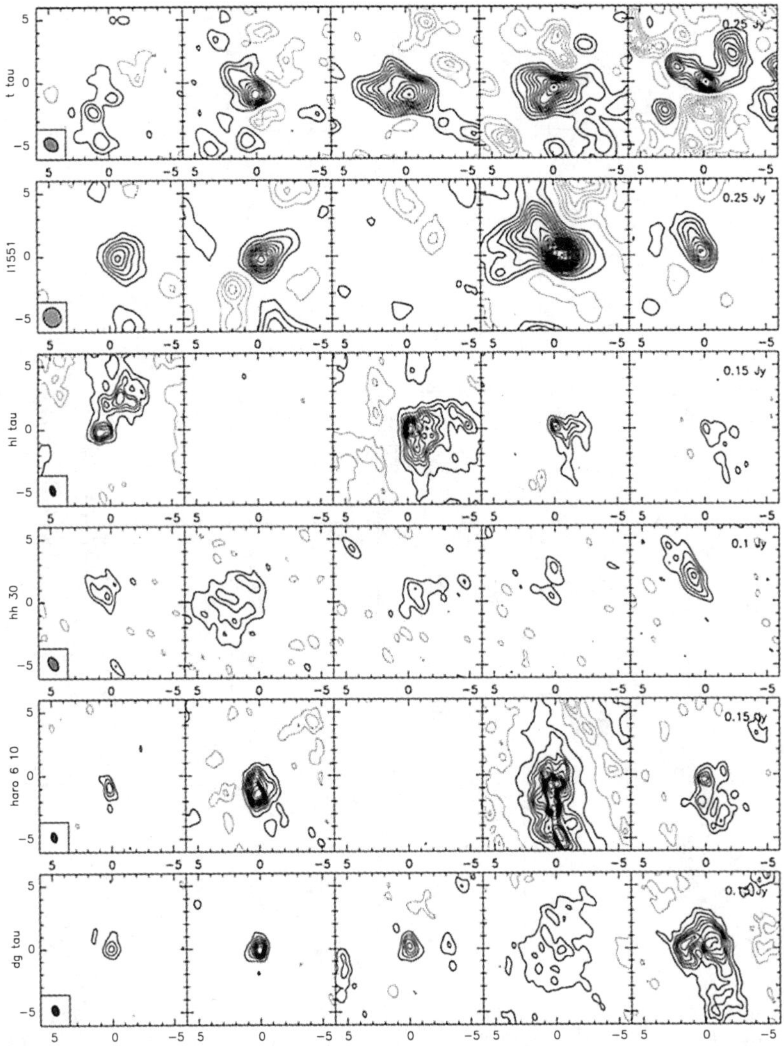

Figure 2. ^{12}CO $J = 2 \rightarrow 1$ channel maps of a few embedded YSOs in Taurus–Auriga, observed with the IRAM interferometer. The *dashed contours* are negative; the CO emission is extended and resolved, in several cases, most of the flux is missing. Confusion with the CO cloud can be seen in all sources as "blank" channels at the velocities where the $COJ = 2 \rightarrow 1$ emission from the cloud is optically thick. Moreover, the emission features from the objects do not show the characteristic pattern of rotation. Indeed, the CO emission is dominated by the outflow motions of the sources.

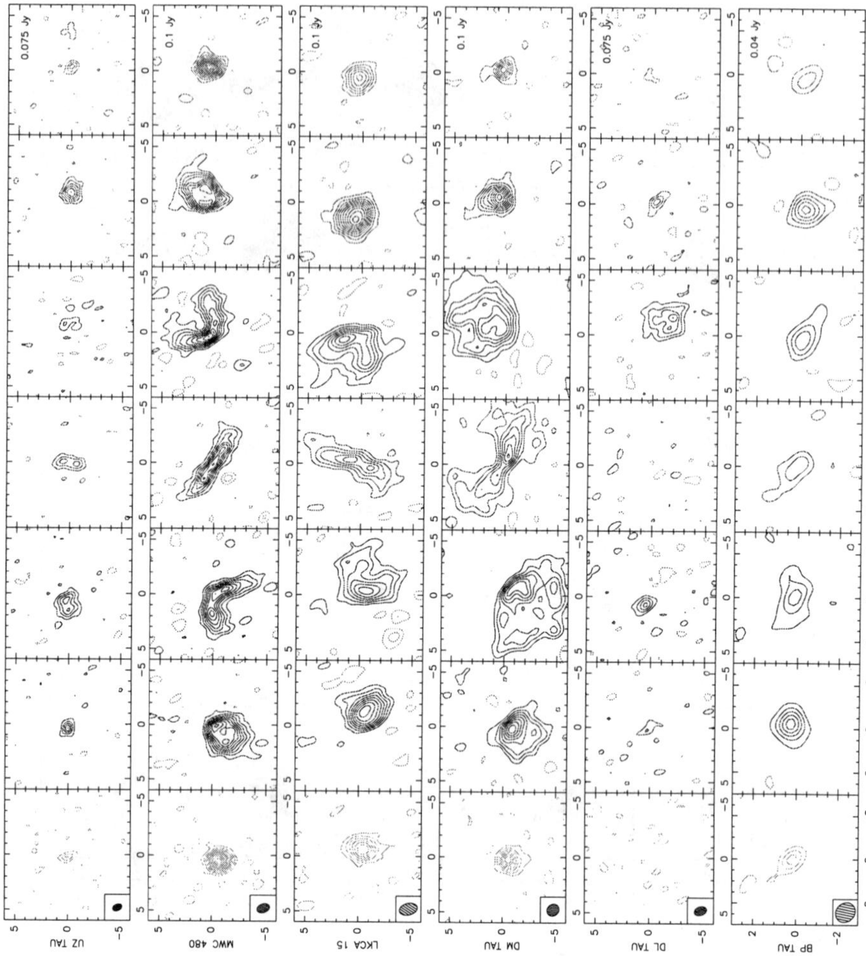

Figure 3. ^{12}CO $J = 2 \rightarrow 1$ channel maps of a few Class II objects in Taurus–Auriga, observed with the IRAM interferometer. In all cases, the CO emission shows the typical pattern of Keplerian disks orbiting the central stars. The central channel corresponds to systemic velocities (the "peanut" features). These sources are free of confusion with the surrounding CO cloud except for DL Tau. In the latter case, the CO $J = 2 \rightarrow 1$ emission from the cloud, optically thick at the DL Tau systemic velocity, hides the characteristic "peanut" pattern.

rotation around the central star. With an outer radius of 2–5″, the gas disks shown here also appear larger than the canonical value of $r_{out} = 100$ AU. Even if current mm arrays are biased toward the larger features, a significant fraction of T Tauri stars have large disks with typical outer radii of 300–1000 AU (Mundy et al. 2000).

GM Aur has been imaged from the mm to optical. A Keplerian CO disk around this YSO was first reported by Koerner et al. (1993). *HST* observations (Stapelfeldt et al. 1997) reveal a reflection nebula inclined by about 60° along the line of sight. Figure 4 (*top panel*) is a montage of IRAM observations showing three channels of the ^{12}CO $J = 2 \rightarrow 1$ line and the thermal dust emission at 1.3 mm (from Dutrey et al. 1998). The apparent size of the dust disk at 1.3 mm is smaller than those of the CO disk. The dust emission, very optically thin in the outer part of the disk, is resolved by the mm array and escapes its threshold of detection (see also Dutrey 2000 for details of mm array sensitivity compared to dust and gas distributions in disks). Comparison between the resolved IRAM observations and the *HST* data shows that the position angle, the inclination angle and the size of the disk agree. Indeed, CO emission and dust scattering the stellar light at optical wavelengths trace the *same material*: a Keplerian disk orbiting the T Tauri star and seen nearly edge on.

3.3. Binary Stars

It is now commonly recognized that most stars form in binary or multiple systems. Understanding the properties of the close gas and dust around multiple systems is therefore of prime importance with respect to planetary formation. Moreover, material around binary systems of separation of about 0.2–2″ are very suitable for study with current instruments either in the mm or at NIR and optical wavelengths.

GG Tau GG Tau is a CTTS binary of separation 0.26″ on the plane of sky. The binary is surrounded by a very massive (0.15 $M_{\odot}$) ring of gas and dust which was first detected by mm interferometry (Dutrey, Guilloteau, & Simon 1994) then in NIR scattered light (Roddier et al. 1996). More recent mm images obtained with the IRAM interferometer by Guilloteau, Dutrey,& Simon (1999) fully resolve the central hole (see Figure 7 from Guilloteau et al.). The disk appears to be in Keplerian rotation about a central binary of mass $\sim 1.3\ M_{\odot}$, the inner radius of the ring ($R_{in} \simeq 180$ AU) is in perfect agreement with predictions of tidal truncation. Moreover, the size of the central hole, the P.A. and the inclination angle of the disk determined from both the mm and the NIR data are in very good agreement. Recent polarization images at 1 μm confirm the geometry of the ring (Duchêne et al. , this volume, p. 169). These data clearly demonstrate that the dust observed in the thermal mm regime and in the scattering regime is at least partially mixed, and that the CO emission also originates from *the same dust and gas disk*.

UY Aur UY Aur is a CTTS binary of separation $\sim 0.89''$ on the plane of sky. Using the IRAM array, Duvert et al. (1998) have observed in ^{13}CO $J = 1 \rightarrow 0$ and $J = 2 \rightarrow 1$ a circumbinary disk in Keplerian rotation around the stars, while the continuum emission at 1.3 mm peaks on the binary itself and is associated

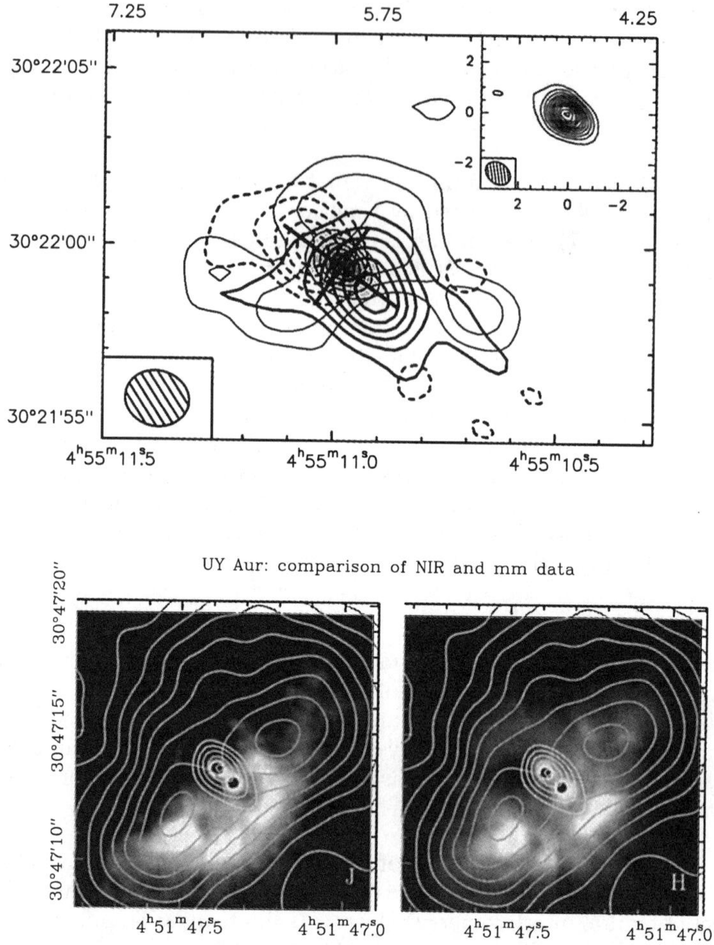

Figure 4. Gas and dust around single and binary stars: examples. *Top:* Single star GM Auriga (Dutrey et al. 1998, Figure 1). Three channels of ^{12}CO $J = 2 \rightarrow 1$ emission in the GM Aurigae disk superimposed on the thermal dust emission at 1.3 mm (grayscale). The continuum image is also shown in the insert. The cross gives the location of the major and minor axes of the disk. As expected for rotation, the velocity gradient of the CO emission is along the disk major axis. *Bottom:* Binary star: UY Aurigae (Close et al. 1998, Figure 3). Superposition of the ^{13}CO $J = 2 \rightarrow 1$ integrated-area map on the circumbinary reflection nebula in the J and H bands. The reflection nebula traces material producing forward scattering. As expected for a circumbinary disk truncated by tidal effect, the CO emission is not centrally peaked. Contrary to the dust scattering the stellar light, the thermal dust emission at 1.3 mm (white contours) originates from the inner disks.

with the inner circumstellar disks. Further NIR observations from the CFHT using the adaptive optics of the University of Hawaii and performed by Close et al. (1998) reveal a reflection nebula, which confirms the existence of the circumbinary disk, tidally truncated. As in the case of GM Aur, the dust which scatters the stellar light originates from the same location as the CO emission (Figure 4). The non-detection of the circumbinary material in continuum at mm wavelengths is well explained by the fact that, contrary to the GG Tau disk, this one is not massive. The optically thin mm emission is also resolved by the mm array and escapes its threshold of detection. Recently, Potter et al. (2000) have reported a resolved polarization image at 1.2 μm of the reflection nebula (their Figure 1*a*); these observations are in agreement with a large flattened disk of dust surrounding both stars (their Figure 4).

When comparing the morphology observed in the mm with those seen in NIR and optical, the previous examples show a wide variety of behavior. When confusion is not a limitation, these examples suggest that, due to opacity effects, disks are preferentially imaged in the NIR and optical when they are nearly edge on. For example, the DM Tau disk, tilted by $\sim 30°$ from pole on (see CO channel map, Figure 3), is only observed in the mm range.

In the same way, the non-detection of the circumbinary disk surrounding UY Aurigae in continuum at 1.3 mm can be explained as a result of interferometry on material having a low surface brightness (see also Dutrey 2000 for details).

4. Physical Properties of Class II Objects

Contrary to Class I objects, Class II sources allow quantitative studies because they are less embedded in their parent clouds. This is particularly true at mm wavelengths.

4.1. CO Modeling from mm Arrays

Mm arrays are now routinely providing maps of CO lines at angular resolutions of around ~ 0.5–$2''$ (e.g., MWC 480: Mannings et al. 1998; DM Tau: Guilloteau & Dutrey 1998; GM Aur: Dutrey et al. 1998). Some of these data possess both high angular and high spectral resolutions (~ 0.2–0.1 km s^{-1}), which are required for a quantitative analysis of the disk properties.

Guilloteau & Dutrey (1998) have demonstrated that a proper comparison between a standard disk model and a resolved CO image (by χ^2 minimization on disk model parameters to evaluate error bars) allows the quantification of physical parameters such as the temperature distribution. This kind of modeling also allows the retrieval of the mass of the central star from the dynamical mass of CO, independently of optical properties. This appears to be a powerful method for constraining PMS evolutionary tracks (Simon, Dutrey, & Guilloteau 2000).

Up to now, only a dozen T Tauri disks have been analyzed in this way. Dutrey et al. (2000) have applied this method to a sample of about 8 CTTSs using CO $J = 2 \rightarrow 1$ images from the IRAM interferometer. They find that: i) CO disks are in Keplerian rotation with $V(r) \propto [r/(100 \text{ AU})]^{-0.50\pm0.05}$, ii) the kinematic pattern indicates that the turbulence is small, of order ~ 0.1 km s^{-1}, and iii) the kinetic temperature is consistent with stellar heating in flared disks with $T(r) \propto r^{-0.62-0.68}$.

Such methods are very promising and should be applied in the near future to large samples of stars. There is, however, a major limitation on current mm analysis. Because of beam dilution, which is very important for the small collecting areas of current mm arrays, it is not possible to study the gas properties in the inner part of disks (Dutrey 2000). The results quoted above are significant for the outer disks or $r \geq 30$–60 AU at the distance of Taurus. As a comparison with our Solar System, we are sampling today the equivalent properties of gas disks outside the Kuiper Belt.

Moreover, the ^{12}CO $J = 1 \rightarrow 0$ line being strongly optically thick, the density profile (and hence the disk mass) cannot be measured from these kinds of data.

4.2. Direct Estimate of the Gas-Disk Mass

Determining disk masses of protoplanetary disks still remains very difficult due to many hidden parameters. Tracers can be either i) the thermal dust emission in the mm domain (the most usual method), or ii) molecular species with optically thin emission lines. In the first case, as discussed in Section 2, a proper knowledge of the dust absorption coefficient is required, together with the gas-to-dust ratio. In the second case, both the molecular properties (line strength and excitation) and abundance relative to H_2 must be known.

Molecular-line excitation offers an alternative way to constrain the outer-disk masses, independent on any assumption about the dust properties or molecular abundances. The basic idea (Dutrey et al. 1997) is to constrain the local H_2 density by measuring molecular lines of various critical densities, a by-product of the method being a direct measurement of the molecular abundance with respect to H_2. This method was applied by Dutrey et al. (1997) to the GG Tau and DM Tau disks where they detected HCO^+, CN, CS, HCN, HNC, H_2CO, and C_2H. For the DM Tau disk. They concluded that CO is depleted by a factor of about ~ 5–8 from the standard abundance found in TMC 1 and other molecules up to factors ~ 100 (HCN and H_2CO). Moreover the H_2 mass of the DM Tau disk is about $\sim 0.004\ M_\odot$, still seven times smaller than the mass derived from the dust emission ($\sim 0.03\ M_\odot$). This disagreement could imply that the gas-to-dust ratio differs from the canonical value of 100, and/or that the dust emissivity differs from the values quoted by Beckwith et al. (1990). Nonetheless, both continuum and molecular data suffer from different observational biases and a deeper analysis is required to resolve the discrepancy. However, this method associated with mapping of the dust distribution should in the long term provide estimates of the gas-to-dust ratio.

5. Summary and Future Prospects

It is now clear in most examples, that the CO lines and dust emission observed either in the thermal or scattering regimes arise from the *same dust and gas disk*, even if the vertical dust distribution and mixing with the gas inside the disk is not known.

Understanding the gas and dust distributions on the scale at which planetary formation occurs requires not only proper modeling of the disk properties but also observations of high angular resolution and sensitivity, which are now becoming available. Imaging the inner part of protoplanetary disks on the scale of our Solar System (30–50 AU) is beyond the scope of current instruments at mm or optical/NIR wavelengths. However, the next generation of interferometers, such as the Keck and the VLTI in the optical or ALMA at mm wavelengths (see also the contribution by Guilloteau, this volume, p. 645) will provide the first constraints on the initial conditions leading to planetary formation.

Acknowledgments. I warmly thank S. Guilloteau, M. Simon and F. Gueth for a fruitful collaboration. L. Close, G. Duchêne, N. Grosso, F. Ménard, D. Padgett, D. Potter, and K. Staperlfeld are acknowledged for providing figures and/or material for this review.

References

Adams, F. C. et al. 1987, ApJ, 312, 788

Beckwith, S. V. W. et al. 1990, AJ, 99, 924

Burrows, C. J. et al. 1996, ApJ, 473, 437

Close, L. et al. 1998, ApJ, 499, 883

Dutrey A. 2000, in ASP Conf. Ser., Science with ALMA (San Francisco: ASP), in press

Dutrey, A., Guilloteau, S., & Simon, M. 1994, A&A, 286, 149

Dutrey, A. et al. 1996, A&A, 409, 493

Dutrey, A., Guilloteau, S., & Guélin, M. 1997, A&A, 317, L55

Dutrey, A., et al. 1998, A&A, 338, L63

Dutrey, A. et al. 2000, in preparation

Duvert, G., et. al. 1998, A&A, 332, 867

Grosso, N. 1999, PhD Thesis

Guilloteau, S., & Dutrey, A. 1998, A&A, 339, 467

Guilloteau, S., Dutrey, A., & Simon, M. 1999, A&A, 348, 57

Koerner, D. et al. 1993, Icarus, 106, 2

Malbet, F. et al. 1998, ApJ, 507, L149

Mannings, V. et al. 1996, MNRAS, 267, 361

Mannings, V. et al. 1998, Nat, 388, 555

Mundy, L. et al., 2000 in Protostars & Planets IV, ed. V. Mannings, A. P. Boss, & S. S. Russell (Tucson: University of Arizona Press), p. 355

Pollack, J. et al. 1994, ApJ, 421, 615

Potter, D. et al. 2000, ApJ, 540, in press

Roddier, C. et al. 1996, ApJ, 463, 326

Simon, M., Dutrey, A., & Guilloteau, S. 2000, ApJ, submitted

Stapelfeldt, K. et. al. 1997, in Proc. IAU Symp. 182, Herbig–Haro Flows and the Birth of Low Mass Stars, ed. B. Reipurth & C. Bertout (Ddordrecht: Kluwer), 355

Stapelfeldt, K. et al. 1999, ApJ, 516, L95

Disks, Planetesimals, and Planets
ASP Conference Series, Vol. 219, 2000
F. Garzón, C. Eiroa, D. de Winter, and T. J. Mahoney, eds.

Images and Physical Parameters of Accretion Disks

François Ménard

Canada–France–Hawaii Telescope Corporation, PO Box 1597, Kamuela, HI 96742, USA[1]

Karl Stapelfeldt

MS 183-900, Jet Propulsion Laboratory, Pasadena CA 91109, USA

Gaspard Duchêne and Gilles Duvert

Laboratoire d'Astrophysique de l'Observatoire de Grenoble, CNRS/UJF, BP 53, F-38041 Grenoble cedex 9, France

John Krist

Space Telescope Science Institute, 3700 San Martin Dr., Baltimore, MD 21218, USA

Deborah Padgett

IPAC, California Institute of Technology, Pasadena, CA 91125, USA

Abstract. In this contribution we present optical and/or near-infrared high resolution images of the disks around the stars UY Aur, HK Tau/c, HV Tau C, and TW Hya. Results from scattered-light models are discussed and parameters of the disks are estimated.

1. Introduction

The current paradigm for star formation includes the presence of an accretion disk around the central object. The disk is probably the mediator for most of the mass accretion, but it also supplies the power for mass loss and helps regulate the angular-momentum evolution of the central object as it grows.

From indirect evidence, the fraction of stars with disks is high. *IRAS* detected excess infrared emission in more than 80% of low-mass young stars. This excess is interpreted as dust thermal emission and is thought to be a good tracer of disks. This fraction is somewhat lower at 1.3 mm, where $\sim$ 50% of the stars are detected in Taurus. This number is similar to the statistics on near-infrared excesses.

Gaining direct information on these disks is more difficult. The disks are expected to be small, with radii of order 100–500 AU, and much fainter than

[1]See http://www-laog.obs.ujf-grenoble.fr/~menard/

the central star. Therefore, orientation is expected to play a key role in the detectability of the objects: a pole-on disk will appear symmetrical and therefore difficult to isolate from the PSF (also symmetrical) while an edge-on disk should produce a dark lane, easily identified. We used *HST* to survey about 100 young stars and preliminary results indicate a detection rate of ~10% (Stapelfeldt et al. 2000).

This is a low number, but the positive results allow for a finer estimation of disk parameters than indirect observations do. A comparison of the observations with models of the scattered light allows a detailed study of the disk structure. In the next section we will present four representative results of our campaign.

2. Results: Three Different Disk Configurations

Observations tell us that binary stars are the norm rather than the exception in star-forming regions. One may intuitively imagine that a nearby companion may have a significant effect on the disk structure when the separation between the stars is comparable to the size of the disks. Numbers are difficult to give with certainty, but for separations of order 1–100 AU we expect the disks around individual stars to be perturbed, i.e., to have their outer and colder parts truncated by tidal and/or direct interactions. Observations of the dust thermal emission of large samples of young stars support this view. Osterloh & Beckwith (1995), Jensen, Mathieu, & Fuller (1996) and Dutrey et al. (1996) found that the (sub)millimeter thermal flux was on average smaller in binaries in that 1–100 AU separation range than for singles.

On the contrary, wider binaries have millimeter fluxes similar to those of single stars. In that case the separation is probably large enough to allow for two "normal" circumstellar disks to be present. More puzzling are the spectroscopic binaries. Their separations are small, below 1 AU, and once again their millimeter fluxes appear similar to those of single stars. Possibly, in that case, the two circumstellar disks are mostly absent and a common circumbinary disk exists whose flux is undistinguishable from that of circumstellar disks. Discovery and observation of a larger sample of spectroscopic binaries is needed to ascertain this question.

2.1. Circumbinary Disks

So far only two circumbinary disks, or rings rather, have been imaged at optical and near-infrared wavelengths. They are found around GG Tau and UY Aur. GG Tau has been covered extensively elsewhere in this volume (Duchêne et al., p. 169; Dutrey, p. 128). In this section we will focus on the disk found around UY Aur.

UY Aur is a binary consisting of two classical T Tauri stars of spectral types K7 and M2 (Duchêne et al. 1999) separated by $0''.9$. Both stars have a near-infrared excess, suggesting that they each possess circum*stellar* disks. Interesting to us, circum*binary* molecular CO gas was discovered by Duvert et al. (1998), in Keplerian rotation about the central stars. The thermal emission from the dust in the disk is not detected however, but Close et al. (1998) secured near-infrared images of the scattered light (by dust) in the same circumbinary ring using adaptive optics at the Canada–France–Hawaii Telescope. The discovery

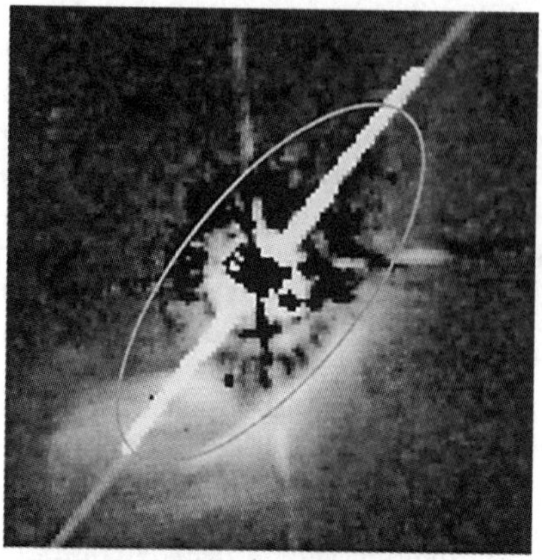

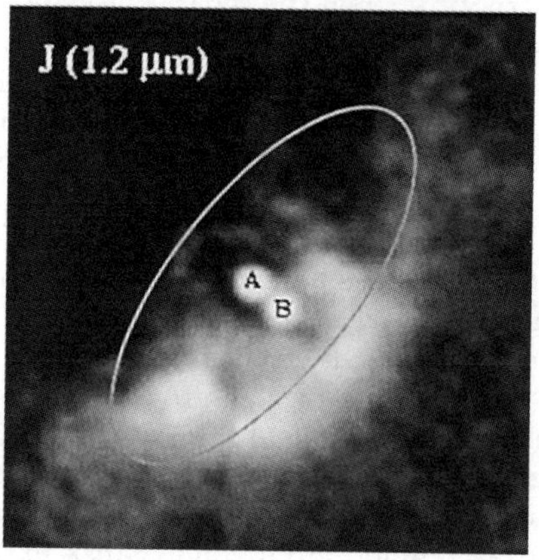

Figure 1. *Top panel: HST* image of the binary UY Aur obtained with WFPC2 (F606W filter). The PSF has been subtracted from the image. Imperfections in the PSF subtraction can be seen as negative (*black*) regions close to the central stars. The primary is saturated and the CCD "bleeding" trail is seen in diagonal, as the image was rotated. *Bottom panel:* Near-infrared *J*-band adaptive-optics image of UY Aur. This is the first image of the circumbinary ring in scattered light. The image is deconvolved. Artifacts are also visible close to the center, due mostly to the diffraction spikes. North is *up* and east to the *left* in both images. Also, the best-fit ellipse to the circumbinary ring is overplotted.

near-infrared adaptive optics image is presented in Figure 1*b*. Figure 1*a* presents an *HST*/WFPC2 F606W image.

In the *HST* image, two components are identified in the circumstellar environment: a bright and narrow extended filament, and a more diffuse, broad patch of nebulosity located at the south-east end of the filament. Both components appear to be overlapping. Artifacts are present, diffraction spikes are visible roughly along the cardinal directions, and a bright linear trail, the "bleeding" of the CCD along columns, emanates from UY Aur A. The "bleeding" trail is tilted because the image has been rotated. Negative zones (in *black* in Figure 1*a*) are also located close to the stars on the east, north, and west sides. They result from slight inaccuracies in the PSF-subtraction process. These artifacts do not affect our detection of the circumbinary ring.

The same basic structures are seen around UY Aur in the near-infrared adaptive-optics image, Figure 1*b*. Noticeably however, the nebulosity appears more clumpy than in the *HST* image. This is partly because of artifacts (diffraction spikes) but also because the SE broad patch seen with *HST* is not resolved as sharply in the AO image.

The filament seen to the SW of the binary in Figure 1 is at a position angle similar to the rotating gas disk detected in ^{13}CO(2 $\rightarrow$ 1) with the IRAM Plateau de Bure interferometer, namely 135° (Duvert et al. 1998). It also extends over roughly the same distance about the central objects, $r \sim 800$ AU, indicating that the dust and gas components are well mixed in the disk. Close et al. (1998) assumed that it traced the *inner edge* of the circumbinary ring around UY Aur and estimated its inclination to $i = 42°$ by ellipse fitting. Their fit included the SE broad nebulosity patch.

A new fit made on the filament alone gives $i = 64°$. This is obtained without taking the broad patch into account. We estimate the range of acceptable values at $59° < i < 65°$. *This is significantly more edge on* than the value obtained by Close et al. (1998). The result of the fit, an ellipse, is superimposed on Figures 1*a* and *b*).

The nature of the broad patch is unclear at present, but it can be shown that it is coincident with a peak of CO emission that is independent of the Keplerian disk discovered by Duvert et al. (1998). This new result supports our suggestion that there are two *independent* components in the circumbinary environment of UY Aur.

Close et al. (1998) measured the contrast between the nebulosity on the NE and SW sides of UY Aur; respectively, the back and front sides of the circumbinary disk. The contrast increases up to a value of ~ 8 at shorter NIR wavelengths, at 1.2 μm. The *HST* images of Figure 1 show the contrast to be at least 15 (or 3 mag) at 0.6 μm, in the optical. Ménard & Bertout (1999) and Ménard et al. (2000) argue that the larger value of inclination naturally explains the large contrast between the back (NE) and front (SW) sides of the disk. The large contrast is a natural consequence of the inefficiency of ISM-sized spherical grains at backscattering photons at large angles. Small grains are naturally forward-throwing, and large disk inclinations naturally lead to large contrasts when illuminated by a central source. The inclination of $\sim 45°$ estimated by Close et al. (1998) *cannot* explain the large contrast (~ 15) observed in the optical, at 0.6 μm.

Furthermore, based on observed kinematics and position, and assuming $i = 42°$, Duvert et al. (1998) concluded that some ^{13}CO $J = 2 \rightarrow 1$ molecular gas was located interior to the circumbinary ring. The "gap" did not appear empty, as would be expected if cleared by gravitational interaction with the inner binary. A larger inclination and the vertical density distribution of the disk naturally combine to explain the *apparent* detection of gas within the gap as a projection effect, without the need to invoke the presence of material inside the ring. However, one should note that the presence of material within the "gap" is not ruled out by our observations.

2.2. Circumstellar Disks in Binary Systems

Figure 2*a* (*left panel*) shows the $2\overset{''}{.}4$ separation binary HK Tau. Both stars in the system are classical T Tauri stars of spectral types M1 and M2 for the primary and secondary (Duchêne et al. 1999). We interpret the dark lane on the secondary's image (to the south) as a disk seen edge on.

The classical T Tauri star HK Tau was imaged quasi-simultaneously with *HST*/WFPC2 in the optical and with adaptive optics at the Canada–France–

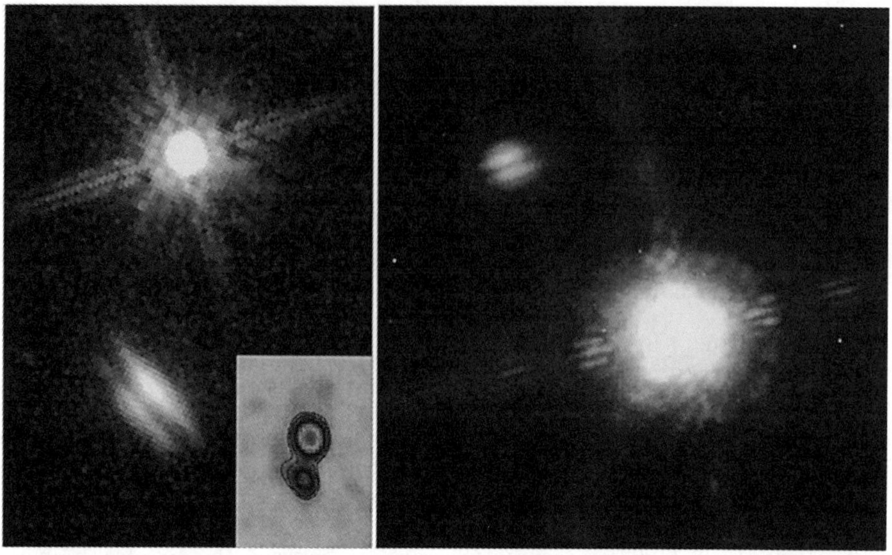

Figure 2. Images of two circumstellar disks in binary systems. *Left: HST I*-band image of the HK Tau binary system (from Stapelfeldt et al. 1998). HK Tau/c, to the south, is nebulous and extended. The insert is a map of the dust thermal emission at 1.4 mm obtained with the IRAM Interferometer (from Ménard et al. 2000). *Right:* CFHT/PUEO Adaptive Optics image of HV Tau. The companion, 4″ to the north-east, is also nebulous and extended. From Stapelfeldt et al. (2000). In both cases no central star is detected and the dark lane is interpreted as a disk seen edge on. North is *up* and east to the *left* on all images.

Hawaii Telescope in the near infrared in 1998. Both methods complement each other and provide the large wavelength coverage necessary for accurate parameter estimation. This method should also yield interesting results on the edge-on disk seen in HV Tau C (Figure 2*b*, *right panel*; from Stapelfeldt et al. 2000).

The HK Tau results are interesting because they were the first to show a *circumstellar* disk around one component of a binary system. The *HST I*-band image (F814W) is repeated in Figure 3 (*left panel*). From the images, one can again directly extract a few geometrical parameters. For example, the disk has a very similar extent at all wavelengths from 600 nm to 2200 nm; namely, $\sim 1''\!.5$ or 210 AU in diameter. This tells us that the edges are likely to be fairly sharp. This is one of the smallest disks resolved so far. It is interesting to note that the disk around HV Tau C is also very small (see Figure 2*b*, *right panel*). Because the projected separation of the two components of HK Tau ($2''\!.4$, or 330 AU) is only three times larger than the disk radius, it is interesting to ask whether there is a causal link between the small size of the disk and the presence of a companion. It is too early to conclude, but systems like HK Tau and HV Tau contain the answer to that question.

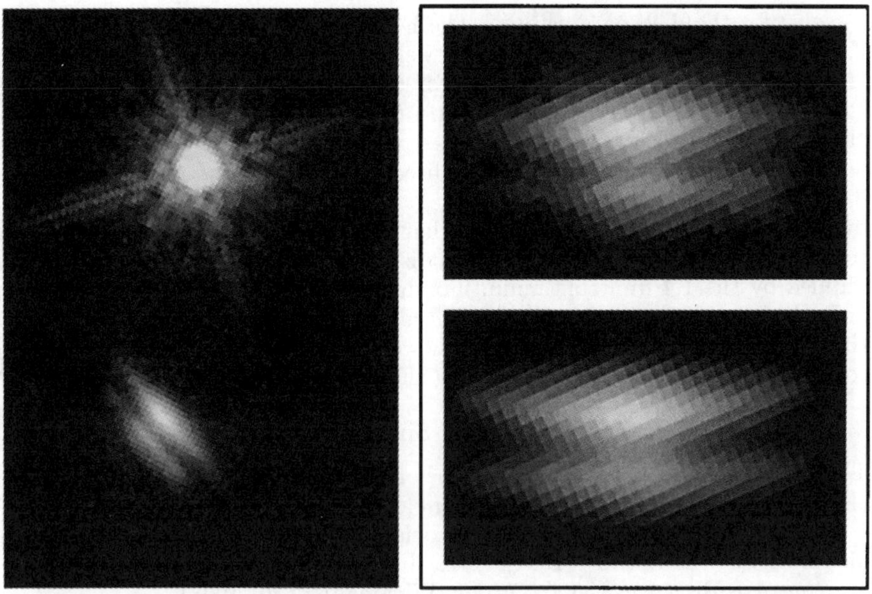

Figure 3. *Left: HST I*-band image of the HK Tau binary system. North is *up* and east to the *left* in this panel only. *Right:* Best-fit model of the scattered light compared to the F814W image from *HST*. The *top panel* shows the observations, the *bottom panel*, the best fit model from single-scattering numerical simulations. The field of view is $1''.6 \times 1''.0$. Both images are from Stapelfeldt et al. (1998).

Close inspection of Figure 3 reveals that the north-west nebula on HK Tau/c is slightly brighter than the south-east one (topmost nebula, *right panel*). Numerical simulations of the scattered light by a disk suggest $i = 5^\circ \pm 1^\circ$. The central star is not seen either, even at 2.2 μm where the extinction is ten times smaller than at 550 nm. Given our detection limit, we estimate that the extinction $A_V > 50$ magnitudes in the plane of the disk, in line with the idea these disks are optically thick.

The nearly edge-on orientation is very favorable for studying the scale height of the dust distribution in the disk plane. This possibility was discussed and tested extensively by Burrows et al. (1996) in the case of HH 30. The model presented for HK Tau/c in Figure 3 was obtained for single scattering by spherical homogeneous dust grains (Mie scattering) distributed in the disk with a parameterized surface density $\Sigma(r) = \Sigma_0(r/r_0)^p$ and Gaussian vertical density profile $\rho(z) = \rho_0 \exp[-z^2/H(r)^2]$. This vertical profile is valid for vertically isothermal, hydrostatic, non-self-gravitating disks. A parameterized scale height $H(r) = H_0(r/r_0)^\beta$ was further assumed. Following the numbers extracted above, we adopted an outer radius of 105 AU. A χ^2 minimization procedure was used to solve for the disk parameters[2]. The normalization H_0 of the scale height is 3.8 AU at $r = 50$ AU. This number is well constrained. The disk of HK Tau B is significantly flatter than HH 30's. For HH 30, $H_0 = 15.5$ AU at $r = 100$ AU. Presented differently, at 50 AU the ratio $H/r = 0.08$ for HK Tau/c but 0.12 for HH 30.

The mass of an edge-on disk is directly related to the "thickness" of the dark lane. For HK Tau/c, we estimate a total disk mass of $\sim 6 \times 10^{-5}\ M_\odot$ in dust and gas if the grains have interstellar properties. This is small with respect to the minimum-mass solar nebula. It is also small with respect to the mass implied by the 1.4 mm continuum flux density (40 mJy) detected by Osterloh & Beckwith (1995). The expected disk mass from this mm flux is roughly 100 times larger. Ménard et al. (2000) showed, however, that $\sim 2/3$ of that flux comes from the primary, partly solving the issue (see Figure 2*a*, *insert*).

2.3. Circumstellar Disks around Single Stars

TW Hydrae, at 10–20 Myr, is one of the oldest classical T Tauri stars known. It also lies outside of any known molecular cloud, a rather unusual occurrence. It is a member of the small TW Hydrae association, a group of young stars recently identified by Webb et al. (1999). It is the nearest group of young stars based on *Hipparcos* distances (56.4 ± 7.0 pc, Wichmann et al. 1998).

The CTTS nature of TW Hya is confirmed by the presence of significant near- and mid-infrared excess, large and variable Hα equivalent width (Rucinski & Krautter 1983), and periodic photometric variability (Mekkaden 1998), suggesting the presence of hot spots from accretion (Herbst & Koret 1988).

The signs of active accretion activity and the proximity of TW Hya make it a prime target for disk imaging. Figure 4 was obtained with *HST*/WFPC2 (Krist et al. 2000). Although the raw image appears symmetric, a careful subtraction reveals unambiguously that a smooth circular halo of reflected light is present in excess of the usual *HST* PSF. The azimuthal symmetry of the nebulosity discovered suggests that we are looking at a disk seen very close to pole on, $i = 0°$.

Various tests, described in Krist et al. (2000) and summarized in Figure 4, were performed to insure the reality of our detection. Errors in the scaling factor between the image and the reference PSF to subtract do not match the intensity profile we detect. Also, the size of the halo is similar in both filters used (F606W and F814W), being slightly larger in the short filter. On the contrary, PSF artifacts are expected to induce the opposite behavior since the PSF scales with the wavelength. The detection of a similar halo in the near infrared by Weinberger et al. (1999; see also this volume, p. 329) also supports our findings. All these results give us confidence that the halo we detect, 300 times fainter in integrated brightness than TW Hya itself, is real.

To image the disk of TW Hya, we saturated, on purpose, the core of the PSF. This, and the PSF subtraction process, prevent us from imaging the disk inside $0''\!.5$ (~30 AU) from the center. Because the brightness profile is regular from this position outward, we can rule out the existence of a large ($R > 30$

[2]Stapelfeldt et al. (1998) present a discussion of the results from various models and discuss the reliability of the best solution, presented here.

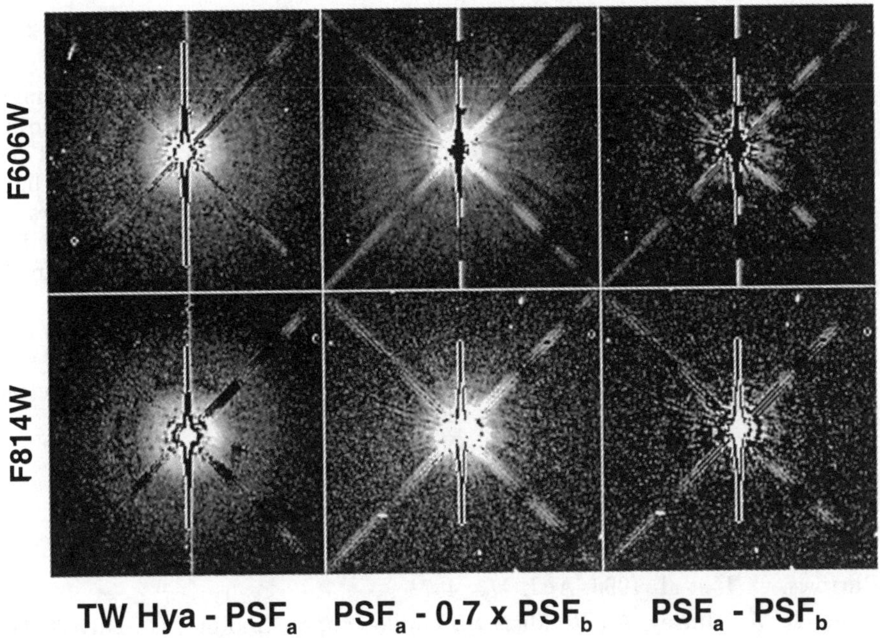

Figure 4. The disk in TW Hydrae observed at two wavelengths. PSF subtraction comparison of TW Hya and a reference star with and without subtraction errors. *Left:* The reference star HD 17637 is subtracted from TW Hya following our standard procedure and normalization. *Middle:* Images of HD 17637 with another PSF (DN Tau) subtracted with a 30% normalization error added on purpose. This error is evident from the dominantly positive residuals in the diffraction spikes and from the presence of radial spikes, also seen in the raw *HST* PSFs. *Right:* Image of HD 17637 subtracted by DN Tau with the correct, standard normalization factor.

AU) inner hole, as expected. Furthermore, the surface-brightness profile appears to be independent of azimuth. Assuming ISM-like particles are responsible for the scattered light, this suggests an inclination very close to pole on. A large inclination, i.e., $i > 15°$, would result in a measurable contrast between the front and back faces of the disk. A similar effect was discussed in Section 2.1.

Models of the scattered light around TW Hya also yield a dust mass in the disk of 33 Earth masses. This is comparable to the 45 Earth masses of dust estimated by Weintraub, Sandell, & Duncan (1989) from 800 μm photometry. The similarity between these independent results suggests that the population of small grains in the disk has not been significantly depleted yet, by an age of 10–20 Myr (Krist et al. 2000). Kastner et al. (1997) estimated $3 \times 10^{-5}\ M_\odot$ of CO gas, suggesting a significant depletion of the CO component however. Estimates of the total disk mass are therefore difficult to perform reliably at this stage. Assuming a normal gas-to-dust ratio of 100, the total disk mass would be $9 \times 10^{-3}\ M_\odot$.

3. Concluding Remarks

The previous sections have shown that a diversified multi-wavelength and multi-technique approach to estimate the disk parameters is not only possible nowadays, but also important to make full use of the observational data. Optical and near-infrared high-resolution and high-contrast imaging of the scattered light completes very well the information obtained by millimeter interferometers on the molecular gas and dust thermal emission. The efforts made in the last few years now give us estimates of parameters for a dozen disks or so (see Ménard & Bertout 2000 for a recent review). These are seen at a wide range of inclination angles. This is important for removing the coupling between parameters and getting a better sense of the reliability of our estimations.

We would like to thank the Organizing Committee for a faultless meeting and the editors for their patience in the preparation of these proceedings.

References

Burrows, C. J. et al. 1996, ApJ, 473, 437

Close, L. M., Dutrey, A., Roddier, F., Guilloteau, S., Roddier, C., Duvert, G., Northcott, M., Ménard, F., Graves, J. E., & Potter, D. 1998, ApJ, 499, 883

Duchêne, G., Monin, J.-L., Bouvier, J., & Ménard, F. 1999, A&A, 351, 954

Dutrey, A., Guilloteau, S., Duvert, G., Prato, L., Simon, M., Schuster , K., Ménard, F. 1996, A&A, 309, 493

Duvert, G., Dutrey, A., Guilloteau, S., Ménard, F., Schuster, K., Prato, L., & Simon, M. 1998, A&A, 332, 867

Herbst, W., & Koret, D. L. 1988, AJ, 96, 1949

Jensen, E. L. N., Mathieu, R. D., & Fuller, G. A. 1996, ApJ, 458, 312

Kastner, J. H., Huenemoerder, D. P., Schulz, N., Zuckerman, B., Weintraub, D. A., Gaidos, E. J., & Negoro, H. 1997, BAAS, 191.9206

Krist, J. E., Stapelfeldt, K. R., Ménard, F., Padgett, D. L., & Burrows, C. J. 2000, ApJ, in press

Mekkaden, M. V. 1998, A&A, 340, 135

Ménard, F., & Bertout, C. 1999, NATO Science Series, Series C, Vol. 540, Proc. NATO/ASI, The Origins of Stars and Planetary Systems, ed. C. J. Lada & N. D. Kylafis (Dordrecht: Kluwer), 341

Ménard, F. et al. 2000, ApJ, in preparation

Osterloh, M., & Beckwith, S. V. W. 1995, ApJ, 439, 288

Rucinski, S. M., & Krautter, J. 1983, A&A, 121, 217

Stapelfeldt, K. R., Krist, J. E., Ménard, F., Bouvier, J., Padgett, D. L., & Burrows, C. J. 1998, ApJ, 502, L65

Staplefeldt, K. R., et al. 2000, ApJ, in preparation

Webb, R. A., Zuckerman, B., Platais, I., Patience, J., White, R., Schwartz, M. J., & McCarthy, C. 1999, ApJ, 512, L63

Weinberger, A. J., Schneider, G., Becklin, E. E., Smith, B. A., & Hines, D. C. 1999, BAAS, 194, 69.04

Weintraub, D. A., Sandell, G., & Duncan, W. D. 1989, ApJ, 340, L69

Wichmann, R., Bastian, U, Krautter, J., Jankovics, I., & Rucinski, S. M. 1998, MNRAS, 301, 39L

Discussion

Keith Horne: In HK Tau, star A, could you find or put upper limits on scattered light from the disk by subtracting the PSF? Since star A has more dust mass than star B, why don't you see its disk in scattered light?

François Ménard: It's probably a question of size of the disk, and orientation too. HK Tau B's, being edge on, is oriented very favorably for detection. HK Tau A is possibly oriented differently, and we do not have enough contrast in our images, close to the center of the PSF, to see the disk in scattered light.

Disks, Planetesimals, and Planets
ASP Conference Series, Vol. 219, 2000
F. Garzón, C. Eiroa, D. de Winter, and T. J. Mahoney, eds.

The Optical Constant of Crystalline Spinel

Hiroki Chihara and Akira Tsuchiyama

Department of Earth and Space Science, Osaka University, 1-1, Machikaneyama, Toyonaka, Osaka, 560-0043, Japan

Chiyoe Koike and Hisato Sogawa

Kyoto Pharmaceutical University, Misasagi, Yamashina, Kyoto, 607-8412, Japan

Abstract. For natural and synthetic crystalline spinel ($MgAl_2O_4$), the mass absorption spectra of the fine particles and the reflectance of the bulk crystals were measured in the mid- and far-infrared regions. The absorption spectra showed the prominent peaks at 14.7, 19.7, and 32.4 μm. The reflectance also showed three features at 13.8, 18.5, and 32.4 μm. From the reflectance, the optical constants (n, k) were derived on the natural spinels.

1. Introduction

ISO SWS/LWS observations revealed the existence of the considerably large amount of solid-state materials such as crystalline silicate and water ice in the circumstellar envelope of oxygen rich young and evolved stars (Waters et al. 1996; Waelkens et al. 1996; Whittet et al. 1996; Malfait et al. 1998). These revolutionary results allowed us to study astronomy by means of materials science. The discovery of crystalline silicate especially provided us with a fundamental clue to understand planet formation within the framework of thermodynamic equilibrium condensation (Grossman & Larimer 1974). The precise laboratory data of high-temperature condensation materials are essential to solving the puzzle of grain formation in circumstellar disks and/or the early stage of protoplanetary disks.

Magnesium aluminum spinel ($MgAl_2O_4$) is one of the high-temperature condensation species. Spinel is condensed at relatively high temperature 2378 K $\sim$ 2408 K (Alper et al. 1962; Roy & Hastert 1983) in the early stage of the thermodynamic condensation sequence (Tielens et al. 1998) and occurs in CAIs (calcium and aluminum inclusions) in the primitive carbonaceous chondrite. The structure of spinel is face-centered cubic, and its chemical composition is expressed as $MgO \times m\ Al_2O_3$, where m varies between 0.9 and 4 in general.

Tropf & Thomas (1991) reviewed the optical constants for this species. In an astronomical context, it was discussed as a candidate for the carrier of the 13 μm feature in a recent paper (Posch et al. 1999) from the observational point of view. In the field relevant to astrophysics, however, few accessible optical constants of this species exist. The precise dataset in a wide wavelength range is required. In this paper, we report on the measurement of the optical quantities of crystalline spinel in the mid- and far-infrared regions for stoichiometric natural samples and a non-stoichiometric synthetic sample.

2. Experimental

The measured samples in this paper were two natural crystalline minerals, white spinel from Burma and pink spinel from Sri Lanka, and a spinel synthesized by the Verneuil method. The natural samples are almost stoichiometric, but the pink spinel includes a very small amount of Cr ($\leq 0.5\%$) which is responsible for the pink coloration. The synthetic spinel was slightly Al_2O_3 rich ($m \approx 2$).

Mass-absorption spectra and reflectance were obtained with a Jasco-FT/IR-350 spectrometer (Kyoto Pharmaceutical Univ.) at a resolution of 0.5 cm^{-1} and 1.0 cm^{-1}, respectively. For the mass-absorption spectra, the bulk crystals were crushed and ground in a WC mortar. These fine particles, less than 1 μm in size, were dispersed in a polyethylene and KBr matrix for far-IR and mid-IR measurements, respectively. For measuring the reflectance, the bulk crystals were cut and polished. The areas of the cross-section of the surface are above 10 mm^2. The polarizer was not used because the optical properties of spinel are isotropic due to its cubic structure. The wavelength ranges were $6 \sim 50$ μm for the mass-absorption spectra and $1.5 \sim 100$ μm for the reflectance.

3. Results

The mass-absorption spectra and reflectances for all the samples are shown in Figures 1 and 2, respectively. The natural samples (white and pink spinel) have three strong absorption bands at 14.8, 19.7, and 32.4 μm. The broad bands around 40 μm are artifacts due to the interference fringes. No significant discrepancy between the two natural samples is detected. Therefore, the effect of Cr in pink spinel can be neglected. In the reflectance, there are also three prominent features at 13.7, 18.5, and 32.4 μm. Moreover, small features and shoulders occur at 12.0, 17.0, and 20.8 μm. A small band at around 7 μm is detected in the absorption spectra, but it is not detected in the reflectance.

For the synthetic spinel, however, four broad bands are detected at 12.1, 14.6, 18.8, and 31.8 μm in the mass-absorption spectrum. The reflectance features are also different from those of the natural samples; intermediate features are detected at 11.9, 14.0, and 18.4 μm. In the case of the non-stoichiometric spinel, the lattice defect tends to occur because of Al_2O_3 enrichment. It is generally difficult to synthesize spinel stoichiometrically. Therefore, the lattice structure of synthetic spinel is apt to be amorphous. As a result, the absorption spectrum and the reflectance of synthetic spinel show broad features. The peak positions are summarized in Table 1 for all the samples.

For the natural crystalline samples, the reflectances were fitted with the multiple oscillator model (e.g. Bohren & Huffman 1983) in order to obtain the optical constants, (n, k). The measured reflectances and the fitted results plotted in Figure 2a and b for the pink and the white spinel, respectively. From the fitted curves, the optical constants are derived in the wavelength range between 7 μm and 90 μm. Figures 3a and b show the optical constants (n, k) for the pink and the white spinel, respectively. Our results are roughly consistent with the optical constants in Tropf & Thomas (1991). However, our dataset shows the features more clearly and sharply because the measurements were carried out at high resolution, Moreover, the strength of the prominent features is approximately 10% greater, and the some small features are detected at 12.0, 17.0, and 20.8 μm in our data.

Table 1. The band positions for all samples (μm). "A" and "R" mean the "absorption spectra" and the "reflectance", respectively. The prominent peaks are underlined

Pink spinel		White spinel		Synthetic spinel	
A	R	A	R	A	R
<u>32.4</u>	<u>32.4</u>	<u>32.4</u>	<u>32.4</u>	31.8	32.3
	20.8		20.7		
<u>19.6</u>	<u>18.4</u>	<u>19.8</u>	<u>18.6</u>	18.8	18.4
17.3	17.0	17.3	17.1		
<u>14.7</u>	<u>13.7</u>	<u>14.8</u>	<u>13.8</u>	14.6	14.0
13.1	12.0	13.2	12.1	12.1	11.9
11.4		11.3		11.3	11.2
7.2		7.2		7.2	

References

Alper, A. M., McNally, R. N., Ribbe, P. G. & Doman, R. C. 1962, J. Am. Ceram. Soc., 45, 264

Bohren, C. F. & Huffman, D. R. 1983, Absorption and Scattering of Light by Small Particles, (New York: Wiley & Sons), 244

Grossman, L. & Larimer, J. W. 1974, Rev. Geophys Spca. Phys, 12, 71

Malfait, K., Waelkens, C., Waters, L. B. F M., Vandenbussche, B., Huygen, E., & Graauw, M. S. 1998 A&A, 332, L25

Posch, T., Kershebaum, F., Mutschke, H., Fabian, D., Dorschner, J., & Hron, J. 1999, A&A, 352, 609

Roy, D. W., & Hastert, J. L. 1983, Proc. SPIE, 400, 37

Tielens, A. G. G. M., Waters, L. B. F. M., Molster, F. J. & Justtanont, K. 1998, Ap&SS, 255, 415

Tropf, W. J. & Thomas, M. E. 1991, Handbook of Optical Constants of Solids II, ed. E. D. Palik (San Diego: Academic Press), 883

Waelkens, C. et al. 1996, A&A, 315, L245

Waters, L. B. F. M. et al. 1996, A&A, 315, L361

Whittet, D. C. B. et al. 1996, A&A, 315, L357

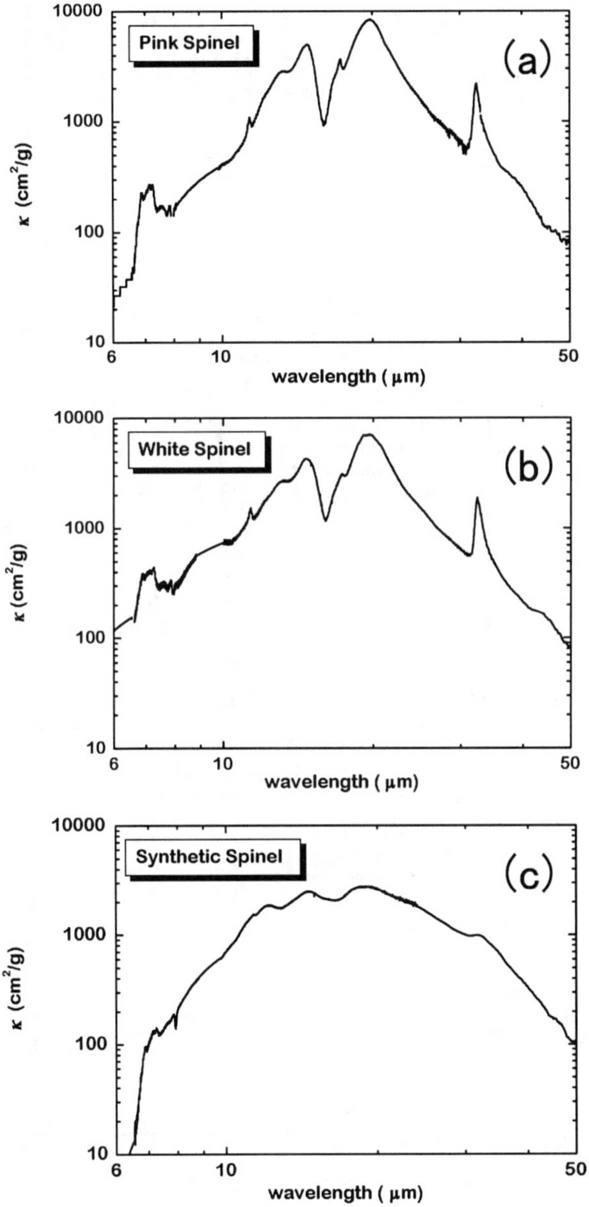

Figure 1. The mass-absorption spectra for all the samples. (a) Pink spinel, (b) white spinel, and (c) synthetic spinel.

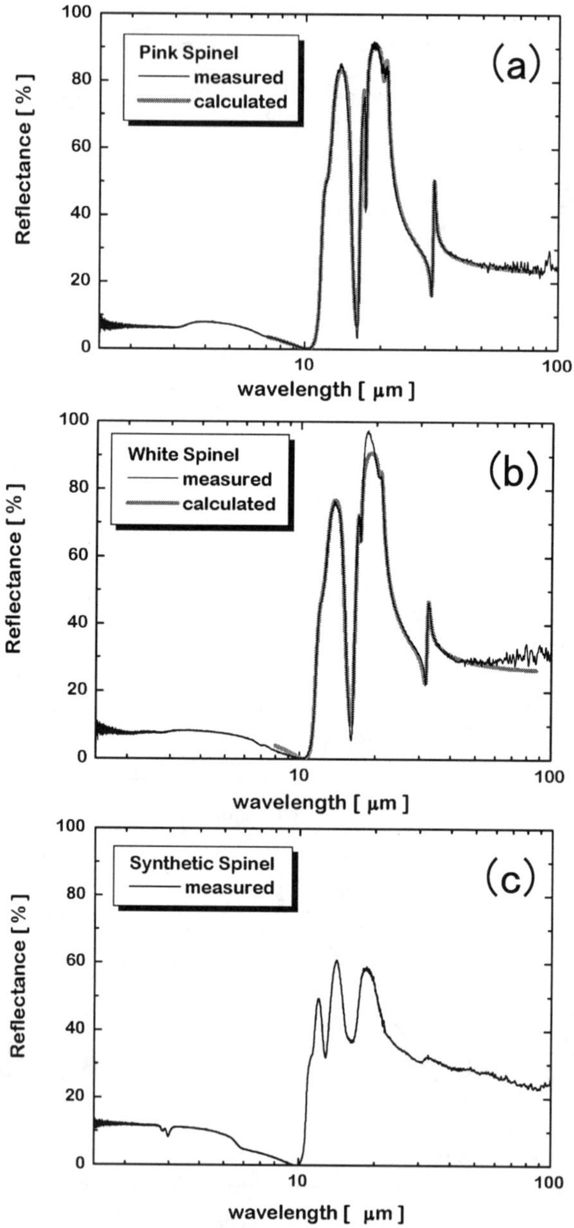

Figure 2. The reflectances for all the samples. For the natural crystalline spinels, the fitted curves are superposed. The *solid lines* indicate the measured spectra and the *gray lines* indicate the fitted curves. (a) Pink spinel, (b) white spinel, and (c) synthetic spinel.

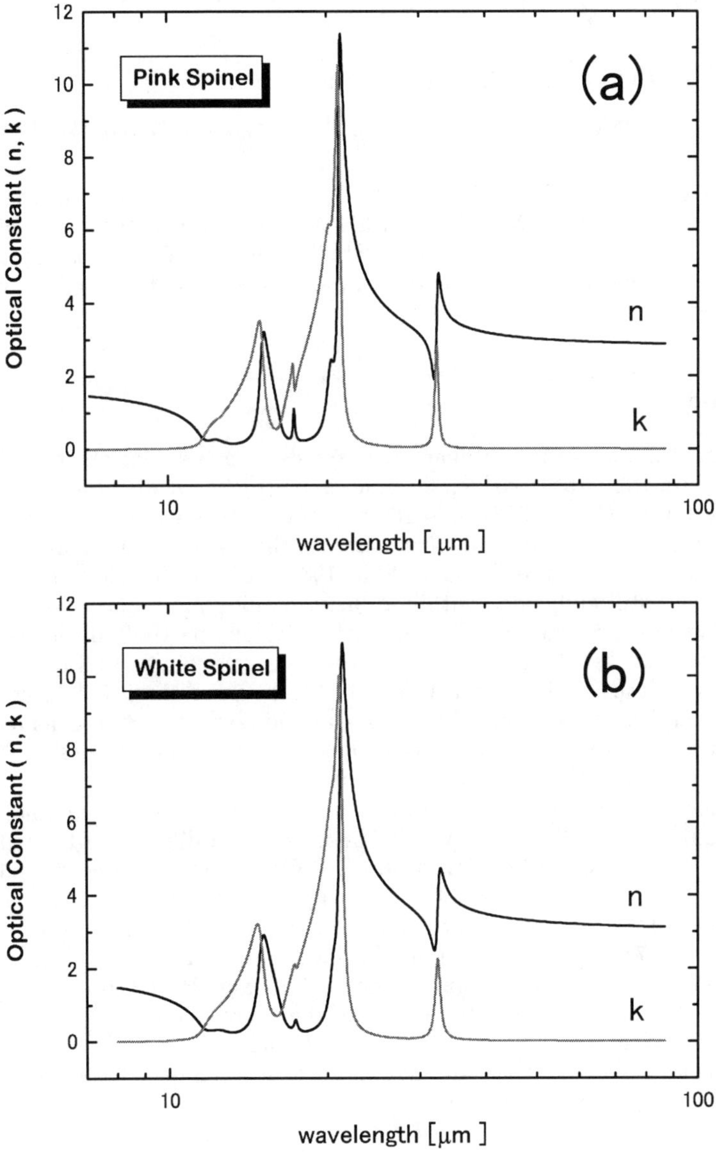

Figure 3. The optical constants (n, k) for the natural spinels. (a) Pink spinel and (b) white spinel.

Disks, Planetesimals, and Planets
ASP Conference Series, Vol. 219, 2000
F. Garzón, C. Eiroa, D. de Winter, and T. J. Mahoney, eds.

HST/NICMOS Observations of YSOs: HH 30, HL Tau, and L1551 IRS5

Angela Cotera, Erick Young, and Hua Chen

Steward Observatory, University of Arizona, Tucson, AZ 85721, USA

Abstract. We have obtained *HST*/NICMOS Camera 2 images of HH 30, HL Tau and L1551 IRS5, with a nominal FWHM resolution of ~ 20 AU. Here we present the F110W, F160W, and F204M images, discuss the observed morphology of the regions, and present preliminary results.

1. Introduction

HH 30, HL Tau, and L1551 IRS5 are prototypical young stellar objects (YSOs), located within the L1551 dark cloud first cataloged by Lynds (1962), at a distance of $\sim$ 140 pc (Elias 1978) to $\sim$ 160 pc (Cohen, Beiging, & Schwartz 1982). These objects represent three different classes of the low-mass pre-main-sequence YSOs as defined by Adams, Lada, & Shu (1987). L1551 IRS5 is the prototype Class I object, still embedded within a dusty infalling molecular envelope. HL Tau is between Class I and Class II, with the envelope partially dispersed, such that the star can be observed at infrared wavelengths but is still obscured at optical wavelengths. HH 30 is a Class II object where the infalling envelope has dispersed, leaving an optically visible star: however, HH 30 is oriented at a P.A. $\sim 84°$ (Burrows et al. 1996), such that the star is entirely obscured by a circumstellar disk, which is then seen in absorption.

We have obtained *HST*/NICMOS using Camera 2 with a nominal plate scale of $0''.0755 \pm 0''.005$ arcsec/pixel. The images are diffraction limited with measured FWHMs of $0''.140$ at 1.104 μm (F110W filter) to $0''.154$ at 2.033 μm (F204M filter). Assuming a distance of $\sim$ 140 pc, this corresponds to an angular resolution of 20 AU. We selected the filter combination F110W, F160W ($\lambda_{\rm eff}$ = 1.593 μm), F187N ($\lambda_{\rm eff}$ = 1.871 μm), F204M, and F212N ($\lambda_{\rm eff}$ = 2.121 μm), to investigate the NIR continuum emission, Paα and H_2 features within the region. No Paα has been detected in any of the sources, but faint H_2 has been detected coincident with the jet axis in HH 30 and in L1551 IRS5. Below, we briefly discuss the observations and data reduction techniques used and present preliminary results for each region.

2. Observations and Data Reduction

The observations of HH 30 were made on 1997 Sept 29, HL Tau on 1997 Nov 23, and L1551 IRS 5 was observed on 1997 Nov 19. F110W, F160W, F204M, F187N, and F212N images were taken sequentially at each of the dither positions for each source. For HH 30 there were four dithers on source, HL Tau is the result of two on-source dithers, the peak of L1551 IRS 5 also has two dither positions. Total integration times for HH 30 were 64 s at F110W and F160W, 112 s at F204M, 144 s in F187N and 160 s at F212N. For HL Tau and L1551 IRS5 times were 32 s at F110W, F160W and F187N, and 80 s at F204M and F212N. Dark frames were taken at the end of each set of observations. Darks for F110W, F160W, and F187N were created using observed dark frames and the IRAF routine NICSKYDARK, part of the NICRED data reduction software (McLeod 1997). Darks for F204M and F212N used observed dark frames and data frames. The flats used were those produced by Marcia Rieke (private communication). The data was reduced using the IRAF data reduction package NICRED. The IDL routines NICMOSAIC and NICSTIKUM were use to shift and add the images respectively (Stobie et. al. 1999). The combined images were then flux calibrated using the values derived by Marcia Rieke (private communication).

3. HH 30

The F160W image of HH 30 is presented in Figure 1. Although at first glance HH 30 appears roughly symmetric, it is in fact asymmetric about both the jet axis, and the axis of the dust midplane. These observed asymmetries are also wavelength dependent. The predicted decrease in the width of the dust lane with increasing wavelength is seen, but there is also an asymmetric, wavelength-dependent difference in the degree of flaring at either end of the disk. In addition, the peak of the reflected emission in both of the lobes moves further away from both the dust midplane and the axis defined by the star + jet as seen in *HST*/WFPC2 F675W images (Burrows et al. 1996) as a function of wavelength. The peak in the F204M filter image is actually very close to the jet axis for both the upper and lower lobes. In the F110W image, on the other hand, the peak is significantly offset.

Since there is no sharp edge to the disk apparent in the images, to quantify the wavelength-dependent differences, the location of the minimum emission in the disk and the maximum emission in the reflection nebula have been determined for F110W, F160W, and F204M. We note that the north-north-east (upper) lobe can be traced to much larger distances from the star than the lower lobe. This is consistent with the effects of a disk not perfectly edge on, but with an orientation of $\sim 84°$ as proposed by both Burrows et. al. (1996) and Wood et. al. (1998). The total disk thickness is taken as the the difference between the determined maxima of the upper and lower nebular lobes. For the F110W and F160W images, the disk thickness increases more rapidly on the upper right side than at the lower left. At F110W the minimum disk thickness is $0.''66$, F160W is $0.''58$, and F204M is $0.''53$: at 140 pc this corresponds to $\sim$ 92 AU, $\sim$ 81 AU, and $\sim$ 74 AU, respectively.

We do not detect the star in the F204M filter, but we can still use the observed flux at the presumed location of the star to estimate the extinction to the star. If we can determine the minimum and maximum magnitudes for a star we would have been able to distinguish in our images, we can derive an upper and lower limit on the extinction to the star itself. There is an important caveat to this approach: we have no way of knowing what the actual unextinguished magnitude of the star is. We therefore assume a K-band magnitude ~ 9, which

Figure 1. F160W image of HH 30. Clearly seen are the disk in absorption and the asymmetry between the north-north-east (*upper*) and south-south-west (*lower*) lobes.

is typical for Class II stars in the Taurus–Auriga Cloud (Kenyon & Hartmann 1995). We have also determined that the magnitude difference between ground-based K and F204M is ~ 0.03 for typical main-sequence stars, considerably less than the error introduced by having to assume an intrinsic K-band magnitude.

Using an artificially created PSF (see below), we then add or subtract a star scaled to a known magnitude in the F204M filter. If we add a PSF star with F204M magnitude ~ 18.9, the star is detectable above the background. If we subtract a PSF of magnitude ~ 19.7, we can detect a hole in the emission. From this we derive a lower limit on the extinction of $A_{\mathrm{F204M}} \sim 9.9$ and an upper limit of ~ 10.7, which corresponds to an $80 \lesssim A_V \lesssim 86$. This is considerably more than the lower limit of $A_V \sim 24$ mag derived by Burrows et. al. (1996), but the same as the value of $A_V \sim 80$ suggested by Stapelfeldt & Moneti (1999).

4. HL Tau

The F204M image of HL Tau is presented in Figure 2. The analysis of HL Tau has been hampered by the fact that at the longer wavelengths, the star is no longer enshrouded by the molecular envelope. This results in the F204M and F212N images being dominated by the star, and in particular the NICMOS point spread function (PSF). Previously, using high-resolution adaptive-optics J, H and K images of HL Tau, Close et al. (1997) reported detecting a ~ 150 pc accretion disk around the central star. At our resolution, this corresponds to a ~ 12 pixels, but the diffraction pattern dominates the emission to ~ 10

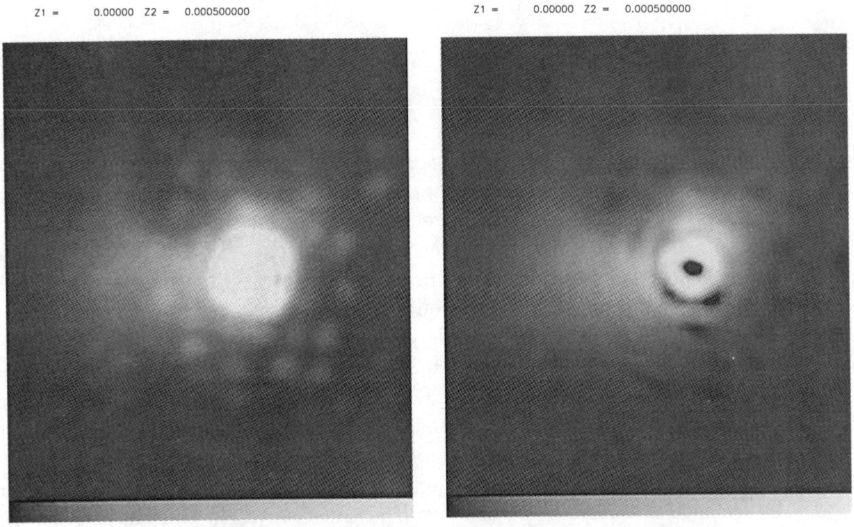

Figure 2. *Left:* F204M image of HL Tau. *Right:* F204M image minus PSF produced by TinyTim.

pixels. Therefore, before we can address the science within the dataset, we have to remove the NICMOS PSF.

During the development stage of *HST*/NICMOS the TinyTim (Krist & Hook 1997) program was modified to simulate the NICMOS PSF. Version 5.0b was released on 2000 April 23 and includes, among other updated information, the effects of the cold mask and color corrections. By adjusting the input parameters to a best match of the conditions when the images were observed, a good (although not yet perfect) PSF can be created. In particular, the position of the point source on the array, and the focus (including the telescope "breathing") must be properly determined in order to create the best TinyTim PSF. Best-estimate TinyTim PSFs have been created for each observed position for each filter. The PSFs for each dither position have to be created separately and then combined in the same manner as the data. For all of the stars, we have assumed the spectral colors of an M1.5V star since spectra of pre-main-sequence stars are unavailable.

Also shown in Figure 2 is the F204M PSF-subtracted image. As can be seen, although a significant amount of the stellar flux within the diffraction pattern has been removed, at the level of a possible accretion disk we still have excess flux. Part of the problem may be that at F204M, the star at HL Tau is not a true point source but is in fact extended. The FWHM of the best-model PSF is $0\farcs153$, while the FWHM of the source is $0\farcs186$. Even when the model PSF is significantly defocused, we are unable to match the source FWHM. At F160W and F110W, the stellar flux is significantly less, and we are better able to subtract off the stellar flux. In the F110W and F160W PSF-subtracted image we see possible excess emission at the orientation proposed for the disk by Close et al. (1997), but additional work needs to be done to determine if this is a significant feature, and if it is detectable at F204M.

5. L1551 IRS5

A contour map of the central region of the F160W image of L1551 IRS5 is presented in Figure 3. In the image, we can see faint emission from the well known jet, and the source resolves into a "C"-like reflection nebulae with two distinctive ridges, not unlike HL Tau. This is consistent with near-infrared observations of L1551 IRS5 with image sharpening obtained by Lucas & Roche (1996), although the C is much better defined in the *HST*/NICMOS images. Our images also reveal a pointlike knot on the lower portion of the C. The source is not a point source, but is extended significantly beyond the model PSF for this image; however, the emission is very spherically symmetric as it extends above the ridge of emission with which it coincides.

Previously, the outflow source of L1551 IRS5 has been placed at the cusp of the "C" (e.g. Davis et al. 1995; Lucas & Roche 1996), although details of the astrometric registration were not provided. It has also been suggested at various times that L1551 IRS5 is in fact a protobinary system (e.g. Rodriguez et al. 1998). Initial attempts to use the *HST* header world coordinates to align the radio and NICMOS images were not successful because the *HST* Guide Star coordinates for the reference star were not accurate. Using the ACT (Astrographic Catalog—Tycho) catalogue available from the US Naval Observatory, we were able to correct the Guide Star coordinates, including precession to the NICMOS observation epoch. The coordinates presented in Figure 3, are the result of this correction. Overlaid on the F160W contours, are the positions for the 7 mm radio sources of Rodriguez et al. (1998) precessed to J2000: source A, RA (J2000) $4^{h}\ 31^{m}\ 34\overset{s}{.}142\pm0\overset{s}{.}002$ Dec. (J2000) $+18^\circ\ 8'\ 5\overset{''}{.}12\pm0\overset{''}{.}03$, source B, RA (J2000) $4^{h}\ 31^{m}\ 34\overset{s}{.}142 \pm 0\overset{s}{.}002$ Dec. (J2000) $+18^\circ\ 8'\ 4\overset{''}{.}80 \pm 0\overset{''}{.}03$. The northern radio source (source A) is intriguingly close to the pointlike source in the NICMOS image, but the results should be considered preliminary at this point.

References

Adams, F. C., Lada, C. J., & Shu, F. H. 1987, ApJ, 312, 788

Burrows, C. J. et. al. 1996, ApJ, 473, 437

Cohen, M., Bieging, J. H., & Schwartz, P. R. 1982, ApJ, 253, 707

Close, L. M., Roddier, F., Northcott, M. J., Roddier, C., & Graves, J. E. 1997, ApJ, 478, 766

Davis, C. J., Mundt, R., Eislöffel, J., & Ray, T. P. 1995, AJ, 110, 766

Elias, J. H. 1978, ApJ, 224, 857

Kenyon, S. J., & Hartmann, L. 1995, ApJS, 101, 117

Krist, J. E., & Hook, R. N. 1997, in The 1997 HST Calibration Workshop with a New Generation of Instruments, 192

Lynds, B. T. 1962, ApJS, 7, 1

Lucas, P. W., & Roche, P. F. 1996, MNRAS, 1996, 280

McLeod, B. 1997, in The 1997 HST Calibration Workshop, ed. S. Casertano et al., 281

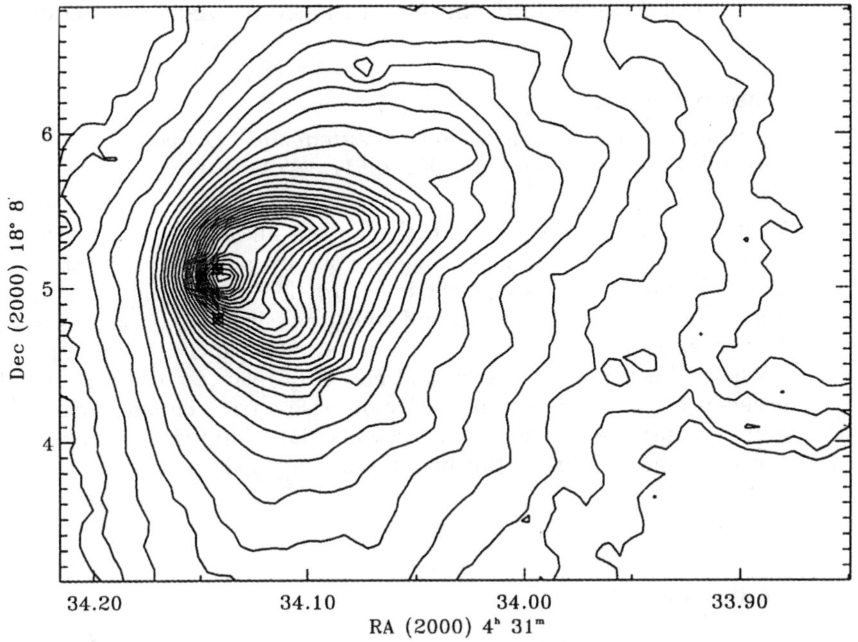

Figure 3. Contours of the F160W image of L1551 IRS 5 overlaid with the location, with errors, of the peak 3 cm radio emission from Rodriguez et. al. (1998). The astrometry represents *preliminary* efforts to align results correctly at different wavelengths.

Rodriguez, L. F., D'Alessio, P., Wilner, D. J., Ho, P. T. P., Torrelles, J. M., Curial, S., Gómez, Y., Lizano, S., Pedlars, A., Cantó, J., & Raga, A. C. 1998, Nat, 395, 355

Stapelfeldt, K., & Moneti, A. 1999, in The Universe as Seen by ISO, ed. P. Cox & M. F. Kessler, ESA-SP 427, 521

Stobie, E., Lytle, D., Barg, I., & Ferro, A. 1999, in Proc. NICMOS and the VLT Workshop held in Pula, Sardinia, Italy May 26-27 of 1998, ed. W. Freudling, & R. Hook, 77

Wood, K., Kenyon, S. J., Whitney, B., & Turnbull, M. 1998, ApJ, 497, 404

Disks, Planetesimals, and Planets
ASP Conference Series, Vol. 219, 2000
F. Garzón, C. Eiroa, D. de Winter, and T. J. Mahoney, eds.

Methanol Masers and the Circumstellar Disk Hypothesis

James M. De Buizer

University of Florida Department of Astronomy, 211 Space Science Research Building, Gainesville, FL 32611, USA

Abstract. It is not certain whether or not all massive stars form circumstellar disks. The earliest stages in the life of a massive star, when such disks would form, are poorly understood. Recent radio observations of methanol maser emission around young massive stars strongly suggest the existence of disks around these objects. This paper contains a review of what is known about masers and how they relate to massive stars. It discusses the hypothesis that methanol masers trace circumstellar disks around young massive stars and presents the mid-infrared observations that may lend support to this hypothesis.

1. Introduction

Massive stars are extremely important astronomical objects on both the smallest and largest scales. From the moment they form photospheres, massive stars are violently changing their environment. They produce copious amounts of photons with wavelengths shorter than the Lyman continuum limit and thus ionize the interstellar gas around them. They can heat material out to large distances in the clouds from which they formed and in the process enrich the interstellar medium with molecules through evaporation of dust-grain mantles. They may have powerful outflows which shock and churn the interstellar medium. Massive stars live fast and furious, and end their lives just as spectacularly. They explode as supernovae, releasing as much energy in the process as they had produced over their entire lives. Supernovae are invaluable alchemists. They produce the heavy elements that are otherwise impossible to create by normal stellar fusion processes. The force of their explosions spread these heavy elements throughout large distances in a galaxy, enriching the interstellar medium and thereby affecting later generations of stellar chemistry. These supernovae generate disruptive shock waves, which may in turn spark the formation of the next generation of stars. In this way, massive stars are responsible for the creation and distribution of elements in a galaxy and are therefore ultimately responsible for the chemical building blocks necessary for the creation of other stars, planets, and life as we know it.

Two of the most fundamental questions in astronomy are: *How do stars form?* and *How do planets form around these stars?* Much has been done both observationally and theoretically to understand the phenomena associated with these questions. However, despite the importance of massive stars, most of these studies have focused on stars that are similar to our Sun in mass or smaller. It still remains unclear exactly how massive stars form. Studies have led to a four-stage hypothesis for the formation of low-mass stars: 1) the formation of a core of material within a molecular cloud, 2) the inside–out collapse of the core creating a protostar, 3) the development of an accretion disk and bipolar outflow, and 4) the end of infall, revealing a newly formed star and a circumstellar disk. It is from these disks that planets are assumed to form. While this scenario has worked well to explain star formation for low=mass stars, it is not clear if it is applicable to massive star formation. We still do not know if massive stars form by accretion or by stellar mergers. Consequently, we do not know if the formation of circumstellar disks and outflows are a general characteristic of massive star formation. Furthermore, young massive stars are often associated with masers. We are only just now beginning to understand how masers of different species relate to the physical processes that occur during the formation and earliest evolutionary stages of massive stars.

2. Masers and Their Environment

Masers are naturally occurring molecular lasers that exist only under certain conditions and in very confined regions near newly forming stars. They are usually detected as extremely intense and narrow radio spectral lines. Interferometric mapping has revealed that maser emission is emitted from many individual "maser spots". Typically found in groups, maser spots each have their own well-defined velocity, and each comes from a region with a projected size on the order of 1 AU. There are four conditions necessary for the creation and continued existence of a maser. First, the interstellar medium through which the maser emission passes must be rich in molecular material. Second, a majority of the population of molecules involved must be in an excited state. This is referred to as a "population inversion", which is produced by either radiative or collisional processes. We refer to any mechanism which creates a population inversion as a "pump". Third, there needs to be a large column density of material through which the maser emission passes. Large column densities lead to large maser gains. And fourth, there needs to be velocity coherence of the molecules. Maser photons will interact only with molecules whose transition frequency has not been Doppler-shifted outside the linewidth of the maser transition.

There are several different molecules which exhibit maser action. The first observed astronomical maser was found by Weaver et al. (1965) in the 1665 MHz mainline transition of the hydroxyl (OH) radical. Shortly thereafter, Cheung et al. (1969) discovered the 22 GHz water maser, and Ball et al. (1970) detected the first methanol (CH_3OH) maser at 834 MHz. While water and hydroxyl masers have been well studied since their discovery, only in the last decade has the nature of the methanol maser begun to be seriously explored. Observations have led to the ideas that hydroxyl and water masers are associated with infall (Reid et al. 1980; Forster & Caswell 2000), outflow (Felli, Palagi, & Tofani 1992; Claussen et al. 1998), and circumstellar disks (Brebner et al. 1987; Slysh et al. 1999), as well as the density-enhanced regions between the shock and ionization fronts that lie close to, and are created by, a massive star (Cook 1966; Gaume & Mutel 1987). Since there is evidence supporting each of these scenarios, it is likely that water and hydroxyl masers trace a variety of stellar phenomena. As we will see in this paper, methanol maser excitation may be linked to several different stellar processes as well.

3. The Methanol Maser

One of the most exciting hypotheses for the location of masers is in circumstellar disks. As we have discussed, there are a few instances where water and hydroxyl masers are thought to trace disks. Recent observations of 6.7 and 12.2 GHz methanol masers have led to the hypothesis that they preferentially lie in circumstellar accretion disks around massive stars (Norris et al. 1993, 1998; Phillips et al. 1998). These studies have shown that groups of methanol maser spots tend to be distributed in lines or arcs in about 40% of the sites of methanol maser emission. Furthermore, in a limited number of cases, there exists a well defined increase or decrease in velocities measured across the linear distribution of maser spots, indicative of rotation.

Circumstellar disks are excellent locations for maser emission. There is a high density of molecular material, plenty of radiation from the central star for pumping the molecules, and, because the knots of material in the disk are caught up in the rotation, there is also velocity coherence. It is believed that if methanol maser spots form in rotating linear patterns, then the simplest explanation is that they exist in, and delineate, edge-on circumstellar disks. Why only edge-on disks? A primary reason would be that there is greater column density through the plane of the disk, which allows for more amplification, which in turn creates easily detectable methanol masers.

The circumstellar disk hypothesis is controversial and has its opponents. Some claim that linear distributions of masers can be caused by a variety of phenomena. Moreover, since there are only a small fraction of maser sites displaying velocity gradients along the spots, rotation may not be a general feature of methanol masers. These opponents advance the shock or outflow hypotheses. Some authors believe that shocked material associated with the expanding regions of ionized gas created by massive stars (also known as UCHII regions) can have just the right density for the existence of masers (e.g. Menten et al. 1988). If this is the case, the masers could be pumped radiatively or by the dissipation of the relative kinetic energies of the shocked and unshocked gas. Furthermore, there would be collective motion of the material for velocity coherence. The same holds true for the outflow hypothesis. If the masers were created by the shock associated with an outflow impinging on the interstellar medium, a similar circumstance could arise. The methanol masers may exist in the outflowing molecular material near the star as well, where the material can be radiatively pumped, and where there is velocity coherence.

4. Mid-Infrared Observations

The high-spatial-resolution radio observations of the individual methanol masers were primarily performed in the southern hemisphere, and until recently there had been no complementary data at other wavelengths. This has slowed progress in figuring out which of the excitation scenarios discussed in the last section are responsible for the methanol maser emission. It is difficult to ascertain what

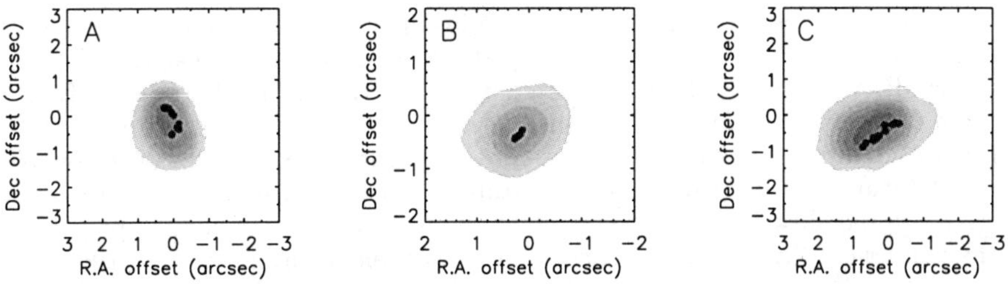

Figure 1. Gray-scale contours of the three resolved mid-infrared sources from De Buizer et al. (2000) with methanol masers overplotted as filled black circles. In each case the mid-infrared dust emission is elongated at a similar position angle as the associated linear maser distribution: *A)* G309.92+0.48, *B)* G323.740-0.263, *C)* G339.88-1.26.

stellar process maser emission is tracing when the locations of the associated massive stars are with respect to the maser spots are not known. Only in recent years have researchers looked for and imaged the radio continuum from the UCHII regions produced by massive stars. It was discovered that many sites of methanol masers have no detectable radio continuum emission (e.g. Walsh et al. 1998). Imaging at another wavelength was needed to find the associated young stellar sources. Optical imaging is not useful because massive stars evolve so quickly that they stay heavily embedded in their birth clouds, even after they reach the ZAMS, and are thus totally obscured visually during these formative years. However, because infrared radiation is much less affected by extinction than visible radiation, infrared images can probe through the cool obscuring dust. Furthermore, mid-infrared images (5–25 μm) can trace the radiation from the $\sim$ 200 K dust close to the stellar sources associated with the methanol masers, and could even detect circumstellar disks around these stars if they exist. The first mid-infrared survey of massive stars associated with maser emission was conducted by De Buizer, Piña, & Telesco (2000). The main thrust of this work was to locate the stellar sources associated with the methanol masers and determine what stellar process or processes methanol maser emission is tracing.

Twenty-one sites of methanol maser emission were imaged by De Buizer et al. (2000) at 10.46 and 18.06 μm, ten of which have their masers distributed in a linear fashion. Only in three of the ten sites were mid-infrared sources resolved at the location of the linearly distributed methanol masers. However, all three resolved sources are elongated in their thermal dust emission at the same position angles as their methanol maser distributions (Figure 1).

Modeling of methanol masers by Sobolev, Cragg, & Godfrey (1997) has led to the conclusion that methanol masers can be effectively pumped via mid-infrared photons from warm dust near stellar sources. De Buizer et al. (2000) find that the physical size of the regions containing methanol masers seems to be smaller than the resolved mid-infrared-emitting regions of the associated stellar sources. Moreover, the mid-infrared source peaks coincide with the methanol masers, a necessary condition if the masers are tracing disks.

All three resolved and elongated mid-infrared sources have mid-infrared companions which lie at approximately the same position angle as the methanol maser distributions. These companions are on the order of 10 000 AU from their counterparts, and are not visible in Figure 1. While they may not be gravitationally bound binaries, the two stars might be coeval. Wide binaries could form via the growth of an instability in the outer parts of a massive circumstellar disk (Bonnell 1994). In this formation scenario, the spin axes of the stars would always be aligned, and consequently circumstellar disks would be coplanar to the position angle of the binary. Recent observational evidence suggests that binaries with circumstellar disks may preferentially form in such a way as to produce this coplanarity (Monin, Menard, & Duchêne 1998; Jensen 2000).

Are these elongated mid-infrared sources circumstellar disks? I will summarize the evidence, which is circumstantial, but convincing. First, the mid-infrared peaks are coincident with the methanol masers. Second, the dust elongations seen in the mid infrared are at the same position angle as the linearly distributed methanol maser spots. Third, there is theoretical evidence that methanol masers require mid-infrared photons for pumping. All three maser distributions are contained within these mid-infrared elongations, and circumstellar disks radiate copious amounts of mid-infrared photons. Fourth, the resolved sources may be binary in nature. These pairs of sources lie at similar position angles as the methanol masers, and there is observational evidence that wide binaries with disks tend to be coplanar. And finally, one of the three sources has a well defined velocity gradient along the maser spots in the linear distribution, which may be caused by rotation. The simplest way to explain the nature of these elongated mid-infrared sources is that they are indeed circumstellar disks.

5. Planets and Disks around Massive Stars

The notion that planets form and exist around massive stars may be evidenced by the first ever discovery of extrasolar planets, detected around the pulsar PSR 1257+12 (Wolszczan & Frail 1992). These planets could have formed before or after the supernova explosion that left the pulsar we see today. Though the current idea is that planets form after a supernova, this scenario is not without its problems. Although controversial, there is some theoretical evidence that planets could survive a supernova explosion (Postnov & Prokhorov 1992). This would imply that the planets can form much earlier in the lifetime of a massive star, just as the case for low-mass stars. It is believed that accretion disks eventually become planet-building disks in the low-mass star formation paradigm. Therefore, the first step towards resolving the question of whether or not young massive stars can form planets is to first find out if they form circumstellar disks.

Though the data is still limited, the results of De Buizer et al. (2000) suggest that circumstellar disks do indeed form around young massive stars. Moreover, methanol masers may trace circumstellar disks more often than any other maser species. However, De Buizer et al. (2000) conclude that there are several methanol maser groups in their survey which most probably do not trace circumstellar disks. They find that some linearly distributed methanol masers may best be described as existing in shocks or outflows. Nevertheless, outflows are indirect evidence for the existence of circumstellar disks around massive stars because the disks serve as a collimator of stellar winds.

All of these results show that methanol masers are important tools which aid in the search for circumstellar disks around massive stars. Whether the masers directly trace disks, or indirectly tell us of the existence of a disk from tracing an outflow, they provide important collaborative evidence for the existence of circumstellar disks around massive stars. These studies also give us much needed insight into the early evolution of massive stars and suggest that these stars may indeed form in a similar fashion as their low-mass counterparts.

References

Ball, J. A., Gottlieb, C. A., Lilley, A. E., & Radford, H. E. 1970, ApJ, 162, L203

Bonnell, I. A. 1994 MNRAS, 269, 837

Brebner, G. C., Heaton, B., Cohen, R. J., & Davies, S. R. 1987, MNRAS, 229, 679

Cheung, A. C., Rank, D. M., Townes, C. H., Thornton, D. D., & Welch, W.J. 1969, Nat, 221, 626

Claussen, M. J., Marvel, K. B., Wootten, A., & Wilking, B. A. 1998, ApJ, 507, L79

Cook, A. H. 1966, Nat, 210, 611

De Buizer, J. M., Pina, R. K., & Telesco, C. M. 2000, ApJ, submitted

Felli, M., Palagi, F., & Tofani, G. 1992, A&A, 255, 293

Forster, J. R., & Caswell, J. L. 2000, ApJ, 530, 371

Gaume, R. A., & Mutel, R. L. 1987, ApJS, 65, 193

Jensen, E. L. N., in preparation

Menten, K. M., Reid, M. J., Moran, J. N., Wilson, T. L., Johnston, K. J., & Batrla, W. 1988, ApJ, 333, L83

Monin, J.-L., Menard, F., & Duchêne, G. 1998, A&A, 339, 113

Norris, R. P. et al. 1998, ApJ, 508, 275

Norris, R. P., Whiteoak, J. B., Caswell, J. L., Wieringa, M. H., & Gough, R. G. 1993, ApJ, 412, 222

Phillips, C. J., Norris, R. P., Ellingsen, S. P., & McCulloch, P. M. 1998, MNRAS, 300, 1131

Postnov, K.A., & Prokhorov, M. E. 1992, A&A, 258, L17

Reid, M. J., Haschick, A. D., Burke, B. F., Moran, J. M., Johnston, K. J., & Swenson, G. W. 1980, ApJ, 239, 89

Slysh, V. I., Val'tts, I. E., Migenes, V., Fomalont, E., Hirabayashi, H., Inoue, M., & Umemoto, T. 1999, ApJ, 526, 236

Sobolev, A. M., Cragg, D. M., & Godfrey, P. D., 1997, A&A, 324, 211

Walsh, A. J., Burton, M. G., Hyland, A. R., & Robinson, G. 1998, MNRAS, 301, 640

Weaver, H., Williams, D. R. W., Dieter, N. H., & Lum, W. T. 1965, Nat, 208, 29

Wolszczan, A., & Frail, D. A. 1992, Nat, 355, 145

Disks, Planetesimals, and Planets
ASP Conference Series, Vol. 219, 2000
F. Garzón, C. Eiroa, D. de Winter, and T. J. Mahoney, eds.

Circumbinary Disks around T Tauri stars: *HST*/NICMOS Near-Infrared Images and Polarimetric Maps

Gaspard Duchêne

Observatoire de Grenoble, Université Joseph Fourier, BP 53, 38041 Grenoble Cedex 9, France

Joel Silber

Department of Physical Sciences, University of Hertfordshire, College Lane, Hatfield, Hertfordshire AL10 9AB, UK

François Ménard

Canada-France-Hawaii Telescope Corporation, PO Box 1597, Kamuela HI 96743, USA

Tim Gledhill

Department of Physical Sciences, University of Hertfordshire, College Lane, Hatfield, Hertfordshire AL10 9AB, UK

Abstract. We have obtained new near-infrared images of both GG Tau and UY Aur circumbinary disks with the polarimetric modes of NICMOS aboard the *Hubble Space Telescope*. The 1 μm intensity map of GG Tau reveals a more complete elliptical shape than in previous ground-based images. Its eastern and western sides are definitely asymmetric. Our image strongly supports the ring geometry proposed by Guilloteau, Dutrey, & Simon (1999) on the basis of their millimeter interferometry images: a geometrically thick and sharply edged ring surrounding an empty gap around the binary. Around UY Aur, we identify structures that are in excellent agreement with the optical images of Ménard et al. (1999), which confirms that the inclination of the system to the line of sight is about 60 degrees. We also find tentative new structures closer to the stars.

1. Introduction

The presence of circumstellar disks around T Tauri stars has been suspected for a long time, but it is only very recently that these disks were directly detected, using high-angular millimeter imaging. These images revealed extended gas structures which appeared to be in Keplerian rotation around the central object. Among the rare detections so far, two disks were found around binary T Tauri stars: GG Tau (Dutrey, Guilloteau, & Simon 1994) and UY Aur (Duvert et al. 1998). The separations of the binaries are $0\farcs25$ and $0\farcs89$ respectively, which correspond to projected physical separations of 35 and 125 AU at the distance of the Taurus star-forming region (140 pc).

In both cases, light scattered off the surface of the disks has been detected afterwards with adaptive optics imaging at near-infrared wavelengths. Roddier et al. (1996) found that the GG Tau ring has a clumpy appearance and that several radial spokes of material extend from the ring onto the central stars. The ring is brighter in its northern part, but is detected in all directions. They interpret this brightness difference as being due to the scattering geometry. The UY Aur case is very different, as Close et al. (1998) only detected the disk on one side of the binary. Furthermore, they found evidence that a "spiral arm" splits from the main disk and gets closer to the star. Deconvolution processes were applied in both studies to retrieve the highest spatial resolution allowed by adaptive-optics devices, and this may lead to some artifacts in the final images.

More recently, the first visible wavelength images of UY Aur were obtained by Ménard et al. (1999) at 600 and 800 nm with *HST*/WFPC2. The PSF-subtracted images revealed a more complicated structure that was found by Close et al. (1998): a large "clump" appears to be independent of the disk itself. If true, this implies that the inclination of the system to the line of sight is larger than was first thought (about 60° instead of about 40°).

To improve our knowledge of these two circumbinary disks, we have performed new observations at 1 μm and 2 μm of these systems with *HST*/NICMOS. We used the polarimetric modes, and we obtained both intensity and polarization maps, which do not need to be deconvolved. The GG Tau polarization maps are the first ever obtained of this system, while Potter et al. (1998) already presented a deconvolved J-band polarization map of UY Aur which revealed a nice centro-symmetric pattern. Polarization maps are powerful tools for investigating the dust-grain properties and the geometry and structure of the disks.

In Section 2, we summarize our observations and data processing steps, and the maps of both systems are presented and commented in Section 3. Section 4 describes some implications of our results on the properties of these disks.

2. Observations and Data Processing

The 1 μm and 2 μm images were obtained with Camera 1 and Camera 2 respectively, providing pixel scales of $0\farcs043$ and $0\farcs075$. Both binaries were observed through the three polarizers at each wavelength, during three 96 s exposures for each filter. The regular NICMOS data reduction pipeline proved to be unsatisfactory, and we had to re-reduce all the data, with particular care concerning the so-called "pedestal effect", to obtain final images where the sky level is flat all over the detector.

To allow clear detections of the disks, it is mandatory to remove the bright stellar point spread functions (PSFs). We first tried TinyTim PSFs, but it appeared that their match with the real ones is quite poor, so we turned to a "natural" star, i.e., a bright single star observed through the same filters. The diffraction-spike subtraction, though imperfect, is quite good, and the optical ghosts induced by some polarizers are naturally removed. Some residuals in the core of the PSFs, however, are still large, and nothing can be detected with certainty in the inner $0\farcs5$ at 1 μm. At 2 μm, some fringing can be seen at

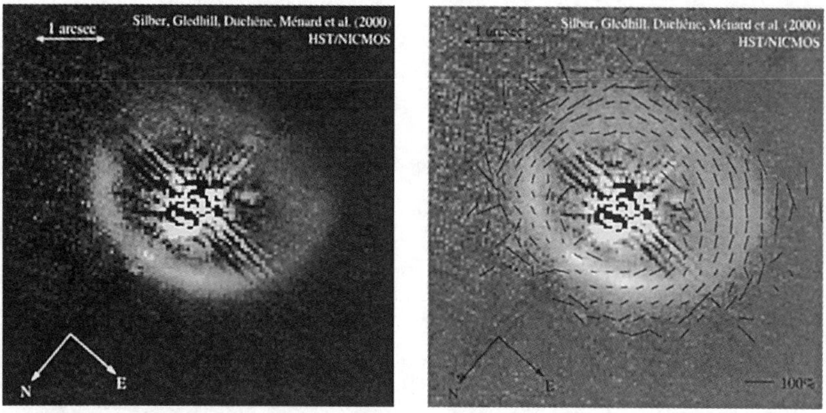

Figure 1. PSF-subtracted intensity (*left*) and polarization (*right*) maps of GG Tau at 1 μm.

separations as large as 3″.5. No deconvolution process was applied to our images, which allows an easier interpretation.

3. Results

3.1. GG Tau

The new 1 μm image of the GG Tau ring is presented in Figure 1. Its overall geometry is in good agreement with Roddier et al.'s (1996) images, though with a higher signal-to-noise ratio. However, there are some noticeable features. First, the ring does not appear clumpy in our image. This property was probably an artifact introduced by the deconvolution process applied to the adaptive-optics images. Fitting an ellipse to the ring, we find a semi-major axis, a position angle and an inclination in excellent agreement with the millimeter results of Guilloteau et al. (1999). It is noticeable, however, that this ellipse is not centered on the center of mass of the binary. Our image does not allow us to confirm the existence of the spokes of material discovered by Roddier et al. (1996), because of the large PSF-subtraction residuals inside the ring. Finally, a significant east–west asymetry in the northern part of the ring is seen in our intensity map.

The polarization vectors are strikingly well organized in a centro-symmetric pattern, which is symmetric about the semi-minor axis of the ring. The brightest part of the ring, which is closest to the observer, displays a lower polarization level than the faintest side, typically 20% as opposed to 50–60%.

At 2 μm, the disk is too close to the stars, and the large subtraction residuals prevent us from obtaining a clear image of the ring. However, we calculated the polarization map at this wavelength and, though the image is strongly dominated by the unpolarized stellar fluxes, a centro-symmetric pattern is found in the polarization vectors, with a typical level of 5–10%, indicating that the intrinsic polarization level of the light scattered by the ring is high.

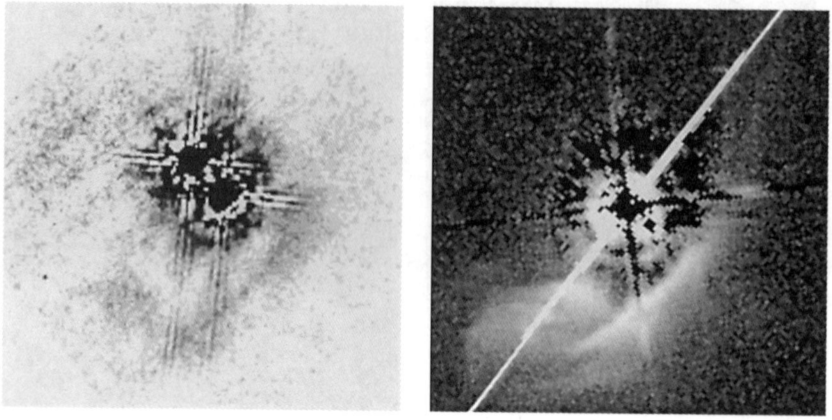

Figure 2. Intensity maps of UY Aur at 1 μm (this paper) and at 600 nm (from Ménard et al. 1999). Image sizes are $7''\!.7$ and $12''$, respectively. North is *up*, east to the *left*. The noisy area located to the east of the binary in the 1 μm image is due to a very low response of the detector.

3.2. UY Aur

The morphology of the UY Aur circumbinary disk in our new 1 μm image is in good agreement with Ménard et al.'s (1999) optical image, though the former suffer from a poor signal-to-noise ratio. As an be seen in Figure 2, the disk appears as an unresolved arc to the south-west of the binary at both wavelengths, while a bright clump to the south-east appears to be unrelated to this structure. Noticeably, the bright arc seems to widen to the west of the binary in our image. This can be interpreted as the arc breaking into two separate arcs: the disk itself remaining at least $2''\!.5$ away from the stars, and an inner arc, getting closer to the stars. The latter would correspond to the "spiral arm" described by Close et al. (1998). A second feature in our map is an inner arc which is much brighter than the disk itself. It lies about $1''$ to the south-west of the secondary. This may be a PSF artifact, but it can be seen using both a natural or a TinyTim-built PSF. Furthermore, its coincidence with a similar arc detected in the WFPC2 images is suggestive (see Figure 2). At 2 μm, the only area which is clearly separated from the PSF residuals is found to the south-east of the binary. It can be traced up to $5''$ away from the stars. The back side of the disk remains undetected at both wavelengths, which provides strong constraints on the dust-grain properties. We note, however, that the small arc seen to the north-east of the primary at 1 μm can be seen in all individual images.

The polarization pattern at 2 μm is well organized in its south-eastern part, with all vectors well aligned, in a fashion consistent with centro-symmetry. The typical polarization level is about 40 per cent. At 1 μm, however, the picture is very different: though the vectors are basically aligned in the south-western part of the disk with a typical level of 20%, again in a more or less centro-symmetrical pattern, they are quite randomly oriented in the south-eastern clump. Whether this is due to our reduction pipeline, to the low signal-to-noise ratio, or reflects

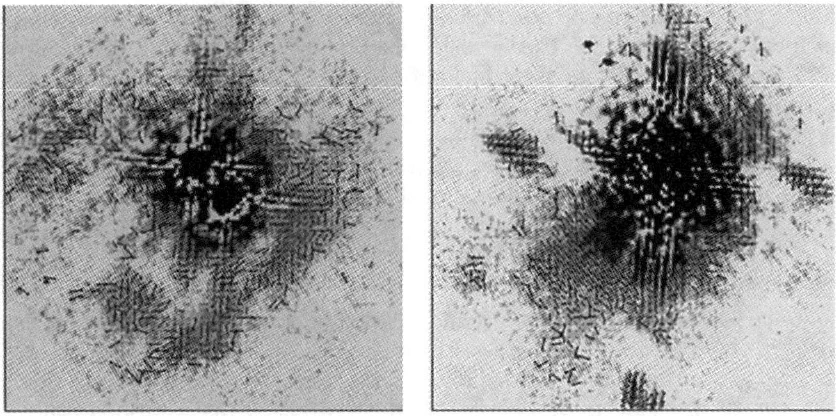

Figure 3. Polarization vectors superimposed on the intensity maps of UY Aur at 1 μm (*left*) and 2 μm (*right*). Image sizes are $7''.7$ and $13''$, respectively. Orientation of both images as in Figure 2.

the intrinsic pattern is unclear. It should be noted that Potter et al. (1998) found a very different behavior. Their data reduction process, however, included a deconvolution step, whose impact on the polarization is unknown.

4. Implications and Open Questions

4.1. GG Tau

As already pointed out by Guilloteau et al. (1999), the shift between the apparent center of the ring and the center of mass of the binary is naturally explained by a thick-ring geometry. This is related to the fact that, in Mie theory, forward scattering is strongly favored. Hence, most of the light scattered towards the observer comes from the upper part of the disk's inner edge, whose projection on the sky is not symmetric about the physical center of the ring. The quantitative model proposed by Guilloteau et al. (1999) is a ring with a half-thickness of 60 AU at its inner radius (180 AU) and a very sharp edge. The observed location of the ring appears to be in excellent agreement with the prediction from this model. Roddier et al. (1996) suggested that the thickness-to-radius ratio was about one tenth at the inner edge of the ring, which seems incompatible with our results.

The origin of the east–west asymetry is unclear. It may be due to the presence of two illuminating stars, or be related to the slight asymmetry found in the millimeter-wavelength image, which itself may reveal internal structures differences.

4.2. UY Aur

Fitting ellipses to the large south-western, unresolved arc both at 1μm and 600 nm yields very similar figures and, especially, an inclination to the line of sight of about 60° in both cases. This is much larger than the 42° estimated by Close et al. (1998). The reason for that discrepancy is that they assumed that the south-east clump belongs to the disk, which now seems unlikely. Duvert et al. (1998) pointed out that a larger inclination (60–70°) is in better agreement with the millimeter observations.

If the inner arc close to the secondary star proves to be real, it will represent a challenge for theoretical studies, as most of these predict that structures at such a location should be highly unstable due to interaction with the binary system.

4.3. Monte Carlo Modeling

The polarization levels, as well as the front-to-back side flux ratios, are tightly linked to the disk geometry and to the dust grain properties. Our next step in this study is to run Monte Carlo simulations to investigate these properties. For instance, we may determine whether the dust-grain size distribution in the disks is compatible with that of interstellar grains. In principle, we will also constrain the amount of "flaring" in both disks, as well as their geometrical height and optical depth.

References

Close, L., Dutrey, A., Roddier, F., Guilloteau, S., Roddier, C., Northcott, M., Ménard, F., Duvert, G., Graves, J., & Potter, D. 1998, ApJ, 499, 833

Dutrey, A., Guilloteau, S., & Simon, M. 1994, A&A, 286, 149

Duvert, G., Dutrey, A., Guilloteau, S., Ménard, F., Schuster, K., Prato, L., & Simon, M. 1998, A&A, 332, 867

Guilloteau, S., Dutrey, A., & Simon, M. 1999, A&A, 348, 570

Ménard, F., Stapelfeldt, K., Krist, J., Duvert, G., Padgett, D., & Burrows, C. 1999, BAAS, 194, 6811

Potter, D., Close, L., Roddier, F., Roddier, C., Graves, J., & Northcott, M. 1998, in ESO Conf proceedings, 56, Astronomy with Adaptive Optics, ed. D. Bonaccini (Garching: ESO), p. 353

Roddier, C., Roddier, F., Northcott, M., Graves, J., & Jim, K. 1996, ApJ, 463, 326

Disks, Planetesimals, and Planets
ASP Conference Series, Vol. 219, 2000
F. Garzón, C. Eiroa, D. de Winter, and T. J. Mahoney, eds.

From Aromatic Molecules to Carbon Dust Grains

C. Moutou and A. Léger

ESO, Alonso de Cordoba 3107, Santiago 19, Chile

L. Verstraete

IAS, CNRS, Université Paris Sud, F-91405 Orsay, France

W. Schmidt

Institut für PAH-Forschung, Flurstrasse 17, 86926 Greifenberg, Germany

K. Sellgren

Astronomy Dept., Ohio State University, 140 West 18th Avenue, Columbus OH 43210, USA

Abstract. Recent analysis of *ISO* spectra highlights the process of grain formation. In several environments of the interstellar medium, a new emission band at 16.4 μm has been detected. Comparison with laboratory experiments suggests that the dominant carriers are large aromatic molecules (PAHs) containing pentagonal rings. The implications of this result for the formation routes of carbon grains are discussed.

1. Introduction

Interstellar dust plays an important role in the star formation process. Indeed, charged grains have been shown to be more effective than ions in coupling the gas to the magnetic field. As a consequence, the collapse of a dense molecular core is significantly slowed down by the presence of dust (via ambipolar diffusion) whereby it also alleviates the magnetic flux dissipation problem (see Ciolek & Mouschovias 1993, and references therein). Moreover, the structure of protostellar and/or protoplanetary objects is best probed in the infrared and/or submillimeter range, where extinction is low. For instance, the presence of disks has been inferred first from the spectral energy distribution (in the observations of Cohen & Schwartz 1983, as analyzed by Adams & Shu 1985) then from the spatial structure (e.g., Koresko et al. 1993; Ménard et al. 1993) observed in the infrared. The analysis of such observations requires a detailed knowledge of the dust emissivity in the infrared and submillimeter (in this latter case also to deduce the mass of the proto-object).

Dust grains are a mixture of silicates and carbonaceous material which condense in the expanding atmospheres of evolved stars and are then expelled into the interstellar medium. The radii of these grains range from a few Å to a few tens of micrometers (Désert, Boulanger, & Puget 1990; Li & Greenberg 1997; Dwek et al. 1997). During their lifetime in the interstellar medium, the physics and chemistry of dust grains is governed mostly by starlight and shocks (Jones, Tielens, & Hollenbach 1996). In dense, shielded clouds an ice mantle forms on top a refractory core. We focus here on the population of smallest grains which produce the 3–20 μm emission bands of the interstellar medium. It is well known that these bands arise from vibrational modes in aromatic bonds. Among the possible carriers of these bands are large aromatic molecules or polycyclic aromatic hydrocarbons (PAHs). In fact, models of dust formation in stellar atmospheres (Goeres & Sedlmayr 1992; Goeres 1996; Cherchneff 1995) show the importance of PAHs in the gas-phase carbon chemistry. Interstellar emission spectra obtained with the *Infrared Space Observatory* (*ISO*) are shown and discussed here. These recent observations relate interestingly to theories of carbon dust formation. Section 2 presents the observations, Section 3 describes results obtained in laboratory experiments on PAHs. Section 4 discusses the implications for the formation of carbon grains.

2. Observations

Short Wavelength Spectrometer (SWS) complete spectra (de Graauw et al. 1996) of three bright galactic sources—NGC 7023, M17-SW, and the Orion Bar—have been obtained with *ISO* in a $14'' \times 27''$ beam. The data reduction was carried out with SWS-IA and is described in Moutou et al. (1999). The high spectral resolution (see Figure 1 *left*) allows a detailed spectroscopic analysis. All three spectra show the strong features characteristic of aromatic bond vibrational modes between 3 and 13 μm. Hydrogenated aromatic molecular clusters such as PAHs show these features. Excited by stellar light, PAHs are small enough to undergo large temperature excursions. In the framework of the PAH model, hot (a few 100 K), aromatic molecules produce the observed mid-IR features via transitions in the fundamental vibrational modes of the C–H and C–C bonds (see reviews by Salama 1999 and Tielens et al. 1999). The 3-to-13 μm features have been extensively studied in the past and we focus here on a new feature visible only with *ISO* spectrometers.

Indeed, the spectra of NGC 7023, M17-SW, and the Orion Bar exhibit an emission band at 16.4 μm (Figure 1 *left*). This new feature, first mentioned by Boulanger et al. (1998) and Tielens et al. (1999), is spatially correlated to the other PAH features. The band is centered at 609 cm^{-1} and has an FWHM of 5.8 cm^{-1} (0.16 μm).

3. Laboratory Spectra

We have used existing laboratory spectroscopic databases of PAHs to assign this new band. The low-temperature absorption spectra of many PAHs exhibit a band around 16.2 μm (Moutou et al. 1996; Schmidt et al. unpublished; Hudgins & Sandford 1998).

Due to anharmonic couplings, which depend on the temperature, this band should appear redshifted in the interstellar spectrum, where the emission originates from hot gaseous species (Joblin et al. 1995). The expected redshift

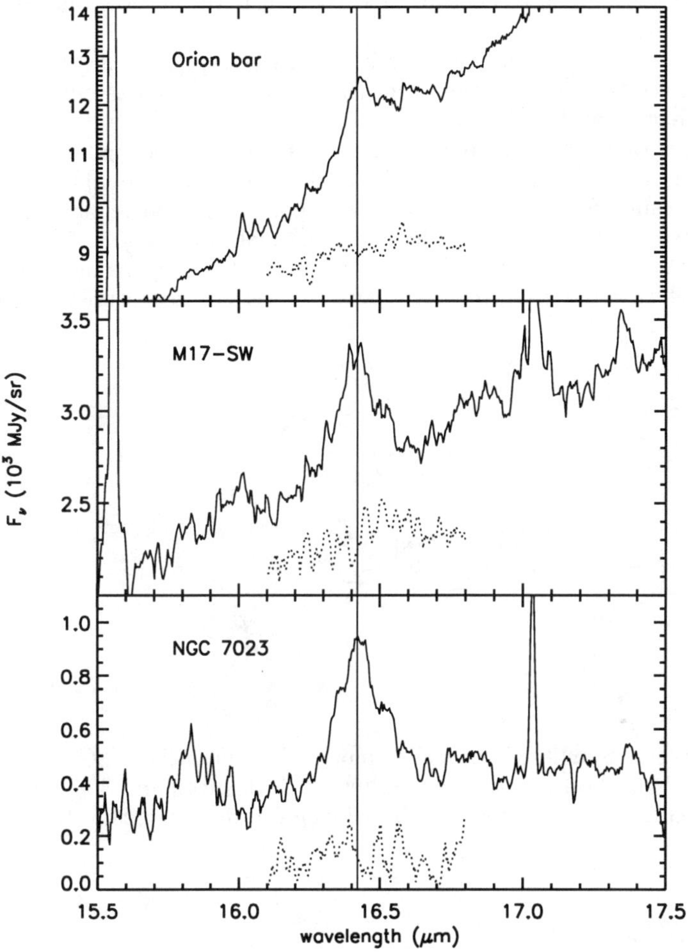

Figure 1. *Left:* The 16.4 μm feature in the *ISO*-SWS01 spectra (*solid lines*) with $\lambda/\Delta\lambda = 900$ of the Orion Bar, M17-SW, and NGC 7023. The dotted lines show the noise estimate as the halved difference between two independent spectra. The unresolved bright line at 15.5 μm in Orion and M17-SW is [Ne III] while the 0-0 $S(0)$ rotational H_2 lines is visible in NGC 7023 and M17-SW at 17 μm. *Right:* Histograms (in bins of width 0.13 μm) of PAH absorption bands in the 16 μm region measured in the laboratory (Moutou, Léger, & d'Hendecourt 1996; Schmidt et al. unpublished; Hudgins & Sandford 1998a–c). They do not take into account the absolute values of the absorption cross-section, but are merely presented as a percentage of the band presence in each PAH family. The *dashed line* shows the position of the emission band in the *ISO* spectrum. A redshift of 0.16 μm has been applied to all laboratory bands, in order to take into account the expected temperature effects ($T \approx 300$ K).

would place the emission band at 16.4 μm, precisely the position observed in the astronomical spectra.

Using laboratory data, we show in Figure 1 (*right*), separated into seven PAH families representing various molecular structure, the percentage of family members that show the 16.4 μm band.

We find that the contribution of the PAHs containing pentagons is the largest; hence, it could possibly dominate the interstellar emission at 16.4 μm. This band results from a global vibrational mode of the carbonaceous skeleton, as illustrated in the case of the fluoranthene molecule (Figure 2, J. Brunvoll and S. Cyvin, private comm.). Finally, we note that pentagons have no spectral signature which could be inconsistent with the rest of the observed IR-spectrum (Moutou et al. 2000).

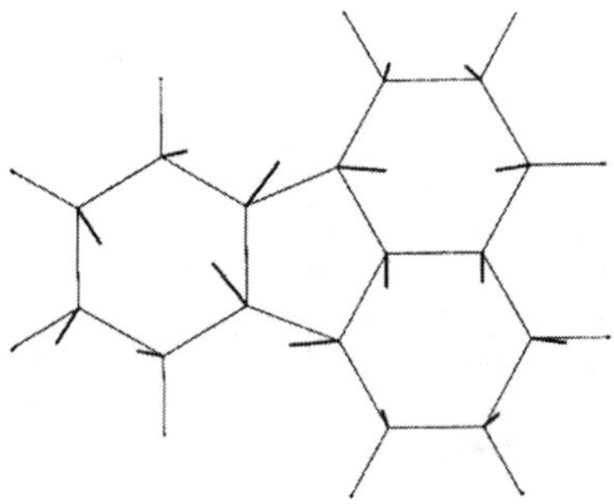

Figure 2. Simulated molecular motion in fluoranthene ($C_{16}H_{10}$) at the origin of the 16 μm mode (courtesy of J. Brunvoll and S. Cyvin): it looks as if the central five-membered ring rotates inside the structure.

4. Discussion

Pentagonal rings tend to bend a molecular plane composed of pure hexagonal rings. The molecule then becomes three dimensional. The chemical growth of aromatic compounds, from simple planar PAHs to small carbon grains, apparently requires such an intermediate evolutionary state where molecules of ~ 100 carbon atoms are curved. Modeling the carbon chemistry in carbon-star shells, Goeres & Seldmayr (1992) predict the formation of large aromatic molecules containing pentagons because pentagons offer the best trade-off between entropy and energy in the process of carbon dust growth. Figure 3 shows the prediction of this model based on carbon nucleation (Goeres 1996); three main steps are required from carbon chains to monocyclic rings, which finally aggregate to form polycyclic aromatic clusters. The role of pentagonal rings in this sequence is fundamental, as they allow the transition to three-dimensional grains which are more thermodynamically stable. Kroto & McKay (1988) proposed an alternative growth sequence to produce carbon grains involving pentagonal rings.

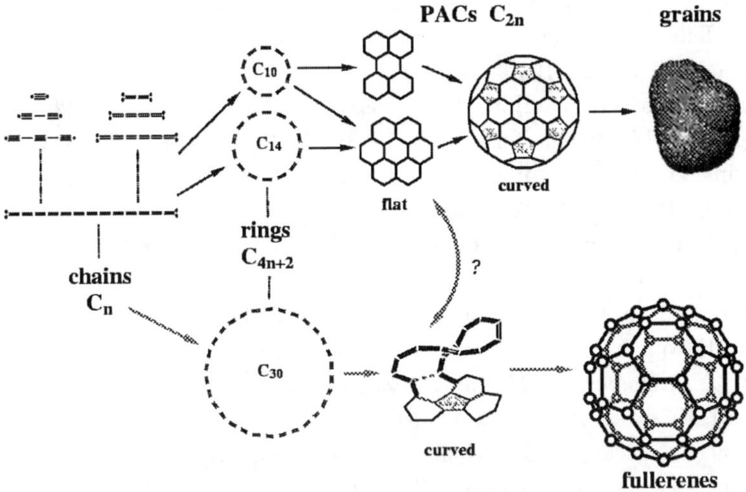

Figure 3. Sketch of the carbon chemistry network possibly occurring in the atmospheres of R CrB stars. Three main structural types are seen: carbon chains, monocyclic rings, and polycyclic aromatic carbon clusters (PACs). The role of five-membered rings in the curvature of planar molecules is shown by the evolution towards larger dust grains (courtesy of Goeres 1996).

Finally, interactions between molecular clusters and grains are also suggested by the behavior of interstellar absorption bands in circumstellar environments: the band profiles observed may be explained by the sticking of their molecular carrier onto grain mantles in dense and cold clouds around forming stars (Krełowski and Greenberg 1999).

Acknowledgments. The authors wish to thank S. Cyvin, J. Brunvoll, P. Klaeboe, and A. Goeres for their interest in this work and their fruitful comments.

References

Adams, F. C., & Shu, F.H. 1985, ApJ, 296, 655

Boulanger, F., et al. 1998, A&A, 339,194

Cherchneff, I. 1995, ApSS, 224, 379

Ciolek, G. E., & Mouschovis, T. C. 1993, ApJ, 418, 774

Cohen, M., & Schwartz, R. D. 1983, ApJ, 265, 877

Désert, F. X., Boulanger, F., & Puget, J. L. 1990, A&A, 237, 215

Dwek, E., et al. 1997, ApJ, 475, 565

Goeres, A., & Sedlmayr, E. 1992, A&A, 265, 216

Goeres, A., 1996, in ASP conf. Ser., 96, Hydrogen-Deficient Stars, ed. C. S. Jeffery & U. Heber (San Francisco: ASP), p. 69

de Graauw T., et al. 1996, A&A, 315, L49

Hudgins, D. M., & Sandford, S. A. 1998a, J. Phys. Chem. 102, 329

Hudgins, D. M., & Sandford, S. A. 1998b, J. Phys. Chem. 102, 344

Hudgins, D. M., & Sandford, S. A. 1998c, J. Phys. Chem. 102, 835

Joblin, C. 1995, A&A, 299, 835

Jones, A. P., Tielens, A. G. G. M., & Hollenbach, D. J. 1996, ApJ, 469, 740

Koresko, C. D., et al. 1993, AJ, 105, 1481

Krełowski, J., & Greenberg, J. M. 1999, A&A, 346, 199

Kroto, H. W., & McKay, K. 1988, Nat, 331, 328

Li, A., & Greenberg, J. M. 1997, A&A, 323, 566

Ménard, F., et al. 1993, ApJ, 414, L117

Moutou, C., Léger, A., & d'Hendecourt, L. 1996, A&A, 310, 297

Moutou, C., et al. 1999, A&A, 347, 949

Moutou, C., et al. 2000, A&A, 354, L17

Salama, F. 1999, in Solid IS Matter: the ISO Revolution, ed. L. d'Hendecourt, C. Joblin, & A. P. Jones, p. 65

Tielens, A. G. G. M., et al. 1999, in The Universe as seen by ISO, ed. P. Cox & M. F. Kessler (Noordwijk: ESA), p. 579

Disks, Planetesimals, and Planets
ASP Conference Series, Vol. 219, 2000
F. Garzón, C. Eiroa, D. de Winter, and T. J. Mahoney, eds.

The Circumstellar Environment of the TTS SU Aurigae

J. M. Oliveira[1] and B. H. Foing

Solar System Division, ESA Space Science Department, ESTEC/SO, Postbus 299, 2200 AG Noordwijk, The Netherlands

J. Th. van Loon

Institute of Astronomy, Madingley Road, Cambridge CB3 0HA, UK

Abstract. In this contribution we investigate how a disk determines the physical and geometrical properties of the circumstellar environment of the T Tauri star SU Aurigae. Our model of the spectral energy distribution of this star includes the central young star, a flat black disk and a diffuse envelope. We also describe the inner interaction region between the disk and the star, the magnetosphere, by analyzing the accretion and wind signatures in several spectral lines observed during the MUSICOS 96 campaign.

1. Introduction

Young stellar objects have extremely complex circumstellar environments: material in an infalling envelope surrounding the central star falls onto a circumstellar disk, dissipating angular momentum; magnetic fields can then channel the disk material onto the star by way of accretion funnels (magnetospheric accretion), causing hot shock regions on the stellar photosphere.

Classical T Tauri stars (CTTSs) are low-mass pre-main-sequence stars. These stars are active, with broad energy distributions and Balmer and metallic (permitted and forbidden) line emission. CTTSs are class II objects, i.e., the systems include a central young star and an accretion disk. The presence of the disk was inferred from the IR and UV excesses in the spectral energy distributions and it can also account for some of the photometric variability and optical veiling; for some stars the disk has also been imaged directly.

Magnetospheric accretion can explain the photospheric hot spots and the He I spectral line, as a shock region at the stellar surface results in the release of kinetic energy. Spectroscopic evidence of magnetospheric accretion are the redshifted absorption components on several spectral lines. The spectra of T Tauri stars also present blueshifted absorption features that are signatures of stellar or disk winds.

SU Aurigae is either classified as a CTTS (e.g., Bouvier et al. 1993) or as the prototype of a separate class of young stellar objects (Herbig & Bell 1988).

---

[1]Centro de Astrofísica da Universidade do Porto, Rua das Estrelas s/n, 4150 Porto, Portugal.

The star's spectral type is G2 (Herbig 1960), earlier than most CTTSs, and it is an exceptionally fast rotator, $v \sin i \sim 66$ km s^{-1} (Hartmann & Stauffer 1989). Its photometric period has been rather difficult to determine, but it is usually considered to be close to 3 days. From our dataset, we favor a somewhat shorter period of 2.5–2.8 days (Oliveira et al. 2000).

In this contribution we describe the SU Aur system from two different viewpoints: i) using the spectral energy distribution to characterize the larger scale circumstellar environment and ii) probing the disk–star interaction by analyzing the inner magnetosphere, using results from our multi-site spectroscopic campaign on this target.

2. The Outer Circumstellar Environment: Disk and Envelope

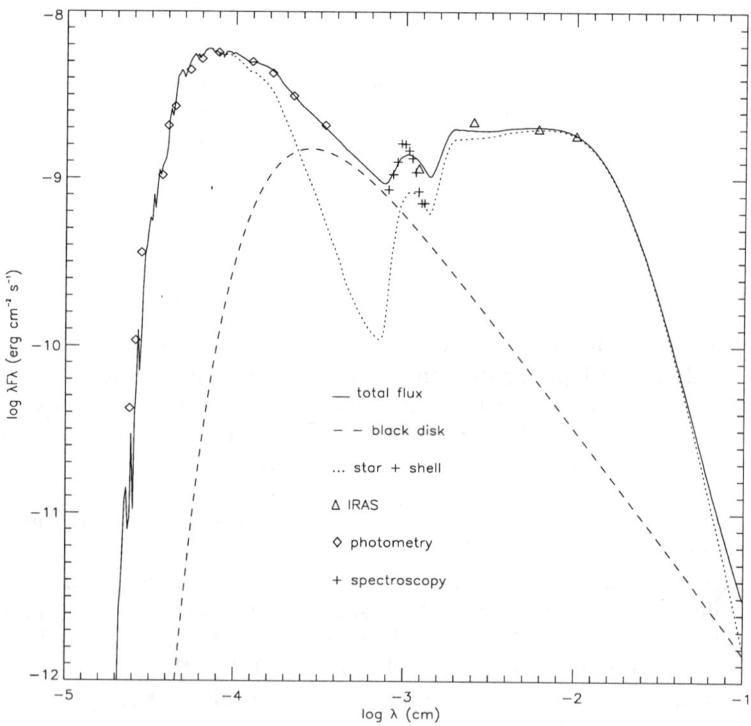

Figure 1. SED for SU AUR: ◇ represent photometric (Bertout, Basri, & Bouvier 1988), +, spectroscopic (Hanner, Brooke, & Tokunaga 1998) and △ IRAS measurements. The resulting model is plotted (*full line*) as well as the star + shell (*dotted line*), computed with the MoDust code, and the flat black disk (*dashed line*) contributions. The fit obtained is quite good, with exception of the 12–20 μm range (silicate features and detailed treatment of chemistry is required) and the UV excess (shock region emission).

Accretion disks can explain two of the most important peculiarities of the spectral energy distribution (SED) of the CTTSs: the UV and IR excesses. The dust grains in the disk are heated either radiatively by the central star or by accretion heating (via the loss of gravitational energy as the gas spirals through the disk to the star) and they emit in the IR. As the visual extinction is relatively low for most CTTSs the dust distribution has to be flattened, as was proved recently by *HST* images of several CTTSs. The UV excess can originate at the shock regions that form where the material accretes from the disk onto the stellar surface, causing the release of large amounts of kinetic energy.

The SEDs of some CTTSs seem to display more far-IR excess than expected from a disk. Several methods have been used to model these so called flat spectra: adding to the system an accreting envelope component (Calvet et al. 1994), or considering a flared disk, i.e., a disk whose vertical height varies with the radial distance (Kenyon & Hartmann 1987).

We review here the SED of SU Aur using photometry from the literature and redetermining the *IRAS* fluxes. This SED is relatively flat in the far IR. It seems to have two contributions: a cool and a hot dust component. We find that the cool component can be well represented by a spherical dust shell at some distance from the star, whilst a disk closer to the star can account for the hot component. We modeled this disk as a flat, black disk (Beckwith 1999). The fit obtained is quite good (within 0.05 in units of $\log \lambda F_\lambda$), with the exception of the silicate features at 10 and 18 μm, that demand a deeper study into the dust properties in the disk and envelope. Our model also does not fit the UV excess properly: the emission at those wavelengths is believed to be related to the accretion energy released at the footpoints of the accretion funnels at the stellar surface. The resulting fit to the SED is shown in Figure 1 and the main model parameters are given in Table 1.

Table 1. The model parameters for the SED of SU Aur

Star		Shell		Disk	
distance	150 pc	R_{in}	200 R_*	R_{in}	3 R_*
L_*	9.5 $L_\odot$	R_{out}	10^6 R_*	R_{out}	3×10^4 R_*
R_*	3.4 $R_\odot$	ρ_{in}	5×10^{-17} (g cm^{-3})	inclination	60°
T_{eff}	5500 K				

3. The Inner Circumstellar Environment: Magnetosphere and Spectral Line Variability

Shu et al. (1994) proposed a magnetospheric model where the accretion disk is disrupted at a few stellar radii from the star by a stellar dipole magnetic field. At the truncation point, the ionized disk material is loaded either onto inner closed magnetic field lines, thereby accreting onto the stellar photosphere, or it is loaded onto outer open magnetic field lines that can drive a disk-wind flow.

Johns & Basri (1995) suggested that in the case of SU Aur the rotational and magnetic axes are misaligned, causing the observed periodicity (~ 3 days) in the wind and accretion flows and their anti-phase behavior.

We used the data set obtained in the MUSICOS 96 campaign to characterize the variability timescales of several lines. In Figure 2 we show the average and variance profiles of Hα, Hβ, Na I D and He I D3. Using the cross-correlation function, we searched for time lags between the variability of these spectral lines. We have found that the variations in the red wing of Hβ are out of phase with the variations in the blue wing of Hβ by slightly more than half of the rotational period (Figure 3*a*). Thus, as expected in an oblique magnetosphere, the accretion funnel and the wind components are out of phase.

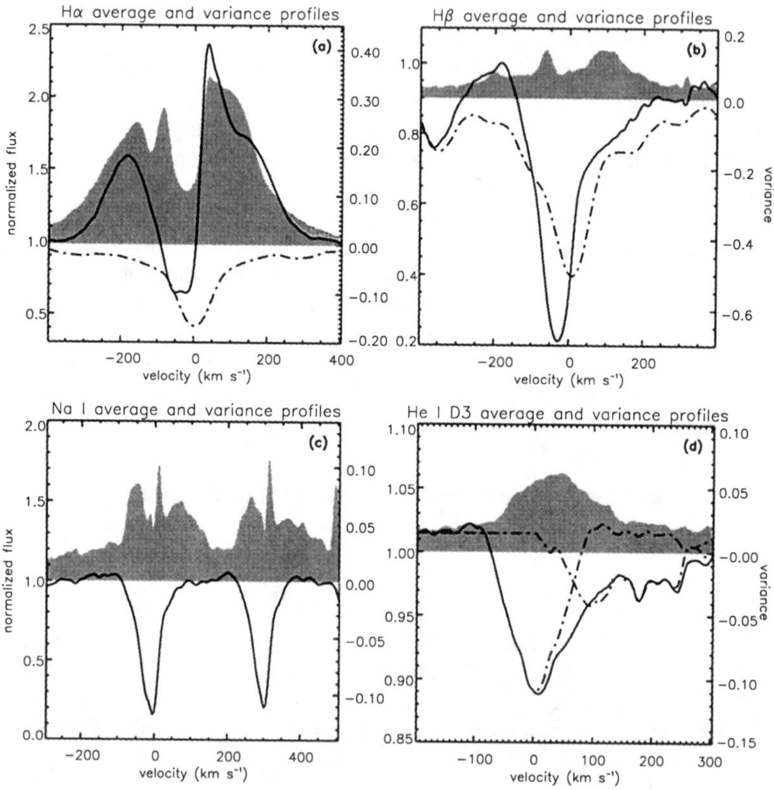

Figure 2. The average (line plots) and variance (*shaded*) of Hα (*a*), Hβ (*b*), Na I D (*c*), and He I D (*d*). The *y*-axis range on the left side is for the average and on the right side for the variance.

The He I D3 line is in absorption in SU Aur, contrary to what is observed for most of the other CTTSs. Still, we have also decomposed this profile into two components, one centered on the rest velocity and another centered at about 80 km s^{-1}. We have found that there might be a small time lag between these components, suggesting that the first component might originate at the surface and the redshifted component in the accretion column. We have also found that the variability in the He I D3 and the red wing of Na I D are almost in synchrony (Figure 3*c* and *d*) and it propagates in a few hours to the absorption components in the red wings of Hβ and Hα (Figure 3*b*).

We thus observe variability that propagates from the inner magnetosphere outwards. Such behavior cannot be caused by non-steady accretion nor by simple occultation effects in the oblique magnetospheric model. We propose that this can occur only if the magnetic field lines are azimuthally distorted due to differential rotation between the star and the disk (Oliveira et al. 2000). In this situation, the time symmetry for ingress/egress of occultation events is broken and then the variability of the higher-energy lines precedes that of the lower-energy lines.

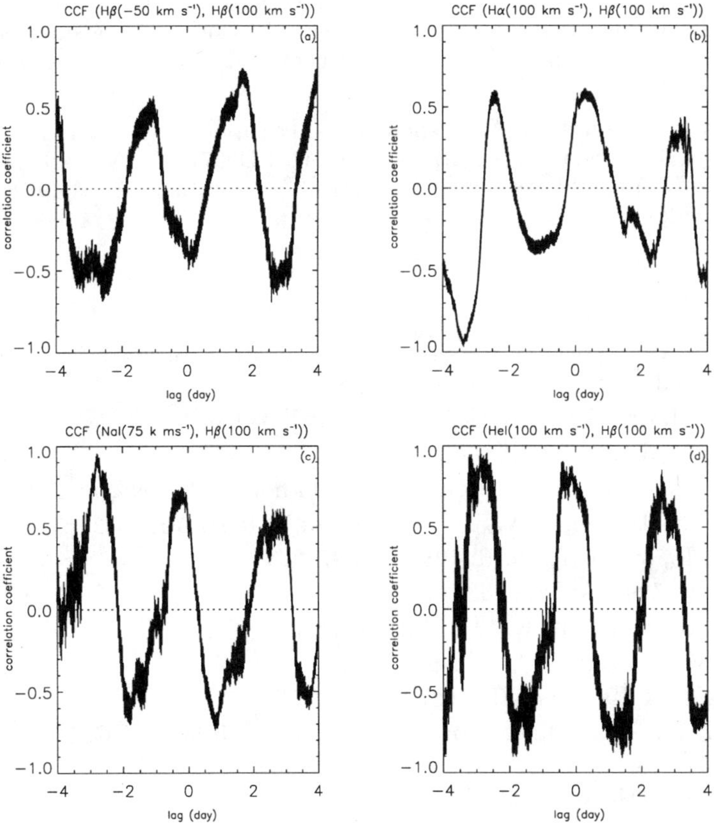

Figure 3. Cross-Correlation as a function of time-lag. The quasi-periodic shape of the CCF reflects the periodicity of the spectral lines. The variability in He I D3 and Na I D slightly precedes the variability in Hβ; the variability in Hβ precedes the variability in Hα. Thus, the observed variability seems to propagate from the inner magnetosphere to the outer regions.

4. Conclusions

The enhanced activity observed in SU Aur can be explained by the presence of an accretion disk and spherical dust shell. We describe the SED of SU Aur as including the star, a flat black disk (a hot dust component to account for the near-IR excess emission) and a diffuse envelope (a cold dust component to account for the far-IR excess). The model is rather simple but it is in very good agreement with the observations available.

Probing the inner magnetosphere, we observe a time-lagged behavior of the variability of spectral lines with different excitation conditions that is not compatible with the standard magnetospheric models, even when the magnetic dipole is inclined with relation to the rotation axis. The behavior we observe requires the magnetic field lines to be azimuthally distorted probably due to differential rotation in the star + disk system. Obviously, more stars need to be observed quasi-continuously to allow a generalization of the results presented here.

Acknowledgments. JMO acknowledges the support of the Fundação para a Ciência e Tecnologia (Portugal) under grant BD9577/96. We acknowledge Alex de Koter from the Astronomical Institute, Amsterdam for the use of his MoDust code.

References

Beckwith, S. V. W. 1999, in Physics of Star Formation and Early Evolution II, ed. C. J. Lada & N. D. Kylafis, in press

Bertout, C., Basri, G., & Bouvier, J. 1988, ApJ, 330, 350

Bouvier, J. et al. 1993, A&A, 272, 176

Calvet, N., Hartmann, L., Kenyon, S.J, & Whitney, B. A. 1994, ApJ, 434, 330

Hanner, M. S., Brooke, T. Y., & Tokunaga, A. T. 1998, ApJ, 502, 871

Hartmann, L., & Stauffer, J. R. 1989, AJ, 97, 873

Herbig, G. H. 1960, ApJS, 4, 337

Herbig, G. H., & Bell, K. R. 1988, LicOB, 1111, 1

Johns, C. M., & Basri, G. 1995, ApJ, 449, 34

Kenyon, S. J., & Hartmann, L. 1987, ApJ, 323, 714

Oliveira, J. M, Foing, B. H., van Loon, J. Th., & Unruh, Y. C. 2000, A&A, submitted

Shu, F. et al. 1994, ApJ, 429, 781

Part 3

The Pre-Main Sequence and Main Sequence

Disks, Planetesimals, and Planets
ASP Conference Series, Vol. 219, 2000
F. Garzón, C. Eiroa, D. de Winter, and T. J. Mahoney, eds.

Disks and Planets in Binary Systems

Wilhelm Kley and Andreas Burkert

Max-Planck-Institut für Astronomie, Königstuhl 17, D-69117 Heidelberg, Germany

Abstract. The star formation process in molecular clouds usually leads to the formation of multiple stellar systems, mostly binaries. Remaining disks around those stars may be located around individual stars (circumstellar disks) or around the entire binary system (circumbinary disk). We shall briefly review the present observational evidence for both types of disks in binary stars, in particular the properties of circumbinary disks.

We then present recent results of the theoretical modeling of the collapse and fragmentation of gravitationally unstable molecular cloud cores and their implications for binary and disk formation, and discuss the dynamical influence of the binary companions on disk truncation and gap formation. The presence of binaries may have profound influence on the process of planet formation as well. We present results on the stability and evolution of orbits of planets in disks around binaries.

1. Observational Properties

This contribution deals primarily with theoretical aspects; however, for completeness, we summarize first some basic observational features of disks and planets in binary stars. An excellent recent review on this topic has been given at PPIV by Mathieu et al. (2000).

1.1. Binary Stars

The majority of stars is found in binary or multiple systems. To quantify the occurrence of multiplicity one usually defines the *binary frequency* (BF) as

$$\text{BF} = \frac{\text{number of multiples}}{\text{total number of systems}}$$

The seminal study of Duquennoy & Mayor (1991) found for field stars (not members of a stellar cluster) having an age larger than 1 Gyr a binary frequency of BF = 60%. The distributions of the orbital elements of binary stars such as a mass ratio, q, periods, P, or eccentricities, e, are typically very broad and they have mean values of

$$\langle q \rangle = 0.3$$

$$\langle \log(P[\text{days}]) \rangle = 5.0$$

$$\langle e \rangle = 0.5$$

For pre-main-sequence stars one finds a bimodal binary frequency. Dense clusters in giant molecular clouds, such as the Trapezium Cluster in Orion (with a stellar density of over 2000 stars per pc^3), tend to have a binary frequency similar to the field value, (60%). In low-stellar-density clusters (such as the Taurus or Ophiuchus T Tauri associations) the binary frequency is much higher than the field value, about 80–100% (Ghez, Neugebauer, & Matthews 1993; Leinert et al. 1993). This difference in the BF may be the result of several mechanisms, the most important of which are (see Mathieu et al. 2000): a) observational effects due to incompleteness, which may lead to a wrong determination of the BF; b) evolutionary effects, which may change the BF with time; c) more frequent stellar encounters, in higher density clusters, which may lead to the disruption of systems and a lowering of the BF, and d) the BF, which may just depend on the environment, such as the temperature or density of the cloud core.

From these results one may infer that most of the field stars originated in dense clusters.

1.2. Disks in Binaries

The classical signatures for the presence of disks around stars are an excess in IR to mm radiation, Balmer and forbidden lines, polarization, and optical veiling (see Hartmann, this volume, p. 95).

Disks in binary star systems may be divided into two main configurations 1) In the case of a *circumbinary disk* the complete binary system is surrounded by a disk which may have an inner clearing. 2) If the two disks surround each component of the binary individually, the situation is similar to that for single stars and one refers to this configuration as *circumstellar disks*

The *mm flux* from systems gives an indication for the mass of the disk. For binaries with larger separations ($d > 100$ AU) and for spectroscopic binaries ($d < 1$ AU) the mm flux is similar to the field value. This indicates that in the first case (large separations) the two stars are surrounded individually by circumstellar disks behaving just as disks around unperturbed single stars. In the second case (very close stars) the system is surrounded by one circumbinary disk which, when observed from far away, resembles a disk around a single star.

In binaries with intermediate orbital separations (d between 1 and 100 AU) the mm flux is mostly undetected, which may indicate that in this case the binary system has neither extended circumstellar disks, as a consequence of the gravitational disturbance of the companion, nor luminous circumbinary disks. However, there are classical T Tau signatures for disks with $d = 10$–100 AU.

Additionally, speckle and mid-IR observations give evidence for circumstellar disks around the stars HK Tau, HR 4796A, and GG Tau3.

The most prominent observational example of a circumbinary disk is presented by GG Tau, where the disk has been directly imaged in the CO 1.3 mm line (Guilloteau, Dutrey, & Simon 1999), in the J Band at 1.2 μm (Roddier et al. 1996), and in the optical. The central binary system consists of two stars with a total mass of 1.3 $M_\odot$ at a separation of $d = 45$ AU, which are surrounded by a disk with an inner radius of $r_{\rm disk} = 180$ AU. The inferred mass of the disk (ring) is about 0.12 $M_\odot$. Additional systems with imaged circumbinary disks are UY Aur and BD+31°643.

In several spectroscopic binaries (e.g., UZ Tau, DQ Tau, or GW Ori) mm observations have indicated disk masses of about 0.03 to 0.05 $M_\odot$ At the same time there is evidence for inner disk clearing (lack of near-IR emission), but there are signs for hot gas near the stars which may indicate the flow of material through the cleared inner disk (leaking gaps).

An interesting example of the simultaneous occurrence of circumstellar and circumbinary disks in a quadruple system is given by UZ Tau (Jensen, Koerner, & Mathieu 1996).

1.3. Planets in Binaries

Among the 32 known extrasolar planets around main-sequence stars three (16 Cyg B, 55 Cnc, and τ Boo) are definitely known to be in binary systems with relatively wide orbital separations (see Table 1). Additionally, in one system (Gl 86) there is some evidence for a stellar companion with a distance that may be as close as 10 AU. It has been argued that the high eccentricity of 16 Cyg B may be caused be the presence of a companion star. However, as the ratio of distance, d, of the secondary to the semi-major axis, a, of the planets falls in the range

Table 1. The Properties of known planets in binary stars

Parameter	16 Cyg B	55 Cnc	τ Boo	Gl 86
Binary d (AU)	700	1150	240	> 10
$M \sin i$ (M_J)	1.5	0.84	3.87	4
Planet a (AU)	1.72	0.11	0.046	0.11
Planet e	0.63	0.05	0.02	0.04

$$\frac{d}{a} = 400 - 10\,000,$$

the companion has basically no influence on the eccentricity of the planet.

2. Star Formation

The observational evidence strongly suggests that binary formation is closely related to the star formation process. Stars form in dense molecular cloud cores (Myers 1985, 1987) with scales of order 0.1 pc, masses in the range of a few solar masses, number densities of order 10^4–10^5 cm^{-3} and temperatures of 10 K. In general, pre-stellar cores appear to be elongated and irregular. They are centrally condensed with Gaussian or power-law density distributions (Ward-Thompson et al. 1994; Henriksen, André, & Bontemps 1997). Rotation has been detected in about 50% of all cases with values for the angular velocity ranging between $\Omega \approx 10^{-13}$ and 10^{-15} s^{-1} (Goodman et al. 1993).

The initial conditions for protostellar collapse could be strongly affected by magnetic fields, which can support the cores against collapse. In magnetically subcritical cores, ambipolar diffusion (Shu et al. 1993) leads to central contraction and the formation of an r^{-2}-density distribution at which point the core experiences an inside-out collapse. If the initial conditions are produced in this manner, numerical simulations have shown that the core would not fragment, leading to the formation of single stars (Myhill & Kaula 1992) Binary formation

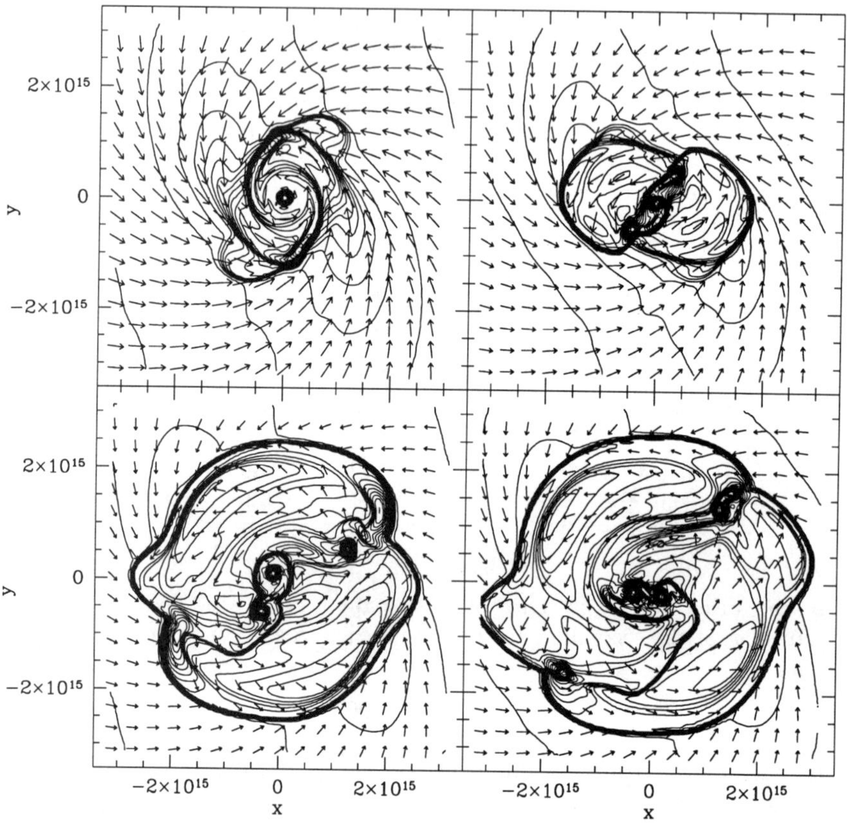

Figure 1. Evolution of a collapsing gas cloud with an initial power-law density distribution. Logarithmically spaced contours of equal density and velocity vectors in the equatorial plane are shown for the innermost region (outer cloud radius: 5 $\times 10^{16}$ cm). The maximum density is 10^{-10} g cm^{-3}, the maximum velocity is 3×10^5 cm s^{-1}.

results from the fragmentation of magnetically supercritical cores. A wide variety of (often idealized) initial conditions has been explored numerically, adopting three-dimensional hydrodynamical codes; for example, a sphere (constant density or centrally condensed) in uniform rotation with a density perturbation in the form of an $m = 2$ mode (Burkert & Bodenheimer 1993, 1996; Truelove et al.

1997), or elongated cylindrical, rotating clouds (Boss 1996; Bonnell et al. 1992). Most of these calculations have treated the isothermal collapse phase. Radiation has been included in some simulations adopting simplified approximations (Boss 1993; Myhill & Kaula 1992).

In general, the resulting fragmentation process occurs in various modes. The most simple cases are non-linear initial density perturbations which lead to fragmentation into wide binaries with separations of order 100–1000 AU during the early collapse phase (Bonnell et al. 1992; Monaghan 1994). In this case, the result depends strongly on the adopted initial conditions.

Figure 1 shows the collapse of a centrally condensed gas cloud with a linear $m = 2$ density perturbation (Burkert, Bate, & Bodenheimer 1997). In this case, the perturbation does not have enough time to grow and become non-linear during the early gravitational collapse phase. The gas instead settles into a rotationally supported, partly self-gravitating protostellar disk. The subsequent evolution of the disk depends critically on its temperature (Bate & Burkert 1997; Boss et al. 2000). If the transition from the optically thin to the optically thick regime occurs before disk formation, fragmentation will be inhibited. Otherwise strong spiral arms form around a central condensation which continually wrap up, interact and reform because of differential rotation. Ultimately, density maxima appear at the ends of the arms, leading to a binary around the central object. As higher-angular-momentum material falls in, spiral arms extend beyond the triple system which rapidly breaks up into a hierarchical triple. Further fragmentation occurs in the outer parts of the disk. The multiple system will ultimately break up into single stars and eccentric binaries with wide period distributions.

3. Influence of Companions on Disks

Let us first consider the case of a disk surrounding one star (circumprimary disk) which is orbited by another secondary star. The presence of a companion excites tidal waves in the disk which carry angular momentum and energy (see, for example, Lin & Papaloizou 1993; Terquem, this volume, p. 19). If the disk lies inside the orbit of the companion the wave will carry negative angular momentum because the companion has a smaller orbital velocity than the disk material. Dissipation of the wave, for example through shock waves, will slow down the matter in the disk and lead to an inward drift of the material. This truncates the outer edge of the disk.

In case of a binary star within a circumbinary disk, the torques created by the binary lead to angular-momentum transfer from the binary's orbit to the disk, and eventually to an inward truncation or clearing of the inner disk (gap formation).

The location of this (inner and outer) truncation radius is determined by the balance of viscous (gap-closing) and gravitational (gap-opening) torques. In linear theory the gravitational potential is expanded into a Fourier series, and for each component the effect on the disk is determined. For the general case of a companion on an eccentric orbit this analysis has been performed for circumprimary and circumbinary disks by Artymowicz & Lubow (1994).

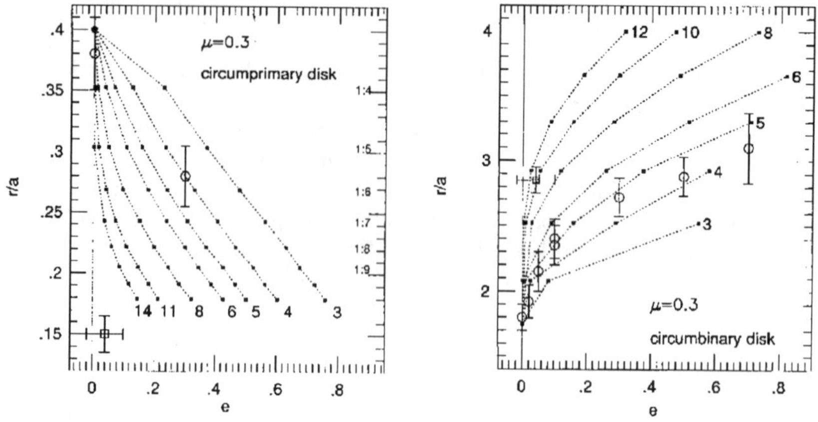

Figure 2. The truncation radius of the disk in units of the semi-major axis of the binary for a mass ratio $q = M_2/M_1$, and an eccentricity $e = 0.5$. Different values of the viscosity are labeled by the logarithm of the Reynolds number in the disk. Adapted from Artymowicz & Lubow (1994).

Their main results are displayed in Figure 2. The label $\mu = 0.3$ refers to the reduced mass $\mu = M_2/(M_1 + M_2)$. In case of the circumprimary disk a larger viscosity (lower Reynolds number) leads to a larger truncation radius r_t while for a circumbinary disk it reduces the truncation radius. For typical values of protostellar disks Re $\approx 10^5$, and for typical binary parameter (see above, $q = 0.5$, $e = 0.5$) we obtain for the circumprimary disk $r_t/a = 0.17$, and for the circumbinary disk $r_t/a = 3.0$.

4. On the Formation of Planets in Binaries

In this final section we consider the restrictions that the presence of a secondary companion has on the efficiency of planet formation. The secondary acts as a perturber for the disk and alters directly the environment in which planets form.

4.1. Planetesimal Disk

The influence a secondary has on the growth process from planetesimals to planetary embryos was studied by Heppenheimer (1978) and Whitmire et al. (1998). They considered a planetesimal disk at runaway phase where collisions and subsequent merging between different particles lead to a rapid growth. The collisions are non-destructive only if the relative velocities, $U_{\rm rel}$, are smaller than a critical value $U_{\rm crit} = 100$ m s^{-1}.

They integrated a four-body system consisting of two stars and two planetesimals using a symplectic mapping procedure and an integration time over a few 10^4 yr. For different physical parameter, varying the semi-major axis of the binary stars, their eccentricities, mass ratios, and the initial separation in semi-major axis, $\Delta a_0 = 0.001$–0.01 AU, of the planetesimals, they investigated

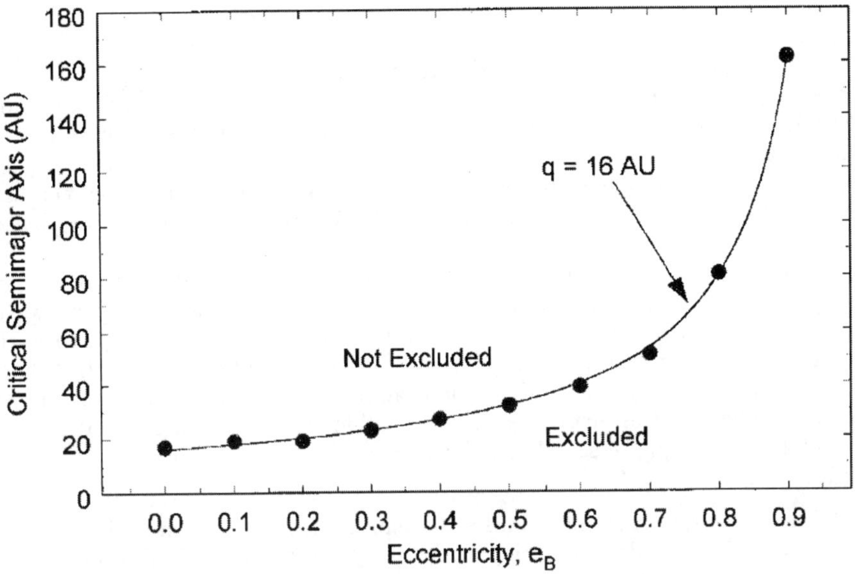

Figure 3. Critical semi-major axis for which collisions between planetesimals are too large to allow for particle growth. The mass of the central star was 1 $M_\odot$, the mass-ratio $q = 1.0$, and the initial semi-major axis of the planetesimals was 1 AU. Adapted from Whitmire et al. (1999).

whether $U_{\rm rel}$ was exceeding $U_{\rm crit}$ and deduced a critical semi-major axis below which the disturbing companion does not allow for sticking (growing) collisions. Their main results are summarized in Figure 3, where the critical semi-major axis is plotted for an equal-mass binary versus eccentricity. If the separation of the binary falls below the curve, the induced relative velocities are too large and planetesimals cannot grow. The results are normalized to the initial mean separation, $\bar{a} = 1$ AU, of the planetesimals. The results of the numerical computations are given by the black dots, and the solid curve gives the location where, for the given eccentricity, the minimum separation (periastron) of the stars is exactly 16 AU.

The result implies that for a one-solar-mass star the minimum distance to an orbiting companion, such that planetesimals at 1 AU can grow to larger bodies, must always be larger than 16 AU. The critical semi-major axis scales only weakly, $a_{\rm crit} \propto (\bar{a}/{\rm AU})^{0.8}$, with distance from the star.

4.2. The Case of L 1551

Another path of studying the formation of planets in binary stars was taken by Nelson (2000), who studied the interaction of two binary stars, each of which is surrounded by its own circumstellar disk. The physical motivation is based on the radio (VLA) observations of the system L 1551 (Rodriguez et al. 1998), which consists of a binary system of two half-solar-mass stars separated by 50 AU.

This system is modeled numerically by solving the full hydrodynamic equations for the two disks using the method of smoothed particle hydrodynamics (SPH), where the continuum equations are modeled by an ensemble of interaction particles.

Each disk is modeled by 60 000 particles. The model includes the self-gravity of the disks, an ideal equation of state, dissipative heating, and radiative cooling.

During the evolution, the initially axisymmetric disks (with respect to their central stars) are strongly perturbed at the time of periapsis. Spiral waves are induced in them and they are tidally truncated. At times of apoapsis the spiral features tend to disappear and the disks become more axisymmetric again. These periodic changes of the distortions alter the internal structure of the disk as well. The question arises as to what influence these changes have on the planet formation process. Planet formation is believed to proceed along two different lines:

a) Through gravitational collapse of a spiral structure. Instability is given when the so called Toomre parameter, Q, which measures the importance of pressure versus gravity, becomes (locally) smaller than unity. Then pressure forces are not sufficient to prevent gravitational collapse.

b) Through coagulation of solid material and subsequent gas accretion.

As can be inferred from Figure 4, the Q parameter is always much greater than 1, indicating a gravitationally stable disk. On the other hand, the temperatures in the disk during periapsis are so high that even the most abundant species (water ice) cannot condense to form the seeds for further growth of the planetesimals.

Hence, from the computations one may conclude that, at least for these chosen parameters of the disks, planet formation is inhibited in binary stars. One has to keep in mind, however, that the heating and cooling of the disk is still treated approximately and the calculations are only two-dimensional. For larger orbital separations, of a few hundred AU, say, the effects of the secondary become smaller and planet formation will not be affected.

4.3. Stability of Orbits

An important prerequisite for the formation of planetesimals is the long-term stability of their orbits. Dvorak (1986) and Holman & Wiegert (1999) have studied the restricted three-body problem, in which the planet is treated as a test particle in the potential of two orbiting stars. As for this problem no analytical description of the orbital evolution of the planet exists, they perform numerical integrations using a symplectic mapping or Bulirsch–Stoer integrator. The total integration time covers more than 10^4 orbits of the planet. The mass ratio of the stars, their eccentricity, and semi-major axis are varied. The critical distance, $a_{\rm crit}$, a planet must have to be on a stable orbit is determined as a function of these orbital parameters of the binary star.

Two different configurations of orbits have been studied. Following the designation of Dvorak (1986), the first are planetary or P-type orbits which are

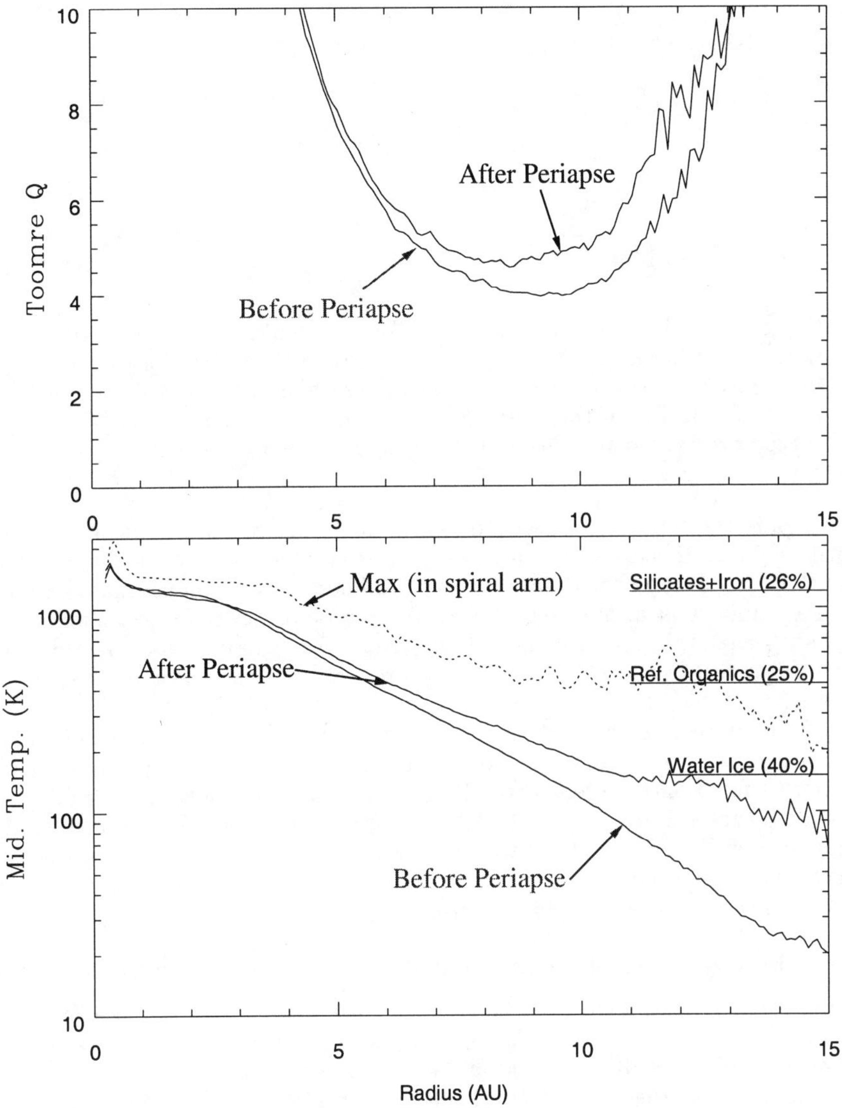

Figure 4. Azimuthally averaged Toomre parameter, Q (*top*), and the midplane temperature in the disk versus radius. Indicated are the vaporization temperatures of various species.

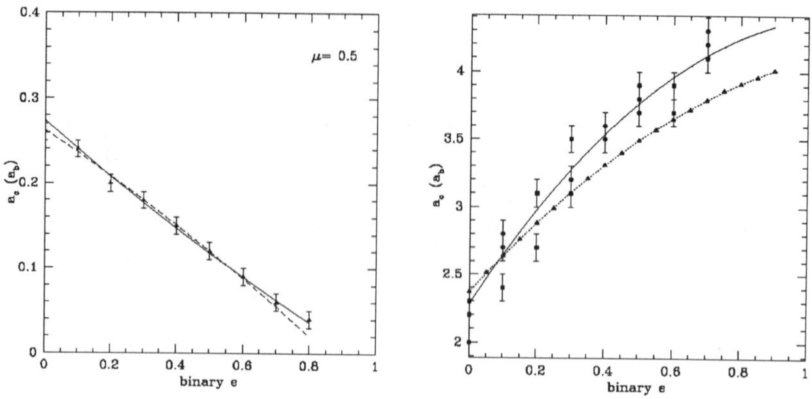

Figure 5. Critical semi-major axis for an equal-mass binary. After Holman & Wiegert (1999). *Left:* For S-type orbits. Separations smaller than the *solid curve* (which is a fit the numerical data points) are stable. *Right:* Evolution of the mass (*bottom*) and semi-major axis (*top*) for P-type orbits. Separations above the *solid curve* are stable.

well outside the binary. The second type studied are S-type orbits, where the planet orbits near one of the stars, with the second star to be considered as a perturber. For P-type orbits there exist a minimum, $a_{\rm crit}$, a planet must have to be on a stable orbit around the binary. For S-type orbits there is a maximum, $a_{\rm crit}$, such that the perturbations of the perturber remain finite. The S-type orbits are certainly more relevant to the properties of the observed extrasolar planets (see Table 1).

The main results are displayed in Figure 5, where, for an equal-mass binary, the critical distance (in units of the semi-major axis of the binary-star system) is given as function of eccentricity. For an average eccentricity of $e = 0.5$ one finds for S-type orbits $a_{\rm crit} \approx 0.14$. All observed extrasolar planets have semi-major axes well below this limit. For P-type orbits the minimum distance a planet must have from the binary is $a_{\rm crit} \approx 3.6$. This value lies beyond the inner truncation radius of a circumbinary disk ($r_{\rm gap} = 3.0$).

4.4. The Evolution of an Embedded Planet in a Binary Star

Another approach to analyzing the influence of a companion on the formation of a massive planet has been taken by Kley (2000), who studied the evolution of a massive planet still embedded in a protoplanetary disk.

A planet with the mass 1 $M_{\rm Jup}$ is placed initially on a circular orbit around a 1 $M_\odot$ star at a distance of $a_{\rm Jup} = 5.2$ AU. The surrounding protostellar disk has a mass of $M_{\rm d} = 0.01\ M_\odot$ within 1–20 AU, a surface density profile $\Sigma(r) \propto r^{-1/2}$, and a Reynolds number of 10^5. The *secondary* has a given fixed mass of $M_2 = 0.5\ M_\odot$ and an eccentricity of $e_2 = 0.5$. The semi-major axis, a_2, of the secondary is varied from about 50 to 100 AU for different models. To minimize initial disturbances, the secondary is placed at apastron in the beginning.

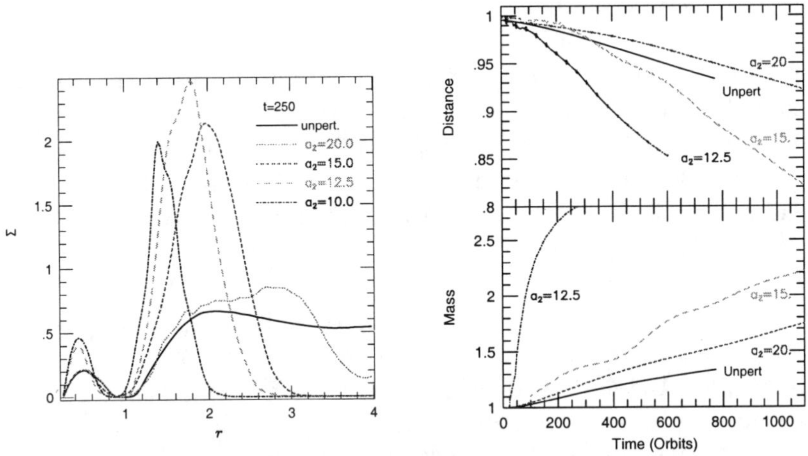

Figure 6. *Left:* Azimuthally averaged surface density for models with varying distance (a_2) of the secondary, given in units of the planetary distance (5.2 AU), after 250 orbits of the planet. The *solid black line* indicates the initial surface density profile. *Right:* Evolution of the mass (*bottom*) and semi-major axis (*top*) of the planet. The evolution of an unperturbed planet (i.e., no secondary star), which is described in detail in Nelson et al. (2000) is given by the *solid lines* (denoted "Unpert").

During the evolution, the planet may accrete material from the disk and increase its mass. At each time-step, the gas density in the inner half of the Roche lobe is reduced by a given fraction. Initially, the disk is axisymmetric with respect to the primary star and has a gap imposed to speed up the computations (Kley 1999; Lubow, Seibert, & Artymowicz 1999).

In the presence of the secondary, the disk becomes truncated at the outer radius. At the same time the planet will truncate the disk from the inner side such that the disk material will be confined to a narrow ring (see Figure 6). The values obtained for the outer truncation radius agree favorably with the value $r_r/a = 0.17$ as obtained above. The very steep density and pressure gradient near the planet lead to an increased mass accretion rate onto the planet, and to a faster inward migration, cf. Figure 6.

The reduced migration time may then also reduce the growth of a protoplanet and planet formation will again be inhibited by the presence of a companion.

References

Artymowicz, P., & Lubow S. H. 1994, ApJ, 421, 651

Bate, M. R., & Burkert, A. 1997, MNRAS, 288, 1060

Bonnell, I., Arcoragi, J.-P., Martel, H., & Bastien, P. 1992, ApJ, 400, 579

Boss, A. P. 1993, ApJ, 410, 157
Boss, A. P. 1996, ApJ, 468, 231
Boss, A. P., Fisher, R. T., Klein, R. I., & McKee, C. F. 2000, ApJ, 528, 325
Burkert, A., & Bodenheimer, P. 1993, MNRAS, 264, 798
Burkert, A., & Bodenheimer, P. 1996, MNRAS, 280, 1190
Burkert, A., Bate, M. R., & Bodenheimer, P. 1997, MNRAS, 289, 497
Duquennoy, A., & Mayor, M. 1991, A&A, 248, 485
Dvorak, R. 1986, A&A, 167, 379
Ghez, A. M., Neugebauer, G., & Matthews, K. 1993, AJ, 106, 2005
Goodman, A., Benson, P., Fuller, G., & Myers, P. 1993, ApJ, 406, 528
Guilloteau, S., Dutrey, A., & Simon, M. 1999, A&A, 348, 570
Henriksen, R., André, P., & Bontemps, S. 1997, A&A, 323, 549
Heppenheimer, T. A. 1978, A&A, 65, 421
Holman, M. J., & Wiegert, P. A. 1999, AJ, 117, 621
Jensen, E. L. N., Koerner, D. W., & Mathieu, R. D. 1996, AJ, 111, 243
Kley, W. 1999, MNRAS, 303, 696
Kley, W. 2000, in preparation
Leinert, Ch., Zinnecker, H., Weitzel, N., Christou, J., Ridgway, S. T., Jameson, R., Haas, M., & Lenzen, R. 1993, A&A, 278, 129
Lin, D. N. C., & Papaloizou, J. C. B. 1993, in Protostars and Planets III, ed. E. H. Levy & J. I. Lunine (Tucson: Univ. Arizona Press), p. 749
Lubow, S. H., Seibert, M., & Artymowicz, P. 1999, ApJ, 526, 1001
Mathieu, R. D., Ghez, A. M., Jensen, E. L. N., & Simon, M. 2000, in Protostars and Planets IV, ed. V. Mannings, A. P. Boss & S. S. Russell (Tucson: University of Arizona Press), p. 703
Monaghan, J. J. 1994, ApJ, 420, 692
Myers, P. C. 1985, in Protostars and Planets II, ed. D. C. Black & M. S. Matthews (Tucson: Univ. Arizona Press), p. 81
Myers, P. C. 1987, in Star Forming Regions, ed. M. Peimbert & J. Jugaku (Dordrecht: Reidel), p. 33
Myhill, E., & Kaula, W. M. 1992, ApJ, 386, 578
Nelson, A. 2000, ApJ, submitted
Nelson, R. P., Papaloizou, J. C. B., Masset, F., & Kley, W. 2000, MNRAS, in press (astro-ph/9909486)
Roddier, C., Roddier, F., Northcott, M. J., Graves, J. E., & Jim, K. 1996, ApJ, 463, 326
Rodriguez, L. F., D'Alessio, P., Wilner, D. J., Ho, P. T. P., Torrelles, J. M., Curiel, S., Gomez, Y., Lizano, S., Pedlar, A., Canto, J., & Raga, A. C. 1998, Nat, 395, 355
Shu, F. H., Najita, J., Galli, D., Ostriker, E., & Lizano, S. 1993, in Protostars and Planets III, ed. E. H. Levy & J. I. Lunine (Tucson: Univ. Arizona Press), p. 3

Truelove, J. K., Klein, R. I., McKee, C. F., Holliman, J. H., Howell, L. H., & Greenough, J. A. 1997, ApJ, 489, L179

Ward-Thompson, D., Scott, P. F., Hills, R. E., & André, P. 1994, MNRAS, 268, 276

Whitmire, D. P., Matese, J. J., Criswell, L., & Mikkola, S. 1998, Icarus, 132, 196

Discussion

Hans Deeg: The work by Artymowicz you presented showed that the inner boundary of a circumbinary disk is about 2.7 times the binary separation, with a concentration of material just beyond the inner boundary. Could one therefore deduce that planets might preferably form near the inner boundary, say at three times the separation?

Willy Kley: Mass is probably accumulating near the inner edge of the disk. However, the presence of tidal effects of the binary or the dynamics of planetesimals and protoplanets may hinder the formation process. Thus, I would expect planets to form at larger radii, which may have moved to the edge by migration.

William Borucki: the paper by Duquennoy and Mayor showed 60% of F- and G-type stars were binary systems. Fisher and Marcy found that K and M stars had a lower frequency. Does your model indicate the reason for the lower frequency?

Willy Kley: To my knowledge the two results are identical.

Alan Boss: A comment on the semantics of the "minimum-mass solar nebula". A 0.01 $M_\odot$ disk is often called the "minimum-mass solar nebula", but it is actually the *minimum* estimate of the "minimum-mass solar nebula". Weidenschilling (1977) estimated the "minimum-mass solar nebula" to lie in the range of 0.01 $M_\odot$ to 0.1 $M_\odot$. An 0.1 $M_\odot$ disk is then the maximum estimate of the "minimum-mass solar nebula".

Willy Kley: In the future I will refer to the mass just in absolute units instead of using the phrase "minimum-mass solar nebula".

Lee Hartmann: You mentioned that 0.1 of the core mass was in the fragments at the end of the calculations by Bodenheimer et al., but collapse has not finished in these simulations, and the fragments should grow further.

Willy Kley: That's true, at this stage only very little of the mass has accumulated into cores, and the further evolution depends also on the rotation of the cloud, as has been seen in the results of M. bates. However, the mass ratio (q) may give and indication of the expected final value of q.

Disks, Planetesimals, and Planets
ASP Conference Series, Vol. 219, 2000
F. Garzón, C. Eiroa, D. de Winter, and T. J. Mahoney, eds.

Infall, Accretion, and the Spectroscopic Evidence for Planetesimals

C. A. Grady

NOAO/STIS and Goddard Space Flight Center, Code 681, NASA's GSFC, Greenbelt, MD 20771, USA

Alcione Mora

Departamento de Física Teórica, C-XI, Universidad Autónoma de Madrid, Cantoblanco, 28049 Madrid, Spain

Dolf de Winter

TNO/TPD-Space Instrumentation, Stieltjesweg 1, PO Box 155, 2600 AD Delft, The Netherlands

EXPORT

Abstract. The presence of planetesimals in protoplanetary disks can be spectroscopically detected via the presence of infalling, circumstellar gas that is episodically present, departures from the composition of the star in the infalling material, and via the presence of small, warm grains which are below the radiation-pressure blow-out radius, and hence which require rapid replenishment either via transport from the outer portions of the disk, or by the disintegration of larger bodies. The evidence for such activity is reviewed.

1. Introduction

The last few years have seen rapid advances of our knowledge of exoplanetary systems, including the steadily expanding inventory of planets around nearby, solar-type stars (Mayor & Queloz 1995; Marcy & Butler 2000; Mayor & Udry, this volume, p. 441). Imaging studies of protoplanetary disks are now available in the millimeter (Beckwith & Sargent 1996; Mannings & Sargent 1997, 2000), mid-IR through optical (McCaughrean, Stapelfeldt, & Close 1999; Jayawardhana et al. 1998; Koerner et al. 1998; Schneider et al. 1999; Silverstone et al. 1998; Grady et al. 1999, 2000a; Weinberger et al. 1999, 2000). These studies have demonstrated that after ~ 4 Myr single or widely separated Herbig Ae/Be (HAeBe) disks have an architecture similar to that of the solar system, with largely cleared inner zones, and dust rings with sharp inner edges. Data for the younger star AB Aur ($t \sim$ 2–4 Myr, Grady et al. 1999), show striations in the circumstellar dust, but no indication of central clearing or cleared annuli. The systems older than ~ 4 Myr, therefore, are making the transition to debris disks similar to β Pic (Heap et al. 2000). Crisp inner edges to the disks indicate that systems like HR 4796A harbor shepherding bodies which may be either massive planets or brown dwarfs (Weinberger et al. 1999; Kenyon et al. 1999; Goldreich & Tremaine 1980).

Theories of planet formation predict that a necessary step in planet formation is the assembly of large numbers of 1–10 km-sized bodies (Lissauer 1997; Pollack et al. 1996). Such bodies are expected to reflect the composition of the refractory portion of the protoplanetary disk in which they formed. By virtue of their surface area, such bodies are expected to be exceedingly difficult to detect compared to the small dust grain population in the disk, unless the bodies are perturbed into star-grazing orbits, producing cometary activity. Contemporary Sun-grazing comets in our Solar System are known to be preferentially detected prior to perihelion and to be detectable in scattered light in optical coronagraphic images of the Sun (Biesecker et al. 1999), via their H I emission (Raymond et al. 1998), and emission from refractory species such as Na I (Evans & McKim Malville 1967) either close to the star, or when impacting onto another large body (e.g., S-L/9, Costa et al. 1997).

2. Observational Motivation

The first indication that systematically redshifted, circumstellar absorption features could be detected toward stars known to have circumstellar disks was provided by β Pic (Kondo & Bruhweiler 1985; Hobbs et al. 1985). The infalling gas in this system reaches velocities of up to 400 km s^{-1} (Lagrange-Henri et al. 1987, 1989), a substantial fraction of the free-fall velocity onto the star, is detectable in a wide range of the higher oscillator strength transitions of the more abundant refractory elements, C I (Roberge et al. 2000; Vidal-Madjar et al. 1994), but not in the Balmer series of hydrogen (Ferlet & Vidal-Madjar 1994). Infalling gas is known to be episodically present in the spectrum of β Pic (Petterson & Tobin 1999), with low velocity infall features (20–40 km s^{-1} relative to the star) persisting in the line of sight for 3–4 days, significantly longer than the stellar rotation period. The infall features in the β Pic system have been interpreted as the gaseous comae of star-grazing planetesimals (Beust & Morbidelli 2000; Beust et al. 1996).

Redshifted absorption features are not limited to β Pic, but are also seen in the spectra of a number of HAeBe stars (Graham et al. 1992; Welty et al. 1992; Grady et al. 1996, 2000a). UV surveys of the HAeBe stars demonstrated that infall velocities up to 400 km s^{-1} are routinely seen. Infalling material is detectable in transitions of a wide range of refractory, semi-refractory, and nominally volatile (C, N, O) species (Grady et al. 1996). Many of the UV line profiles exhibited flat-bottomed absorption well above the instrumental scattered-light level, compatible with saturated absorption which does not fully obscure the stellar disk. The bulk of the UV data were obtained sporadically and thus cannot be used to assess the frequency of infall "events" in HAeBe star spectra, except to note that the ease of detection in temporally sparse datasets implies a high frequency for the activity. Demonstrably episodic infall is seen in stars such as 51 Oph, which, while young, has an IR excess more typical of a debris disk (Grady & Silvis 1993), and in other A-shell stars with at most modest IR excesses (Mora et al., this volume, p. 411). Among the HAeBe stars with UV data, clearly episodic infall has been detected only toward the near-ZAMS star, HD 100546 (Grady et al. 1997).

3. Accretion vs. Star-Grazers

The IR spectral energy distributions of many pre-main-sequence stars (Hillenbrand et al. 1992; Malfait et al. 1998) are characterized by prominent excesses which can be best fitted by active accretion disk models (Hartmann, this volume, p. 95). Accretion manifests in the optical and UV spectra of T Tauri stars in the form of continuum veiling of the stellar photospheric spectrum (Folha & Emerson 1999; Chelli 1999; Batalha & Basri 1993; Hartigan et al. 1989; Bertout et al. 1988, 1989), the presence of UV excesses (Gomez de Castro & Fernandez 1996; Herbig & Goodrich 1986), and redshifted absorption features (Edwards et al. 1994). The detection of UV excesses associated with HAeBe stars has been more problematic since the higher stellar $T_{\rm eff}$ require observations at $\lambda \leq 1300$ Å to firmly detect excess emission. Optical continuum veiling and the presence of forbidden emission features are easier to assess. Several studies have shown that signatures of active accretion for both HAeBe and T Tauri stars are correlated with signatures of mass loss, both via strong stellar winds and the presence of forbidden emission features compatible with the presence of more collimated outflows (Mundt & Eislöffel 1998; Corcoran & Ray 1997, 1998).

The observational challenge, then, is to identify under what conditions infalling gas can be associated with accretion of primordial disk material, and under what conditions does the infalling gas contain a significant fraction of material which has been processed through planetesimal-sized bodies.

4. Criteria for Distinguishing between Star-Grazers and Accretion

If the available data on young T Tauri stars and β Pic are any indication it should be feasible to distinguish between accretion and star-grazing events on the basis of:

- Composition of the infalling material;
- Correlation between the infall activity and other signatures of high-rate mass accretion;
- Frequency of infall events and persistence of material with a stable profile in the line of sight.

4.1. Composition of the Infalling Material

The simplest criterion for distinguishing between star-grazing events and accretion of primordial disk material is the composition of the infalling gas. The infall profiles studied by Edwards et al. (1994) demonstrated that high-velocity infall in classical T Tauri spectra was equally conspicuous in the higher Balmer series lines and in species such as Na I, after normalization of the profiles by photospheric spectra derived from observations of diskless (e.g., weak T Tauri) stars. Sorrelli, Natta, & Grinin (1996) and Natta et al. (2000) have carried out simulations of infalling material under conditions appropriate to an early-type Herbig Ae star. They find that material of solar composition will exhibit infall profiles in the higher Balmer lines (indicating densities $n \sim 10^{8-12}$ cm^{-3}) and unsaturated, redshifted absorption in transitions of refractory elements such as Ca II and Na I. In contrast, infall of cometary-composition material (metallicity $m \sim 500$) does not result in simultaneous detection of strong H I absorption and *unsaturated* absorption of refractory species such as Na I. This suggests that high-resolution optical spectra spanning the higher Balmer lines, Ca II, and Na I will be critical in distinguishing systems with accretion from those where infall is dominated by planetesimal infall.

4.2. Correlation with Accretion/Outflow Signatures

The presence of forbidden emission lines in the spectrum of an HAeBe star, together with veiling of the photospheric spectrum, suggests that the star is continuing to actively accrete material from its circumstellar disk and may be driving a circumstellar outflow. Forbidden optical emission lines are known to be present in the spectra of some HAeBe stars (Böhm & Catala 1995; Corcoran & Ray 1997), and should be more conspicuous when the star is heavily embedded or occulted. Such conditions are achieved at minimum light for the UXORs (Grinin, this volume, p. 216).

Signatures of outflows are not limited to the optical spectrum. Several UXORs, including the class prototype, UX Ori (Grady et al. 1995) have low-resolution UV spectra obtained at both minimum and maximum optical light. While the maximum-light spectra show the heavy line blanketing characteristic of circumstellar Fe II absorption in the near UV, at minimum light the absorption spectrum is replaced by a featureless, blue continuum with prominent Mg II and Fe II emission. The minimum light spectra are effectively coronagraphic and are dominated by circumstellar material in any outflow. Similarly, optical coronagraphic spectra and/or imagery, covering the wavelengths of Hα, [O I], and/or [S II] can reveal the presence of accretion-driven outflows in stars showing no forbidden emission features in their optical spectra (Grady et al. 2000b).

4.3. Short-Term Variability: Features Persisting in the Line of Sight

For T Tauri stars, high-velocity infall features are believed to originate at the point that the stellar magnetic field disrupts the disk, and material from the disk free-falls onto the stellar photosphere (Hartmann, this volume, p. 95). Such disruption of the disk is believed to occur at the corotation point, and, if the accretion footprint is localized on the star, should result in rotational modulation of the infall signatures. Such features have been observed in association with some T Tauri stars (Gomez de Castro & Fernandez 1996). The periodic presence of infall features with variability on a timescale of days is likely to represent the detection of accretion rather than star-grazer activity. If the accretion footprint is not localized, no clear periodicity in the infall frequency should be observed. The signature here would be variable rate, but ubiquitous infall signatures. On the other hand, the persistence of infall features at low velocity (a few tens of km s^{-1}) in the line of sight for more than one stellar rotation period suggests the presence of material partially occulting the stellar disk and lying beyond the corotation radius. With rotation periods of 1–2 days, establishing the presence of such features in the spectra of HAeBe stars will require synoptic studies lasting longer than a week.

4.4. Long-Term Variability: Episodic Infall

A distinctive characteristic of the infall activity seen toward β Pic is the episodic nature of the infall, with long-term fluctuations in the level of infall activity on timescales of years and potentially consistent with the orbital period of a giant planet at 4 AU (Beust & Morbidelli 2000). The ease of detection of infall activity in the sparse UV and optical datasets considered by Grady et al. (1996) suggests that infall activity is rather routinely present in the spectra of HAeBe stars that are inferred to have inclinations close to equator-on to the line of sight, based on anticorrelated polarimetric and photometric variability (Grinin, this volume, p. 216). The lone example showing clearly episodic infall is HD 100546, which also has one of the older evolutionary ages (van den Ancker et al. 1997, 1998).

5. Comparison with the LPIT'98 Sample

Identification of infall activity that can be linked to active accretion from the disk, or to star-grazing solid bodies should ideally rest on several criteria. The UXORs with polarimetric data indicating that we view the stars close to equator-on are ideal for simultaneous assessment of infall activity and searches for outflow signatures. A suite of data including optical high-resolution spectra supported by lower-resolution data sampling Hα and the stronger forbidden lines associated with outflows, and with sufficient photometric and polarimetric data to identify the degree of obscuration of the star, such as is provided by the LPIT'98 dataset (Eiroa et al., this volume, p. 3; Mora et al., this volume, p. 411) is ideal for assessing the nature of the infall activity, and vastly superior to previous optical studies limited to Na I and Hα, or to the UV.

5.1. Global Sample Characteristics

The LPIT'98 dataset for the UXORs contains no spectra obtained at or close to optical minimum light, with all of the observations within 1 magnitude of maximum light. The LPIT'98 dataset, therefore, forms a homogeneous dataset for searches for forbidden emission features. Infall signatures are seen in all of the UXORs in the sample with more than one spectrum. A number of the LPIT'98 program stars have been included in previous optical surveys for forbidden emission-line features, and cover the observed range in [O I] equivalent width, and by inference mass-loss rate seen in the Corcoran & Ray (1997) sample.

5.2. Stars with Prominent Accretion Signatures

The two HAeBe stars in the LPIT'98 sample with the strongest forbidden emission features in the Corcoran & Ray (1997) sample are KK Oph and VV Ser (Figure 1). INT spectra of VV Ser (de Winter et al., this volume, p. 356) show [O I]6300, 6346 Åemission, with the λ 6300 emission reaching 1.2 times the local continuum with $W_{[\mathrm{O\,I}]\lambda}6300 = 0.45$ Å. The INT data provide no hint of Hα absorption: instead, the line is entirely filled in with emission in a type III P Cygni

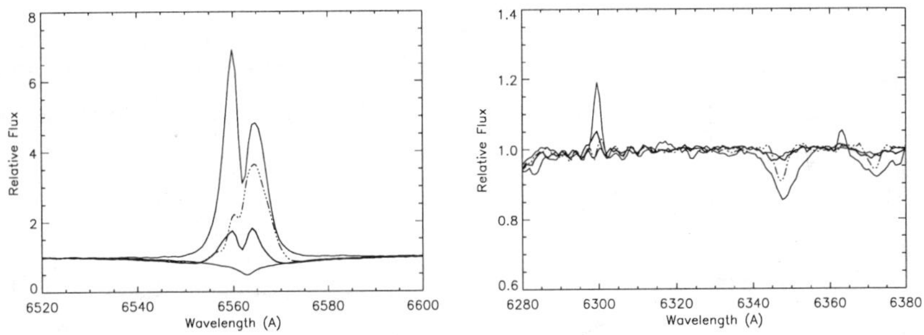

Figure 1. Portions of intermediate-resolution spectra obtained with the INT for VV Ser (*solid line*), SV Cep, (*solid bold line*), HD 163296 (*long-dashed line*), and the A-shell star HR 10 (*short-dashed line*). *Left:* Hα. *Right:* [O I] 6300, 6347 Å.

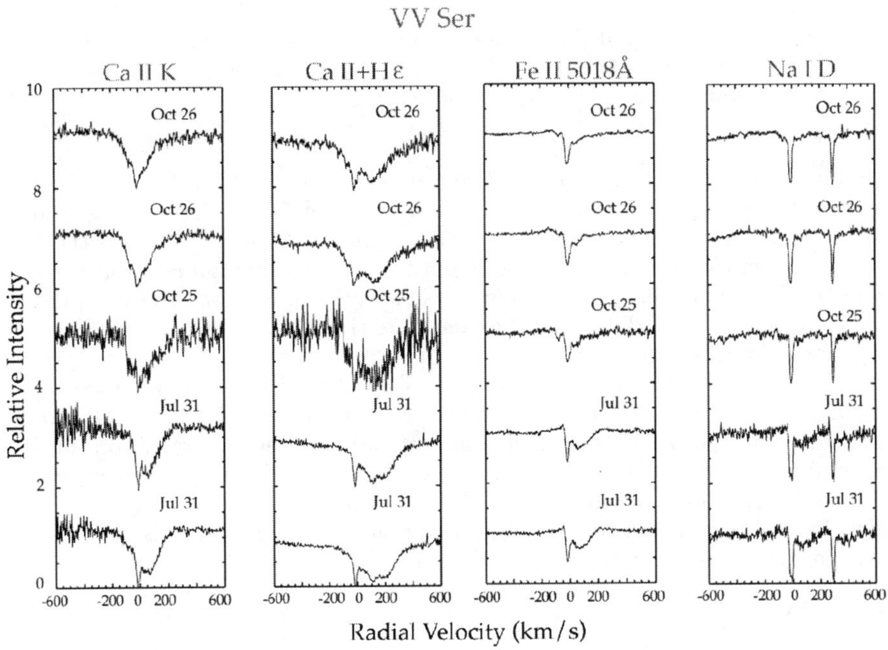

Figure 2. Selected line profiles for VV Ser from the WHT observations. Infalling material is visible in Ca II, H I, Fe II, and Na I.

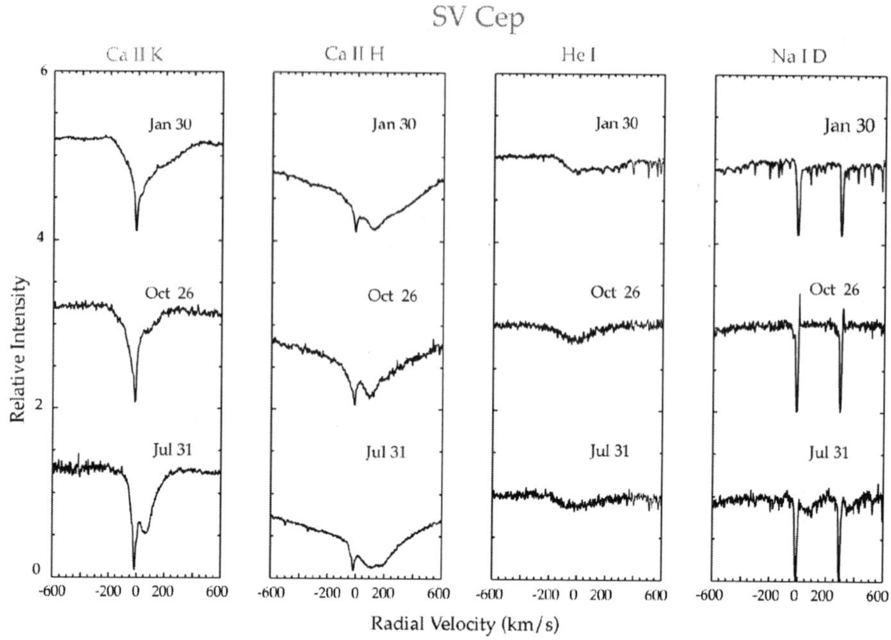

Figure 3. Selected line profiles from WHT UES spectra of SV Cep obtained as part of the LPIT'98 campaign. Infalling gas profiles are conspicuous in Ca II, H I, He I, and Na I.

profile reaching 6–7 times the local continuum level. He I is represented by an inverse P Cygni profile with the emission reaching 6–7% of the local continuum level. The Na I D lines are clearly inverse type I P Cygni profiles in the INT data, with emission reaching up to 5% of the local continuum level on the blue wing of the circumstellar absorption. Infall is visible in the Na I lines and in Si II $\lambda\lambda$ 6347, 6371. WHT data from the same date (Figure 2) show Na I profiles with saturated high-velocity absorption reaching 200 km s^{-1} and a covering factor of ≈ 0.2. The higher Balmer series lines show high-velocity infall traceable to +400 km s^{-1} through H6 and peak covering factors of $\sim 70\%$. The Ca II profiles are similar to the H I features. Infall features are also seen in transitions of Fe II including λ 5018 (Figure 1). The presence of veiling, the large covering factor, high-velocity H I absorption, and the [O I] emission all suggest that, in this star, we view solar-composition, infalling material. The implication is that we are viewing accretion of unprocessed disk material.

5.3. SV Cep

INT spectra for SV Cep (Figure 1) show weaker inverse P Cygni features in He I λ 5876 and no net emission in Na I and Si II. [O I] 6300 Å emission is present, with an equivalent width of 0.1 Å. [O I] λ 6363 is not detected. Hα emission while present, does not fill in the photospheric absorption and reaches only 1.8 times the local continuum with a P Cygni type III profile. WHT spectra for

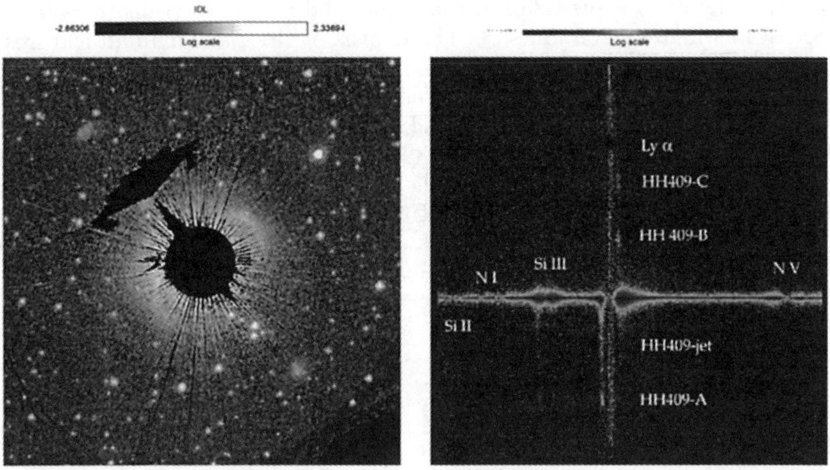

Figure 4. left) 20″ × 20″ detail of the STIS coronagraphic image of HD 163296 (Grady et al. 2000). FUV long-slit spectrum of HD 163296 showing the HH knots and the jet in Lyα and Si III.

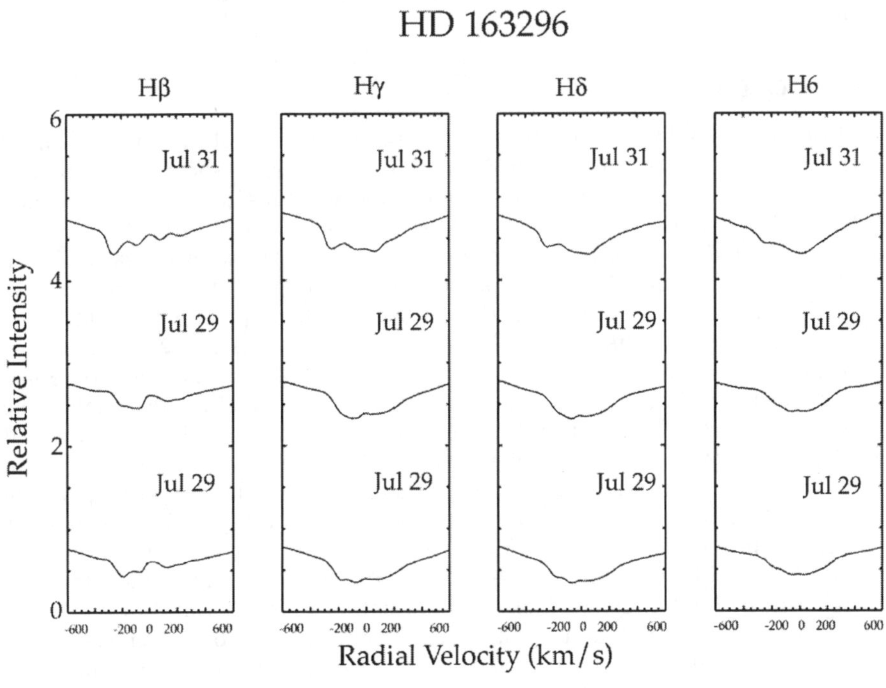

Figure 5. Selected Balmer line profiles for HD 163296.

SV Cep (Figure 3) show infall in the higher Balmer series transitions reaching velocities as high as 200 km s^{-1}. The depths of the Balmer infall features are similar for Hβ through Hϵ suggesting optically thick and saturated absorption covering at most 60% of the stellar disk. Ca II is similar to the H I profiles. He I λ 5876 absorption can be traced to +400 km s^{-1}. Na I infall features can be traced to at most 200 km s^{-1}. The silicate emission profile seen in *ISO* SWS data (Palacios et al., this volume, p. 255) closely resembles GEMS and the interstellar silicate feature, suggesting minimal grain growth and processing of the warm dust in this disk (Waters et al. 2000). The combination of infall profiles dominated by H I and forbidden emission features in tandem with the silicate data suggests that we view accretion of unprocessed disk material for this star also.

5.4. HD 163296

HD 163296 is an $\approx$ 4 Myr-old Herbig Ae star which is viewed, unlike the UXORs, at $i \sim 60°$. In contrast to the UXORs, which all show Na I in absorption, Na I and He I are both in emission. No net [O I] emission is seen in the INT data (Figure 1), however, coronagraphic observations made with *HST* (Grady et al. 2000) reveal a bipolar outflow with HH objects radiating in [O I], [S II] and Hα (Figure 4). High-resolution echelle data obtained during the LPIT'98 campaign show infall in the higher Balmer lines (Figure 5). Together with the *HST* data, these observations indicate that this star is still actively accreting material from its disk.

5.5. CQ Tau

CQ Tau is a nearby, A8–F2 Herbig Ae star with an evolutionary age of 10 Myr (van den Ancker et al. 1998). Unlike VV Ser, it shows no forbidden emission features at maximum light and only modest Hα emission, reaching only $\approx$ 1.5 times the local continuum level (Kozlova et al. 2000). The LPIT'98 data are limited to three INT spectra from 1999 January with similar emission levels in Hα, and no discernible photospheric veiling. Na I routinely shows redshifted absorption features (Kozlova et al. 2000), frequently with saturated absorption in the infalling material and absorption depths of 0.2–0.6 times the local continuum. Infall to +300 km s^{-1} is typical. The WHT spectrum, obtained on 1999 October 25 (Figure 6), shows no indication of infalling material in the Balmer lines, a result confirmed by comparable-resolution echelle data obtained with the Hamilton Echelle Spectrograph on the Shane 3 m at Lick Observatory in 1999 November (Quirrenbach, private communication). The lack of infall in the higher Balmer series is compatible with cometary gas composition, in contrast to UXORs like VV Ser. CQ Tau does have some features in common with the other UXORs: infall is routinely observed, and there are striking night-to-night changes in the infall profiles. This suggests that accretion is still going on in the CQ Tau system, although at a sufficiently low rate that obvious signatures of a collimated outflow are not seen in *HST* STIS optical coronagraphic imaging. The data for this star imply that there is significant chemical evolution in the composition of the protoplanetary disk by the time the star approaches the ZAMS.

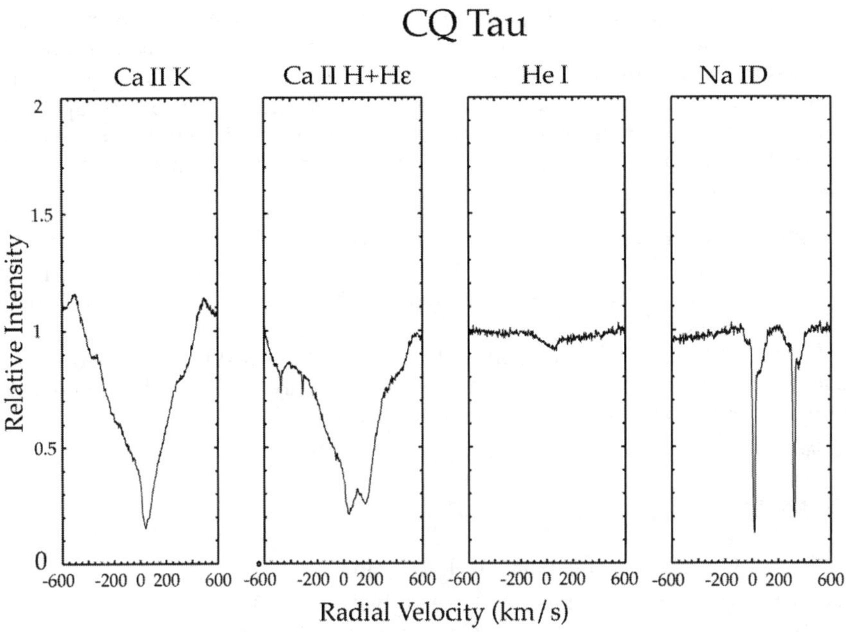

Figure 6. WHT data for CQ Tau from 1998 October 25, showing infall in Na I and not in H I.

6. Discussion

The LPIT'98 dataset has revealed that infall activity seen toward most HAeBe UXORS differs from that seen toward β Pic and A-shell stars, e.g., HR 10.

- Like the T Tauri stars (Edwards et al. 1994), infall activity in most of the LPIT'98 UXORs is seen in the higher Balmer series lines. As noted by Natta et al. (2000), the presence of infall in these transitions indicates infall of stellar rather than cometary-composition gas.
- These stars typically show other signatures of active accretion in the form of veiling of the photospheric spectrum at Hα, forbidden emission lines such as [O I] 6300, 6347, and inverse P Cygni profiles in He I in at least some of the LPIT'98 observations.
- Striking night-to-night variability in the infall activity is seen with no indication, in the LPIT'98 data of periodicity or of episodic infall among the UXORs with other signatures of accretion.
- However, not all UXORs are accreting stellar-composition material. One UXOR in the LPIT'98 sample, CQ Tau, shows infall in refractory species such as Na I and in Hα, but not in the higher Balmer series.

The two Herbig Ae stars with either episodic infall (HD 100546) or infall of material which appears to be cometary in composition (CQ Tau) are among the older, nearby pre-main-sequence stars. Both stars have evolutionary ages of 10 Myr (van den Ancker et al. 1997, 1998), making them contemporaries of candidate young planetary systems such as HR 4796 A, and only a factor of two younger than β Pic itself. These data suggest that the composition of the protoplanetary disks are changing with time, and that future searches for signatures of planetesimal-dominated infall concentrate on the near-ZAMS objects.

Acknowledgments. Support for this study was provided by by NASA Guaranteed Time Observer (GTO) funding to the STIS Science Team in response to NASA A/O OSSA-4-84 through the Hubble Space Telescope Project at Goddard Space Flight Center, and through inter-agency transfer of funds to The National Optical Astronomy Observatories. NOAO is operated by the Association of Universities for Research in Astronomy (AURA), Inc., under co-operative agreement with the National Science Foundation.

References

Batalha, C. C., & Basri, G. 1993, ApJ, 412, 363

Beckwith, S. V. W., & Sargent, A. I. 1996, Nat, 383, 139

Bertout, C., Basri, G., & Bouvier, J. 1988, ApJ, 330, 350

Bertout, C., Bouvier, J., & Basri, G. 1989, in Modelling the Stellar Environment: How and Why? ed. P. Delache (Gif-sur-Yvette: Editions Frontières), p. 189

Beust, H. et al. 1996, A&A, 310, 181

Beust, H., & Morbidelli, A. 2000, Icarus, 143, 170

Biesecker, D. et al. 1999, DPS, 31.1404

Böhm, T., & Catala, C. 1995, A&A, 301, 155

Chelli, A. 1999, A&A, 342, 763

Corcoran, M., & Ray, T. P. 1997, A&A, 321, 189

Corcoran, M., & Ray, T. P. 1998, A&A, 331, 147

Costa, R. D. D. et al. 1997, ApJ, 485, 380

Edwards, S. et al. 1994. AJ, 108, 1056

Evans, C., & McKim Malville, J. 1967, PASP, 79, 310

Folha, D. F. M., & Emerson, J. 1999, A&A, 352, 517

Goldreich, P., & Tremaine, S. 1980, ApJ, 241, 425

Gomez de Castro, A., & Fernandez, M. 1996, MNRAS, 283, 55

Grady, C. A. et al. 2000a,in Protostars and Planets IV, ed. V. Mannings, A. P. Boss, & S. S. Russell (Tucson: University of Arizona Press), p. 613

Grady, C. A. et al. 2000b, ApJ, in press

Grady, C. A. et al. 1999, ApJ, 523, L151

Grady, C. A. et al. 1997, ApJ, 483, 449

Grady, C. A. et al. 1996, A&AS, 120, 157

Grady, C. A. et al. 1995, A&A, 302, 472

Grady, C. A., & Silvis, J. M. S. 1993, ApJ, 402, L61
Graham, J. A. 1992, PASP, 104, 677
Hartigan, P. et al. 1989, ApJS, 70, 899
Heap, S. R. et al. 2000, ApJ, in press
Herbig, G. H., & Goodrich, R. W. 1986, ApJ, 309, 294
Hillenbrand, L. A. et al. 1992, ApJ, 397, 613
Hobbs, L. et al. 1985, ApJ, 293, L29
Jayawardhana, R. et al. 1998, ApJ, 503, L79
Kenyon, S. J. et al. 1999, ApJ, 524, L119
Koerner, D. et al. 1998, ApJ, 503, L83
Kondo, Y., & Bruhweiler, F. C. 1985, ApJ, 291, L1
Kozlova, O. V. et al. 2000, Astron. Rep., 44, 36
Lagrange-Henri, A. M. et al. 1987, A&A, 173, 289
Lagrange-Henri, A. M. et al. 1989, A&A, 215, L5
Lissauer, J. J. 1997, Nat, 386, 18
Malfait, K. et al. 1998, A&A, 331, 211.
Mannings, V., & Sargent, A. I. 1997, ApJ, 490, 792
Mannings, V., & Sargent, A. I. 1998, ApJ, 529, 391
Marcy, G., & Butler, P. 2000, PASP, 112, 137
Mayor, M., & Queloz, D. 1995, Nat, 378, 355
McCraughrean, M. J., Stapelfeldt, K. R., & Close, L. M. 1999, Protostars and Planets IV, ed. V. Mannings, A. P. Boss, & S. S. Russell (Tucson: University of Arizona Press), p. 485
Mundt, R., & Eislöffel, J. 1998, AJ, 116, 860
Natta, A. et al. 2000, ApJ, in press
Petterson, O. K. L., & Tobin, W. 1999, MNRAS, 304, 733
Pollack, J. B. et al. 1996, Icarus, 124, 62
Raymond, J. C. et al. 1998, ApJ, 508, 410
Roberge, A. et al. 2000, ApJ, in press
Schneider, G. et al. 1999, ApJ, 513, L127
Silverstone, M. D. et al. 1998, AAS, 193, 7316
Sorelli, C., Natta, A., & Grinin, V. P. 1996, A&A, 309, 155
van den Ancker, M. E. et al. 1997, A&A, 324, L33
van den Ancker, M. E. et al. 1998, A&A, 330, L145
Vidal-Madjar, A. et al. 1994, A&A, 290, 245
Vidal-Madjar, A., & Ferlet, R. 1995 in Circumstellar Dust Disks and Planet Formation, ed. R. Ferlet & A. Vidal-Madjar (Paris: Editions Frontiéres), p. 7
Waters, L. B. F. M. et al. 2000, in ASP Conf. Ser., 196, Thermal Emission Spectroscopy and Analysis of Dust, Disks, and Regoliths, ed. M. L. Sitko, A. L. Sprague, & D. K. Lynch (San Francisco: ASP), p. 3
Weinberger, A. et al. 1999, ApJ, 524, L119

Weinberger, A. et al. 2000, ApJ, in press
Welty, A. D. et al. 1992, AJ, 103, 1673

Discussion

Keith Horne: What is the jet velocity in HH 409?

Carol Grady: The approaching jet is moving with a radial velocity of -400 ± 50 km s^{-1} (dictated by the slit width: $0.02'' \leftrightarrow 100$ km s^{-1} if the jet fills the aperture). The receding jet is moving at +400 km s^{-1}. We have no sensitivity to material with velocities $< | \pm 200$ km $s^{-1}|$, due to absorption by Lyα in the foreground interstellar medium. Our preliminary proper motion data suggest motions of $0.6''$/yr in the plane of the sky (~ 300 km s^{-1}).

Alan Boss: What is known about the evolutionary state of AB Aurigae?

Carol Grady: Mario van den Ancker and collaborators (1997, 1998) estimated the age as ~ 2 Myr using the *Hipparcos* data and the Palla & Stahler (1994)[1] models. If the D'Antona & Mignatelli models are used the estimated age, 154 Myr, is in good agreement with AB Aur's common proper motion companion, SU Aur. Both estimates have uncertainties due to the uncertainties in the parallax data of ~ 2 Myr. However, as Lee Hartmann has pointed out, the evolutionary models have their own large uncertainties. In any event, AB Aur is more luminous than objects such as HD 163296 or HD 100546, and is therefore presumably younger.

Lee Hartmann: The ages of A stars are uncertain; in many clusters there are systematic differences between A- and M-star ages that are probably due to uncertainties in the birthline positions of intermediate-mass stars.

Carol Grady: If we assume AB Aur and SU Aur are coeval, the Ae star is ~ 4 Myr old (e.g., older than predicted by the Palla & Stahler model, not younger, as Lee indicates, for the Orion objects). Clearly, a better handle on ages of intermediate-mass PMS stars is needed.

Lee Hartmann: Is there a systematic difference in the IR excesses of A stars with β Pic-type infall and A stars with T Tauri-like infall?

Carol Grady: The Herbig Ae/Be stars with more volatile gas-rich infall have larger IR excesses, more typical of the less embedded classical T Tauri stars. HD 100546, which shows episodic infall and composition departures from solar abundances (Grady et al. 1997), has a NIR SED that shows signs of central clearing, although not as much as for β Pic. This star is apparently on the ZAMS and may be transitional between the younger—if you believe the ages—(larger IR excess) objects and the systems which appear to be young planetary systems and /or debris disks.

---

[1]Palla, F., & Stahler, S. W. 1994, in ASP Conf. Ser., 62, The Nature and Evolutionary Status of Herbig Ae/Be Stars, ed. P. S. The, M. R. Perez, & E. P. J. van den Heuvel (San Francisco: ASP), 391.

Bill Herbst: Your interpretation of the infall as relatively unprocessed disk material, linking it to the CTTS accretion, is consistent with our (Herbst & Shevchenko)[2] view that the large-amplitude variability has a common source: variable accretion. The fact that the infall events are found preferentially in the UXORs (at maximum light), while the spectrum becomes shell-like (at minimum light) suggests that accretion is the principal luminosity source at maximum.

Carol Grady: No! The optical and UV spectra are "shell-like" (e.g., with large amounts of line blanketing due to wind and infalling material) within ~ 1.5 magnitudes of maximum light. At minimum light in the UV the spectra of 3 stars with infall, large-amplitude light variability (UX Ori, BF Ori, and CQ Tau) go to a faint, flat continuum, which is bluer than the stellar spectrum at maximum light and emission in MG II and Fe II. This is consistent with detection of a bipolar outflow (lines) and light scattered by the disk when the star is effectively occulted (continuum). If we were detecting enhanced accretion at maximum light, the spectrum should be bluer (more accretion luminosity) and the photosphere should be most visible at minimum light. Neither effect is seen. Geometrical occultation of the stars by material in a possibly flared disk is a better match to the data, with major occultations (deep minima) comparatively infrequent. This point is addressed in more details in Vladimir Grinin's paper (this volume, p. 216).

Antonella Natta: A comment: if one observes redshifted absorption in the higher Balmer lines, the chemical composition of the infalling gas cannot be "cometary"; very probably, we are seeing infall of original gas.

---

[2]Herbst, W., & Shevchenko, V. S. 1999, AJ, 118, 1043.

Disks, Planetesimals, and Planets
ASP Conference Series, Vol. 219, 2000
F. Garzón, C. Eiroa, D. de Winter, and T. J. Mahoney, eds.

Photopolarimetric Activity of Pre-Main-Sequence Stars

V. P. Grinin

Astronomical Institute of St. Petersburg University, 198904, Bibliotechnaya pl., 2, Petrodvorets, St. Petersburg, Russia

Abstract. A review of the main properties of the photometric and polarimetric variability of young stars is presented. Special attention is given to stars with large-amplitude photopolarimetric variability caused by variable circumstellar (CS) extinction. The most photometrically active objects of this group (whose prototype is the Herbig Ae star UX Ori) are those observed directly through their CS disks. Therefore observations of these stars give us direct information on the dust distribution in CS disks and are a means of searching for the protoplanets.

1. Introduction

Optical light and polarimetric variability are among the brightest manifestations of the youth of stars and are sources of information about the physical processes in the neighborhoods and on the surfaces of young stars. The results obtained in this field are discussed in the proceedings of the conference on Herbig Ae/Be stars (Thé, Pérez, & van den Heuvel 1994), in a few chapters of the book *Protostars and Planets IV* (Mannings, Boss, & Russell 2000) and also in the reviews papers by Pérez & Grady (1997) and Waters & Waelkense (1998). In this review I summarize the main properties of the photopolarimetric activity of young stars and discuss the different mechanisms suggested for their explanation. The accent is on the photopolarimetric activity of young stars connected with their CS disks.

2. Variability of Spotted Stars

The simplest mechanism of light variability is realized in the case of rotating spotted stars. Since the optical radiation of cool stars is sensitive even to small temperature perturbations, this type of variability is frequently observed in the cool T Tauri stars (later than K0). It was first observed in classical T Tauri stars (CTTSs) by Schaefer (1983), Rydgren & Vrba (1983), Bouvier, Bertout, & Bouchet (1986) and then in the weak T Tauri stars (WTTSs) by Vrba, Herbst, & Booth (1988), Walker (1990), and many others. The spots can be cool or hot. Cool spots are similar to the magnetic spots on the Sun. Hot spots are caused by the gas accretion onto the star along the magnetic lines.

2.1. Cool Spots

The rotational modulation of brightness caused by cool spots is usually observed in WTTSs. The reason is very simple: these stars do not show the large-amplitude variability caused by gas accretion or by variable CS extinction, which is more typical of CTTSs (and mask the low-amplitude rotation modulation due to the spots). The observational data connected with this type of activity are discussed in the papers by Herbst et al. (1994), Stassun et al. (1999), and Herbst, Rhode, & Hillenbrand (2000), and can be summarized as follows:

1. The periods of light variations range from a few days to about twenty days. Long-term observations show that the periods can change slowly from year to year. Such changes indicate that the rotation of stars is differential, and that spots can migrate on the stellar surface as on the Sun in the different phases of the activity cycle.

2. The amplitudes of periodic brightness variations are usually small: 0.1–0.3 stellar magnitudes or less. The largest amplitude, $\Delta V \approx 0.6$, has been observed in the WTTS V410 Tau (Vrba et al. 1988). Its photometric period ($P = 1.872^{\rm d}$) is stable on the timescale of the order of seven years, although the shape of the light curve changes slightly from year to year (Petrov et al. 1994).

3. The light variations are accompanied by color changes: the star becomes redder when fainter. An exception is the U band (and sometimes the B band), in which the additional sources of emission (active regions above the magnetic spots or flares) can be important (Rydgren & Vrba 1983; Vrba et al. 1988).

In the case of V410 Tau the effective spot area is so large (about 40% of the stellar surface) that the rotation of the star evokes not only periodic light variability but also strong periodic changes in the radial velocity of the star itself with an amplitude of about 10 km s^{-1} (Welty & Ramsey 1996). This value is thousands of times larger than expected in the case of a planet or brown dwarf orbiting the star. Thus, the surface inhomogeneity of cool stars can mimic radial-velocity variations similar to those expected in the case of planet orbiting a star.

2.2. Hot Spots

The photometric activity caused by the hot spots dominates in CTTSs. The star DF Tau was the first in which this type of activity was recognized and attributed to gas accretion onto the star along the magnetic lines (Grinin 1980). The estimations show that the temperature of spots can reach 7000–10 000 K and their effective area can cover a few per cent of the stellar disk. Since gas accretion is the essentially a non-stationary process the parameters of hot spots are unstable and therefore the rotation modulation of stellar fluxes is a quasi-periodic function of time (Herbst et al. 1986; Vrba et al. 1986; Bouvier & Bertout 1989; Bertout et al. 1996).

The most important consequences of hot-spot emission are as follows:

1. It is a source of "veiled" continuum which masks the absorption lines in the spectra of CTTSs. Synchronous photometric and spectroscopic observations of CTTSs show that the brightness variations of stars correlate well with the veiling of their spectra (see Chelli et al. 1999, and references therein).

2. The rotation of the spotted star causes the periodic illumination of the CS dust envelope. As a result, a periodic component of polarized radiation can exist in the intrinsic polarization of TTSs (Wood et al. 1996; Stassun et al. 1999).

3. In the case of young binaries with eccentric orbits periodical modulation of the accretion rate is possible (Artymowicz & Lubow 1996), which evokes the periodical brightening of such stars at their periastron passages (Mathieu et al. 1997).

2.3. Flares of T Tauri Stars

Magnetic fields on the surface of T Tauri stars and non-stationary gas accretion are the natural sources of the flares and rapid light fluctuations observed frequently in these stars. The observational properties of such events are summarized in the papers by Gahm et al. (1995) and Gullbring et al. (1996). There is a growing log of data indicated that the most of the observed flares of TTS are caused by inhomogeneous matter infall controlled by magnetic fields.

3. FUORs

FU Ori-type stars (FUORs) belong to a very sparse group of young stars (about ten objects). Their unusual light curves and some other characteristics have been the the subject of numerous discussions. The FUOR phenomenon is a very short (tens years) and probably repeated phase in the life of young stars. If the interpretation of this phenomenon adopted now (see Hartmann, Kenyon, & Hartigan 1993, and references therein) is correct, then in the case of FUORs we observe not the star itself but the accretion disk around it. The high luminosity of the disk demands the corresponding accretion rate (10^{-5} $M_\odot$ yr^{-1} or more). The energy released in a typical FUOR event is about 10^{45}–10^{46} erg. More detail information about these objects is given in the above review by Hartmann et al.

4. Pulsations of Young Stars

According to Marconi & Palla (1998) low-amplitude non-linear pulsations can exist in very young stars of intermediate mass. The typical range of periods is about 0.1–0.3 days. In this phase of evolution young stars become visible but are still surrounded by a large amount of CS matter. Therefore, the variable CS extinction and variable emission of accreting gas can mask the low-amplitude periodic variations of the stellar flux. Nevertheless, the results of the observations by Marconi et al. (2000) seem to be very promising. If confirmed, they will give an independent test of the physical structure and evolutionary status of these

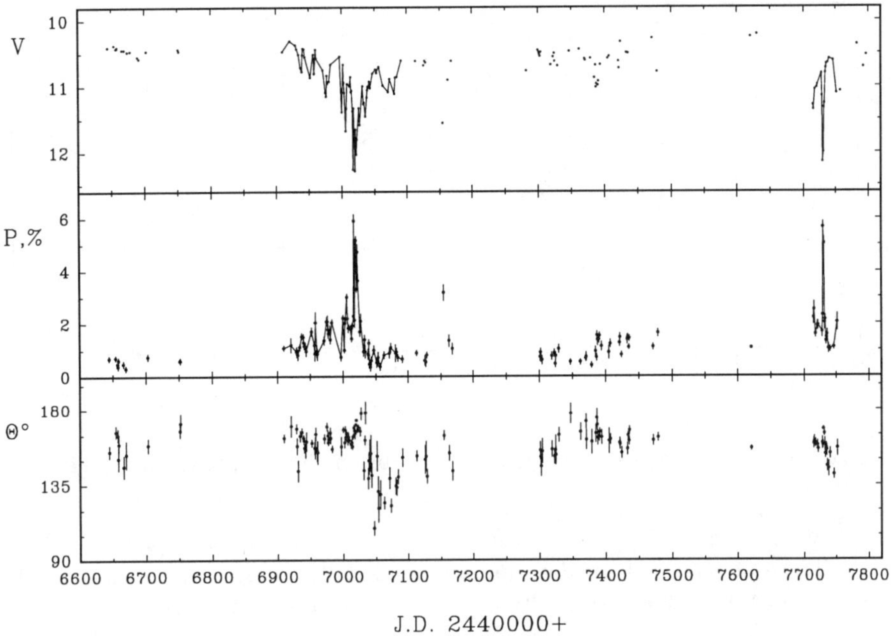

Figure 1. Variability IN brightness, polarization degree, and position angle of WW Vul; from Grinin et al. (1988) and Berdjugin et al. (1992).

young objects and, in turn, a method for estimating poorly known parameters such as the stellar mass and age.

It should be noted that stellar pulsations have to be present also in the variability of spectral lines, and there are attempts to find them (see Baade & Stahl 1989a). However, they are not so convincing as the photometric results by Marconi et al. (2000), which are probably connected with the contamination of photospheric spectra by the variable shell absorption components.

5. UX Ori-Type Stars

Many young stars reveal irregular variations in brightness with amplitudes of about 1–2 stellar magnitudes. Polarimetric observations show that the linear polarization of young stars is also variable in many cases on different timescales (Bastien 1988; Kopatskaya & Shulov 1978; Gullbring & Gahm 1996; Yudin & Evans 1998; Beskrovnaya et al. 1998). In its most spectacular form, irregular variability is observed in a small group of the young stars known in the literature as the UX Ori-type stars or UXORs. In the catalogue of variable stars by Kukarkin et al. (1982) they are denoted as the stars with non-periodic Algol-type minima. At the present time this subclass includes a few tens OF objects. Among them are Herbig Ae/Be (HAeBe) stars, CTTSs, and some others. The observational properties of UXORs are described in the review papers by Thé (1994) and Grinin (1994). The most important of these are as follows:

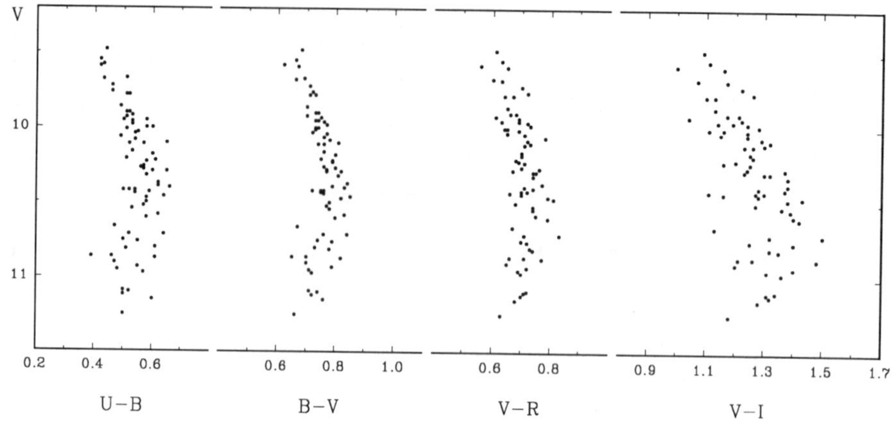

Figure 2. The color–magnitude diagrams of CQ Tau; from Berdjugin et al. (1990).

1. The amplitudes of the deep minima reach 2–3 stellar magnitudes in the V band; their durations are from a few days to a few weeks (Figure 1).

2. The amplitudes of photometric activity depend on the spectral type of the stars: Herbig Ae (HAe) stars are usually more active then Herbig Be stars (Finkenzeller & Mundt 1984; Thé 1994; Herbst & Shevchenko 1999), and most UXORs are the HAe stars.

3. Spectral observations made at the different brightness states do not show the variations of the spectral types of stars (Walker 1980; Petrov et al. 1999), although changes in the spectral lines are observed in the deep minima. In particular, the equivalent width of the H_α emission line, $WE(H_\alpha)$, increases when a star fades and the line profile changes from an asymmetric two-component profile in the bright state of the star to a single one in the deep minima (Kolotilov 1977; Grinin et al. 1994; Kozlova, Grinin, & Rostopchina 2000).

4. The brightness variations are also accompanied by changes in the color and linear polarization of stars.

5.1. The "Blueing" Effect

This unusual effect is described in many papers (see, for example, Zaitseva 1986; Gahm et al. 1989; Thé 1994; Eaton & Herbst 1996). and was first discovered by von Götz and Wenzel (1968) in CQ Tau. Observing this star in the deep minima, they found that, beginning from some brightness level, the initial reddening of the star stopped and its $U - B$ and $B - V$ colors become bluer again with a further decrease in the visual flux. Later, similar color variations were observed by Zaitseva (1973) in the deep minima of UX Ori. To the present time the "blueing" effect has been observed in the deep minima of many UXORs (see Figure 2) and is an important common property of their photometric behavior.

Omitting the discussion of the earlier attempts to explain this effect (see Thé 1994) we note here only that model which gives the most natural explanation of the blueing effect as a consequence of the coronagraphic effect produced by CS dust clouds (Grinin 1986), and which has direct observational evidence. This model takes into account that the observed radiation of young stars is the sum of the direct stellar radiation reduced by CS extinction in the CS dust clouds intersected from time to time in the line of sight and the radiation scattered by CS dust:

$$I_{\rm obs} = I_* \cdot e^{-\tau_\lambda} + I_{\rm sc}, \qquad (1)$$

where τ_λ is the current value of the optical thickness of CS dust in the line of sight; the other designations are standard.

The initial reddening of a star is the result of selective extinction of the stellar radiation in the dust cloud (as was suggested by Wenzel 1969). In the deep minima, when the direct radiation of a star is blocked by the CS cloud, the scattered radiation dominates, giving rise to the turnover of color tracks in the color–magnitude diagrams (Figure 2).

5.2. Scattered Radiation in the CS Disk: the Coronagraphic Effect

Since CS dust envelopes around young stars have a disk-like shape the scattered radiation should be polarized. Therefore, if the variability of UXORs is really caused by the screening of a star by opaque CS dust clouds, the linear polarization has to increase when a star fades. This observational test was confirmed by the synchronous polarimetric and photometric observations of UXORs started in 1986 at CrAO. Part of these observations were carried out in cooperation with N. Kiselev's group at the Sunglok Observatory (Tadjikistan), and with A. Okazaki and S. Kikuchi at the Dodajra Astronomical Observatory (Japan). It was shown that the linear polarization of UXORs anti-correlates as rule with the brightness changes, and that its dependence on the stellar magnitudes agrees well with that predicted (see review papers by Grinin et al. 1991; Grinin 1994, and references therein). Similar results for three UXORs were obtained by Kardopolov & Rspaev (1989, 1990) and Kardopolov, Pavlova, & Rspaev (1991), and for the classical HAeBe star R Mon by Matsumura, Seki, & Kawabata (1999).

If the dust cloud intersects the line of sight far from the star, its shadow will be small in comparison with the solid angle 4π. In this case the Stokes parameters of scattered radiation will not change at the moment of eclipse, and the variation in linear polarization has to follow to the simple law

$$\vec{P}_{\rm obs}(\Delta m) = \vec{P}_{\rm is} + \vec{P}_{\rm in} \cdot 10^{0.4\Delta m}, \qquad (2)$$

where $P_{\rm obs}(\Delta m)$ is the observed linear polarization of the star, $P_{\rm in}$ is intrinsic polarization of the star in the bright state, and $P_{\rm is}$ is the interstellar polarization.

The other potentially important mechanism of polarization due to the optical dichroism of aligned non-spherical grains gives the following dependence (Grinin et al. 1988):

$$\vec{P}_{\rm obs}(\Delta m) = \vec{P}_{\rm is} + \vec{a} \cdot \Delta m, \qquad (3)$$

where $\vec{a}$ is the pseudo-vector depending on the optical dichroism of dust in CS clouds.

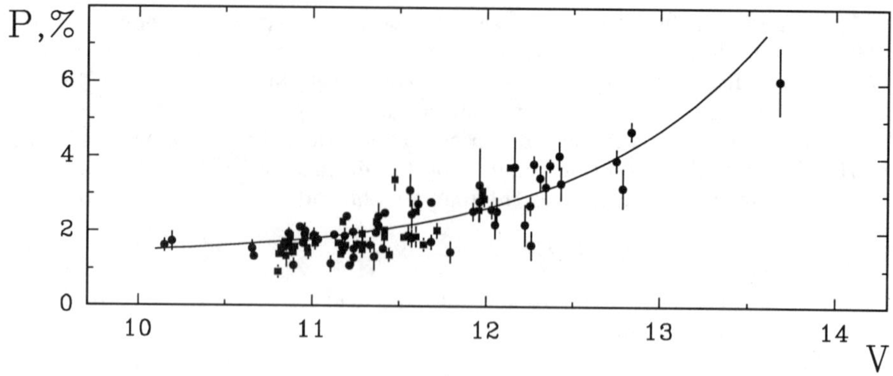

Figure 3. The polarization/magnitude (P/m) dependence of RR Tau in the V band; from Rostopchina et al. (1997). The *thin line* corresponds to the theoretical dependence (2).

The observations confirm the first of these dependences (see Figure 3), although the scatter of the observing points in some cases can be quite large and has a tendency to increase with the accumulation of the data. One of the reasons for this scatter is the fluctuations of the Stokes parameters of the scattered radiation due to the variable illumination of the CS dust caused by the heterogeneous structure of the dust envelopes. This reason is important since the main deposit in the variable CS extinction can produce dust clouds moving in the near vicinities of stars, probably in the region of dust sublimation (Grinin 1994). In the case of T Tauri stars the characteristic size of this region is about of ten stellar radii and CS dust can contact immediately with the stellar magnetosphere and corotate with a star producing the quasi-periodic brightness variations (Gahm et al. 1993; Bouvier et al. 1999; Chelli et al. 2000).

The other cause of the dispersion of observing poins in the polarization/magnitude diagrams, as well as in the color–magnitude diagrams (Figure 2), is the cyclical variability of UXORs (see below).

Since the CS disks of UXORs are not resolved, their scattered radiation (which can be considered as the direct analogy of solar zodiacal light) plays the role of the natural limitation of the amplitudes of the brightness variability, which cannot exceed the value:

$$\Delta m_{\rm max} = 2.5 \cdot \log \frac{I_* + I_{\rm sc}}{I_{\rm sc}}. \tag{4}$$

The typical intensities of the scattered radiation followed from the interpretation of the blueing effect are about 0.03–0.1 (in units of I_*) in the visible (Grinin 1986)[1], which explains why we do not observe the very deep minima of UXORs with amplitude 5 or more stellar magnitudes.

[1]These values are thousands of times larger than in the older protoplanetary disk of β Pictoris (Smith & Terrile 1984).

The independent observational evidence in favor of the variable CS extinction model should also be mentioned; i.e., the anti-correlation between the changes of stellar brightness and the equivalent widths of the forbidden lines [O I] 6300/6343, which have been recently observed in RR Tau (Grinin et al. 2000). These lines are formed in the outer parts of CS gas envelope, and the CS dust clouds that screen the star from an observer are not able to screen this more extended region. A similar coronagraphic effect is also the reason for the above-mentioned variations of WE(H_α) and transformation of the H_α line profile in the deep minima (Grinin & Tambovtseva 1995).

5.3. Evidence for the Edge-on Orientation of CS Disks

There are the reasons to believe that the violent photopolarimetric activity of UXORs is connected with the"optimal" orientation of their CS disk-like dust envelopes (Grinin et al. 1991). The main argument in favor of this suggestion is the high linear polarization systematically observed in deep minima (up to 5–8%),which indicates the preferential orientation of their CS disks relative to the observer: edge on or at a small inclination to the line of sight. It is also supported by the statistical analysis of the H_α line profiles observed in the spectra of HAeBe stars wirh different level of photometric activity (Grinin 1992; Grinin & Rostopchina 1996): the accretion line profiles are typical for UXORs, and the stellar wind signatures are more frequently observed in the photometrically quiet stars. Such a dependence agrees qualitatively with the expected velocity field in the magnetospheric accretion models.

A similar result was obtained independently by Grady et al. (1996). These authors have noted also that the rotation velocities of UXORs are systematically larger than those in the photometrically quiet stars of the same spectral types. This result was not confirmed by Herbst & Shevchenko (1999), who have analyzed the rotation velocities of about twenty HAeBe stars. Recently, however, Grinin & Kozlova (2000) have shown that the more homogeneous statistics based only on the Herbig Ae stars give the weak but statistically real correlation between $v sin i$ and ΔV. Such a correlation is an additional argument in support of the preferential orientation of UXORs relative to the observer.

It should be noted also that, according to estimates by Natta et al. (1997) based on the observations of the HAe stars at millimeter wavelengths, the masses of CS disks do not correlate with the amplitudes of the visual variability of UXORs and the positions of these stars on the HR diagram are practically the same as those of the low-activity HAe stars. This result is an important indirect indication that the "optimal" orientation of CS disks relative to the line of sight is probably the main condition of the large-amplitude photometric activity of UXORs.

6. The Long-Term Photopolarimetric Activity of UXORs

Due to the specific orientation of CS disks mentioned above the photometric monitoring of UXORs gives immediate information about the variability of the dust column density in the line of sight. In combination with polarimetric observations this creates the basis for the study of the inner structure of CS disks.

6.1. The Activity Cycles

There have been several attempts to find the periodicity of the appearance of the Algol-type minima of UXORs but none of these has been successful. The existence has also been noticed of the wave-like variations of their bright state on a timescale from a few to ten or more years (Shevchenko et al. 1993a,b; Minikulov et al. 1993; Friedemann et al. 1993). There are several cases of suspected periodicity although the density of the time coverage of the observations is usually only a few cycles (Figure 3), and the periodic properties of the light curves need the confirmations.

Recently, we summarized all photometric observations of UXORs, including our own data, and discussed the possible origin of the activity cycles (Grinin, Rostopchina, & Shakhovskoy 1998; Rostopchina, Grinin, & Shakhovskoy 1999; Shakhovskoy, Grinin, & Rostopchina 2000). The existence of such cycles implies the existence of large-scale structures of gas and dust in the CS disks of young stars. Such structures could be caused by gravitation perturbations in CS disks connected with the presence of low-mass components (see the SPH simulations by Artymowicz & Lubow 1996), or long-lived vortices in the protoplanetary disks orbiting a star.

The maximum duration of the cycle (about 22–25 years) is observed in the light curve of CQ Tau (Minikulov et al. 1993; Shakhovskoy et al. 2000), which indicates the existence of global deviations of the CS disk from axial symmetry on a characteristic scale of about 10 AU. Another indication to the non-axially symmetric distribution of CS dust comes from observations of the very high intrinsic polarization (up to 10–15%) in the deep minima of some UXORs (see, for example, Rostopchina et al. 2000a). These observations show that the axially symmetric models of CS disk-like envelopes that are frequently used for modeling CS dust around UXORs (Voshchinnikov & Grinin 1991; Voshchinnikov 1998) have limited possibilities of description of the observations and should be revised.

6.2. Indications of Comet-Like Activity

Gahm & Greenberg (19834) were probably the first to assume that the photometric activity of young stars can be partly caused by extinction in the dust coma of proto-comets. Later, a similar idea was used in several papers to interpret the variability of young stars (see Pérez & Grady 1997, and references therein), including the cyclic variability of UXORs (Shevchenko et al. 1993). It is not very easy, however, to distinguish the extinction in the dust coma of a proto-comet and in the primitive CS dust clouds intersecting the line of sight. Therefore, we note here only a few of the most interesting observations in support of this idea.

The first of these is the detection of the spectroscopic signatures of crystalline silicates in the infrared (IR) spectra of some HAeBe star similar to those observed in comets in the Solar System (Waelkense et al. 1998; Malfait et al. 1998; Sitko et al. 1999). Another indication comes from the observations by Hutchinson, Albinson, & Barrett (1994) and Sitko et al. (1994) of the optical and IR variability of a few HAeBe stars. Both groups have independently observed the same type of variability: an increase in the IR and a decrease in the visible fluxes. Such a variability indicates the appearance of dust clouds in the

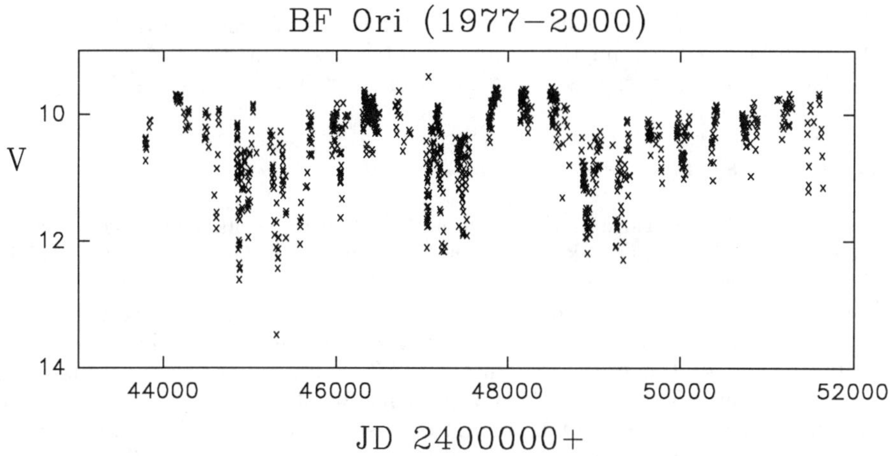

Figure 4. Example of the cyclic variability of BF Ori; from Grinin et al. (1998) and from the Crimean data base (the last two observing seasons).

line of sight in the immediate vicinity of the star (Sitko et al. 1994),which is difficult to explain without assuming comet-like activity. The same conclusion also follows from the analysis of the photopolarimetric behavior of some UXORs (Grinin 1994).

Several spectroscopic observations are also known of accretion events in UXORs that are similar in many details to those observed in β Pictoris (see Pérez & Grady 1997). However, their interpretation is not unique (Sorelli, Grinin, & Natta 1996; Natta, Grinin, & Tambovtseva 2000).

6.3. Flare-Like Events on HAeBe Stars

Analysis of long-term photometric observations of Herbig Ae/Be stars shows that in the light curves of some of these there occur flare-like events, in which the star becomes suddenly brighter and then returns to its normal state. The duration of such events is probably quite short and does not exceed two nights (see Rostopchina et al. 1997, and references therein). In comparison with T Tauri stars where the flare-like events have a clear physical basis (see Section 2) the "flares" on hot stars are quite enigmatic. According to the interpretation of Rostopchina et al. (1997) the observed "flares" on RR Tau were the result of a decrease in dust extinction in the line of sight due to formation of short-lived "holes" free of dust in the neighborhood of the star. This suggestion is based on the behavior of the color indices of the star during the "flares": they were in agreement with the reddening law caused by CS extinction. Approximately the same behavior of the color indices has been observed in the "flares" on BF Ori by Koval'chuk & Pugach (1991) and T Ori (Herbst & Shevchenko 1999).

Zaitseva & Lyutij (1997) assumed that the optical radiation of the accretion disk can give some deposit in the photometric variability of WW Vul in the deep minima. However, this seems rather improbable since the main part of the accretion energy is realized in the inner layers of accretion disks near the stellar surface (Hartmann 1999), and when CS dust clouds screen the star from an observer they screen this part of the accretion disk at the same time.

7. Conclusion

The photopolarimetric investigations of PMS stars give us a knowledge of the physical processes that take place on the surface of the young stars and in their neighborhoods. In particular, the variability of young spotted stars gives important information on the rotation of PMS stars and the location of the magnetic spots on their surfaces. Observations of stellar pulsations are a way to investigate of the inner structure of PMS stars and provide estimates of the stellar mass. The photopolarimetric behavior of UXORs is sensitive to the presence of the relatively small amount of dust in the line of sight. The further accumulation of photopolarimetric observations of these stars will permit us to investigate the dust distribution around young stars in more detail. It is a unique method for studying the thin structure of protoplanetary disks perturbed by protoplanets or by secondary components in the young binary systems.

Acknowledgments. I am thankful to A. Rostopchina and D. Shakhovskoy for their preparation of the figures and discussions, and to M. Marconi for useful comments. This work has been supported in part by grant RFFI 99-02-18520.

References

Artymowicz, P., & Lubow, S. H. 1996, ApJ, 467, L77

Baade, D., & Stahl, O. 1989a, A&A, 209, 255

Baade, D., & Stahl, O. 1989b, A&A, 209, 268

Bastien, P. 1988, in Polarized Radiation of Circumstellar Origin, ed. G. V. Coyne et al. (Vatican City State/Tucson: Vatican Observatory/University of Arizona), p. 303

Berdjugin, A. V., Berdjugina, S. V., Grinin, V. P., & Minikulov, N. Kh. 1990, AZh, 67, 812

Berdjugin, A. V., Grinin, V. P., & Minikulov, N. Kh. 1992, Bull. Crimean Astrophys. Obs., 86, 69

Bertout, C., et al. 1996, AJ, 112, 2159

Beskrovnaya, N. G., Pogodin, M. A., Yudin, R. V., Franco, G. A. P., Vieira, S. L. A., & Evans, A. 1998, A&AS, 127, 243

Bouvier, J., & Bertout, C. 1989, A&A, 211, 99

Bouvier, J., Bertout C., & Bouchet, P. 1986, A&A, 158, 149

Bouvier, J., et al. 1999, A&A, 349, 619

Chelli, A., et al. 1999, A&A, 345, L9

Eaton, N. L., & Herbst, W. 1996, AJ, 110, 2369

Finkenzeller, U., & Mundt, R. 1984, A&AS, 55, 109

Friedemann, C., et al. 1993, A&A, 277, 184

Gahm, G. F., Fischerström, C., Liseau, R., & Lindroos, K. R. 1989, A&A, 211, 115

Gahm, G., & Greenberg, J. M. 1984, in Asteroids, Comets, Meteors, ed. C.-I. Lagerkvist & H.Rikman (Uppsala: Astronomiska Observatoriet), p. 375

Gahm, G. F., et al. 1993, A&AS, 100, 371

Gahm, G. F., Loden, K., Gullbring, E., & Hardstein, D. 1995, A&A, 301, 89

Grady, C., et al. 1996, A&AS, 120, 157

Grinin, V. P. 1980, Astrophysics, 16, 147

Grinin, V. P. 1986, unpublished poster paper presented at IAU Symp. 122, Circumstellar Matter (see Sov. Astron. Lett., 1988, 14, 27)

Grinin, V. P. 1992, A&A Trans., 3, 17

Grinin, V. P. 1994, in ASP Conf. Ser., 62, The Nature and Evolutionary Status of Herbig Ae/Be Stars, ed. P. S. Thé, M. R. Pérez, & E. P. J. van den Heuvel (San Francisco: ASP), p. 63

Grinin, V. P. Kiselev, N .N., Minikulov, N. Kh., & Chernova, G. P. 1988, Pis'ma v Astron. Zn., 14, 514

Grinin, V. P., Kiselev, N. N., Minikulov, N. Kh, Chernova, G. P., & Voshchinnikov, N. V. 1991, Ap&SS, 186, 283

Grinin, V. P., & Kozlova, O. V. 2000, Astrophysics, accepted

Grinin, V. P., & Rostopchina, A. N. Astron. Rep., 40, 171

Grinin, V. P., Rostopchina, A. N., & Shakhovskoy, D. N. 1998, Astron. Lett., 24, 802

Grinin, V. P., & Tambovtseva, L. V. 1995, A&A, 293, 396

Grinin, V. P., Thé, P. S., de Winter, D., Giampapa, M., Rostopchina, A. N., Tambovtseva, L. V., & van den Ancker, M. 1994, A&A, 292, 165

Grinin, V. P., et al. 2000, in preparation

Gullbring, E., & Gahm, G. F. 1996, A&A, 308, 821

Gullbring, E., et al. 1996, A&A, 307, 791

Hartmann, L. 1999, New Astron. Rev., 43, 1

Hartmann, L., Kenyon, S., & Hartigan, P. 1993, in Protostars and Planets III, ed. E. H. Levy & J. I. Lunine (Arizona: University of Arizona Press), p. 497

Herbst, W., & Shevchenko, V. S. 1999, AJ, 118, 1043

Herbst, W., et al. 1986, ApJ, 310, L71

Herbst, W., et al. 1994, AJ, 108, 1906

Herbst, W., Rhode, K., & Hillenbrand, L. A. 2000, AJ, 119, 261

Hutchinson, M. G., Albinson, J. S., & Barrett, P. 1994, A&A, 285, 883

Kardopolov, V. I., Pavlova, L. A., & Rspaev, F. K. 1991, AZn, 68, 565

Kardopolov, V. I., & Rspaev, F. K. 1989, Kinem. Fiz. Neb. Tel., 5, 50

Kardopolov, V. I., & Rspaev, F. K. 1990, AZh, 76, 1253

Kolotilov, E. A. 1977, Afz, 13, 33

Kopatskaya, E. N., & Shulov, O. S. 1978, Tr. Astron. Obs. Leningrad Univ., 34, 94

Koval'chuk, G. U., & Pugach, A. F. 1991, Kinem. Fizika Neb. Tel., 7, 33

Kozlova, O. V., Grinin, V. P., & Rostopchina, A. N. 2000, ARep, 44, 36

Kukarkin, B. V., et al. 1986, General Catalogue of Variable Stars (Moscow: Nauka)

Malfait, K., et al. 1998, A&A. 332, L25

Mannings, V., Boss, A. P., & Russell, S. S. (eds.) 2000, Protostars and Planets IV (Tucson: University of Arizona Press)

Marconi, M., & Palla, F. 1998, ApJ, 507, L141

Marconi, M., et al. 2000, A&A, in press

Mathieu, R. D., Stassun, K., Basri, G., Jensen, E. L. N., Johns-Krull, C. M., Valenti, J. A., & Hartmann, L. W. 1997, AJ, 113, 1841

Matsumura, M., Seki, M., & Kawabata, K. S. 1999, AJ, 117, 429

Minikulov, N. Kh., Rakhimov D. Yu., Volchkova, N. A., & Pikhun, A. I. 1993, Afz, 36, 55

Natta, A., Grinin, V. P., Mannings, V., & Ungerechts, H. 1997, ApJ, 491, 885

Natta, A., Grinin, V. P., & Tambovtseva, L. V. 2000, ApJ, in press

Pérez, M., & Grady, C. 1997, Sp. Sci. Rev., 82, 407

Petrov, P. P., et al. 1994, A&ASS, 107, 922

Petrov, P. P., et al. 1999, A&A, 341, 553

Rostopchina, A. N., et al. 2000, AZh, accepted

Rostopchina, A. N., et al. 1997, A&A, 327, 145

Rostopchina, A. N., Grinin, V. P., & Shakhovskoy, D. N. 1999, Astron. Lett., 25, 291

Rostopchina, A. N., Grinin, V. P., & Shakhovskoy, D. N. 2000, ARep, submitted

Rydgren, A. E., & Vrba, F. J. 1983, ApJ, 267, 191

Shaefer, B. E. 1983, ApJ, 266, L45

Shakhovskoy, D. N., Grinin, V. P., & Rostopchina, A. N. 2000, Astron. Lett., submitted

Shevchenko, V. S., et al. 1993a, Ap&SS, 202, 121

Shevchenko, V. S., et al. 1993b, Ap&SS, 202, 137

Sitko, M. L., et al. 1994, ApJ, 432, 753

Sitko, M. L., et al. 1999, ApJ, 510, 408

Smith, B. A., & Terrile, R. J. 1984, Science, 226, 1421

Sorelli, C., Grinin, V. P., & Natta, A. 1996, A&A, 309, 155

Stassun, K. G., Mathieu, R. D., Mazeh, T., & Vrba, F.J. 1999, AJ, 117, 2941

Stassun, K., & Wood, K. 1999, ApJ, 510, 982

Thé, P. S. 1994, in ASP Conf. Ser., 62, Nature and Evolutionary Status of Herbig Ae/Be Stars, ed. P. S. Thé, M. R. Pérez, & E. P. J. van den Heuvel (San Francisco: ASP), p. 23

Thé, P. S., Pérez, M. R., & van den Heuvel, P. J. (eds.) 1994, ASP Conf. Ser., Vol. 62, The Nature and Evolutionary Status of Herbig Ae/Be Stars (San Francisco: ASP)

von Götz, W., & Wenzel, W. 1968, Mitt. Vernd. Sterne, 5, 2
Voshchinnikov, N. V. 1998, ARep, 42, 54
Voshchinnikov, N. V., & Grinin, V. P. 1991, Afz, 34, 181
Vrba, F., et al. 1986, ApJ, 306, 199
Vrba, F., Herbst, W., & Booth, J. L. 1988, AJ, 96, 1032
Waters, L. B. F. M., & Waelkense, C. 1998, ARA&A. 36, 233
Waelkense, C., et al. 1996, A&A, 315, L245
Walker, M. F. 1980, PASP, 92, 66
Walker, M. F. 1990, PASP, 102, 726
Welty, A. D., & Ramsey L. W. 1996, AJ, 110, 336
Wenzel, W. 1969, in IAU Coll. 4, Non-Periodic Phenomena in Variable Stars, ed. L. Detree, p. 61
Wood, K., Kenyon, S., Whitney, B. A. & Bjorkman, J. E. 1996, ApJ, 458, L79
Yudin, R. V. & Evans, A. 1998, A&AS, 131, 234
Zaitseva, G. V. 1973, Variable Stars, 19, 63
Zaitseva, G. V. 1986, Afz, 25, 471
Zaitseva, G. V., & Lytij, V. M. 1997, Pis'ma v Astron. Zn. J., 23, 1

Discussion

Mario van den Ancker: You showed a plot of the anti-correlation between the optical and infrared brightness of two UXORs. But moving a clump of dust in and out of our line of sight should not affect the infrared excess, which comes from the entire circumstellar environment, which is always seen because it is in emission. Can you comment on this?

Vladimir Grinin: If the dust cloud moves along a highly eccentric orbit it can approach the near vicinity of the star and give (under certain conditions) a sizable contribution to the near-infrared flux. This means that the results of Hutchinson et al. (1994) and Sitko et al. (1994) indicate that such highly eccentric orbits do really exist.

Carol Grady: For systems like CQ Tau and BF Ori, what binary separation and magnitude difference would you predict if the clustering of the large-amplitude light variations is due to perturbations on a circumbinary disk?

Vladimir Grinin: For CQ Tau we predict periodic perturbations in the CB disk with $P \approx 25$ yr (Shakhovskoy, Grinin, & Rostopchina 2000, in press) and for BF Ori with $P \approx 5.5$ or 11 yr (Grinin, Rostopchina, & Shaklovskoy 1998). For the other UXORs, the characteristic time of the cyclical variability is about a few years. This means that the separation between the components is in the range 3–10 AU. Unfortunately, we still do not know the magnitude difference between the primary and secondary components because we still do not know their mass ratio.

Sebastian Wolf: What are the typical optical depth, size, and distance from the central stars of the clouds or density perturbations in a circumstellar disk that can be derived from the intensity variations in the optical and near-infrared wavelength range?

Vladimir Grinin: We can speak only about the dust clouds which can produce the sizable screening effect. Their characteristic dimensions have to be not less than 3–5 stellar radii and the optical thickness should be enough (τ ~3–5 or more) in order to explain the observed optical variability. The rapid photometric and polarimetric variations which were observed in some episodes show that the dust clouds can reach the nearest vicinity of stars (~ 2 AU or less). Such clouds can in principle give the sizable variability of the near-infrared fluxes.

Patrick Godon: Your are using simulations of the gas (SPH) to interpret observations of the dust particles. Did you check that the modeling is self-consistent?

Vladimir Grinin: Yes, under the present conditions the dust particles can be modeled using the gas flow. (for micron-size particles, the dust acts as a tracer in the gas flow.)

François Ménard: Herbig Be stars do not show (as others, if at all) the large polarimetric variability that the Herbig Ae stars (or UXORs) do. Do you hav, in your large database, such large polarimetric events for later-type stars?

Vladimir Grinin: Yes, we have the same results as for the UXOR-type stars for the classical T Tauri stars BM And (Grinin, Kolotilov, & Rostopchina 1995, A&AS, 112, 457) and CO Ori Grinin et al., in press). DF Tau does not show the anti-correlation between the linear polarization and stellar brightness.

Disks, Planetesimals, and Planets
ASP Conference Series, Vol. 219, 2000
F. Garzón, C. Eiroa, D. de Winter, and T. J. Mahoney, eds.

Spectroscopic Monitoring of PMS and Vega-Type Stars

Alcione Mora

Departamento de Física Teórica C-XI, Universidad Autónoma de Madrid, Cantoblanco 28049 Madrid, Spain

Dolf de Winter

TNO/TPD-Space Instrumentation, Stieltjesweg 1, PO Box 155, 2600 AD Delft, The Netherlands

R. Ferlet

CNRS, Institut d'Astrophysique de Paris, 98bis Bd. Arago, 75014 Paris, France

C. A. Grady

NOAO/STIS, Goddard Space Flight Center, Code 681, NASA/GSFC, Greenbelt, MD 20771, USA

EXPORT

Abstract. High- and intermediate-resolution spectroscopy of pre-main-sequence (PMS) and Vega-type stars was the main contribution to the EXPORT observing project. Roughly 80 stars were observed at intermediate resolution while high-resolution spectra were obtained for about 50 objects. A major characteristic of the data relies on the spectroscopic monitoring, with monitoring timescales of hours, days, and months for many of the observed stars. Details of these observations and a few examples of our results are presented in this contribution.

1. Introduction

Planets are thought to be an end-product of the evolutionary process starting from the protostellar disks associated with protostars. In this scenario, the dissipation timescale for the disk or circumstellar (CS) material must be overruled by the protoplanetary formation timescale. A first step leading to planets must therefore be the formation of large solid bodies. The formation and evolution of protoplanetary disks and the study of planetesimals as possible planet precursors motivated the EXPORT project (Eiroa et al., this volume, p. 3) to devote intensive observational attention to PMS and Vega-type stars. Clues suggesting the importance of these objects for (proto)planetary-system formation are the different observational results for PMS and MS stars that have been interpreted in terms of circumstellar disks harboring solid bodies (Grady et al. 2000; Lagrange, Backman, & Artymowicz 2000). A more specific example is β Pic, which is the best studied case of an MS star with a dust and gas disk, where transient absorption components in the spectral lines of several chemical species could be explained in terms of kilometer-size solid bodies acting as star-grazers, also known as the falling evaporating bodies (FEB) hypothesis (e.g., Beust et al. 1998, and references therein). In addition, a significant fraction of PMS stars show β Pic-like spectroscopic events. It has been suggested that these stars, whose best studied member is UX Ori, are β Pic system progenitors.

The EXPORT spectroscopic monitoring program aimed to study the variable gaseous circumstellar activity on different timescales of stars of different evolutionary ages. High-resolution spectroscopy allows us to analyze the kinematics of transient spectroscopic gaseous events in detail, while intermediate-resolution spectroscopy provides better time resolution plus the possibility of observing faint objects. It is important to point out that these observations were complemented with simultaneous optical photo- and polarimetry and near-IR photometry. With such a unique dataset information on possible simultaneous changes or correlations in the gaseous and dusty structures of circumstellar disks can be achieved

2. EXPORT Spectroscopic Observations

2.1. Intermediate Resolution

Intermediate-resolution spectra were taken with the 2.5 m Isaac Newton Telescope (INT) equipped with the Intermediate Dispersion Spectrograph (IDS). Observations were carried out during 16 nights distributed over four observing runs, approximately four nights per run, in 1998 May, 1998 July, 1998 October, and 1999 January. This distribution allowed us to obtain long-term (months) and short-term (hours, days) spectroscopic monitoring. The quality of the nights was good, even photometric. Spectra of many stars were obtained each night when observable.

Two different instrumental setups were used. An EEV CCD was used in 1998 May; the spectral range covered from 5854 Å to 6728 Å, with a resolution of $\Delta\lambda \sim 0.707$ Å per pixel and a resolving power $\sim$ 4500 at 6300 Å. An upgraded EEV CCD was used during the other three observing runs. Now the wavelength range covered 5712 to 6812 Å, the resolution was $\Delta\lambda \sim 0.474$ Å per pixel and the resolving power $\sim$ 6600 at 6300 Å. The slit width was always set at 1.0 arcsecond.

During our observations 81 stars were observed for which 507 individual spectra were collected. Table 1 lists the observations and the number of spectra obtained per object. Exposure times ranged from one minute for the brightest objects, e.g., HR 10, $V = 6.23$ mag, and resulted in a signal-to-noise ratio (SNR) $\sim$ 280, to half an hour for the faintest, e.g., VY Mon, $V \sim 14.0$ mag, SNR $\sim$ 95. Figure 1 presents a histogram of the number of exposures per star in our sample. Around 50% of the stars were observed at least eight times, while only 11 out of the 81 stars were observed only once; the later ones are those apparently showing 'pure' photospheric spectra.

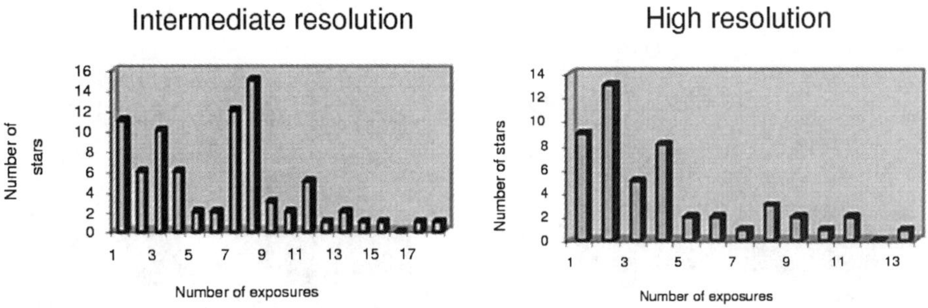

Figure 1. Number of observations per star.

2.2. High Resolution

High-resolution spectra were obtained at the 4.2 m William Herschel Telescope (WHT) equipped with the Utrecht Eechelle Spectrograph (UES). In this case, the overall observing time consisted of 12 nights distributed over the same runs as the INT observations: two nights in 1998 May and October and four nights in 1998 July and 1999 January. In the May and October runs four nights were fully dedicated to observing 51 Peg and τ Bootis (see Harris et al., this volume, p. 454)

UES was set to provide a wavelength coverage from 3800 to 5900 Å, a spectral resolution of 49 000 and a slit width of 1.15 arcseconds projected on the sky. The dataset consists of 198 spectra for 49 stars. The observed objects and the number of UES spectra obtained are shown in Table 1. Integration times range from 10 minutes for the brightest objects (e.g., HR 10, $V = 6.23$ mag, SNR $\sim$ 2400) to 45 minutes for the faintest (e.g., VV Ser, $V \sim 11.9$ mag, SNR $\sim$ 14). The histogram of the number of UES spectra per star is also shown in Figure 1. In this case, the frequency of observations per star is considerably lower than in the case of the INT spectra, due to the exposure times needed to obtain good-quality UES spectra. Nevertheless, the most interesting objects, as indicated by the INT observations, were observed as frequently as possible.

3. Observational Results

The sample selection contained 81 objects of different masses and ages, 57% of which are PMS stars and 43% considered MS stars. PMS stars consist of HAeBe, HAeBe/ZAMS, T Tauri and UXOR stars. UXORs are a subgroup of HAe stars and T Tauri stars which share photometric and spectroscopic characteristics with the prototype of the class, UX Ori. We have singled them out because it has been suggested that UXORs could be the progenitors of Vega-type/β Pic systems. The MS group consists of Vega-type stars, A-shell stars and post-T Tauri (PTT) stars. PTT stars were taken from Lindroos's list (Lindroos 1986) of visual binaries composed of an early-spectral-type star (primary) and a late-type star thought to be in the PTT phase. Because of the MS-like photospheric spectra of these stars, we introduce this group into the MS stellar sample, though

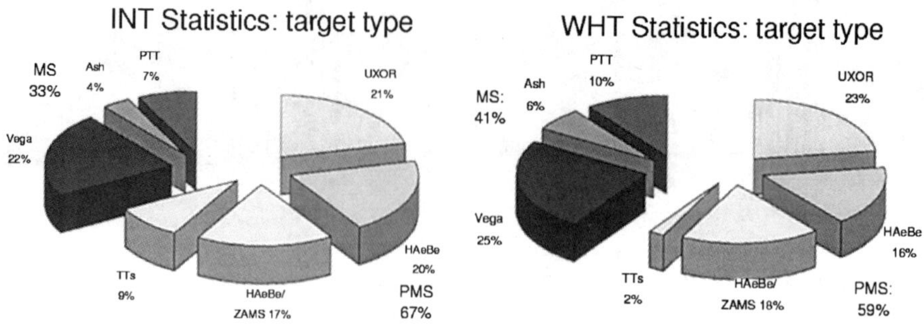

Figure 2. Statistics of target types.

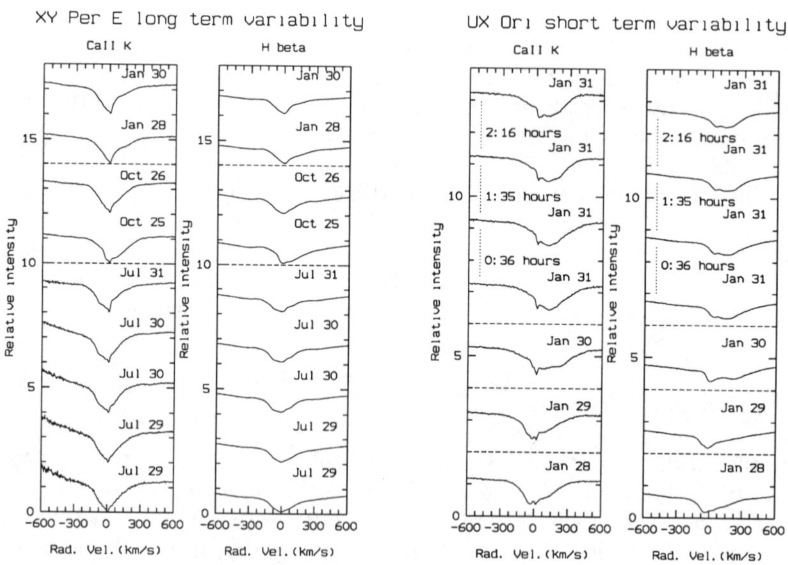

Figure 3. Examples of long- and short-term variations. The interval in hours between consecutive exposures is shown for the UX Ori short-term variations on 1999 January 31.

Table 1. List of observed stars. The number of high- and intermediate-resolution exposures is given

Name	Type	# WHT	# INT	Name	Type	# WHT	# INT
17 Sex	Ash	4	3	HR 26	PTT	2	2
24 CVn	Ash	4	3	HR 26-B	PTT	-	2
49 Cet	Vega	2	4	HR 419	PTT	1	-
51 Oph	Vega	8	8	HR 1847	Vega	-	7
AS 442	HAeBe	-	5	HR 1847-B	Vega	-	7
BD+31 643	Vega	5	8	HR 2174	Vega	1	7
BD+31 643-B	Vega	-	2	HR 2174-B	Vega	-	2
BF Ori	UXOR	4	9	HR 4757	PTT	3	4
BH Cep	HAeBe	8	17	HR 4757-B	PTT	2	4
BO Cep	T Tauri	-	13	HR 5422	PTT	2	4
BM And	T Tauri	-	11	HR 5422-B	PTT	-	4
CO Ori	UXOR	3	8	HR 9043	Vega	2	2
CQ Tau	UXOR	1	8	KK Oph	UXOR	5	10
CW Tau	T Tauri	-	7	KK Oph-B	UXOR	-	3
DK Tau	UXOR	-	7	λ Boo	Vega	13	8
DK Tau-B	UXOR	-	1	LkHα 200	T Tauri	-	6
DR Tau	UXOR	2	8	LkHα 234	HAeBe	-	9
HD 23362	Vega	1	1	LkHα 262	T Tauri	-	4
HD 23680	Vega	-	1	LkHα 262-B	T Tauri	-	2
HD 31293	HAeBe	1	-	MWC 297	HAeBe/ZAMS	-	1
HD 31648	HAeBe/ZAMS	4	7	NV Ori	HAeBe	1	8
HD 34282	HAeBe/ZAMS	2	8	PX Vul	HAeBe	-	9
HD 34700	Vega	2	7	R Mon	HAeBe	-	3
HD 58647	HAeBe/ZAMS	3	3	RR Tau	UXOR	3	8
HD 109085	Vega	4	7	RY Ori	T Tauri	1	7
HD 123160	Vega	6	8	RY Tau	UXOR	4	7
HD 141569	HAeBe/ZAMS	11	10	SV Cep	HAeBe	7	14
HD 142666	HAeBe/ZAMS	6	12	T Ori	UXOR	1	7
HD 142764	Vega	2	6	UX Ori	UXOR	10	15
HD 144432	HAeBe/ZAMS	4	11	UX Ori-N	UXOR	-	1
HD 150193	HAeBe	-	3	V1685 Cyg	HAeBe	-	1
HD 158352	HAeBe/ZAMS	-	5	V1685 Cyg-B	HAeBe	-	1
HD 163296	HAeBe/ZAMS	9	11	V1686 Cyg	UXOR	-	8
HD 179218	HAeBe/ZAMS	-	3	V346 Ori	HAeBe	-	8
HD 190073	HAeBe/ZAMS	2	8	V350 Ori	UXOR	-	7
HD 199143	HAeBe/ZAMS	2	8	VV Ser	HAeBe	11	18
HD 203024	HAeBe/ZAMS	-	1	VX Cas	HAeBe	3	11
HD 203024-B	HAeBe/ZAMS	-	1	VX Cas-B	HAeBe	-	1
HD 233517	Vega	4	3	VY Mon	HAeBe	-	3
HD 233517-B	Vega	-	1	WW Vul	UXOR	8	13
HK Ori	UXOR	1	8	XY Per E	HAeBe	9	11
HR 10	Ash	2	3	Z CMa	HAeBe	2	-

few of the secondaries show Li I absorption—which is an indication of youth. Figure 2 shows the statistics of the observed stars of each group with both INT and WHT data. Only one T Tauri was observed with the WHT because of the faintness of the selected stars of this subgroup.

One of the main aims of the observations is to monitor transient events that could be indicative for the presence of orbiting solid bodies in the circumstellar dust disks. Our spectra show the presence of blue- and redshifted absorption components in several lines of many of the stars observed for their PMS and Vega-type natures. Some examples are shown in other EXPORT contributions in this volume. As examples we show the results for XY Per E and UX Ori. XY Per E was observed during three different runs, while UX Ori was observed in two. Figure 3 shows the CaII K and Hβ line profiles as observed with the WHT on those dates. The spectra in Figure 3 show remarkable changes on timescales of hours, days, and months and they demonstrate the feasibility of detailed monitoring of the evolution of transient events in terms of their kinematics and strength.

Finally, it is well known that Hα emission is a good indicator of the presence of gas in the circumstellar environment. In the case of the PMS group most of the stars show, as expected, Hα in emission, and for 79% of these objects the line is clearly strongly variable. Only in three cases of the HAeBe/ZAMS subgroup Hα is pure absorption, though with a complex profile, HD 158352, HD 199143, and HD 203024. In the MS group only three stars show Hα emission: 51 Oph, HR 1847B, and HD 34700.

Acknowledgments. A. Mora is partly supported by Spanish grant ESP 98-1339.

References

Beust, H., et al. 1998, A&A, 338, 1015

Grady, C. A., et al. 2000, in Protostars and Planets IV, ed. V. Mannings, A. P. Boss, & S. S. Russell (Tucson: University of Arizona Press), p. 613

Lagrange, A. M., Backman, D. E., & Artymowicz, P. 2000, in Protostars and Planets IV, ed. V. Mannings, A. P. Boss & S. S. Russell (Tucson: University of Arizona Press), p. 639

Lindroos, K. P. 1986, A&A, 156, 223

Discussion

Carlos Eiroa: The amount of data collected by the EXPORT collaboration is huge. It is our opinion that these data can be of interest to astronomers outside the EXPORT collaboration. Therefore we will make the data available to the community once we find the best way.

Mario van den Ancker: Can you make a clear spectroscopic distinction between the main-sequence stars in your sample, e.g., HD 141569 and the Herbig Ae stars?

Alcione Mora: Yes, the PMS stars are much more active and/or peculiar than the MS stars.

William Herbst: I believe that you are grouping the stars as UXORs or HAeBes according to the amplitude of variation. Is that correct? Do you agree that there is no fundamental difference between the A-type UXORs and the HAeBe stars?

Alcione Mora: The UXORs are HAeBes or T Tauris, but we have separated them because the UXORs may be the progenitors of the β Pic-like stars. We agree that there is no possible evolutionary difference.

Disks, Planetesimals, and Planets
ASP Conference Series, Vol. 219, 2000
F. Garzón, C. Eiroa, D. de Winter, and T. J. Mahoney, eds.

Optical Photopolarimetry and Near-Infrared Photometry of Pre-Main-Sequence and Main-Sequence Objects

René D. Oudmaijer

Department of Physics and Astronomy, E C Stoner Building, Leeds LS2 9JT, UK

J. K. Davies

Joint Astronomy Centre, 660 N. A'ohoku Place, University Park, Hilo, Hawaii 96720, USA

C. Eiroa and J. Palacios

Dpto. Fisica Teorica, Facultad de Ciencias, Universidad Autonoma de Madrid, 28049 Cantoblanco. Madrid, Spain

F. Garzon

Instituto de Astrofísica de Canarias, E-38205 La Laguna, Tenerife, Spain

D. de Winter

TNO/TPD-Space Instrumentation, Stieltjesweg 1, PO Box 155, 2600 AD Delft, The Netherlands

EXPORT

Abstract. This contribution discusses the optical *UBVRI* broad-band photopolarimetry of the EXPORT sample obtained at the 2.5 m Nordic Optical Telescope and the near-infrared*JHK* photometry obtained at the 1.5 m Carlos Sánchez Telescope. The data, consisting of multi-epoch photopolarimetry of more than 70 pre-main-sequence and main-sequence stars, comprises the largest set of its kind.

1. Introduction

One of the main projects within the EXPORT proposal concerned the study of the circumstellar gas and dust around pre-main-sequence (PMS) and main-sequence (MS) objects thought to be accompanied by planets, or by planets in the process of formation (see Eiroa et al, this volume, p. 3). The data concerned are semi-simultaneous spectroscopy, photometry, and polarimetry. In simple terms, the spectroscopic data probe the gas closest to the stars, while the photopolarimetry probes gas and dust further out from the stars.

Although studies of individual objects have been carried out in the recent past, a large number of objects has not been studied in a systematic and statistical manner. This is mainly due to the large amounts of telescope time needed. Rectifying this omission was made possible by the allocation of the La Palma International Time to the EXPORT project. This contribution discusses the optical photopolarimetric and near-infrared (NIR) photometric data acquisition, details of the spectroscopic effort are found in Mora et al (this volume, p. 231).

The photopolarimetric data is of use in two ways. In the first place, the data can be used to flux calibrate the spectroscopic data. This enables both changes in emission line equivalent width and energy output to be traced. Secondly, combined polarimetry and photometry can reveal a great deal about the geometry of the circumstellar material. Stellar light that is scattered off circumstellar dust grains will result in the light being polarized. In the case of spherically symmetric shells, or face-on circular disks, no net polarization will be observed, as all polarization vectors cancel each other out. On the other hand, in the case of deviation from circular symmetry on the sky, a net polarization will be observed (see also Grinin, this volume, p. 216). However it is hard to decide, based on single epoch polarimetry, whether a polarized object has a disk-like dust configuration. This is because the separation between interstellar and circumstellar polarization is not trivial. A comparison with the so-called Serkowski law, which describes the normal behavior of interstellar polarization has to be made, but this method is not always conclusive. Multi-epoch polarimetry can resolve this question quite straightforwardly. Any variations in polarization must be due to circumstellar effects because interstellar dust clouds are not expected to vary within the epochs under consideration.

Polarimetric variability has been observed in several intermediate-mass PMS stars and is commonly referred to as the "UX Ori" phenomenon (Grinin et al. 1994). As an aside, the fact that the dust polarization of the UXORs is variable implies that the stars are surrounded by flattened dust structures. Interestingly, this could long ago have settled the question whether the intermediate-mass Herbig Ae/Be PMS objects are surrounded by disks or not. Instead a long debate over this issue has ensued (see, for example, Pezzuto et al 1997; Oudmaijer & Drew 1999 for overviews).

2. Observations and Data Reduction

The optical data were obtained using the TurPol photopolarimeter mounted on the 2.5 m Nordic Optical Telescope (NOT), La Palma, Spain. This instrument observes the light through a diaphragm, then, using a beam-splitter, the $UBVRI$ photons are recorded simultaneously on photomultipliers. Approximately 50% of the 16 observing nights, during four epochs in 1998 May, July, and October, and 1999 January, were photometric. Almost 70 targets were observed altogether, most of the objects more than four times in two epochs. The target acquisition was guided by the progress of the INT spectroscopy to ensure near-simultaneity. The photometry is accurate to within 0.03 mag, an error which is dominated by systematic effects rather than photon shot noise, as on average millions of photons were recorded to reach an accuracy of 0.05–0.10% in V-band polarization.

The NIR photometry was obtained during the same observing epochs as the NOT observations. In the May and October runs a NIR photometer was used, while in July and January a NIR camera was used. Sixty objects were observed during 13 nights, which were at least partly photometric. The resulting accuracies were about 0.05 mag with the single-channel photometer and of the order of two times worse in the case of the 2D NIR camera

3. Results

3.1. Near-Infrared Photometry

Of the 60 objects that were observed, 42 were PMS stars, massive Herbig Ae/Be stars and intermediate-mass T Tau stars. The remainder were stars beyond that phase: A- to B-type main-sequence, (Vega-type) objects and naked post-T Tau stars. The main results are discussed in a poster (see Mora et al., this volume, p. 411). Analysis of this dataset is still in progress, but the global characteristics of the sample can be summarized as follows. The majority of the "evolved" objects do not show evidence for circumstellar (hot) dust in the NIR bands; their spectral energy distributions up to 2 μm are still photospheric. In contrast, the majority of the PMS objects show excess emission, due to hot dust. Half of these latter objects are photometrically variable.

3.2. Optical Photopolarimetry

The main global property that we discuss here is the presence of polarimetric variability—which confirms the presence of a flattened structure around the objects. The presence of variability was investigated in the data using relatively simple statistical methods. The main result for the more than 15 Vega-type objects is that these systems are hardly polarized at all and are not variable to within the error bars.

The 15 objects that were classified as UXORs all showed significant variability, both on short (days) and long (months) timescales. This was not necessarily unexpected as UXORs by definition show this type of variability. Interestingly, around half of the remaining Herbig Ae/Be stars do show variability, albeit of mostly lower amplitudes.

In most cases, the polarimetry is anti-correlated with the photometry: the objects show larger polarization when they are fainter. This is generally interpreted in terms of aspherically distributed dust clouds orbiting the star. These give rise to more extinction in the line of sight, while the scattered light will not be blocked. This results in an increase of its relative contribution to the total observed light, giving rise to larger polarization (e.g., Grinin et al 1994).

Arguably, the distinction between "UXORs" (Herbig Ae/Be stars showing large variations in photometry and polarization) and Herbig Ae/Be stars becomes less pronounced based on results like these. Indeed, this large dataset indicates that the majority of the observed Herbig Ae/Be stars show evidence for non-spherical envelopes. Although there may have been a certain personal bias in the selection of the targets, it should be noted that most of the observable (northern) known Herbig Ae/Be stars have been observed in this project. Considering that the number of Herbig Ae/Be stars only number about one hundred (Thé et al. 1994), we speculate at this point that perhaps *all* massive PMS objects have aspherical structures. This is in line with the random orientations of any disks in the line of sight. If the disks are circular, then those disks oriented face on would not show any polarization/polarization variability. Inclined disks would give detectable polarizations. Of course, we need to analyze the data further, but such a conclusion, albeit tentative, illustrates the importance of observing a large sample of PMS stars to be able to reach general conclusions on the properties of this evasive class of object.

References

Grinin, V. P., et al. 1994, A&A, 292, 165

Oudmaijer, R. D., & Drew, J. E. 1999, MNRAS, 305, 166

Pezzuto, S., et al. 1997, ApJ, 485, 290

Thé, P. S., et al., 1994, A&AS, 104, 315

Disks, Planetesimals, and Planets
ASP Conference Series, Vol. 219, 2000
F. Garzón, C. Eiroa, D. de Winter, and T. J. Mahoney, eds.

ISO Observations of Pre-Main-Sequence and Vega-Type Stars

M. E. van den Ancker

Harvard-Smithsonian Center for Astrophysics, 60 Garden Street, MS 42, Cambridge, MA 02138, USA

Abstract. I present an overview of the results obtained by the *Infrared Space Observatory* (*ISO*) on circumstellar material in pre-main-sequence (PMS) stars. Results obtained for embedded YSOs, Herbig Ae/Be systems and T Tauri stars are reviewed and their connection to the disks around Vega-type systems is discussed. Although the gas contents of the PMS environment will also be discussed briefly, this review will mainly focus on the composition, mineralogy and evolution of dust in these systems, and the results will be compared to those found in other classes of objects, including Solar System comets.

1. Introduction

One of the most exciting developments in the study of pre-main-sequence and young main-sequence stars is the advent of space-based infrared observatories. The first infrared space mission, *IRAS*, discovered the presence of excess infrared emission above photospheric levels in many young stellar objects (YSOs) due to the presence of circumstellar disks or envelopes containing small dust particles. More surprisingly, *IRAS* also discovered infrared excesses in several nearby main-sequence stars. These Vega-type stars (after the prototype Vega; Aumann et al. 1984) have infrared excesses that usually start at longer wavelengths (typically > 25 μm) than in YSOs, indicative of much cooler dust temperatures. A particularly intriguing case is the A-type star β Pictoris, exhibiting both an infrared excess and a circumstellar disk seen in scattered light, which may form the evolutionary link between YSOs and Vega-type stars. Thirteen years after the *IRAS* mission, its successor saw first light: the *Infrared Space Observatory*[1] (*ISO*; Kessler et al. 1996). Unlike the all-sky *IRAS* mission, *ISO* could do pointed observations and offered multiple modes of operation. It gave us our first glimpse of the sky at wavelengths longer than 100 μm and revealed the richness of the infrared spectrum.

In this review, I will discuss the results obtained by *ISO* on pre-main-sequence and Vega-type stars, focusing on the the evolution of circumstellar dust. In Section 2 I will briefly introduce the capacities of the instruments on

[1] Based on observations with *ISO*, an ESA project with instruments funded by ESA Member States (especially the PI countries: France, Germany, the Netherlands, and the United Kingdom) and with the participation of ISAS and NASA.

board *ISO*. The next section is devoted to a qualitative discussion of the infrared spectra of young stars and outlines a possible evolutionary scenario for circumstellar dust. Section 4 deals with the originating region of the infrared radiation from pre-main-sequence stars, whereas Section 5 reviews the new picture of Vega-type stars that *ISO* had given us and their relation to young stars. In the last section I will summarize the main conclusions of this review and will look forward to the new questions that may be answered by future infrared missions.

2. The *Infrared Space Observatory*

ESA's *Infrared Space Observatory* consisted of a 60 cm cryogenically cooled telescope with four focal-plane instruments, operating at wavelengths from 2.4 to 240 microns. Launched in 1995 November, it collected more than 26 000 scientific observations before it finally ran out of liquid helium in 1998 April, after a hugely successful mission that lasted eight months longer than expected.

The first of the four instruments, the infrared camera (ISOCAM; Cesarsky et al. 1996) covered the 2.5–18 μm band with two different 32×32 pixel detectors. Each detector had several discrete band-pass filters as well as circular variable filters (CVFs) with a resolution of about 35. The second focal-plane instrument, the photopolarimeter (ISOPHOT; Lemke et al. 1996), covered the largest wavelength range of the *ISO* instruments (2.5–240 μm). Its scientific capabilities included multi-filter and multi-aperture photometry, polarimetry, imaging, and low-resolution ($R \approx 100$) spectrophotometry in the 2.5–4.9 and 5.8–11.6 μm wavelength ranges.

The other two instruments on board *ISO* were two complementary spectrometers. The short-wavelength spectrometer (SWS; de Graauw et al. 1996) consisted of a scanning spectrograph with a spectral resolution ranging from 1200 to 2400 covering the wavelength range of 2.4–45 μm. By inserting Fabry–Pérot filters, the resolution of the instrument could be enhanced by a factor 20 in the wavelength range from 11.4 to 44.5 μm. Similar to SWS, the *ISO* long-wavelength spectrometer (LWS; Clegg et al. 1996) consisted of a scanning spectrograph in the 43–198 μm range, with a resolution of either 150–200, or 6800–9700 in Fabry–Pérot mode.

3. Spectral Evolution of Pre-Main-Sequence Stars

3.1. Massive YSOs

The *ISO* mission allowed us for the first time to explore those parts of the infrared spectrum that are hidden from view by the Earth's atmosphere. Because of this, major steps forward have been made in our understanding of the star formation process. In the standard picture of star formation (Shu, Adams, & Lizano 1987) four distinct phases can be distinguished. First, a number of dense cores develop within a molecular cloud. At some point, the dense cores collapse to form embedded protostars which accumulate material by accretion from the surrounding cloud. As the embedded protostars accrete mass, they also lose mass due to a strong stellar wind, which eventually leads to the dispersion of the reservoir of mass from the protostellar envelope and the end of accretion. The former protostar, now a pre-main-sequence star surrounded by a circumstellar disk, begins to contract slowly, increasing its central temperature until hydrogen ignition takes place and it has become a stable star on the main sequence.

Spectra taken with all four focal-plane instruments on board *ISO* have revealed a large variety of properties of gasses, dust, and ices around YSOs that can be well interpreted in the scenario outlined above. The first stages of the spectral evolution of pre-main-sequence objects are illustrated in Figure 1, where we show *ISO*-SWS spectra of three massive YSOs, NGC 7538 IRS9, AFGL 2591, and S106. The infrared spectrum of NGC 7538 IRS9 (Whittet et al. 1996) consists of a smooth continuum with broad, strong absorption bands due to silicates and ices. Frozen H_2O, CO, CO_2, CH_4, CH_3OH, and a band around 6.8 μm whose carrier remains unidentified are all commonly detected in the lines of sight towards embedded YSOs and may be regarded as a strong indicator of youth.

The infrared spectrum of AFGL 2591 (van der Tak et al. 1999) already shows striking differences with that of NGC 7538 IRS9. Although the massive young star is embedded to about the same degree as NGC 7538 IRS9, as evidenced by the similarity of the depth of their 10 μm silicate profile, the relative prominence of the ice absorption of some species, such as H_2O and CO_2 is diminished, whereas the absorption bands of CO, CH_4, and CH_3OH have completely vanished. In contrast to NGC 7538 IRS9, higher resolution *ISO*-SWS spectra of AFGL 2591 show a large number of sharp absorption lines due to gas-phase CO, CO_2, and H_2O (Helmich et al. 1996; van Dishoeck et al. 1996). The temperatures derived from these gaseous molecules amount to a few 100 K, well above the evaporation temperature of most ices.

The difference between the infrared spectra of NGC 7538 IRS9 and AFGL 2591 can easily be understood by *assuming* that the former is at an earlier evolutionary stage than the latter. Whereas the NGC 7538 IRS9 inner envelope consists of mostly unprocessed interstellar material, AFGL 2591 has already heated the inner region of its envelope to above the evaporation temperature of most ices and has formed a so-called "hot core" region. By analyzing gas/ice abundances in a larger sample of YSOs, van Dishoeck et al. (1999) found a good correlation between gas temperature and gas/ice abundance, showing that this process of the evaporation of ice mantles and the heating of the YSO's envelope is more general than illustrated here by the two examples of NGC 7538 IRS9 and AFGL 2591, and may in fact be a good indicator of age for embedded YSOs.

A second difference between the spectra of NGC 7538 IRS9 and AFGL 2591 in Figure 1 is the presence of emission lines due to H_2, [Si II] and [S I] in AFGL 2591, which appear absent in NGC 7538 IRS9. In particular the detection of [S I] 25.25 μm is significant, since this line can only be formed with detectable intensity in regions of shocked gas (van den Ancker et al. 2000b). This shocked gas is most probably the result of the interaction of AFGL 2591's outflow with the surrounding cloud and traces the clearing of the circumstellar envelope of a young star.

Compared to NGC 7538 IRS9 and AFGL 2591, the spectrum of S106 (van den Ancker et al. 2000a), the third massive YSO shown in Figure 1, looks very

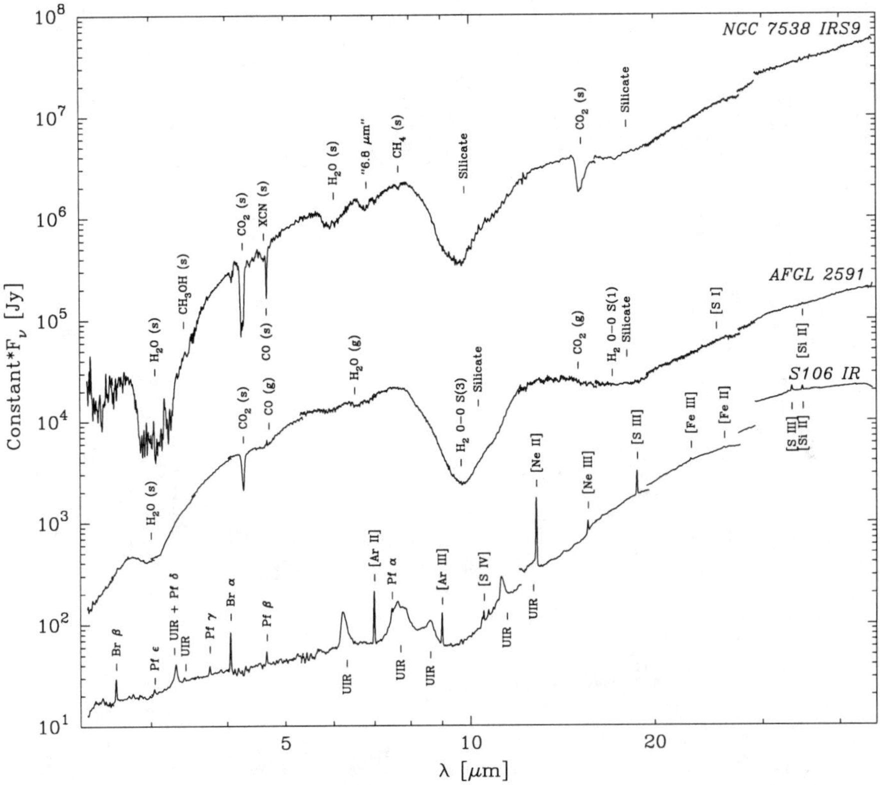

Figure 1. *ISO*-SWS spectra of the heavily embedded massive YSO NGC 7538 IRS9 (Whittet et al. 1996), the intermediately embedded massive YSO AFGL 2591 (van der Tak et al. 1999), and the bipolar nebula S106 (van den Ancker et al. 2000a), illustrating the spectral evolution of high-mass YSOs.

different. It consists of a smooth[2] continuum with a large number of sharp, strong emission lines due to recombination in the Brackett and Pfund series of H I, and due to numerous fine-structure transitions of ionic species. In addition to this, the unidentified infrared (UIR) emission bands, often attributed to polycyclic aromatic hydrocarbons (PAHs)[3], are present at 3.3, 3.4, 6.2, 7.6, 7.8, 8.6, 11.3, and 12.7 μm. The central star in S106 has developed a strong stellar

[2]The "jumps" seen at discrete wavelengths in the spectrum are due to the fact that the SWS uses entrance apertures of different sizes for different wavelength regions, so for a source that is not point-like, such as S106, one may receive more flux where such a change in aperture occurs.

[3]Note however that ISOCAM CVF images did not retrieve the dependence of the relative strength of the UIR bands on UV field expected for PAHs (Uchida et al. 2000, and references therein).

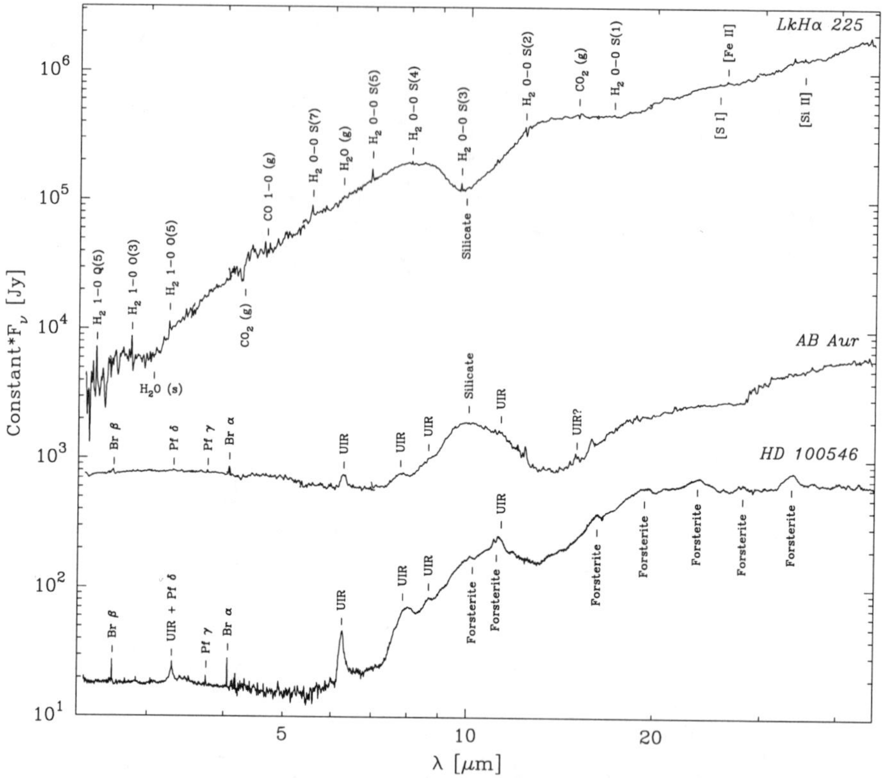

Figure 2. *ISO*-SWS spectra of the embedded intermediate-mass YSO LkHα 225 (van den Ancker et al. 2000b), and the Herbig Ae stars AB Aur (van den Ancker et al. 2000c), and HD 100546 (Malfait et al. 1998), illustrating the spectral evolution of intermediate-mass pre-main-sequence stars.

wind, as evidenced by the recombination lines, which has cleared most of the its natal cloud. The high-ionization species in the spectrum show that it is now surrounded by a substantial H II region. The UIR emission as well as emission due to H_2 and other gas-phase molecules indicate that a so-called photodissociation region (PDR; see Hollenbach & Tielens 1999 for a comprehensive review), a mostly neutral region whose surface is photoelectrically heated under the influence of UV radiation, is present as well. The presence of this PDR emission shows that some regions of mostly neutral material, perhaps the remnants of the highest-density clumps in the original cloud, remain.

3.2. Intermediate-Mass Stars

So far our discussion of YSO spectral evolution has focused on massive objects. The infrared spectrum of these objects is typically dominated by their wide circumstellar environment and in fact evidence for the existence of protoplanetary disks around massive YSOs is weak. We now focus on intermediate-mass

(2–10 $M_\odot$) pre-main-sequence stars, the Herbig Ae/Be (HAeBe) stars. In Figure 2 we again show the *ISO*-SWS spectra of three of these, the embedded intermediate-mass YSO LkHα 225, and the two relatively isolated Herbig Ae/Be systems AB Aur and HD 100546.

Although there are some indications that deeply embedded intermediate-mass YSOs may have a different method of thermally processing their circumstellar ices (c.f. the case of Elias 29, Boogert et al. 2000), the overall spectra of embedded intermediate-mass YSOs appear similar to those of high-mass YSOs. For example, the spectrum of LkHα 225 (van den Ancker et al. 2000b) shown in Figure 2 resembles that of many moderately embedded massive YSOs. Silicate absorption, as well as absorption due to water ice are visible and gas-phase CO, CO_2, and H_2O are present, indicative of a relatively evolved, warm core. Strong shock-excited emission lines of H_2, [S I], [Fe II], and [Si II] are present, again most probably related to the presence of an outflow from this star.

However, the next stage in the evolution of an intermediate-mass YSO, represented here by the well-studied Herbig Ae star AB Aur (van den Ancker et al. 2000c), yields a very different spectrum from that in the corresponding stage in the life of a massive star. The lines from the Pfund and Brackett series of H I, indicative of a strong wind, are still present, but the spectrum lacks the multitude of atomic fine-structure lines seen in the spectrum of S106, indicating that the central star of AB Aur does not emit sufficient UV photons to create a substantial H II region.

Several UIR bands can be seen in the spectra of AB Aur and HD 100546 shown in Figure 2. An analysis of all spectra of Herbig Ae/Be stars found in the *ISO* data archive shows that the UIR bands are present in about 50%, without a strong preference for a certain spectral type. Whereas undoubtedly in some cases the observed UIR emission arises in filamentary PDR structures included in the rather large beam of the *ISO* spectrometers (e.g., $20'' \times 14''$ for the SWS), ground-based imaging as well as ISOCAM CVF data show that more commonly the UIR emission arises close to the star. At present it is not clear why some Herbig Ae/Be stars show this compact UIR emission, whereas others with apparently very similar properties do not. However, the absence of UIR bands in some Herbig Ae/Be stars does show that models depending on the presence of very small dust grains to explain the observed near-infrared excess in Herbig Ae/Be stars will not be successful in all cases.

The answer to the question why only some HAeBes show UIR emission may come from the observation that a few objects (e.g., AB Aur, Figure 2) exhibit strong UIR bands due to C–C stretching modes, whereas the bands due to C–H bonds are weak or absent, suggesting that either the PAHs have been stripped of most of their hydrogen atoms, or that they are rather large (> 100 C atoms). The presence of the rare UIR bands at 3.43 and 3.53 μm in a small number of Herbig stars (most noticeably HD 97048 and Elias 1; Van Kerckhoven et al. 1999) may reflect an even further chemical processing stage, in which nanodiamonds, not unlike those found in meteorites in our own solar system, have been formed (Guillois¡ Ledoux, & Reynaud 1999).

Although C-rich dust is certainly present, the general appearance of the infrared spectrum of most Herbig Ae/Be stars is determined by O-rich dust particles. Broad emission features in the 8–12 and 17–19 μm ranges due to silicate O–Si–O bending and stretching modes are found in a large fraction of Herbig Ae/Be stars. In some more embedded objects they are seen in absorption. At longer wavelengths, emissions due to water ice and hydrous silicates have been reported in some Herbig stars (Malfait et al. 1998, 1999). A broad spectral feature ranging from 14 to 38 μm was observed by van den Ancker et al. (2000c) in the Herbig Ae/Be stars AB Aur and HD 163296, which could due to a blend of the 28 μm resonance of iron oxide with the 19 μm resonance of amorphous silicates. Interestingly, iron or iron oxide at $\sim$ 1000 K can also naturally explain the near-infrared excess observed in virtually all HAeBes.

The fact that the silicate features are seen in emission suggests that the emitting material is optically thin at 10 μm in the objects where this is the case. This puts strong constraints on the distribution of dust and its temperature structure. However, millimeter continuum observations imply that the masses contained in these disks can be substantial, leading some authors to suggest that the observed silicate emission could arise in an extended optically thin region above the dense midplane of the disk (Natta et al. 1999; Bouwman et al. 2000.

In most stars the 9.7 μm silicate emission appears broad and smooth, such as in AB Aur. It is essentially the inverse of the very smooth silicate features seen in absorption towards all embedded YSOs (e.g., AFGL 2591 and LkHα 225 in Figures 1 and 2). Comparison with laboratory determinations of optical constants of silicates shows that the lattice structure of these silicates must be amorphous, i.e., relatively unordered. Although most common, not all Herbig stars show the very smooth silicate emission profiles seen in AB Aur. An especially spectacular example is that of the Herbig B9 star HD 100546 (Waelkens et al. 1996; Malfait et al. 1998), shown as the bottom curve in Figure 2. Here we see that the 8–12 μm silicate feature shows a lot of sub-structure, and many broad bands with strengths up to that of the underlying continuum appear in emission at the longer wavelengths of the SWS wavelength range. These sharper features are indicative of silicates in which the lattice structure is more crystalline. Note that detailed modeling of the dust composition shows that even though the crystalline silicates dominate the appearance of the infrared spectrum, the bulk of the silicates is still in the amorphous form (Malfait et al. 1999). Comparison with lab data shows that the peaks seen in the HD 100546 spectrum must be due to the mineral forsterite (Mg_2SiO_4). This is exactly the same material as is found in comets in our own Solar System (Crovisier et al. 1997), suggesting that the processing of circumstellar dust in the primitive Solar System bears similarity to that in HAeBes.

3.3. T Tauri Stars

The SWS and LWS were sufficiently sensitive to allow the study of massive YSOs at distances up to several kiloparsecs, as well as the nearest intermediate-mass stars, but they lacked the sensitivity to study even the nearest T Tauri stars, low-mass pre-main-sequence stars which may represent a better analog of the protosolar nebula than the Herbig Ae/Be stars. However, the spectroscopic modes of the more sensitive ISOPHOT and ISOCAM instruments, although limited to wavelengths smaller than 18 μm, also allowed valuable information on the dust composition in the nearest young stars of solar mass to be gathered. Based on ISOPHOT data, Natta, Meyer, & Beckwith (2000) show that the amorphous silicate emission features exhibited by most Herbig Ae stars are also found around all nine T Tauri stars studied in the Chamaeleon I dark cloud. Possibly due to the higher requirements with respect to signal to noise and spectral resolution, crystalline silicates still await their first detection in a low-mass YSO. Although a detailed study of the dust composition of T Tauri objects awaits more sensitive mid-IR spectroscopic instruments, such as the infrared spectrograph on board NASA's *Space Infrared Telescope Facility* (*SIRTF*), there are at the time of writing no indications that the spectral evolution sketched in the previous subsection for intermediate-mass YSOs would not equally apply to solar-mass pre-main-sequence stars.

4. Disks and Envelopes around Young Stars

The *IRAS* mission detected infrared radiation, believed to originate in circumstellar disks or envelopes, from many young stellar objects. However, its poor spatial resolution inhibited a comprehensive study of young clusters. Higher spatial resolution at wavelengths not hindered by interstellar extinction allowed *ISO*'s camera to provided a new complete census of nearby star-forming regions, resulting in a doubling of the known number of YSOs in each (Olofsson et al. 1999). The new cluster members are mainly at the low-luminosity end of the luminosity distribution, indicating that the initial mass function continues to rise slowly (with index −0.2) until brown dwarf masses.

Preliminary results of an ISOPHOT study of 97 T Tauri stars in five young clusters have been reported by Robberto et al. (1999). They found the fraction of stars with 60 μm excesses to decrease from around 60–70% in Taurus, ρ Oph and the R CrA region to below 20% in the older Cham I and TW Hydrae associations. More than 80% of the stars in these clusters that exhibit near-infrared excess emission also show far-infrared excesses. In contrast, only one star that has no near-infrared excess has one at 60 μm. No 60 μm excesses were detected in the oldest clusters studied by Robberto et al., suggesting that for low-mass stars the dust in disks between 0.3 and 3 AU disappears on timescales of $\sim$ 10 Myr.

A long-standing problem in studies of pre-main-sequence stars has been the question of whether the observed infrared excesses arise in a relatively stable circumstellar disk or in an infalling envelope. This question can be answered by studying the spatial extent of sources in the infrared. Ábrahám et al. (2000) observed seven Herbig Ae/Be stars at mid-infrared wavelengths with ISOPHOT and concluded that at $\lambda < 25$ μm the observed emission mainly arises from a compact area, whereas at longer wavelengths it is spatially extended. At wavelengths longer than 100 μm, the emission is never dominated by the Herbig Ae/Be stars, but by dust cores of about $1'$ in size. These results show that at least in these cases the observations exclude a "disk-only" geometry and a (not necessarily spherical) envelope needs to be present as well. They do not exclude the possibility that in some cases only a disk remains. Further study of a sample that also contains Herbig stars located in relative isolation is required to provide the definitive answer to this problem.

Although the infrared continuum will be dominated by the circumstellar dust, the gas in the disk will also produce spectral signatures in the infrared.

The pure rotational lines of H_2, all located in the mid infrared, are especially interesting since they directly probe the dominant gas component in the disk and hence allow us to measure disk masses without relying on an essentially unknown CO/H_2 ratio. For most YSOs observed with *ISO* the observed H_2 appears too warm (typically 500 K) to be associated with the circumstellar disk: it probably arises in the loose stellar surroundings and is associated with either PDRs or with shocked gas in outflows (e.g., van den Ancker et al. 2000a,b). However, van Dishoeck et al. (1998) report the detection of the H_2 0–0 S(0) and S(1) lines toward the isolated Herbig Ae star HD 163296. The implied temperature of only 150 K, the disappearance of the H_2 emission off source, and the absence of ro-vibrational H_2 emission in ground-based imaging all argue in favor of an interpretation in which the H_2 lines observed in HD 163296 originate in the circumstellar disk. The total observed gas mass in this object is 0.007 $M_{\odot}$. The value of 35 for the gas-to-dust ratio in the HD 163296 disk that is derived by combining this gas mass with the dust mass in the HD 163296 computed by Bouwman et al. (2000) is surprisingly close to the canonical value of 100 for the gas/dust ratio in the interstellar medium.

5. Vega-Type Stars

One of the most unexpected results of the *IRAS* mission was the discovery of an infrared excess due to the presence of circumstellar dust around several nearby main-sequence stars (Aumann et al. 1984). Spectacular images showing an edge-on disk around the relatively young main-sequence star β Pictoris (Smith & Terrile 1984) raised the question whether these disks are a remnant of the star formation process, or represent the extrasolar equivalents of the Kuiper Belt in our own Solar System. However, the *IRAS* survey and subsequent ground-based follow-up research lacked the sensitivity to determine what fraction of nearby stars show the Vega phenomenon.

Using the *ISO* photometer, Dominik et al. (1999) performed a deep survey on a volume-limited sample of nearby main-sequence stars of spectral types A, F, G, and K. They found infrared excesses in approximately 21% of nearby stars. However, Vega-type phenomena appear much more common in A- and F-type than in later-type stars. This need not necessarily reflect a different disk mass, but may be due to a mixture of disk illumination and time evolution. Based on the same dataset, Habing et al. (1999) report a distinctly bimodal distribution of the frequency of dust disks with age: more than 50% of stars younger than 300 Myr have infrared excesses at 25, 60, or 170 μm, while most stars older than 400 Myr do not. Ninety percent of the disks disappear when the star is between 300 and 400 Myr old. These timescales are similar to those derived for the final clean-up phase of our own Solar System. This result clearly shows that in most cases the Vega-type phenomenon represents remnants of the star formation process, and that the vicinity of the Sun hosts a considerable population of relatively young stars.

However, not *all* debris disks decay after 400 Myr. In about 10 per cent of the cases circumstellar dust persists for longer than 400 Myr. At present is is unclear what causes these disks to live longer than most. A hint to the answer may come from the detection of a 60 μm excess in ρ^1 Cnc (Dominik et al. 1998). Around this 4 Gyr-old star both a planet and a disk are present and it is not inconceivable that the presence of a third body may be required to keep a circumstellar disk stable on long timescales.

Soon after their discovery it was realized that the timescale for the removal of dust particles around Vega-type stars is much smaller than their age: radiation pressure will push the small dust particles out of the system, whereas Poynting–Robertson drag will bring the larger ones in to where they are evaporated. Therefore they must be re-supplied, at least for some fraction of the stellar lifetime. Possibilities that have been speculated upon to provide the mechanism for the replenishment of small dust grains include collisions between larger bodies or the evaporation of extra-solar comets.

Indications of the origin of the dust seen in the Vega-type systems may come from infrared spectroscopy. Solar System comets show a number of distinctive spectral signatures in the infrared due to the presence of crystalline silicate dust (e.g. Crovisier et al. 1997). Ground-based spectroscopy of the β Pic system in the 10 μm atmospheric window seemed to suggest a high degree of similarity of the dust seen in this system to that in comets (Telesco & Knacke 1991; Knacke et al. 1993). However, Pantin, Waelkens, & Malfait (1999) report an absence of the expected mid-infrared crystalline silicate features in the *ISO*-SWS spectra they obtained of β Pic. At present it is unclear whether comets need to be discarded as the source of dust in Vega-type systems, or that the in many ways unusual β Pic disk has a different composition from most Vega-type stars. Infrared spectroscopy of fainter, more typical, Vega-type stars with *SIRTF* will be able to clarify this issue within the coming years.

6. Concluding Remarks

The *Infrared Space Observatory* has yielded great new insights in the the evolution of disks into planetesimals and planets. For the first time we are able to analyze the chemical composition and evolution of circumstellar dust, and only after the *ISO* mission has it been realized how important solid-state resonances are for the interpretation of infrared energy distributions. In fact the flux in all four *IRAS* bands, centered at 12, 25, 60, and 100 μm, can be dominated by solid-state emission features, demonstrating the necessity of infrared spectroscopy to interpret the energy distributions of pre-main-sequence objects.

One of the most exciting *ISO* results has been the discovery of crystalline silicates, such as found in our own Solar System, in some Herbig Ae/Be stars. The fact that in embedded YSOs the silicates are invariably amorphous shows that for intermediate-mass stars, the transition from amorphous to crystalline must occur in the HAeBe phase of evolution. The mechanism responsible for this transition is still unclear. Laboratory experiments show that crystallization of silicates may be achieved by either a slow annealing of warm grains, or by heating the dust grains to temperatures above $\sim$ 1000 K. Since the crystalline dust in Herbig Ae/Be stars is observed to have much lower temperatures, substantial mixing in the disk would be required in this scenario. However, a study of *ISO* spectra of post-AGB stars by Molster et al. (1999) shows that in these objects crystallization occurs at much lower temperatures over long timescales. They suggest that the same low-temperature annealing process at work in these stars may also be responsible for the crystallization of silicates in young stars. If proven, this same process will be responsible for the creation of crystalline silicates in Solar System comets, a conclusion which ultimately may alter our view of the origin of our own Solar System.

ISO has also left us with a feeling for the richness and variety of the infrared spectra of young stars, resulting in the qualitative picture of spectral evolution sketched in Section 3. However, a quantitative conformation of this picture awaits bigger samples and in particular its extension towards solar-mass young stars. Indeed, van den Ancker et al. (2000c) remark upon the fact that stellar ages as derived from the position in the Hertzsprung–Russell diagram do not show a one-to-one correlation with degree of crystallinity. The fact that in a typical young cluster only half of the pre-main-sequence objects shows an infrared excess (e.g. Beckwith et al. 1990), suggests that this may be due to greatly differing timescales for disk evolution from object to object. The lack of correlation between submm spectral index, an indicator for grain growth, and degree of crystallinity noted by these authors is more puzzling. Apparently, these two processes are not necessarily coupled, suggesting that processes other than stellar mass and age, perhaps the environment of the star, are important.

The discovery that the Vega-type phenomenon is common in relatively young main-sequence stars, whereas it is rare in stars older than 400 Myr has considerably strengthened the connection between pre-main-sequence and Vega-type stars. However, the final proof of an evolutionary link between these two categories of objects still remains to be made and most probably will be made soon: the last of NASA's great observatories, *SIRTF*, is scheduled to be launched in the near future and, with its greatly enhanced sensitivity, promises to continue the path of discoveries that has been explored by *ISO*.

References

Ábrahám, P., Leinert, C., Burkert, A., Henning, T., & Lemke, D. 2000, A&A 354, 965

Aumann, H. H., Gillett, F. C., Beichmann, C. A., de Jong, T., Houck, J., Low, F. J., Neugebauer, G., Walker, R., & Wesselius, P. R. 1984, ApJ, 278, L23

Beckwith, S. V. W., Sargent, A. I., Chini, R. S., & Güsten, R. 1990, AJ, 99, 924

Boogert, A. C. A., Tielens, A. G. G. M., Ceccarelli, C., Boonman, A. M. S., van Dishoeck, E. F., Keane, J. V., Whittet, D. C. B., & de Graauw, Th. 2000, A&A, in press

Bouwman, J., de Koter, A., van den Ancker, M. E., & Waters, L. B. F. M. 2000, A&A, in press

Cesarsky, C. J., et al. 1996, A&A, 315, L32

Clegg, P. E., et al. 1996, A&A, 315, L38

Crovisier, J., & et al. 1997, Science, 275, 1904

de Graauw, Th., et al. 1996, A&A, 315, L49

Dominik, C., et al. 1999, in The Universe as Seen by ISO, ed. P. Cox & M. F. Kessler, ESA SP-427, p. 203

Dominik, C., Laureijs, R. J., de Muizon, M., & Habing, H. J. 1998, A&A, 329, L53

Guillois, O., Ledoux, G., & Reynaud, C. 1999, ApJ, 521, L133

Habing, H. J., Dominik, C., de Muizon, M., Kessler, M. F., Laureijs, R. J., Leech, K., Metcalfe, L., Salama, A., Siebenmorgen, R., & Trams, N. 1999, Nat, 401, 456

Helmich, F. P., et al. 1996, A&A, 315, L173

Hollenbach, D. J., & Tielens, A. G. G. M. 1999, Rev. Mod Phys., 71, 173

Kessler, M. F., et al. 1996, A&A, 315, L27

Knacke, R. F., Fajardo-Acosta, S. B., Telesco, C. M., Hackwell, J. A., Lynch, D. K., & Russell, R. W. 1993, ApJ, 418, 440

Lemke, D., et al. 1996, A&A, 315, L64

Malfait, K., Waelkens, C., Waters, L. B. F. M., Vandenbussche, B., Huygen, E., & de Graauw, Th. 1998, A&A, 332, L25

Malfait, K., Waelkens, C., Bouwman, J., de Koter, A., & Waters, L. B. F. M. 1999, A&A, 345, 181

Molster, F. J., et al. 1999, Nat, 401, 563

Natta, A., Meyer, M. R., & Beckwith, S. V. W. 2000, ApJ, in press

Natta, A., et al. 1999, A&A, 350, 541

Olofsson, G., et al. 1999, in The Universe as Seen by ISO, ed. P. Cox & M. F. Kessler, ESA SP-427, p. 459

Pantin, E., Waelkens, C., & Malfait, K. 1999, in The Universe as Seen by ISO, ed. P. Cox & M. F. Kessler, ESA SP-427, p. 385

Robberto, M., Meyer, M. R., Natta, A., & Beckwith, S. V. W. 1999, in The Universe as Seen by ISO, ed. P. Cox & M. F. Kessler, ESA SP-427, p. 195

Smith, B. A., & Terrile, R. J. 1984, Science, 226, 1421

Shu, F. H., Adams, F. C., & Lizano, S. 1987, ARA&A, 25, 23

Telesco, C. M., & Knacke, R. F. 1991, ApJ, 372, L29

Uchida, K. I., Sellgren, K., Werner, M. W., & Houdashelt, M. L. 2000, ApJ, 530, 187

van den Ancker, M. E., Bouwman, J., Wesselius, P. R., Waters, L. B. F. M., Dougherty, S. M., & van Dishoeck, E. F. 2000c, A&A, 358, 339

van den Ancker, M. E., Tielens, A. G. G. M., & Wesselius, P. R. 2000a, A&A, in press

van den Ancker, M. E., Wesselius, P. R., & Tielens, A. G. G. M. 2000b, A&A, 355, 194

van der Tak, F. S., van Dishoeck, E. F., Evans, N. J., Bakker, E. J., & Blake, G. A. 1999, ApJ, 522, 991

van Dishoeck, E. F., et al. 1999, in The Universe as Seen by ISO, ed. P. Cox & M. F. Kessler, ESA SP-427, p. 437

van Dishoeck, E. F., et al. 1996, A&A, 315, L349

van Dishoeck, E. F., Thi, W. F., Blake, G. A., Mannings, V., Sargent, A. I., & Mundy, L. G. 1998, in ISO's View on Stellar Evolution, ed. L. B. F. M. Waters et al. (Ap&SS 255, 77)

Van Kerckhoven, C., Tielens, A. G. G. M., & Waelkens, C. 1999, in The Universe as Seen by ISO, ed. P. Cox & M. F. Kessler, ESA SP-427, p. 421

Waelkens, C., et al. 1996, A&A, 315, L245

Whittet, D. C. B., et al. 1996, A&A, 315, L357

Discussion

Antonella Natta: What is the basis for your statement that the crystalline silicates can be produced at low temperature?

Mario van den Ancker: This statement comes from the fact that in evolved systems (AGB and post-AGB stars), crystalline silicates are not observed in systems where the disk has a temperature above or at the glass temperature, whereas in much older systems, in which the temperature of the circumstellar silicates is only a few hundred K, they are present. No mechanism is known that will heat the disks in between. therefore, in these systems, crystallization of silicates can occur, on longer timescales ($\sim 10^6$–10^7 yr), at low temperatures. In analogy with this, we suggest that the same mechanism is responsible for crystallization of silicates in the disks of young stars.

Keith Horne: The laboratory work of Hollenbeck and Nuth (1998) suggests that the annealing time for Mg-rich olivine smokes becomes essentially infinite for $T < 1000$ K. what mechanism(s) do you propose to account for the presence of forsterite features in some Herbig Ae/Be stars?

Mario van den Ancker: Currently, the mechanism is not clear. But observations show that annealing does occur in circumstellar disks with $T \ll 1000$ K. Possibly, irradiation by cosmic rays could anneal olivines over long timescales, or some other mechanism that we do not yet understand.

Robert Fisher: Do you see a way to easily distinguish between a HAeBe star and a Vega-like star in the spectra?

Mario van den Ancker: I think the distinction between the two is mainly a matter of semantics. There definitely are stars that have already arrived at the ZAMS and are spectroscopically indistinguishable from HAeBes. But maybe it would be better to *define* Vega-type stars as having dusty disks without Hα emission, in which case the two would be easily distinguishable observationally.

Disks, Planetesimals, and Planets
ASP Conference Series, Vol. 219, 2000
F. Garzón, C. Eiroa, D. de Winter, and T. J. Mahoney, eds.

ISO-SWS Observations of EXPORT Targets

J. Palacios

Universidad Autónoma de Madrid, Facultad de Ciencias, Campus de Cantoblanco, E-28043 Madrid, Spain

P. R. Wesselius

SRON-Groningen, Landleven 12, 9747 AD Groningen, The Netherlands

EXPORT

Abstract. We present a first analysis of the *ISO*-SWS spectra of the pre-main-sequence and Vega-type stars observed during the 1998 La Palma International Observing Time program. PMS stars generally show excesses in the whole SWS spectral range, while the excess in Vega-type stars normally appears at wavelengths longer than 10 μm. PMS stars normally show silicate emission, while most of the Vega-type stars do not. The observed silicate profiles are similar to those observed in cometary silicates. The SWS results, together with EXPORT ground-based data, could reflect the evolution of circumstellar disks from PMS to Vega-type stars.

1. Introduction

Most of the EXPORT 1998 International Time of the Canary Islands Observatories was devoted to optical spectroscopy and near-IR and optical photopolarimetry of a large sample of main-sequence (MS) and pre-main-sequence (PMS) objects. Such observations are very suitable for the study of physical properties of the protoplanetary gaseous and dusty disks around these stars. In addition, the mid IR provides a lot of information on the dusty component of the disks, due to the expected grain temperatures and to the solid-state spectral features found in this wavelength regime. In this respect, the Short Wavelength Spectrometer (SWS, de Graauw et al. 1996) on board on the *Infrared Spatial Observatory* (*ISO*[1], Kessler et al. 1996) has obtained very interesting results. Thus, in order to complement the EXPORT ground-based data we decided to analyze the available SWS spectra of the EXPORT targets.

Among the several solid-state features present in the SWS spectral range, the 10 μm silicate band is one of the best studied and it appears in many differ-

[1]Based on observations with *ISO*, an ESA project with instruments funded by ESA Member States (especially the PI countries: France, Germany, the Netherlands, and the United Kingdom) and with the participation of ISAS and NASA.

ent astrophysical scenarios. It is also known that the band profile allows us to determine the degree of processing of the dust particles. Thus, the comparison of silicate features observed towards young stars with those observed in Solar System comets is very useful for studying the dust properties in young protoplanetary systems, because comets are supposed to have a chemical composition very close to that of the primitive protosolar nebula. In fact, the discovery of comet-like silicate emission around some young stars (Telesco & Knacke 1991) is a major supporting argument for the existence of stellar systems in which the very first steps towards planet formation are probably taking place. The coincidence of these features with optical transient spectroscopic events, which have been interpreted as due to the evaporation of solid bodies around Vega-type and some PMS stars (e.g., Grady et al. 1999) can lead to new insights into the formation and evolution of protoplanetary disks.

2. Available Data and Reduction

The *ISO* post-mission archive was surveyed for SWS spectra of the EXPORT targets. Only SWS01 observations have been considered in our first approach. This observing mode provides a spectrum for the full grating instrumental set-up, designed to produce a spectrum for the whole wavelength range covered by the SWS detectors (3 to 45 μm) at a nominal resolution of around 1000.

The retrieved datasets are shown in Table 1. One third of the stars are classified as Vega-type in the literature, and most of the PMS stars are HAeBe stars close to the ZAMS (van den Ancker et al. 1997, 1998).

The reduction of the SWS spectra was performed using the SWS Interactive Analysis package (IA3[2]). Special attention was devoted to dark-current calculation and subtraction due to the low flux levels of many of the objects. A description of the reduction procedure, close to the standard one, is detailed by Palacios et al. (2000).

3. The Mid-IR Spectra

The SWS spectral energy distribution of each star in Table 1 has been compared with a blackbody with the effective temperature of the star and normalized to the observed flux around 3 μm. Two distinct types of behavior are observed. Most of the Vega-type stars show SEDs compatible with a photosphere in the shorter SWS spectral range ($\sim$ 10 μm), and the Vega excess typically appears beyond 20–30 μm. SEDs of the PMS stars are flatter than blackbodies and their slopes are not compatible with a photosphere even at 3 μm. Considering the SWS SED, the Vega-type stars 51 Oph and HD 34700 should be classified as PMS objects, since both stars show a remarkable excess at the shorter wavelengths. On the other hand, the star HD 141569, which is among the oldest HAeBe stars according to *Hipparcos* data (van den Ancker et al. 1997), could be classified as a possible Vega-type star, since the SWS SED at wavelengths shorter than

[2]IA3 is a joint development of the SWS consortium. Contributing institutes are SRON, MPE, KUL, and the ESA Astrophysics Division.

Table 1. Individual *ISO*-SWS observations of stars included in the EXPORT ground-based program

Object		Sp. type	Type	TDT number
HR 10		A2 V	Vega	37802001
BD+31°643		B5	Vega	65201414
HD 31293	AB Aur	A0 V	HAeBe/ZAMS	68001206
MWC 480	HD 31648	A3	HAeBe/ZAMS	83501201
HD 34282		A0	HAeBe/ZAMS	83301240
HD 34700		G0 V	Vega	66302638
HD 109085	GL 9411	F2 V	Vega	24002304
λ Boo	HD 125162	A0 V	Vega?	35101303
HD 141569		A0 V	HAeBe/ZAMS	62802937
HD 142666		A7 Ve	HAeBe/ZAMS	10402952
				44901283
HD 144432		A5 V	HAeBe/ZAMS	45000284
HD 150193	MWC 863	A1 Ve	HAeBe	8200444
51 Oph	HR 6519	B9 Ve	Vega	10703103
HD 163296		A1 Ve	HAeBe/ZAMS	32901191
MWC 297		O9	HAeBe/ZAMS	70800234
HD 179218		B9e	HAeBe/ZAMS	32301321
WW Vul		A0e	HAeBe	17600305
V 1686 Cyg	LkH$_\alpha$ 224	G2 V		85800502
SV Cep		A0e	HAeBe	28800703
HR 9043		A5 V	Vega	40400904

10 μm is practically photospheric; this is also probably the case for HD 34282, although this star is too faint for a reliable classification to be made.

With respect to the presence of the silicate feature, the band is not observed in any of the objects in Table 1 with a photospheric SWS SED, i.e., the Vega-type stars. The opposite is true for the PMS objects, where the feature appears in emission in most of the cases. In this group, MWC 297 shows the silicate band in absorption, while no feature is observed in V 1686 Cyg, whose SED shows increasing flux densities with wavelength. This star is located inside an active young stellar cluster(van den Ancker 1999).

We have made a first comparison of the silicate emission observed towards the stars in our sample and other astronomical silicates. Figure 1 shows the normalized SWS 7–14 μm spectra of our objects and the Trapezium (Hanner, Brooke, & Tokunaga 1995) and Comet Kohoutek (Merrill 1974) silicates. The emission in the Trapezium is believed to be characteristic of amorphous silicates in a dense interstellar environment, such as that found in the direction of the Orion molecular cloud, while the comet emission is an example of crystalline silicates and should be a good sample of dust material processed while the central object evolves. For comparison, we have removed the underlying dust emission in our stars and all spectra have been normalized at 9.7 μm. The agreement is

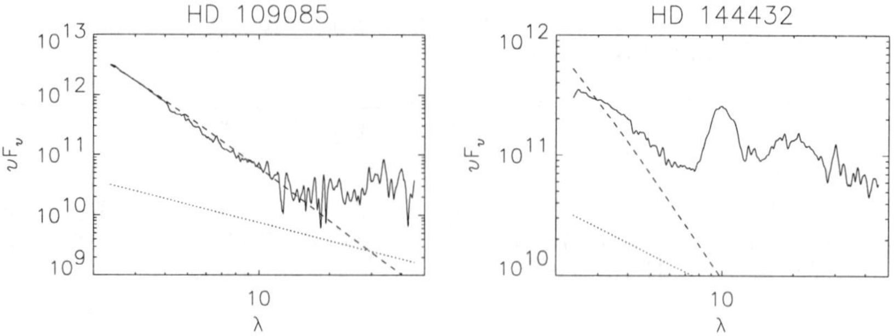

Figure 1. SWS Spectral Energy Distributions for two sample object: a Vega-type star (left) and a PMS object (right). The dashed line represents a blackbody fitted to the 3 μm flux. The dotted line marks a 0.25 Jy flux level.

very poor for the Trapezium silicates; however, it is in general very satisfactory in the case of the cometary silicates. In particular, the observed silicate emissions of AB Aur and HD 163296 are quite similar to cometary silicates.

4. Comparison with EXPORT Ground-Based Results

The EXPORT ground-based data contain optical spectroscopic and photopolarimetric information on the gas and dust properties of the circumstellar environment. It is an interesting exercise to compare these results with the SWS data in order to see if any kind of general trend or relationship exists.

The ground-based results for the sample of SWS stars are summarized in Table 2. A *blank* in any of the columns 2 to 6 means that there is no information on the corresponding star, while an *open circle* means that the behavior of the star is apparently normal, i.e., unpolarized, photospheric near-IR colors and SWS SED up to $\sim$ 10 μm, photospheric spectra, and no silicate feature. The second column indicates if the object is polarized in the *UBVRI* bands (*black dot*); a *question mark* in this column means that the object is probably polarized, with a very low polarization degree[3]. Infrared photometry is shown in the third column; *black dots* mean a near-IR excess. The near-IR colors of BD +31°643 are compatible with a reddened star of its spectral type. *Black dots* in the fourth column identify non-photospheric spectra with spectral lines indicating gaseous circumstellar activity. The last two columns indicate the presence of an SWS excess at wavelengths shorter than $\sim$ 10 μm and of a silicate feature.

The results in Table 2 suggest an evolution from the PMS objects to Vega-type stars. As expected, most of the PMS stars have a circumstellar component, as shown by the gas and dust signatures in the columns of Table 2. In contrast, Vega-type stars usually do not show any observational properties revealing the

[3] HD 163296 could show variable polarization at very low levels.

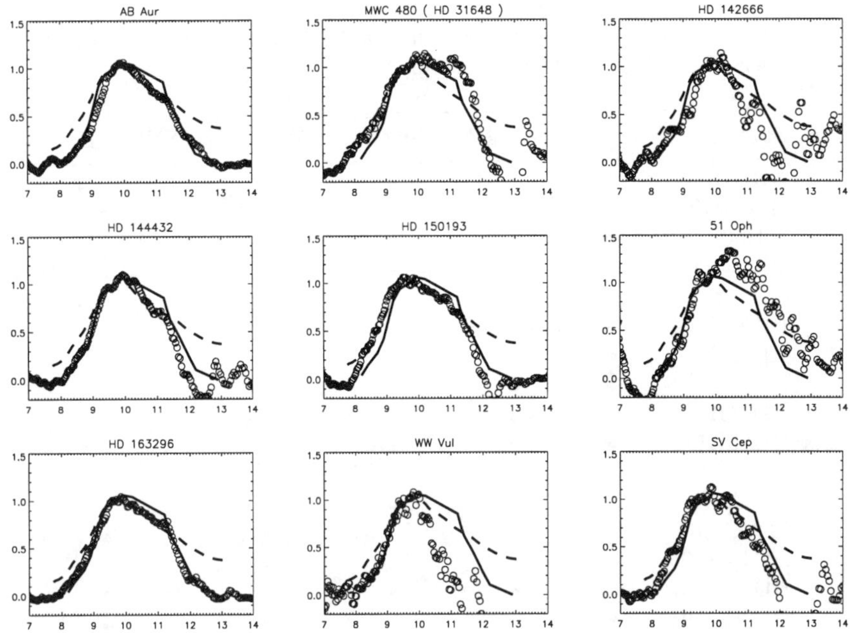

Figure 2. Normalized 7–14 μm spectra of the objects with silicate emission features. The Trapezium (*dashed line*) and Comet Kohoutek silicates (*continuous line*) are superimposed.

presence of dust around the stars. However, in at least two cases, HR 10 and HR 9043, the optical spectra indicate circumstellar gas. Five objects are probably at intermediate evolution phases, three Vega-type stars—BD +31°643, HD 34700, and 51 Oph—and two HAeBe/ZAMS stars—HD 34282 and HD 141569. All these stars are polarized and show some of the properties of PMS objects in Table 2.

5. Conclusions

The *ISO*-SWS data of the PMS and MS stars studied by EXPORT show a good correlation with the results obtained from the ground and well indicate the presence of dust. It is interesting to point out that the observed silicate profiles in the sample of stars are similar to the profiles observed in Solar System comets. Apparently, dust particles in the protoplanetary disks around young stars experience similar physical processes to those prevailing in the formation phase of the Solar System.

Acknowledgments. The data were analyzed with the support of the Dutch *ISO* Data Analysis Centre of the Space Research Organization Netherlands, Groningen, the Netherlands. Discussions with the SWS staff members were very useful. J. Palacios is partly supported by Spanish grant ESP 98-1339.

Table 2. Properties derived from EXPORT ground-based and SWS observations (see description in text)

Star	Pol	Near IR	Spectra	SWS-SED	SWS-silicates
HR 10	○	○	●	○	○
HD 109085	○		○	○	○
λ Boo	○	○	○	○	○
HR 9043	○	○	●	○	○
BD+31°643	●	reddened	○	?	○
HD 34700	●	●	●	●	○
51 Oph	●	●	●	●	●
HD 34282	●	●	○	?	○
HD 141569	●	○	●	○	○
AB Aur			●	●	●
MWC 480	●	●	●	●	●
HD 142666	●	●	●	●	●
HD 144432	●	●	●	●	●
HD 150193	●	●	●	●	●
HD 163296	?	●	●	●	●
MWC 297	●	●	●	●	abs
HD 179218	?	●	●	●	●
WW Vul	●	●	●	●	●
V 1686 Cyg	●	●	●	●	○
SV Cep	●		●	●	●

References

de Graauw, T., et al. 1996, A&A, 315, L49

Grady, C., et al. 1999, in Protostars and planets IV, ed. V. Mannings, A. P. Boss, & S. S. Russell (Tucson: University of Arizona Press), p. 613

Hanner, M. S., Brooke, T. Y., & Tokunaga, A. T. 1995, ApJ, 438, 250

Kessler, M. F. et al. 1996, A&A, 315, L27

Merrill, K. M. 1974, Icarus, 23, 566

Palacios, J., et al. 2000, in ISO beyond the Peaks, in press

Telesco, C. M., & Knacke, R. F. 1991, ApJ, 372, L29

van den Ancker, M. E. 1999, PhD Thesis, University of Amsterdam

van den Ancker, M. E., de Winter, D., & Tjin A Djie, H. R. E. 1998, A&A, 330, 145

van den Ancker, M. E., et al. 1997, A&A, 324, L33

Disks, Planetesimals, and Planets
ASP Conference Series, Vol. 219, 2000
F. Garzón, C. Eiroa, D. de Winter, and T. J. Mahoney, eds.

ISO Observations of Isolated Herbig Ae/Be Stars

Koen Malfait[1,2] and Christoffel Waelkens[2]

[1] *Faculteit Wetenschappen, KULAK, E. Sabbelaan 53, 8500 Kortrijk, Belgium*

[2] *Instituut voor Sterrenkunde, KULeuven, Celestijnenlaan 200B, 3001 Heverlee, Belgium*

Abstract. We discuss spectra of isolated Herbig Ae/Be stars that have been obtained with the spectrometers on board the *Infrared Space Observatory* (*ISO*). This satellite opened up a new field in stellar astronomy—astro-mineralogy—enabling detailed studies of the composition of the dust particles present in the circumstellar dust disks that surround Herbig Ae/Be stars, among others. The detection of large quantities of crystalline dust, contrary to what was expected, is one of the most remarkable results of *ISO*, and was therefore called the crystalline revolution!

1. Introduction

Herbig Ae/Be stars are pre-main-sequence stars of intermediate mass (2–10 $M_{\odot}$). As a newly formed star continues to accrete material from the molecular cloud it collapsed from, the conservation of angular momentum prohibits the direct infall of all gravitationally attracted material onto the rotating (proto)star, and the infalling material settles into an accretion disk. The circumstellar dust disks that surround Herbig Ae/Be stars are thus a natural by-product of the star formation process.

When the star has gathered enough mass (through direct or disk accretion), a bipolar outflow is developed which removes the surrounding material, except from the more resistant disk. It is believed that the disk will finally dissipate before or just after the star has reached the main sequence. However, in 1983, *ISO*'s precursor, *IRAS*, detected a number of main-sequence stars that still exhibit an infrared excess, caused by a cool circumstellar dust disk. The most famous of these Vega-type stars are its prototype Vega, β Pictoris, and Fomalhaut (e.g., Aumann et al. 1984). They are found to be older than the computed lifetimes of a PMS disk, which directly implies ongoing replenishment processes inside the disks. By looking at the chemical composition of a circumstellar disk at different stages of the pre-main-sequence and early main-sequence evolution, it is our final aim to be able to comprehend the processes going on in these disks. Here, we will focus on the dust component of the disks. For infrared brightness reasons, we opted for the Herbig Ae/Be stars instead of the lower-mass T Tauri stars.

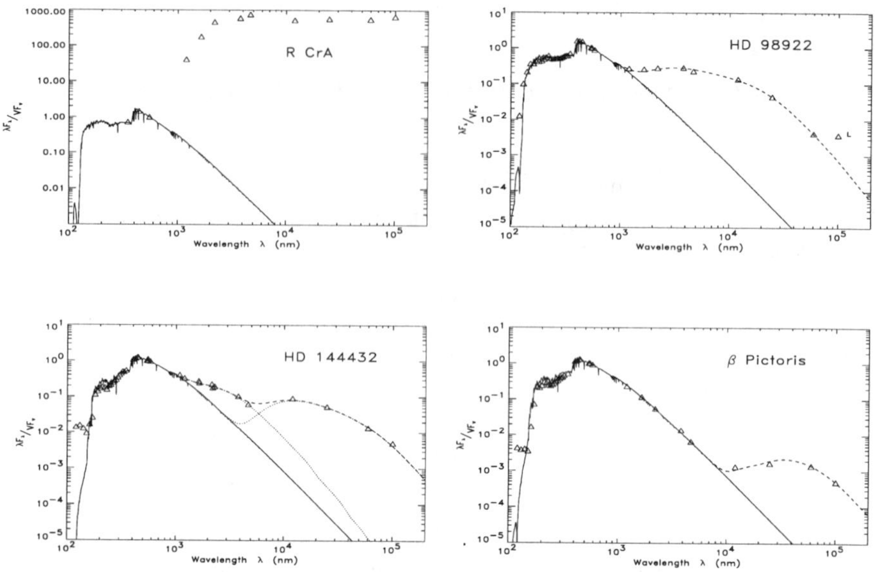

Figure 1. Spectral energy distributions (Malfait et al. 1998a)

2. Pre-*ISO* Observations

To get a better insight in the evolution from very young embedded Herbig Ae/Be stars towards β Pictoris-like objects, which are believed to be evolutionary successors of the Herbig Ae/Be stellar class, we have studied the overall spectral energy distribution of a sample of stars by combining *IRAS* infrared photometry and UV and optical measurements (Malfait, Bogaert, & Waelkens 1998a). The youngest, most embedded objects are heavily reddened and exhibit a huge IR excess, mainly caused by the remnant radiation of the clump from which the star formed. The Vega-type stars, on the other hand, are not situated in a star-forming region, have cold circumstellar dust disks, and thus only exhibit far-IR excesses. In Figure 1, we show the spectral energy distribution (SED) of four objects. R CrA is an embedded object with a huge IR excess. HD 98922 on the other hand is more evolved, from which the filamentary dust has been blown away, and where we look at a star–disk system. We modeled the IR excess with an optically thin dust model, which is shown in *dashed* and *dotted* lines. Contrary to what was expected, the infrared excess was not found to disappear gradually from the shortest wavelengths towards the longer ones. In almost half of the SEDs of the isolated Herbig Ae/Be stars, which are believed to be more evolved that classical Herbig Ae/Be stars, an apparent dip at approximately 10 μm is observed (e.g., HD 144432). For completeness, the SED of β Pictoris is included as well.

A possible explanation for the dip in the SED is the splitting of the disk into a hot and a cold part. Such a splitting up of the disk could be related to the formation of a planet, which is thought to sweep away its dusty environment and create a hole or empty ring in the disk. However, other explanations, such as a different disk geometry are also proposed. If, for example, a dust disk has an optically thick central region, the dust which is shielded remains cold, creating distinct hot and cold dust components (Bouwman et al., in preparation). Also, other processes such as the selective removal of the dust particles which radiate at these specific mid-IR wavelengths should be considered (Meeus et al., in preparation). In addition, if particles grow to sizes of several microns, they will no longer emit efficiently at shorter wavelengths. In order to discriminate between these different explanations, a study of the spectral dust fingerprints is needed, by detailed analysis of *ISO* spectroscopy.

3. *ISO* Observations

3.1. Why Isolated Objects?

Isolated objects have already left the star-forming region they originated from, or have been formed in small dense regions. The main advantage of studying them is that the flux which we receive is certainly emitted by the star–disk system, as no emitting loose surroundings are present. The splitting up of their SED demonstrates that effective dust processing is taking place during this evolutionary stage.

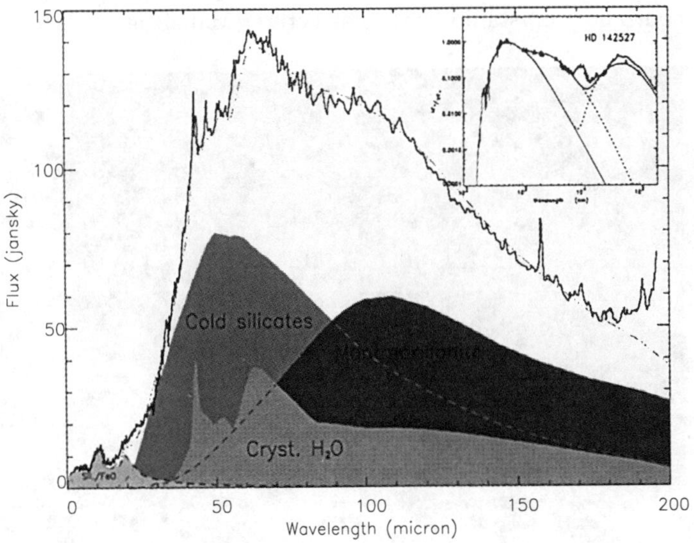

Figure 2. Dust modeling of the *ISO* SWS-LWS spectrum of HD 142527. In the inset, the spectral energy distribution of the star–disk system is shown (Malfait et al. 1999).

3.2. Selected Objects

Here, we concentrate on the results of our detailed analysis of HD 142527, HD 100546, and HD 179218, which have respective ages of 10^5, 10^7, and 10^5 yr, estimated from their *Hipparcos* parallaxes and stellar evolutionary models (van den Ancker, de Winter, & Tjin A Djie 1998). The determination of these ages is somewhat uncertain, but it is evident that HD 100546 is much older than HD 142527 and HD 179218, which both have similar ages. A wider analysis of all the spectra that we have obtained (± 30 objects) will be published elsewhere (Meeus et al., in preparation).

HD 142527 In Figure 2, we have combined the SED of HD 142527 with the *ISO* spectrum and its model fit. The optically thin dust model described by Malfait et al. (1998a) fits the photometric points rather well (with a two-disk model), but was unsuccessful in fitting the *IRAS* 12 μm point. As can be seen from the *ISO* spectrum, the mismatch results from the presence of strong silicate emission at the appropriate wavelength. This again demonstrates that broad-band photometry is not sufficient for a detailed understanding of the disks. Radiative optically thin modeling using the MODUST code developed by Bouwman et al. (in preparation) reproduces the cold part of the spectrum ($T < 60$ K) within the errors, using 25% of amorphous silicates, 60% of crystalline H_2O-ice, and 15% of layer-lattice hydrous silicates (we used the optical constants of montmorillonite—Koike & Shibai 1990—though similar species might produce an equally good fit). Such hydrous silicates have also been found in interplanetary dust particles or IDPs (Sandford & Walker 1985). The hot dust, on the other hand, appears to be composed of metallic Fe, FeO, and silicates, with both an amorphous and a crystalline lattice structure (Malfait et al. 1999).

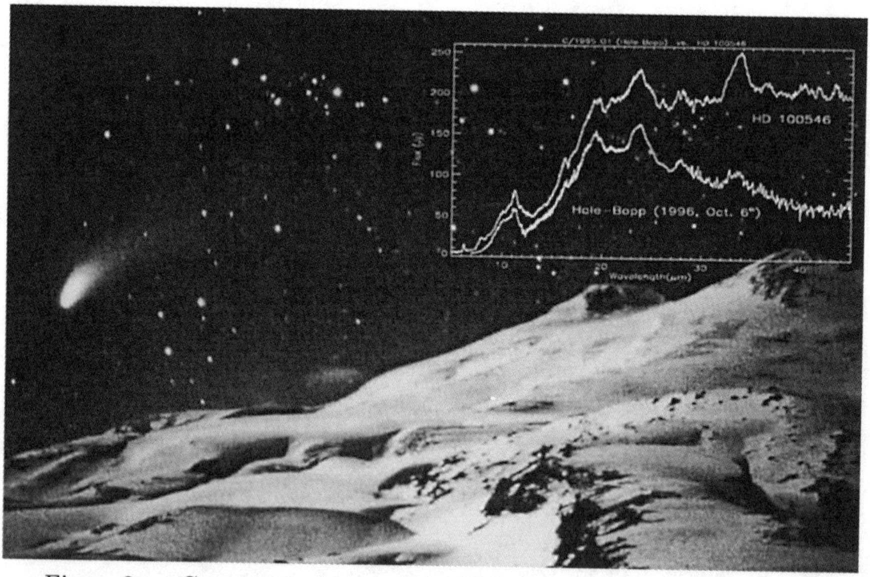

Figure 3. Comparison between the *ISO*-SWS spectra of Comet Hale–Bopp (Crovisier et al. 1997) and HD 100546.

HD 100546 HD 100546 with an estimated age of $\sim 10^7$ yr is located close to or on the ZAMS. Again, a cold and a hot dust component can be distinguished. The dust is mainly O-rich, but also evidence for PAHs is available. Some 55% of the cold dust is composed of mainly crystalline H_2O ice, while the rest consists of silicates, of which 10% has a crystalline structure. It is important to note that these crystalline silicates (mainly Mg_2SiO_4) have temperatures below 250 K.

A fortunate coincidence was that the brightest comet of the last decades appeared during *ISO*'s lifetime; when we compare the spectrum of HD 100546 with that of Hale–Bopp, the similarity is striking (Figure 3). This indicates that HD 100546 is surrounded with cometary dust, as is the case for β Pictoris. The infall of comet-like bodies onto isolated Herbig Ae/Be stars, which has been argued for by Grady et al. (1996), is thus confirmed. However, a temperature conflict emerges: the crystalline H_2O ice and the crystalline silicates are now too cold to allow crystallization on a reasonable timescale. The similarity with Hale–Bopp, which originates from the Oort Cloud, where crystallization is also unexpected, enables the following reasoning: the dust which is crystalline to date might have been crystallized before it was ejected to a larger distance (and colder temperature) by gravitational effects due to the presence of giant planets (Fernandez 1978). This implies that the formation of an Oort Cloud and/or Kuiper Belt might be going on in the environment of HD 100546 and—by inference—that giant planets are already present there (Malfait et al. 1998b).

HD 179218 HD 179218 has an age similar to that of HD 142527, but its spectrum is similar to that of HD 100546, which is a factor of 100 times older (see Figure 4). The spectrum is again dominated by cold crystalline silicates. Clinoenstatite is present despite the very strict formation conditions it requires, among which, again, is a high temperature. It is thus clear that no direct correlation between spectrum and stellar age is present, though an obvious evolutionary trend is present using a larger sample (from amorphous to crystalline dust, from optically thick to optically thin dust—Malfait et al. in preparation).

4. Discussion and Conclusions

Isolated Herbig Ae/Be stars are interesting study objects because they are located in the evolutionary stage between classical Herbig stars and Vega-type objects. *ISO* has confirmed this general evolutionary scenario, though there seems to be no clear correlation between age and dust-spectrum characteristics. Dust processing occurs during this evolutionary stage, changing the lattice structure from amorphous to crystalline. The presence of crystalline material links these objects with our own Solar System. The similarities between both indicate that we might be witnessing planet formation in the environment of these young stars. Similar studies of T Tauri stars can evidently result in the detection of new solar systems like ours.

Important questions on the cosmic dust cycle can be raised as well, since crystalline silicates appear not only in the disks of young objects, but also around some evolved objects and have on the other hand not been found in the ISM so far. Waters et al. (1998) studied the circumstellar dust around the evolved binary HD 44179 and its reflection nebula the Red Rectangle. The *ISO* spectrum of this system can be divided into a C-rich part, at $\lambda < 15$ μm, while O-rich dust dominates at longer wavelengths. As can be seen from the additional images

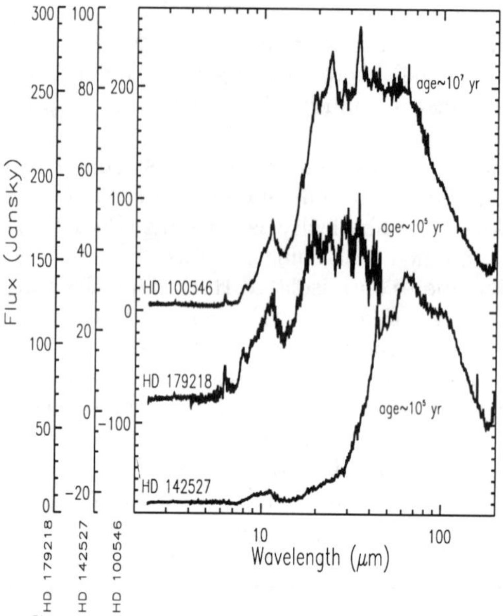

Figure 4. Comparison of selected stars.

shown in Figure 5, the PAH and C-rich emission originates from the present outflow, while the O-rich dust is located in a long-lived circumbinary disk that probably formed while the central mass-losing star was still O-rich. Molster et al. (1999) found that the degree of crystallization in this and similar stable disks surrounding evolved stars is larger than in regular outflows of O-rich matter. Since the early temperature evolution (the cooling) in both processes is similar, they conclude that time plays an important role in the crystallization process. Although no processes are known to date which can explain the crystallization at low temperatures, we should no longer rule out this possibility. Since the circumstellar dust disk has similar fingerprints to those of young stellar objects, planet formation (or at least dust grain growth, which has been argued for by Jura, Balm & Kahane in 1995) around evolved objects might occur as well.

Acknowledgments. KM is a postdoctoral fellow of the Fund of Scientific Research, Flanders. KM and CW also wish to thank the *ISO* group of the Instituut voor Sterrenkunde of the K. U. Leuven and collegues from the Astronomical Institute of the University of Amsterdam, as well as *ISO*'s instrument-dedicated teams.

References

Aumann, H., et al. 1984, ApJ, 278, 23

Crovisier, J., et al. 1997, Science, 275, 1904

Fernandez, J. 1978, Icarus, 34, 173

Grady, C., et al. 1996, A&AS, 120, 157

Jura, M., Balm, S., & Kahane, C. 1995, ApJ, 453, 721

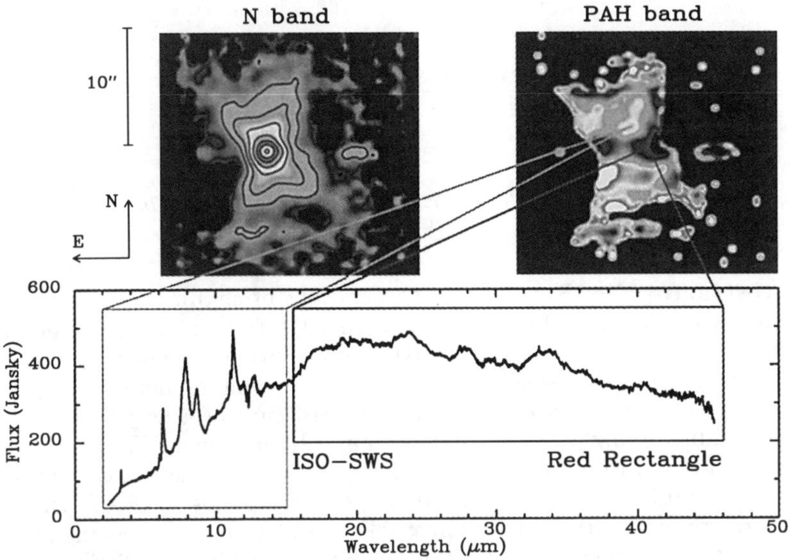

Figure 5. The Red Rectangle (Waters et al. 1999).

Koike, C., & Shibai, H. 1990, MNRAS, 246, 332
Malfait, K., Bogaert, E., & Waelkens, C. 1998a, A&A, 331, 211
Malfait, K., 1998b, A&A, 332, L25
Malfait, K., et al. 1999, A&A, 345, 181
Molster, F., et al. 1999, Nat, 401, 563
Sandford, S., & Walker, R. 1985, ApJ, 291, 838
van den Ancker, M., de Winter, D., & Tjin A Djie, H. 1998, A&A, 330, 145
Waters, L., et al. 1998, Nat, 391, 868

Discussion

Antonella Natta: Are there submm and mm data for these stars?

Koen Malfait: Yes, but we have not included these points in the modeling so far. The points do seem to indicate grain particle growth. A combination of the optical and infrared spectra and submm points is studied by Meeus et al., and will be published soon!

Antonella Natta: What is the mass of dust in the disks?

Koen Malfait: We found masses that range to 0.1 $M_\odot$ with the assumption of a typical gas-to-dust ratio of 100.

Lee Hartmann: How did you get age estimates?

Koen Malfait: The age estimates have been derived by van den Ancker et al. (1998), using the parallax measurements performed by the *Hipparcos* satellite.

Disks, Planetesimals, and Planets
ASP Conference Series, Vol. 219, 2000
F. Garzón, C. Eiroa, D. de Winter, and T. J. Mahoney, eds.

Mid-Infrared Observations of Disks

Charles M. Telesco

University of Florida, Department of Astronomy, PO Box 112055, Gainesville, FL 32611-2055, USA

Abstract. Mid-IR (5–30 μm) observations of circumstellar disks sample dust located within a few hundred AU of the stars, a region that is particularly relevant to the birth and evolution of planetary systems. While planets themselves are currently too faint to detect directly, a planet can perturb a disk, altering its structure in a way that may be detectable by high-angular-resolution mid-IR imaging. With the largest ground-based telescopes, the angular resolution at these wavelengths is typically $0\farcs3$, which for a star at 20 pc, corresponds to 6 AU. One can therefore probe structures comparable in size to potential planetary orbits. Multi-wavelength mid-IR imaging also provides a realistic basis for estimating the dust temperature, emission efficiency, and number density as a function of position within disks. Among other things, this information bears on the particle size distribution and, consequently, on the dynamical behavior of the disk particles in response to radiation pressure and other forces. In this review, I describe the role that mid-IR astronomy can play in the exploration of circumstellar disks and I present an assessment of some of the most recent mid-IR imaging of disks and their environments. I emphasis Vega-type disks and their (presumed) progenitors, the Herbig Ae/Be stars. Finally, I touch on the capabilities that the newest generation of large ground-based telescopes will provide for exploration of disks in the mid-IR.

1. Introduction

While the discovery of circumstellar debris disks by *IRAS* made disk research serious business, the modern era of mid-IR observations of circumstellar disks was ushered in with the publication of the beautiful image of β Pictoris by Lagage & Pantin (1994). This 12 μm image showed how the new generation of detector arrays could reveal directly fundamental disk structure, the kind that must underlie any reasonable formulation of disk evolution. For example, their density map, derived from the image, showed unambiguously the central cleared zone, or hole, for which there had previously been only indirect evidence based on the spectral energy distribution. Detection of the central hole in β Pic has fueled speculation about the existence of one or more planets there.

Mid-IR images like this play an important role in considerations of disk structure and the connection between planets and disks. As the new 8-to-10 meter telescopes continue to come on line, we will see a flood of mid-IR observations, partly because the detectors are getting better, but also because, in the background-limited regime that characterizes broad-band mid-IR imaging, the sensitivity increase with telescope aperture is enormous; the integration time it takes to achieve a given signal-to-noise ratio on a point source is inversely proportional to the fourth power of the aperture diameter. Here I present highlights that I hope illustrate the potential power of mid-IR observations of disks.

2. The Mid-IR as a Probe of Disk Properties

The mid-IR spectral region spans the wavelength range 5–30 μm. The long-wavelength end is traditionally defined by the atmospheric cut-off (which is site-dependent), beyond which the atmosphere is opaque out to the submillimeter region. The short-wavelength end is defined roughly by the transition to different types of detectors. The mid-IR radiation that we see from disks is thermal radiation from dust heated directly or indirectly by starlight and/or (for the densest, youngest disks) viscous accretion. Where heating by starlight dominates, we expect a relatively simple monotonic fall-off of the dust temperature with distance from the star. Figure 1 illustrates this trend for astronomical silicate Mie spheres of different diameters around an A0V star. Generally, in astrophysical environments the dust we see in the mid-IR is hotter than about 100 K, which Figure 1 indicates is attained within a hundred or so AU of an A star. This scale is comparable to the size of our Solar System, and so the mid-IR is a particularly useful probe of circumstellar regions relevant to the exploration of the origin and evolution of planets. Of course, lower-mass stars will not heat the dust as affectively as OBA stars, but the mid IR is still a useful, if more indirect, probe of the disk properties even if the disk is unresolved.

One other practical advantage of working in the mid-IR is that, from high, dry sites like Mauna Kea, there are two excellent atmospheric windows, one near 10 μm and one near 20 μm, that permit observations to be made with large ground-based telescopes. The diffraction limits of an 8 m telescope like Gemini or the VLT are $0\farcs2$ and $0\farcs5$ arcsec at 10 and 20 μm, respectively. Thus, we can resolve details in disks that have scale sizes of the order of 5–10 AU, clearly useful for our exploration of planetary disks.

3. Massive Pre-Main-Sequence Stars

The environments of young massive stars often appear to be incredibly complex. This is partly because massive stars evolve more quickly than lower-mass stars, so there is less time for them to remove or leave their natal clouds. They are also much more luminous than lower-mass stars, and they illuminate a much larger part of their neighborhood, which may not only include disks but enormous envelopes and adjacent molecular clouds. Detailed imaging is needed to avoid confusion and misinterpretation of the observations of such regions. Because of this complexity, progress in understanding the nature of disks around young massive stars has been slow.

One intriguing type of source associated with massive stars and possibly their disks is the methanol maser, particularly those of the so-called Class II sources coincident with ultra-compact H II regions (Menton 1991). Norris et al.

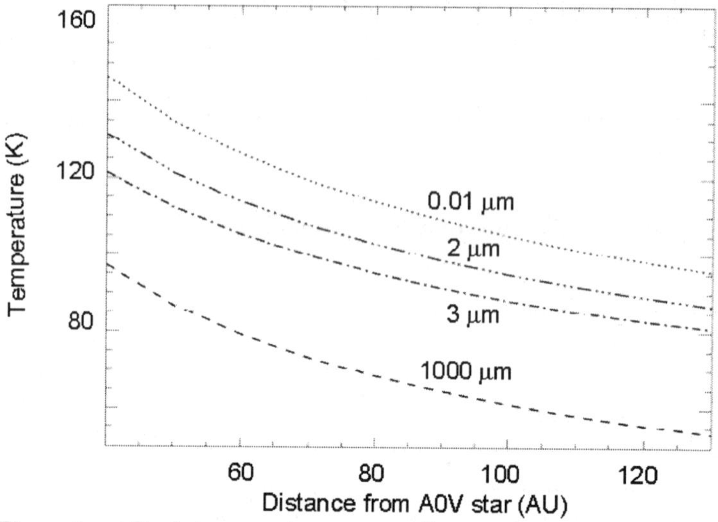

Figure 1. Dust temperature versus distance from an A0V star, from Wyatt (1999). Particles are Mie spheres composed of astronomical silicates (Draine & Lee 1984). Curves are labeled with particle diameters.

(1993) and Walsh et al. (1998) have found that 6.7 and 12.2 GHz maser spots occur in linear or roughly linear distributions for about 1/3 of the methanol maser sources. These maser distributions often span several thousand AU (Norris et al. 1993). The Doppler shifts of the maser spots vary systematically across some of the distributions in a way indicating that the masers may trace a nearly edge-on circumstellar disk undergoing Keplerian rotation (Norris et al. 1993). In a mid-IR imaging survey of 21 methanol maser sites, De Buizer et al. (this volume, p. 162) detected eight out of the ten sites that have linearly distributed maser spots. Three of these eight mid-IR sources were resolved and elongated along the same orientation as the maser distributions, although the masers span regions much smaller than the mid-IR-emitting regions (De Buizer, Piña, & Telesco). The maser sources for this sample are embedded in the mid-IR-emitting sources, which supports the view presented by Sobolev, Cragg, & Godfrey (1997) that the masers are pumped by mid-IR photons. Most importantly for the topic of disks, the fact that, at least for some sources, the maser and mid-IR distributions have the same global elongation suggests that, in mid-IR images, we may be able to see circumstellar disks for which the masers provide critical complementary dynamical information.

The class of Herbig Ae/Be stars illustrates well the role that detailed mid-IR imaging can play in clarifying the nature of disks and the environments of early-type stars. Herbig Ae/Be stars are considered to be the progenitors of main-sequence B and A stars, i.e., the higher-mass (2–10 solar masses) analogs to T Tauri stars, which are pre-main-sequence solar-mass stars. A variety of arguments strongly support the idea that the excess IR emission from T Tauri stars originates in circumstellar disks. By analogy, then, it has seemed reasonable to make the same assumption about the excess IR emission in Herbig Ae/Be

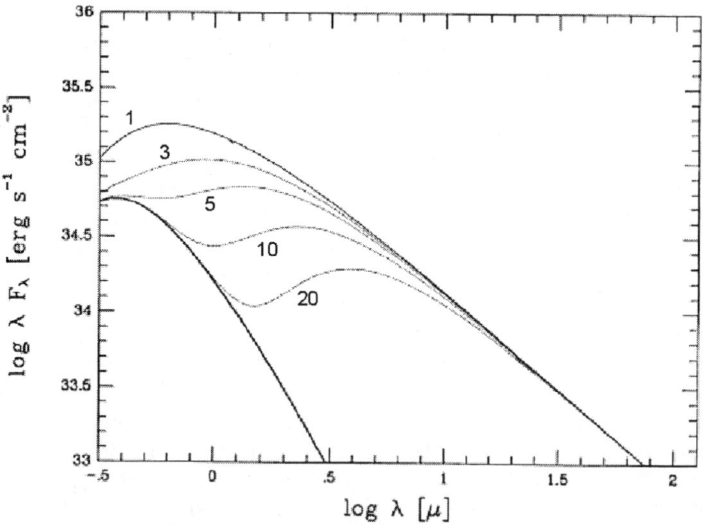

Figure 2. SEDs expected for optically thick disks around an A0 star (Hillenbrand et al. 1992). The *heavy solid curve* represents the stellar SED, and the lighter curves represent the SEDs for disks with central holes of various radii, in units of the stellar radius, as indicated.

stars (e.g., Lada & Adams 1992; Hillenbrand et al. 1992). The spectral energy distributions (SEDs) of many of these sources are key evidence in support of this hypothesis: longward of ~2 μm one often sees the shape $\lambda F_\lambda \sim \lambda^{-4/3}$ expected for a heated optically thick disk. Model SEDs for optically thick disks (Figure 2) indicate that the expected SED is seen at mid-IR and longer wavelengths even if there are modest central holes in the disk.

Hillenbrand et al. (1992) considered the SEDs for a sample of 47 Herbig Ae/Be stars, with low-angular-resolution *IRAS* photometry playing an important role in the determination of the longer-wavelength parts of the SEDs. Thirty of these stars had SEDs ("Group I") expected for an optically thick disk, 11 had considerably more IR emission, particularly longward of 10 μm, than expected and so were classified as disk/envelope candidates ("Group II"), and six had small excesses, all of which could be attributed to ionized gas ("Group III"). Even before recent high-angular-resolution IR imaging of Herbig Ae/Be stars, clear observational evidence at far-IR wavelengths was found for the expected large envelopes in several Group II objects, but, surprisingly, several Group I stars were also found to have large envelopes (e.g., Natta et al. 1993; Evans & Di Francisco 1995). These Group I envelopes had been anticipated by several studies, particularly that of Hartmann, Kenyon, & Calvet (1993), who showed that models using plausible dust properties and incorporating envelopes could match the Group I SEDs at least as well as the optically-thick disk models.

High-angular-resolution mid-IR imaging can play a decisive role in clarifying the relative roles of disks and envelopes in generating the IR emission from Herbig Ae/Be stars, as recent observations by Polomski et al. (this volume, p.

416) illustrate. In their mid-IR imaging survey with arcsec or better resolution, Polomski et al. (in preparation) have observed the Group I Herbig Ae/Be star AS 310, for example, an early B star possibly located at ~2.5 kpc (see, for example Pezzuto, Strafella, & Lorenzetti 1997). Complex structure clearly unassociated with a circumstellar disk is evident out to more than 10″ (25 000 AU) from the emission peak, although the compact emission near the peak may arise in an unresolved disk. However, the compact emission within a 1″ radius of the peak represents only ~12% of the total 10 μm flux in the image, so obviously any attempt to model the disk structure must carefully exclude most of the extended emission, which is impossible without these kinds of detailed images. The multi-wavelength images by Polomski et al. (this volume, p. 416) also permit the determination of dust temperatures and optical depths throughout this complex region. They show, for example, that, while the maximum column density of emitting dust coincides with the emission peak, the dust temperature is highest relatively far away from the source, an indication that very small grains may contribute significantly to the mid-IR emission. Having pointed out the complexity of the environments of Herbig Ae/Be star, I want to emphasize that most sources seem to be much less complex than AS 310. More than half of the sources in the Polomski et al. sample are unresolved, and all of their emission could arise in disks. Perhaps the source WL 16, which shows what, with admirable restraint, DeVito & Hayward (1998) refer to as an 'extended oval nebula,' is an example of this pure disk structure.

4. Lower-Mass Pre-Main-Sequence Stars

OBA stars brightly illuminate their environments, so that dust temperatures at several hundred, and in some cases several thousand, AU in the disks and envelopes are sufficiently high that they can sometimes be spatially resolved in the mid-IR. The lower temperatures and luminosities of solar-mass stars means that grains at a given distance have lower temperatures compared to OBA stars, and thus the emitting regions will be harder to resolve. For example, assuming the dust emission efficiency varies as λ^{-2}, the radius at which the dust around a G0V star attains a given temperature is ~ 16% the radius at which that same temperature is attained around an A0V star. The result is that our chances of resolving T Tauri disks are generally not very good. Nevertheless, mid-IR observations of them can provide useful information.

Details of the disk structure are evident in the SED, and the SED can be modeled to assess the optical depth of various parts of the disk. For example, Skrutskie et al. (1990, their Figure 4) compare the 1–100 μm SEDs of the T Tauri stars AA Tau and V819 Tau to that expected for an optically thick reprocessing disk (i.e., where the IR emission is entirely due to absorbed stellar radiation). The entire SED of AA Tau lies above the model SED, indicating an optically thick disk, possibly with an envelope. Interestingly, while the SED of V819 Tau longward of 30 μm indicates that the disk is optically thick, it is too low at shorter wavelengths, nearly equal to the photospheric values. V819 Tau appears to have a central hole in its disk (cf. Figure 2), the hole being large enough to include the disk region that would otherwise (as in the case of AA Tau) contribute most of the mid-IR emission.

Thus, even unresolved mid-IR observations are useful because the appearance of central holes may affect the mid-IR flux level that can accompany a critical stage in the disk evolution, perhaps as planet formation comes to fruition. This kind of information permits estimation of disk evolutionary timescales, as demonstrated beautifully by Skrutskie et al. (1990) and Hartigan et al. (1990). They show for large samples of T Tau stars that there is little or no mid-IR excess for those stars without Hα emission or veiling of photospheric absorption lines; i.e., when there is a halt to the accretion (which gives rise to the optical and near-IR emission near the disk–star boundary), the mid-IR emission from the inner disk effectively disappears, too. These studies support a consistent picture for the clearing of the inner disk regions. Furthermore, by comparing the number of transition T Tau stars (i.e., those intermediate between the two groups with and without mid-IR emission, veiling, and line emission) to the number with thick disks, Skrutskie et al. (1990) estimate a "transition time" of about 0.3 Myr. This timescale may correspond to the most rapid phase of central clearing associated with planet formation. However, planets may not be the only way to clear out the inner region of a disk and suppress the mid-IR flux. In their study of the (possibly) very young (< 0.1 Myr) triple T Tauri system UZ Tau, Ghez et al. (1994) show that nearly all of the mid-IR excess is associated with the single eastern component, which is $3\farcs6$ from the western component consisting of a binary with separation $0\farcs35$. Since the eastern star and the dominant western star are of nearly the same mass, Ghez et al. (1994) speculate that their disks should be equally prominent in the mid-IR unless the binary companion has had an affect on the disk of the western system. One obvious possibility is that the western star's close companion has cleared out the central, mid-IR zone of the disk, perhaps as described by Artymowicz & Lubow (1994).

As the above examples suggest, identifying and surveying coeval stellar clusters in the mid IR can be key to our understanding disk evolution. The fractions of stars with excesses at various wavelengths in clusters of different ages can tell us the timescales for significant evolution of those disk zones dominating emission at a particular wavelength; the mid-IR regime probes the inner $\sim$100 AU zone inhabited by planets. The recently identified TW Hydrae association emphasizes this point. This possible group (Kastner et al. 1997; but cf. Jensen, Cohen, & Neuhäuser 1998) of about 15 stars is $\sim$ 50 pc away and has an age of $\sim$ 10 Myr (see below). The TW Hydrae association is the nearest young group and consists of both pre-main-sequence and main-sequence stars. Jayawardhana et al. (1999a,b) showed that only four of these stars have detectable mid-IR flux. Three of these (HD 98800, Hen 3-600, and HR 4796A) have disks with central holes, whereas TW Hya itself has an optically thick disk. Therefore, Jayawardhana et al. (1999b) conclude that by 10 Myr most of the TW Hydrae stars have cleared their inner, mid-IR-emitting disks of dust through coagulation into larger bodies and/or by accretion onto the stars, thus providing crucial observational constraints on the planet-formation timescale (e.g., Podosek & Cassen 1994). Some of the stars in this association are main-sequence stars, which we now consider.

5. Main-Sequence Stars: Vega-Type (Debris) Disks

As everyone here well knows, it was the *IRAS* detection of the IR excess associated with dust around Vega and several other stars that energized this field of study. There was previously no real inkling that such a normal main-sequence star would have significant dust around it. Incorporation into a disk seemed the only viable way that the dust could endure. Confirmation of this hypothesis came with the beautiful optical coronagraphic image by Smith & Terrile (1984) of scattered light from β Pic's edge-on disk. Disks around main-sequence stars are called Vega-type disks or debris disks, the latter designation indicating that generally these systems are too old for the dust to be primordial, as discussed further below. Consistent with this is the observation that their dust masses are smaller than 0.01 solar masses, which is roughly the minimum mass required for the formation of a planetary system like our own (see, for example, Strom, Edwards, & Skrutskie 1993). Thus, any planet-building phase in these disks is probably completed. More than a hundred debris disks are known, with the lists accumulated by Sylvester et al. (1996) and Mannings & Barlow (1998) being among the most current published ones.

One of the biggest challenges has been to determine the detailed structure of a large enough sample of disks to be able to draw general conclusions. Disks around B and A stars offer the best chance of resolution because of the higher stellar temperatures and luminosities. Since the first mid-IR resolution of β Pic in 1988 (Telesco et al. 1988), no other debris disks have been resolved in the mid-IR until very recently, and that list is still very small: β Pic, HR 4796A, and HD 141569 are (to my knowledge) the only unambiguous examples. I will comment mainly on the latter two objects, since β Pic has been reviewed extensively elsewhere (Artymowicz 1997).

Its *IRAS* excess first indicated that HR 4796 contains a disk. Because those observations showed no discernible emission from dust hotter than $\sim$ 110 K, Jura (1991) concluded that there is no dust closer to the star than about 40 AU. Planets were considered as one possible cause of the hole (e.g., Jura et al. 1995). HR 4796 is a binary for which the secondary has a reasonably well determined age of 8 Myr. This age is consistent with the idea that any planets there may be nearly fully formed by now. The disk was first resolved independently by Jayawardhana et al. (1998) and Koerner et al. (1998). Both images, at $\sim$ 20 μm, clearly show what appears to be a nearly edge-on disk with a maximum extent of $3''$ (200 AU at 67 pc). They also give clear evidence of the predicted central hole in the disk, with the higher-resolution Keck data indicating a hole radius of about 55 AU (Koerner et al. 1998). Subsequent 10 and 18 μm images from Keck are shown in Figure 3 (Telesco et al. 2000). At 18 μm one sees three peaks: the central one is primarily stellar photospheric emission, and the two outer ones are the lobes of the edge-on disk. Even before removing the starlight, the hole is apparent unambiguously as brightness minima on each side of the stellar peak. Much of the 10 μm emission is seen as extended wings coinciding with the 18 μm lobes.

I want to make two points about HR 4796A's disk. Both of these are debatable but will no doubt be decided by mid-IR imaging and spectroscopy in the near future. First, we can derive an estimate of the color temperature, and therefore a characteristic size, of the mid-IR-emitting dust particles from the

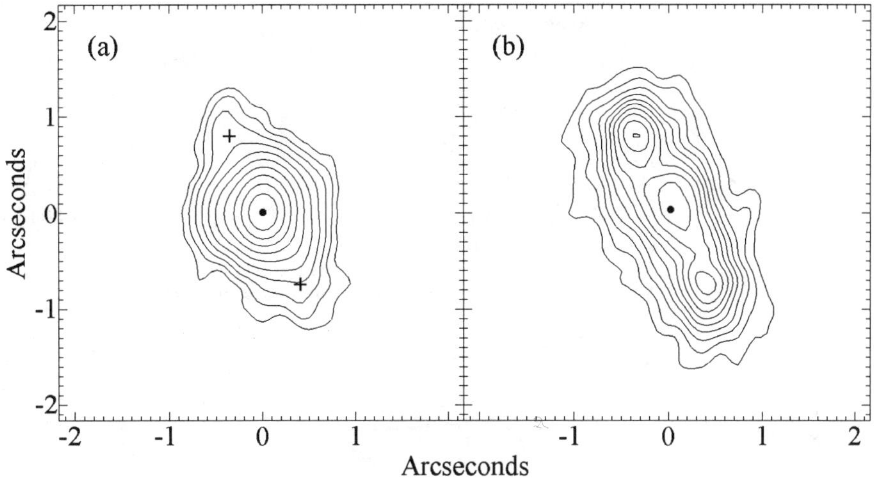

Figure 3. 10.8 μm (*left*, logarithmic contour spacing) and 18.2 μm (L*right*, linear contour spacing) images of HR 4796A obtained at Keck (Telesco et al. 2000).

multi-wavelength images; the details are presented in Wyatt et al. (1999) and Telesco et al. (2000). If the particles are assumed to be Mie spheres composed of astronomical silicates, their diameter is roughly 2–3 μm. We can then calculate β, the ratio of the radiation pressure force to the gravitational force on a particle; we derive the value $\beta \gg 1.7$ for the dust in HR 4796A. If the particles are fluffier, β is smaller. If this value is accurate, it tells us that the mid-IR-emitting particles should be blown out of the system by radiation pressure in a few hundred years, and therefore that they cannot be primordial. The particles in this debris disk do indeed appear to be "debris," presumably recently created collisional fragments or cometary ejecta.

The second point I want to make is explored more fully by Wyatt in another presentation at this meeting (this volume, p. 289; see also Wyatt et al. 1999). The mid-IR images shown in Figure 3 indicate a marginally significant asymmetry in the lobe brightness, the north-eastern lobe being roughly 5% brighter than the south-western lobe. Until now we have ignored the fact that HR 4796 is a binary system, with the secondary (B) located $7\rlap{.}''7$ (516 AU in projection) to the south west of A. Secular perturbations on the disk by B might result in this brightness asymmetry. Actually, we have no information about the orbit, but Wyatt et al. (1999) find that simple solutions consistent with the observed stellar locations can result in a shift of the disk toward the apocenter, as indicated schematically in the cartoon in Figure 4. With such a shift, the inside of the hole on the companion's pericenter side is closer to the star and therefore warmer and more mid-IR-emissive ("pericenter glow"). The shift can be very small, in fact only a few AU, but it can lead to a noticeable effect. A planet in the hole could lead to the same effect. Whether or not the asymmetry is

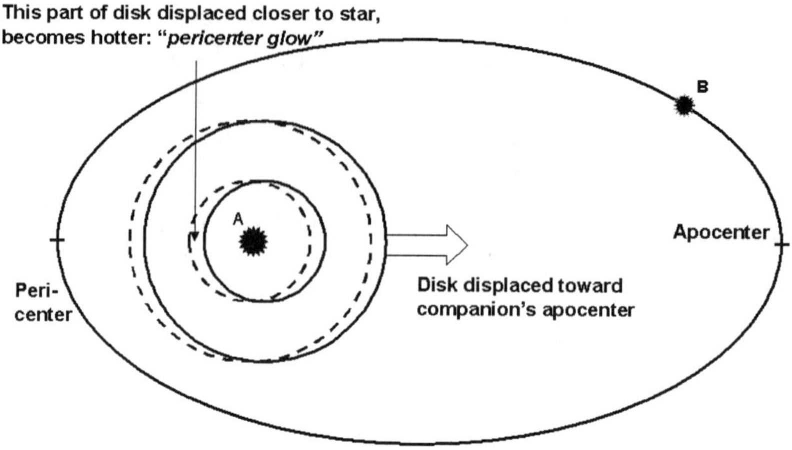

Figure 4. Cartoon showing geometry associated with pericenter glow conjectured to account for the possible brightness asymmetry in the HR 4796A disk.

confirmed, this analysis demonstrates a way in which perturbations by a stellar or planetary companion can have an observable effect on a disk.

HD 141569 is the last of the three disks to be resolved in the mid-IR. Its classification as a debris disk is somewhat ambiguous, however, since some authors classify it as a Herbig Ae/Be star (see comparison in Fisher et al. 2000). We will consider it here with the debris disks, because it does appear to be on the main sequence (e.g., Weinberger et al. 1999). However, HD 141569 may represent a transitional object between the Herbig Ae/Be and debris-disk classes. HD 141569 is only slightly resolved at 10.8 and 18.2 μm (Fisher et al. 2000; Fisher et al., this volume, p. 369), no doubt partly due to its great distance (99 pc). The faintest discernible emission appears to extend out to about 100 AU from the star, and so the overall extent of the mid-IR-emitting part of the disk is comparable to the mid-IR-emitting regions in both β Pic and HR 4796A.

Comparison of the mid-IR images of HR 4796A to the NICMOS coronagraphic images by Weinberger et al. (1999) and Augereau et al. (1999) reveal that the near-IR emission, which is thought to be starlight scattered by the dust in the disk, is much more extended than the mid-IR emission (Fisher et al. 2000). In fact, all of the detected mid-IR radiation originates within the central region unobserved in the NICMOS image because of the coronagraphic spot. It seems likely that the dust seen by NICMOS is cooler than the mid-IR-emitting dust and emits the far-IR radiation detected by *IRAS*. This relationship between the mid-IR emitting region and the larger-scale scattered-light region, which may also emit the far-IR flux, is reminiscent of the situation encountered for β Pic; e.g., compare the mid-IR image of β Pic (Lagage & Pantin 1994) to the R-band images presented by Kalas & Jewitt (1995). It is in stark contrast to the case of HR 4796A, for which the NICMOS image (Schneider et al. 1999) shows a scattered-light disk that coincides with the mid-IR-emitting disk; there is no evidence for disk extension outside of the narrow mid-IR-emitting torus in HR 4796A. Of these three objects, only HR 4796A unambiguously has a companion (but see Weinberger et al. 1999 regarding HD 141569). One may speculate that the disk in HR 4796A has been tidally truncated by the companion star, whereas the outer disks in β Pic and HD 141569 remain relatively unperturbed.

6. The Emerging Importance of Mid-IR Spectroscopy

Mid-IR spectroscopy of circumstellar disks is a rapidly emerging area of study from which we will be able to learn about dust size, temperature, composition, and mineralogy, all of which relate to the origin and evolution of dust grains and the disks themselves. The best known of the mid-IR spectral diagnostics for dust is the 10 μm silicate feature, a broad lattice resonance feature that has been observed for years in many other astrophysical environments. Some modest progress has been made so far in demonstrating its value for disk exploration. It was discovered in β Pic's disk through broad-band filter photometry (Telesco & Knacke 1991). Dust grains much larger that 10 μm would be optically thick and not expected to display the feature, and therefore just the detection indicates that there are dust grains in the disk that are a few micrometers or less in size. As mentioned above, smaller grains cannot survive for long in the disk, so they must be continuously created, perhaps as they are in our own Solar System. The feature in β Pic was found to be broader than the one attributed to amorphous silicates in the general interstellar medium (Telesco & Knacke 1991; Knacke et al. 1993); it closely resembles the feature in some comets, most notably Halley and Levy, which supports the notion that comets may be a significant source of dust in β Pic. Interestingly, the 10 μm silicate feature associated with the Herbig Ae/Be star WW Vul looks like the ISM feature (Hanner et al. 1998); i.e., that dust has not undergone the kind of processing that the β Pic disk particles have. Because WW Vul is a young star, perhaps we are seeing part of the "pristine" natal envelope or an immature disk?

β Pic's feature has a peculiar bump near 11.3 μm that is probably due to crystalline olivine, which perhaps suggests an annealing phase in the dust's formation or in its transition from an interstellar origin. That this can tell us something important about the mineralogical history of the disk particles seems undeniable, but teasing out those secrets is a challenge. Wooden et al. (1999), among others, are making significant progress in establishing the basis for interpreting the feature. By monitoring cometary spectra as a function of heliocentric distance, they have shown that the feature shape changes because of the different temperature dependences of various minerals (Wooden et al. 1999). The *ISO* SWS/LWS spectrum of the Herbig Ae/Be star HD 100546 (Malfait et al. 1998) is a beautiful demonstration of the diagnostic power of the infrared region: the features between 8 and 100 μm are numerous and, taken together, apparently unique. The point to emphasize here is that many of these features are accessible through the 10 and 20 μm atmospheric windows with the new generation of large ground-based telescopes. Indeed, the three main constituents (forsterite, amorphous olivine, and FeO) that Malfait et al. (1998)

identify in HD 100546 are most evident in the mid IR. Great progress therefore seems imminent.

7. Comments on the Future

All the new big telescopes have acquired, or are acquiring, mid-IR imagers and spectrometers. The signal-to-noise ratio achievable in a given integration time at mid-IR wavelengths varies inversely as the fourth power of the aperture diameter, and thus many projects that are impossible on smaller telescopes can be carried out in a few hours or a few nights on, say, Gemini or the Gran Telescopio Canarias. The projected broad-band mid-IR sensitivity (1σ, 1 h of chopped integration) on these telescopes is typically 50–100 μJy for a point source. This means that sources like β Pic or HR 4796A could be detected, if unresolved, out to 1 kpc with a signal-to-noise ratio of five. Since there are nearly 1500 L*Hipparcos* main-sequence A stars within only 100 pc, we cannot help but be struck by the vista that is now opening to us in the mid IR.

References

Artymowicz, P. 1997, AREPS, 25, 175

Artymowicz, P., & Lubow, S. 1994, ApJ, 421, 651

Augereau, J., Lagrange, A., Mouillet, D., & Menard, F. 1999, A&A, 350, L51

Draine , B., & Lee, H. 1984, ApJ, 285, 89

De Buizer, J., Pina, R., & Telesco, C. 2000, ApJ, in press

DeVito, B., & Hayward, T. 1998, ApJ, 504, L43

Evans, N., & Di Francisco, J. 1995, Rev. Mex. Astron. Astrofis., Ser. Conf., 1, 187

Fisher, R., Telesco, C., Pina, R., Knacke, R., & Wyatt, M. 2000, ApJ, 532, L141

Ghez, A., et al. 1994, ApJ, 434, 707

Hanner, M., et al. 1998, ApJ, 502, 871

Hartigan, P., et al. 1990, ApJ, 354, L25

Hartmann, L. Kenyon, S., & Calvet, N. 1993, ApJ, 407, 219

Hillenbrand, L., Strom, S., Vrba, F., & Keene, J. 1992, ApJ, 397, 613

Jayawardhana, R., et al. 1998, ApJ, 503, L79

Jayawardhana, R., et al. 1999a, ApJ, 520, L41

Jayawardhana, R., et al. 1999b, ApJ, 521, L129

Jensen, E., Cohen, D., & Neuhäuser, R. 1998, AJ, 116, 414

Jura, M. 1991, ApJ, 383, L79

Jura, M., et al. 1995, ApJ, 445, 451

Kalas, P., & Jewitt, D. 1995, AJ, 110, 794

Kastner, J., Zuckerman, B., Weintraub, D., & Forveille, T. 1997, Science, 277, 67

Knacke, R., et al. 1993, ApJ, 418, 440

Koerner, D., Ressler, M., Werner, M., & Backman, D. 1998, ApJ, 503, L83

Lada, C., & Adams, F. 1992, ApJ, 393, 278

Lagage, P., & Pantin, E. 1994, Nat, 369, 628

Malfait, K., et al. 1998, A&A, 332, L25

Mannings, V., & Barlow, M. 1998, ApJ, 497, 330

Menton, K. 1991, in ASP Conf. Ser., 16, Proc. Third Haystack Observatory Meeting, Atoms, Ions, and Molecules: New Results in Spectral Line Astrophysics, ed. A. D. Haschick.& P. T. P. Ho (San Francisco: ASP), 119

Natta, A., et al. 1993, ApJ, 406, 674

Norris, R., Whiteoak, J., Caswell, J., Wieringa, M., & Gough, R. 1993, ApJ, 412, 222

Pezzuto, S., Strafella, F., & Lorenzetti, D. 1997, ApJ, 485, 290

Podosek, F., & Cassen, P. 1994, Meteoritics, 29, 6

Schneider, G., et al. 1999, ApJ, 513, L127

Skrutskie, M., et al. 1990, AJ, 99, 1187

Smith, B., & Terrile, R. 1984, Science, 226, 1421

Sobolev, A., Cragg, D., & Godfrey, P. 1997, A&A, 324, 211

Strom, S., Edwards, S., & Skrutskie, M. 1993, in Protostars and Planets III, ed. E. Levy & J. Lunine (Tucson: University of Arizona Press), 837

Sylvester, R., Skinner, C., Barlow, M., & Mannings, V. 1996, MNRAS, 279, 915

Telesco, C., Becklin, E., Wolstencroft, R., & Decher, R. 1988, Nat, 335, 51

Telesco, C., et al. 2000, ApJ, 530, 329

Telesco, C., & Knacke, R. 1991, ApJ, 372, L29

Walsh, A., Burton, M., Hyland, A., & Robinson, G. 1998, MNRAS, 301, 640

Wooden, D., et al. 1999, ApJ, 517, 1034

Weinberger, A., et al. 1999, ApJ, 525, L53

Wyatt, M. 1999, PhD Thesis, University of Florida

Wyatt, M., Dermott, S. F., Telesco, C. M., Fisher, R. S., Grogan, K., Holmes, E. K., & Piña, R. K. 1999, ApJ, 527, 918

Discussion

Glenn Schneider: The peak of the flux density isophotes in your 18 μm images appear to be interior to the locations of the ansal flux density peaks in the NICMOS 1.1 μm images by $\sim$ 5 AU. Can you comment on this?

Charles Telesco: We found that about half of this offset is accounted for by the fact that the angular resolution of the mid-IR images is lower than that of the NICMOS images. That is, if we convolve the narrow-disk model inferred from NICMOS with our PSF, the lobes shift toward the star. We think the resto of the difference is due to the fact that there is a more gradual fall-off with radius of the surface density as you move from the density peak into the central hole than implied by the NICMOS model. Our dynamical models suggest that such a gradual fall-off would result from the fact that the particle orbits are slightly elliptical, and therefore particles have a slight radial excursion during each orbit. These radial excursions "soften" the edge of the central cleared zone. We discuss these issues in our February 2000 *Astrophysical Journal* article on HR 4796A.

Disks, Planetesimals, and Planets
ASP Conference Series, Vol. 219, 2000
F. Garzón, C. Eiroa, D. de Winter, and T. J. Mahoney, eds.

ADONIS Detection and Models of the Pre-Main-Sequence Disk of HD 100546

E. Pantin and P. O. Lagage

DSM/DAPNIA/SAp CE Saclay 91191 Gif sur Yvette Cedex, France

C. Waelkens and K. Malfait

Instituut voor Sterrenkunde, Celestijnenlaan 200B, B-3001 Leuven, Belgium

Abstract. We report the first resolved images of the circumstellar dust disk around the pre-main-sequence star HD 100546[1] in J and K_s. These near-infrared images obtained with the adaptive optics system ADONIS reveal a disk extending 2 arcsec from the star. Data modeling shows that the disk has a "β Pic-like" structure with a density peaking at a distance of 40 AU from the star.

1. Introduction

The discovery of infrared excesses around an important fraction of main-sequence stars (Aumann et al. 1984; Plets & Vynckier 1999) and attributed later to the presence of dust grains orbiting the star geometrically arranged in a disk has triggered a lot of studies because the latter may be linked to planetary formation. For a long time, the best example, the disk around the star β Pictoris, was the only one to be resolved both in the visible (Smith & Terrile 1984) and in the mid-infrared range (Lagage & Pantin 1994). Recent discoveries of "debris" circumstellar disks around relatively "old" pre-main-sequence stars such as HD 141569 (Weinberger et al. 1999; Augereau et al. 1999a) or HR 4796A, thanks to high-resolution observations in the visible/near-infrared range (Schneider et al. 1999; Augereau et al. 1999b) or mid-infrared images (Koerner et al. 1998; Jayawardhana et al. 1998) have shown the possibility of observing and studying precursors to main-sequence dust disks. These so-called "baby β Pic" dust disks are the denser precursors to main-sequence debris disks. *IRAS* and *ISO*/SWS observations have shown that they usually produce a huge infrared excess (typically 250 times the infrared excess produced by the β Pic disk); some of them, as HD100546 for instance, showing prominent signatures of crystalline water ice (Waelkens et al. 1996; Malfait et al. 1998), are particularly interesting targets in the visible/near infrared because of high particle albedo. Observing this class of disks at various stages of evolution will provide clues to a better un-

[1]Based on observations obtained at the ESO 3.6 m Telescope, La Silla, Chile (proposal 61D-0540).

derstanding of their mode of formation and dissipation. We report here on the first adaptive-optics images of the disk around HD 100546, an isolated "old" pre-main-sequence star. Using a scattering model by dust grains, we derive the morphology of the disk outwards of 10 AU and show evidence for a density maximum around 40 AU.

2. Observations

We observed the HD100546 star ($m_V = 6.7$) using the SHARPII+ camera coupled with the adaptive-optics system ADONIS, and mounted on the ESO 3.6 m Telescope at La Silla, Chile. A pre-focal-optics coronagraph (Beuzit et al. 1997) was used to reject the direct starlight and increase the integration time in each elementary exposure. The mask size we used has a diameter of $0.74''$. J (1.25 μm) and K_s (2.15 μm) exposures were performed on nights 23 and 24 of June, 1999. The seeing was quite variable, but had decreased to a value of $0.7''$ at the time of the J and K_s observations. The point spread function (PSF) was frequently monitored thanks to interlaced observations of two reference stars, HD 97218 and HD 101713. These reference stars were used later in the reduction process to remove the wings of the star PSF and for photometry calibration. Observations of empty fields were also performed to estimate and remove the background flux which can be important in the K_s band.

3. Reduction Procedure

First, standard reduction techniques were applied to the data: bias subtraction, flat-field correction. For each filter, we got a set of HD 100546 observations and corresponding PSFs. In spite of the use of a coronagraph mask, the disk surface brightness is still dominated by the star emission at any distance from the star. In order to get the disk image, one has to remove the starlight wings numerically. The straightforward subtraction of a scaled PSF to the HD 100546 image gives poor results because of slight shifts between the reference and object images, uncertainties in the references fluxes, and residual backgrounds (ADONIS bench emission, different airmasses). We developed a specific method to get an optimum subtraction. For each HD 100546 image (Obj) and its corresponding PSF (Psf), we have to find three parameters: a shift $(\delta x, \delta y)$ between the two images, a scaling factor, R, and a residual background, Bg. These parameters are estimated by minimizing the following error function:

$$J = \sum |\mathrm{Obj} - S(\mathrm{Psf}/R), \delta x, \delta y) - \mathrm{Bg}|, \tag{1}$$

where the S function stands for image shift. The sum is performed on a set of pixels (typically 1000) located in a region of the images where no disk emission is expected, and from which inaccurate ones are excluded (bad pixels, area with diffracted light from coronagraph support). The functional minimum is found using a zeroth-order minimization algorithm called the Powell method (Press 1993). In order to evaluate the errors in the subtraction process, test the stability of the PSF, and make sure that the disk obtained is not an artifact created by the process, the same subtraction process was applied to the two different PSFs.

Figure 1 shows the resulting images in the J and K_s filters of the disk and the corresponding residuals between the two reference stars. The position of the star under the mask, and therefore the disk center, is also carefully determined in this procedure thanks to offset observations of the reference stars.

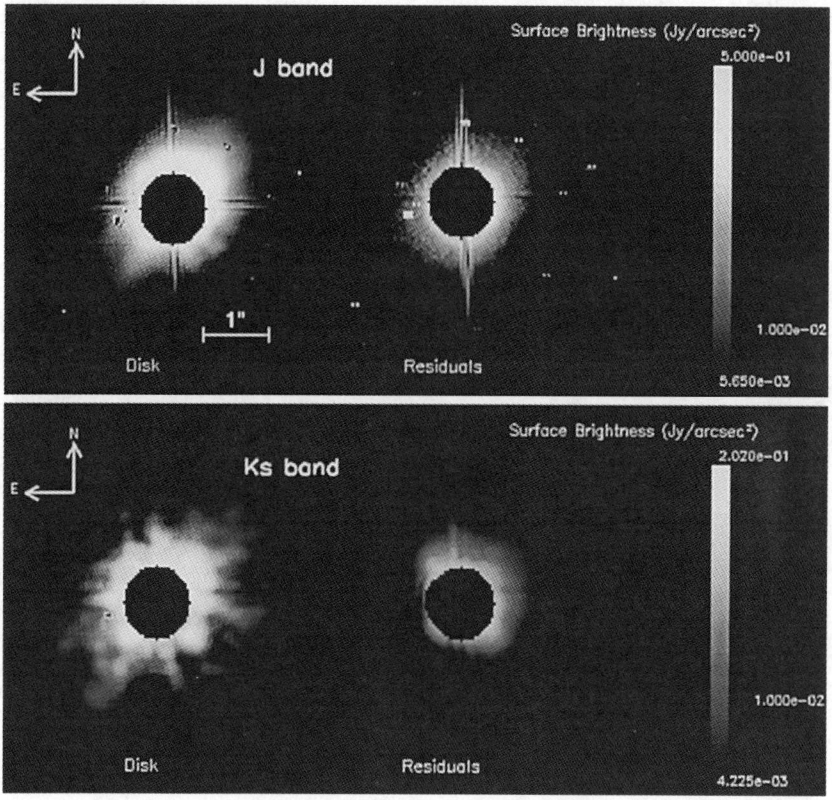

Figure 1. *Upper left:* The disk seen in the J band; north is to the *top* and east to the *left*. The pixel scale is 0.035 arcsec/pixel. *Upper right:* The corresponding residuals obtained when applying the reduction method described in the text. *Lower left:* the disk in the K_s band. *Lower right :* The residuals in the K_s band.

4. Results

Since the dataset in the J band is more accurate (because of a larger integration time, better observing conditions, and a better sensitivity of the whole system) than is the K_s band, the geometrical parameters and the disk model are based on the J-band image. We checked in a second step that the results are compatible with the K_s-band data.

4.1. Disk Geometrical Parameters

Given a disk inclination with respect to the line of sight, which is $\approx 45°$, we expect to observe some deviations from the perfect-ellipse case, due to non-isotropic scattering of the starlight by particles relatively big with respect to the wavelength of observation. One simple way to reduce this effect is to symmetrize the disk, i.e., construct an isotropic-like virtual disk by averaging the original disk image and its symmetrical image with respect to the major axis. Best-ellipse fitting on these symmetrized images gives the following results:

- a) Disk position angle on the sky: 37 ± 5 degrees with respect to the east;
- b) Ellipse major-to-minor axial ratio: 1.2, implying a disk inclination of 50 ± 5 degrees with respect to the line of sight (or from the edge-on case);
- c) Disk extension: ≈ 2.0 arcsec from the star corresponding to a distance to the star around 200 AU (assuming a distance to the Earth of 103 parsecs (van den Ancker et al. 1998) and a limiting sensitivity of 5 mJy/arcsec^2.

4.2. Photometry and Surface Brightness

From the final reduced images, we deduce the following photometric values :

- a) Disk flux in J band: 0.27 ± 0.05 Jy outside the the coronagraph mask and a maximum surface brightness of 0.5 ± 0.1 Jy/arcsec^2. From images of HD 100546 and corresponding reference star images (HD 101713 and HD 97218) obtained when shifting the targets outside the mask, we can deduce a total of scattered flux in the J band around 1 Jy. This result is compatible with the infrared excess computed thanks to a Kurucz model of the star (B9 V spectral type). Consequently, the disk flux outside the coronagraph mask represents roughly 30% of the total disk emission.
- b) K_s band: 0.13 ± 0.05 Jy outside the coronagraph mask and a maximum surface brightness of 0.17 ± 0.05 Jy/arcsec^2.

4.3. Model and Observational Fit

In order to derive the physical parameters of the disk, we built a numerical model of starlight scattering by an optically thin disk made of dust particles. Assuming that no multiple scattering is occurring in the disk, and since the exact physical properties of the particles is poorly known (composition, shape, size distribution), we chose to use a global, single-parameter, scattering phase function. The most practical one, although not realistic, is the Henyey–Greenstein phase function (Henyey & Greenstein 1941). The disk is assumed to be seen at an inclination derived above, and has a proper thickness using the same parameterization used by Artymowicz, Burrows, & Paresce (1989) in the case of β Pic. The disk midplane density is described by a scattering area which follows a series of broken power laws. We start with a first guess deduced from the radial surface brightness and we iterate on the parameters until we get a satisfactory fit along the four directions defined in Figure 3.

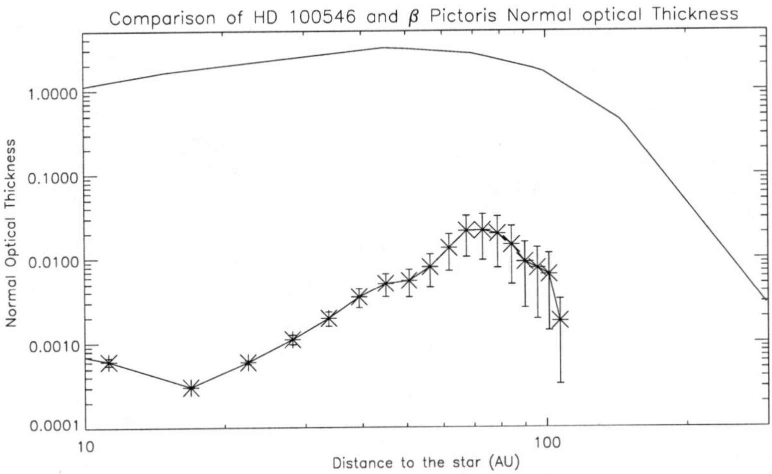

Figure 2. The normal optical thickness of the disk deduced from model fitting of the data. We have overplotted (*stars*) the normal optical thickness found in the case of the disk around β Pictoris (Pantin, Lagage, & Artymowicz 1997)

4.4. Fit Results

The best-fit model provides the following parameters:

- The Henyey–Greenstein phase function parameter found is $g = 0.2$, corresponding to dominant spherical particles size around 0.1 micron;
- A radial normal optical thickness fitted by a series of broken power laws whose exponents are: 0.6 in the range [10, 45] AU, -0.4 in the range [45, 70] AU, -1.35 in the range [70, 100] AU, and -5 beyond 100 AU (see Figure 2);
- An exponentially decreasing vertical profile and an opening angle around 0.1 radian;
- A total scattering area of 10^{29} cm^2 and a dust mass of 0.02 $M_\oplus$;
- An approximate disk surface density at the maximum around 3×10^{14} particle/m^2 assuming single-sized particles of 0.1 μm, or an equivalent maximum scattering area of the order of the unity/m^2.

5. Discussion and Conclusion

The radial surface-brightness distribution shows that the disk density peaks at a distance of 40 AU from the star corresponding to the edges of the planetary zone in our Solar System. The normal optical thickness found is more than 100 times

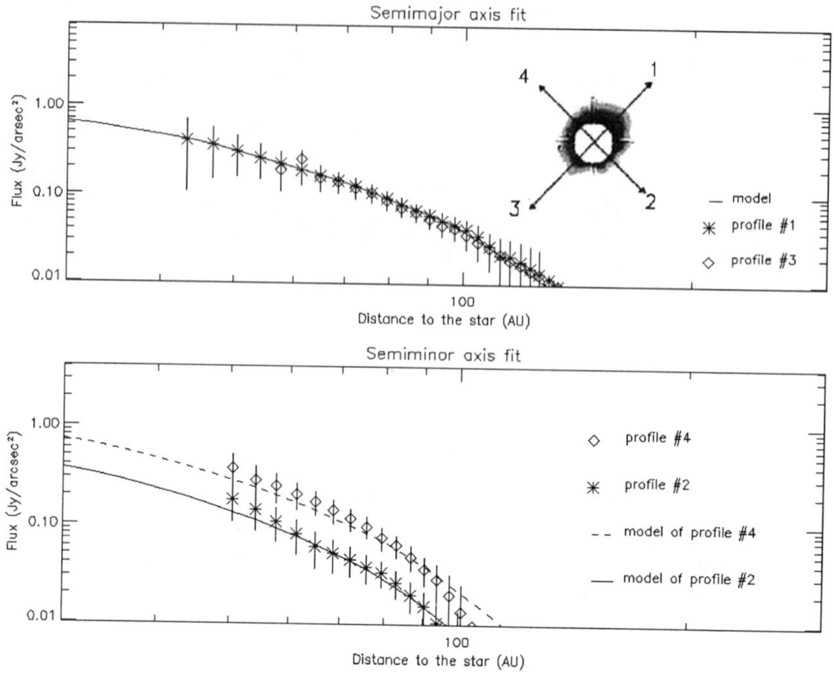

Figure 3. Fit of the data profiles. *Upper plot:* The modeled profiles (*plain lines*) and (*stars* and *diamond*) the data profiles 1 and 3 defined on the disk image inserted. *Lower plot:* Model of profile 4 (dominant forward scattering, *dashed line*) and model of profile 2 (*plain line*).

bigger than in β Pic. Simultaneously, the presence of sub-micron particle sizes, well below the radiation pressure cut-off in sizes, suggests that huge quantities of very small particles are produced continuously by evaporation or collisions of planetesimals. We are probably witnessing an analog of the young Kuiper Belt whose planetesimals are undergoing violent agitation. If the dust particles are the result of collisions, a high rate is required to produce the observed disk. This implies that gravitational perturbations occur frequently in this disk, maybe because of giant planets. If the disk is replenished by planetesimal evaporation, as suggested by Lecavelier des Etangs (1998) in the case of the β Pic disk, 10^9 planetesimals whose orbits have recently been perturbed by planet migration would be required.

Thanks to coronagraphic adaptive-optics observations we could detect the disk surrounding the PMS star HD 100546. This detection of a disk, which is believed to represent a class of precursors to β Pic-like disks, adds a new clue to our understanding of the formation and evolution of these debris disks. Other imaging observations of disks at various stage of evolution are necessary to sketch more closely the possible scenario of disk creation and evolution. This should be achieved by next-generation instruments such as VLT NAOS (adaptive optics) and VISIR (mid-infrared) instruments, or the NGST.

Acknowledgments. We would like to express our warm gratitude to the persons who helped in achieving these observations: D. Le Mignant, F. Marchis, O. Marco, V. Meriño, E. Wenderoth, and the whole 3.6 m team.

References

Augereau, J. C., Lagrange, A. M., Mouillet, D., Papaloizou, J. C. B., & Grorod, P. A. 1999b, A&A, 348, 557

Augereau, J. C., Lagrange, A. M., Mouillet, D., & Ménard, F. 1999a, A&A, 350, 51

Artymowicz, P., Burrows, C., & Paresce, F. 1989, ApJ, 337, 494

Aumann, H. H., et al., 1984, ApJ, 278, L23

Beuzit, J.-L., Mouillet, D., Lagrange, A.-M., & Paufique, J. 1997, A&AS, 125, 175

Henyey, L. G., & Greenstein, J. L. 1941, ApJ, 93, 70

Jayawardhana, R., et al. 1998, ApJ, 503, L79

Koerner, D. W., Ressler, M. E., Werner, M. W., & Backman, D. E. 1998, ApJ, 503, L83

Lagage, P. O., & Pantin, E. 1994, Nat, 369, 628

Malfait, K., Waelkens, C., Waters, L. B. F. M., Vandenbussche, B., Huygen, E., & de Graauw, M. S. 1998, A&A, 332, L25

Pantin, E., Lagage, P. O., & Artymowicz, P. 1997, A&A, 327, 1123

Plets, H., & Vynckier, C. 1999, A&A, 343, 496

Press, W. H. 1993, Numerical Recipes (Cambridge: Cambridge University Press)

Schneider, G., et al., 1999, ApJ, 513, L127

Smith, B. A., & Terrile, R. J. 1984, Science, 226, 1421

van den Ancker, M. E., de Winter, D., & Tjin A Djie, H. R. E. 1998, A&A, 330, 145

Waelkens, C., et al. 1996, A&A, 315, L245

Weinberger, A. J., et al. 1999, ApJ, 525, L53

Discussion

Bernard Foing: Based on your different images at 10 and 20 μm, is it justified to take ratio images to derive particle properties, as mentioned in C. Telesco's paper, even though the emission does not originate from the same region? How, then, can T, the particle size, and density distribution be derived as functions of distance?

Pierre-Olivier Lagage: In the case of a disk composed of a thin ring of matter, as discussed by C. Telesco, it is justified to take ratio images at 10 and 20 μm to derive particle properties.

In the case of the disk around β Pictoris, it is not justified; the 10 and 20 μm emitting regions are different. What we have done is to assume a grain size distribution and composition (silicate, as suggested by the 10 μm features) in order to derive the temperature of the grains, and then to retrieve from the 10 μm image the grain spatial density distribution. Then we made predictions on how the disk should appear at 20 μm and compared the predictions with the observations presented here.

Disks, Planetesimals, and Planets
ASP Conference Series, Vol. 219, 2000
F. Garzón, C. Eiroa, D. de Winter, and T. J. Mahoney, eds.

Pericenter Glow: A Signature of Hidden Planets in HR 4796?

M. C. Wyatt

UK Astronomy Technology Centre, Royal Observatory, Blackford Hill, Edinburgh, EH9 3HJ, UK

S. F. Dermott and C. M. Telesco

Department of Astronomy, University of Florida, 211 Bryant Space Science Center, Gainesville, FL 32611, USA

Abstract. Mid-IR observations of the disk of dust around the main sequence A0 star HR 4796A show an asymmetry which may be indicative of gravitational perturbations from an unseen planetary system. This paper describes the *pericenter glow* mechanism by which a planetary system could be causing the observed asymmetry and discusses the implications of the observations for the existence of planets hiding in the HR 4796 disk.

1. Introduction

The disk of dust around the A0 V star HR 4796A provides an important opportunity for studying the planetary formation process, since the system's age, ~ 10 Myr, places it at an evolutionary epoch when any planetary system should be substantially formed. The mid-IR and near-IR emission from this disk has been shown to be concentrated in two lobes, one either side of the star (Schneider et al. 1999; Telesco et al. 2000). This indicates that the disk is being observed nearly edge on, and that its inner region, ~ 40 AU in radius from the star, is almost completely devoid of dust. Like the dust disks found around the main-sequence stars β Pictoris, Fomalhaut, and Vega, its optical depth implies that the collisional lifetime of the 1–100 μm-sized dust that is seen is much shorter than the age of the system, implying that this dust must be continually replenished (Backman & Paresce 1993). The most likely source of dust in HR 4796 is collisions between a population of larger bodies which, like the objects in the Kuiper Belt in the Solar System, formed due to the aggregation of interstellar dust grains in the protoplanetary nebula, but which did not manage to grow to sufficient size to develop into a planet. This picture is supported by the existence of the Solar System-sized central cavity which implies that, like the Solar System, bodies that are large enough to be able to clear a cavity (i.e., planets) were able to form there. However, the existence of such planets is, as yet, purely speculative.

It is known from observations of the disk of dust in the Solar System, the zodiacal cloud, that the gravitational perturbations of a planetary system to the orbits of particles in a debris disk can cause asymmetries in that disk's structure (Dermott et al. 1998). Such asymmetries may be present in the observed structure of the β Pictoris dust disk (Kalas & Jewitt 1995). Observations of the HR 4796 disk also show an interesting asymmetric feature (Telesco et al. 2000): the mid-IR lobes appear to be of unequal brightness. In a recent paper, Wyatt et al. (1999) modeled these observations and showed that such a lobe brightness asymmetry is to be expected if there is another body orbiting HR 4796A that is on an eccentric orbit, since the long-term effect of the gravitational perturbations from this body would force the center of symmetry of the disk to be offset from the star in a direction away from the pericenter of the body's orbit, thus causing the dust near this forced pericenter to glow.

The dynamics behind the pericenter glow phenomenon is described in Section 2 and the pertinent results of the Wyatt et al. model are presented in Section 3. The implications of the model on the existence of planets hiding in the cavity region is discussed in Section 4.

2. Pericenter Glow

The long-term gravitational perturbations from a planetary system to the orbits of any particles in that system are called secular perturbations (for a good review of the dynamics involved, see Murray & Dermott 1999). Their effect on a disk particle's orbit can be demonstrated using Figure 1*a*. A particle on an orbit described by the eccentricity vector SD has an eccentricity, e, given by the length of the vector and a pericenter orientation (relative to an arbitrary direction), $\tilde{\omega}$, given by the angle that vector makes with the x-axis. The action of secular perturbations is to decompose the particle's eccentricity vector into forced (SC) and proper (CD) components. The forced component is that imposed on the particle's orbit by the planetary system, which is the same for all particles orbiting at the same distance from the star, and the proper component precesses anticlockwise about the forced component on a circle with a radius determined by the particle's intrinsic orbital properties.

Consider the break-up of one planetesimal in the disk, the orbit of which is described by the eccentricity vector SD in Figure 1*a*. Initially, all collisional fragments have very similar orbits to that of the original planetesimal and so have eccentricity vectors that are almost coincident. The secular perturbations of any planets in the system then make these fragments' proper elements precess about the forced elements. However, since their orbits are slightly different, due to the velocity dispersion imparted in the collision, they precess at slightly different rates, which means that after a few precession timescales their eccentricity vectors are no longer coincident but randomly distributed about the *dotted circle* in Figure 1*a*. There is evidence that such an evolution has occurred in the asteroid belt, where there are families of asteroids the members of which orbit at the same distance from the Sun, and which have eccentricity vectors spread around circles centered on the forced elements known to be imposed on their orbits by the Solar System's secular perturbations; each family is thought to have been created by the break-up of a much larger asteroid. Figure 1*b* shows the eccentricity vectors of the asteroids in the Eos, Themis, and Koronis asteroid families.

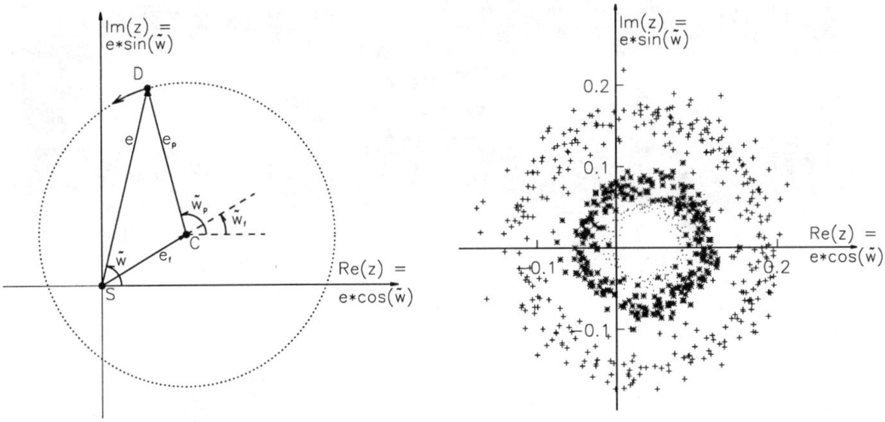

Figure 1. *a)* The evolution of the eccentricity vector of a debris disk particle due to the secular perturbations from a planetary system (see text for discussion). *b)* The eccentricity vectors of asteroids in the Themis (+), Eos (*), and Koronis (.) families.

It can be shown by considering the geometry of each of the orbits in the eccentricity vector distribution of Figure 1*b* that the disk of material comprising one family's members forms a uniform torus that is rotationally symmetric, not about the star, but about a point that is offset from the star in a direction away from the forced pericenter direction, $\tilde{\omega}_f$, by an amount that is proportional to the forced eccentricity imposed on the family's orbits, e_f (Wyatt et al. 1999). Since the offset direction, $\tilde{\omega}_f$, and amplitude, e_f, is the same for all families at the same distance from the star, a relatively narrow disk formed by the break-up of Kuiper Belt-like objects should appear rotationally symmetric about a point offset from the star, the location of which is determined solely by the secular perturbations imposed on the disk by any planets in the system, which may themselves be unseen. Such an offset would cause material on the forced pericenter side of the disk to be closer to the star than that on the forced apocenter side, thus causing that material to be hotter, and therefore brighter, a phenomenon we call pericenter glow. This is the phenomenon that we think may be causing the HR 4796 disk's lobe brightness asymmetry.

3. The HR 4796 Disk Model

It was the aim of the modeling to see whether the pericenter glow phenomenon could indeed be responsible for the brightness asymmetry, and if so, to find how large a forced eccentricity, e_f, (and hence an offset) would have to have been imposed on the disk to be causing the observed 5% asymmetry. The essence of the modeling is a program called SIMUL which was developed by Stan Dermott and his Solar System Dynamics group at the University of Florida for the study of the detailed observed structure of the zodiacal cloud. This program creates a three-dimensional model of the distribution of dust in a system given the

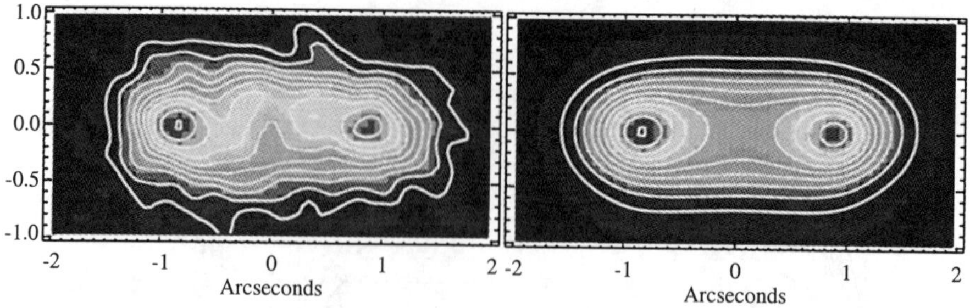

Figure 2. 18 μm images of the HR 4796 disk: the Telesco et al. (2000) observation (*left*), and the Wyatt et al. (1999) model of this observation (*right*). The contours in both of these images are linearly spaced at 58, 93, 128, 163, 198, 234, 304, 339, and 374 mJy/arcsec^2.

distribution of the orbital elements of the constituent particles, thus providing a unique way of linking a disk's observed structure with the physical and dynamical evolution of its constituent particles.

The distribution of orbital elements of the dust particles in the HR 4796 disk model was taken from a combination of our understanding of the consequences of secular perturbations (e.g., the distribution of their eccentricity vectors was assumed to be circularly symmetric about a forced eccentricity similar to Figure 1*b*), and inferences from the observations (e.g., the distribution of their semi-major axes was assumed to follow a power law between inner and outer bounds). The details of the model can be found in Wyatt et al. (1999); it is only the asymmetry that will be discussed here. It was found that having a forced eccentricity imposed on the disk does indeed cause the disk to have an offset center of symmetry, and that this could cause the disk's lobes to appear to have asymmetric brightnesses. The forced eccentricity that was needed in the model to cause the 5% asymmetry was found to depend on the orientation of the forced pericenter to the line of sight; for the majority of the possible forced pericenter orientations, it was found that the observations could be matched with a forced eccentricity of 0.02, while for the remaining orientations a higher forced eccentricity was needed. The Telesco et al. observation and the final Wyatt et al. model of that observation (which has $e_f = 0.02$) are shown in Figure 2. This figure shows that as well as fitting the asymmetrical heights of the lobes, the model also gives an extremely good fit to the lobes' horizontal and vertical structure.

4. ... Hidden Planets in HR 4796?

What can we conclude from the modeling about whether there are any planets hiding in the HR 4796 disk? What the modeling explicitly shows is that if $e_f = 0.02$ is imposed on the particles in the disk's lobes at $a \approx 62$ AU, then, depending on the orientation of the disk to our line of sight, a 5% asymmetry could be observed. However, it does not specify the origin of e_f, which could come from one or more massive perturbers either interior or exterior to the disk.

If we assume that there is a planetary system inside the disk cavity that is causing the hole, then it would be the outermost planet of that system that affects the outer disk most. By analogy with the Kuiper Belt, the inner edge of which is at the 2:3 resonance with Neptune, the orbit of this outermost planet could have $a_{\rm pl} = 62 * (2/3)^{2/3} = 47$ AU. This means that to cause $e_{\rm f} = 0.02$ at 62 AU its orbit only needs to have $e_{\rm pl} > 0.023$. Furthermore, assuming that the timescale for this asymmetry is the same as the secular precession timescale, which is inversely proportional to the mass of the perturbing planet, then the age of the system implies that this planet need only have a mass of $> 10\ M_\oplus$ to be causing the asymmetry. This leads to the conclusion that if any planets did form in the disk, then unless they were born with a negligible eccentricity (which could, for example, be possible due to the interaction of the planet with a massive disk, Ward & Hahn 1998), or an adverse observing geometry prevents it, the lobes would inevitably have asymmetric brightnesses.

In this system, however, we may not need to invoke hidden planets to explain the forced eccentricity in the lobes because there is another perturber in the system, the binary companion HR 4796B. The orbit of this binary is unknown. Observations in 1931 put the projected separation and position angle of the two stars at [$6\rlap{.}''9$, 226°], while observations in 1992 put them at [$7\rlap{.}''7$, 225°] (Jura et al. 1993). If both of these observations were absolute, then the two stars cannot be bound, since it can be shown that for bound orbits, the motion of the star on the sky in $''$/yr:

$$v_{\rm obs} < 2\pi\sqrt{2(M_A + M_B)/\rho'' d^3}, \qquad (1)$$

where $M_A + M_B \approx 2.56$ is the total mass of the binary pair in solar masses, $d = 67.1$ is the distance to the stars in pc, and ρ'' is the observed projected separation in $''$; i.e., the change in 61 years should have been less than $0\rlap{.}''6$. The most likely reason for the discrepancy is observational error and the two stars are bound; however, a preliminary search through archival observations was unable to determine whether this is indeed the case. If, for pedagogical reasons, we assume that the semi-major axis of the orbit is equal to its most statistically likely value[1] of $a_B = 697$ AU, then the orbit needs an eccentricity of $e_B > 0.18$ for the binary to be causing the asymmetry. Such a high eccentricity is to be expected for a binary this wide (Duquennoy & Mayor 1991) and would be consistent with a high apparent motion on the sky. In other words, the lobe asymmetry is also likely to be an inevitable consequence of the presence of the binary companion. Thus it appears that it would have been rather surprising if an asymmetry had not been observed in this disk!

It is most likely that both HR 4796B and the outermost planet of a planetary system are perturbing the disk. In this case it is difficult to make any quantitative statements about such a planet without first determining the orbit of HR 4796B. However, we do know that if the planet is very massive (e.g., $> 0.1\ M_{\rm Jup}$, where $M_{\rm Jup} = 10^{-3}\ M_\odot$ is the mass of Jupiter), then the magnitude and orientation of the asymmetry will be determined by the orbit of the planet, while it will be determined by the orbit of HR 4796B if the planet is much

[1] $\log a'' = \log \rho'' + 0.13$, Duquennoy & Mayor (1991).

smaller than this. We also know that if the two perturbers have different orbital planes, and the mass of the planet were $\sim 0.1\ M_{\rm Jup}$, then the disk could appear warped (Wyatt et al. 1999). Clearly, if the orbit of HR 4796B was known, we would be able to determine its effect on the disk, and we would be able to set some quantitative constraints on any planets in the system; a detailed study that determines the orbit of HR 4796B is therefore warranted.

Of course, the asymmetry may not be a consequence of pericenter glow. Without detouring from the planetary theme, we can think of at least one mechanism for forming such an asymmetry: Many of the particles in the disk could be trapped in the 2:3 resonance of the outermost planet in the system, just as many Kuiper Belt objects are trapped in that of Neptune. This would give the disk clumpy structure that would follow the planet in its orbit. If the asymmetry is caused by such a clump that is currently residing in one of the lobes, then in half a planetary orbital period, ~ 100 years, the asymmetry will be reversed. It would not, however, take 100 years to ascertain whether this was the case or not; such moving structure could be detected with current technology in ~ 3 years, since this is the time it would take for a clump (which could be detected in the residuals of a high-S/N observation once a smooth model has been subtracted) to move 1 pixel on the Telesco et al. image.

In conclusion, because of the existence of HR 4796B, which could be causing the asymmetry, the fact that HR 4796 is a 10-Myr-old debris disk with a hole about the size of the Solar System remains the best evidence for planets in this system. However, the modeling has shown that the observed asymmetry could have been caused by the perturbations of a 10 $M_\oplus$ planet and has shown the observations that are necessary if we want to be able to use the disk observations to make any quantitative statements about the planetary system.

References

Backman, D. E., & Paresce, F. 1993, in Protostars & Planets III, eds. E. H. Levy, & J. Lunine (Tucson: Univ. Ariz. Press), 1253

Dermott, S. F., Grogan, K., Holmes, E. K., & Wyatt, M. C. 1998, in Proc. Exozodiacal Dust Workshop, ed. D. E. Backman et al., 59

Duquennoy, A., & Mayor, M. 1991, A&A, 248, 485

Jura, M., Zuckerman, B., Becklin, E. E., & Smith, R. C. 1993, ApJ, 418, L37

Kalas, P., & Jewitt, D. 1995, AJ, 110, 794

Murray, C. D., & Dermott, S. F. 1999, Solar System Dynamics (Cambridge: Cambridge University Press)

Schneider, G., Smith, B. A., Becklin, E. E., Koerner, D. W., Meier, R., Hines, D. C., Lowrance, P. J., Terrile, R. J., Thompson, R. I., & Reike, M. 1999, ApJ, 513, L127

Telesco, C. M., Fisher, R. S., Piña, R. K., Knacke, R. F., Dermott, S. F., Wyatt, M. C., Grogan, K., Holmes, E. K., Ghez, A. M., Prato, L. A., Hartmann, L. W., & Jayawardhana, R. 2000, ApJ, 530, 329

Ward, W. R., & Hahn, J. M. 1998, AJ, 116, 489

Wyatt, M. C., Dermott, S. F., Telesco, C. M., Fisher, R. S., Grogan, K., Holmes, E. K., & Piña, R. K. 1999, ApJ, 527, 918

Discussion

Glenn Schneider: What is the sensitivity of the model in changing the initial constraints of the brightness anisotropy of the ansal flux density? At shorter wavelengths (1.1–1.6 μm) we see an $\sim$ 10% difference in ansal flux, whereas at 18 μm Telesco sees half of this. Does this change the anticipated location of the perturber?

Mark Wyatt: No. The first point to mention is that we expect to see different brightness asymmetries at different wavelengths, and that it would certainly be a useful constraint on the model for it to be able to reproduce the asymmetries observed at all wavelengths. As for the effect of a different brightness asymmetry (in the mid IR) on the location of the perturber, if the 18.2 μm asymmetry had been 10% instead of 5%, this would have implied a forced eccentricity imposed on the disk of greater than $\sim$ 0.04 (rather than 0.02); it would, however, have no effect on the discussion of the radial location of the perturber, which cannot be inferred from the modeling.

Thomas Henning: Is there any reliable estimate of the gas content of the disk?

Mark Wyatt: There is an upper limit to the mass of gas in the system of between 1 and 7 $M_\oplus$ (Greaves et al. 2000). For comparison, the mass of the dust disk has been estimated to be between 0.1 and 1 $M_\oplus$ (Jura et al. 1998).

Hubertus Klahr: Did you estimate the mass reservoir (5 $M_\oplus$) one needs in order to replenish the dust over 10 Myr?

Mark Wyatt: If the dust we are seeing is indeed blown out of the system by radiation pressure, then the mass of emitting dust ($\sim 1.4 \times 10^{-3}$) and the blow-out timescale ($\sim$ 370 yr) imply a mass-loss rate of 4×10^{-6} $M_\oplus$/yr. Assuming a constant mass-loss rate, this implies that the disk is currently 40 $M_\oplus$ less massive than it was at its birth. To make any more quantitative statements, one needs to consider the physics of grain–grain collisions. Of course, we still do not know whether or not the particles we are seeing are really being blown out of the system by radiation pressure.

Disks, Planetesimals, and Planets
ASP Conference Series, Vol. 219, 2000
F. Garzón, C. Eiroa, D. de Winter, and T. J. Mahoney, eds.

Submillimeter Studies of Dusty Debris around Nearby Stars

J. S. Greaves and W. S. Holland

Joint Astronomy Center, 660 N. A'ohōkū Place, University Park, Hilo, Hawaii 96720, USA

Abstract. Imaging of remnant disks of dusty debris around main-sequence stars gives valuable clues to the planetary formation process. Submillimeter observations of such disks are relatively new, but with technological advances we now have the capability to directly image the dust distributions, with negligible signal from the star itself. This review discusses recent results for the nearby stars whose disks were first detected in the far-infrared by *IRAS*—in particular, the "Big Four" prototypes: Vega, Fomalhaut, β Pic, and ϵ Eri. The disks appear very diverse, and some surprising features were uncovered, including dominance of grains of particular sizes, and clumping within disks that may trace perturbations by planets. Generally, the amounts of dust detected are very small (of the order of the mass of the Moon), and the grains need to be continually replenished, for example from collisions of larger bodies. The decline of dust mass with stellar age allows us to establish a rough timescale of $\lesssim$ 10 Myr for the formation of rocky planets and the solid cores of gas giants. Finally, we discuss future submillimeter observations that will reveal the diversity of disks around far more nearby stars, and probe scales as small as a few AU.

1. Introduction

"Vega-excess" stars were discovered serendipitously during the *IRAS* mission (Aumann et al. 1984), when emission above photospheric levels was seen around otherwise normal main-sequence stars. This excess, generally seen at 60 and 100 μm wavelengths, was attributed to orbiting dust particles heated by the central stars. Thus the question was raised: is this the debris left over from the formation of planets?

Follow-up observations have been hampered by the inaccessibility of far-infrared wavelengths from the ground and the faintness of scattered light from the disks in the near-infrared and optical. Until recently, only β Pictoris had been studied extensively (e.g., Kalas & Jewitt 1996), since it is an edge-on system and so the disk appears very clearly in coronagraphic data. However, technology has advanced significantly recently at wavelengths that have moderate transmission from the ground and are nearer the peak of the thermal emission from warm dust; namely, the mid infrared and submillimeter. In this review, the emphasis is on recent results for the nearest "Vega-excess prototypes", obtained with a new submillimeter camera on the James Clerk Maxwell Telescope.

2. Overview of Submillimeter Observations

Submillimeter imaging data reveal the *large-scale* dust distribution, as much as hundreds of AU from the star. The downside is that spatial resolution is low with single-dish telescopes, but 8–15″ beam sizes on the JCMT correspond to only 25–50 AU for the closest star in our survey. The submillimeter is well matched to the cool dust spectrum, which rises into the far infrared, and the emission is optically thin, allowing accurate mass estimates. Also, since we obtain *direct* images of the thermal grain emission, we eliminate the uncertainties involved in modeling scattered light. Finally, unlike at shorter wavelengths, stellar photospheric emission is very weak and can be subtracted with submJy errors. Thus we can be confident that we are really studying the structure of the disk, not any residuals from subtraction of stellar light.

In the last decade, pioneering submillimeter and millimeter observations were made with single-element detectors. For example, Zuckerman & Becklin (1993) mapped Vega and Fomalhaut using the photometer on the JCMT; very time-consuming deep maps totaling 7–9 points did detect emission at 800 μm, and also gave an impression of the size of the dusty regions. However, some controversies emerged: for example, towards Vega some emission was detected offset to the north-east of the star, and towards ϵ Eridani, the fluxes appeared somewhat inconsistent with those at 1300 μm (Chini et al. 1991), obtained using slightly different beam sizes. Both of these anomalies later turned out to be due to real source structure, and it is impressive that the earlier observers deduced this possibility, even with their very limited data.

2.1. New Instrumentation

The data presented here were obtained with the one of the first submillimeter cameras, the Submillimetre Common-User Bolometer Array (SCUBA), on the James Clerk Maxwell Telescope in Hawaii. The Mauna Kea site has excellent transmission in the submillimeter, and the 15 m JCMT is the largest telescope dedicated to this regime. SCUBA has been in use since 1997, and offers simultaneous imaging at 850 and 450 μm over a 2.3 arcmin field of view, as well as other wavelengths and observing modes in the 350 to 2000 μm atmospheric windows (Holland et al. 1999).

The results discussed here are primarily from 850 μm images, with some data from the 450 μm maps (which are noisier due to poorer atmospheric transmission). The observing mode used was "jiggle-mapping", which uses small offsets of the telescope's secondary mirror to fill in a grid of points that are spaced by half a beam width (Nyquist-sampled) at both wavelengths. The data are calibrated using a skydip (a series of measurements of the sky brightness at different elevations) to measure atmospheric opacity, and a map of a standard source to calculate the flux conversion factors (Jy per measured volt in a map pixel).

2.2. Survey Methodology

With the joint advantages of improved single-pixel sensitivity (an order of magnitude better than the previous single-bolometer instrument) and a multi-pixel capability, it was possible for the first time to undertake an imaging survey of nearby Vega-type stars. The goals were to obtain high-resolution maps of the dust emission, with beam sizes small enough to potentially resolve "holes" swept out by the formation of planetary systems, and to study the evolution of dust masses and grain sizes at different stellar ages.

The data have been modeled (Dent et al. 2000) using a full 2D continuum radiative transfer model for thin disks. Input parameters include the stellar luminosity, temperature and distance, and the 2D density structure of the disk and its inclination to the line of sight. For simplicity a single representative grain size is used, and the emissivity is dependent on a critical wavelength of the order of a, the grain radius. At short wavelengths, the emission follows a blackbody spectrum, while longwards of the critical wavelength the emissivity depends on a power law with spectral index β. As shown by Dent et al., varying just these parameters gives excellent fits both to the images and spectral energy distributions (SEDs).

Numerical simulations (Liou & Zook 1999; Liou et al. 2000) have also been performed, initially for the Solar System and ϵ Eri, to study the perturbing effects of planets on the disks. These simulations include one or more giant planets plus dust grains acting as test particles. The dust is assumed to be released from asteroid collisions and then operated on by Poynting–Robertson (radiation) drag and the stellar wind. Output consists of model images as a function of planet mass, location, observer's viewing angle, etc., and perturbations within the dust disk can be matched to observed structures.

Table 1 lists some properties of the four observed stars along with a summary of other relevant data, including imaging at other wavelengths and estimates of stellar inclinations from rotational velocities.

Table 1. Stellar properties: type, luminosity, distance from the Sun, estimated age, and summary of observations relevant to disks. *ISO* data have been published recently by Walker & Heinrichsen (2000); for other sources see Holland et al. (1998) and Greaves et al. (1998).

	Fomalhaut (α Piscis Austrini)	β Pictoris	Vega (α Lyrae)	ϵ Eridani
Stellar type	A3 V	A5 V	A0 V	K2 V
Luminosity ($L_\odot$)	13	9	60	0.3
Distance (pc)	7.7	19.3	7.8	3.2
Estimated age (Myr)	200	~ 20	350	500–1000
Previous data	*IRAS* *ISO* close to edge on?	*IRAS* *ISO* optical near infrared edge on	*IRAS* *ISO* close to face on?	*IRAS* *ISO* close to face on?

3. Results

Figures 1–4 show the 850 μm images of each of the four stars (Holland et al. 1998; Greaves et al. 1998), which are now discussed individually.

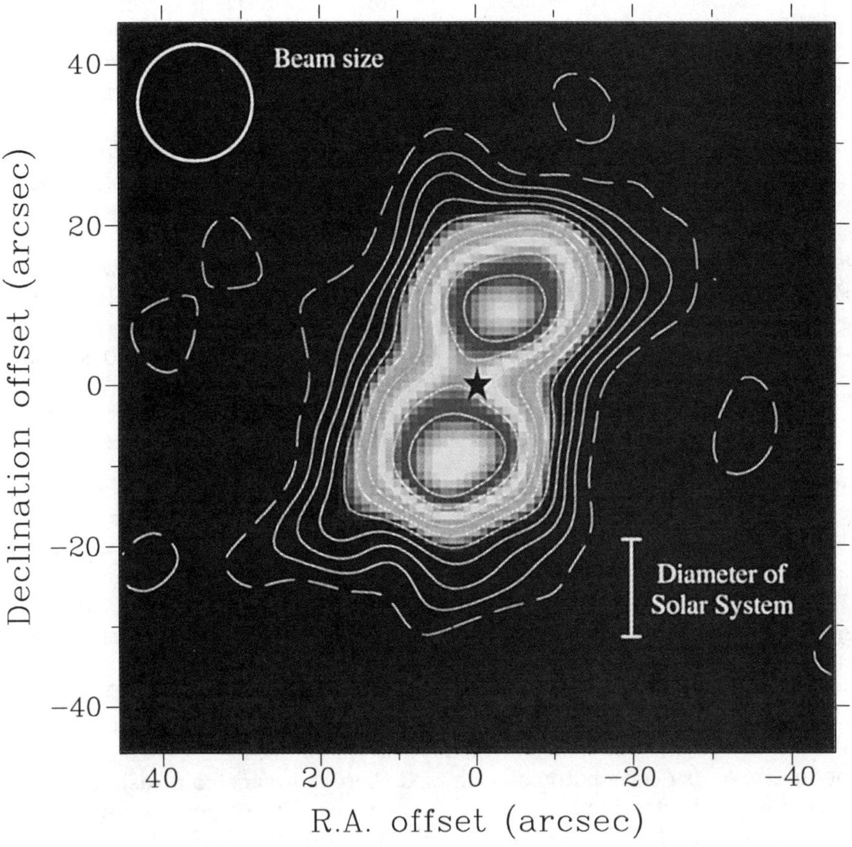

Figure 1. 850 μm image of Fomalhaut. Contours are 2σ (*dashed*), 3σ, 4σ, ...; also marked are the size of the telescope beam and the Solar System if viewed by the observer at Fomalhaut's distance.

3.1. Fomalhaut

Towards Fomalhaut, two peaks are seen, both with fluxes of 28 mJy and offset symmetrically from the star by about 10 arcsec (80 AU). The image structure is consistent with an edge-on torus, with dimensions of approximately 315 by 140 AU (after deconvolving from the 15 arcsec beam). There is a very distinct dip between the two peaks, towards the star, which we interpret as a cavity with much less dust emission. (The cavity must actually be *more* empty than it appears, since the peaks are smeared out by the beam, and tend to artificially fill in the hole.) One possibility is that this cavity has been cleared out by dust coagulating into planetesimals; since large "rocks" do not emit significantly in the submillimeter compared to the same mass in dust grains, this will reduce the central flux.

The SED for Fomalhaut shows a well-defined peak in the far-infrared, which can be fitted with a dust opacity index $\beta = 1.1$ and a dust temperature of 40 K (see Dent et al. 2000). The deduced representative grain size is 30 μm, and thus grain replenishment is required, as otherwise Poynting–Robertson drag would have forced grains this large into the star at a 200 Myr stellar age. The excellent fit obtained for a single temperature suggests the grains lie in a ring between a limited range of radii. A range of locations, or of grain sizes (with consequently different temperatures), would broaden the FIR peak more than is observed.

The model image corresponds to the data for disk inner and outer radii of 70 and 140 AU, and a dust depletion of one hundred times lower density in the inner hole. This depletion cannot be explained by P–R drag, as although this moves grains in from the outer disk, dust then piles up near the star with a $1/r$ density profile that is not seen. Grain mantles could sublime, thus reducing the emission cross-sections, but this is also an unlikely explanation for the cavity as the grains are colder than the sublimation temperatures of roughly 100 K. Thus the most likely explanation for the cavity is planetesimal formation.

3.2. β Pic

Towards β Pic, the dust is elongated along P.A. = 32°, close to the optical major axis. Emission extends out to at least 250 AU radius from the star, but we do not resolve the central cavity (Lagage & Pantin 1994). Radiative-transfer modeling (Dent et al. 2000) shows that both the submillimeter image and the SED can be fitted using a dust distribution taken from optical data, and thus the scattering and emitting grains must be co-spatial. The characteristic grain size is $a = 10$ μm, with a spectral index of $\beta = 0.8$.

Beyond the main dust disk, there is also a secondary "blob" of emission located 700 AU to the south-west of the star. This is a 7 σ detection, and has no counterpart on the opposite side of the disk (in spite of the 2 σ contour here on the image, later deep photometry showed there is in fact no measurable flux towards this position). Some possible explanations for the south-west feature include a fragmented disk—but this is unlikely on only one side of the star; a background unresolved galaxy—but in this case, the flux of 18 mJy would make it almost the brightest known in SCUBA extragalactic surveys, and it would have to be coincidentally close to the β Pic disk plane; or a second dust-enshrouded source. Although the disk is asymmetric on smaller scales and a fly-by has been suggested (Kalas et al. 2000) that could fragment the dust distribution, the discrete nature of the blob suggests that this is an unlikely explanation, and a second source is more plausible. However, we need to invoke an accreting (i.e., optically hidden) planet or brown dwarf, at the late time of $\sim$ 20 Myr and large separation of at least 700 AU.

3.3. Vega

Towards Vega, the dust forms an approximately circular structure, but the peak emission of 16 mJy is offset about 70 AU to the north-east of the star. This is in fact consistent with the point-by-point map made many years earlier by

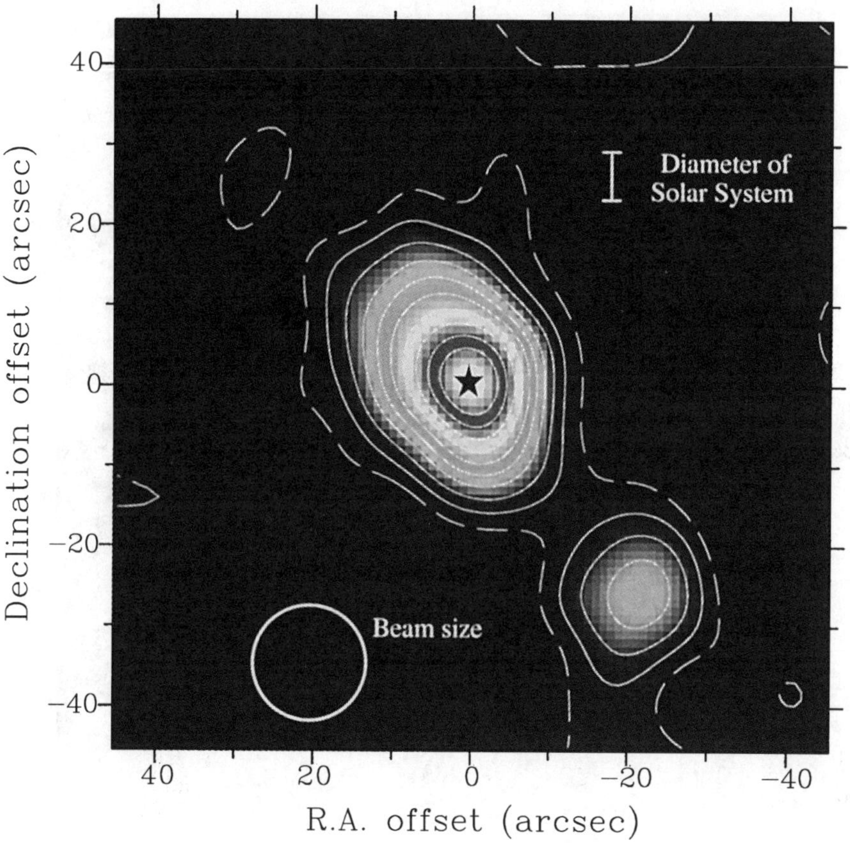

Figure 2. 850 μm image of β Pictoris, annotated as in Figure 1.

Zuckerman & Becklin (1993). After removal of the photospheric contribution, there is a small dip towards the star and a fainter secondary peak to the south-west. This gives the image a superficial resemblance to Fomalhaut. However, an interpretation using an edge-on torus is problematic, as the low rotational velocities from stellar lines suggest Vega is viewed close to pole on, and the outer contours are roughly circular as would be expected for a face-on disk, coplanar with the star.

The asymmetry could be explained by a body in orbit, which perturbs the dust (e.g., Gorkavyi et al., this volume, p. 51). No companion is seen in 2.2 μm images down to a limiting magnitude of 16, so any such body must still be dust-enshrouded. A more remote possibility is that dust has been liberated by the collision of two major planetesimals, but this would be a rare event given the dust mass involved (Table 2), and would not explain the fainter secondary peak to the south-west of the star.

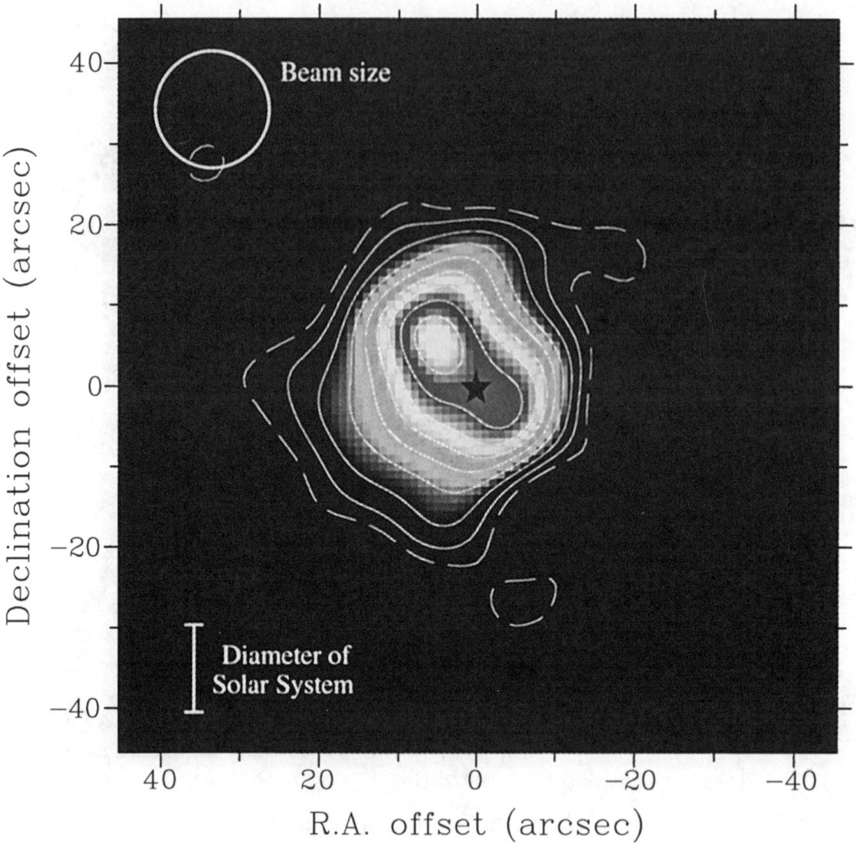

Figure 3. 850 μm image of Vega, annotated as in Figure 1.

3.4. ϵ Eridani

This star differs from the other three in being closer in type to the Sun—K2 compared to the Solar G2, whereas Vega, β Pic, and Fomalhaut are much more luminous A-type stars. The dust will be cold (Table 2) since stellar heating is not very effective from a K star, and the emission is detectable only because the star is extremely close to us, at a distance of 3.2 pc. Even so, the peak emission at 850 μm is only 8 mJy in the brightest parts of the image.

A ring of dust emission is seen around the star, with a central cavity that could again be explained by clearing due to planetesimal formation. The ring is very similar to the Kuiper Belt in size (Greaves et al. 1998), with a cavity equivalent in scale to Neptune's orbit and most of the dust lying between 35 and 75 AU from the star. Thus we could be looking at an analog to the Solar System at an age about 10–20% of that of the Sun. The mass of the ϵ Eri ring is lower than for the Kuiper Belt, but we detect only the dust component with SCUBA and miss any large "rocks", whereas the KB mass (estimated at 0.03–0.4 $M_{\oplus}$) is mainly in comets. As expected if dust is progressively absorbed into bigger

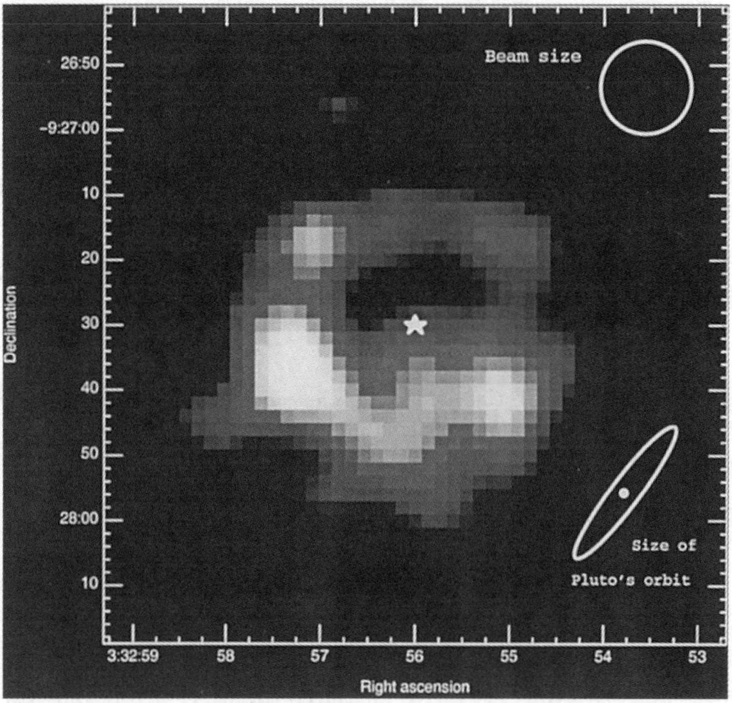

Figure 4. 850 μm image of ϵ Eridani, annotated similarly to Figure 1.

bodies or radiatively expelled, the present-day mass of dust in the KB is tiny. Values of the order of 10^{-5} $M_{\oplus}$ (see discussion in Teplitz et al. 1999), compared to the ϵ Eri dust mass of around 10^{-3} $M_{\oplus}$, suggest a timescale longer than a Gyr for elimination of the majority of the dust. Interestingly, the partially cleared appearance of the ϵ Eri cavity is consistent with Solar System history in that the Earth was heavily bombarded up to about 600 Myr, which falls in the estimated age range of ϵ Eri.

There are significant variations in brightness around the ring, which could be due to perturbation by a planet. Simulations of the young Solar System (Liou & Zook 1999) show that Jupiter had most effect in clearing a central cavity while Neptune caused strong perturbations on the outer dust. Bright spots appear in the simulated dust distribution that are due to mean motion resonances. At least one of these is suggested by the ϵ Eri image, in particular the bright spot to the south-east and the "trail" along the southern side of the ring. Modeling by Liou et al. (2000) suggests the presence of a Jupiter-like planet in an orbit of about 40 AU radius, although the image has not yet been reproduced exactly.

Table 2. Modeled dust properties: disk size, inclination to the plane of the sky, mass and characteristic grain temperature and size (Dent et al. 2000).

Star	Inner, outer radii (AU)	Inclination (degrees)	M (dust) ($M_{\rm lunar}$)	T (dust) (K)	grain size (μm)
Fomalhaut	100,140	70	1.4	40	100
β Pic	20,—	> 80	3.0	85	10
Vega	80,120	5	0.2	80	70
ϵ Eri	50,80	20	0.07	35	30

This inhomogeneity, together with the cleared cavity, implies the presence of two (or more) member planetary system.

4. Disk and Grain Evolution

Figure 5 shows a plot of estimated dust mass versus stellar age, where it is assumed that all the grains are fairly large (thus masses are in effect lower limits). There is a roughly linear trend of decreasing mass between about 10 Myr and a few Gyr, although a steeper decline may be required prior to this, for consistency with the dust fluxes measured towards emission-line stars. After 10 Myr, the slope indicates a decline of dust mass at a rate of about $t^{-1.2}$— surprisingly, somewhat slower than the t^{-2} decline of far-infrared dust opacity found from *ISO* observations of Vega-excess stars (Spangler et al. 1999).

Since Figure 5 shows that total dust masses fall below the mass of the Earth at a few Myr, terrestrial planets must form on timescales shorter than this. However, there is some dispersion in the evolution around different stars, as Jayawardhana et al. (2000) have recently detected about 10^{-3} $M_\oplus$ of dust towards 55 Cnc, a G8 star with a known planet; in contrast, our own Kuiper Belt has perhaps 1% of this dust mass although both stars are a few Gyr old.

By adopting submillimeter imaging, we have to some extent biased our view of the grain properties towards these stars. For example, some warmer grains closer to the stars might be more prominent in mid-infrared imaging. However, modeling shows that the cavities are real, since large numbers of such central warm grains would significantly broaden the SEDs. For Fomalhaut and ϵ Eri, the grain density must drop by $\gtrsim$ 90% at the inner edges of the cavities (Dent et al. 2000). Also, the models indicate that the grains are mostly large compared to those found in the interstellar medium, with sizes of tens of microns. This is supportive of the idea of grains accumulating into larger bodies, as part of the picture of planetesimal and planet formation, but since grains approaching 100 μm in size will spiral into the star under P–R drag, we also require erosive processes. For example, asteroid collisions could continually generate more large grains, although the minimum requirement is simple grinding down of "gravel" a few millimeters across (Dent et al. 2000).

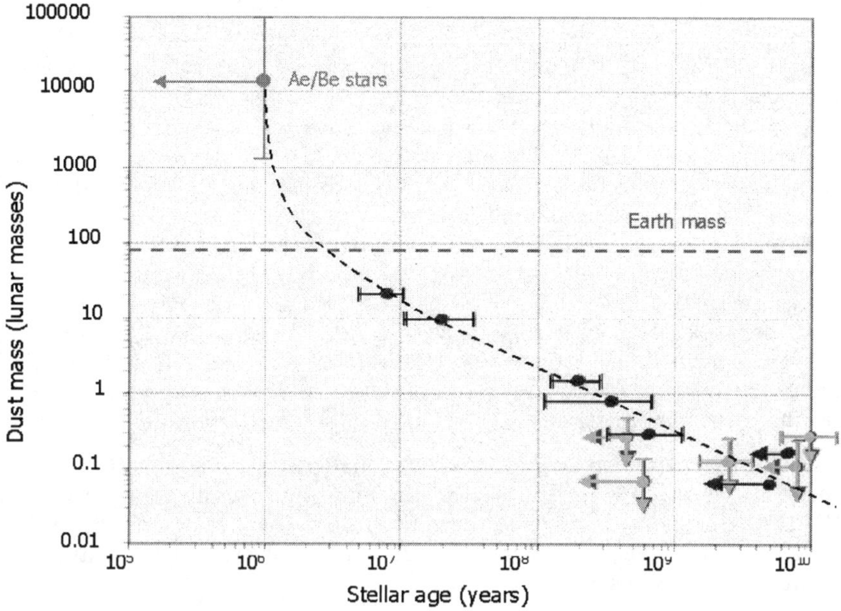

Figure 5. Plot of dust masses against stellar age, for the "Big Four" and other nearby Vega-excess stars. The point for Herbig Ae/Be stars is from data in Mannings (1994).

4.1. Ongoing Work

Deeper imaging of these four stars is continuing, particularly to improve the 450 μm data that are obtained simultaneously with the 850 μm results discussed above. These will have the advantage of a factor of two higher resolution, or 25 AU at ϵ Eridani. Remnant dust disks are also being sought around stars with known planets, with a first submillimeter detection of 55 Cnc (Jayawardhana et al. 2000). We have also begun a search for the molecular gas component needed to form gas giants, concentrating on younger stars than the four prototypes, since gas appears to disperse very early (Zuckerman, Forveille, & Kastner 1995; Dent et al. 1995; Greaves, Coulson,& Holland 2000).

5. Conclusions

General properties of the dust disks include similarity in extent to the Kuiper Belt, with cleared cavities roughly the diameter of the Solar System. This is in itself suggestive of the formation of planets, and is supported by modeling showing that accumulation has proceeded at least as far as making large grains. More directly, asymmetries towards Vega and ϵ Eridani suggest perturbations due to low-mass companions such as planets. The detection of a dusty "blob" several hundred AU from β Pic warns that our understanding of companion formation may be incomplete. Finally, the big difference between dust masses around the Sun and 55 Cnc shows that not all planetary systems may evolve in the same way.

In the future, higher resolution submillimeter studies will be possible with the SMA and ALMA interferometers; beam sizes of 0.1 arcsec would allow us to probe scales less than 10 AU for Vega, Fomalhaut, and ϵ Eridani. Also, second-generation submillimeter cameras will greatly extend the range of disk surveys: for example, SCUBA-2 is already being planned for the JCMT and will be more sensitive, as well as having instantaneous full sampling of the image plane. Images such as that of ϵ Eri could be obtained in only 30 minutes, and there are at least another 20 stars within 25 pc that are too faint for SCUBA, but are known to have dust disks from *IRAS* excesses. Coupled with this, data from mid-infrared imaging and optical/near-IR coronagraphy promises to revolutionize our "picture gallery" of extrasolar planetary systems in formation.

Acknowledgments. We thank our collaborators, who contributed many of the ideas discussed here: Harold Butner, Iain Coulson, Bill Dent, Walter Gear, Ray Jayawardhana, Tim Jenness, Jer-Chyi Liou, Geoff Marcy, Chris McCarthy, Gerald Moriarty-Schieven, Ian Robson, Helen Walker, Dolores Walther, Rich Webb, and Ben Zuckerman.

References

Aumann, H. H., Beichman, C. A., Gillett, F. C., de Jong, T., Houck, J. R., Low, F. J., Neugebauer, G., Walker, R. G., & Wesselius, P. R. 1984, ApJ, 278, L23

Chini, R., Krügel, E., Kreysa, E., Shustov, B., & Tutukov, A. 1991, A&A, 252, 220

Dent, W. R. F., Greaves, J. S., Mannings, V., Coulson, I. M., & Walther, D. M. 1995, MNRAS, 277, L25

Dent, W. R. F., Walker, H. J., Holland, W. S., & Greaves, J. S. 2000, MNRAS, in press

Greaves, J. S., Coulson, I. M., & Holland, W. S. 2000, MNRAS, 312, L1

Greaves, J. S., Holland, W. S., Moriarty-Schieven, G., Jenness, T., Dent, W. R. F., Zuckerman, B., McCarthy, C., Webb, R. A., Butner, H. M., Gear, W. K., & Walker, H. J. 1998, ApJ, 506, L133

Holland, W. S., Robson, E. I., Gear, W. K., Cunningham, C. R., Lightfoot, J. F., Jenness, T., Ivison, R. J., Stevens, J. A., Ade, P. A. R., Griffin, M. J., Duncan, W. D., Murphy, J. A., & Naylor, D. A. 1999, MNRAS, 303, 659

Holland, W. S., Greaves, J. S., Zuckerman, B., Webb, R. A., McCarthy, C., Coulson, I. M., Walther, D. M., Dent, W. R. F., Gear, W. K., & Robson, I. 1998, Nat, 392, 788

Jayawardhana R., Holland, W. S., Greaves, J. S., Dent, W. R. F., Marcy, G. W., Hartmann, L. W., & Fazio, G. G. 2000, ApJ, in press

Kalas, P., Larwood, J., Smith, B. A., & Schultz, A. 2000, ApJ, 530, L133

Kalas, P., & Jewitt, D. 1996, AJ, 111, 1347

Lagage, P. O., & Pantin, E. 1994, Nat, 369, 628

Liou, J.-C., Zook, H. A., Greaves, J. S., & Holland, W. S. 2000, in 31st Lunar & Planetary Science Conf., abstract #1416

Liou, J.-C., & Zook, H. A. 1999, AJ, 118, 580

Mannings, V. G. 1994, MNRAS, 271, 587

Spangler, C., Silverstone, M. D., Becklin, E. E., Hare, J., Zuckerman, B., Sargent, A. I., & Goldreich, P. 1999, in The Universe as seen by ISO, ed. P. Cox & M. F. Kessler, ESA-SP 427, p. 405

Teplitz, V. L., Stern, S. A., Anderson, J. D., Rosenbaum, D., Scalise, R. J., & Wentzler, P. 1999, ApJ, 516, 425

Walker, H. J., & Heinrichsen, I. 2000, Icarus, 143, 147

Zuckerman, B., & Becklin E. E. 1993, ApJ, 414, 793

Zuckerman, B., Forveille, T., & Kastner, J. H. 1995, Nat, 373, 494

Discussion

William Borucki: Are your measurements of the dust distribution surrounding β Pictoris consistent with dust being expelled from the original nebula that formed the β Pic system?

Jane Greaves: It's possible that the extra dust emission to the SW could have been expelled—the recent data on companion stars by Barrado y Navasués et al. could suggest a close passage of another star dragging some material off the β Pic disk.

Alan Boss: A comment on 55 ρ^1 Cnc. The latest data from Paul Butler and his colleagues show that the second companion must be rather long-period (20 to 30 years or more) and fairly massive (many Jupiter masses).

Jane Greaves: Yes—that is very interesting for our group, as such a planet may clear out the inner edge of the dust disk, which we saw in our 850 μm photometry.

Paul Ho: Has CO been looked for in the other systems besides 55 Cnc?

Jane Greaves: We have looked briefly at 51 Peg, but needed a broader bandpass to get an accurate spectral baseline. Such a project is planned in the future, concentrating on the stars with planets around 0.05 AU.

Disks, Planetesimals, and Planets
ASP Conference Series, Vol. 219, 2000
F. Garzón, C. Eiroa, D. de Winter, and T. J. Mahoney, eds.

Models of β Pictoris and Disks around Main-Sequence Stars

Alain Lecavelier des Etangs

Institut d'Astrophysique de Paris, CNRS, 98 bis bld Arago, F-75014 Paris, France

Abstract. Disks around main-sequence stars are probably related to the presence of young planetary systems in a phase of strong activity and show that the planetary systems are still active and evolve after their formation. The present review focuses on β Pictoris, which is an archetype of these circumstellar disks, presenting almost all the features observed separately on other similar stars.

Here we discuss models of the properties and dynamics of the dust, as well as the issue of the nature and origin of this disk that must be produced by collision or by evaporation of Kuiper Belt-like objects. We emphasize the scenario in which the β Pictoris disk is a gigantic multi-cometary tail supplied by slowly evaporating bodies like Chiron. This evaporation could be due to the formation and the migration of an outer planet like Neptune trapping bodies in resonance. The issues of asymmetries and warp of the inner disk can also be interpreted as due to one (or several) perturbing planet(s).

Although apparently more difficult to interpret, the gas disks detected through stable and variable spectroscopic signatures provide accurate information and give the opportunity of the most detailed modeling of the circumstellar activity. In particular, the β Pictoris spectroscopic variability is now well explained in many details by the evaporation of cometary objects close to the star.

Examples of stars with similarities to β Pictoris are presented.

1. Introduction

The infrared excess "Vega-like" stars and their circumstellar dusty environment were discovered by *IRAS* in the 80's (Aumann et al. 1984). After the *IRAS* discovery, "disks, planetesimals, and planets" really became the subject of interest when few months later Smith & Terrile (1984) published the first image of a disk in the history of astronomy: the β Pictoris disk, seen edge on from the Earth. Since that date, the question has been whether β Pictoris is really a "star" harboring a (young?) planetary system.

Other images have been published giving information on the dust distribution and properties (Lecavelier et al. 1993; Golimowski et al. 1993; Lagage & Pantin 1994; Mouillet et al. 1997a; Heap et al. 2000, Figure 1), and the puzzling asymmetries (Kalas & Jewitt 1995). Asymmetries are surprising in such a

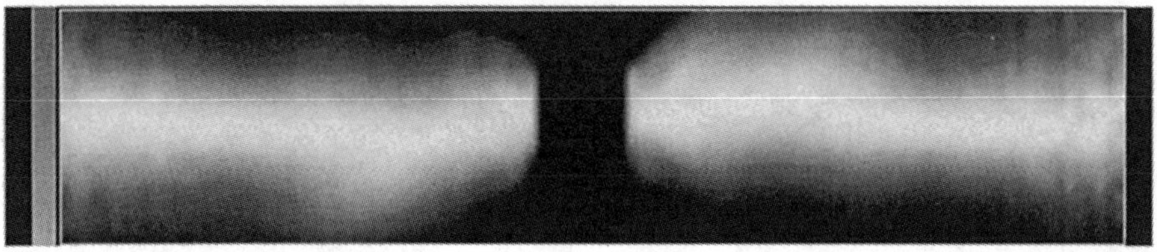

Figure 1. The β Pictoris disk imaged by STIS/*HST* (Heap et al. 2000).

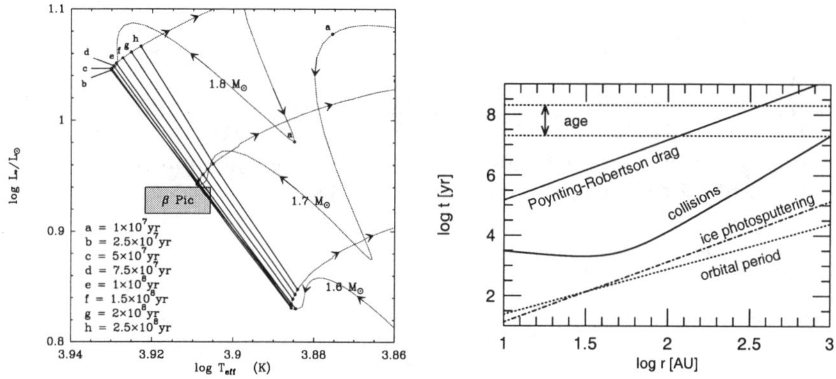

Figure 2. *a)* HR diagram with the estimated error box of β Pictoris. The evolutionary tracks are shown for 1.6, 1.7 and 1.8 $M_\odot$ (Beust et al. 1998). *b)*. Dust lifetime versus distance to β Pictoris for different destruction processes (Artymowicz 1997).

system where Keplerian relative motion should remove any asymmetry on very short timescale. The disk morphology and the inferred spatial distribution of the dust have been analyzed in great details (Artymowicz, Burrows, & Paresce 1989; Kalas & Jewitt 1995). Among these infrared excess stars, β Pictoris has a very peculiar status because, in addition to the dust disk, β Pictoris shows a gas disk seen through absorption lines together with an intense spectroscopic activity. In short, β Pictoris is presenting a whole set of features also seen individually around some other stars. It is, however, the only known case where all the features described here are simultaneously present around the same star.

Other reviews on this system and other similar main-sequence stars have been published recently (Artymowicz 1997; Vidal-Madjar et al. 1998; Lagrange, Artymowicz, & Backman 2000), each of them focusing on different aspects. The present paper is divided into two parts dedicated to dust disks (Section 2.) and gas disks (Section 3.). It focuses on β Pictoris, while similar stars are discussed in the last subsections (Sections 2.5. and 3.2.).

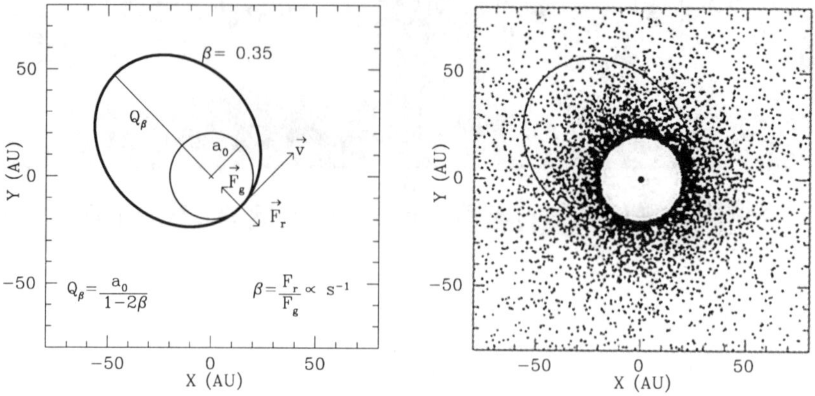

Figure 3. *a)* Orbit of a dust grain (*thick line*, $\beta = 0.35$) produced by a parent body on a circular orbit of radius a_0 (*thin line*). *b)* Set of dust grains with a range of sizes and produced at different positions in the orbit of the parent body. This is a disk!

2. Dust Disks

2.1. Lifetime and Production of Dust

A key point concerning disks around main-sequence stars is the dust lifetime shorter than the age of these systems (Figure 2). In the β Pictoris case, dust particles are destroyed by collisions between grains which produce submicronic debris quickly eliminated by the radiation pressure. This phenomenon dominates the Pointing–Robertson (PR) effect within the whole disk. The PR effect is therefore irrelevant to the physics of this disk. Anyway, PR effect should be considered in the other less dense disks where collision timescale is much larger, as, for example, in the ϵ Eri dust ring (Lecavelier et al. 1996b).

As the dust lifetime is very short, one must consider that the observed dust is continuously resupplied (Backman & Paresce 1993). It is generally thought that the β Pictoris disk is a substantial disk of colliding planetesimals (Backman & Paresce 1993; see also Backman et al. 1995 for an analogy with collision in the Solar System Kuiper Belt). Artymowicz (1996) proposed that the exponential proliferation of submicronic grains on hyperbolic orbits colliding and producing new grains, which can again collide, can start a significant avalanche of grain outflow. Grain destruction by internal sand-blasting and collision avalanche limits the total dust mass. In that model, particles are in eccentric, and even hyperbolic, orbits, which then drive into the outer part of the disk any asymmetric perturbation produced in the inner disk. This may explain asymmetries and some morphological properties, but a detailed modeling of this promising idea has still to be done (see also Krikova, Krivov, & Mann, this volume, p. 387).

2.2. Dust Production by Evaporation, an Alternative to Collisions

Here we advocate for an alternative scenario. Two mechanisms can continuously resupply the disk in dust: collision or evaporation of kilometer-sized parent bodies. In both cases, because of the radiation pressure, the particles ejected from the parent bodies follow very eccentric orbits whose eccentricities are related to the grains size (Figure 3*a*). If we assume a zeroth-order model of a narrow ring of bodies producing dust, the particles are then distributed on a disk-like structure presenting three morphological similarities with the β Pictoris disk (Figure 3*b*). First, the central region of the disk is empty of dust, its limit corresponds to the inner radius of the parent bodies' orbits. Second, the width of this zeroth-order model disk increases with the distance to the central star, r, because the distribution of the particles' inclinations are the same as those of the parent bodies. Finally, the dust density decreases with r; moreover, this density distribution follows a power law. Consequently, if seen edge on from the Earth, the radial brightness profile along the mid-plane of this disk follows also a power law: $F(r) \propto r^{-\alpha}$, with $\alpha \sim 5$ (Lecavelier des Etangs et al. 1996a).

The slope of this zeroth-order model is at the upper limit of the range of or steeper than the observed slope of the power law in the β Pictoris disk ($\alpha \sim 4$, Kalas & Jewitt 1995; $\alpha \sim 5$, Heap et al. 2000). To explain this distribution, it is possible that the dust is produced by collisions of Kuiper Belt-like objects spread over a wide range of distances. But an alternative model is also possible in keeping the assumption that the parent bodies remain in a narrow ring close to the star. For example, the disk can be produced by the evaporation of parent bodies of size $\gtrsim 10$ km and located at several tens of AU from the central star (Lecavelier et al. 1996a). Several arguments are in favor of this alternative scenario in the case of the β Pictoris disk. First, it is obviously easy to explain any asymmetry even at large distances, because a planet in the inner disk (in an eccentric orbit) can have influence on the distribution of nearby parent bodies, and this non-axisymmetric distribution is projected outward by the particles on very eccentric orbits. The observed CO/dust ratio is an independent argument in favor of this scenario (see below). Finally, the connection between the inner radius of the disk and evaporation limit is a direct consequence of this scenario because the periastron distances of the particles are similar to the periastron of the parent bodies. A hypothetical planet at this limit is no longer needed to explain the presence of the inner void in the disk.

A Necessary Source of CO An important characteristic of the β Pictoris disk is the presence of cold CO and C I (Vidal-Madjar et al. 1994; Roberge et al. 2000). CO is cold with a typical temperature of less than 30 K, which corresponds to the temperature of CO evaporation. For instance, with an albedo of 0.5 this temperature corresponds to an evaporating body located between 100 and 200 AU. CO and C I are destroyed by UV interstellar photons and have lifetime shorter than the star age ($t_{\rm CO} \sim t_{\rm C\,I} \sim 200$ yr). A permanent replenishment mechanism must exist. From the column density of CO ($N_{\rm CO} = 2 \times 10^{15}$ cm^{-2}), the present mass of CO is estimated to be $M_{\rm CO} \approx 7 \times 10^{20}$ kg. The known photodissociation rate of CO, $\tau_{\rm CO} = 2 \cdot 10^{-10}$ s^{-1} (van Dishoeck & Black 1988) gives a corresponding supplying rate of CO $\dot{M}_{\rm CO} = M_{\rm CO}\tau_{\rm CO} \approx 10^{11}$ kg s^{-1}.

We can also estimate the supplying rate of dust needed, $\dot{M}_\mathrm{d} = M_\mathrm{d}/t_\mathrm{d} \approx 10^{11}$ kg s^{-1}, where M_d is the mass of the dust disk ($M_\mathrm{d} \sim 10^{23}$ kg), t_d is the dust life-time ($t_\mathrm{d} \approx 10^4$ yr). It is very interesting to see that the dust/CO supply rate is consequently $\dot{M}_\mathrm{d}/\dot{M}_\mathrm{CO} \approx 1$. This is very similar to the ratio in the material supplied by evaporation in the Solar System. This provide *independent* evidence that the β Pictoris dust disk can be supplied by *orbiting evaporating bodies.*

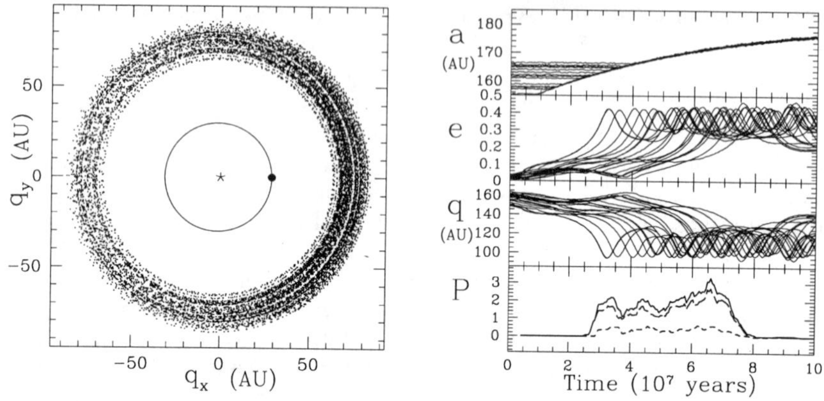

Figure 4. *a)* Spatial distribution of the periastron of a set of bodies perturbed by a planet on a slight eccentric orbit : $e_\mathrm{p} = 0.05$ (Lecavelier 1998). *b)* Example of evaporating bodies trapped in the 1:4 resonance with an outward migrating planet. *Top three panels:* Orbital parameters: semi-major axis (a), eccentricity (e), and periastron (q) of 19 bodies trapped in the resonance. Although the semi-major axes of the bodies increase, their periastra decrease. *Bottom panel:* Dust production by these trapped bodies (P) as a function of time. The production of dust starts when the periastron distance is small enough for the CO to evaporate. Then, it stops when the parent bodies are exhausted and have no more volatiles. Consequently, the dust production is transient and is large only when the bodies trapped in the resonance are passing through the evaporation limit. With even a small eccentricity of the migration planet ($e_\mathrm{p} \lesssim 0.05$), the production rate is not axisymmetric and is larger in the direction of the periastron of the perturbing planet (*long dashed line*) than in the opposite direction (*short dashed line*) (Lecavelier, Vidal-Madjar, & Ferlet 1998).

β Pictoris a Transient Phenomenon? It might be difficult to imagine that this CO production can last for several 10^7 yr. A total mass of several Earth mass of CO should have been evaporated! ($\dot{M}_\mathrm{CO} \times 10^7$ yr $\times\, \tau_\mathrm{CO} = 2\ M_\oplus$). This gives evidence that either the present mass of CO is overestimated, or that β Pictoris is a transient phenomenon. There is in fact no reason to believe that the β Pictoris system has always been as dusty as is observed today. Of course, the idea that this disk is not transient is a consequence of any model of collisional erosion from asteroid to dust. But with other scenarios, we can easily imagine that a

particular phenomenon occurred recently, and that the density of the β Pictoris disk must be significantly smaller during the quiescent phase of simple collisional erosion, during which the density is similar to the characteristic density of the more common Vega-like stars.

The Asymmetry In contrast to production by a set of collisional bodies at very large distances where planets have no influence, if the dust is produced by a narrow ring of orbiting evaporating bodies, these bodies must be close to the star where the planetary perturbations can be important. In this case, the asymmetry can simply be due to an eccentric orbit of the perturbing planet. For instance, a planet in an eccentric orbit causes a modulation of the precessing rate of the periastron of the evaporating bodies. When the periastron of the parent body is roughly aligned with the periastron of the perturbing planet, the precession rate is slower. Consequently, the number density of bodies with about the same longitude of periastron as the planet is larger (See Figure 4*a* for bodies between 70 and 90 AU and perturbed by a planet with a mass $M_{\rm p} = 3 \cdot 10^{-4}\, M_{\odot}$, a semi-major axis $a_{\rm p} = 30$ AU, and an eccentricity $e_{\rm p} = 0.05$). Such an effect would obviously cause an asymmetry in a disk *if it is produced by evaporation* of bodies with a distribution of periastron perturbed in this way. As the dust is mainly produced at the periastron of the parent bodies and principally observed during the apoastron, the part of the disk at 180 degrees from the periastron of the perturbing planet should be more dense.

Planet Migration and Perturbation on Parent Bodies Fernández (1982) proposed that the accumulation and scattering of a large number of planetesimals causes the migration of the outer planets during their formation. In this model, Neptune moved outwards by several AU (Fernández & Ip 1996). Malhotra (1993, 1995) showed that this can explain the particular orbit of Pluto and some Kuiper Belt objects. In short, Pluto was trapped in the orbital commensurability moving outward during the expansion phase of Neptune's orbit. In the β Pictoris system with a typical age of few 10^7 yr, if an outer planet is moving outward a fraction of bodies can be trapped in resonances. Their semi-major axes and eccentricities increase significantly and the net result is a decrease in their periastron (Figure 4*b*). The trapped bodies can evaporate. This could explain the putative slow evaporation of a large number of bodies and produce a significant dust disk (Lecavelier des Etangs 1998).

2.3. Inner Warp

Concerning the inner part of the β Pictoris disk, the most intriguing properties is certainly the so-called "warp", which consists of a change in the inclination of the mid-plane of the disk inside about 80 AU. This effect has been independently discovered by Burrows et al. (1995) with *HST* and Mouillet et al. (1997a) with adaptive optics on the ESO 3.6 m Telescope. It is now confirmed by the very nice images done with STIS aboard *HST* (Heap et al. 2000, Figure 1). This warp has been explained by Mouillet et al. (1997b) by the presence of a planet in an inclined orbit with the same inclination as the tilt of the inner disk. This planet causes a precession of the inclination of planetesimals around the plane of the perturbing planet. The orbits of the planetesimals keep the same angle to the plane as the planet but after a given time the longitudes of the ascending nodes are spread over 360 degrees: the disk opening half-angle is thus increased to the angle between the plane of the planet and the initial plane of the disk of planetesimals. The precession timescale is larger for planetesimals at larger distances, therefore the warp is spreading outward up to the distance where the precession of the inclination has made at least one rotation since the beginning of the system. Given that at a distance r the precession frequency is $w_{\rm p} = -(3GM_{\rm p}D_{\rm p}^2)/(4\Omega r^5)$, where $M_{\rm p}$ and $D_{\rm p}$ are the mass and the distance of the perturbing planet, $|w_{\rm p}t| > 1$ is valid up to the distance $\sim$ 80 AU if we have $M_{\rm p}D_{\rm p}^2 \approx 2 \cdot 10^{-3} M_*(10\ {\rm AU})^2(t/10^7\ {\rm yr})^{-1}$ (Mouillet et al. 1997b). It is noteworthy that if the age of the system is $t \sim 2 \cdot 10^7$ yr (Barrado y Navascués et al. 1999), then a Jupiter-mass planet at 10 AU and inclined at 5 degrees to the disk plane can easily explain the observed warped.

2.4. Rings in the Outer Disk

The last interesting and puzzling property of the dust disk has been recently discovered by Kalas et al. (2000), who have shown the presence of seven over-densities (arc-like structure) between 500 and 785 AU in the north-east part of the disk. These features are not visible in the south-west part. They thus require a perturbation by a body in a very eccentric orbit. Kalas et al., have proposed that this could be due to a recent encounter with a neighboring star about 90 000 years ago.

2.5. Other Dust Disks

Many other disks have been imaged in the last three years. In the optical, Kalas & Jewitt (1997) have shown that a disk-like structure is surrounding BD+31°643. Here, the dust grains are probably very small, as concluded from the blue color of the disk. The dust spatial distribution can be explained by a dust outflow with a steady-state equilibrium between the strong radiation pressure and the drag by the gas of the cluster in which BD+31°643 is embedded (Lecavelier des Etangs et al. 1998).

The majority of the available images of dust disks around main-sequence stars have been obtained either in the infrared through the scattered light (e.g., Augereau et al. 1999) or the thermal emission at 10 or 20 microns (e.g., Koerner et al. 1999), or in the submillimetric (Holland et al. 1998; Greaves et al. 1998) Many of these new discoveries are discussed in the present book. All these nice images, already obtained or still to be done in the forthcoming months with the new devices, should help to understand if the β Pictoris disk is an exception or simply at a particular stage of the common evolution of planetary systems.

3. Gas Disks

3.1. The β Pictoris Circumstellar Gas

Stable Gas Component Because the β Pictoris disk is seen nearly edge on, high-resolution spectroscopy has revealed the presence of circumstellar gas through absorption lines. Because one absorbing component is seen identically in all observations and at the same radial velocity as the star (20 km s^{-1}), it is named the "stable" absorption gas component. This stable gas is composed of small

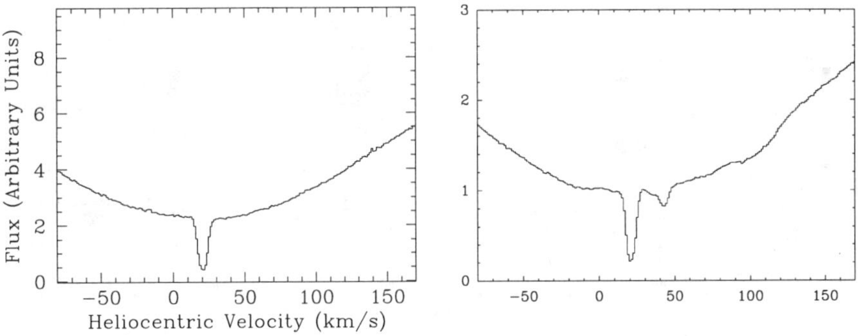

Figure 5. *a)* Ca II spectrum of β Pictoris. Here only the stable component is detected. *b)* This second spectrum shows—in addition to the stable gas at 20 km $^{-1}$ heliocentric velocity—two redshifted components, one at low velocity and one broad absorption line at high velocity (> 100 km s^{-1}).

amounts of neutral sodium and iron, as well as large amounts of singly-ionized species such Ca II, Fe II, Mg II, Mn II, and Al II. The overall composition seems to be close to solar (Lagrange et al. 1997). Two very peculiar elements were detected: C I and the CO molecule with a short lifetime (Section 2.2.). On the other hand, the OH molecule was not detected with a relatively tight upper limit (Vidal-Madjar et al. 1994).

Variable Components In addition to this stable component, variable absorption features have been detected and surveyed for more than 15 years (Figure 5). Slowly variable features are most of the time confined to one or two components redshifted by 10 to 30 km s^{-1} relative to the star. Although such structures seem to be changing in both velocity and strength (by about ± 10 km s^{-1} and large factors in strength), they nevertheless remain very comparable during a few consecutive hours, often from one day to the next and even sometimes over weeks.

Some other components present strong variability, in particular in the Mg II and Al III lines. They are also observed in the Ca II and Fe II line as weak and broad absorptions spread over few tens of km s^{-1}. The changes are observed on a very short timescale (hours or even less). These features are mostly strongly redshifted, with shifts that could reach 300 to 400 km s^{-1}. These highly varying features were detected only in ionized species, including highly (over-)ionized ones like Al III and C IV completely unexpected in such a relatively "cool" stellar environment. The rapid changes make these features difficult to track. Sometimes they are there and later on they are just gone.

Interpretation of the Spectral Variations: the Star-Grazing Comet Scenario A detailed development of the star-grazing comet scenario can be found in the English translation of the "director thesis" of Beust (1998), which can be found on the IAP web site (www.iap.fr).

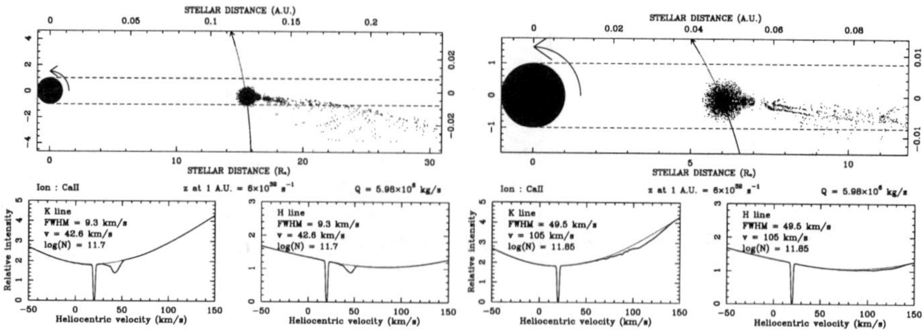

Figure 6. *a)* Simulation of Ca II absorption lines produced by a FEB at a large distance. The line is sharp and at low redshift. *b)* Simulation of a FEB close to the star. The radial velocity is larger, and, because of the stronger radiation pressure, the line is broader.

The presence of strongly redshifted ionized gas is in fact very challenging, since the very high radiation pressure should expel it very quickly (the ratio β of radiation to gravitational forces is $\gg 1$ for Ca II and several other ions). The absorption lines are almost always redshifted, although blueshifted ones would have been expected. The gas must then be injected with very high inward radial velocity. There is only one simple way to produce this situation, namely the evaporation of grains moving toward the star. Radiation pressure still acts on the grains. Thus, they must also be injected with high velocity, through evaporation from more massive bodies for which the gravitation is much larger that the radiation force. This model has been developed in great detail (Beust et al. 1989, 1990, 1991) and can be summarized by 'evaporation of star-grazing comets' (Figure 6). Indeed, the strong variability in the circumstellar lines of the ionized elements such as Fe II, Al III, C IV, and Mg II is now attributed to the evaporation of kilometer-size, "cometary-like" bodies falling toward the star: this is the falling evaporating bodies (FEB) scenario. The over-ionized variable species Al III and C IV cannot be produced by photoionization, but Beust & Tagger (1993) showed that they can be formed by collisional ionization in the coma surrounding these falling evaporating bodies.

Among the different phenomena that can be explained by this model, we would like to stress on the observation of abnormal ratios in doublet lines. For example, although the Mg II doublet has an intrinsic oscillator strength ratio of two, the measured ratio is exactly one, even for unsaturated lines (Vidal-Madjar et al. 1994). This proves that the absorbing gas cloud is optically thick (ratio equal to one) but does not cover the total stellar disk (lines unsaturated). This behavior was expected as a test of the FEB model, which produces clumpiness of the absorbing clouds (Beust et al. 1989). Similar ratios have also been detected in redshifted absorptions toward other "β Pictoris-like stars" (Section 3.2.).

With several hundreds of FEBs per year, the frequency is several orders of magnitude higher than that of Sun-grazing comets in the Solar System. Planetary perturbations are thought to be the basic process. Direct scattering by close encounters with a massive planet does not seem to be efficient unless the planet eccentricity is very high (Beust et al. 1991b). Levison et al. (1994) proposed a model involving the eccentricity pumping effect of a secular resonance with two planets. Beust & Morbidelli (1996) proposed a more generic model based on a mean-motion resonance with a single massive planet on a moderately eccentric orbit ($e \simeq 0.05$). Indeed, a test particle trapped in the 4:1 resonance with such a planet becomes star-grazing after $\simeq$ 10 000 planetary revolutions. This model explains not only the preferred infall direction but also the radial velocity–distance relation observed in the FEBs (Figure 7*a*).

Models of the Stable Component The origin and even the stability of the stable component is still unclear. In the stable component, ions like Ca II and Fe II have the same radial velocity as the star, in spite of their large β ratio ($\beta \gg 1$). A residual stable gas is needed to support the radiation pressure of these ions. This idea has been quantitatively investigated (Lagrange et al. 1998; Beust et al. 1998) by assuming that a permanent massive torus of gas, for instance H I, is present around β Pictoris. The observed ions are supposed to be produced relatively close to the star and blown away. When encountering the gas torus, they are slowed down to very small radial velocities relative to the star. The detected "stable" component may correspond to the slowed down ions passing through the gas torus (Figure 7*b*). These ions are then again blown away from the star. The possible emission lines of Fe II at $\sim$ 100 km s^{-1} observed with *HST* at 10 AU from the star could thus be explained by the outflow outside this torus (Lecavelier des Etangs et al. 2000). The composition of this required stable gas unaffected by radiation pressure could be H I, H_2, or even O I and C II the final products of the CO dissociation. Only an upper limit of $M_{\rm HI} < 10^{28}$ g (0.5% the mass of Jupiter) has been reported (Freudling et al. 1995). This limit corresponds to a column density of about 10^{19} cm^{-2} , while 10^{18} cm^{-2} would be enough to support the radiation pressure on Fe II (Lagrange et al. 1997; Beust et al. 1998).

3.2. β Pictoris-Like Disks

Circumstellar gas signatures similar to those of β Pictoris are also seen around some other main-sequence and pre-main-sequence stars. It should, however, be noted that in the rare cases where a Ca II (or, for example, Fe II) line at the star radial velocity has been detected toward other stars like HR 10, these stars have been identified because they also present spectral variability, or the detection of over-ionized species, or redshifted absorption lines, which are sometimes shown to be optically thick. This lack of detection of only circumstellar absorption at the systemic velocity may be due to possible confusion with the interstellar medium. The first main-sequence star discovered to have a very similar spectroscopic behavior to β Pictoris is HR 10 (Lagrange et al. 1990). This star shows variable redshifted or blueshifted absorption lines (Lagrange et al. 1991; Welsh et al. 1998), and a central component seems relatively stable. Very highly excited levels of Fe II have been detected, proving that the gas is circumstellar. Redshifted, optically thick lines of Mg II have also been detected (Lecavelier et al. 1997). Again, this shows that small clouds of excited gas are falling toward the star.

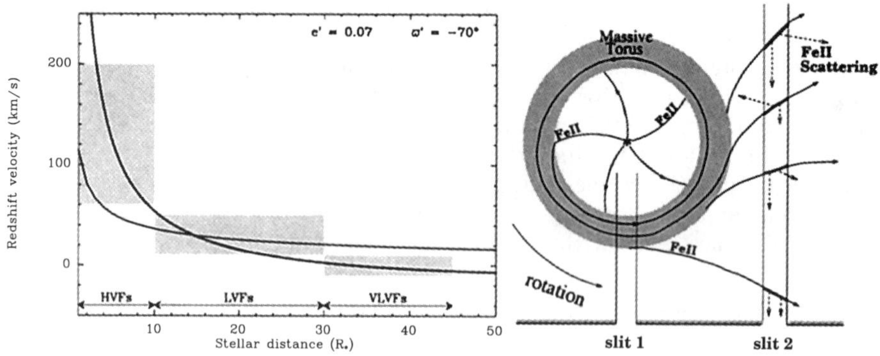

Figure 7. *a)* Radial velocity of an absorbing component versus distance to the star. The *gray boxes* represent the observations. The *thin line* gives the result if all the FEBs had the same longitude of periastron, and the *thick line* is for FEBs produced by a 4:1 resonance with a planet at 10 AU (Beust & Morbidelli 1998). *b)* Sketch of the dynamics of Fe II ions trapped in a massive torus when observed at zero radial velocity by absorption spectroscopy of the starlight (slit 1), and at large radial velocity through emission lines (slit 2) (Lecavelier et al. 2000).

Two other stars deserve detailed analysis because circumstellar dust is present simultaneously with the gas. 51 Oph presents a dusty infrared excess due to cold dust, silicate emission features, absorption lines by over-ionized species (Grady & Silvis 1993) and by Fe II at excited levels, abnormal Mg II ratio, and finally a possible detection of C I in the circumstellar gas. On the other hand, HD 100546 shows many β Pictoris-like spectroscopic features, and *ISO* has shown the presence of silicate dust. Links between these gas and dust features have still to be understood.

Many other stars show redshifted absorption lines (HR 2174, 2 And, etc.). All these detections of redshifted absorption lines raise the question of the explanation for the quasi-absence of blueshifted events. In the β Pictoris case, it is believed that the orbits of the star-grazing comets always present about the same angle to the observer. However, when observed on several stars, some blueshifted orientation should be expected. Given the very impressive fit between the observations and the FEB model, this point is certainly not enough to reject this model. It is more likely that the β Pictoris FEB phenomenon is somehow particular and that other stars present real "infall" on the star or their evaporating bodies may be generally destroyed before they reach the periastron (Grinin et al. 1996). In this last case, the process needed to put these bodies in very eccentric orbits on less than one orbital timescale remains to be found.

In conclusion, despite many searches β Pictoris seems to remain an exception in the general picture of the circumstellar disk. In any case, even if it is a unique case in the solar vicinity, its complexity and subsequent beauty make β Pictoris a first-class subject for observations, modeling, and discoveries.

Acknowledgments. Figures 1, 2, 5, 6, and 7*a* have been reproduced with the kind permission of P. Artymowicz, H. Beust, and S. Heap. Many thanks are due to R. Ferlet and A. Vidal-Madjar for many fruitful discussions.

References

Artymowicz, P., Burrows, C., & Paresce, F. 1989, ApJ, 337, 494

Artymowicz, P. 1996, in The Role of Dust in the Formation of Stars, ed. H. U. Käufl & R. Siebenmorgen (Berlin: Springer), p. 137

Artymowicz, P. 1997, Ann. Rev. Earth Planet., 25, 175

Augereau J. C., Lagrange A.-M., Mouillet D., & Ménard F. 1999, A&A 350, L51

Aumann, H. H., et al. 1984, ApJ, 278, L23

Backman D.E., & Paresce F. 1993, in Protostars and Planets III, ed. E. H. Levy & J. Lunine (Tucson: University of Arizona Press), p. 1253

Backman, D. E., Dasgupta, A., & Stencel R. E. 1995, ApJ, 450, L35

Barrado y Navascués, D., Stauffer, J. R., Song, I., & Caillault, J.-P. 1999, ApJ, 520, L123

Beust, H., et al. 1998, A&A, 338, 1015

Beust, H., Lagrange-Henri, A.-M., Vidal-Madjar, A., & Ferlet, R. 1989, A&A, 223, 304

Beust H., & Morbidelli A. 1996, Icarus, 120, 358

Beust H., Lagrange A.-M., Crawford I.A., et al. 1998, A&A, 338, 1015

Beust, H., & Morbidelli, A. 1996, Icarus, 120, 358

Beust, H., & Morbidelli, A. 2000, Icarus, 143, 170

Beust, H., & Tagger, M. 1993, Icarus, 106, 42

Beust, H., Vidal-Madjar, A., Ferlet, R., & Lagrange-Henri, A.-M. 1990, A&A, 236, 202

Beust, H., Vidal-Madjar, A., Ferlet, R., & Lagrange-Henri, A.-M. 1991a, A&A, 241, 488

Beust, H., Vidal-Madjar, A., & Ferlet, R. 1991b, A&A, 247, 505

Burrows, C. J., Krist, J. E., Stapelfeldt, K. R., & WFPC2 Investigation Definition Team 1995 , AAS, 187, 3205

Fernàndez, J. A. 1982, AJ, 87, 1318

Fernàndez, J. A., & Ip, W. H. 1996, Planet. Space Sci., 44, 431

Freudling, W., et al. 1995, A&A, 301, 231

Golimowski, D. A., Durrance, S. T., & Clampin, M. 1993, ApJ, 411, L41

Grady, C. A., & Silvis, J. M. S. 1993, ApJ, 402, L61

Greaves, J. S., et al. 1998, ApJ, 506, L133

Grinin, V., Natta, A., & Tambovtseva, L. 1996, A&A, 313, 857

Heap S. R., et al., 2000, ApJ, in press

Holland W. S., et al. 1998, Nat, 392, 788

Kalas, P., & Jewitt, D. 1995, AJ, 110, 794

Kalas, P., & Jewitt, D. 1997, Nat, 386, 52

Kalas, P., Larwood, J., Smith, B. A., & Schultz, A. 2000, ApJ, 530, L133

Lagage, P. O., & Pantin, E. 1994, Nat, 369, 628

Lagrange-Henri, A.-M., Beust, H., Ferlet, R., Vidal-Madjar, A., & Hobbs L. M. 1990a, A&A, 227, L13

Lagrange-Henri, A.-M., et al., 1990b, A&AS, 85, 1089

Lagrange, A.-M., et al. 1998, A&A, 330, 1091

Lagrange, A.-M., Artymowicz, P., & Backman, D. E. 2000, in Protostars and Planets IV, ed. V. Mannings, A. P. Boss, & S. S. Russell (Tucson: University of Arizona Press), p. 639

Lecavelier des Etangs, A., et al. 1993, A&A, 274, 877

Lecavelier des Etangs, A., Vidal-Madjar, A., & Ferlet, R. 1996a, A&A, 307, 542

Lecavelier des Etangs, A., Scholl, H., Roques, F., Sicardy, B., & Vidal-Madjar, A. 1996b, Icarus, 123, 168

Lecavelier des Etangs, A. 1997, A&A, 325, 228

Lecavelier des Etangs, A. 1998, A&A, 337, 501

Lecavelier des Etangs, A., Vidal-Madjar, A., & Ferlet, R. 1998, A&A, 339, 477

Lecavelier des Etangs, A., et al. 2000, A&A, 356, 691

Levison, H. F., Duncan, M. J., & Wetherill, G. W. 1994, Nat 372, 441

Malhotra, R. 1993, Nat, 365, 819

Malhotra, R. 1995, AJ, 110, 420

Mouillet, D., Lagrange, A.-M., Beuzit, J. L., & Renaud N. 1997a, A&A, 324, 1083

Mouillet, D., Larwood, J. D., Papaloizou, J. C. B., & Lagrange, A. M. 1997b, MNRAS, 292, 896

Roberge, W. G., et al. 2000, ApJ, 538, 904

Smith, B. A., & Terrile, R. J. 1984, Sci, 226, 1421

van Dishoeck, E. F., & Black, J. H. 1988, ApJ, 334, 771

Vidal-Madjar, A., et al. 1994, A&A, 290, 245

Vidal-Madjar, A., Lecavelier des Etangs, A., & Ferlet, R. 1998, Planet. Sp. Sci., 46, 629

Welsh, B. Y., Craig, N., Crawford, I. A., & Price, R. J. 1998, A&A, 338, 674

Discussion

Lee Hartmann: The reality of the Kalas–Jewett disk in IC 348 is questionable. There is a very large scattered light structure (an arc) passing through the position of the "disk", and my guess is that this is simply some interstellar feature. Would collisions make a shallower surface density distribution in your simple model when the particles originate at only one radius?

Alain Lecavelier: As far as I know, the disk discovered by Kalas & Jewitt has not been directly confirmed or disproved. It is noteworthy that it is consistent with the polarimetric measurements made by Andersson & Wannier (1997, ApJ, 491,

L103). Clearly, a confirmation is still needed. In the simple model presented (Lecavelier et al. A&A, 339, 477) the collision between particles should not change the result because we consider very small particles for which the radiation pressure is dominant as for the debris of the particle collisions.

Hubertus Klahr: Even a tiny amount of gas will reduce the efficiency of the Poynting–Robertson effect. Do you have an idea what the amount of bulk gas (H_2, He, etc.) might be in your observed systems?

Alain Lecavelier: Even in the best known case (Beta Pictoris), the amount of bulk gas is in fact poorly constrained. Concerning the total quantities, we have an upper limit on the H I content (Freudling et al. 1995), and *ISO* should give at least interesting constraints on H_2. In the line of sight to the star, absorption spectroscopy is very powerful, and the *detected* CO column density allows to us show that CO emission must be very difficult to detect at millimeter wavelengths. FUSE will also give at least a strong constraint on the H_2 column density in the very near future.

Alan Boss: For Beta Pictoris, are the properties of the planet inferred to lead via resonances to the Ca II absorption features (redshifted only) consistent with those of the unseen planet inferred to cause the inner warp of the disk? Can they be the same planet?

Alain Lecavelier: Yes, these two inferred planets can be the same object. This object can also be the same as the planet which may have caused an occultation of the star in 1981 (Lecavelier et al. 1995, A&A, 299, 557; 1997, A&A, 328, 311).

Disks, Planetesimals, and Planets
ASP Conference Series, Vol. 219, 2000
F. Garzón, C. Eiroa, D. de Winter, and T. J. Mahoney, eds.

Dust–Gas Interaction in the Early Solar Nebula: the Role of Coherent Structures

Paolo Tanga

Osservatorio Astronomico di Torino, Strada Osservatorio 20, 10025 Pino Torinese, Torino, Italy

Patrick Michel

Observatoire de la Côte d'Azur, UMR 6529 Cassini, B. P. 4229, F-06304 Nice Cedex 4, France

Abstract. The dynamics of dust particles in a gas flow exhibits complex behavior even in very simple situations. This suggests that the particle distribution in the primordial Solar Nebula could have been far from homogeneous. Vortices capable of trapping small planetesimals could play a role in the process of accretion and thus in the subsequent phase of purely gravitational evolution.

1. Introduction

The problem of planetesimal formation in the primordial Solar Nebula is strongly coupled to that concerning the character of the turbulence in the gas disk. While it is widely accepted that small-scale turbulence can be generated by friction between the gaseous and the dust components due to systematic drifts (Cuzzi, Dobrovolskis, & Champney 1993), the presence of larger-scale phenomenology is still under debate. Several possible non-linear mechanisms have been proposed for sustaining turbulence in rotating disks, but the existence and properties of large-scale structures, though probable in analogy with other rotating flows, is still matter of debate.

Turbulence in planetary formation is strictly bound to the more general problem of the influence of the gas component on the motion of solid particles of different sizes. The simplest approach considers turbulence statistically, as a source of random motion. This approximation is valid only when the typical length-scales of the turbulent motion are not larger than particle sizes.

Current planetesimal-formation models take into account dust friction with the gas disk and the subsequent orbital decay (Weidenschilling 1977, 1997). In a few cases it was attempted to investigate the effects of segregation into two and three dimensions (Cuzzi, Dobrovolskis, & Hogan 1996; Squires & Eaton 1991).

In the domain of turbulent transport in geophysical flows, several theoretical and numerical investigations have been made into the dispersion of passive tracers. We recall here recently obtained results about tracer behavior in the presence of coherent structures, first in the case of simple 2D models in a non-rotating frame (Section 2), then in the case of disks, where dust trapping in vortices is observed (Section 3). The subsequent accretion and gravitational evolution of the vortex-trapped particles is briefly discussed.

2. Transport Properties of Dust in 2D Flows

In recent years research into non-perfect tracers in two-dimensional flows has been developed by the investigation of the Lagrangian dynamics of particles having a density different from that of the advecting flow (Crisanti et al. 1992). Reasonable assumptions lead to the following equation of motion for an advected particle (Maxey & Riley 1983):

$$\frac{d^2\mathbf{x}}{dt^2} = \delta \frac{D\mathbf{u}}{Dt} - \mu \left(\frac{d\mathbf{x}}{dt} - \mathbf{u} \right), \tag{1}$$

in which $\mathbf{u}$ is the velocity field of the advecting flow, while δ is the ratio between the fluid density and the particle density. The term on the LHS is simply the acceleration of the particle placed at the $\mathbf{x}$ coordinates. The first term on the RHS represents the acceleration exerted by the fluid over a hypothetical fluid particle in the same position as the impurity. The second term is the Stokes drag, and μ is the friction parameter (inverse of the *stopping time*, τ). The substantive derivative D/Dt is the derivative computed along the trajectory of a hypothetical fluid particle at the position of the impurity.

In the following, we will show how the value of δ is critical in changing the tracer dynamics. Even if in the case of planetesimals and disks we will deal only with particles having a much higher density than the fluid ($\delta \ll 1$), the case of $\delta > 1$ will also be briefly addressed, for a better illustration of the relevant processes.

In order to investigate the role of coherent structures over tracer diffusion, some numerical investigations have been made with a simple, non-divergent, and stationary two-dimensional flow (Crisanti et al. 1992; Tanga & Provenzale 1994) described on the (x, y) plane by the stream function:

$$\Psi(x,y) = A\left(\cos x + \cos y\right), \tag{2}$$

in which A is a constant. It should be noted that in the case of fluid particles ($\delta = 1$) the system given by (1) is integrable, and its solution corresponds to a regular, close trajectory. On the other hand, it can be shown analytically that the centers of the vorticity cells become attractive fixed points for $\delta > 1$ ("light" particles) and repulsive fixed points for $\delta < 1$ ("heavy" particles). As a result, particles less dense than the fluid have trajectories spiraling toward the center of vortices. On the contrary, heavier particles are ejected from vortices and diffuse while preferentially remaining close to the separatrices. Both types of behavior are illustrated in Figure 1.

The friction parameter, μ, just determines the timescales of ejection or trapping, without changing the qualitative behavior.

It is thus clear that even a stationary, regular flow can produce complex, chaotic dynamics of the Lagrangian tracers. For this reason the inference of a dust distribution from a knowledge of the Eulerian flow (and *vice-versa*) remains a largely unresolved issue.

In the case of a much more complex situation the numerical simulation of a fully developed, forced two-dimensional turbulent flow was adopted as advecting flow. It was obtained by the pseudo-spectral numerical integration of the

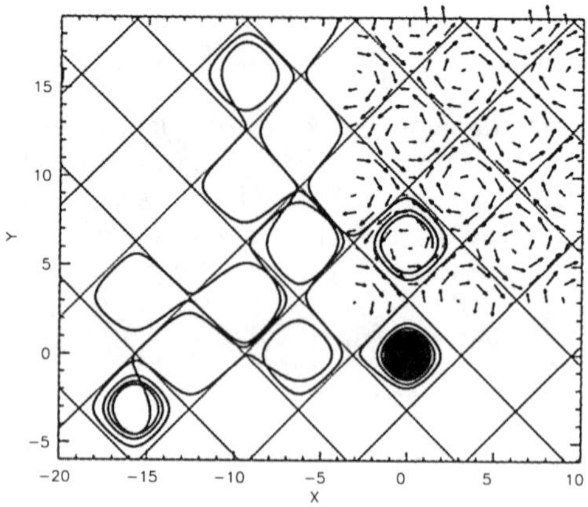

Figure 1. The velocity field defined by (2) divides the plane into advection cells. Velocity vectors of fluid motion are shown in part of the field. The *heavy lines* represent particle trajectories. The particle with $\delta = 1.1$ is shown to go toward the center of the cell close to (0,0), while the other one ($\delta = 0.9$) diffuses along the separatrices. In this case it was assumed that $\mu = 1$.

vorticity equation on a 128^2 grid, by the code previously adopted by Elhmaïdi, Provenzale, & Babiano (1993) for a Lagrangian study of fluid particles. The evolving flow field shows several vortices of a typical size corresponding to the forcing scale (wavenumber $k_I = 10$), as well as a background field of small vorticity fluctuations and low kinetic energy.

The trajectories of 1000 particles were integrated on timescales an order of magnitude longer than the eddy turnover time and their distribution compared to the topology of the flow. The results clearly show that after sufficiently long times all *heavy* particles have been ejected by the vortices and populate the background. Conversely, *light* particles have been efficiently concentrated by the vortices, and a few have been released to the background during interactions between vortices.

3. Transport Properties in a Rotating Disk

It is well known that global rotation induces a two-dimensionalization of turbulence and promotes the inverse energy cascade, as observed both in nature and in laboratory experiments (see, for example, Dubrulle & Valdettaro 1992). The macroscopic effect is thus the formation of large-scale vortices. The importance of global rotation relatively to fluid motion scales is usually quantified by the Rossby number $Ro = U/(2L\Omega)$, where U and L are the typical velocity and length scales of the flow, while Ω is the angular velocity. For $Ro < 1$ rotation dominates, and the emergence of large structures is observed.

While turbulence allows us to justify, for example, the angular momentum transport, its generation and characteristics in accretion disks are not clear. Since the Keplerian shear is linearly stable to infinitesimal perturbations, other mechanisms are also investigated and are the subject of an open debate.

Despite these uncertainties, it is still interesting to give a clear picture of the mechanisms influencing dust transport in presence of vortices i n the disk horizontal plane. In order to do this, a modification of Equation (1) is required. In a frame rotating with angular velocity Ω, it can be written as:

$$\frac{d^2\mathbf{x}}{dt^2} = \delta\frac{D\mathbf{u}}{Dt} - \mu\left(\frac{d\mathbf{x}}{dt} - \mathbf{u}\right) - 2\,\Omega \times \left(\frac{d\mathbf{x}}{dt} - \delta\mathbf{u}\right). \tag{3}$$

The last two terms on the RHS represent the net effect of Coriolis acceleration. The effect of the central gravitational attraction is assumed to be balanced by the centrifugal term. This approximation holds as long as the radial displacements of the particles are small (for a complete discussion, see Tanga et al. 1996).

The following expression, written in polar coordinates (r,θ) and representing a standing Rossby wave (corotating with the flow) on a two-dimensional disk, was chosen in order to reproduce the simple conditions of the stream function (2):

$$\Psi(r,\theta) = A\cos\left(Q\ln\left(\frac{r}{l_t}\right)\right)\cos(m\theta). \tag{4}$$

With appropriate choices of Q, m, and l_t, the size of advection cells was fixed to about 0.01 AU at 5 AU from the Sun, in agreement with the expected vortex dimension in the turbulent disk (Dubrulle 1993; Dubrulle & Valdettaro 1992) when a standing-wave solution is sought.

In fact, the integration of (3) in the flow (4), with Ω corresponding to the local Keplerian rotation rate at 5 AU from the Sun, clearly shows a completely new phenomenology. Using standard parameters of the minimum-mass Solar Nebula, the density ratio was fixed at $\delta = 10^{-8}$, and different simulations were made with stopping times ranging from $1/\mu = 1$ yr to $1/\mu = 500$ yr.

As also suggested by Barge & Sommeria (1995), it was found that anticyclonic vortices are capable of efficiently trapping dust particles, while cyclonic vortices eject them. The timescale of trapping increases with μ, but the behavior is always observed, as long as the parameter A which determines the typical eddy velocity scale, remains sufficiently small. In fact, in this case, inside anticyclonic vortices the Coriolis acceleration efficiently counteracts the centrifugal force associated with the circular vortex motion. However, for sufficiently large values of A, the centrifugal term is always larger and the particle is ejected. Interestingly, this condition for dust trapping corresponds to $Ro \leq 1$, i.e., to the regime of the rotation-dominated fluid motion.

The process of capture (with $\delta = 0$) has also been tested by Bracco et al. (1999), who demonstrated the possibility of generating vortices from strong random perturbations. Anticyclonic vortices are observed to survive for several revolutions, the lifetime being mainly determined by viscous dissipation. In that case the integrations were made with a two-dimensional non-compressible model, but, as shown by Godon & Livio (1999), anticyclonic vortices also survive when compressibility is taken into account.

In both cases dust particles are efficiently trapped inside vortices (Figure 2), showing again that the process is rather robust. The same phenomenology is observed in the simulations by Nauta (2000). In particular, as expected, it seems clear that the flow satisfies the condition $Ro < 1$, thus allowing trapping to take place.

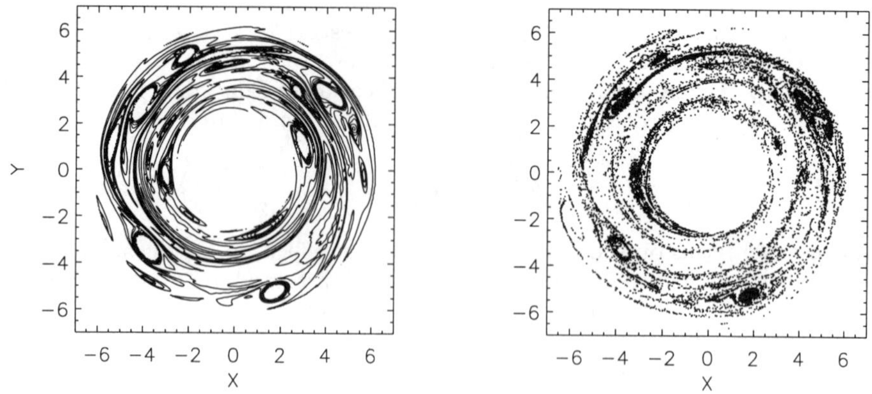

Figure 2. Vorticity isolines and dust particles with $\mu = 10$ from the simulations of Bracco et al. (1999) at $t = 20$ yr.

3.1. Gravitational Evolution

All previous simulations were made neglecting gravitational interactions or encounters between particles, whereas, from the point of view of planetesimal growth, these processes are essential.

In order to investigate the gravitational evolution of a disk of planetesimals not uniformly distributed in the disk, we used the N-body tree-code described by Richardson (1993). Planetesimal positions and velocities were initialized using the results of integrations of dust particles trajectories by Bracco et al. (1999), at the time $t = 20$ when trapping in vortices of various sizes has already happened.

In this first test we decided to reproduce the surface density of the nebula at 5 AU. Given the small number of particles (around 6000 were extracted in the range 4.5–5.5 AU) and the assumed density of 1.5 $g\ cm^{-3}$, their resulting diameter is rather large (800 km). Thus, they cannot be considered to coincide with the real, much smaller particles trapped inside vortices, but nevertheless they could represent the qualitative result of the concentration and accretion of small grains inside vortices. They were all obtained assuming a velocity dispersion along z, $\sigma_z = 0.01$ AU, initialized as described in Richardson (1993). In future investigations, other values of dispersion along z will be considered.

In Figure 3, the first stages of gravitational evolution are shown. A preliminary qualitative analysis shows that the initial particle clusters trapped inside vortices could last for several rotations. Of course, the characteristics of velocity dispersion are strictly determined by the velocity field inside the vortices, and different conditions must be simulated before reaching a solid conclusion. The persistence of self-gravitating clusters of planetesimals could also be important for theories of cometary nuclei formation (Weidenschilling 1995). Quantitative analyses are under way and a complete study will be published elsewhere.

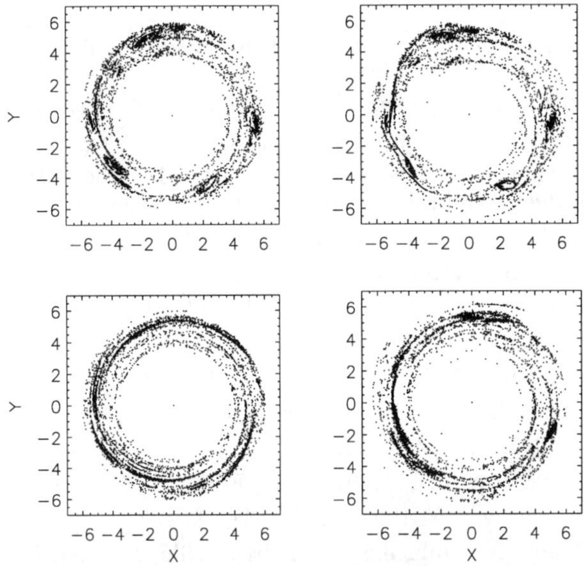

Figure 3. Results of gravitational evolution at $t = 2.4$ yr (*top*) and $t = 14.3$ yr (*bottom*). The scales are in AU. Initial positions and velocities were extracted from the simulations of Bracco et al. (1999) at $t = 20$ for the *right-hand panels*. *Left panels:* Circular orbits with Keplerian velocities.

4. Conclusions

Even though the existence of vortices on protoplanetary disks is not proved, it is nevertheless clear that the homogeneous distribution could be an approximation rather far from reality. As is widely observed in several situations, the Lagrangian dynamics of advected particles can give rise, even in very simple situations, to highly inhomogeneous distributions. The presence of vortices could lead to the segregation of particles of different sizes. Thus, besides the growth speed-up of small planetesimals in the 1–10 cm range (Tanga et al. 1996), the effects of segregation could propagate to later evolutionary stages. Further investigations taking into account both Lagrangian dynamics and gravitation are necessary to improve our knowledge of the early stages of planetesimals accretion.

Acknowledgments. This work was supported in part by the *Programme National de Planétologie* of INSU, France. We acknowledge A. Bracco and collaborators for providing data for gravitational simulation and for Figure 2, Ch. Froeschlé, A. Provenzale, and D. Richardson for very useful discussions.

References

Barge, P., & Sommeria, J. 1995 A&A 295, L1

Bracco, A., Chavanis, P. H., Provenzale, A., & Spiegel, E.A. 1999 Phys. Fl., 11, 2280

Crisanti, A., Falcioni, M., Paladin, G., Provenzale, P., Tanga, P., & Vulpiani, A. 1992 Phys. Fl. A, 4, 1805

Cuzzi, J. N., Dobrovolskis, A. R., & Champney, J. M. 1993 Icarus, 106, 102

Cuzzi, J. N., Dobrovolskis, A. R., & Hogan, R. C. 1996, in Chondrules and the Protoplanetary Disk, ed. R. Hewins, R. Jones, & E. R. D. Scott (Cambridge: Cambridge University Press), p. 35

Dubrulle, B., & Valdettaro, L. 1992 A&A, 263, 387

Dubrulle, B. 1993, Icarus,106, 59

Elhmaïdi, D., Provenzale, A., & Babiano, A. J. 1993, Fl. Mech., 257, 533

Godon, P., & Livio, M. 1999, ApJ, 523, 350

Maxey, M. R., & Riley, J. J. 1983, Phys. Fl., 26, 883

Nauta, M D. 2000, PhD Thesis, Utrecht University

Richardson, D. 1993, MNRAS, 261, 396

Squires, K. D., & Eaton, J. K. 1991, Phys. Fluids A, 3, 1169

Tanga, P., & Provenzale, A. 1994, Physica D, 76, 202

Tanga, P., Babiano, A., Dubrulle, B., & Provenzale, A. 1996, Icarus, 121, 158

Weidenschilling, S. J. 1977, MNRAS, 180, 57

Weidenschilling, S. J. 1995, Icarus, 116, 433

Weidenschilling, S. J. 1997, in ASP Conf. Ser., 122, From Stardust to Planetesimals, eds. Y. Pendleton & A. G. G. M. Tielens (San Francisco: ASP), p. 281

Discussion

Patrick Godon: The calculations of Bracco et al. are incompressible and therefore the dust particles do not drift radially inwards. When compressibility is taken into account, the particles drift radially inwards and fill in the void created by the cyclonic vorticity perturbations.

Paolo Tanga: The details of the dust distribution depend on poorly known parameters, so we preferred to study particle trapping as an isolated phenomenon, even if it is clear that it will couple to other effects in realistic simulations. Anyway, even the meaning of realistic here is questionable, since we cannot really say that we are capable of building a Keplerian disk numerically. It is certain, instead, that we cannot capture all the details. So, despite these limitations, the fundamental think to remember is the deficiency of models relying upon homogeneous particle distributions and simple assumptions on the velocity dispersion.

Hubertus Klahr: Please remind us which particle sizes are affected by this concentration mechanism!

Paolo Tanga: In principle, if vortices are sufficiently long-lived, a "particle" of any size can be trapped. In practice, we dealt with small planetesimals in a range of stopping times going from 2 to 500 years. At a distance of about 5 AU, assuming (badly known) nebula parameters, that corresponds roughly to particles in the 1 cm to 1 m range of sizes.

Disks, Planetesimals, and Planets
ASP Conference Series, Vol. 219, 2000
F. Garzón, C. Eiroa, D. de Winter, and T. J. Mahoney, eds.

NICMOS Imaging of Dusty Circumstellar Disks

A. J. Weinberger and E. E. Becklin

UCLA, 405 N. Hilgard Ave., Los Angeles, CA 90095-1562, USA

G. Schneider

Steward Obs., U. Arizona, 933 N. Cherry Ave., Tucson, AZ 85721, USA

Abstract. Using the coronagraph in the Near Infrared Camera and Multi-Object Spectrometer (NICMOS) instrument aboard the *Hubble Space Telescope*, we have imaged circumstellar disks around stars of age $\sim$ 8 Myr. These dusty disks vary in physical size by over an order of magnitude and exhibit radial anisotropies in their brightness distributions which may be indicative of dynamical confinement or sculpting of the disk particles by unseen planetary bodies. We highlight results from three stars: HD 141569, HR 4796A, and TW Hydrae.

1. Introduction

If planets are to form, their cores must begin to build before disks are destroyed; therefore, to understand planet formation it is important to know how disk dust particle size and composition evolves. Disks at ages of $\sim 10^7$ yr are thought to exist at a critical juncture in evolution between when most of the original circumstellar material has disappeared and when embryonic bodies formed in the protostellar disk begin to collide and generate secondary debris.

We observed 19 disk candidates in our NICMOS program selected on the basis of known *IRAS* IR excesses with $L_{\rm IR}/L_* \gtrsim 10^{-3}$. Using the NICMOS coronagraph, and taking advantage of the angular resolution and relative stability of the HST/NICMOS PSF, we pushed our investigations into previously and otherwise unreachable sensitivity ranges by reducing the background light at $1''$ to $\sim 10^{-6}$ of the peak stellar intensity. We imaged disks in about a third of our candidates with a wide range of sizes, shapes and surface brightnesses. Stars which are all $\sim 10^7$ yr old, including HD 141569 (Weinberger et al. 1999), HR 4796A (Schneider et et al. 1999), and TW Hya (Weinberger et al. 2000b) have radically different disk morphologies suggesting that dynamical influences are already at work, and that planets may already be forming at this age.

2. Data Analysis

In the reduced and calibrated NICMOS images, scattered and diffracted light from the star still dominates the disk flux levels at radii of 1–5$''$, so in order to see any disk which may be present, a PSF must be subtracted. The detailed character of the *HST* PSF changes on multi-orbit timescales; hence some star subtractions produce much lower residuals than others. In our coronagraphic program, 15 stars were observed with the F110W filter and ~ 50 stars of spectral type A–K were observed with the F160W filter. Targets which do not show extended disk emission form a library of possible PSFs.

Each PSF star brighter than the target was registered using cubic convolution interpolation to the target, and the PSF that minimized the subtraction residuals in the diffraction spikes and in annuli between $0.''3$ and $4''$ was chosen. The magnitudes of the target and PSF were extrapolated from their H-band magnitudes, and the uncertainty in their flux ratio completely dominates the uncertainty in the photometry of the observed disk. A change of a few per cent in the subtracted PSF flux can make an order of magnitude larger difference in the integrated disk flux density. All targets were observed at two telescope rolls, and the two PSF-subtracted images were rotated to a common orientation and averaged. Regions of the image obscured by the diffraction spikes in one roll were replaced where possible by uncorrupted regions in the other roll.

3. HD 141569

HD 141569, classified as B9.5Ve (Jaschek & Jaschek 1992), is a Herbig Ae/Be star at a distance of 100 pc. Our full results on HD 141569 are presented elsewhere (Weinberger et al. 1999, 2000a). We also direct the reader to the NICMOS coronagraphic results of Augereau et al. (1999 and this volume, p. 335).

Scattered-light imaging at 1.1 μm shows an elongated disk that is clearly evident at radii from ~ 1.6 to $4''$ (160 to 400 AU) from the star (Figure 1*a*). The disk major axis position angle is $356 \pm 5°$ and ratio of the minor to major axes of 0.63 ± 0.04. If the disk is intrinsically circular, the axial ratio implies an inclination of $51 \pm 3°$ from face on. The azimuthally averaged disk surface brightness peaks at 0.3 mJy arcsec^{-2} at a radius of 185 AU and falls off to larger and smaller radii. At 250 AU an annulus of lower surface brightness, or a "gap" in the disk, can be seen.

The eastern side of the disk is 1.5 ± 0.2 times brighter than the western side. This difference is not an artifact of the PSF subtraction and can be explained by dust particles which preferentially scatter in the forward direction. The brighter side, therefore, is inclined toward us, and the ratio of forward-to-backward scattering implies an asymmetry parameter, g, of 0.11 in the analytic phase function of Henyey & Greenstein (1941).

The optical depth to visual absorption (τ) implied by the far-infrared excess is 8.4×10^{-3}; if this is also the absorption optical depth at 1.1 μm, then the albedo of the scatterers is ~ 0.3. However, about 1/3 of the mid-infrared luminosity arises from a region within 150 AU of the star (Fisher et al. 2000). So, the total infrared τ overestimates the optical depth seen in scattered light; in the region at $\sim 1.''9$ (190 AU), the albedo is probably more like 0.4. This is similar to the albedos of mineral mixtures and ices found in ISO spectra of Herbig Ae/Be disks (Malfait et al. 1998).

We have investigated two stars close to HD 141569 on the plane of the sky (hereafter HD 141569B and C) via optical spectroscopy and near-infrared

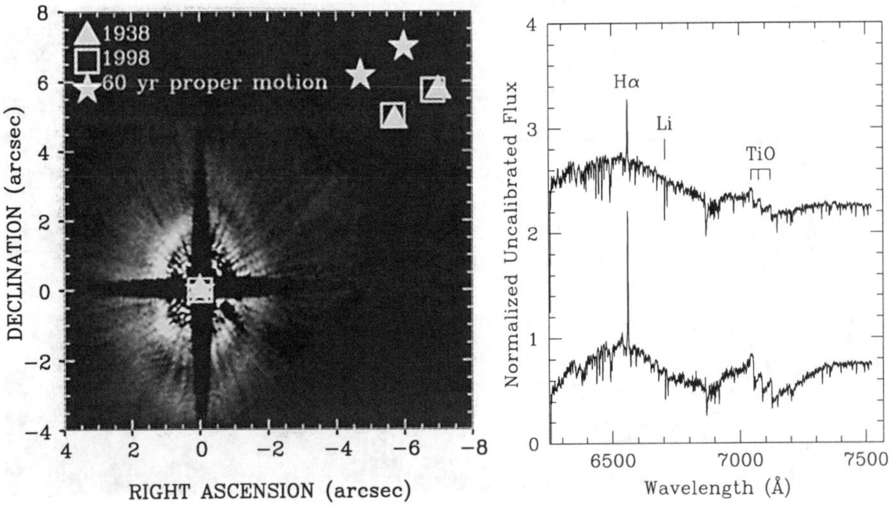

Figure 1. *Left:* NICMOS 1.1 μm image of HD 141569 and the relative locations of the two companion stars in 1938 (*filled triangles*) and 1998 (*open squares*). If B and C were stationary background stars, the proper motion of A would predict a vastly different location for B and C in 1998 (*filled stars*). *Right:* Low-resolution spectra of HD 141569 B and C show that they are young M-type stars.

imaging. Our spectra from the Keck Telescope show that they are of type M2 and M4, respectively and show Hα in emission and Li in absorption (Figure 1*b*). Furthermore, they have radial velocities consistent with that of A. Using astrometry from NICMOS and from historical measurements of Rossiter (1955), we see that the three stars have had the same proper motion over the last 60 years (Figure 1*a*). Therefore, they almost certainly form a triple star system, although they may not be bound. From their X-ray flux, lithium absorption and location on pre-main-sequence evolutionary tracks, the companions appear to be 5 Myr old. HD 141569 A is now one of the few stars with a circumstellar disk that has a well-determined age.

The inferred size of the grains in the HD 141569A disk, <5 μm (Fisher et al. 2000), implies that radiation pressure blows them away in at least an order of magnitude less time than our derived stellar age. Thus, they must be continuously regenerated, probably through collisions of larger bodies.

4. TW Hya and HR 4796A

Both TW Hya (K7V) and HR 4796A (A0V) are young ($\sim$ 8 Myr-old) members of the eponymous "TW Hya Association" at a distance of $\sim$ 60 pc (Webb et al. 1999). At about the same age as HD 141569, these three stars show the variety of evolutionary influences at work on circumstellar disks over short timescales.

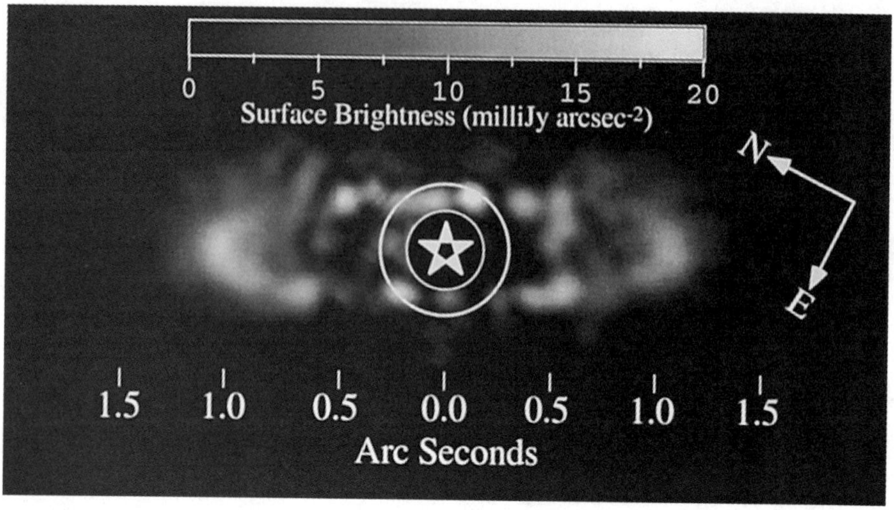

Figure 2. NICMOS 1.1 μm image of HR 4796A after reprocessing to remove background due to PSF residuals. The outer white circle is the size of the coronagraphic hole, and the annulus is the width of the PSF. Thus the filled circle shows the region where the PSF is fully occulted. Compare to Figure 1*b* of Schneider et al. (1999).

NICMOS images of HR 4796A at 1.1 μm and 1.6 μm show a ring at a position angle of 26.8° with a semi-major axis of 70 AU (1.″05). The disk appears to be abruptly bounded both interior and exterior to the mid-ring. A one-dimensional deconvolution along the major axis using a NICMOS PSF sets an upper limit to the ring width of 14 AU. The north-east ansa is brighter than the south-west ansa by $\sim$ 15%. The observed flux ratios of the disk to the central star are 0.0014 at 1.1 μm and 0.0024 at 1.6 μm indicating somewhat red reflection. This implies a mean particle size that is large compared to these wavelengths. Assuming $\tau_{\rm dust} \sim L_{\rm disk}/L_*(\lambda)$, there is reasonable agreement between the scattered stellar flux in the near infrared and that which is absorbed in the visible and reradiated in the thermal infrared (Koerner et al. 1998; Telesco et al. 2000).

Since the time of the published images in Schneider et al. (1999), we have reprocessed the data to model and remove the PSF subtraction residuals. The background at the location of the ring was inferred by interpolating the background inside and outside the ring over a spatial scale of $\sim$ 5 pixels. This corrects for the local non-uniformity of the subtraction residuals. The result is that the ring can now be seen nearly all the way around the star (Figure 2). Most if not all of the lumpiness in the image is likely to be induced by incomplete and imperfect modeling of the PSF residuals and should not be interpreted as real azimuthal discontinuities, or clumps, in the ring.

Unlike the tightly confined ring of HR 4796A, the T Tauri star TW Hya is surrounded by a broad, face-on disk (Figure 3). From 0.″7 (40 AU) where the disk begins to dominate the PSF subtraction residuals, to 3.″1 (175 AU), the

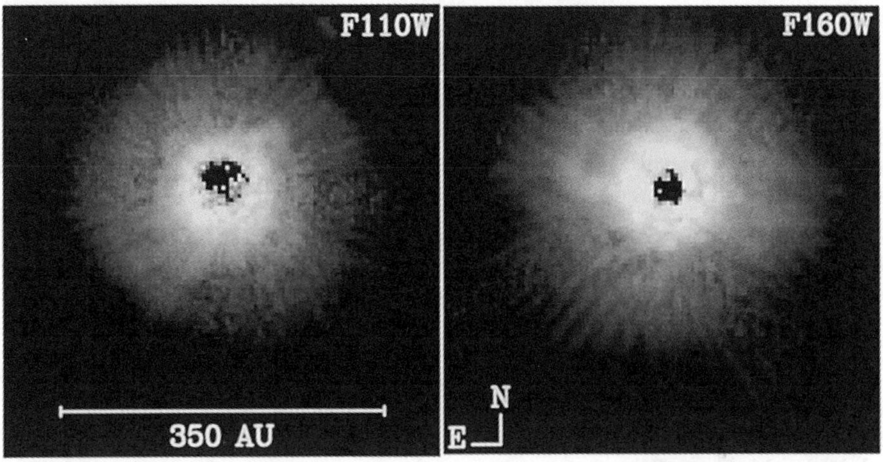

Figure 3. NICMOS PSF-subtracted images of TW Hya at 1.1 and 1.6 μm.

disk is visible at both 1.1 and 1.6 μm. The surface brightness declines smoothly with radius, peaking close to the star at 2–4 mJy/arcsec^2. The disk isophotes appear to be circular and assuming the disk is intrinsically circular, it is therefore viewed face on. Between radii of 40 AU and 135 AU, the disk is well fit by a single power law, $r^{-2.6}$, at both wavelengths. Between 135 and 175 AU, the disk falls off more sharply. Visible light WFPC images also show the disk with approximately the same extent and surface brightness profile (Krist et al. 2000).

The average color of the disk is, within the uncertainties, exactly that of the star. Since the two observations were taken at different telescope orientations, the PSF features do not appear in the same place in the images from the two filters, so no color map can be made. The ratio of the total scattered light to the flux density from the star is 0.02 at both wavelengths, but the mid-/far-infrared $\tau = 0.2$, so the disk must be optically thick. For single scattering by a geometrically thin, optically thick disk, the flux density per square arcsec goes as r^{-3} (Whitney & Hartmann 1992). This is steeper than the $r^{-2.6}$ we measured and clearly does not fit our data. Furthermore, the thin-disk model does not reproduce the observed amount of scattered light at the distances from the star at which it is observed, but is low by a factor of $\sim 10^4$. The TW Hya disk therefore must be flared so that the outer parts of the disk are able to intercept and scatter stellar photons.

The amount of scattered light at any given radius is determined not just by the surface density of scatterers but also by the extinction along the line of sight from the star to that part of the disk. At $\sim$ 80 AU from the star, the disk scattered light is somewhat lower over a region of $\sim$ 20 AU. This depression in the amount of scatter could be caused by a small bump in the disk that creates a "shadow," or by a decrease in the surface density of the disk at that location. Either way, it may indicate dynamical effects.

In conclusion, we see that by an early age, 5–10 Myr, stellar disks can have dramatically different morphologies. The disks presented here span sizes of 70–400 AU. TW Hya's disk is optically thick, whereas HD 141569 and HR 4796A's are thin; the destruction of gas and grains is expected to be faster in the luminous, massive, early-type stars. However, HD 141569 and HR 4796A are similar in spectral type, age, and amount of infrared excess, yet their disks differ greatly in size and character.

Acknowledgments. This work is supported in part by NASA grant NAG 5-3042, and based on observations with the NASA/ESA *Hubble Space Telescope*, obtained at the Space Telescope Science Institute, which is operated by the Association of Universities for Research in Astronomy, Inc., under NASA contract NAS5-26555. This work has also made use of the W. M. Keck Observatory, which is operated as a scientific partnership between the California Institute of Technology, the University of California, and the National Aeronautics and Space Administration and was made possible by the generous financial support of the W. M. Keck Foundation.

References

Augereau, J. C., Lagrange, A. M., Mouillet, D., & Menard, F. 1999, A&A, 350, 51

Fisher, R. S., Telesco, C. M., Piña, R. K., Knacke, R. F., & Wyatt, M. C. 2000, ApJ, 532, 141

Henyey, L. G., & Greenstein, J. L. 1941, ApJ, 93, 70

Jaschek, C., & Jaschek, M. 1992, A&AS, 95, 535

Koerner, D. W., Ressler, M. E., Werner, M. W., & Backman, D. E. 1998, ApJ, 503, L83

Krist, J. E., Stapelfeldt, K. R., Menard, F., Padgett, D. L., & Burrows, C. J. 2000, ApJ, in press

Malfait, K., Waelkens, C., Waters, L. B. F. M., Vandenbussche, B., Huygen, E., & de Graauw, M. S. 1998, A&A, 332, L25

Rossiter, R. A. 1955, Publ. Michigan Obs., 11, 1

Schneider, G., et al. 1999, ApJ, 513, 127

Telesco, C. M., et al. 2000, ApJ, 530, 329

Webb, R. A., Zuckerman, B., Platais, I., Patience, J., White, R. J., Schwartz, M. J., & McCarthy, C. 1999, ApJ, 512, L63

Weinberger, A. J., Becklin, E. E., Schneider, G., Smith, B. A., Lowrance, P. J., Silverstone, M. D., Zuckerman, B., & Terrile, R. J. 1999, ApJ, 535, L53

Weinberger, A. J., Rich, R. M., Becklin, E. E., Zuckerman, B., & Matthews, K. 2000a, ApJ, submitted

Weinberger, A. J., Becklin, E. E., Schneider, G., & Zuckerman, B. 2000b, in preparation

Whitney, B. A., & Hartmann, L. 1992, ApJ, 395, 529

Disks, Planetesimals, and Planets
ASP Conference Series, Vol. 219, 2000
F. Garzón, C. Eiroa, D. de Winter, and T. J. Mahoney, eds.

Dust Disks around Old Pre-Main-Sequence Stars: *HST*/NICMOS2 Scattered-Light Images and Modeling

Jean-Charles Augereau, Anne-Marie Lagrange, and David Mouillet

Laboratoire d'Astrophysique de l'Observatoire de Grenoble, Université J. Fourier, 414 rue de la piscine, BP 53, 38041 Grenoble Cedex 9, France

François Ménard

CFH Telescope Corporation, PO Box 1597, Kamuela, HI 96743, USA

Abstract. We present recent near-infrared detections of circumstellar disks around the two old PMS Herbig stars, HD 141569 and HD 100546, obtained with *HST*/NICMOS2. They reveal extended structures larger than 350–400 AU in radius. While the HD 100546 disk appears as a continuous disk down to 40 AU, the HD 141569 environment seems more complex, split into at least two dust populations. As a convincing example, the full modeling of the disk surrounding HR 4796, another old PMS star, is detailed and confronted with more recent observations.

1. Introduction

The presence of massive disks with large gas-to-dust ratios around T Tauri stars is now well established (see a recent review by Dutrey 1999). Most recent near-IR and millimeter resolved images reveal optically thick disks in Keplerian rotation around single or multiple systems (e.g., GG Tau, Guilloteau, Dutrey, & Simon 1999). The dust appears to be the direct remnant of the primordial nebula, and Weidenschilling & Cruzzi (1993) have shown from a theoretical point of view that planetesimals may form during this stage.

It is also known that a large fraction of main-sequence (MS) stars are surrounded by material heated by their central star (for a recent review, see Lagrange, Backman, & Artymowicz 2000). Among these Vega-like stars, the β Pic circumstellar disk was the only one imaged till recently. Submillimetric observations resolved for the first time dust emission around a few isolated MS stars (e.g., Greaves et al. 1998). Vega-like stars bear optically thin disks with small gas-to-dust ratios, the smallest grains being continuously replenished by collision and/or evaporation among larger bodies (Backman et Paresce 1993). We also have direct (e.g., 55 Cnc, Butler, Marcy, & Williams 1997) or indirect (β Pic: e.g., Beust & Morbidelli 2000) evidence for planets in these systems.

To understand better the full evolutionary scenario leading to planetary-system formation, one must answer the following questions: what are the time-scales for dissipating disks? How does they depend on the spectral type and multiplicity? What are the dynamically dominant processes? How do the gas and dust couple? What is the origin and nature of the dust? And finally, how does this correlate with planet formation? This requires studying intermediate-age circumstellar disks which represent the missing link between embedded young disks and evolved ones. Full spectral energy distributions (SEDs) give precious constraints on the chemical composition of the grains (e.g., HD 100546, Malfait et al. 1998 with it ISO) but only qualitative information on the dust distribution itself. Resolved data are needed on these objects.

In Section 2, we present *HST*/NICMOS2 images of two Herbig (Vega-like?) stars: HD 141569 and HD 100546. We then discuss in Section 3 how we can infer valuable constraints on the dust composition and distribution as well as on the dynamics of the system by combining all available data (SED and multi-λ images) through a consistent model. The case of HR 4796 will be detailed.

2. *HST*/NICMOS2 observations of HD 141569 and HD 100546

2.1. Targets and Coronagraphic Data

HD 141569 is a B9.5 Ve star older than 10 Myr and 99^{+9}_{-8} pc away (*Hipparcos* measurements, van den Ancker et al. 1998). This star was selected due to its *IRAS* infrared (IR) excess and its small intrinsic polarization (Yudin & Evans 1998), which are both clues to the presence of an extended optically thin disk. Moreover, the ^{12}CO $J = 2 \rightarrow 1$ detection by Zuckerman, Forveille, & Kastner (1995) ensures that gas is present at a level consistent with a young MS star.

The Herbig B9 V star HD 100546, associated with the dark cloud DC 296.2-7.9, shows a strong IR excess peaking at about 25 μm (*IRAS*) due to a large amount of circumstellar material. The photometric, polarization and spectroscopic events (e.g., van den Ancker et al. 1998; Yudin & Evans 1998; Grady et al. 1997) remain uncertain but are characteristic of Herbig Ae/Be stars. HD 100546 is 103^{+7}_{-6} pc away and 10 Myr old according to van den Ancker et al. (1998) and appears to be in a less evolved state than HD 141569.

We have performed, $\lambda = 1.6$ μm coronagraphic images of HD 141569 and HD 100546 with the *HST*/NICMOS2 camera. During the *same* orbit (so as to avoid PSF variations), a close star free of known circumstellar material so far and with similar spectral type and magnitude was observed. The reduction procedure consists in subtracting this reference star carefully scaled to the star of interest. The determination of the scaling factor is critical since a small change can significantly modify the photometry or in the worst case be responsible for the presence of unrealistic features. It is found by azimuthally averaging the division between the star of interest and the reference-star images.

2.2. Results for HD 141569

The final reduced image of HD 141569 shown in Figure 1 (*left*) is interpreted as an inclined dust ring peaking at 325 ± 10 AU from the star, which scatters a small fraction (a few 10^{-3}) of the stellar light. We derive from least-squares ellipse fitting (Figure 1, *right*) a position angle (P.A.) of $355.4^\circ \pm 1^\circ$ and a disk inclination from edge on of $37.5^\circ \pm 4.5^\circ$, assuming that the disk is axisymmetric with respect to the star. The radial surface brightness indicates that the outer edge of the ring steeply decreases with distance from the star following an $r^{-6.87\pm0.14}$ radial power law between 360 AU and 420 AU.

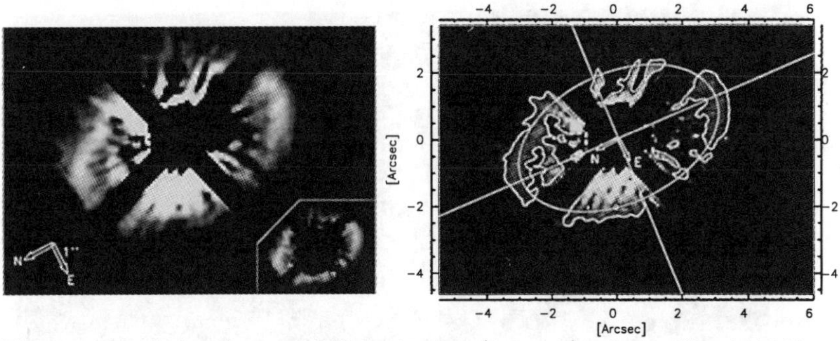

Figure 1. *Left:* Scattered-light image (1.6 μm) of the HD 141569 circumstellar disk on a logarithmic scale (from Augereau et al. 1999b). In the *bottom right-hand corner* we show the same disk where most of the unrealistic bright patterns close to the star have been removed to highlight the resolved annular structure. *Right:* Ellipse fitting.

The presence of an inclined extended disk-like structure around HD 141569 has been confirmed independently by Weinberger at al. (1999) through *HST*/ NICMOS2 1.1 μm images. The interpretation of the inner part of disk (below $\sim$ 240 AU along the major axis) is controversial. They found that most of the scattered light comes from a rather peaked region at $\sim$ 200 AU from the star. Scattering properties cannot account for the difference between our 1.6 μm and 1.1 μm images. At 1.6 μm, bright areas which appear inside the inclined annulus are interpreted as unrealistic residues from PSF subtraction. This interpretation is supported by the fact that the ratio of HD 141569 (before subtraction) to the reference star clearly shows a lack of significant detected circumstellar material inside the main resolved shape (see Figure 2 of Augereau et al. 1999b).

2.3. Results for HD 100546

The reduced image of HD 100546 shown in Figure 2 also reveals a large elliptical circumstellar structure centered on and bright close to the star. This observation seems consistent with preliminary results from Pantin & Lagage (2000), obtained with the adaptive-optics system ADONIS (ESO). Contrary to that of HD 141569, the HD 100546 disk does not exhibit a ring-like morphology but rather a continuous surface-brightness distribution from the very close edge of the coronagraphic mask ($\sim$ 40 AU) up to $\sim$ 350 AU. Isophotal ellipse fitting constrains the P.A. of the disk to $161^\circ \pm 5^\circ$ and provides an upper limit of $51^\circ \pm 3^\circ$ for the inclination of the disk with respect to the line of sight. The measured flux density of 84 $\pm$ 8 mJy leads to a lower limit of $10^{-2} \pm 15\%$ for the scattered-to-photospheric flux, assuming that 55% to 60% of the total flux (star + disk) at 1.6 μm is due to thermal emission from very hot unresolved grains.

Basic radial power laws fit the measured surface brightness profile very well with indexes -2.92 ± 0.04 in the radial range 40–250 AU and -5.5 ± 0.2 further than 270AU. This implies a surface density roughly proportional to r^{-1}, coinciding with results for less massive young stars (Dutrey et al. 1999, and references therein). Subtracting a synthetic disk, supposed to exhibit an axisymmetric surface brightness as described above and inclined at 51° towards P.A. = 161°,

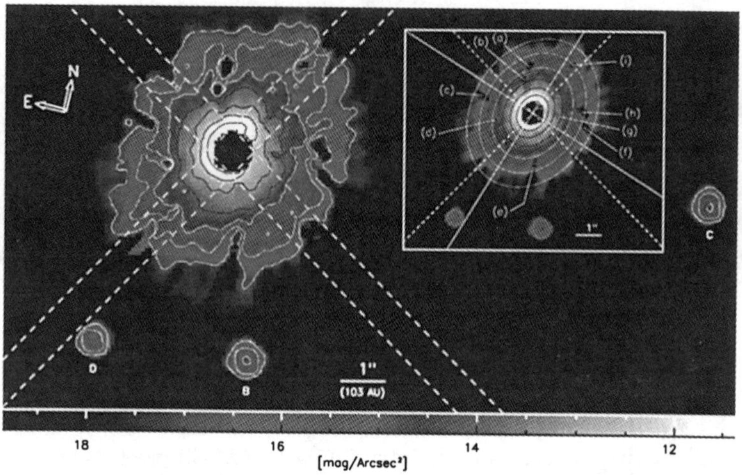

Figure 2. Scattered-light image (1.6 μm) of the HD 100546 circumstellar disk (from Augereau et al. 2000). The lowest contour level is 17 mag/arcsec^2. The isophote spacing is 1 mag/arcsec^2 in increasing order. The *small panel* in the *upper right-hand corner* highlights both the fix patterns, (*a*)–(*i*), blurring the image and the result of isophotal ellipse fitting. *Plain lines:* major and minor axes of the disk; *dashed lines:* position of spider arms.

from the observed image reveals a NE–SW brightness asymmetry: i.e., a flux excess up to 1.1 mag/arcsec^2 at 0.5–0.6″ collimated in a direction coinciding well with the precise minor axis of the disk. Anisotropic scattering dust grains properties can explain this effect (e.g., Figure 12 of Augereau et al. 1999a).

3. Modeling of Dust Disks around Old PMS Stars

3.1. Brief Overview of Model Assumptions

We developed an optically thin disk model able consistently to reproduce scattered and thermal observations, as well as SEDs. The dust distribution is described by a smooth combination of radial power laws. The vertical distribution is assumed to follow an exponential shape with height depending on the distance from the star as a radial power law. Grain optical properties form a key step in the modeling. We assume that grains are porous aggregates made of a silicate core coated by an organic refractory mantle (Greenberg 1986). Vacuum is assumed to fill the holes due to porosity which can be partially taken up by H_2O ice if the grain temperature falls below 110–120 K. We adopt two types of grains: amorphous "ISM-like grains" with porosity of about 50% and crystalline "comet-like grains" with large porosities ($\sim$ 95%). Dust optical properties are computed using Mie theory. They depend on the complex index of refraction of the mixture and on the grain size, *a*. We adopt the Maxwell–Garnett effective-medium theory to compute these indexes and assume a collisional grain size distribution following an $a^{-3.5}$ power law between $a_{\rm min}$ and $a_{\rm max}$. A more detailed description of model assumptions can be found in Augereau et al. (1999a).

3.2. An Example: The HR 4796 Disk

HR 4796 is a wide binary A0V star showing an infrared excess twice that of β Pic and 5 to 15 times younger (8 ± 2 Myr according to Stauffer, Hartmann, & Barrado y Navascués 1995) and thus traces an evolutionary stage prior to that of β Pic. We used the above model to reproduce all available observations of the HR 4796 circumstellar system before 1999 (Augereau et al. 1999a). We find that two dust populations are required to account for the full SED compatible with thermal and scattered-light resolved images.

The first population forms a sharp, cold ring peaking at 70 AU from the star and made of "ISM-like grains". It is responsible for the 25 μm–1 mm dust continuum emission (see Figure 11 of Augereau et al. 1999a). This implies grains larger than $a_{\rm min} = 10$ μm, which exactly corresponds to the blow-out size limit for grains produced by larger bodies in circular orbits. Also, such large bodies must be present either to reproduce millimeter measurements or to resupply the ring with smallest grains. Indeed, the 10 μm grains have very short timescales under collisions ($\sim$ 1000 yr) compared to the stellar age.

The second hot-dust population is poorly constrained but would rather be composed of large crystalline grains at 9–10 AU from the star. This population is necessary to fit the 10 μm measurements, which cannot be reproduced with a single ring at 70 AU as sharp as required by scattered-light images (Schneider et al. 1999). Interestingly, Telesco et al. (2000) later marginally resolved the HR 4796 disk in the N band ($\lambda = 10.8$ μm, $\Delta\lambda = 5$ μm) and showed that a significant fraction of this emission comes from the outer ring. To check the consistency of proposed model with these observations, we convolve with the observed PSF from Telesco et al. (2000) a simulated disk at 10.8 μm. We also simulate a second disk at 12.8 μm, since the N band is wide and since the disk emission across this band varies considerably (Figure 11 of Augereau et al. 1999a). Figure 3 shows that the outer ring detection is predicted by the model at a level fully consistent with the observations given the spectral width of the N band. This implies in particular that the relative fractions of flux coming from the inner and outer disks are correctly predicted.

4. Concluding Remarks

Further, and especially dynamical, modeling, is necessary to improve our knowledge of disks at different stages. In the case of HR 4796, Wyatt et al. (1999), Kenyon et al. (2000), and Klahr & Lin (2000) have proposed models to explain either the ring-shaped structure or brightness asymmetries. Further studies of HD 141569 and HD 100546 will allow us to show which dynamical processes are generic or particular to such old PMS stars. But very few observations of these circumstellar environments are available at present. As a new example, SED fitting of the HD 141569 disk using the radial dust distribution derived from the 1.6 μm images has predicted the presence of an inner dust population (Augereau et al. 1999b), recently confirmed by Fisher et al. (2000).

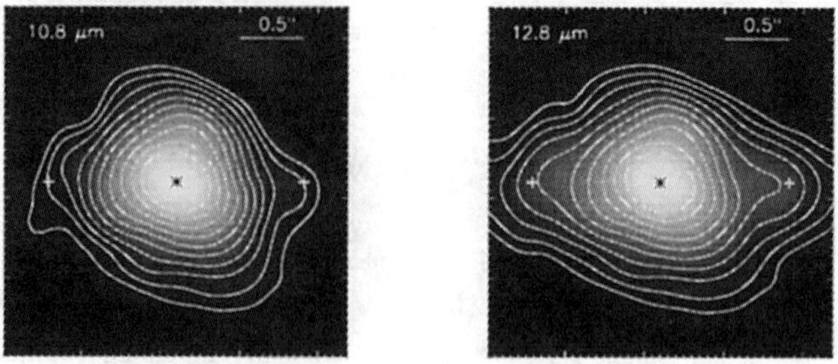

Figure 3. Simulated thermal images of the HR 4796 disk convolved with the N-band observed PSF from Telesco et al. (2000). To allow direct comparison, the contours levels are strictly the same as those of Figure 1a of these last authors. *White crosses* indicate the position of the outer ring.

Acknowledgments. We thank Mark Wyatt for his suggestion to check the consistency of the HR 4796 model with Telesco et al.'s (2000) 10 μm observations and Charles Telesco for kindly providing the corresponding observed PSF.

References

Augereau, J. C., et al. 1999a, A&A, 348, 557

Augereau, J. C., et al. 1999b, A&AL, 350, 51

Augereau, J. C., et al. 2000, A&A, submitted

Backman, D., & Paresce, F. 1993 in Protostars & Planets III, ed. E. H. Levy & J. Lunine (Tucson: University of Arizona Press), p. 1253

Beust, H., & Morbidelli, A. 2000, Icarus, 143, 170

Butler, R. P., Marcy, G. W., & Williams, E. 1997, ApJ, 474, 119

Dutrey, A. 1999 in NATO-ASI Series C, Planets outside the Solar System, Theory and Observations, ed. J.-M. Mariotti & D. Alloin (Dordrecht: Kluwer), p. 13

Fisher, R. S., et al. 2000, ApJ, 532, L141

Grady, C. A., et al. 1997, ApJ, 483, 449

Greaves, J. S., et al. 1998, ApJ, 506, L133

Greenberg, J. M. 1986 in Light on Dark Matter, ed. F Israel (Dordrecht: Reidel), p. 177

Guilloteau, S., Dutrey, A., & Simon, M. 1999, A&A, 348, 570

Kenyon, S. J., Wood, K., Whitney, B. A., & Wolff, M. 1999, ApJ, 524, L119

Klahr, H. H., & Lin, D. N. C. 2000, ApJ, submitted

Lagrange, A. M., Backman, D., & Artymowicz, P. 2000 in Protostars & Planets IV, ed. V. Mannings, A. P. Boss, & S. S. Russell (Tucson: University of Arizona Press), p. 639

Malfait, K., et al. 1998, A&A, 332, L25

Pantin, E., & Lagage, P. O. 2000, A&A, submitted

Schneider, G., et al. 1999, ApJ, 513, L127

Stauffer, J. R., Hartmann, L. W., & Barrado y Navascués, D. 1995, ApJ, 454, 910

Telesco, C. M., et al. 2000, ApJ, 530, 329

van den Ancker, M. E., de Winter, D., & Tjin A Djie, H. R. E. 1998, A&A, 330, 145

Weidenschilling, S. J., & Cruzzi, J. N. 1993 in Protostars and Planets III, ed. E. H. Levy & J. Lunine (Tucson: University of Arizona Press), p. 1031

Weinberger, A., et al. 1999, ApJ, 525, L53

Wyatt, M. C., et al. 1999, ApJ, 527, 918

Yudin, R. V., & Evans, A. 1998, A&AS, 131, 401

Zuckerman, B., Forveille, T., & Kastner, J. H. 1995, Nat, 373, 494

Disks, Planetesimals, and Planets
ASP Conference Series, Vol. 219, 2000
F. Garzón, C. Eiroa, D. de Winter, and T. J. Mahoney, eds.

The Evolution of the IR Excesses in Vega-Like Stars

David Barrado y Navascués

Departamento de Física Teórica, C-XI, Universidad Autónoma de Madrid, Cantoblanco, E-28049 Madrid, Spain

Inseok Song and J.-P. Caillault

Department of Physics and Astronomy, University of Georgia, Athens GA 30602, USA

John R. Stauffer

Harvard–Smithsonian Center for Astrophysics, 60 Garden St., Cambridge MA 02138, USA

Abstract. We have estimated the ages of a sample of Vega-like stars of spectral types A, F, G, and K. We show that there is a correlation between the IR excesses and the age for the early A types, as opposed to what happens for the late types.

1. The Vega Phenomenon

The β Pic type stars show IR excesses associated with the presence of circumstellar dusty disks (e.g., Backman & Paresce 1993). The prototypes are β Pic, Vega, and Fomalhaut. These circumstellar dusty disks are assumed to be formed from T Tauri disks or a subproduct of planet formation. Knowledge of the ages of β Pic stars is therefore one of the keys to understanding the formation and evolution of their disks. However, there is only a handful of estimates of the ages of these systems. Recently, several searches have been conducted in open clusters, which have accurated ages. But so far, no final conclusion has been reached about the presence of IR excesses in members of these clusters (see Angelova, Backman, & Stauffer 1998; Abrahám et al. 1998). All three prototypes are A stars, and estimates from their post-ZAMS evolution give approximate ages of 100, 200, and 400 Myr for β Pic, Fomalhaut, and Vega, respectively (Backman & Paresce 1993).

2. Estimating the Ages

Although the three prototypes have early spectral types, the Vega-like phenomenon is not circumscribed to hot stars. Actually, stars such as ϵ Eri have strong IR excesses, despite their late spectral types (K2 V in the case of ϵ Eri). Therefore, we have taken advantage of the properties of each spectral range to estimate the ages. Essentially, we have used four methods:

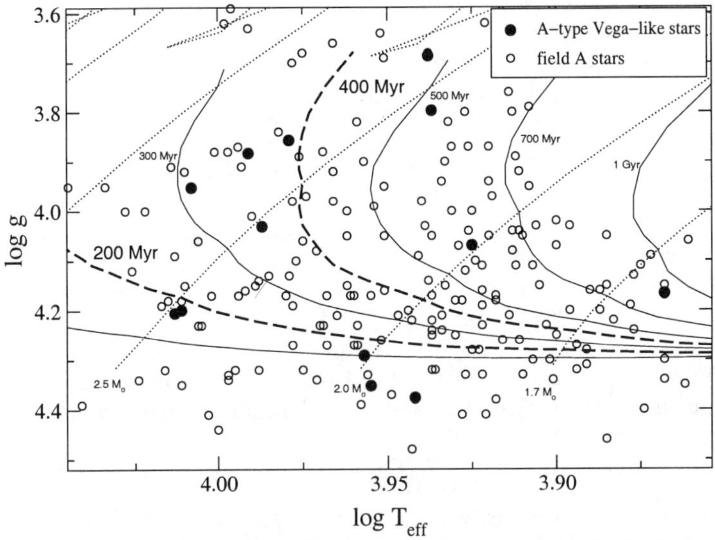

Figure 1. Gravity against effective temperature, as derived from Strömgren photometry. Several isochrones and evolutionary track from Schaller et al. (1992) are included as *dash-dotted* and *dotted lines*, respectively.

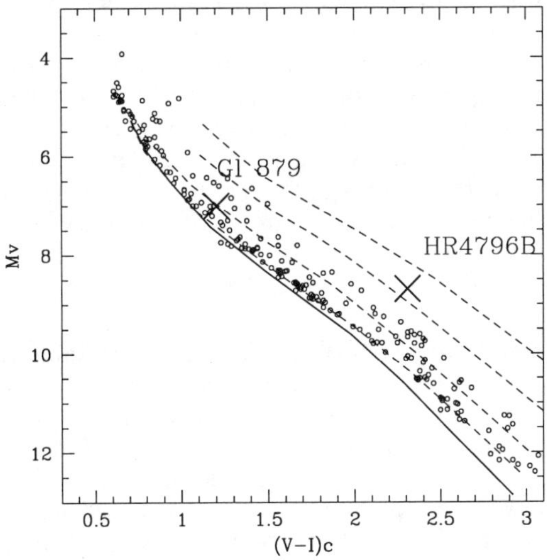

Figure 2. Color–magnitude for GJ 879, a physical companion of Fomalhaut, and HR 4796B, a physical companion of HR 4796A. The isochrones are from D'Antona & Mazzitelli (1994).

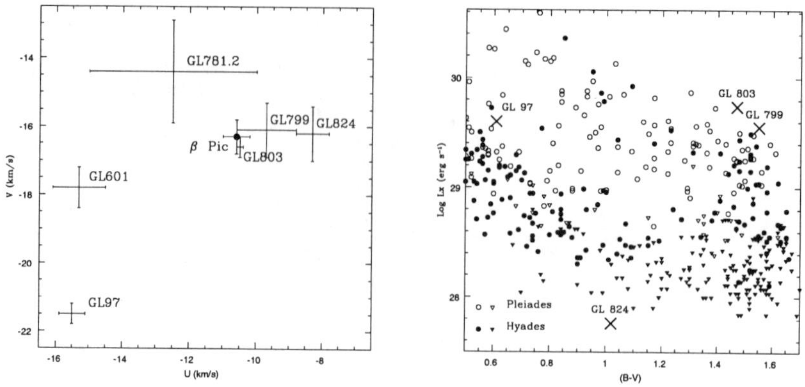

Figure 3. Proposed members of the β Pic moving group. *a)* Components of the Galactic velocity. *b)* Coronal activity against $(B-V)$ color.

- Strömgren photometry of A-type stars.—We derived $T_{\rm eff}$ and gravities based on Strömgren photometry ($uvby\beta$ filters). The comparison between the Vega-like values, those corresponding to open-cluster stars of different ages, and theoretical isochrones allow us to estimate the ages of the former. Figure 1 illustrate this technique. In this case, we show *bona fide* field Vega-like A stars, nearby field A stars, and the isochrones and evolutionary tracks of Schaller et al. (1992). See Song et al. (2000b).

- Properties of low-mass stars physically associated with A-type stars.—Most stars have physical companions. We have searched for visual companions to Vega-like stars, trying to verify if they are indeed physically associated with the primary. This is the case of Fomalhaut and GJ879, a K5 Ve star; and HR 4796A and B. For these coeval systems, we have used several age-dependent properties of these secondary components, such as their location on color–magnitude diagrams (Figure 2), lithium abundance, and activity, to estimate the age of the primary. See Stauffer et al. (1995).

- Stellar kinematic groups (SKGs).—We have been able to identify SKGs, as strictly defined by Soderblom & Mayor (1993), which contain Vega-like stars. One of them is the Castor moving group, which could include Vega and Fomalhaut. If both stars are members of the groups, they would be coeval. The other SKG is the β Pic moving group, which shares its motion with two late-type, very active stars (GJ799 and GJ803), as shown in Figures 3*a* and *b*. See Barrado y Navascués (1998) and Barrado y Navascués et al. (1999).

- Properties of late-type Vega-like stars.—-Some Vega-like stars are in fact late-type stars. Therefore, it is possible to study their rotation rate, stellar activity, and lithium abundance to compare their values with models and those corresponding to members of open clusters, and to estimate an age for them. See Figures 4 & 5 and Song et al. (2000a).

We have applied the same techniques to field stars without IR excesses and verified that our age-dating methods are not biased.

Table 1. Ages and spectral types of Vega-like stars

Name	Sp. type	Age (Myr)	Reference
HR 4796A	A0 V	8±2	Stauffer et al. (1995)
Vega	A0 V	200±100	Barrado y Navascués (1998)
βUMa	A1 V	300±100?	Abrahám et al. (1998)
Fomalhaut	A3 V	200±100	Barrado y Navascués et al. (1997)
Beta Pic	A5 V	20±10	Barrado y Navascués et al. (1999)
HD 10647	F9 V	3500	Decin et al. (2000)
HD 22484	F9 IV–V	5250	Decin et al. (2000)
HD 41700	G0 IV–V	<4300	Decin et al. (2000)
HD 73752	G3/5 V	600–2000	Song et al. (2000)
HD 128400	G5 V	1100±600	Song et al. (2000)
HD 20794	G8 V	7250	Decin et al. (2000)
HD 69830	K0 V	600–2000	Song et al. (2000a)
HD 53143	K1 V	965±200	Song et al. (2000a)
HD 53143	K1 V	980	Decin et al. (2000)
HD 67199	K1 V	2000±200	Song et al. (2000a)
ϵ Eri	K2 V	730±200	Song et al. (2000a)
HD 221354	K2 V	1500±1000	Song et al. (2000a)
HD 98800	K5 V	7±5	Song et al. (2000a)

3. The Ages of β Pic Stars and the Evolution of the IR Excesses

Figure 6 displays the ratio between the luminosity from the dust and bolometric luminosities for three different samples: (i) early-type Vega-like stars (*solid circles*), including Vega, Fomalhaut, β Pic, and HR 4796A; (ii) F and G Vega-like stars (*squares*). Note that stars from Decin et al. (2000) appear as open symbols. (iii) K-type Vega-like stars (*solid diamonds*), including ϵ Eri and HD 98800.

The figure seems to indicate that there is a clear trend between the IR excess and the age for early-type Vega-like stars, as pointed out by several authors (Holland et al. 1998; Barrado y Navascués et al. 1999; Song et al. 2000a,b). This could indicate that an evolutionary effect is present, and that the presence of debris disks is a normal characteristic in early-type stars. This disk would evolve with time, disappearing on a timescale of hundreds of millions of years. However, the situation is different for stars having spectral type F or later.

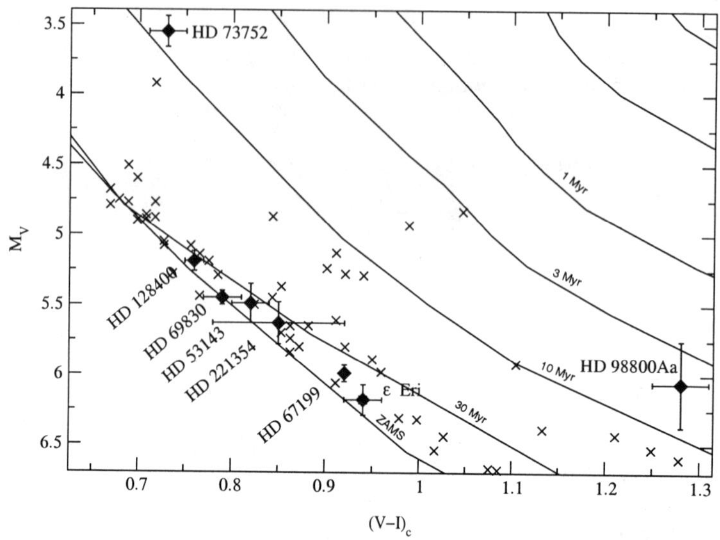

Figure 4. Color–magnitude diagram for late-type Vega-like stars.

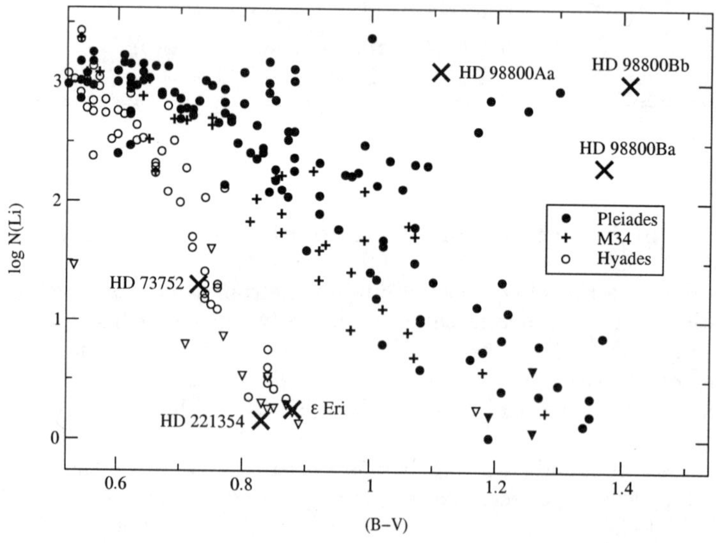

Figure 5. Lithium abundances for Vega-like stars of late spectral type.

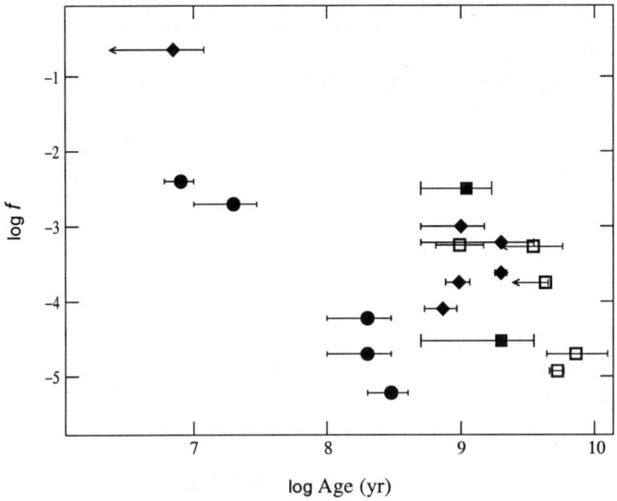

Figure 6. Fractional IR luminosity arising from the dust against age.

Except for HD 98800, there are no F-, G-, or K-type Vega-like stars having ages in the range ~ 10–500 Myr. Since we have been able to estimate ages in this range for stars without IR excesses, we have rejected the possibility of introducing a bias due to our age-dating techniques. This situation might suggest that the timescale for the evolution of the debris disks is much longer in late-type than in early-type stars.

References

Ábrahám, P., Leinert, C., Burkert, A., Lemke., D., & Henning, T. 1998, A&A, 338, 91

Angelova, A. M., Backman, D. E., & Stauffer, J. R. 1998, AAS, 192.5320

Backman, D. E., & Paresce, F. 1993, in Protostars and Planets III, ed. E. H. Levy & J. Lunine (Tucson: University of Arizona Press), p. 1253

Barrado y Navascués, D., Stauffer, J. R., Hartmann, L., & Balachandran, S. 1997, ApJ, 475, 313

Barrado y Navascués, D. 1998, A&A, 339, 831

Barrado y Navascués, D., Song, I., Stauffer, J. R., & Caillault, J.-P. 1999, ApJ, 520, L123

D'Antona, F., & Mazzitelli, I. 1994, ApJS, 90, 467

Decin, G., et al. 2000, A&A, in preparation

Holland, W. S., et al. 1998, Nat, 392, 788

Schaller, G., Schaerer, D., Meynet, G., & Maeder, A. 1992, A&AS, 96, 269

Soderblom, D. R., & Mayor, M. 1993, AJ, 105, 226

Song, I., Caillault, J.-P., Barrado y Navascués, D., & Stauffer, J. R., 2000a, ApJ, in press

Song, I., Caillault, J.-P., Barrado y Navascués, D., & Stauffer, J. R. 2000b, ApJ, submitted

Stauffer, J. R., Hartmann, L., & Barrado y Navascués, D. 1995, ApJ, 454, 910

Disks, Planetesimals, and Planets
ASP Conference Series, Vol. 219, 2000
F. Garzón, C. Eiroa, D. de Winter, and T. J. Mahoney, eds.

G Dwarfs and the Vega Phenomenon

G. Decin, M. Malfait, and C. Waelkens

Instituut voor Sterrenkunde, KULeuven, Celestijnenlaan 200B, B-3001 Leuven, Belgium

C. Dominik

Sterrenkundig Instituut 'Anton Pannekoek', Kruislaan 403, NL-1098 SJ Amsterdam, The Netherlands

M. Mayor

Observatoire de Genève, CH-1290 Sauverny, Switzerland

Abstract. In this paper, we present and discuss photometric infrared measurements taken with ISOPHOT at 60 μm for a sample of 30 G dwarfs. This sample is scrutinized for radial-velocity variations by the Geneva group, aiming at the detection of extrasolar planets.

In our sample, five stars display an infrared excess at 60 μm. These five stars are probably older than 3 Gyr, from which we suggest that the disks around cool stars may survive longer than those among earlier-type objects. For two stars in our sample, the fractional luminosity of the disk is significantly higher than for typical Vega-type stars, and approaches the exceptional value observed for β Pic.

To investigate the correlation between an infrared excess and planets/companion stars, we combine our results with the first results of the CORALIE survey. None of the radial-velocity variables detected till now displays an infrared excess. That the stars for which an infrared excess is found lack a companion cannot, however, be concluded at the present stage.

1. Introduction

Since the *IRAS* space mission detected in 1983 an excess of infrared radiation for Vega (Aumann et al. 1984), Fomalhaut, and β Pictoris (Gillett 1986), and since Smith & Terrile (1984) imaged the dust disk surrounding β Pictoris in the optical for the first time, the study of this so-called "Vega phenomenon" has received a substantial amount of interest, opening new possibilities for studying other planetary systems than our own. The interest in planetary material around other stars received new stimuli with the detection of Jupiter-like planets around solar-type stars (Mayor & Queloz 1995; Marcy & Butler 1995, 1996).

It rapidly became clear that the dusty disk observed in Vega-like stars need replenishment to survive. This replenishment is widely attributed to collisions of small-sized bodies, comets or planetesimals and evaporation of larger bodies, suggesting that the process towards the formation of major planets was to some extent hampered in such systems. Moreover, gravitational perturbations by giant planets are thought to remove a large fraction of the remaining dust, comets and planetesimals from the inner system on short timescales. Whether an anti-correlation exists between the presence of a planet and the presence of a dusty disk is currently unclear. Surveys for these phenomena indicate that the presence of both an infrared excess and planets is not uncommon.

2. The Program Stars

The Geneva group has selected a large sample of G dwarfs on which accurate radial-velocity measurements were carried out with CORAVEL and CORALIE. From this sample of G dwarfs, a subsample is compounded according to the following criteria. First, the stars with a visual magnitude higher than 7.5 mag are rejected in order to optimize the signal-to-noise ratio of the results. A second criterion is, of course, their detectability by *ISO*. In the next step the stars affected by a high infrared background (> 0.5 Jy/pixel) and cirrus (> 0.04 Jy/pixel) were removed from the sample, as well as multiple systems and objects which overlap with other programs. Finally, we picked the brightest stars at 60 μm in order to increase the chance of detection. The photospheric flux of the fainter objects of our sample can be lower than 30 mJy, which is the sensitivity limit of ISOPHOT at this wavelength. For these faint objects, disks can be found only if the total flux at 60 μm exceeds the detection limit. This resulted in 34 stars observed by the *ISO* mission. A complete list with the physical parameters of the observed stars can be found in Decin et al. (2000).

3. ISOPHOT Data

3.1. *ISO* Observations

Observation mode AOT PHT37 was first opted for. This consists of a C100 staring observation on source and four staring observations of the background around the target. Because no flat-fielding was possible, we switched after three observations to the AOT PHT22 observation mode and required 3×3 mini-maps with the C100 detector at 60 μm. The step size used was $46''$ and the integration time per raster position was 128 s. The observation of HD 45184 was omitted because of a solar flare event during revolution 722. In total, a usable C100 mini-map reduction was thus carried out for 30 G dwarfs.

3.2. Data Reduction

Our data reduction is performed with PIA v7p3(e) (Gabriel et al. 1997). Long-term responsivity drifts are modeled by determining the background baseline (eyeballing method). Since there is no standard method of extracting the flux from the mini-map observations, a method is developed based on the point spread function fractions related to the ISOPHOT C100 (Laureijs 1999). The method is based on the assumption that the point source is positioned in the center of the mini-map. First, the flat-field correction factors are calculated for

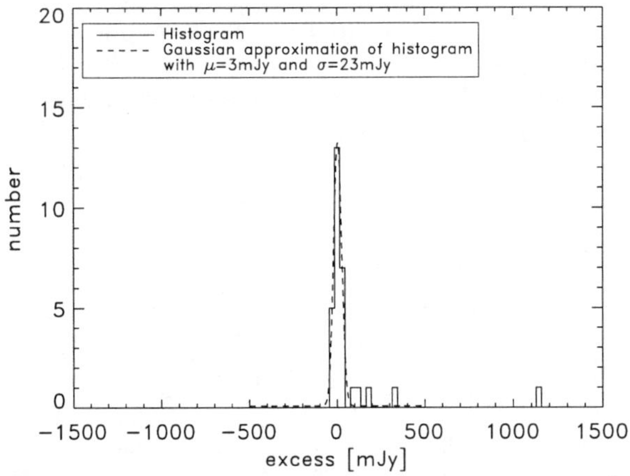

Figure 1. Histogram of the differences between the measured flux density and that predicted at 60 μm for our data sample.

each pixel by assuming that 1) the raster-position-averaged flux of pixels 1, 3, 7, and 9 is the same, as these pixels cover a similar area on the sky relative in distance to the source position, 2) pixels 2, 4, and 8 have the same raster-positioned flux, and 3) each raster position of pixel 5 equals the average of the flat-field-corrected fluxes of the other pixels covering this sky position. After this correction, the source flux and the background flux, together with their respective relative statistical errors, can be calculated using the psf fractions (Decin et al. 2000). Because pixel 6 of the C100 camera shows a dark signal which is a factor 3–7 higher than for the other pixels in the C100 array, this pixel is omitted in the calculations.

4. Observational Results

4.1. Quality Assessment of the Results

Since the sky signal of our sources is not of the same order of magnitude as the FCS signal, the default responses have to be applied. However, these default responses can be uncertain by up to several tens of per cent (Laureijs & Klaas 1999). Taking into account all other effects, the overall error can reach some 30%. Therefore, the reliability of the end result was checked by (1) comparing the measured fluxes with the predictions for the photosphere and (2) comparing the background fluxes with those measured by *COBE*/*IRAS*.

4.2. Results of the Data Reduction

In Figure 1 the histogram of the differences between measured and predicted flux densities—i.e., the excesses—is shown. The distribution can be split into two components: a very narrow distribution around zero plus a wing of positive excesses, i.e., cases where the measurements show more flux than is produced by the stellar photosphere. A Gaussian curve with average $\mu = 3$ mJy and dispersion $\sigma = 23$ mJy is plotted in Figure 1. The value of the σ confirms the estimate of the sensitivity of ISOPHOT at 60 μm. We conclude that a circumstellar disk is present when there is an excess at 60 μm amounting more than $\mu + 3\sigma = 72$ mJy. This criterion results in a list of five stars having an IR excess (see Table 1).

Table 1. The excess stars at 60 μm

HD	Values for log(age/1 yr)			$F_{\nu,\mathrm{src}}^{\mathrm{PHT}}$	$F_{\nu,\mathrm{src}}^{\mathrm{pred}}$	$F_{\nu,\mathrm{src}}^{\mathrm{exc}}$	$\log(\tau_{60})$
	min	median	max	mJy	mJy	mJy	
10647	0.00	9.54	9.76	1188	34	1154	–3.27
20794	9.64	9.86	10.1	264	156	108	–4.70
22484	9.66	9.72	9.77	196	118	78	–4.93
41700	0.00	$<$ 9.63	9.65	197	15	182	–3.75
53143	8.81	8.99	9.17	341	18	323	–3.25

Another quality check is a comparison between the background measured by ISOPHOT and that measured by *COBE*/*IRAS*. Figure 2 reveals a clear correlation between both fluxes. The questions arise as to whether 1) this correlation is real or induced by a systematic error, or 2) this correction has to be applied to the flux measurements on source. In this case, the amount of stars with an IR excess at the 3σ level decreases to 3.

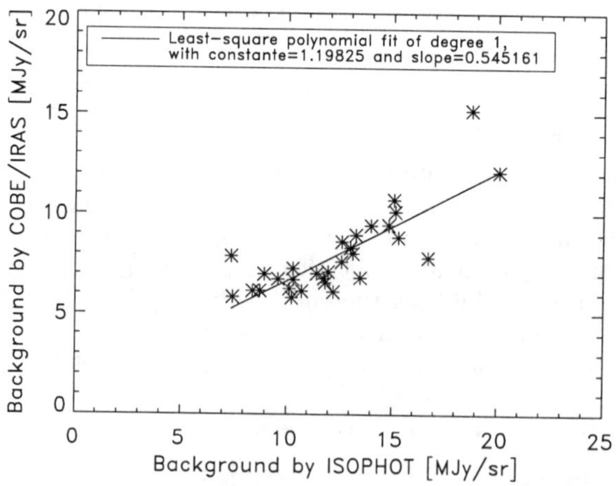

Figure 2. Comparison between the background measured by ISOPHOT and that estimated by *COBE*/*IRAS*.

5. Discussion

5.1. Survival of the Remnant Dust Disk

In our sample, 17% of the stars have an IR excess in good agreement with the fraction of stars displaying the Vega phenomenon, 13 ± 10% (Plets & Vynckier 1999). Habing et al. (2000) found that 60% of the main-sequence stars with spectral classes between A and K, and younger than 400 Myr have a disk, but only 8% of those older than 400 Myr. From their survey they conclude that the A stars in general possess a disk upon their arrival on the main sequence, and that these disks are dispersed within about 400 Myr. They also suggest that the disks around F, G, and K stars decay in a similarly short time after arrival on the main sequence. On the other hand, our stars are most probably older than 3 Gyr, and still 17% display an IR excess at 60 μm. Also HD 207129, a G2 V star, and ρ^1 Cancri, a G8 V star, are surrounded by a disk and are older than 3 Gyr (Jourdain de Muizon et al. 1999; Dominik et al. 1998). It is currently unclear why the timescale of the removal of colliding planetesimals around these old G dwarfs is longer than for most other stars with a disk. However, if the relation between the background measured by ISOPHOT and by *COBE/IRAS* is correct and a correction has to be applied, the amount of G dwarfs displaying the Vega phenomenon decreases to 10%, which is close to the 8% found for the stars older than 400 Myr and having an IR excess in the sample of Habing et al. (1999, 2000). In that case, our results fit well with the results of their survey.

5.2. Fractional Luminosity

The IR excess is believed to be caused by thermal emission of dust particles in a disk around the star. Backman & Gillett (1987) introduced the fractional luminosity, τ, defined as $\tau \equiv F_{\rm bol}^{\rm exc}/F_{\rm bol}^{\rm photosp} = L^{\rm dust}/L^{\rm bol}$. This parameter is linked with the mass of the disk and can be estimated from observations (see Habing et al. 2000). In Figure 3 the fractional luminosity, τ_{60}, derived from the 60 μm data, is plotted with respect to the stellar age, both for our sample as for the sample of Habing et al. (2000). In the sample of Habing et al. (2000) β Pic was an exception in having a value of τ greater than 10^{-3}. The fractional luminosities of HD 10647 and HD 53143 fill the gap between β Pic and the sample of Habing et al. (2000). Both stars are located at approximately the same distance as β Pic, but are cooler, less massive, and older. HD 10647 and HD 53143 appear to be true main-sequence stars and thus do not support the suggestion that massive disks appear only around pre-main-sequence stars. Note that β Pic is at this moment the only "warm" main-sequence star known with such a high luminosity.

Again, these results have to be evaluated against the hypothesis of the background. The logarithm of the fractional luminosities of HD 10647, HD 41700, and HD 53143 would then decrease to −3.54, −4.03, and −3.43, respectively. The highest values still put these stars in the gap between β Pic and the other stars, but the discrepancy is less marked.

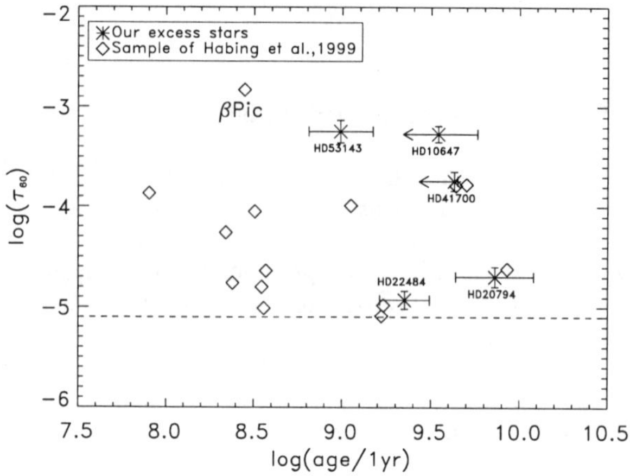

Figure 3. Fractional luminosities as a function of stellar age.

5.3. Coexistence of a Companion Star or Planet with a Circumstellar Dust Disk: First Results

The goal of this survey was to initiate an investigation into the (anti)-correlation between the presence of an IR excess and the occurrence of planets or companion stars. Our sample is included in the CORALIE survey, which started recently. Eight stars in our sample have already been detected as being double. None of these stars has a 60 μm infrared excess. For the other stars the results of the CORALIE survey are still not known. At the present stage, no clear correlation has been found between dust disks and planets from our sample. However, based on the limited sample on which both phenomena have already been investigated, the conclusion that one phenomenon prohibits the presence of the other would be premature. A scenario in which a planet is located close to the central object, not affecting that part of the circumstellar dusty disk located further out, is not wishful thinking. The completion of the CORALIE survey will hopefully provide more definite results.

Acknowledgments. The ISOPHOT data presented in this paper was reduced using PIA, which is a joint development by the ESA Astrophysics Division and the ISOPHOT consortium. GD is supported by project IUAP P4/05 financed by the Belgian Federal Scientific Services (DWTC/SSTC). KM is Postdoctoral Fellow of the Fund for Scientific Research, Flanders (Belgium).

References

Aumann, H., et al. 1984, ApJ, 278, L23

Backman, D. E., & Gillett, F. C. 1987, in Cool Stars, Stellar Systems and the Sun, ed. J. Linsky & R. E. Stencel (Berlin: Springer), p. 340

Decin, G., et al. 2000, A&A, in press

Dominik, C., et al. 1998, A&A, 329, L53

Gabriel, C., et al. 1997 in ASP Conf. Ser., 125, Astronomical Data Analysis and Software Systems VI, ed. G. Hunt & H. E. Payne (San Francisco: ASP), p. 108

Gillett, F. 1986 in Light on Dark Matter, ed. F. Israel (Dordrecht: Reidel), p. 61

Habing, H. J., et al. 1999, Nat, 401, 456

Habing, H. J. 2000, in preparation

Jourdain de Muizon, M., et al. 1999, A&A, 350, 875

Laureijs, R. J. 1999, Point spread function fractions related to the ISOPHOT C100 and C200 arrays

Laureijs, R. J., & Klaas, U. 1999, ISOPHOT error budgets: derive SPD processing steps

Marcy, G. W., & Butler, R. P. 1995, AAS, 175, 70.04

Marcy, G. W., & Butler, R. P. 1996, ApJ, 464, L153

Mayor, M., & Queloz, D. 1995, Nat, 378, 355

Plets, H., & Vynckier, C. 1999, A&A, 343, 496

Smith, B. A., & Terrile, R. J. 1984, Sci, 226, 1421

Disks, Planetesimals, and Planets
ASP Conference Series, Vol. 219, 2000
F. Garzón, C. Eiroa, D. de Winter, and T. J. Mahoney, eds.

Spectro- and Photopolarimetric Monitoring of the HAeBe Star VV Serpentis

D. de Winter

TNO-TPD/Space, Stieltjesweg 1, PO Box 155, 2600 AD Delft, The Netherlands

A. Mora, C. Eiroa, and J. Palacios

Dpto. Física Teórica C-XI, Facultad de Ciencias, Universidad Autónoma de Madrid, Cantoblanco, 28049 Madrid, Spain

R. D. Oudmaijer

Department of Physics and Astronomy, E C Stoner Building, Leeds LS2 9JT, UK

F. Garzón

Instituto de Astrofísica de Canarias, E-38205 La Laguna, Tenerife, Spain

EXPORT

Abstract. The Herbig Ae star VV Serpentis was monitored intensively during the EXPORT 1998 observing campaign. Variations on timescales of days and months were detected in the spectroscopy and photometry, while the object was polarimetrically stable within the error bars. The photopolarimetric behavior is consistent with the presence of an edge-on dust disk with obscuring cloud structures. The spectroscopic changes indicate accretion activity in this 3 $M_\odot$ system, which is a few million year old. Although the star is remarkably similar to earlier observed UXORs, its differences from this group are most interesting for the study of CS disks around pre-MS objects.

1. Introduction

Planetary systems are thought to originate in the circumstellar (CS) material around young stars. If this simple statement is realistic, the study of the latter should give key information on the formation of planets. Studies of relatively young MS stars have revealed disks with inhomogeneous structures (e.g., Backman & Paresce 1993) in a few cases. Evidence for more solid bodies in the CS disks of pre-MS and MS stars has also been reported (e.g., Grady et al. 2000).

Information on the outer regions of young disks can be obtained by imaging and photopolarimetry (see several contributions in this volume), while the gaseous regions up to the stellar photosphere can be investigated by spec-

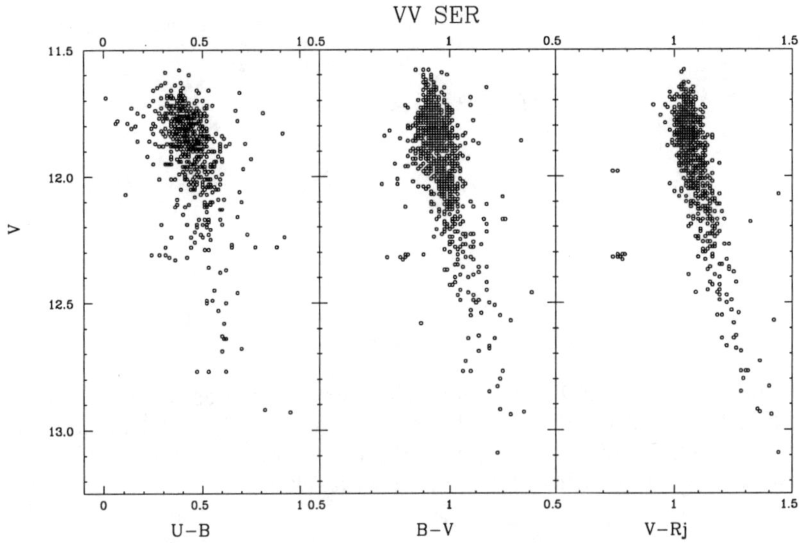

Figure 1. The photometric behavior of VV Ser presented in color–magnitude diagrams. Besides EXPORT results, we added monitoring data of Shevchenko et al. (1993 and private communication).

troscopy. Many young MS and pre-MS stars are found to be highly variable in their spectroscopic line structure. If these lines are non-photospheric, their variations are a powerful tool for studying the gaseous disk. Indeed, spectroscopic monitoring has revealed solid disk contents for a number of pre-MS and MS stars, such as star-grazers and other orbiting structures. Good examples are the HAeBes UX Ori (Grinin et al. 1994) and BF Ori (de Winter et al. 1999), and the Vega-type star β Pic (Beust et al. 1998).

For such reasons we introduced the monitoring of young intermediate-mass stars as one of the main projects within the EXPORT proposal (Eiroa et al., this volume, p. 3). Here we report the monitoring results for VV Ser, an object quite similar to UX Ori and BF Ori (see Table 1).

2. EXPORT Observations of VV Ser

Details of the EXPORT observing nights and conditions can be found in the contributions of Eiroa et al. (this volume, p. 3), Mora et al. (this volume, p. 231 and p. 411) and Oudmaijer et al. (this volume, p. 238). The following monitoring results were obtained for VV Ser.

UBVRI Polarimetric Observations We obtained 12 polarimetric observations of VV Ser: 1998 May 14–16, July 28–31, and October 22–25. The light of VV Ser is significantly polarized and the variations behave as follows: U = 1.40–1.85%, B = 1.42–1.91%, V = 1.48–2.20%, R= 1.41–1.80%, I = 1.05–1.62%.

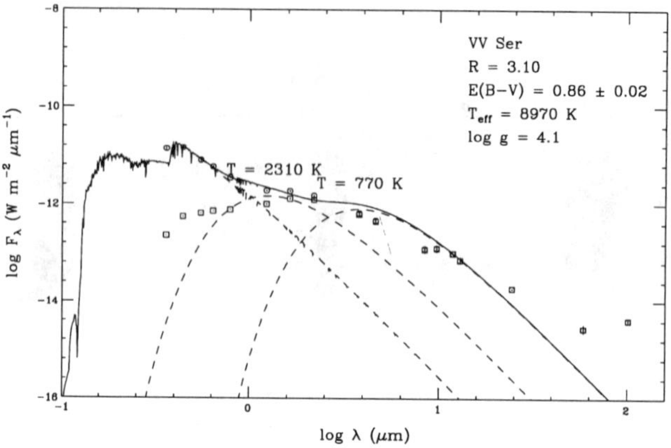

Figure 2. The SED of VV Ser. The photometry was taken from the EXPORT results and those of Berrilli et al. (1992) and Hillenbrand et al. (1992). For the astrophysical parameters see Table 1. The IR excess was fitted with two appropriate Planck curves.

A roughly constant position angle of 73–86 degrees was observed throughout. The peak-to-peak variations of the polarization are only slightly larger than the individual error bars and, as such, statistically speaking we cannot be certain that the object was variable during our observations.

UBVRI Photometric Observations VV Ser was observed seven times on 1998 May 14–15, July 28, and October 22–25. During all these observations VV Ser was near maximum light (see Figure 1). Besides October 24, when all the magnitudes were about $0\overset{m}{.}2$ brighter, VV Ser remained at the following magnitudes: $U = 13\overset{m}{.}23$–$13\overset{m}{.}27$, $B = 12.76$–12.85, $V = 11\overset{m}{.}85$–$11\overset{m}{.}94$, $R = 11\overset{m}{.}11$–$11\overset{m}{.}23$, $I = 10\overset{m}{.}33$–$10\overset{m}{.}54$. This means that the amplitude increases for longer wavelengths.

JHK Photometric Observations We collected nine *JHK* observations for VV Ser on: May 29, July 27–30, and October 22–25. In all three pass-bands VV Ser was variable: $J = 8\overset{m}{.}70$–$9\overset{m}{.}06$, $H = 7\overset{m}{.}47$–$7\overset{m}{.}74$, $K = 6\overset{m}{.}31$–$6\overset{m}{.}51$. Variations are observed between the different runs as well as on consecutive nights.

Intermediate-Resolution INT Spectroscopic Observations VV Ser was observed during 1998 May 15–17, July 28–31, and October 24–28, three, ten, and five times, respectively. The first run covered the 5854–6728 Å region with a 0.707 Å/pixel resolution. The other two runs 5712 to 6812 Å with 0.474Å/pixel.

High-Resolution WHT Spectroscopic Observations Using the UES, we obtained 11 spectra covering 3800 to 5900 Å at a resolving power of 49 000. The spectra were taken during the same observing periods as the INT observations.

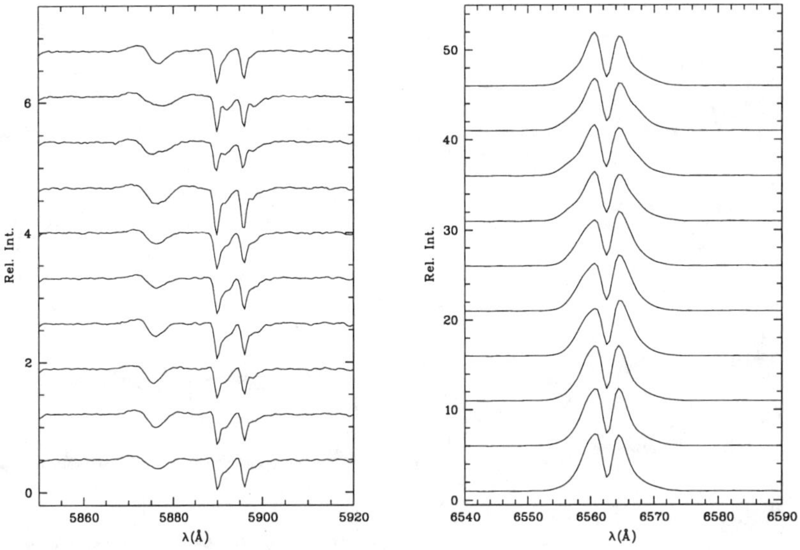

Figure 3. INT intermediate-resolution spectral monitoring results for VV Ser. Shown are regions around the He I, Na I D, and Hα lines. Spectra were taken on July 28 (*top*), and three spectra, with two-hour intervals, on 1998 July 29, 30, and 31 (*bottom*).

3. Photometric and Polarimetric Results

A comparison of our optical photometry with known monitoring results (e.g., Shevchenko et al. 1993) shows that our data fit well previous observations (see Figure 1) that VV Ser was close to maximum light and in a state of relative quiescence. The color–magnitude diagrams show that VV Ser is more reddened at lower light levels, probably due to extra extinction. Such behavior is well known for UXORs (Grinin et al. 1994). The UXOR turn-around phenomenon, getting bluer again in deep minima, may be present only in the $U-B$ color. The polarization variations (if any) are smaller than observed in UXOR-type objects. For UXORs the light behavior is usually explained by dust clouds, located in a disk orbiting the central star (Grinin et al. 1994).

In the infrared, VV Ser has much redder colors than normal for an A-type star. Indeed, the spectral energy distribution (SED) of VV Ser (see Figure 2) shows a large near- to far-IR excess. Such an excess is indicative of a large amount of CS dust. When this is combined with the observed high polarization, we infer that it is distributed in a disk (Oudmaijer et al., this volume, p. 238).

The photometric IR variations of VV Ser decrease at longer wavelengths. This could be due to suppression of the direct stellar light (low in the far red) by CS dust cloud obscurations. However, the amplitude in the visual photometry was only $0^{\rm m}_{.}14$ to $0^{\rm m}_{.}30$ in the red and even $0^{\rm m}_{.}36$ in J andso increases with longer wavelength. It might be that light from the disk, being redder than the star (see

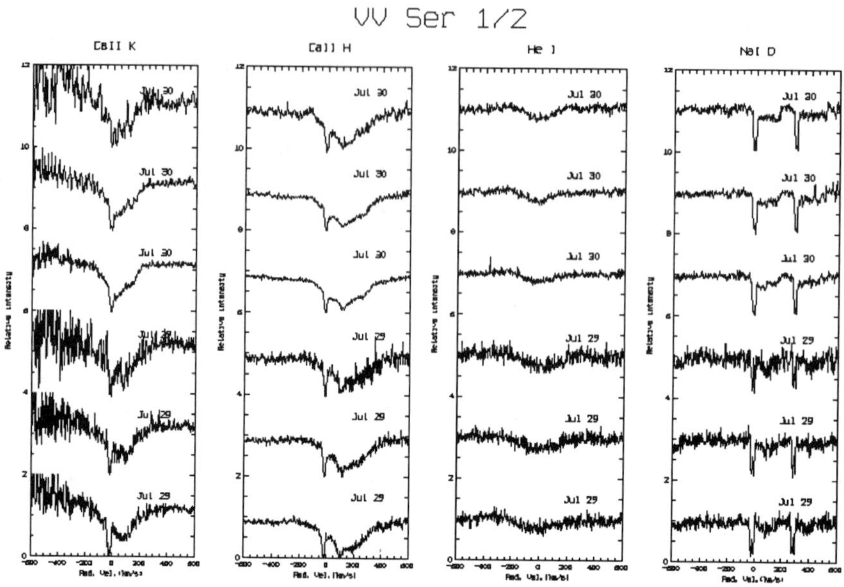

Figure 4. WHT high-resolution spectral monitoring results for VV Ser. Shown are regions around the Ca II, He I, and Na I D lines. Observing dates are indicated. The spectra were taken simultaneously with the INT spectra (see Figure 3.

Figure 2), is obscured. At even longer wavelengths this contribution gets lower as this radiation comes from outer disk regions and is then less affected by local extinction. More detailed analysis is needed, especially since, for the blue colors, our amplitude variations are close to the observational errors.

In Table 1 we compare the astrophysical parameters of VV Ser with those of two UXORs: UX and BF Ori. Although their similar astrophysical parameters, the $v \sin i$ as well the A_V, are significantly higher for VV Ser.

4. Spectroscopic Monitoring Results

Our intensive monitoring allows us to study VV Ser's spectroscopic behavior on timescales of hours, days, and months. The spectrum of VV Ser is a combination of an A2 photospheric spectrum enriched with shell lines (e.g., Si II and Fe II) and emission at Hα, Hβ, and [O I]. A more detailed study of the spectra reveals the presence of many variable lines on any of the monitoring timescales. Here, we present only a few monitoring results.

Of interest are the variations in the He I, Na I D, and Hα lines (see, for example, Figure 3 for the 1998 July dataset). Reversed P Cygni is seen in He I,

Table 1. Astrophysical parameters of VV Ser and the UXORs UX Ori and BF Ori. The data were collected from Thé, de Winter, & Pérez (1994), van den Ancker[1] et al. (1997), and Solano[2] et al. (this volume, p. 422)

Name	V_{max}	A_V	Sp. T.	$\log T_{eff}$	D (pc)	$v \sin i$ (km s^{-1})
BF Ori	$9^m.7$	$0^m.3$	A5 III	3.908	430	100(10)[1]–34[2]
UX Ori	$9^m.5$	$1^m.1$	A3 III	3.934	460	70(6)[1]–189[2]
VV Ser	$11^m.6$	$2^m.8$	A2 III	3.953	440	200(20)[1]

accompanied by redshifted absorption components (RACs) in the sodium lines. Note the covariation of the RACs with the red He I absorption wing at similar velocities. Hα is double-peaked in emission with the red peak the strongest, peak intensities are reversed only on the third night. Such a reversal, as well as RACs in the sodium lines and the presence of (variable) He I in A-type stars, is known for UXORs. Similar line behavior is studied and explained for UX and BF Ori (Grinin et al. 1994; de Winter et al. 1999). In their model, the RACs are due to evaporating star-grazers and He I is present when these bodies are rich in volatiles or cause heated (shock) regions. However, additional effects of stellar rotation and orbiting bodies are needed to fully explain the more complex line variations, especially Hα, that are also visible in our dataset.

RACs are also visible in the Fe II and Ca II lines (see Figure 4). The RACs in these lines reach velocities up to 250 km s^{-1}. The He I variations are detected up to similar velocities. Of interest are the RACs detected in Hγ and Hδ, with covariations and similar velocities to those of the Na I D RACs.

5. Conclusions

Spectrometric and photopolarimetric monitoring of VV Ser on time scales from hours up to months revealed variations almost identical to those known for the UXORs UX Ori and BF Ori. It is therefore proposed that VV Ser has an edge-on CS disk with star-grazer bodies. Also in astrophysical parameters, VV Ser is a twin to these objects. The main differences are its higher $v \sin i$, A_V, and lack of scattered light in deep minima as attested by the non-detection of polarimetric variations. This might tell us more on the disk properties of pre-MS objects in which (proto)planet formation might be active.

References

Backman, D. E., & Paresce, F. 1993, in Protostars and Planets III, ed. E. H. Levy, J. I. Lunine, & M. S. Matthews (Tucson: University of Arizona Press), p. 1253

Berrilli, F., Corciulo, G., Ingrosso, G., Lorenzetti, D., Nisini, B., & Strafella, F. 1992, ApJ, 398, 254

Beust, H., Lagrange, A.-M., Crawford, I. A., Goudard, C., Spyromilio, J., & Vidal-Madjar, A. 1998, A&A, 338, 105

de Winter, D., Grady, C. A., van den Ancker, M. E., Pérez M. R., & Eiroa, C. 1999, A&A, 343, 137

Grady, C. A. et al. 2000, in Protostars and Planets IV, ed. V. Mannings, A. P. Boss, & S. S. Russell (Tucson: University of Arizona Press), p. 613

Grinin, V. P., Thé, P. S., de Winter, D., Giampapa, M., Rostopchina, A. N., Tambovtseva, L. V., & van den Ancker, M. E. 1994, A&A, 292, 165

Hillenbrand, L. A., Strom, S. E., Vrba, F. J., & Keene, J. . 1992, ApJ, 397, 613

Shevchenko, V. S., Grankin, K. N., Ibragimov, M. A., Mel'nikov, S. Yu., & Yakubov, S. D. 1993, Ap&SS, 202, 121

Thé, P. S., de Winter, d., & Pérez, M. R. 1994, A&AS, 104, 315

van den Ancker, M. E., Thé, P. S., Tjin A Djie, H. R. E., Català, C., de Winter, D., Blondel, P. F. C., & Waters, L. B. F. M. 1997, A&A, 324, L33

Disks, Planetesimals, and Planets
ASP Conference Series, Vol. 219, 2000
F. Garzón, C. Eiroa, D. de Winter, and T. J. Mahoney, eds.

The UXOR-Type Star RY Ori

C. Eiroa

Dpto. Física Teórica, Universidad Autónoma de Madrid, 28049 Madrid, Spain

F. Garzón

Instituto de Astrofísica de Canarias, E-38205 La Laguna, Tenerife, Spain

L. F. Miranda

Instituto de Astrofísica de Andalucía, Apartado de Correos 3004, 18080 Granada, Spain

EXPORT

Abstract. We present the observations of the UXOR-type star RY Ori carried out by EXPORT. A first analysis of the simultaneous optical and near-infrared data suggests that the optical and near-IR variabilities of pre-main-sequence stars can follow different trends when observed at the same time.

1. Introduction

RY Ori (HBC 436) is a young, highly variable PMS star of the Orion population classified as a F6/F8 star. It was studied by Bibo & Thé (1991) in the *ubvy* system and found to have a Δy variability of 1.85 mag in the range 11.15–3.00. Bibo & Thé observed a color reversal in the color–magnitude diagram similar to that observed in UX Ori and noticed that the star spends most of its time at mean brightness and shows rapid activity. The star has been studied at near-IR wavelengths by Cohen (1973) and by Glass & Penston (1974). IRAS 05296-0251 is associated with RY Ori in the SPC catalog. Recently, Yudin & Evans (1998) have carried out polarimetric observations and reported a large polarization in the range $p \approx 4$–6% when the star was near minimum ($V \approx 13.7$ mag).

During the EXPORT observing runs RY Ori showed spectroscopic and very prominent photometric and polarimetric variability. In particular, the optical and near-IR variabilities follow different patterns. This contribution presents a first analysis of the photopolarimetry results.

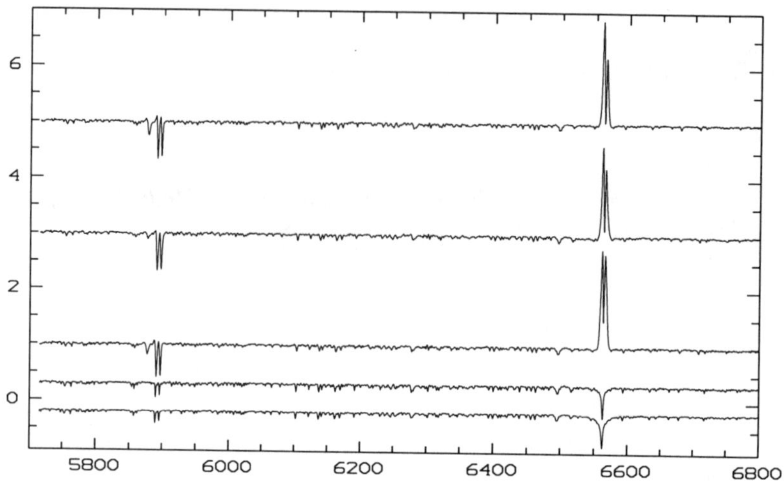

Figure 1. 1999 January INT spectra of RY Ori. *Top to bottom:* The spectra correspond to JD(+2400000) 51208.50, 51209.51, and 51210.52. The fourth spectrum from the top corresponds to an F6 V standard star and the spectrum at the very bottom is an F6 IV star.

2. Observations and Results

RY Ori was observed with the INT during the nights of 1998 October 25–28 and 1999 January 29–31. Details of the observations are given in other EXPORT contributions in this volume. Figure 1 shows the INT January spectra. The spectra correspond very well to an F6 V/IV star. In addition to the photospheric spectrum, RY Ori shows variability in some lines. He I 5875 and Na I D 5890 appear relatively strong in absorption and also show a blueshifted emission in some of the spectra. Hα is a strong double-peaked emission; the red component is always the faintest. Figure 2 shows He I, Na I D, and Hα with some detail (1998 October and 1999 January).

One spectrum was taken with the WHT in the night 1999 January 30/31 (JD 2451209.50). Figure 3 shows the Ca II H and K and the Na I D lines as observed with the WHT. Procyon (F5 V) is also shown for comparison. The Ca II and Na Ilines are very different in both stars. In particular, Na I redshifted absorption components are observed in the RY Ori spectrum. These components are not distinguished in the lower-resolution INT spectrum taken simultaneously (Figure 2). These components are similar to those observed in other UXORs and β Pic-/Vega-type systems and probably are transient, although this requires further confirmation—we have only one high-resolution spectrum.

Optical photopolarimetry (NOT) and near-IR photometry (CST) were carried out in 1998 October and 1999 January. In particular, the 1998 October run was very good in terms of photometric quality. The star showed a strong variability in 1998 October, while it remained at mean brightness level and prac-

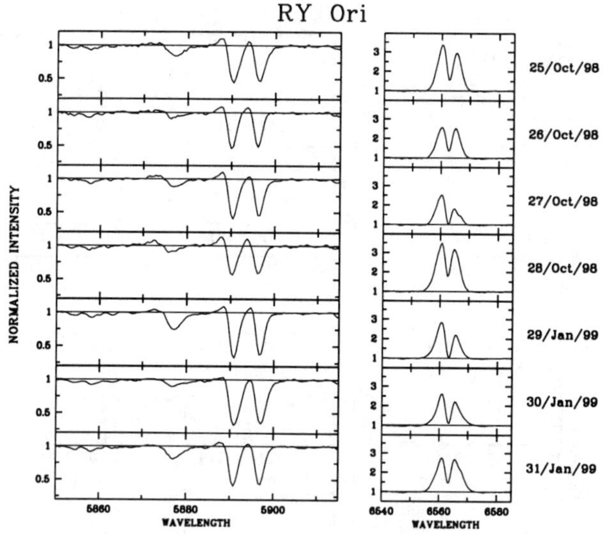

Figure 2. INT spectra of RY Ori around the He I, Na II, and Hα lines.

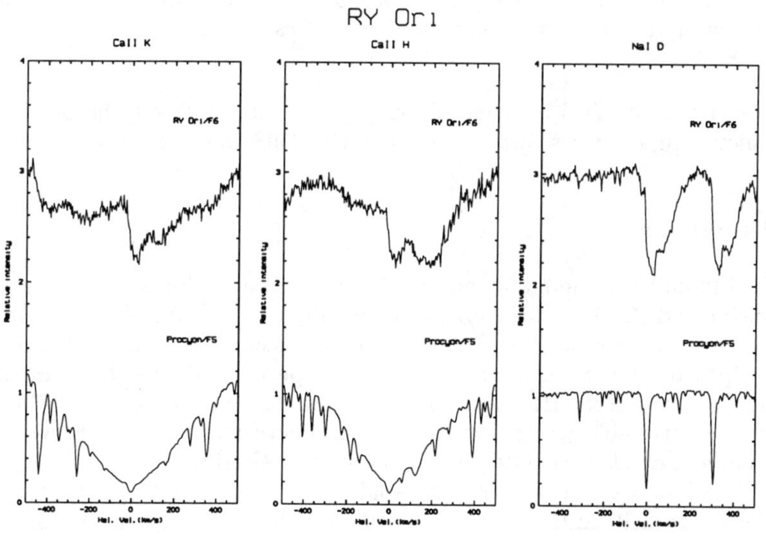

Figure 3. WHT high-resolution Ca II H and K, and Na I D lines from RY Ori (JD 2451209.50). The spectrum of Procyon is also shown for comparison.

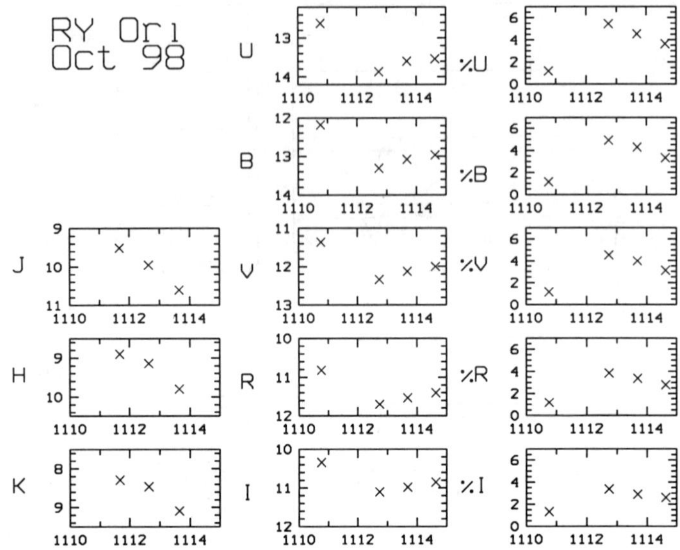

Figure 4. 1998 October photometric and polarimetric light curves of RY Ori. The figure plots magnitudes and polarization degrees versus Julian date (+2450000.0). Note the lack of correlation between the optical and near-IR photometric variabilities.

tically constant in 1999 January. Figure 4 show the observed photometric and polarimetric light curves corresponding to the 1998 October data.

3. Discussion

RY Ori fits into the commonly accepted scheme of a PMS star surrounded by a circumstellar disk. The star shows a photospheric spectrum plus some variable emission and absorption lines that originate in the circumstellar gas and a stellar wind. Optical colors are consistent with an F-type star reddened by an extinction $A_V \sim 1$ mag, while the IR colors indicate a noticeable near-IR excess. The data indicate a noticeable photometric and polarimetric variability. However, *the optical and near-IR variability are not correlated* (Figure 4).

RY Ori shows the largest and rapid near-IR variability of the EXPORT sample: $\Delta J = 1.10$, $\Delta H = 0.90$, $\Delta K = 0.80$ within 48 hours during 1998 October; at optical wavelengths the star also varies significantly. From JD 2451110.76 to JD 2451112.74 the star fades at optical wavelengths ($UBVRI$), V changes from 11.36 mag to 12.34 mag, while the polarization increases from 1.2 to 4.5 %. The amplitude of the variations decreases from U to I. From JD 2451112.74 to JD 2451114.64 the star brightens again and the polarization diminishes; now the amplitude of the variability is similar at all wavelengths while the polarization decreases less with the wavelength. This anti-correlation between optical brightness and polarization degree has previously been observed towards many young stars. The near-IR behavior, however, is very different and the star fades steadily from JD 2451111.67 to JD 2451113.65 all the way. Note that although the near-IR and optical light curves do not exactly overlap in time, the critical points are those from JD 2451112.7 to JD 2451113.7, when the star fades in the near IR but brightens in the optical.

In principle, we would expect the same fading/brightening trend in the optical and near-IR bands. To be sure that the data are real, we have very carefully revised all the 1998 October photometric data. The run was excellent in terms of optical and near-IR photometry. Errors of the standard-star fits and/or errors of the individual RY Ori measurements (as well as of the observed standard stars) are low, normally well below 0.05 magnitudes. Thus, there is nothing indicating that the RY measurements could be wrong. We have analyzed the light curves of other variable stars observed in the same run and found many different types of behavior. In particular, DK Tau, a T Tauri star of spectral type K7, exhibits light curves similar to that of RY Ori; i.e., the optical and near-IR variabilities do not correlate. Taken independently, the variability in both spectral ranges follows usual trends observed in pre-main-sequence stars. Thus, we conclude that the simultaneity of the observations in the optical and in the near-IR demonstrates that the variability in both wavelength regimes does not necessarily follow the same trend at the same time.

Photometric variability of PMS stars is normally attributed to one or several of the following scenarios: 1) Cool stellar spots which rotationally modulate the stellar light. 2) Hot spots caused by a magnetically channeled accretion flow onto the stellar surface. 3) Variable obscuration by dust clumps in the circumstellar environment. 4) Changes in the temperature profile and structure of the circumstellar dust disk. The first three scenarios can produce large amplitude variations, up to several magnitudes, with the amplitude of the variability diminishing with increasing wavelength (although variability greater that $\sim$ 0.2–0.3 mag is difficult to explain in terms of cool spots because it would require a very large coverage of the stellar surface). The fourth scenario would produce greater variability in the near-IR than in the optical, since typical dust-disk temperatures ($T_{\rm D}^{\rm max} \sim 2000$ K) radiate relatively little in the optical. It seems plausible to think that the origin of the observed optical and near-IR variability in RY Ori is a combination of hot stellar spots and/or variable obscuration plus some changes in the disk structure, each mechanism acting with different weights in both the optical and near-IR ranges. A detailed comparison of the observational data with PMS variability models is beyond the scope of this paper but is planned for the near future.

4. Conclusions

Further simultaneous near-IR and optical photometric observations are required to confirm and to characterize much better the observed variability in RY Ori (and also that of DK Tau). In our opinion, the relevant outcome of the data presented here is the real need for simultaneous optical and near-IR observations to provide plausible scenarios in order to explain the variability of pre-main-sequence stars.

Acknowledgments. C. Eiroa is partly supported by the Spanish grant ESP 98-1339-C02.

References

Bibo, E. A., & The, P. S. 1991, A&AS, 89, 319
Cohen, M. 1973, MNRAS, 161, 97
Glass, I. S., & Penston, M. V. 1974, MNRAS, 167, 273
Yudin, R. V., & Evans, A. 1998, A&AS, 131, 401

Disks, Planetesimals, and Planets
ASP Conference Series, Vol. 219, 2000
F. Garzón, C. Eiroa, D. de Winter, and T. J. Mahoney, eds.

Preliminary Results from a Mid-IR Survey of Vega-Like Sources

R. Scott Fisher, Charles M. Telesco, and Robert K. Piña

University of Florida Department of Astronomy, 211 Space Science Research Building, Gainesville, FL 32611-2055, USA

Roger F. Knacke

Penn State University, 139 Behrend Science Building, Station Road, Erie, PA 16563, USA

Abstract. We used OSCIR to conduct a survey of nearly 50 Vega-like sources in the northern and southern sky. Imaging at 10 and 18 μm shows that most of these sources are unresolved at the level we can achieve on a 4 meter-class telescope. However, a subset of the survey sources seem to be marginally resolved, and one of the sources (HD 141569) was definitely resolved at Keck. Here we present some early results from the survey, including new high-resolution observations of SAO 26804 that show that the source is, in fact, unresolved in the mid-IR.

1. Introduction

Vega-like stars are main-sequence stars that exhibit excess radiation at infrared wavelengths. The excess emission is thermal radiation from dust thought to be distributed in a circumstellar disk. The so-called "Vega phenomenon" was first observed around the four main-sequence stars α Lyrae, α PsA, β Pictoris, and ϵ Eridani by *IRAS* (Aumann et al. 1984). Subsequent searches of the *IRAS* Point Source Catalog (PSC) located more than 100 main-sequence stars of most spectral types with similar IR excesses. The first major search of the *IRAS* database by Walker & Wolstencroft (1988, hereafter WW) correlated both the entire SAO catalog and the Catalog of Nearby Stars with the *IRAS* PSC to look for Vega-like sources. WW identify 34 stars of various spectral types that they classify as Vega-like, including 12 stars within 25 pc of the Sun. In 1991, Stencel & Backman used similar but more relaxed constraints to identify 379 stars with Vega-like excesses. These initial searches led to additional photometry and/or spectroscopy of subsets of the newly identified Vega-like sources such as the impressive optical to submm study of 23 sources from the WW list by Sylvester et al. (1996).

To date, there have been approximately a dozen searches of the *IRAS* database looking for main-sequence stars associated with *IRAS* sources. In total the searches have classified $\backsim$ 200 sources as Vega-like. Good agreement between the major searches shows that we are starting to understand how to accurately identify these objects. For example, of the $\backsim$ 150 sources identified by Mannings & Barlow (1998), 34 were included in previous surveys. This classification is not an exact process though. More detailed observations have shown that some of the sources originally identified as Vega-like have probably not yet reached the main sequence, and some may have already left it to start up the giant branch.

Coupled with large-aperture telescopes, new mid-IR camera systems make it possible to directly image disks around a significant number of Vega-like sources for the first time. The sub-arcsecond resolution in the mid IR offered by these cameras gives us the opportunity to resolve sources out to a distance of $\backsim$ 100 pc which significantly increases our chances of finding new disks. Detection of even a few more disks around Vega-like stars will have far-reaching significance in our understanding of these systems and their role in the origin and evolution of planetary systems. In particular, once the characteristics of this type of disk are known, we can start to address the crucial question of how these disks relate to our own Solar System, and to planetary systems in general. Discovering new disks, and placing them in the sequence of disk evolution that the astronomical community is starting to put together were the main goals of our survey. In this paper we discuss the survey and present some of the earliest results from it.

2. The Vega-Like Source Survey

Over the past two years we used front-line infrared instrumentation and large-aperture telescopes to conduct a survey of nearly 50 Vega-like sources in both the northern and southern sky. The majority of the survey was conducted using the 4 meter Blanco Telescope at the Cerro Tololo Inter-American Observatory (CTIO). Parts of the survey were also conducted at the NASA Infrared Telescope Facility (IRTF) and the W. M. Keck Observatory. We imaged each of the survey sources at 10 and 18 μm using the Observatory Spectrometer and Camera for the Infrared (OSCIR), which was developed by the University of Florida Infrared Astrophysics Group. OSCIR uses a 128 $\times$ 128 pixel Si:As blocked impurity band detector produced by Boeing. The plate-scale of the camera ranged from $0\farcs062$/pixel on Keck II to $0\farcs223$/pixel on the IRTF. For the survey observations, we made deep N ($\lambda_0 = 10.8\ \mu$m, $\Delta\lambda = 5.2\ \mu$m) and IHW18 ($\lambda_0 = 18.2\ \mu$m, $\Delta\lambda = 1.7\ \mu$m) images with exposure times of 10 minutes *on-source* ($\sim$ 20 minutes of chopped integration). We felt that this exposure time allowed us to efficiently use the allocated telescope time while giving us a good chance to detect any extended emission associated with the sources.

For this project we wanted to observe a statistically significant number of all the known Vega-like sources while concentrating on those sources we had the best chance of resolving. Our primary source list is that of WW, with Sylvester et al. (1996) and Mannings & Barlow (1998) as secondary lists. A total of 48 sources were observed in the survey, 32 of the 34 WW sources, 19 of the 24 Sylvester sources and six sources from the Mannings and Barlow list. Although Sylvester et al. (1996) obtained mid-IR spectra for 13 of the sources in our sample, prior to our survey most of these sources had never been imaged at high resolution in the mid IR.

Some of the WW sources are listed as luminosity class III or IV, and a few had no luminosity classification at all at the time the WW list was published.

Today only about half of the sources on the WW list sources are known to be on the main sequence (Mannings & Barlow 1998). Since an important part of our research is to place these sources into an overall evolutionary sequence, we need to see these sources in all stages of their evolution. This includes the time immediately before and the time just after they are on the main sequence. It is therefore not unreasonable to make observations of PMS objects and post-main-sequence objects that meet the criteria of the Vega-like surveys to fill in the gaps in the evolutionary sequence.

Perhaps the most important result of the survey is that the vast majority of these Vega-like sources are unresolved on an arcsecond scale. For most of the survey, we measure unresolved source sizes with FWHM in the range of $\sim 0\farcs6$ to $1\farcs0$. This allowed us to probe within 100 AU of most of the sources, and to within 10 AU of the closest ones. We also placed stringent upper limits on the surface brightness of any extended emission around the sources at the few mJy/square arcsecond level. The mid-IR emitting regions of the few known Vega-like disks are all $\sim$ 100 AU in radius and have surface brightnesses of (at least) tens of mJy/square arcsecond (Fisher et al. 2000, and references therein). Therefore, we believe we would have detected similar disks around the survey sources if they existed. Not all of the survey sources were unambiguously unresolved. A subset of the sources seemed to be marginally resolved, and one source (HD 141569) was definitely resolved at 10.8 and 18.2 μm (Fisher et al. 2000). Other sources in the survey distinguished themselves for various reasons that will be discussed in a forthcoming series of publications.

3. SAO 26804 (HD 233517)—A "De-Resolved" Source

The unusual nature of SAO 26804 was first recognized by *IRAS* when it detected a large IR excess at $\lambda > 12$ μm associated with the star. The spectral energy distribution (SED) of SAO 26804 shows that all of the mid- and far-IR emission associated with this source is excess. WW included the object in their list of Vega-like sources and found that a single temperature blackbody with $T_{\rm dust}$ = 105 K fitted the excess well. Sylvester et al. (1996) included SAO 26804 in their follow-up survey and observed the source from the optical to the submm. Skinner et al. (1995) observed SAO 26804 from the near IR to the millimeter and presented an N-band image showing what appears to be a highly inclined disk with a FWHM of $1\farcs5$ on the major axis (Skinner et al. 1995). Skinner et al. (1995) also presented detailed modeling of the dust in the system assuming the star was a main-sequence dwarf.

About the same time as the Skinner et al. (1995) result was published, a paper by Zuckerman, Kim, & Liu (1995) presented a list of $\sim$ 90 luminosity class III stars with far-IR excess emission. Although SAO 26804 was not included in the this list, the Zuckerman et al. (1995) paper was perhaps the first to discuss the idea of a class of giant stars with dust *orbiting* them. Fekel et al. (1996) presented evidence that SAO 26804 is in fact a member of the luminosity class III stars with far-IR excesses, a conclusion that is at odds with the previous classification of the star as Vega-like. Conversely, Miroshnichenko, Bergner, & Kuratoc (1996) argue that SAO 26804 is a dwarf star (luminosity class V), but with a later spectral type (K5–7) than that assumed by Skinner et al. (1995) and

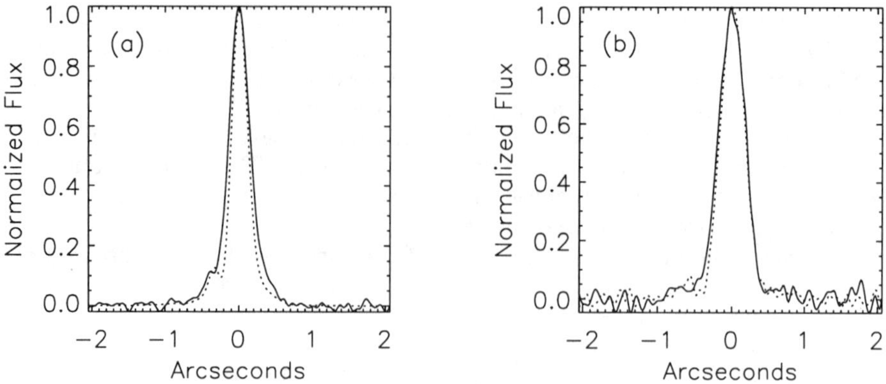

Figure 1. North–south scans through SAO 26804 and the PSF star. In both plots SAO 26804 is the *solid line* and the PSF star is the *dotted line*. *a)* Scans at 10.8 μm. FWHM of SAO 26804 = $0\farcs31$. The small bump seen on the *left* (south) side of the scans is part of the first Airy ring. *b)* Scans at 18.2 μm. FWHM of SAO 26804 = $0\farcs41$.

Sylvester et al. (1996). In 1999, we observed SAO 26804 as part of our Vega-like survey. Our observations have shed new light on the existence of a disk around the star and give information about the nature of the mid-IR-emitting grains around the source, but we cannot yet answer the question of where this star is in its evolution.

3.1. New High-Resolution Imaging from Keck

We obtained diffraction-limited 10.8 μm (N Band) and 18.2 μm images of SAO 26804 using OSCIR on Keck in 1999 May. These observations show SAO 26804 to be an unresolved point source with FWHM of $0\farcs31$ at 10.8 μm and FWHM of $0\farcs41$ at 18.2 μm. This 10.8 μm source size is inconsistent with the N-Band image presented by Skinner et al. (1995). In Figure 1 we show north–south scans of SAO 26804 and a nearby point-spread-function (PSF) star at 10.8 μm and 18.2 μm. Figure 1 shows that there is no appreciable difference between the profile of SAO 26804 and the PSF star along the major axis of the disk reported by Skinner et al. (1995). The profiles of SAO 26804 and the PSF star are virtually identical at 18.2 μm (Figure 1*b*).

Given our FWHM measurements, we know that there is no emission from the source at radii greater than about $0\farcs5$. Translating this angular size into physical units is problematic though. The ambiguity in the luminosity classification of this source leads to a similar uncertainty in its intrinsic luminosity and therefore its distance. Fekel et al. (1996) propose that SAO 26804 is a K2 III star with $L/L_{\odot} = 79.1$; $M/M_{\odot} = 1.15$; $T = 4420$ K; $d = 600$ pc. Using this distance, our images limit the size of any extended emission from the source to a radius of 90 AU at 10.8 μm and 125 AU at 18.2 μm. These limits are consistent with the sizes of the mid-IR-emitting regions of the other resolved Vega-like disks (Fisher et al. 2000, and references therein). Conversely, if we assume that SAO 268094 is a main-sequence dwarf and use the stellar parameters given in

Miroshnichenko et al. (1996) [lum. class = K5–7 V; $L/L_\odot = 0.28$; $M/M_\odot = 0.73$; $T = 4170$ K; $d = 40$ pc] our limits on the radial extent of the dust are 6 AU at 10.8 μm and 8 AU at 18.2 μm. The smaller limits in this case are similar to the proposed size of the circumbinary disk in the HD 98800 system. OSCIR observations of HD 98800 imply that the disk in that system is ~ 2 AU in radius (Prato et al. 2000). Similar to HD 98800, the images presented here show that SAO 26804 is unresolved in the mid IR at the highest resolution possible today, but our work cannot rule out the fact that there is an unresolved disk around the star.

3.2. Modeling the Mid-IR-Emitting Grains

Our model was developed by the University of Florida Dust Dynamics Group and has been successfully used to analyze similar data on the circumstellar disks around HR 4796A (Wyatt et al. 1999; Telesco et al. 2000) and HD 141569 (Fisher et al. 2000). In our models we choose the mid-IR emitting grains to be composed of the astronomical silicates given by Drain & Lee (1984) with $\rho = 2.5$ g cm^{-3}. The choice of silicate grains is consistent with modeling performed by Skinner et al. (1995), where they determine that $\sim 74\%$ of the mass of the dust round SAO 26804 is composed of silicates. With the assumption that the grains act as spherical Mie particles, we can calculate temperatures for dust grains of various sizes. Since the temperature of the dust determines the observed ($F_{10.8\mu\mathrm{m}}/F_{18.2\mu\mathrm{m}}$) flux ratio, we estimate the grain size by matching the observed flux ratio to that calculated for grains of different diameters at different distances from SAO 26804.

We ran two sets of models for SAO 26804. For the first run we modeled SAO 26804 as a K2 III star with parameters shown above. Solutions for the grain sizes at different distances from the star that are consistent with the observed flux ratio range from 7 μm grains 20 AU from the star, to ~ 1 μm grains 60 AU from the source. These grain sizes and distances from the star are consistent with both the limiting sizes set by our images and with estimates for the grain sizes in other Vega-like disks (Fisher et al. 2000, and references therein). The second run of models assumed SAO 26804 was a dwarf as described by Miroshnichenko et al. (1996). In this case the solutions consistent with our observations range from 1 μm grains 2.5 AU from the star to 5 μm grains 1 AU from the source. Again, these grain sizes are consistent with grain size estimates of other Vega-like disks, and the size of the disk is very close to the size of the proposed disk in the HD 98800 system (Prato et al. 2000). Our models also return information on the ratio of the radiation force to the gravitational force acting on a particle (called β) (summarized in Artymowicz 1997). A value of $\beta > 1$ for grains of a given size means that those grains would be removed from the system on essentially a free-fall timescale, which is much less than any assumed age for this system, be it a giant or a dwarf. Values of $\beta < 0.5$ for grains imply that they are gravitationally bound in the system and could remain in the system for long times after their capture or creation. In the first case of our modeling, assuming SAO 26804 is K2 III, $\beta > 1$ for all particles that gave us consistent solutions to our models. This implies that in this case none of the mid-IR-emitting grains around SAO 26804 is primordial, and that the grains have been replenished through some mechanism. One possible scenario for this replenishment of these particles is through the collisions of larger bodies, as is apparently the case for β Pic (Artymowicz 1997) and HR 4796A (Wyatt et al. 1999; Telesco et al. 2000). Modeling of SAO 26804 as a K5 V star produced very different results for the grains near the star. In this case $\beta < 0.2$ for all the grain sizes consistent with our observations. In this case, the grains would be assumed to be gravitationally bound in the system and may have been present since the formation of the star. Since our derived grain sizes are larger than the size of interstellar grains, there may have been aggregation of smaller particles into the micron-sized grains that are emitting the observed mid-IR radiation, or the micron-sized grains may well have been created by the collisions of larger particles in the disk. However, in this case we have no dynamical evidence that these grains have ever been replenished. Although our modeling found grain sizes that are consistent with our observations, until we can either accurately measure the distance to SAO 26804, or unequivocally assign a luminosity class to the source, we will not be able to definitively answer questions about the size of the mid-IR-emitting region around this source. Additionally, the ambiguity in the luminosity (and distance) makes it impossible to accurately determine if the mid-IR grains are primordial, or have been replenished in some way. The remaining questions about the nature of this star make it very difficult to classify, but they also make it worthy of further study.

References

Artymowicz, P. 1997, Ann. Rev. Earth Planet. Sci., 25, 175

Aumann, H. H., et al. 1984, ApJ, 278, L23

Draine, B. T., & Lee, H. M. 1984, ApJ, 285, 89

Fisher R. S., et al. 2000, ApJ, 532, L141

Fekel, F. C., et al. 1996, ApJ, 462, L95

Mannings, V., & Barlow, M. J. 1998, ApJ, 497, 330

Miroshnichenko, A. S., Bergner, K. Y, & Kuratoc, K. S. 1996, A&A, 312, 521

Prato, L., et al. 2000, in preparation

Stencel, R. E., & Backman, D. A. 1991, ApJS, 75, 905

Skinner, C. J., et al. 1995, ApJ, 444, 861

Sylvester, R. J., et al. 1996, MNRAS, 279, 915

Telesco, C. M., et al. 2000, ApJ, 530, 329

Walker, H. J., & Wolstencroft, R. D. 1988, PASP, 100, 1509

Wyatt, M. C., et al. 1999, ApJ, 527, 918

Zuckerman, B., Kim, S. S., & Liu, T. 1995, ApJ, 446, L79

Disks, Planetesimals, and Planets
ASP Conference Series, Vol. 219, 2000
F. Garzón, C. Eiroa, D. de Winter, and T. J. Mahoney, eds.

Dust Distribution in Gas Disks: A Model for the Ring around HR 4796A

Hubertus Klahr and Doug Lin

UCO/Lick Observatory, University of California, Santa Cruz, CA 95064, USA

Abstract. There have been several model analyses of the near- and mid-IR flux from the circumstellar ring around HR 4796A (Augereau et al. 1999; Kenyon & Luu 1999; Telesco et al. 2000; Wyatt et al. 2000). All of them consider only a dust distribution in a gas-free environment. But, in the vicinity of a young star, the possibility that the dust ring is embedded within a residual protostellar gas disk cannot be ruled out. In a gas-rich environment, large sizes ($> 100\ \mu$m) are needed for the particles to survive the radiative blowout. The total dust mass required to account for the IR flux is $< 10^{-2}\ M_\oplus$. The combined influence of gas and stellar radiation may also account for the observed sharp inner boundary and rapidly fading outer boundary of the ring. The pressure gradient induced by a small (10%) amplitude variation in the surface-density distribution of a low-mass gaseous disk is sufficient to modify the rotation speed of the gas. The resulting hydrodynamic drag on modest-sized ($> 100\ \mu$m) particles is adequate to compensate for the turbulent stirring, radiative drag, and radiation pressure such that they remain gravitationally bound to the system. The required surface-density variation of the gas may be induced by 1) the perturbation of a low-mass planet or the binary companion HR 4796B, 2) the photo-evaporation of the disk, or 3) from the variations in the viscous angular-momentum transport and mass-diffusion rate in the disk. We show that the structure of the dust ring is preserved during and after gas depletion, such that similar rings may be common among early-type stars.

1. Introduction

In this paper (see also Klahr & Lin 2000), we examine the persistence of the dust-ring structure around HR 4796A. Since the host star is only 8×10^6 yr old, we hypothesize the existence of some residual gas analogous to that found around several young stellar objects with similar ages (Beckwith et al. 1990; Beckwith & Sargent 1993a,b; Zuckerman, Forveille, & Kastner 1995; Hartmann et al. 1998). An upper limit, 1–7 $M_\oplus$, of gas has been inferred from the mm continuum radiation by Greaves, Mannings, & Holland (2000). (A larger gas mass may exist if a substantial amount of heavy elements is depleted from the gas phase and is stored in particles larger than a few mm.) We explore the interaction between the dust, radiation, and residual gas in the disk. In Klahr

& Lin (2000) we recapitulate the physical interaction between particles, stellar radiation, and residual gas more carefully. We show that the hydrodynamic drag tends to promote the effect of radiation pressure and increase the critical particle size for radiative blowout. We also provided some constraints on the size and total mass of particles in the ring around HR 4796A. We showed that variations in the surface distribution of the gas provide a favorable condition for modest-sized (more than 100 μm) particles to congregate into a dust ring. This ring structure is preserved during and after the depletion of the disk gas. We suggest that these particles provide a supply of the dominant contributors to the IR emission of the dust ring.

2. A Model for the Gas around HR 4796A

For illustrative purpose, we consider a simple model in which a disk of dust particles is embedded in a residual gaseous disk. The only free parameters in this model are those that describe the gas surface density, $\Sigma(R)$, as a function of the disk radius, R. All other gas properties are deduced in a self-consistent way. The physical parameters for the first model are given by the mass ($M_\star = 2.5\ M_\odot$) and luminosity ($L_\star = 35\ L_\odot$) of the central object. Additional model parameters include the location of the maximum in surface density at 70 AU, the width of the inner edge as $\delta R_0 = 20$ AU, and the gas mass around the edge between $R_0 - \delta R_0$ and $R_0 + \delta R_0$ which is taken to be 100 earth masses. (Figure 1). But the mass of gas can also be much smaller. The physical parameters for a second model are given by the width of the inner edge, which is $\delta R_0 = 10$ AU, and the gas mass around the edge, which is only 1 earth mass (Figure 2).

3. Gas as an Amplifier of Inhomogeneities

In every sub-Keplerian disk dust is drifting radially inward. But a local increase of the surface density reverses the pressure gradient and stops the radial particle drift (Figure 3, *left:*).

If there is strong radiation from the central object, an even smaller step in density may stop the grains (Figure 3, *right*).

4. Particle Concentration due to Lack of Radial Pressure Gradient

Wherever the radial pressure gradient vanishes ($\Sigma \sim R^{7/4}$), the disk rotates Keplerianly. Towards these points the dust is drifting with a speed of

$$v_r = \tau_f 2\Omega^* dV^* \quad \text{for} \quad St \ll 1. \tag{1}$$

The resulting radial drift velocity can be used with the continuity equation to compute the global evolution of the dust surface density, Σ_d, due to their orbital migration (Figure 4, *right*). After equilibrium is established, the persistence of the gas is not required. Thus, this scenario only needs the gas to be present for the initial few 10^6 yr in order to produce the ring. Afterwards, it can be evaporated or be viscously diffused, while the dust distribution remains intact.

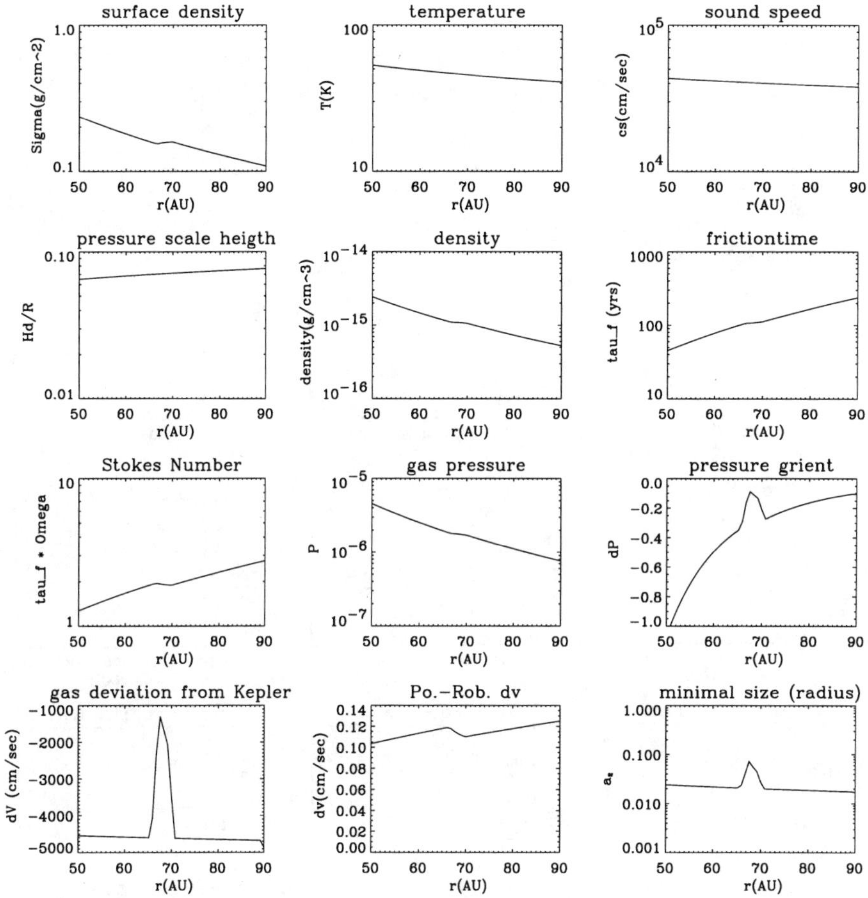

Figure 1. *Model A:* The radial distribution of surface density, density, temperature, friction time for 600 μm-size particles, the resulting Stokes number, pressure scale height, the speed of sound, the pressure, the radial pressure gradient, the deviation from the Kepler rotational profile, the Poynting–Robertson effect, and the size of the smallest gravitationally bound particles.

5. Conclusions

We present a model for the formation of dust rings in circumstellar disks (see Klahr & Lin 2000 for a detailed elaboration). The main difficulty of identifying small particles as the dominant contributors of the IR radiation is the requirement of a non-continuous bimodal size distribution. This problem would be removed if the particles which emit most of the IR radiation were sufficiently large (> 100 μm) to survive the radiative blowout. In this case, a total mass < 0.1 $M_\oplus$ is needed throughout the lifespan of HR 4796A. We suggest that the

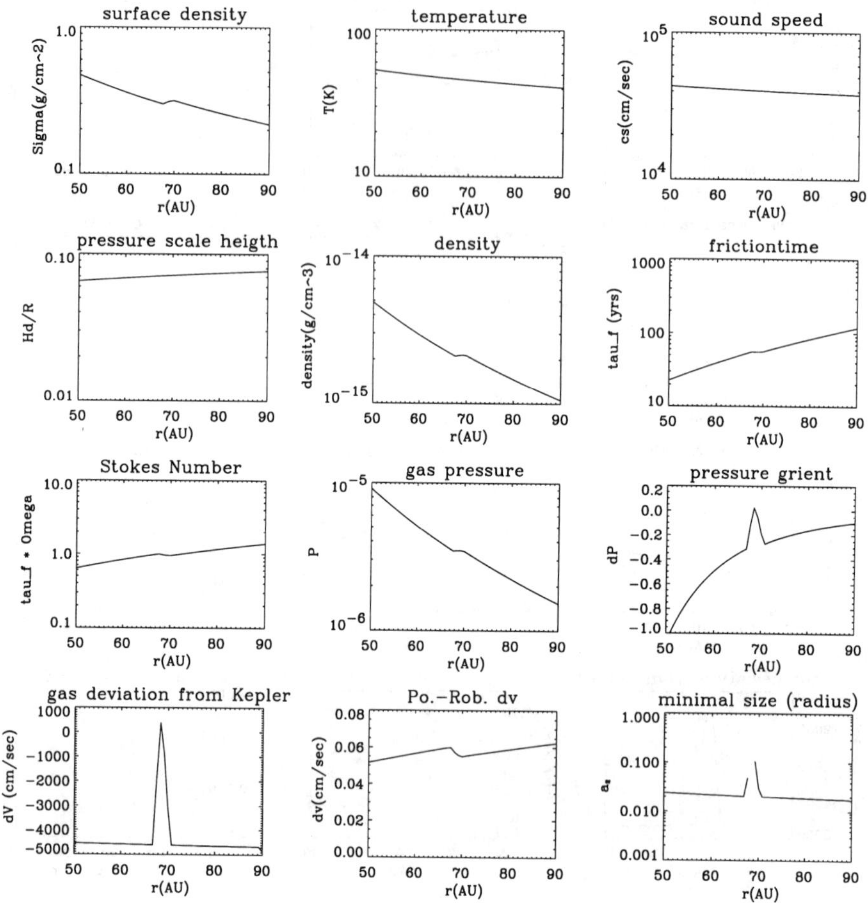

Figure 2. *Model B:* see Figure 1 for explanation.

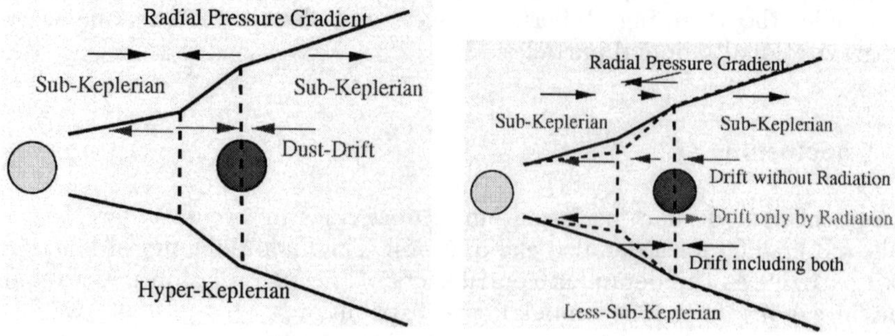

Figure 3. *Left:* Negligible radiation. *Right:* Strong radiation.

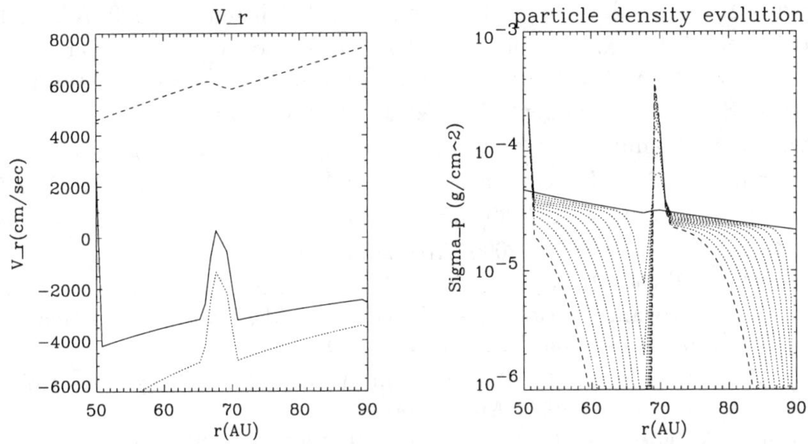

Figure 4. *Left:* Model A, showing the radial-velocity distribution of 600 μ-size particles for the model in Figure 1. The *solid line* gives the effective radial drift. The *dotted line* accounts only for the gas-drag effect, and the *dashed line* denotes only the radiation-pressure effect. *Right:* Model B, showing the evolution of the 600 μm-particle surface-density distribution for the model in Figure 2. The *solid line* gives the initial distribution. The other lines are separated by time intervals of every 10^4 yr.

preservation of these modest-sized particles (whether they are parent particles or the dominant IR emitters) is due to the combined effect of radiation pressure and gas drag. Our model can account for rings with sharply defined edges if they are embedded in a gas disk. A very small amount of inhomogeneity is needed to induce the congregation of the particles. This inhomogeneity may be the result of a planet, but it does not necessarily have to be so. Photo-evaporation, an inhomogeneous initial distribution of mass after the infall, or inhomogeneities in the radial viscous transport of material can also account for such an anomaly in the surface-density distribution. Thus we conclude that dust rings with sharp structures are not necessarily good indicators of embedded planets.

Acknowledgments. We thank D. Trilling and P. Bodenheimer for useful conversations. This work is supported in part through National Science Foundation grant AST-9618548, NASA grants NAG5-4277, 4494, 7515, and a special NASA astrophysics theory program, which supports a joint Center for Star Formation Studies at NASA–Ames Research Center, UC Berkeley and UC Santa Cruz.

References

Augereau, J. C., Lagrange, A. M., Mouillet, D., Papaloizou, J. C. B., & Grorod, P. A. 1999, A&A, 348, 55

Beckwith, S. V. W., Sargent, A. I., Chini, R. S., & Güsten, R. 1990, AJ, 99, 924

Beckwith, S. V. W., & Sargent, A. I. 1993a, in Protostars and Planets III, ed. E. H. Levy & J. I. Lunine (Tucson: University of Arizona Press), p. 521

Beckwith, S. V. W., & Sargent, A. I. 1993b, ApJ, 402, 280

Greaves, J. S., Mannings, V., & Holland, W. S. 2000, Icarus, 143, 155

Hartmann, L., Calvet, N., Gullbring, E., & D'Alessio, P. 1998, ApJ, 495, 385

Kenyon, S. J., & Luu, J. X. 1999, AJ, 118, 1101

Klahr H. H., & Lin, D. N. C. 2000, ApJ, submitted

Telesco, C. M., Fisher, R. S., Pina, R. K., Knacke, R. F., Dermott, S. F., Wyatt, M. C., Grogan, K., Holmes, E. K.,Ghez, A. M., Prato, L., Hartmann, L. W., & Jayawardhana, R. 2000, ApJ, 530, 329

Wyatt, M. C., Dermott, S. F., Telesco, C. M., Fisher, R. S., Grogan, K., Holems, E., & Pina, R. K. 2000, ApJ, in press

Zuckerman, B., Forveille, T., & Kastner, J. H. 1995, Nat, 373, 494

Disks, Planetesimals, and Planets
ASP Conference Series, Vol. 219, 2000
F. Garzón, C. Eiroa, D. de Winter, and T. J. Mahoney, eds.

The Two-Star Photometer at Kourovka Observatory: Design and Noise Analysis

Valerij Kozhevnikov and Polina Zakharova

Astronomical Observatory, Ural State University, Lenin Ave. 51, Ekaterinburg, 620083, Russia

Abstract. We have constructed a two-star photometer that allows observations of a variable star and a nearby comparison star to be made simultaneously. The design of this photometer differs from most of the multichannel photometers described in the literature. During 1996–97 the photometer was employed in observations of the eclipsing binary CM Dra as part of the TEP (Transits of Extrasolar Planets) network (Deeg et al. 1998). Despite the fact that the observations were obtained with a rather small (70 cm) telescope, mostly under unfavorable sky conditions due to the location of Kourovka Observatory in a damp fir forest, the precision for this photometer (5 mmag) turned out to be only slightly lower than the precision for the CCD systems (3–4 mmag) employed by other collaborators of the TEP network. The noise analysis made for the two-star photometer shows that a precision of about 2 mmag can be achieved under excellent photometric conditions and the employment of larger telescopes or in observations of brighter stars. The two-star photometer seems to be a promising instrument for searching for transits of extrasolar planets.

1. The Design

A complexity of most of the multichannel photometers described in the literature is that one or several channels of the photometer must move in order to center the two stars in the diaphragms (see, for example, Belmonte et al. 1991). Since these channels have significant weight, it is difficult to make such a precision mechanism. We offer a new design of the two-star photometer in which all the channels are immovable. A schematic drawing of the photometer is shown in Figure 1. The centering of the two stars in the diaphragms is carried out by rotating the photometer around the optical axis of the telescope and the movement of two rectangular prisms attached to the slide carriage along the optical axis. Besides these prisms, the slide carriage carries another rectangular prism creating an offset field. The movable eyepiece allows a guide star to be chosen in this offset field.

Looking at Figure 1, one might think that this design has a drawback. Due to the movement of the two rectangular prisms along the optical axis, the position of the focal plane changes and cannot match the diaphragm planes. In practice, however, this is of no importance. The small defocusing of stellar

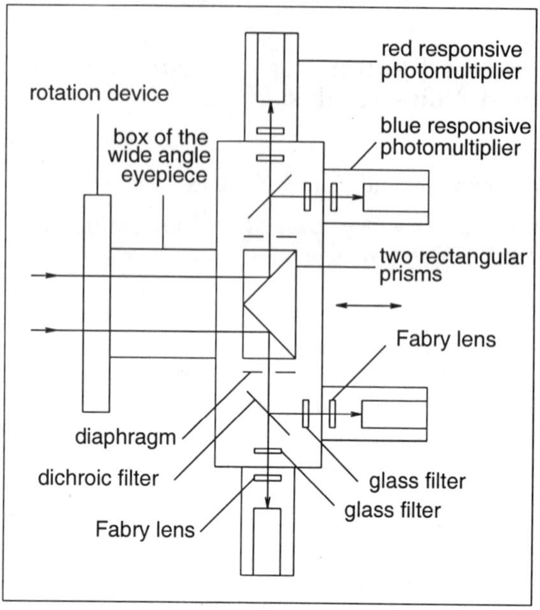

Figure 1. Schematic drawing of the photometer.

images does not prevent stars from being seen, and we can focus the telescope by shifting the secondary mirror. Moreover, a more precise focus can be made with the aid of the guide star.

The advantages of this design are obvious. The photometer has no heavy movable parts. The slide carriage does not undergo significant variable load while the telescope tracks. This allows good centering of the two stars to be maintained in the diaphragms. A third channel, for continuous sky brightness measurements, can be incorporated into such a design. We recently mounted a small rectangular prism in the box of the wide-angle eyepiece. This prism creates one more offset field, which is used in sky brightness measurements.

2. Noise Analysis

As mentioned above, Kourovka Observatory participated in TEP network observations of the eclipsing binary CM Dra with the goal of detecting transits of planets. Nine observatories took part in these observations during 1994–96. With the exception of Kourovka, these observatories employed CCD-equipped telescopes. CM Dra is a very red star of 11 mag in the R band. Because of the absence of suitable comparison stars with similar colors around it differential photometry shows slow airmass-related changes in CM Dra's brightness from differential extinction. In TEP network observations these variations were removed by subtracting a fit that was either linear nor a second-order polynomial to the off-eclipse lightcurve (Deeg et al. 1998). Such variations, as well as nightly averages of the lightcurves obtained at Kourovka Observatory, were also removed by subtraction of a second-order polynomial fit, and these residual lightcurves were adapted to noise analysis for two-star photometry. A total of 25 lightcurves with durations of 3–5 hours turned out to be useful for the analysis. We divided these lightcurves into two nearly equal groups. The first group included the lightcurves obtained on photometric nights. The second group included the lightcurves obtained on non-photometric nights, when the comparison star showed occasional variations from 0.2 to 0.7 mag.

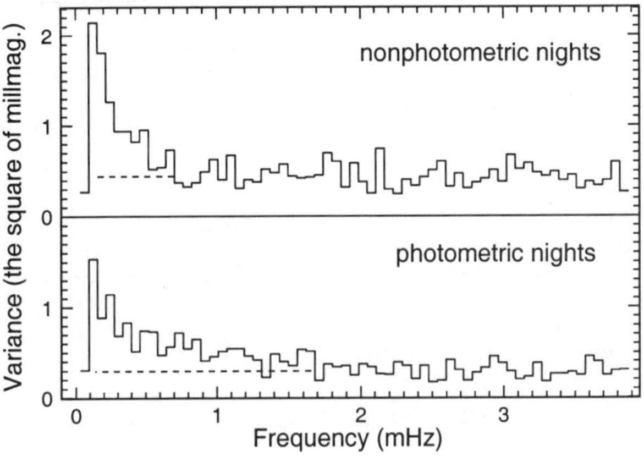

Figure 2. Average power spectra of CM Dra.

We can calculate the noise of these lightcurves and relate the frequency-dependent behavior of the noise to its sources by using power spectra. The discrete form of the Parseval's theorem is

$$\sum_{k=0}^{N-1} |h_k|^2 = \frac{1}{N} \sum_{n=0}^{N-1} |H_n|^2 , \tag{1}$$

where H_n is the discrete Fourier transform of the N points h_k. Since the function h_k is real with a mean value of zero, we can write equation 1 as

$$N\sigma^2 = \frac{2}{N} \sum_{n=1}^{N/2} |H_n|^2 \tag{2}$$

and find the time-series variance, σ^2, adding together all the points of the power spectrum divided by $2/N^2$:

$$\sigma^2 = \frac{2}{N^2} \sum_{n=1}^{N/2} |H_n|^2 = \sum_{n=1}^{N/2} \frac{2}{N^2} |H_n|^2 . \tag{3}$$

The average power spectra for the two groups of lightcurves are shown in Figure 2. These power spectra show $1/f$ noise components (Young et al. 1991; Kjeldsen & Frandsen 1992; Schutt 1993) and white-noise components. We can calculate the white noise and the $1/f$ noise by cutting off the frequency-dependent parts of the power spectra as shown with dashed lines in Figure 2. These noises are $\sigma_{1/f} = 2.7$ mmag, $\sigma_{\rm wn} = 4.1$ mmag on the photometric nights and $\sigma_{1/f} = 2.5$ mmag, $\sigma_{\rm wn} = 4.8$ mmag on the non-photometric nights.

Table 1. The noise in the CM Dra observations

Type of noise	rms (mmag) Phot. nights	Non-phot. nights
$1/f$	2.7	2.5
Photon	3.5	3.7
Scintillation	1.1	1.1
Motion	1.8	2.8
Total	4.9	5.4

The white noise can be written as

$$\sigma^2_{\rm wn} = \sigma^2_{\rm phot} + \sigma^2_{\rm scin} + \sigma^2_{\rm mot}\,, \tag{4}$$

where $\sigma_{\rm phot}$ is the photon noise, $\sigma_{\rm scin}$ is the scintillation noise, and $\sigma_{\rm mot}$ is the noise caused by motion of the stellar images in the diaphragms due to changes in the refractive index of the atmosphere. The photon and scintillation noise are white in the case of one-channel photometry (Young et al. 1991). These noise types are white in the case of two-star photometry because they are also uncorrelated for the two stars. While the image motions are correlated at the angular distance separating the two stars (Young 1974), we can consider the noise due to image motions as white because the power spectra (Figure 2) do not show an appreciable rise at high frequencies. Considering that photon noise is the Poisson noise of the total counts, we can estimate the photon noise of the differential lightcurves. The photon noise in magnitudes for each photometer channel is $\sigma_{\rm p} = 1.08\sqrt{N_{\rm tot}}/(N_{\rm tot} - N_{\rm sky})$, where $N_{\rm tot}$ is the mean total count, and $N_{\rm sky}$ is the mean count in the sky background. In accordance with the observed count rates, the photon noise of the differential lightcurves, $\sigma^2_{\rm phot}$, is equal to 3.5 mmag and 3.7 mmag on the photometric and nonphotometric nights (128 seconds integration times), respectively. We can estimate the scintillation noise according to Young (1967). The scintillation noise in magnitudes for each photometer channel is

$$\sigma_{\rm s} = 1.08\frac{I_{\rm rms}}{I_{\rm dc}} = 0.1D^{-2/3}M^{3/2}\exp\left(\frac{-h}{8000}\right)\Delta t^{-1/2}\,, \tag{5}$$

where D is the telescope diameter in centimeters, M is the airmass for the object under the observed conditions, h is the telescope elevation in meters and Δt is the integration time in seconds. The calculations yield $\sigma_{\rm s} = 0.8$ mmag for our CM Dra observations ($D = 70$ cm, $M = 1.4$, $h = 320$ m, $\Delta t = 128$ s). The scintillation noise of the differential lightcurves is $\sigma_{\rm scin} = \sqrt{2}\sigma_{\rm s} = 1.1$ mmag.

The noise caused by motion of the stellar images can be calculated as

$$\sigma^2_{\rm mot} = \sigma^2_{\rm wn} - (\sigma^2_{\rm phot} + \sigma^2_{\rm scin}), \tag{6}$$

amounting to 1.8 mmag and 2.8 mmag for the photometric and nonphotometric nights, respectively. The noise in the CM Dra observations is given in Table 1.

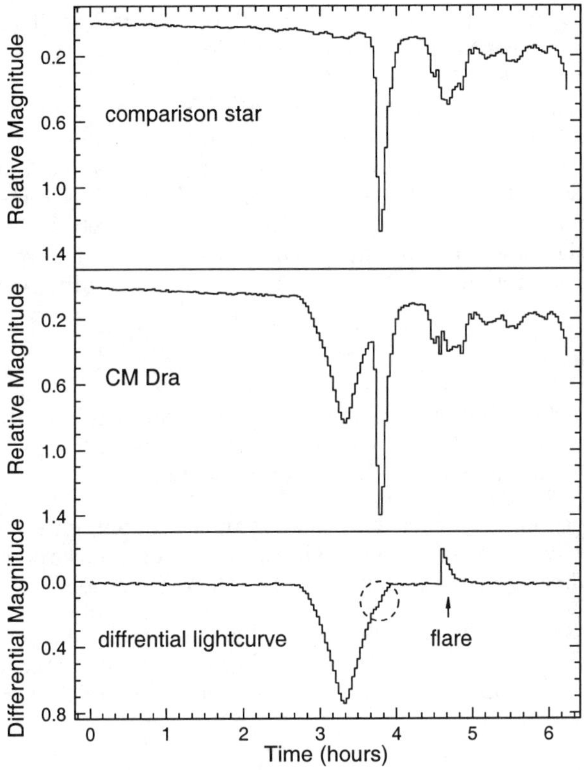

Figure 3. Light-curves obtained on a hazy night.

Two-star photometry reduces $1/f$ noise from the atmosphere because of continuous monitoring of atmospheric transparency. This makes it possible to obtain observations under non-photometric sky conditions. Figure 3 shows the lightcurves obtained on a hazy night. These lightcurves demonstrate the striking capability of the two-star photometer to compensate for the light loss due to thin clouds or haze. Even the occurrence of fog absorbing light by 1.3 mag causes only a barely noticeable distortion of the differential lightcurve. This area is encircled in Figure 3. The flare of CM Dra is hidden in the noise of the lightcurve obtained directly through haze.

We observed CM Dra using a rather small (70 cm) telescope. In order to compare the precision for our photometer with that of the CCD systems used in TEP network observations, we can estimate noise types for this photometer and telescopes with mirrors of 1.0 and 1.3 meters and altitudes of about 3000 meters.

Such telescopes and altitudes seem to be typical of TEP network observations (see Deeg et al. 1998). The counts can be estimated simply if we increase the counts of our CM Dra observations proportionally with the areas of the mirrors of these telescopes. The photon noises must then be equal to 2.2 mmag and 1.8 mmag for mirrors of 1.0 and 1.3 meters, respectively. According to equation 5, the respective scintillation noise must be equal to 0.7 mmag and 0.6 mmag. The possible noise caused by motion of the stellar images can be considered negligible on high mountains with excellent seeing. It is rather difficult to estimate the $1/f$ noise under excellent photometric conditions since the causes of this noise are not fully clear. We can consider two cases when the $1/f$ noise is the same as at Kourovka Observatory and when it is half this value. The total noise for the two-star photometer and the telescopes with mirrors of 1.0 and 1.3 meters must be equal to 3.6 mmag and 3.3 mmag in the first case, and 2.7 mmag and 2.3 mmag in the second case. The typical noise for the CCD systems used in TEP network observations is 3–4 mmag (see Table 3 of Deeg et al. 1998). Thus the precision for the two-star photometer can be the same or higher in comparison with the precision for the CCD systems under similar conditions.

3. Conclusions

1. The two-star photometer constructed at Kourovka Observatory has no heavy movable parts. This simple design allows such a photometer to be built without significant labour requirements.

2. The device for centering the two stars in the diaphragms does not experience significantly variable load while the telescope tracks stars. This allows good centering of the two stars during the observation.

3. Light absorption by thin clouds and haze is largely independent of wavelength. This permits high precision of two-star photometry under nonphotometric conditions even if the program and comparison stars are different in color.

4. Despite the relatively low quantum efficiency of photocathodes, multichannel photometers allow the same or higher precision compared to CCD systems under similar conditions even if the stars are rather faint (11 mag).

References

Belmonte, J. A., Chevreton, M., Mangeney, A., Praderie, F., Saint-Pe, O., Puget, P., Alvarez, M., & Roca Cortes, T. 1991, A&A, 246, 71

Deeg, H. J., et al. 1998, A&A, 338, 479

Kjeldsen, H., & Frandsen, S. 1992, PASP, 104, 413

Schutt, R. L. 1993, PASP, 105, 22

Young, A. T. 1967, AJ, 72, 747

Young, A. T. 1974, ApJ, 189, 587

Young, A.T., et al. 1991, PASP, 103, 221

Disks, Planetesimals, and Planets
ASP Conference Series, Vol. 219, 2000
F. Garzón, C. Eiroa, D. de Winter, and T. J. Mahoney, eds.

Size Distribution of Dust in the Disk of β Pictoris

Natalia A. Krivova[1], Alexander V. Krivov[1], and Ingrid Mann

MPI für Aeronomie, D-37191 Katlenburg-Lindau, Germany

Abstract. Intensive studies of the β Pic dust disk, based on observational data from the visual to the far IR and mm, give a relatively good understanding of the global spatial dust distribution in the disk. In contrast, the size distribution of dust is poorly known—only some constraints on the grain sizes have been put so far. Assuming a continuous replenishment of larger grains from sources in the disk and a collisional production of smaller debris, and considering the dynamics of grains, we derive the size distribution of dust in a broad range from micron-sized grains to tiny nanometer-sized particles. We show that this dynamically justified size distribution explains the observed polarization and colors of the disk much better than commonly assumed single power-law distributions.

1. Introduction

Since the first *IRAS* observations attested to the so-called "Vega phenomenon", considerable study is being given to dust disks around main-sequence stars and, in particular, around β Pictoris. This has yielded a relatively good understanding of the global spatial distribution of dust in the disk (e.g., Artymowicz, Burrows, & Paresce 1989; Kalas & Jewitt 1995; Mouillet et al. 1997) and led to a number of conclusions about the chemical composition of the dust (Telesco & Knacke 1991; Aitken et al. 1993; Knacke et al. 1993; Artymowicz 1997), but its size distribution is as yet scantily known. Models derived from different observational data lead to somewhat controversial inferences (Table 1), especially concerning the content of small submicrometer grains.

Here we assess the size distribution in the β Pic disk from dynamical considerations and then use the derived size distribution to fit observational data. Assuming continuous replenishment of larger grains from sources in the disk and collisional production of smaller debris, we consider then the dynamics of grains under radiation pressure and stellar gravity forces. A more accurate model than that of Krivova, Krivov, & Mann (2000) is used to derive the size distribution of the dust over a wide range. To test the results against observational data, we analyze the polarization of the radiation scattered by dust grains, a valuable source of information about dust properties. We apply different shapes of the size distribution to model the observed spatial variation and wavelength dependence

[1]On leave from the Astronomical Institute, St. Petersburg University, 198904 St. Petersburg, Russia.

Table 1. Previously deduced sizes of the grains in the disk of β Pic

Type of data	Sizes	References
Far-IR + mm data	> 5–$10\ \mu$m	Chini et al. (1991) Zuckerman & Becklin (1993)
Gray colors	> A few μm	Paresce & Burrows (1987) Lecavelier des Etangs et al. (1993)
Vis. + IR images	1–20 μm	Artymowicz et al. (1989)
10 μm sil. feature	$< 10\ \mu$m	Telesco & Knacke (1991)
	$\lesssim 2$–$3\ \mu$m	Aitken et al. (1993)
IR fluxes and scans + vis. data	$\sim 1\ \mu$m	Backman et al. (1992)
10 & 20 μm emission	Subμm	Telesco et al. (1988)
IR to mm data + 10 μm sil. feature	As in Comet Halley + extra-small grains	Li & Greenberg (1998)

of the polarization of the radiation of β Pic and show that the suggested size distribution gives better results than commonly used power-law distributions.

2. Dynamical Evolution of Dust Grains

2.1. Model

We assume the dust disk to consist of two dust populations which, borrowing the terminology from Solar System studies, may be called α- and β-meteoroids. The former are larger grains that move about the star in bound orbits, whereas the latter are smaller particles blown away from the star by stellar radiation pressure. The boundary between the two populations depends on the luminosity-to-mass ratio of the central star and the properties of dust grains. For β Pic and plausible compositions of grains in its disk, the boundary lies at a grain radius of $a_0 \approx 2$ to 3 μm (e.g., Artymowicz 1997), or the mass $m_0 \sim 10^{-10}$ g.

First we postulate the spatial and size distributions of α-meteoroids. We assume that the large grains in the disk are produced by a collisional cascade and the activity of comet-like bodies, so that the number density of α-meteoroids with sizes $[a, a + da]$ at a distance r from the star obeys the power law

$$n_\alpha(r, a)da = n(r)a^{-q}da, \tag{1}$$

with $q \approx 3.5$ being expected for both collisional (Dohnanyi 1969) and cometary (e.g., Greenberg & Hage 1990, Fulle et al. 1995) sources. The spatial part of the distribution $n(r)$ is taken to have the form (Artymowicz & Clampin 1997)

$$n(r) = n_0 \left[(r/r_m)^{-1} + (r/r_m)^{2.7}\right]^{-1}, \tag{2}$$

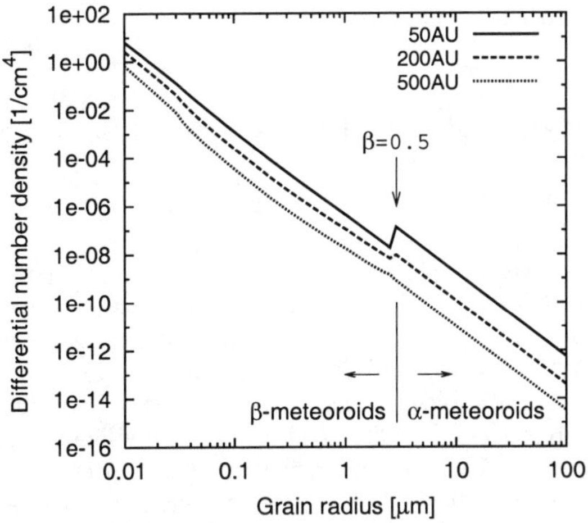

Figure 1. Size distributions of dust at different distances from the star.

where $r_m = 60$ AU and the normalizing constant n_0 is determined from the normal optical depth of the disk $\tau(r)$ given by Artymowicz & Clampin (1997), assuming a vertically uniform disk with the half-opening angle $\epsilon = 7°$.

Next, we assume that α-meteoroids create β-meteoroids through collisional fragmentation. The steady-state number density of β-meteoroids with sizes $[a, a + da]$ at a distance r from the star is

$$n_\beta(r, a)da = \frac{da}{r^2} \int_0^r r_0^2 \, \frac{dn^+}{dt}(r_0, a) \, v^{-1}(r_0, r, a) \, dr_0, \tag{3}$$

where $dn^+/dt(r_0, a)$ is the number of β-meteoroids with sizes $[a, a+da]$ produced per unit time in a unit volume at a distance r inside the disk, and $v(r_0, r, a)$ is the velocity a β-meteoroid of radius a born at a distance r_0 attains at distance r. The β-meteoroid production rate is calculated from the collisional probabilities determined by (1) and from the impact mechanics of colliding particles (Krivov et al., in preparation). Thus a more accurate model is used as compared to Krivova et al. (2000), where all α-meteoroids serving as sources of β-meteoroids were assumed to have a "typical" size. The velocity of β-meteoroids is evaluated from the classical formulae of the "photo-Keplerian" problem.

2.2. Results

The computed size distributions at different distances from the star are depicted in Figure 1. The basic result is that the size distribution of larger grains cannot be simply extrapolated to smaller sizes: the grains with sizes just below the blowout limit are typically somewhat (but not strongly) depleted, and the slope of their size distribution may differ from that of the larger grains. Another result is that the contribution of smaller particles grows with the distance from the star. This is readily understandable: β-meteoroids are blown radially away from the star; thus the β-meteoroids produced at a certain distance from the star contribute to the number density at all larger distances.

3. Polarization and Colors

3.1. Model

To calculate polarization and colors, we represent the dust number density as

$$n(r,z,a) = n(r,a) \exp\left[-\left(\frac{z/r_6}{\zeta(r)}\right)^{\gamma}\right], \quad \zeta(r) = \zeta_6 \left(\frac{r}{r_6}\right)^{\eta}, \tag{4}$$

where $n(r,a)$ is given by (1) for α-meteoroids and (3) for β-meteoroids, and the exponential term describes the dependence on z, the distance from the symmetry plane of the disk (Artymowicz et al. 1989; Kalas & Jewitt 1995). We take $\eta = 1.2$, $\gamma = 1.1$, $r_6 = 6''$, and $\zeta_6 = 0.05$, and assume the viewing angle to be $i = 4°$. The precise choice of the parameters describing the disk geometry is not crucial (Krivova et al. 1999, 2000). We use silicate (Laor & Draine 1993) as inferred from the 10 μm silicate feature (Aitken et al. 1993; Knacke et al. 1993) for Mie calculations of the particle scattering properties and also test other materials. An original numerical code is used to calculate the polarization scans at different wavelengths. The equations are given in Krivova et al. (2000).

3.2. Results

Polarization diagrams are predominantly determined by the dust grain sizes (see discussions in Voshchinnikov & Krügel 1999 and Krivova et al. 2000); this allows us to evaluate the sizes from an analysis of the polarimetric data. Dust composition and structure may only modify the derived values of the sizes, but the general shape of the size distribution still stands.

In Figure 2*a*, the polarization in the *UBVRIH* bands calculated for the derived size distribution and the traditional power-law one a^{-q} ($a_{\min} = 0.005$ μm, $a_{\max} = 100$ μm) is compared with observational data (Gledhill, Scarrott, & Wolstencroft 1991; Wolstencroft, Scarrott, & Gledhill 1995). For the power-law distribution, the index $q = 2.7$ provides a better, yet not satisfactory, agreement with the observational data, which means that small particles contribute less than in the case of the usually assumed $q = 3.5$ (cf. Voshchinnikov & Krügel 1999). For $q = 2.7$ the predicted colors of the disk, which are too blue in the case of $q = 3.5$, are a little bit redder than observed by Paresce & Burrows (1987), but within the observational errors.

Using the distribution derived in Section 2 gives much better results. The colors are also in a good agreement with observations. Still better agreement could be achieved if the sizes of the grains giving the major contribution to the cross-section are shifted to larger values ($\sim$ 5 to 15 μm). This is actually expected from a yet more accurate model (Krivov et al., in preparation).

Figure 2*b* shows the difference in the polarization in two wings. To explain the polarization in the NE wing one should increase the number density of small grains by $\approx$ 20–30 % as compared with the SW wing. There could be different mechanisms leading to such a difference in the sizes. One possibility is that the bombardment of the disk by interstellar dust particles (Artymowicz & Clampin 1997) under plausible conditions could produce more β-meteoroids in one wing (Krivova et al. 2000). The difference could also be explained within the "falling evaporating bodies" scenario under the presence of a planetary perturber (Lecavelier des Etangs, this volume, p. 308).

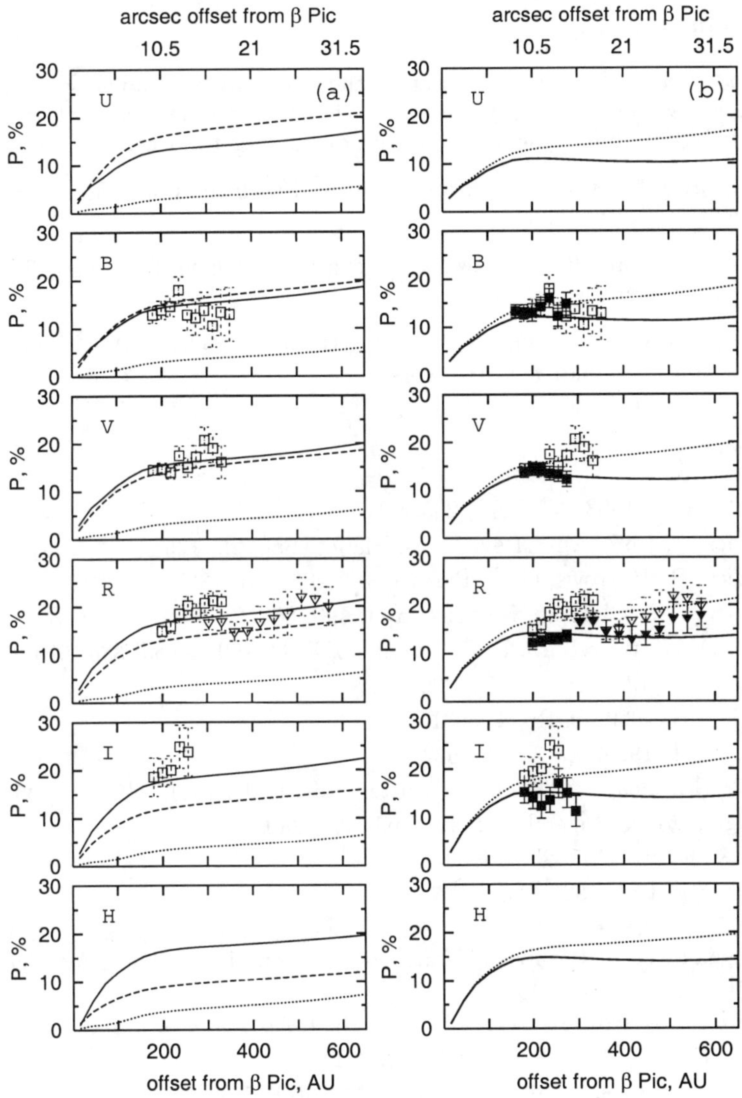

Figure 2. Modeled polarization of the disk of β Pic in $UBVRIH$. Symbols show the observed values from Gledhill et al. (1991, *triangles*) and Wolstencroft et al. (1995, *squares*). *a)* SW wing. *Dotted line*—standard power-law size distribution with the index $q = 3.5$; *dashed line*—the same with the index $q = 2.7$; *solid line*—our derived size distribution (Section 2). *b)* SW wing (*open symbols* for the data and *dotted curve* for the model results) and NE wing (*filled symbols* and *solid line*). About 20% "extra" β-meteoroids, as compared to the distribution in Figure 1, are present in the NE wing.

4. Conclusions

1. Our dynamical model suggests that the grains with sizes below about 2–3 μm are somewhat (but not strongly) depleted, and the slope of their size distribution differs slightly from that of the larger grains.

2. The suggested size distribution agrees well with the polarimetric and colorimetric observations.

3. Different polarizations in two wings can be explained if more small grains are present in the NE wing.

Acknowledgments. The work was partly supported by BMBF. AK is an Alexander von Humboldt fellow at MPAe.

References

Aitken, D., et al. 1993, MNRAS, 265, L41

Artymowicz, P. 1997, Ann. Rev. Earth Planet. Sci., 25, 175

Artymowicz, P., Burrows, C., & Paresce, F. 1989, ApJ, 337, 494

Artymowicz, P. & Clampin, M. 1997, ApJ, 490, 863

Backman, D. E., Witteborn, F. C., & Gillett, F. C. 1992, ApJ, 385, 670

Chini, R., et al. 1991, A&A, 252, 220

Dohnanyi, J. S. 1969, JGR, 74, 2531

Fulle, M., et al. 1995, A&A, 304, 622

Gledhill, T. M., Scarrott, S. M., & Wolstencroft, R. D. 1991, MNRAS, 252, 50

Greenberg, J. M. & Hage, J. I. 1990, ApJ, 361, 260

Kalas, P. & Jewitt, D. 1995, AJ, 110, 794

Knacke, R. F., et al. 1993, ApJ, 418, 440

Krivova, N. A., Mann, I., & Krivov, A. V. 1999, in Meteoroids 1998, ed. W. J. Baggaley & V. Porubčan (Bratislava: Astron. Inst. Slovak Ac. Sci.), p. 291

Krivova, N. A., Krivov, A. V., & Mann, I. 2000, ApJ, 539, in press

Laor, A. & Draine, B. T. 1993, ApJ, 402, 441

Lecavelier des Etangs, A., et al. 1993, A&A, 274, 877

Li, A. & Greenberg, J. M. 1998, A&A, 331, 291

Mouillet, D., Lagrange, A.-M., Beuzit, J.-L., & Renaud, N. 1997, A&A, 324, 1083

Paresce, F. & Burrows, C. 1987, ApJ, 319, L23

Telesco, C. M. & Knacke, R. F. 1991, ApJ, 372, L29

Telesco, C. M., Becklin, E. E., Wolstencroft, R. D. & Decher, R. 1988, Nat, 335, 51

Voshchinnikov, N. V. & Krügel, E. 1999, A&A, 352, 508

Wolstencroft, R. D., Scarrott, S. M., & Gledhill, T. M. 1995, Ap&SS, 224, 395

Zuckerman, B. & Becklin, E. E. 1993, ApJ, 414, 793

Disks, Planetesimals, and Planets
ASP Conference Series, Vol. 219, 2000
F. Garzón, C. Eiroa, D. de Winter, and T. J. Mahoney, eds.

The β Pictoris Dust Disk at 20 Microns

Pierre-Olivier Lagage and Eric Pantin

DSM/DAPNIA/Service d'Astrophysique, CEA/Saclay, Bat 709, F-91191 Gif-sur-Yvette Cedex, France

Abstract. We present 20 μm images of the "debris" disk around the nearby main-sequence star β Pictoris. The observations were made with the Saclay mid-IR camera, CAMIRAS, mounted on the ESO 3.6 m Telescope. The images complement previous 10 μm images and confirm the presence of a void of matter in the inner regions of the disk (< 40 AU), as well as a large asymmetry between the northeast and southwest sides of the disk; the asymmetry peaks at about 65 AU from the central star.

1. Why Observe Debris Disks in the Mid Infrared?

Disks and planets are intimately connected. Planets are made from protoplanetary disks; interactions between newly formed planets and still remaining "protoplanetary" disks can modify the orbit of the planet, which may "migrate" towards the star (see the reviews in this volume by Boss—p. 7—Terquem—p. 19, and references therein). Protoplanetary disks disappear with time, so that main-sequence stars were not expected to be surrounded by a disk. However, collisions between planetesimals or comet sublimation can generate a "second generation" disk; these disks are usually quoted as "debris" disks. The presence of a planet in such a disk can deeply modify the morphology of the disk through gravitational perturbations of the dust orbit (e.g., Roques et al. 1994). Such effects have been observed in the Solar System (Dermott et al. 1994; Reach et al. 1995). Thus, studying the morphology of debris disks is a way to search for "footprints" of planets.

The study of debris disks started with the *IRAS* discovery of an infrared (IR) excess around many main-sequence stars (Aumann et al. 1984; Plets & Vynckier 1999). The best-studied dust disk is that around the star β Pictoris (see reviews by Artymowicz 1997; Vidal Madjar, Lecavelier de Etangs, & Ferlet 1998, and references therein; Lecavelier des Etangs, this volume, p. 322). The outer parts of the dust disk were imaged in the visible soon after the *IRAS* discovery (Smith & Terrile 1984). The inner parts of the disk were imaged ten years later, thanks to the advent of mid-IR cameras (Lagage & Pantin 1994) following the first spatial resolution of the disk in the mid IR with bolometers (Telesco et al. 1988). Both the mid-IR images and recent observations in the visible or near IR (e.g., Mouillet et al. 1997) show features (asymmetry, deficit of matter near the star, warp), which may indicate the presence of planets in the system. Only a few other debris disks have been imaged so far: HR 4796A in the mid IR (Jayawardhana et al. 1998; Koerner et al. 1998; Telesco, this volume, p. 268, and the references therein)) and then in the near IR (Schneider et al. 1999; Augereau et al. 1999a), HD 141569 in the near IR (Weinberger et al. 1999; Augereau et al. 1999b), HR 100546 in the near IR (Pantin et al., this volume, p. 281). The dust disk around the main-sequence stars Vega, α PsA, and ϵ Eri were resolved in the submm (Greaves et al 1998; Holland et al. 1998; Greaves & Holland, this volume, p. 296, and references therein).

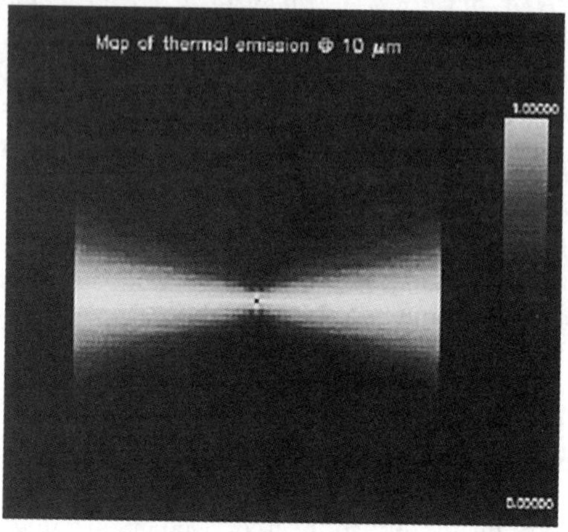

Figure 1. Simulated face-on view of the β Pictoris dust disk regions emitting the 10 μm flux seen by an observer located north in the plane of the disk (up in the image) and thus viewing the disk edge on. The star is located at the center of the image (*dark pixel*). The color scale is linear, (*black* corresponds to a flux divided by more than a factor of 2). *White* corresponds to the largest flux along the line of sight (corresponding to a column in the image); for each line of sight, the flux has been normalized to the largest flux. The pixel resolution of the map is 10 AU. As shown by this map, observations of the 10 μm radiation allow the inner regions of the disk to be probed, contrary to observations in the visible or near infrared (see Figure 2).

The origin of the disk emission observed from the mid-IR radiation (and at longer wavelengths) is dominated by dust thermal emission; at shorter wavelengths, the origin of the emission is dust scattering of the starlight. Observations in the mid IR and in the near IR (or visible) are complementary; the near IR is often more adapted to the study of the outer parts of a disk, when the mid IR is often more adapted to revealing the innermost regions of a disk. Indeed, given that the disk emission often dominates over the star emission in the mid IR, the angular resolution is not limited by the removal of the star contribution, as is the case at shorter wavelengths. Second, dust thermal emission is isotropic when dust scattering is anisotropic (forward scattering). Then, different parts of a disk are probed according to the origin of the observed emission and, in the case of an edge-on disk similar to that of β Pictoris, the scattered emission is dominated by the outer part of the disk (let us say greater than 60 AU from the star), so that the inner parts are barely accessible, whatever the angular

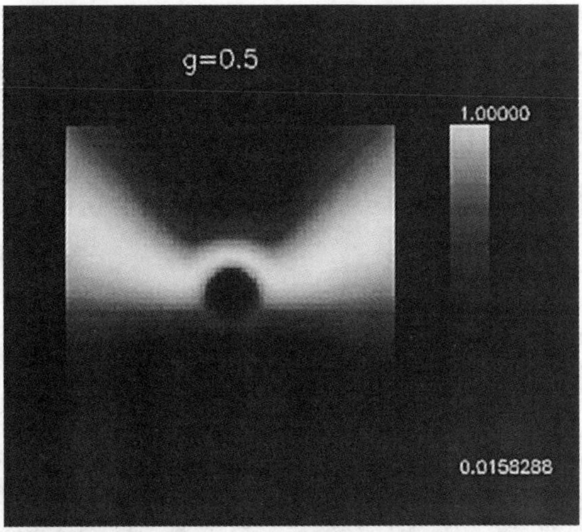

Figure 2. Same as Figure 1, but for the β Pictoris dust regions, which dominate in terms of starlight scattering. The regions probed by the scattering emission are quite different from the regions probed by 10 μm thermal radiation (see Figure 1). In particular the regions close to the star contribute very little to the observed scattered flux. The calculations of the scattered emission have been done with a Henyey–Greenstein phase function coefficient, g, equal to 0.5.

resolution is, how narrow the coronagraph is or how well the star contribution is removed (see Figures 1 and 2).

2. Observations and Results

The observations were performed with CAMIRAS, the Saclay mid-IR camera (Lagage et al. 1992; Galdemard et al. 1998), mounted on the 3.6 m Telescope of the European Southern Observatory (ESO) located at La Silla (Chile). A filter centered at 20.5 μm and with a full half maximum bandwidth of 1.1 μm was used. Observing in the 20 μm atmospheric window requires very low water vapor in the atmosphere. The weather conditions for 20 μm observations were very poor during the first observing run devoted to mid-IR observations of debris disks (1998 February). Only bright objects, such as HD 100546, were observed. No extension was detected for this particular object, which has been resolved thanks to observations in the near IR (Pantin et al. this conference, p. 281). The weather conditions were much better during the second observing run (1998 November), and observations of the β Pictoris dust disk were conducted. The resulting image is shown in Figure 3. The first result is the orientation of the mid-IR disk on the sky, which is confirmed to be 33.5 degrees, while the orientation of the outer parts of the disk are in the 30–31 degree range (Kalas & Jewitt 1995). This is a confirmation that we are not detecting the same regions at 10 μm and in the visible or near infrared, and that the disk is warped.

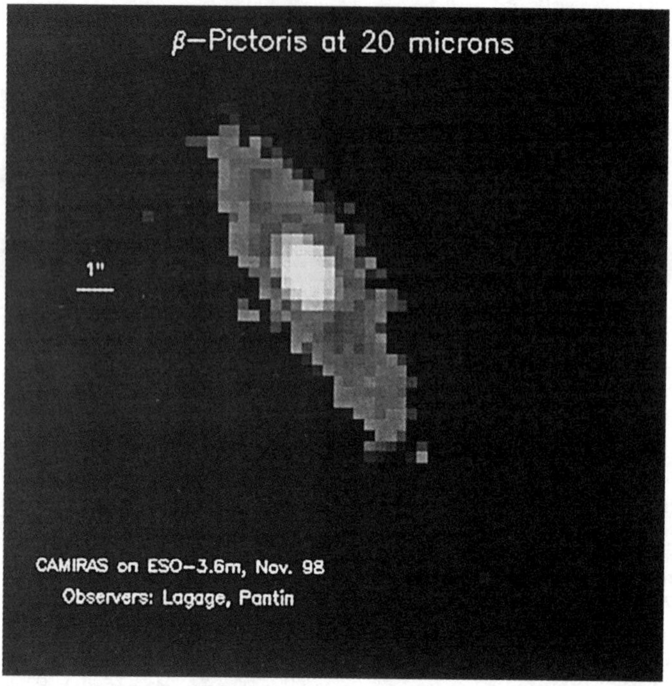

Figure 3. The β Pic dust disk as observed at 20 μm (raw image) with the CAMIRAS Saclay camera mounted on the ESO 3.6 m Telescope in 1998 November. North is up; East is left. The pixel field of view is 0.3 arcsec, which means 5.7 AU at the distance of β Pic (19.3 pc, Crifo et al. 1997). At this wavelength, the star contribution is negligible.

The image was deconvolved from the point spread function (deduced from the observations of the star α Car); then the flux was integrated along the disk thickness. The resulting one-dimensional profile is shown in Figure 4. The profile is compared with that predicted from the density deduced from 10 μm observations (Pantin, Lagage, & Artymowicz 1997). The observed flux is a combination of two dust parameters, temperature and density (integrated along the line of sight). To retrieve the density from the 10 μm observations alone, the temperature had to be modeled assuming a dust grain composition (silicates) and size distribution. The fair agreement found between the predictions and the observations reinforces strongly the conclusions of Pantin et al. (1997).

Another surprising results obtained from the 10 μm observations was the large asymmetry between the northeast and southwest sides of the disk. A large asymmetry is also present in the 20 μm data, as shown in Figure 5. The origin of such a large asymmetry is difficult to assess; one possible explanation may be a dust excess due to the recent collision between two kilometer-sized bodies.

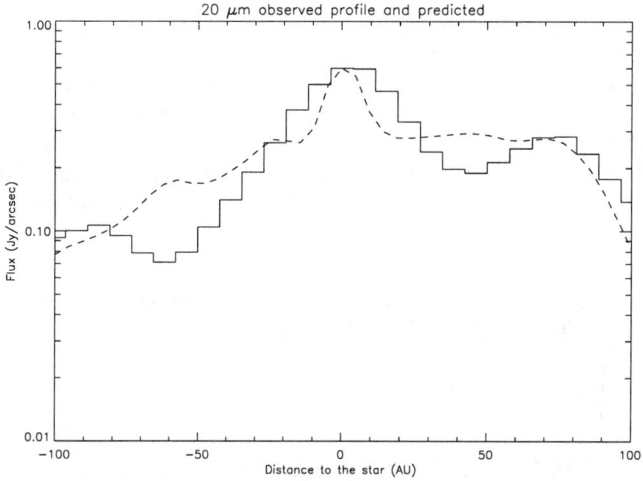

Figure 4. 20 μm flux observed from the β Pictoris dust disk as a function of the distance to the star in astronomical units; (the flux has been integrated over the direction perpendicular to the disk plane, which is seen edge on). A good agreement is found between the observed flux and the predicted flux (*dashed line*); the prediction was made from the dust density profile deduced from previous observations of the disk at 10 μm (Pantin et al. 1997).

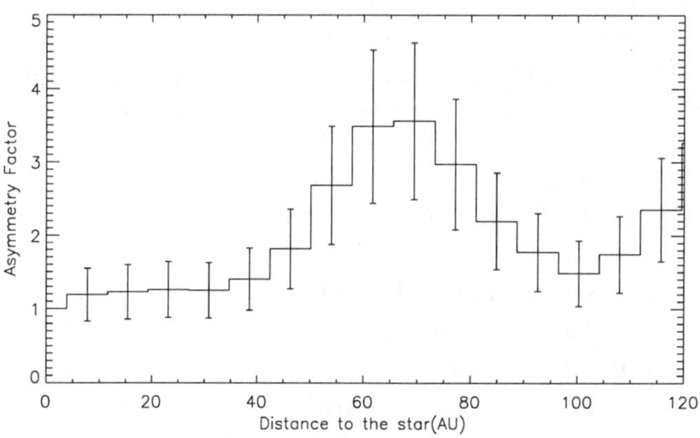

Figure 5. Flux ratio between the southwest and northeast sides of the β Pictoris disk. A large asymmetry is found around 65 AU, confirming previous results obtained at 10 μm.

3. Conclusions and Perspectives

The results deduced from the 20 μm observations presented here (deficit of matter in the inner regions of the disk, large asymmetries) have confirmed that dynamical processes are at work in the β Pictoris system. We can be sure that the next generation of mid-IR cameras and spectrometers to be mounted on large telescopes, such as VISIR at the VLT (Lagage et al. 2000), will scrutinize the β Pictoris disk to further probe the inner parts of the disk, as well as to study several other debris disks. Going from a 3.6 m telescope to an 8 m will substantially increase the angular resolution and dramatically increase the sensitivity.

Acknowledgments. The work reported here is based on observations performed at ESO (proposal numbers 60.D-0699 and 62.L-0246).

References

Artymowicz, P. 1997, Ann. Rev. Earth & Planet. Sci., 25, 175

Augereau, J. C., et al. 1999a, A&A, 348, 557

Augereau, J. C., et al. 1999b, A&A, 350, 51

Aumann, H. H., et al. 1984, ApJ, 278, L23

Crifo, F., et al. 1997, A&A, 320, 29

Dermott, S. F., et al. 1994, Nat, 369, 719

Galdemard, Ph., et al. 1998, SPIE, 3354, 704

Greaves, J., et al. 1998, ApJ, 506, L133

Holland, W. S., et al. 1998, Nat, 392, 788

Jayawardhana, R., et al. 1998, ApJ, 503, L79

Kalas, P., & Jewitt, D. 1995, AJ, 110, 794

Koerner, D. W., et al. 1998, ApJ, 503, L83

Lagage, P. O., et al. 1992, in Progress in Telescope and Instrumentation Technologies, ed. M.-H. Ulrich (Garching: ESO), p. 601

Lagage, P. O., & Pantin, E. 1994, Nat, 369, 628

Lagage, P..O., et al. 2000, SPIE, 4008, in press

Mouillet, D., et al. 1997, MNRAS, 292, 896

Pantin, E., Lagage, P. O., & Artymowicz, P. 1997, A&A, 327, 1123

Plets, H., & Vynckier, C. 1999, A&A, 343, 496

Reach, B., et al. 1995, Nat, 374, 521

Roques, F., et al. 1994, Icarus, 108, 37

Schneider, G., et al. 1999, ApJ, 513, L127

Smith, B. A., & Terrile, R. J. 1984, Sci, 226, 1421

Telesco, C. M., et al. 1998, Nat, 335, 51

Vidal-Madjar, A., Lecavelier des Etangs, A. L., & Ferlet, R. 1998, Planet. Sp. Sci., 46, 629

Weinberger, A. J., et al. 1999, ApJ, 525, L53

Disks, Planetesimals, and Planets
ASP Conference Series, Vol. 219, 2000
F. Garzón, C. Eiroa, D. de Winter, and T. J. Mahoney, eds.

Determination of Spectral Types of the Stars in the EXPORT Sample

Bruno Merín and Benjamín Montesinos[1]

LAEFF-INTA, PO Box 50727, 28080 Madrid, Spain

EXPORT

Abstract. We present the results of the spectral classification of stars included in the EXPORT sample and observed during the 1998–99 International Time campaigns in the Canary Island Observatories. Intermediate-resolution spectra obtained with the Isaac Newton Telescope were used in this work. In addition to the confirmation of some spectral types and changes to previous classifications, we report new results for stars whose spectral type had not been accurately assigned before.

1. Introduction and Motivation for This Work

The determination of spectral types in general, and for the kind of objects we are interested in in particular, is an important and difficult field of work (see Jaschek & Jaschek 1990). Knowledge of all the relevant parameters of stars showing protoplanetary disks, envelopes or any trace of circumstellar material, or attendant planets, is a crucial point for understanding the evolution of these objects and the formation processes of planetary systems.

During the four International Time campaigns, the Isaac Newton Telescope (INT) was devoted to obtain intermediate-resolution spectra of targets of interest for establishing an evolutionary sequence covering objects ranging from those with protoplanetary disks in the pre-main sequence to those similar to β Pic, stars showing the Vega phenomenon and stars with planets already formed.

One of the goals of these observations was the analysis of spectral features and their comparison with standard, normal stars, leading to a revision of the available results on spectral types for some of the targets and to a classification of stars poorly studied previously.

2. Observations

The observations were taken with the INT during the four campaigns awarded to EXPORT in the International Time Programme (1998 May 14–17, July 28–31, and Oct 23–27 and 1999 January 28–31). See Eiroa et al. (this volume, p. 3) for more information on the EXPORT consortium.

---

[1] Also at IAA-CSIC, PO Box 3004, 18080 Granada, Spain.

The Intermediate Dispersion Spectrograph (IDS) was used. The spectral ranges covered were 5854–6728 Å (May run) and 5715–6812 Å (July, October, and January runs), and the resolutions were 0.707 and 0.475 Å/pixel (corresponding to $\lambda/\Delta\lambda = 4500, 6600$ at 6300 Å) respectively. See Mora et al. (this volume, p. 231) for more information on INT EXPORT observations.

3. Tools for Spectral Classification

The spectral classification was performed using spectra of standard stars with well-determined spectral types, taken during the EXPORT runs. A library of high- and intermediate-resolution INT spectra of standard stars (Montes et al. 1997) was also used to determine the spectral type of the cool stars in the sample more accurately.

First of all, an approximate classification was carried out by visual inspection of the spectra. After this, the comparison spectra were broadened using a rotation profile with the $v \sin i$ value of the target star. The formalism given by Gray (1992) was applied. Most of the $v \sin i$ values used here were computed by EXPORT members using high-resolution observations obtained during the four campaigns with the William Herschel Telescope (see Solano et al., this volume, p. 422). The broadened standard spectra were then compared with the target spectra, allowing the determination of the spectral type.

3.1. Cool Stars

For cool stars several lines, in different spectral intervals, have been used, paying attention to the changes in their relative intensity. In Table 1 we list these lines and the interval of spectral types where their intensity changes are significant for discriminating between two spectral types. In addition to these lines, the TiO molecular bands have been also used; these appear around spectral type K7 V and increase their strength as the effective temperature decreases.

3.2. Hot Stars

For hot stars we used the He I lines at 5875 Å and 6678 Å to confirm the B-type assignation when they are clearly photospheric, i.e., when we do not see any changes from one observation to another in the line profiles or equivalent widths. In those cases where variations are apparent, their origin is assumed to be circumstellar and they are therefore not appropriate for classification.

4. Results and Comments

We have very accurately determined the spectral types of 30 stars in the EXPORT sample. Work is in progress to determine the spectral type of the whole EXPORT sample observed with the INT (81 stars).

In Table 2 the spectral types are listed for those stars whose determination is unambiguous and can be considered as highly reliable. Table 3 shows the spectral types for peculiar stars in the EXPORT sample whose assignation could not be accurate either because of the presence of very broadened lines, or because they have many circumstellar and interstellar absorption lines that

Table 1. Lines used for assignation of spectral types for cool stars

Line identification	Interval of spectral types
S II 6102 Å Ca I 6122 Å Fe II 6129 Å	K0 V–K5 V
Fe I 6494 Å Fe I 6546 Å K II 6595 Å	F0 V–F9 V
Na I D_1 D_2 5890,5896 Å Ca I 6162 Å Fe I 6136 Å Fe I 6677 Å Ca I 6494 Å	F0 V–M0 V

mask the photospheric spectrum, or they show a combined spectrum that seems to correspond to a binary system. Sample spectra are shown for HD 199143 in Figure 1.

Regarding the spectral-type suffixes, "e" stands for "emission lines", "v" for "variable spectrum", "p" for "peculiar spectrum" (some non-standard features appear), "n" for very broad or "nebulous" lines, and "m" for "metallic" lines when not expected or stronger than expected.

In the second and third columns we give, for each star, the spectral type and the class of object according to previous studies; the latter can be taken, in general, as a good classification; however, some stars, such as HD 34700, may be a different kind of object following the results of our work. The pre-main-sequence types are abbreviated as follows: TT (T Tauri), HAeBe (Herbig Ae/Be), UXOR (UX Ori-type) and ZAMS (zero age main sequence Herbig Ae/Be). The main sequence types are: MS (main sequence star with planets), Vega (Vega-type), PTT (Post T Tauri) and Ash (A Shell).

Diffuse interstellar bands appear in many spectra and had to be ruled out in performing the visual inspection. These are found at 5780 Å, 5797 Å, 5850 Å, 6196 Å, 6234 Å, 6270 Å, 6284 Å, 6369 Å, 6376 Å, and 6614 Å (Mouton et al. 1999; Schmidt-Kaler 1982).

Almost all the pre-main-sequence stars in the sample present some kind of variations in the spectra. Many show variable Hα in emission. Some of them show some forbidden emission lines ([O I] 6300 Å, [OI] 6363 Å, [S II] 6716 Å, [S II] 6731 Å, [N II] 6583 Å, [N II] 6548 Å and [N II] 6527 Å; see Osterbrock 1989) that may be related to winds.

The stars HD 34700 and HD 233517 show a deeper than expected Li I 6708 Å line in their spectra although they where supposed to be Vega-type main-sequence stars before this study. Variable lines of He I 5877 Å, He I 6678 Å and Na I D_1 D_2 5890,5896 Å that may be produced in the circumstellar disk or

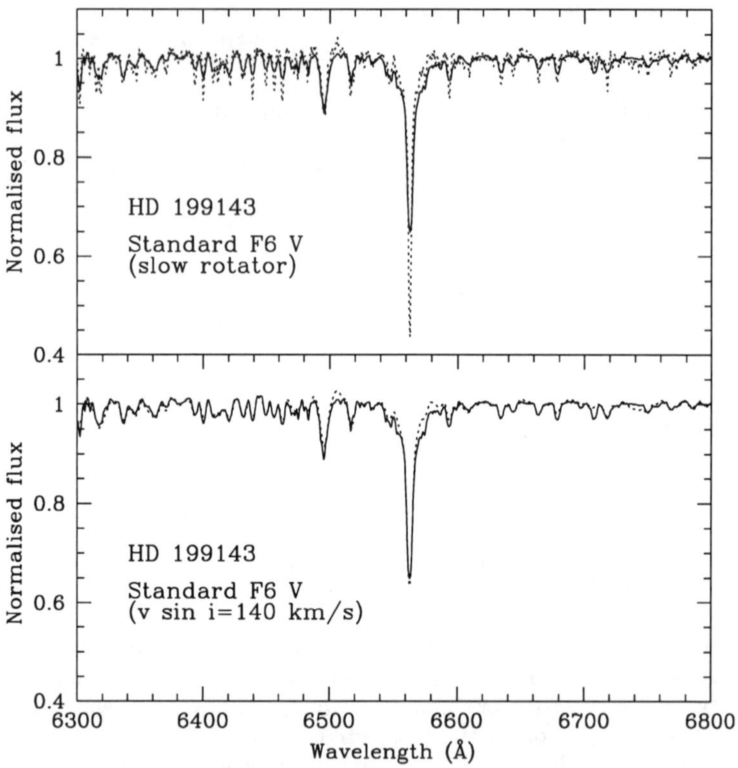

Figure 1. An example of the spectral type determination for the star HD 199143. *Top panel:* The spectrum of the target star (*solid line*) is compared with that of HR 4606, an F6 V standard star with slow rotation (*dotted line*). *Bottom panel:* The standard spectrum has been broadened with a $v \sin i$ = 140 km s^{-1} profile; the agreement is remarkable.

envelope appear in many spectra. In some cases these lines show very peculiar and variable profiles.

The method used here to determine the spectral types of the stars in the EXPORT sample leads to very acceptable results. One important fact we would like to note is that the presence of a number of interstellar lines, and also that variable circumstellar and shell lines in the spectra can lead to erroneous classifications if a small number of observations is used. Therefore, for these kinds of stars, it is highly desirable to work with as many spectra as possible in order to disentangle the nature of each line. In future work we will carry out a deter-

Table 2. Results for the assignation of spectral types

Star	Previous determinations	Type	This study
HD 23362	K2 V, K2	Vega	K5 IIIm
HD 34700	G0 V	ZAMS	G0 IVe
HD 34282	A2 Ve+sh, A0e, A0	HAeBe	A3 Vne
HD 109085	F2 V, F3 V, F2 III-IV	Vega	F2 V
HD 142764	K5	Vega	K7 V
HD 144432	A7 Ve, A9/F0 V, A7 Ve	ZAMS	A9 IVev
HD 199143	F6 V	ZAMS	F6 V
HD 233517	K2, A2	Vega	K0 V
HR 10	A2 V–A5 V, A6 Vn	Ash	A4 Vn
HR 4757 B	K0 V, K2 Ve, K1 V	PTT	K2 V
HR 5422 A	A0 V, B9.2p , B9 Vp	PTT	A0 V
HR 5422 B	K1 V	PTT	G5 V
HR 9043	A5 V, A3 Vn	Vega	A5 Vn
BM And	K5 V	TT	K5 Ve
BH Cep	A/F5 Vea, F5 IVvar	HAeBe	F5 IIIev
24 CVn	A5 V, A4 V, A5.5 V	Ash	A4 V
KK Oph	B3, B,Ae, A5 Ve, A6	UXOR	A8 Ve
T Ori	A3/4e, A3, B9, A5 e	UXOR	A3 IVev
BF Ori	A5/6 IIIe+sh, A5e, A6e	UXOR	A2 IVev
CO Ori	F9/G Ve, F9: e, F8	UXOR	F7 Vev
NV Ori	F0/8 IVe, F4/8 III,V	HAeBe	F6 IIIev
RY Ori	F6/Gep, F8:pe	TT	F6 Ve
UX Ori	A3e, A2/3 IIIe, A1-3 IIIe	UXOR	A4 IVe
51 Peg	G5 V, G2 V, G2.5 V	MS	G2.5 V
DK Tau	M0 V:e, K7 V, K7	UXOR	K5 Ve
RY Tau	F8 V:e, K1 IV, G5e, K7	UXOR	F8 IIIev
PX Vul	F0 V:e, F0, F5	HAeBe	F3 Ve
WW Vul	A3e, A0/3 Ve, A0, A1e	UXOR	A0 Ve
LkHα 200	K1 V, Ke, dK0	TT	K3 Ve
LkHα 234	B5.7e, B5/7, B9/A0e, A7	HAeBe	B5 Vev

mination of the spectral type in the ultraviolet range using observations taken with the *International Ultraviolet Explorer* (*IUE*) by fitting Kurucz atmospheric models. Also, a determination of the absolute magnitude using the results from the *Hipparcos* mission will be very valuable to bracket more accurately the luminosity class of each object.

References

Gray, D. 1992, The Observation and Analysis of Stellar Photospheres (Cambridge: Cambridge University Press)

Table 3. Results for the assignation of spectral types: peculiar stars

Star	Previous determinations	Type	This study
HD 58647	A0 IVe, B9 II–IIIe, Be3	ZAMS	B9 IVep
HD 141569	A0 Ve, B9.5 Ve, B9e	ZAMS	A0 Vev
HD 163296	A3 Vep+sh, A1 V, A0-A2	ZAMS	A1 Vepv
HR 2174	A3 Vn, A1 IV–sh, A3 V	Vega	A2 Vnv
VX Cas	A0/3e, A0, A3, A0ea	HAeBe	A0 Vep
VY Mon	B9/A0e, B8, B-A	HAeBe	A5 Vep
HK Ori	A5, A4, G:ep, B8/A4ep	UXOR	G1 Vepv
VV Ser	A2e, B1–3e:, B1/3e/A2:	HAeBe	A0 Vev
RR Tau	F:e, A2 II–IIIe, B8ea	UXOR	A3 IVev

Jaschek, C., & Jaschek, M. 1990, The Classification of Stars (Cambridge: Cambridge University Press)

Montes, D., et al. 1997, A&AS, 123, 473

Mouton, C., et al. 1999, A&A, 351, 680

Osterbrock, D. 1989, Astrophysics of Gaseous Nebulae and Active Galactic Nuclei (Mill Valley, Ca.: University Science Books)

Schmidt-Kaler, Th. 1982, Landolt-Börnstein New Series, Vol. 2B, Chapt. Physical parameters of the stars (New York: Springer)

Disks, Planetesimals, and Planets
ASP Conference Series, Vol. 219, 2000
F. Garzón, C. Eiroa, D. de Winter, and T. J. Mahoney, eds.

Spectroscopic Observations of BD +31°643, BH Cep, LkHα 234, and LkHα 262

L. F. Miranda
Instituto de Astrofísica de Andalucía, CSIC, Ap. Correos 3004, E-18080, Granada, Spain

C. Eiroa
Departamento de Física Teórica, Facultad de Ciencias, Universidad Autónoma de Madrid, E-28049 Cantoblanco, Madrid, Spain

F. Garzón
Instituto de Astrofísica de Canarias, E-38205 La Laguna, Tenerife, Spain

EXPORT

Abstract. We present spectroscopy of BD +31°643, BH Cep, LkHα 234, and LkHα 262 obtained as a part of the 1998 La Palma International Time Programme on extrasolar planetary systems. We analyze line profiles (mainly Hα, Na I and He I), equivalent widths, the presence of different velocity components, and the spectroscopic variability.

1. Introduction

The La Palma International Time Proposal 1998 has been devoted to the study of planetary systems through the project "Planetary Systems: their formation and properties". EXPORT (EXo-Planetary Observational Research Team) has observed a large number of objects including T Tauri stars, HAeBe, Vega-like stars, and stars for which planets have been identified. In the case of pre-main-sequence (PMS) stars and some main-sequence (MS) stars many observations have demonstrated the existence of circumstellar gas and dust disks (e.g., Beckwith & Sargent 1996; Aumann et al. 1984). In some cases, the presence of solid bodies/planetesimals in both PMS and MS stars has been inferred through transient blueshifted and redshifted absorption components (BACs and RACS) in some metallic lines (Grinin et al. 1994; Crawford, Beust, & Lagrange 1998; Beust et al. 1998). It is thought that planetary systems originate from these disks (e.g., Beckwith & Sargent 1996). Therefore, PMS and MS stars may reveal key information about the properties, formation, and evolution of protoplanetary disks.

Four stars observed by EXPORT have been selected to be analyzed in this contribution: BD +31°643, BH Cep, LkHα 234, and LkHα 262. BD +31°643 is

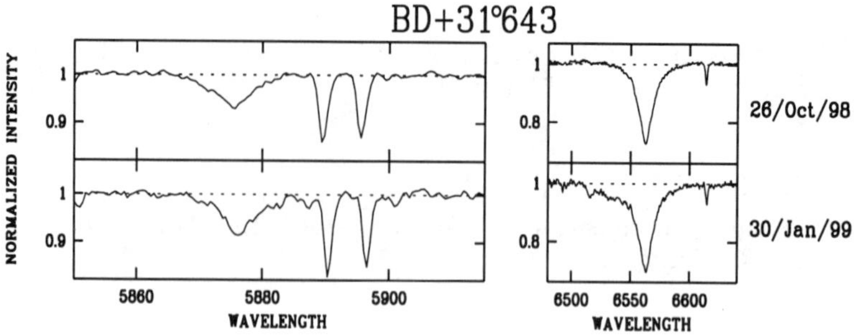

Figure 1. INT spectra of BD +31°643 around the He I, Na I and Hα lines as observed on 1998 Oct 26 and 1999 Jan 30.

a B5 Vega-type star, which is surrounded by a dust disk similar to that of β Pic (Kalas & Jewitt 1997). BH Cep is an F5 IV T Tauri star known by its rapid variations. LkHα 234 is a Herbig Ae/Be star classified as B5–B7 and associated with the NGC 7129 reflection nebula. LkHα 262 is a classical T Tauri star of spectral type around K. These four stars cover a range of spectral types, masses, and different evolutionary status. Thus, it is interesting to analyze them in order to compare the observed phenomenology.

2. Observations

For BD +31°643 the following data have been obtained: INT spectra (1998 October 24–28, 1999 January 29–31), and a WHT–UES spectrum (1998 October 26). For BH Cep we obtained: INT spectra (1998 May 15–17, July 28–31, October 24–28, 1999 January 30), and WHT–UES spectra (1998 May 17, July 31, October 25–26). In the case of LkHα 234 the data are INT spectra (1998 July 28–31, October 24–28). For LkHα 262 we have only INT spectra (1998 October 24,25,28, July 31). A description of the instrumental setup and relevant information on the data can be found in the EXPORT contribution by Mora et al. in this volume (p. 231).

3. BD +31°643

Figure 1 shows the INT spectra around the He I, Na I, and Hα lines obtained on 1998 Oct 26 and 1999 Jan 30. Some variability in the He I profile and equivalent width is observed. The Hα line does not present variations except for an extended blue absorption wing on 1999 Jan 30. The October WHT spectrum has strong Na I absorptions which are interstellar. In addition, weak absorption features might be present around the Na I lines on 1999 Jan 30, coinciding with the presence of the blue Hα wing. New observations are necessary to investigate these weak features in more detail and whether they co-vary with Hα. Strong diffuse interstellar bands are observed at 5780, 5797, 5850,6196, 6270, 6284, 6369, and 6614 Å.

4. BH Cep

Figure 2 shows the He I, Na I, and Hα profiles as observed in the INT spectra. Dramatic changes over timescales of hours, days, and months are observed. However, the changes follow a well-defined sequence during each month. In May and July, the profile is double-peaked, with the blue peak stronger than the red one. The intensity of the peaks changes, being more noticeable in the red component. The heliocentric radial velocity of the blue peak decreases systematically from 1998 May 15 to 1998 Jul 29, with daily values of -220, -198, -160, -150, and -130 km s^{-1}. In October, the Hα profile is completely different from that observed in the previous months. Even though strong changes in intensity are observed, the same absorption and emission components are present during this month. In 1999 January, the profile is similar to that observed on 1998 May 15. The relative intensity of the blue and red emission peaks changes through the four months. In May, July, and January, the blue peak is stronger than the red one while in October, the red peak is the stronger one.

The Na I lines present both blue and red absorption components (BACs and RACs) as well as central emissions, which can be better recognized in the high-resolution WHT spectra (not shown here). The velocities are relatively low with values between $+25$ and $+50$ km s^{-1} for the RACs and -20 and -75 for the BACs. Between 1998 May 15 and 1998 Jul 28, the INT spectra also show a faint blue emission component. Its radial velocity coincides each day with that of the blue peak of the Hα emission (see above), suggesting a relationship between both emissions. A faint blueshifted Na I emission component is also observed during October. The He I line also presents blue emission components and RACs, which appear to be related to the presence of the same components in the Na I line.

5. LkHα 234

The spectra around the He I, Na I, and Hα lines are shown in Figure 3. We note that He I lines are not always observed towards this star. From previous data, a spectral type B8–A3 is obtained (de Winter et al. 2000, in preparation), whereas the present data indicate a spectral type B5 (Merín et al., this volume, p. 399). In our spectra the [O I] 6300,6363 and [S II] 6716,6731 are also exceptionally observed. The Hα profile shows two emission peaks, the red one stronger than the blue one, separated by an absorption reversal, and is the profile always observed for this star. The relative intensity of the peaks changes over timescales of days. However, the radial velocities are rather constant. The red emission peak is observed at a heliocentric radial velocity of $\simeq +50$ km s^{-1} whereas the absorption reversal appears at $\simeq -45$ km s^{-1}. The only velocity variations are observed in the blue emission peak, which is observed at -145 km s^{-1} in July and at -125 to -100 km s^{-1} in October.

The Na I lines exhibit complex and highly variable profiles with absorption and emission components. RACs are clearly seen on 1998 Jul 28 and, probably, on 1998 Jul 31 at a heliocentric velocity of $\simeq +160$ km s^{-1}. Na I emission components are observed in almost all spectra, except in that on 1998 Jul 30. The Na I absorptions are probably interstellar and their variability may be

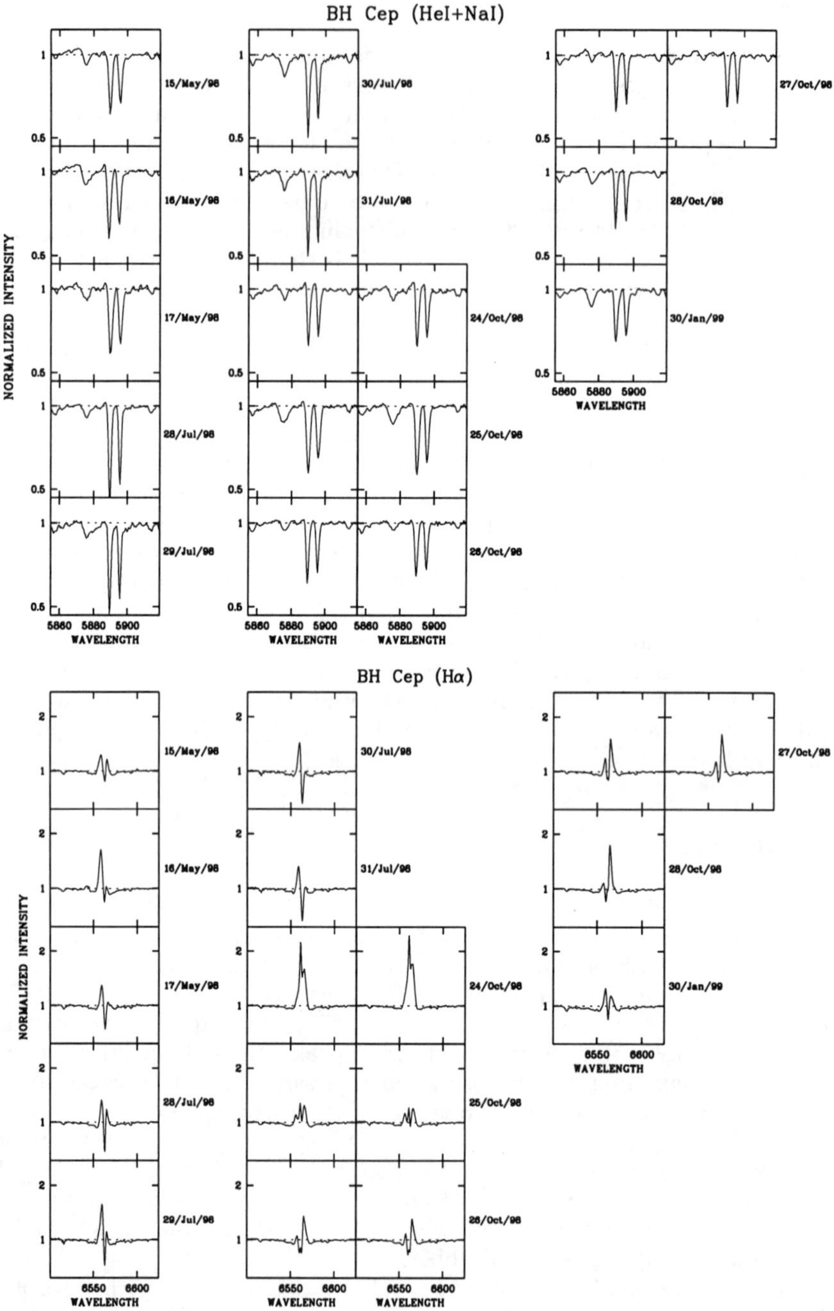

Figure 2. INT spectra of BH Cep around the He I and Na I (*top*) and Hα (*bottom*) lines.

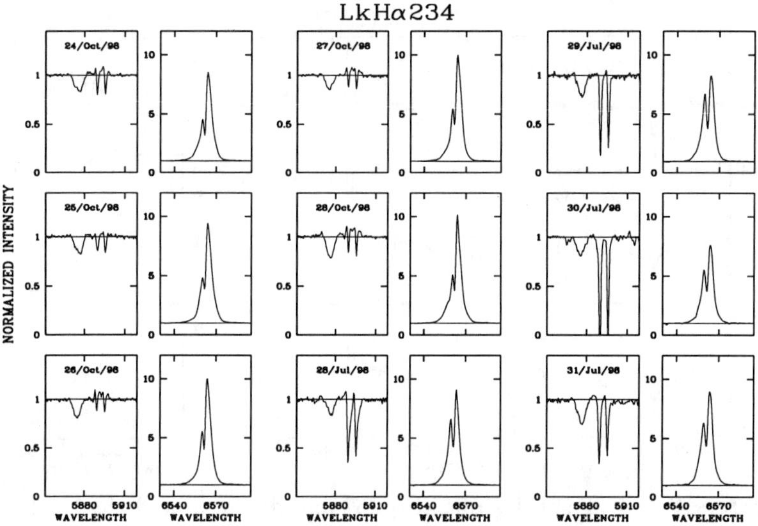

Figure 3. INT spectra of LkHα 234 around the He I, Na I, and Hα lines.

related to that of the emission components. Although the intensity of these emissions varies, their velocity appears approximately constant at $\simeq -70$ and $+95$ km s^{-1}. A correlation between the variations in Na I and Hα is not observed. Variability of the He I line profile and equivalent width is also observed, as well as the possible presence of BACs and RACs. However, the variations do not appear to be correlated with those observed in the Na I and Hα lines.

6. LkHα 262

Figure 4 shows the spectra of LkHα 262. The spectral type changes from about K5 on 1998 Oct 24 to early M on 1998 Oct 25 and 28. Hα, He I 5785, and [O I] 6300,6363 are in emission, as well as the He I 6678 line observed on 1998 Oct 24. The Hα emission is double-peaked, and the equivalent width and radial velocity of the emission peaks and absorption reversal vary. The blue peak is stronger than the red one on 1998 Oct 24 and 28, whereas the contrary is observed on 1998 Oct 25. In the Na I lines, RACs could be present on 1998 Oct 24–25, although the absorption bands prevent definitive conclusions from being drawn.

7. Conclusions

The phenomenology observed in these four stars is varied. Evidence for blue- and/or redshifted transient absorption components is found in BH Cep, LkHα 234, and perhaps in LkHα 262. Similar events in UX Ori and β Pic have been interpreted in terms of solid bodies, although they could also reflect disk inhomogeneities. In our spectra, BD $+31°643$ does not show such clear evidence for blue and red absorption components as those observed in β Pic, which could

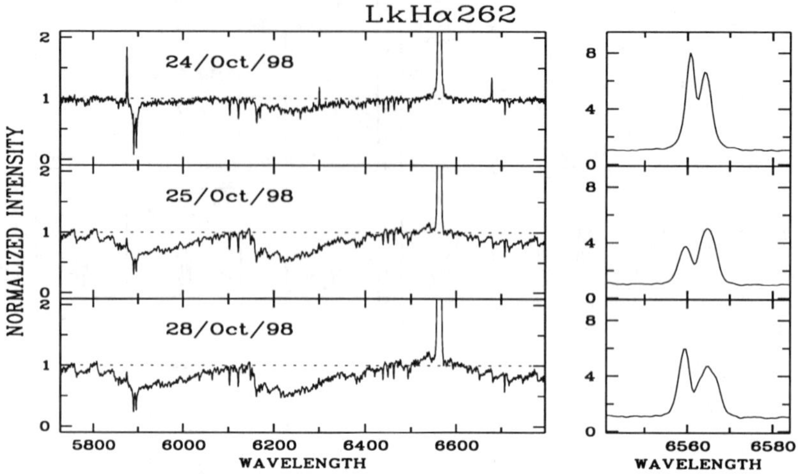

Figure 4. INT spectra of LkHα 262 and detail around the Hα line.

indicate the presence of solid bodies in its disk. In BH Cep, the relationships between some emission components in Na I and Hα suggest that the Hα line may be composed of "circumstellar" and "stellar" contributions. On the contrary, in the Herbig Ae/Be star LkHα 234, Hα and Na I apparently trace different regions with Hα probably related to a stellar wind and Na I to the circumstellar environment.

References

Aumann, H. H., et al., 1984, ApJ, 278, L23

Beckwith, S. V. W., & Sargent, A. I. 1996, Nat, 383, 139

Beust, H., Lagrange, A.-M., Crawford, I. A., Goudard, C., Spyromilio, J., & Vidal-Madjar, A. 1998, A&A, 338, 1015

Crawford, I. A., Beust, H., & Lagrange, A.-M. 1998, MNRAS, 294, L31

Grinin, V. P., The, P. S., de Winter, D., Giampapa, M., Rostopchina, A. N., Tambovtseva, L. V., & van den Ancker, M. E. 1994, A&A, 292, 165

Kalas, P., & Jewitt, D. 1997, Nat, 386, 52

Disks, Planetesimals, and Planets
ASP Conference Series, Vol. 219, 2000
F. Garzón, C. Eiroa, D. de Winter, and T. J. Mahoney, eds.

The Behavior of Vega-Type Stars in the EXPORT Sample

A. Mora, C. Eiroa, and J. Palacios

Dpto. Física Teórica, Universidad Autónoma de Madrid, E-28049 Madrid, Spain

D. de Winter

TNO/TPD-Space Instrumentation, Stieltjesweg 1, PO Box 155, 2600 AD Delft, The Netherlands

R. Ferlet

CNRS, Institute d'Astrophysique de Paris, 98bis Bd. Arago, 75014 Paris, France

F. Garzón

Instituto de Astrofísica de Canarias, E-38205 La Laguna, Tenerife, Spain

R. D. Oudmaijer

The Department of Physics and Astronomy, E C Stoner Building, Leeds, LS2 9JT, W Yorkshire, UK

EXPORT

Abstract. We present a brief summary of the EXPORT observations of a sample of 30 Vega-type systems and post-T Tauri (PTT) stars. The spectroscopic and photopolarimetric data of PTT stars show no evidence of significant amounts of gas and dust in the environment close to these stars, even in those showing a far-infrared excess. On the other hand, our data indicate a rich gaseous and dusty environment around Vega-type stars. Spectroscopic events similar to those observed in β Pic have been detected towards some of the Vega-type stars. In addition, many of the stars are polarized and show colors suggesting the presence of dust very close to the star. At least two of the observed Vega-type stars possibly are pre-main-sequence objects.

1. Introduction

Vega-type objects are main-sequence stars still surrounded by circumstellar disks (Backman & Paresce 1993). The detection of strong variable absorption components in the spectral lines of several elements (e.g., Ca II towards β Pic, the prototype of this class of object) has been explained as due to the evaporation of large solid bodies, i.e., cometesimals, falling onto the star (e.g., Beust et al. 1998). Some shell stars have been found to share these spectral characteristics with β Pic. Thus, the Vega-type/β Pic disks seem to be sites of planet formation, and they even might host planets in some cases. In addition, post-T Tauri stars (PTTs) are young objects very close to the main sequence; in some cases PTTs show far-infrared excesses revealing the presence of dust in the circumstellar environment (Ray et al. 1995). Both types of objects probably represent intermediate scenarios on the road to the formation of planets.

In the EXPORT project we selected a sample of Vega-type stars and PTT candidates to look for optical spectroscopic evidence of circumstellar gas. In addition, optical photopolarimetry and near-IR photometry of these stars are helpful to investigate the characteristics of the circumstellar dust. In this contribution we present the sample of stars we observed and give a short summary of the basic results. A detailed analysis of the whole dataset and of individual objects will be published elsewhere.

2. The EXPORT Sample

Table 1 presents the observed stars. Information on instruments and telescopes are given in other EXPORT contributions in this volume. The type of object (Vega or PTT) and spectral type are given in columns 2 and 3. PTTs are from the Lindroos binary star list, in which the secondary component is thought to be in the post-T Tauri phase, while the primary is an early type star (Lindroos 1986). In some of the PTT stars, we have observed the bright companion only (note that this companion also is referred to as PTT in the table). Columns 4–7 indicate the telescopes used. A blank in these columns means that the object was not observed. A + sign in the NOT and CST columns means that the stars are polarized or reddened or, in the cases of HD 34700 and 51 Oph, have a near-IR excess; a – sign means that the stars are unpolarized or have near-IR colors in accordance with their spectral type. The INT and WHT columns provide information after a quick look on some selected lines of the stellar spectra. A – sign in these columns means that the spectrum of the stars seems photospheric without any obvious non-stellar characteristic. Hα means that the line is observed in emission (normally double-peaked), Li I means that the 6708 Å Li I line is present in the spectrum. In the WHT column we provide information on the Ca II H and K lines only. The symbols mean the following: "cs": "circumstellar", "n": "narrow", "c": "chromospheric", "e": "emission", and "w": "wide".

3. Results

Our data allow us to draw some general preliminary conclusions, although definitive statements must wait for a detailed analysis of the data.

PTT stars in our sample, i.e. the low-mass PTT stars and/or their bright companions, do not show any evidence of the presence of significant amounts of gas or dust in their environment, even though some of them are *IRAS* sources. We do not find any obvious "peculiarity" in any of the PTT stellar spectra. The

Table 1. Observed stars

Object	Class	Sp. T.	INT	WHT	NOT	CST
HR 10	Vega	A4 V	–	Ca II: cs	–	–
HR 26	PTT	B9 Vn	–	–	–	–
HR 26b	PTT	G5	Li I			
HR 419	PTT	B9 V		Ca II: n	–	–
49 Cet	Vega	A3 V	–	Ca II: n	–	–
HR 836	PTT	B6 V				–
BD +31°643	Vega	B5 V	–	–	+	+
HD 23362	Vega	K7	–	–	+	+
HD 23680	Vega	G5	–		+	+
HR 1369	PTT	B9 V			–	–
HD 34700	Vega	G0 III	Hα, Li I	Ca II: n	+	+
HR 1847	Vega	B7	–		+	
HR 1847b			Hα		+	
HR 2174	Vega	A2 Vn	–	Ca II: n	–	–
HR 2147b					–	
HD 233517	Vega	K2 V	Li I	Ca II: c	+	
HD 233517b			–			
17 Sex	Ash	A1	–	Ca II: cs?	–	
HR 4757	PTT	B9.5 V	–	–	–	–
HR 4757b	PTT	K2 V	–	Ca II: e	–	–
HD 109085	Vega	F2 V	Li I weak?	–	–	
24 CVn	Ash	A5	–	–	–	
HD 123160	Vega	K5	–	Ca II: c	+?	
λ Boo	Vega	A0 Vpsh	–	–	–	–
HR 5422	PTT	A0 V	–	Ca II: n	–	–
HR 5422b	PTT	G5 V	–		–	
HD 142764	Vega	K7 V	–	Ca II: c	+	+
51 Oph	Vega	A0 Ve	Hα	Ca II: n	+	+
HD 218396	Vega	A5 V				–
HR 9043	Vega	A5 V	Na: w	Ca II: n	–	–

stars are unpolarized, there is no photometric variability, and their colors are normal, in accordance with their spectral types. Thus, we find no indication of the existence of hypothetical protoplanetary disks around these objects.

Very different is the situation concerning the Vega-type stars of our sample. Most of these stars show Ca II lines which are not purely photospheric. Narrow Ca II absorptions are seen superimposed on the photospheric line, which could originate in the interstellar medium, in a shell, or in a circumstellar gas disk. Particularly interesting is the case of HR 10, where we detected a transient redshifted absorption component superimposed on a "main absorption", also with a velocity change (Figure 1). These data confirm earlier results on this star (Welsh et al. 1998) with respect to the similarity of HR 10 with β Pic.

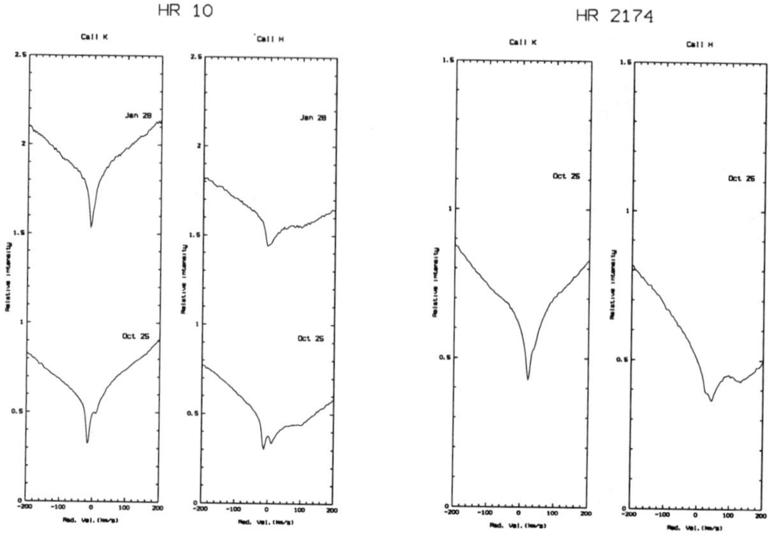

Figure 1. WHT high-resolution Ca II H and K lines of HR 10 (*left panel*) and HR 2174 (*right panel*). Flux units are arbitrary. HR 10 shows a redshifted absorption component in the Oct 25 spectrum which is not seen in the Jan 28 spectrum. In the case of HR 2174, the main absorption in the K line is centered on a velocity of 21 km s^{-1}, and a weak redshifted absorption at velocity $\sim$ 43 km s^{-1} seems to be present. This absorption at 43 km s^{-1} is stronger in the Ca II H than in the K line

Absorption components are also present in the only WHT spectrum we obtained of HR 2174 (Figure 1). Undoubtedly, some of the Vega-type stars deserve further observations aiming at velocity structures in some lines of their spectra.

Some of the observed Vega-type stars show the Li I 6708 Å absorption line. These stars have large *IRAS* excesses. HD 34700, the only one which also was observed in the near IR, shows a weak $H - K$ color excess. It is likely that they are pre-main-sequence stars.

Around 50% of the Vega stars in the sample are polarized. Polarization variability is not obvious. The polarization degree clearly exceeds 1% in several cases. Since the distance to these stars is relatively small, it is likely that at least part of the polarization must arise in the direct vicinity of the stars. The observed stars do not show any clear sign of photometric variability. Four out of 12 Vega stars observed in the near IR are reddened and two have a weak near-IR excess. Since the extinction deduced from the IR colors (and also from the optical ones) is usually around or larger than $A_V \sim 2$ mag and the stars are relatively close, it is plausible to assume that part of the reddening is circumstellar, as also suggested by the polarimetry. In fact, when we have both polarimetric and near-IR data there seems to be a correlation between the polarimetry and the near-IR behavior, in the sense that stars with colors departing from the intrinsic stellar ones are polarized. On the other hand, there is no apparent correlation with the amount of *IRAS* excess, neither with a "peculiar" optical stellar spectrum.

4. Conclusions

Our data show that the hypothetical protoplanetary disks which might exist around PTT stars, as suggested by *IRAS* fluxes, are not prominent enough to show up at optical and near-IR wavelengths, at least in the stars of our sample.

On the other hand, circumstellar gas and dust seem to be detected at optical and near-IR wavelengths in a large fraction of Vega-type stars. This conclusion is based on the spectroscopic and photopolarimetric results. Further studies are needed to elucidate in which cases these signatures come from the presence of gas and dust around the star, or, alternatively, from the interstellar medium. In addition, more information is needed to try to find out the geometrical distribution of the gas and dust when they are circumstellar.

Acknowledgments. The group of the Universidad Autónoma de Madrid is partly supported by Spanish grant ESP 98-1339-C02.

References

Backmann, D. E., & Paresce, F. 1993, in Protostars and Planets III, ed. E. H. Levy, J. I. Lunine, & M. S. Matthews (Tucson: Univ. Arizona Press), 1253

Beust, H., Lagrange, A.-M., Crawford, I. A., Goudard, C., Spyromilio, J., & Vidal-Madjar, A. 1998, A&A, 338, 1015

Lindroos, K. P. 1986, A&A, 156, 223

Ray, T. P., Sargent, A. I., Beckwith, S. V. W., Koresko, C., & Kelly, P. 1995, ApJ, 440, L89

Welsh, B. Y., Craig, N., Crawford, I. A., & Price, R.J. 1998, A&A, 338, 674

Disks, Planetesimals, and Planets
ASP Conference Series, Vol. 219, 2000
F. Garzón, C. Eiroa, D. de Winter, and T. J. Mahoney, eds.

Dust and Disks of Ae/Be Stars

Elisha Polomski, Charles M. Telesco, and Robert K. Piña

Astronomy Department, University of Florida, Gainesville, FL 32611-2055, USA

Abstract. We have conducted a program of mid-IR imaging observations of the intermediate-mass protoplanetary disk candidates, the Ae/Be stars, using the University of Florida mid-IR imager OSCIR at the Infrared Telescope Facility and the Keck Observatory. Our observations have shown that $\sim$ 50% of the mid-IR emission of these objects is unrelated to circumstellar disks. We have discovered large-scale extended emission and multiple companions around many Ae/Be stars. Temperature maps of the extended emission support the hypothesis that some of the mid-IR emission originates in non-equilibrium emission from small grains. In addition, we have constructed revised SEDs which exclude emission from non-disk sources. These SEDs exhibit altered slopes that are not within the ranges seen in simple circumstellar disk models. We present images of several of the objects we have surveyed and discuss their properties. The source of the IR excess of these stars has direct consequences for the presence of circumstellar disks and, ultimately, the formation of planets around intermediate-mass stars.

1. Introduction

Herbig Ae/Be stars were first recognized as intermediate-mass (2–10 $M_{\odot}$) pre-main-sequence stars by Herbig (1960), who found that these stars had many properties in common with the pre-main-sequence T Tauri stars, but were more luminous and of an earlier spectral type. He classified the Herbig stars using the following criteria: stars are of spectral type A or earlier, exhibit emission lines, are located in an obscured region, and illuminate bright nebulosity in their vicinity. More recent studies also characterize these objects as having a large excess of IR flux that is attributed to emission from dust within circumstellar disks (Hillenbrand et al. 1992). Disks have been directly imaged around both T Tauri and Vega-like stars (McCaughrean & O'Dell 1996; Koerner et al. 1998; Fisher et al. 2000). Since disks are believed to be the sites of planet formation, it is essential to know whether Ae/Be stars possess circumstellar disks and can therefore potentially form planets. However, most evidence for disks around Herbig Ae/Be stars is indirect. Blueshifted absorption lines have been interpreted as evidence of accretion disks, polarization vectors imply disk-like geometries, mass estimates from millimeter observations also imply disk geometries, and simple disk models have been fitted to low-resolution photometry (Corcoran & Ray 1997; Bastien & Ménard 1990; Lada & Adams 1992; Hillenbrand et al. 1992).

Other models that may account for the IR excess of Ae/Be stars include spherical dust clouds, embedded companions, or emission from small grains (Hartmann, Kenyon, & Calvet 1993; Natta, Prusti, & Krugel 1993; Li et al. 1994; Brooke, Tokunaga, & Strom 1993). The large beam sizes and low spatial resolutions of previous infrared detectors hindered the direct detection of any spatial structure in the environments of these stars. In this paper we present new high angular resolution mid-IR images of four Herbig stars which illustrate how phenomena other than disks can account for the mid-IR emission.

2. Mid-Infrared Observations

We present mid-IR observations of HD 259431, AS 310, LkHα 234, and MWC 1080. All objects are Hillenbrand group I Ae/Be stars whose SEDs have been previously modeled as circumstellar disks (Hillenbrand et al. 1992; Pezzuto, Strafella, & Lorenzetti 1997). Observations were made between 1996 and 1999 at the IRTF and Keck Observatories with the University of Florida mid-IR camera, OSCIR. We conducted observations in the narrow ($\Delta\lambda = 1$ μm) 4.8, 7.9, 8.8, 9.8, 10.3, 11.7, 12.5, and 18.2 μm filters, as well as the broad ($\Delta\lambda = 5.23$ μm) N filter. We find large-scale dusty envelopes and embedded companions around these stars (Figures 1–3). Small-aperture photometry from the IRTF and higher-resolution data from Keck of MWC 1080, indicate that $\sim$ 50% of the mid-IR emission arises in sources other than circumstellar disks (Table 1).

Hillenbrand et al. defined a 3–10 μm flux index $\alpha = \log[(\lambda F_{\lambda \mathrm{L}})(\lambda F_{\lambda \mathrm{N}})]$ to categorize the SEDs of Ae/Be stars. Circumstellar disk models have mid-IR SEDs of the form $\lambda \mathrm{F}_\lambda \propto \lambda^{-4/3}$, which leads to $\alpha \approx 0.64$. We measured fluxes in 4 and 5$''$ apertures for our 4–12.5 μm and 18 μm observations respectively and constructed revised SEDs. We found that all four stars exhibit α indices less than that expected from a classical geometrically thin, optically thick disk. We present below the highlights of our analysis for each object.

2.1. HD 259431

HD 259431 appears to be a single star with a large cometary-shaped mid-IR dust envelope. Some near-IR observations imply that this object may have one or more companions which are not visible in the mid IR (Ciardi et al. 1999; Corporan 1998; Vieira & Cunha 1994). The mid-IR SED of this object shows an emission feature between 8.8 and 9.8 μm. *IRAS* spectra also show an emission feature in this regime, as well strong emission from polycyclic aromatic hydrocarbon grains (PAHs) at 11.7 μm. We dereddened the SED and derived an α index of 0.29 from the 3.5 and 10.3 μm fluxes. The index for a circumstellar disk is expected to be $\alpha \sim 0.63$ for the same bandpasses. The flatter index suggests that this object may be more like Hillenbrand group II objects whose infrared emission arises from a dust shell. Alternatively, flared disk models are expected to have flatter indices since more surface area of the disk can be heated by the central star (Kenyon & Hartmann 1987). Models with spectral shapes of $\lambda \mathrm{F}_\lambda \propto \lambda^{-2/3}$ have $\alpha \sim 0.32$, quite close to the value we measure for HD 259431.

2.2. AS 310

AS 310 is part of a cluster of young stars (Ageorges et al. 1997; Testi, Palla, & Natta 1998). Our observations show a large dust nebula, an embedded companion 4.7″ north of the primary, and a dust condensation to the NW. We identify the northern companion with Ageorges et al.'s (1997) near-IR source AS 310-C, and the NW condensation may be associated with a double-lobed radio source (Cohen, Bieging, & Schwartz 1982). Our photometry indicates that 75% of the 10.8 μm emission originates in the dust shell and embedded companions. Our revised SED of the primary star has an α index of –0.67 and shows a silicate emission feature also noted by previous authors (Cohen 1974). We calculated a 7.9/18 μm color temperature map of AS 310 assuming $\tau \propto 1/\lambda$ emissivity. This map shows an inverse temperature gradient; lowest temperatures occur at the position of the primary, highest temperatures occur far out in the dust nebula. We suggest that this is due to the influence of strong emission at 7.7 μm from very small grains such as PAHs. These grains may be destroyed by intense radiation near the star, but survive further out in the nebula where they re-radiate absorbed stellar radiation independently of their distance. *IRAS* spectra and 3 μm spectroscopy show emission features that have been attributed to PAH grains, and some observations have indicated that the 3.29 μm PAH feature is extended around AS 310 (Brooke et al. 1993; Volk et al. 1991). Our temperature map shows that PAH emission can have a large contribution to the mid-IR excess and may be the first indication of extended 7.7 μm emission around this object.

2.3. LkHα 234

LkHα 234 has been studied previously in the mid-IR by Cabrit et al. (1997). Here we show higher-resolution observations of disk-like structure associated with the NW companion and the detection of an additional mid-IR companion which is coincident with an H_2O maser to the SW of the primary. The NW companion dominates the SED at $\lambda > 12$ μm and exhibits a silicate absorption feature. This object is also associated with two water masers, a near-IR jet, and a large-scale CO outflow (Tofani et al. 1995; Ray et al. 1990; Mitchell & Matthews 1994). The outflow is oriented approximately perpendicular to the disk-like structure, but higher resolution near-IR and CO observations are needed to confirm the association of these structures with the secondary star and verify the presence of a disk.

An examination of the dereddened fluxes of the primary star shows that it has a featureless mid-IR SED and a flat α index (0.04). Previous authors had cited a silicate emission or absorption feature (Bechis et al. 1978; Berrilli et al. 1987). The index of the secondary remains unknown; this object is not detected at wavelengths shorter than $\sim$ 3.0 μm and a 3.8 μm map shown in Weintraub et al. (1996) does not indicate fluxes for individual stars.

2.4. MWC 1080

MWC 1080 is part of a cluster of young stars and has multiple companions, including an eclipsing binary companion (Shevchenko et al. 1994) and a close (0.7″) companion (Leinert, Richichi, & Haas 1997). Previous observations by Deutsch et al. (1995) showed that MWC 1080 was extended in the mid IR. Our observations resolve the close binary and reveal the presence of a companion 5.2″ east of the star and extended structure similar to two tidal tails. The large-aperture mid-IR flux of this system is due to multiple components; several embedded sources, and dust clouds, which we resolve to determine the shape of the SED for individual sources. Using photometry from Leinert et al. (1997) and our own measurements, we derive indices of 0.67 and 0.02 for the east and west component of the core binary respectively. These two presumably coeval stars have dramatically different SEDs; the index of the brighter component appears similar to that expected from a classical disk, while the companion has a flat index suggestive of a dust envelope.

Assuming a distance of 2500 pc, the close companion and eastern companion orbit at 1750 AU and 13 000 AU respectively. The short-period ($\sim$ 2.9 days) of the eclipsing component indicates a distance of less than 1 AU. Binary, or multiple systems are predicted to form through the fragmentation of a massive circumbinary disk (Bonnell 1998). The interaction between a disk and an inner binary can fragment the disk and induce large scale (1000 AU) spiral structure (Burkert & Bodenheimer 1996). Interactions between the spiral arms may then form a Jeans mass condensation that can collapse to an additional companion, which may then evolve to a wider separation through angular-momentum transfer. The MWC 1080 system exhibits similar structure, with an inner binary, and an outer companion located at the terminus of one of the tidal arms. However, fragmentation processes are expected to occur early in protostellar evolution and it is uncertain whether structure would endure until the stars become optically visible.

Table 1. Mid-IR flux densities (Jy) of Ae/Be stars

	N (4×4″)	*N* (1×1″)	*N* total	18.2 (5×5″)	18.2 total
HD 259431	6.4	...	7.8	5.7	10.3
AS 310	2.8	...	11.8	12.5	48.0
LkHα 234	3.1	...	5.2	3.3	11.5
MWC 1080 IRTF	13.2	...	17.1	11.0	14.8
MWC 1080 A Keck	...	9.5	...	...	...
MWC 1080 B Keck	...	3.0	...	...	...

3. Summary

High spatial resolution observations of four Ae/Be Group I disk candidates show extended dust envelopes and embedded companions. This sample includes objects which show evidence for non-equilibrium grains, disk structure around a binary companion, and tidal structure within a multiple system. On average, 50% of the mid-IR flux previously attributed to circumstellar disks arises in other sources. Revised SEDs show slopes that are inconsistent with classical disk models, but could be accounted for with more sophisticated models. Our observations do not conclusively rule out the presence of disks, but call for a reevaluation of the link between the IR excess and disks.

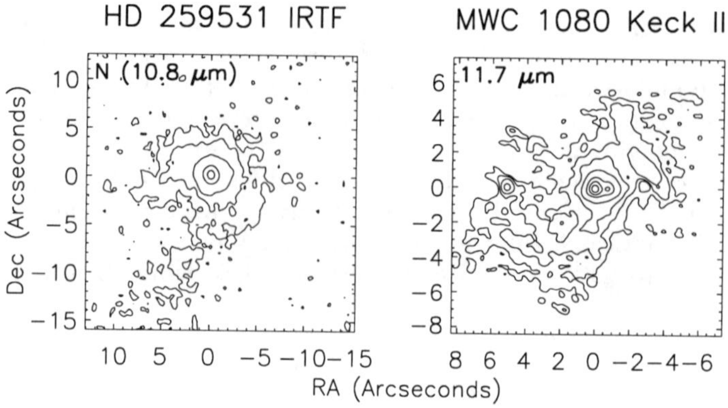

Figure 1. Mid-IR maps of HD 259431 and MWC 1080.

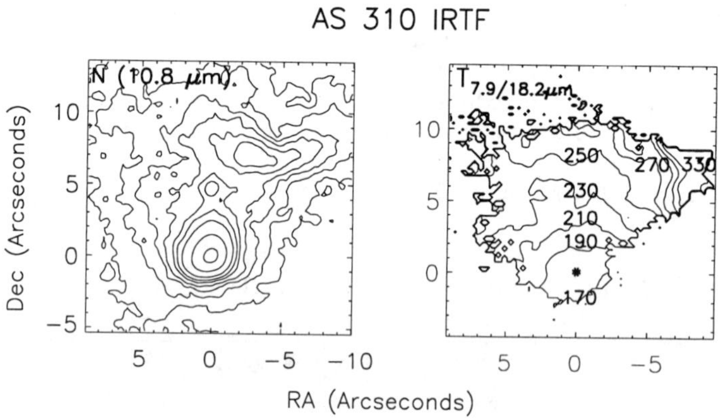

Figure 2. AS 310 10.8 μm map and 7.9/18.2 μm temperature map.

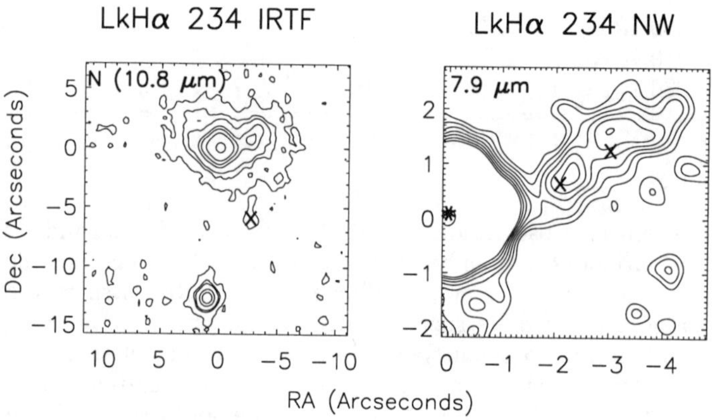

Figure 3. LkHα 234 10.8 μm map and 7.9 μm map of NW companion, "×" indicates H_2O masers.

References

Ageorges, N., Eckart, A., Monin, J.-L., & Ménard, F. 1997, A&A, 326, 632
Bastien, P., & Ménard, F. 1990, ApJ, 364, 232
Bechis, K., Harvey, P., Campbell, M., & Hoffmann, W. F. 1978, ApJ, 226, 439
Berrilli, F., Lorenzetti, D., Saraceno, P., & Strafella, F. 1987, MNRAS, 228, 833
Bonnell, I. A. 1998, in The Origin of Stars and Planetary Systems, ed. C. J. Lada, & N. D. Kylafis (Dordrecht: Kluwer), p. 479
Brooke, T. Y., Tokunaga, A. T., & Strom, S. E. 1993, AJ, 106, 656
Burkert, A., & Bodenheimer, P. 1996, MNRAS, 280, 1190
Cabrit, S., Lagage, P.-O., McCaughrean, M., & Olofsson, G. 1997, A&A, 321, 523
Ciardi, D., Akeson, R., van Belle, G. T., & Lada, E. A. 1999, BAAS, 195, 4905
Cohen, M., Bieging, J. H., & Schwartz, P. R. 1982, ApJ, 253, 707
Cohen, M. 1974, MNRAS, 169, 257
Corcoran, M., & Ray, T. 1997, A&A, 321, 189
Corporan, P. 1998, PhD thesis, Universiteé Joseph Fourier de Grenoble, France
Deutsch, L. K., Hora, J. L., Butner, H. M., Hoffman, W. F., & Fazio, G. G. 1995, Ap&SS, 224, 89
Fisher, R., Telesco, C., Piña, R., Knacke, R., & Wyatt, M. 2000, ApJ, 532, L141
Hartmann, L., Kenyon, S. J., & Calvet, N. 1993, ApJ, 407, 219
Herbig, G. H. 1960, ApJS, 4, 337
Hillenbrand, L. A., Strom, S. E., Vrba, F. J., & Keene, J. 1992, ApJ, 397, 613
Kenyon, S. J., & Hartmann, L. 1987, ApJ, 323, 714
Koerner, D., Ressler, M., Werner, M. W., & Backman, D. 1998, ApJ, 503, L83
Lada, C., & Adams, F. 1992, ApJ, 393, 278
Leinert, C., Richichi, A., & Haas, M. 1997, A&A, 318, 472
Li, W., Evans, N. J., II, Harvey, P. M., & Colome, C. 1994, ApJ, 433, 199
McCaughrean, M., & O'Dell, C. 1996, AJ, 111, 1977
Mitchell, G. F., & Matthews, H. E. 1994, ApJ, 423, L55
Natta, A., Prusti, T., & Krugel, E. 1993, A&A, 275, 527
Pezzuto, S., Strafella, F., & Lorenzetti, D. 1997, ApJ, 485, 290
Ray, T. P., Poetzel, R., Solf, J., & Mundt, R. 1990, ApJ, 357, L45
Shevchenko, V., et al. 1994, in ASP Conf. Ser., Vol. 62, The Nature and Evolutionary Status of Herbig Ae/Be Stars, ed. P. S. Thé, M. R. Pérez, & P. J. van den Heuvel (San Francisco: ASP), p. 43
Testi, L., Palla, F., & Natta, A. 1998, A&AS, 133, 81
Tofani, G., Felli, M., Taylor, G. B., & Hunter, T. R. 1995, A&AS, 112, 299
Vieira, S. L. A., & Cunha, N. C. S. 1994, Inform. Bull. Var. Stars, 4090, 1
Volk, K., Kwok, S., Stencel, R. E., & Brugel, E. 1991, ApJS, 77, 607
Weintraub, D. A., Kastner, J., Gatley, I., & Merrill, K. M. 1996, ApJ, 468, L45

Disks, Planetesimals, and Planets
ASP Conference Series, Vol. 219, 2000
F. Garzón, C. Eiroa, D. de Winter, and T. J. Mahoney, eds.

Rotational Velocities for the EXPORT Sample

E. Solano

INSA-Laboratorio de Astrofísica Espacial y Física Fundamental, Apartado de Correos 50727, E-28080 Madrid, Spain

B. Montesinos

Laboratorio de Astrofísica Espacial y Física Fundamental, Apartado de Correos 50727, E-28080, Madrid, Spain

Instituto de Astrofísica de Andalucía, Apartado de Correos 3004, E-18080, Granada, Spain

A. Mora

Departamento de Física Teórica, Facultad de Ciencias, Universidad Autónoma de Madrid, E-28049 Cantoblanco, Madrid, Spain

EXPORT

Abstract. A homogeneous set of projected rotational velocities of the targets observed during the 1998 La Palma International Time campaigns is presented here. This is part of a project whose main goal is to perform a detailed chemical analysis of these objects and to study their implications in the planetary formation phenomenon.

1. Introduction

A precise determination of rotational velocities is of fundamental importance to understanding the process of planetary formation. Despite this, most of the values published in the literature are based on poor-quality data or are affected by systematic errors (e.g., the combination of observations with different resolutions or the selection of lines broadened by agents other than rotation). These problems are not present in this study, where the same instrumentation and observational configuration have been used for all the stars in the sample and a careful selection of lines, where rotation is the only source of broadening, has been made.

2. The Method

Projected rotational velocities have been measured using the method described in Gray (1992) and based on the relation between $v \sin i$ and the frequencies where the Fourier transform of the rotational profile reaches a relative minimum (Figure 1). If the spectral resolution is reasonably high, the instrumental

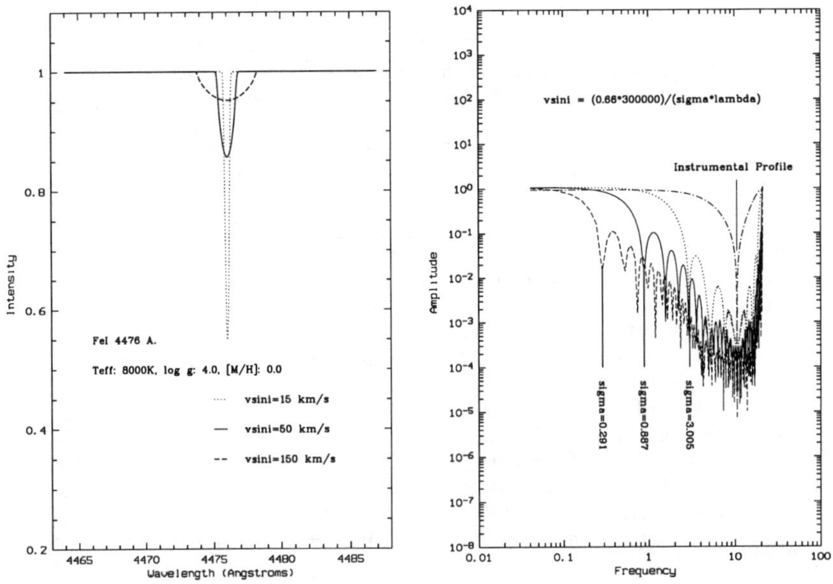

Figure 1. *Left:* Influence of rotational broadening on spectral lines. The synthetic line Fe I 4476 Å is shown for different broadenings. *Right:* Fourier transforms and their relation with rotational velocities.

profile may add, at high frequencies, relative minima to the Fourier transform, thereby not contaminating the transform of the rotational profile (Figure 1, *left*). Projected rotational velocities for our sample of stars are given in Table 1.

Unlike other methods (e.g., Sletteback et al. 1975; Tonry & Davis 1979), the one presented here does not need any a priori calibration and provides direct and independent measurements of $v \sin i$.

The assumptions underlying the method are the following:

- The star is spherical. Stars in our sample rotate far below their break-up velocities.
- The star rotates rigidly. This is not true for late-type stars but latitudinal differential rotation affects only the high-frequency part, whereas $v \sin i$ is calculated from the low-frequency region (Bruning 1981).
- The line profile stays constant along the stellar disk.

3. Macroturbulence

The method used to calculate projected rotational velocities is also valid in determining other sources of broadening-like macroturbulence in slow rotators. The basic assumption is that the observed line profile can be described by the

Table 1. Projected rotational velocities of program stars. The number of lines used in the analysis are given in parentheses. "*" indicates the presence of macroturbulence as additional source of broadening.

Target	Type	$v \sin i$ (km s^{-1})	
		Measured	Adopted
BH Cep	HAe/Be	97 ± 8(8); 97 ± 5(7); 98 ± 6(11)	97
49 Cet	Vega-like	183 ± 9(4); 187 ± 4(3)	186
24 CVn	A shell	168 ± 6(6); 177 ± 5 (3)	171
T Ori	HAe/Be	175 ± 14(9)	175
BF Ori	HAe/Be	34 ± 3(33); 33 ± 6(14); 36 ± 5(10); 44 ± 4(8)	35
CO Ori	UXOR	60 ± 9(21); 68 ± 5(6) ; 61 ± 8 (9)	62
HK Ori	UXOR	28 ± 6(34)	28
NV Ori	HAe/Be	81 ± 8(19)	81
RY Ori	HAe/Be	66 ± 6(19)	66
UX Ori	UXOR	165 ± 4(3); 213 ± 11(3)	189
CQ Tau	HAe/Be	105 ± 5(6)	105
RY Tau	T Tauri	49 ± 7(3); 56 ± 6(8); 57 ± 6(8); 55 ± 10(10)	55
HR 419	B9 V	162 ± 14(7)	162
HR 4227	A-type	152 ± 6(4); 129 ± 8(5)	140
HR 4757B	PTT	5.2 ± 0.7(35)	5*
HR 5422	A-type	7.6±0.7(52); 7.2 ±0.8(45)	7*
HR 9043	β Pic-like	192 ± 22(4); 218 ± 11(3)	213
HD 23362	Vega-like	6± 0.7(25)	6*
HD 31293	HAe/Be	97 ± 20(8)	97
HD 31648	HAe/Be	111 ± 8(8); 87 ± 13(8);101 ± 9(5);96 ± 13(8)	102
HD 34282	HAe/Be	132 ± 15(12); 125 ± 16(6)	130
HD 34700	HAe/Be	48 ± 3(11); 30 ± 8(11)	45
HD 58647	HAe/Be	119 ± 5(3); 117 ± 8(3); 116 ± 12(2)	118
HD 109085	Vega-like	66 ± 5(29); 71 ± 4(9); 67 ± 4(13); 68 ± 3(7)	67
HD 123160	Vega-like	8 ± 1.5(13)	8
HD 141569	HAe/Be	283 ± 10(3)	283
HD 142666	UXOR	74 ± 4(7); 72 ± 4(14); 71 ± 4(21); 75 ± 4(16) 73 ± 3(11); 70 ± 3(13)	72
HD 142764	Vega-like	7 ± 3 (34)	7*
HD 144432	HAe/Be	81 ± 6(10); 85 ± 11(7) ; 88 ± 8(9); 89 ± 9(8)	85
HD 163296	HAe/Be	134 ± 15(7)	134
HD 199143	F-type	159 ± 17(7); 154 ± 9(6)	155
HD 233517	Vega-like	17 ± 3(5); 16 ± 2(6); 15 ± 2(24)	15

following expression:

$$F(\lambda) = T(\lambda) * I(\lambda) * R(\lambda) * M(\lambda), \tag{1}$$

where $*$ means convolution, or, in Fourier space, the product

$$f(\sigma) = t(\sigma)i(\sigma)r(\sigma)m(\sigma), \tag{2}$$

where $F(\lambda)$ is the observed profile (to improve the analysis, an averaged profile of several lines with similar depth of formation has been used), $T(\lambda)$ is the thermal

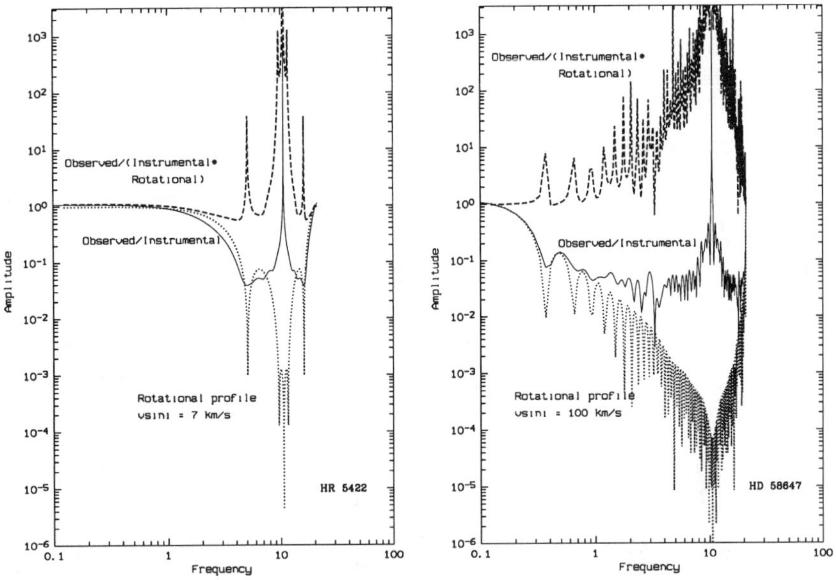

Figure 2. Removal of rotational profile for slow (*left*) and fast rotators (*right*).

profile (negligible for moderate-intensity lines), $I(\lambda)$ is the instrumental profile (measured from calibration spectra), $R(\lambda)$ is the rotational profile, and $M(\lambda)$ is the macroturbulence profile.

Macroturbulence can be estimated by simply removing the other sources of broadening. Figure 2 (*left*) shows how the instrumental broadening-free profile (*solid line*) cannot be fitted by a single rotational profile (*dotted line*), suggesting the presence of an additional source of broadening that can be ascribed to macroturbulence (*dashed line*) comparable, in slow rotators, to rotation. This is not the case for fast rotators, where macroturbulence is negligible and the instrumental broadening-free profile (*solid line*) can be suitably fitted by a single rotational profile (*dotted line*; Figure 2, *right*). Residuals at high frequencies can be ascribed to amplified noise.

4. Estimated Uncertainties

- *Continuum placement:* A displacement of the continuum level may change the line profile and thus distort the shape of the Fourier transform, thereby modifying the position of the relative minima. This is particularly important for high values of $v \sin i$ due to the weakening of the line depth.

- *Sampling frequency:* This is another limiting factor in the calculation of $v \sin i$. Defining this frequency as $\sigma_N = 0.5/\Delta\lambda$ and considering the spectral resolution of our observations, the lowest $v \sin i$ value that can be achieved is ≈ 4 km s^{-1}. Hence, for stars with $v \sin i$ lower than this value

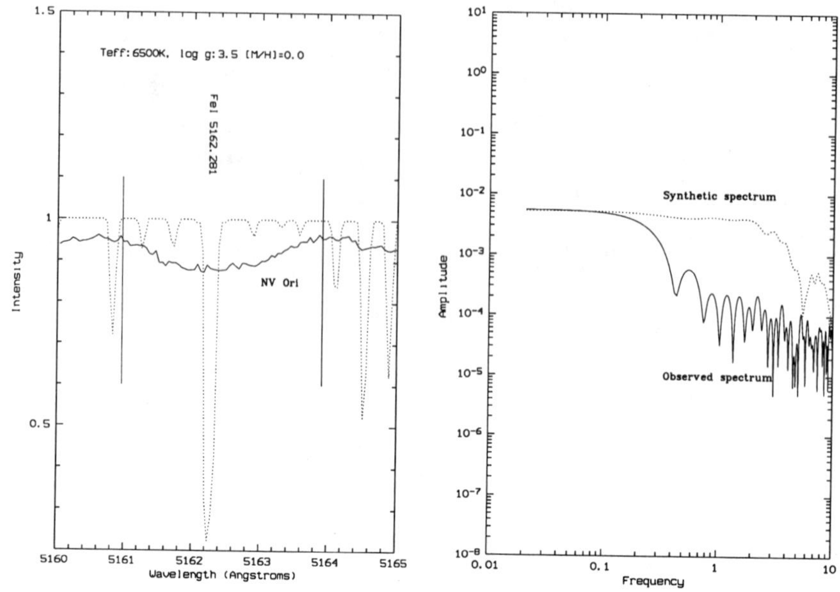

Figure 3. Influence of the intrinsic profile of blended features in the Fourier transform. The observed spectrum (*solid line*) is compared to a synthetic spectrum with similar physical parameters and $v \sin i = 0$ (*dotted line*). Assumed line edges are also marked. The contribution of the intrinsic profile lies at high frequencies not affecting the $v \sin i$ determination.

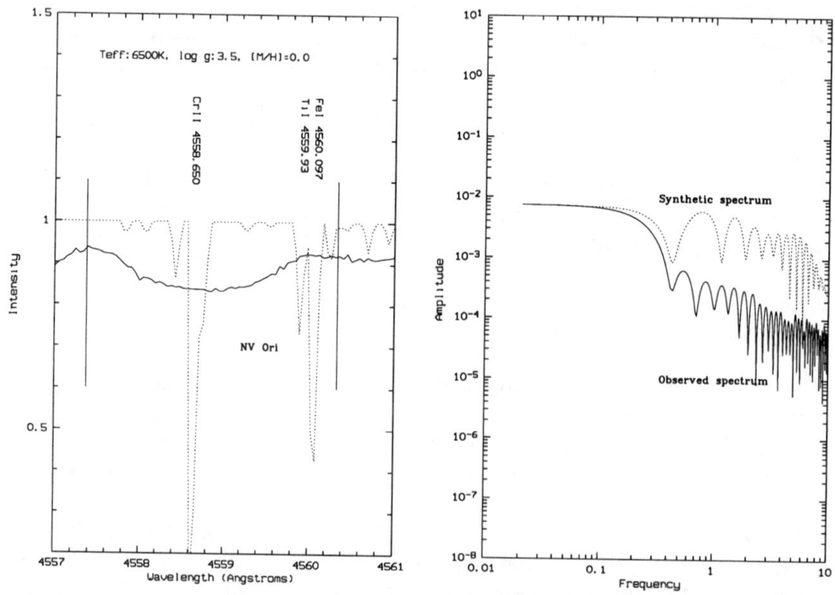

Figure 4. Similar to Figure 3, but in this case the intrinsic profile produces a spurious minimum in the Fourier transform of the observed spectrum that may lead to a wrong $v \sin i$ determination.

it is not possible to determine the rotational velocity but only an upper limit.

- *Blending:* The typically high rotational velocities of most of the stars in our sample make it very difficult to apply the method to isolated lines. Blended features were used instead. A careful inspection using synthetic spectra was performed to avoid contamination in the Fourier transform due to the intrinsic profile of the blended features (Figures 3 and 4).

5. Conclusions

The suitability of the method proposed by Gray (1992) to derive rotational velocities has been demonstrated here. In addition to this, the presence of other sources of broadening, such as macroturbulence, can also be estimated for slow rotators. The characterization of the different type of objects in the EXPORT sample according to their rotational velocities and the implications in the stellar evolution is part of an ongoing study.

References

Bruning, D. H. 1981, ApJ, 248, 274

Gray, D. F. 1992, The Observation and Analysis of Stellar Photospheres (Cambridge: Cambridge Univ. Press), p. 368

Slettebach, A., Collins, G. W., Boyce, P. B., White, N. M., & Parkinson, T. D. 1975, ApJS, 29, 137

Tonry, J., & Davis, M. 1979, AJ, 84, 1511

Disks, Planetesimals, and Planets
ASP Conference Series, Vol. 219, 2000
F. Garzón, C. Eiroa, D. de Winter, and T. J. Mahoney, eds.

Non-LTE Modeling of Accretion Disks in UX Ori-Type Stars

L. V. Tambovtseva

Central Astronomical Observatory Pulkovo, Pulkovskoe shosse 65/1, 196140, St.Petersburg, Russia

V. P. Grinin

Astronomical Institute of St. Petersburg University, Bibliotechnaya pl. 2, 198904, Petrodvorets, St. Petersburg, Russia

Abstract. The diagnostics of accretion-disk parameters are discussed on the basis of non-LTE analysis of the emission spectra of UX Ori-type stars. Line profiles of five hydrogen lines (Hα, Hβ, Hγ, Brγ, and Pa$_{14}$) arising in the envelope of these stars are computed. The accretion disk includes the corotation zone, whose radius is determined from the equation of motion.

An analysis of the theoretical and observed line profiles and equivalent widths permit us to determine the accretion rate: $\approx$ 2–5 $\times 10^{-9}$ $M_\odot$ per yr. The temperature law $T(r) = T(r_*)(r/r_*)^{-1/n}$, with $n \approx$ 2–3, gives the best results where the electron temperature near the star $T(r_*)$ is likely to be between 15 000 and 20 000 K. Such low accretion rates testify in favor of the magnetospheric accretion model. It is clear that such a feeble accretion cannot influence the photometric activity of these stars.

1. Introduction

The numerous accretion features in Herbig Ae/Be stars have been revealed both in the visual (Graham 1992; Grinin et al. 1994; Grady et al. 1996) and ultraviolet (Peréz, Grady, & Thé 1993; Grady et al. 1996) regions. They are usually displayed by young (preferentially Herbig Ae) stars with the non-periodical Algol-type brightness minima, UX Ori being a prototype. A specific feature is violent photopolarimetric activity. The brightness variability is caused by screening of the stars by circumstellar dust clouds that intersect the line of sight from time to time (Grinin 1988) and depends on the orientation of the disks relative to the observer. In particular, it is maximum if the circumstellar disk is oriented edge on or at a small inclination to the line of sight. UX Ori type stars (UXORs) have the above mentioned orientation (Grinin et al. 1991). They all manifest signatures of non-stationary accretion, which resembles both magnetospheric accretion onto T Tau stars (TTSs) and a process of gas infall due to evaporation of comet-like bodies like in β Pic. The goal of our investigations is the theoretical modeling of the emission spectra arising in the envelopes of

UXORs. As often happens in modeling, many parameters are involved the task. In our case the orientation of the system is a well known parameter, a fact that eases our calculations.

2. The Model

Modeling of the accretion process onto young stars is based on the following considerations (Camenzind 1990; Königl 1991; Shu et al. 1994; Hartmann, Hewett, & Calvet 1994): accreting circumstellar (CS) gas forms in a geometrically thin accretion disk; at a certain distance from the star such a disk is destroyed due to interaction with the magnetosphere of the star. The gas then begins to move along the magnetic field lines; in case of TTSs the gas reaches the star in the vicinity of its poles if the axis of the magnetic dipole coincides with the rotational axis of the star. Details of modeling results are to be found in Hartmann et al. (1994). The important result of the theoretical study of accretion in TTSs is the conclusion that part of the accreting matter is rejected by the magnetosphere of the star, forming anisotropic outflows.

Herbig Ae/Be stars have a similar CS envelope structure, but one that is even more complex because these stars are more rapid rotators than TTSs and the situation near the "contact point" of the infalling gas and the magnetosphere is difficult to model. Emission lines form directly in the accretion disk (e.g., Grinin et al. 1994; de Winter 1995). As in TTSs, part of the gas is rejected by the rapidly rotating magnetosphere (Grinin 1998) forming the stellar wind. In the particular case of UXORs, we do not see the wind signatures, owing, apparently, to their specific orientation.

In a recent paper (Tambovtseva et al. 1999) we examined the simplified accretion onto UXORs model with the flat accretion disk and continuous axially symmetric flow. Using a non-LTE calculation, we computed line profiles of three first lines of Balmer series and neutral helium 5876 Å. Here, we complicate the kinematics of the flow in order to approach the magnetospheric accretion model by introducing the corotation region in the disk. We present below a brief description of the model and results of calculations.

3. Calculations

We assume that the CS envelope consists of two zones: the corotation zone, where the motion of accreting gas is controlled by the rotation of the star, and the region of the quasi-Keplerian accretion disk. Figure 1 presents an example of the velocity law adopted for UX Ori. The gas velocity and the corotation radius are determined from the equation of motion. The geometrical thickness, h, of the accretion disk is constant and is an input parameter. The gas density has been derived from the equation of continuity for the disk flow.

The calculation of the line profiles is described in detail by Tambovtseva et al. (1999). The level populations were computed in the Sobolev approximation (numerical code by Grinin & Mitskevich 1990). In the temperature range considered the universal cooling function for the opaque plasma is absent (*ibid.*). Therefore, we accepted the power law for the electron temperature,

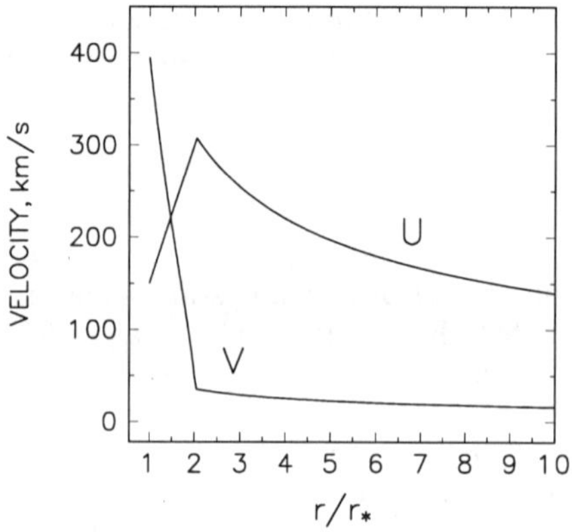

Figure 1. Behavior of the radial (v) and tangential (u) velocities with distance (r) from the star. The radius of corotation $r_{\rm c} = 2r_*$.

$T(r) = T(r_*)(r/r_*)^{-1/n}$, where $T(r_*)$ is the electron temperature near the star and n is the parameter.

We examined several tens of models with different combinations of parameters. Our choices were based on two criteria: i) the width and the intensity of the computed compared to observed profiles, and ii) the close approximation of the theoretical to the observed Balmer decrement. We used observational data obtained with the Nordic Optical Telescope (Grinin et al., in preparation). Figure 2 gives Hα, Hβ, and Hγ line profiles for UX Ori averaged over three close dates during 1995, when the star was in the bright state.

4. Results

One example of the calculated Balmer line profiles is given in Figure 2*a*. Figure 3 presents calculated equivalent widths in the Hα and Pa$_{14}$ (8600 Å) lines, depending on the accretion rate. The model parameters are: $T(r_*) = 10\ 000$ K, 15 000 K, and 20 000 K; $u(r_{\rm c})/v(r_{\rm c}) = 300/50$ (in km s^{-1}); $h/r_* = 0.5$, and $n = 3$. Analysis of numerous hydrogen line profiles leads to the conclusion that the accretion rate is about (2–5)$\times 10^{-9}$ $M_\odot$ yr^{-1}. The other values give the profiles far from those observed and are not able to reproduce the steep Balmer decrement. It should also be noted that a better fit occurs if the accretion disk is slightly inclined (10–15 deg) to the line of sight. After a careful analysis of all profiles computed for the corotation model, we are able to conclude that the radius of corotation has to be $\leq 2r_*$; i.e., the accretion disk has to be quasi-Keplerian practically up to the star.

Summarizing the results of our calculations, we conclude that the accreting envelope is not a continuous formation with the same thickness at all distances from the star. Non-LTE modeling of the neutral helium line implies the presence of very hot ($\geq 17\,000$ K), thin, and dense (10^{12} cm^{-3}) regions in the close vicinity of the star. The helium modeling gives smaller tangential velocities near the star than the that based on hydrogen. This could mean that the rapid rotation of the accreting gas is slowed down by the magnetic field of the star.

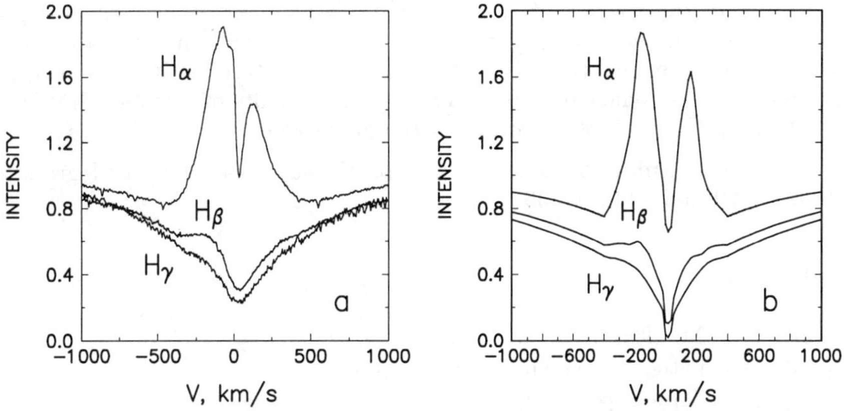

Figure 2. *a)* Observed profiles of Balmer lines for UX Ori. *b)* Theoretical Hα, Hβ, and Hγ line profiles for one of the models considered. The model parameters are: $T(r_*) = 20\ 000$ K, $n = 3$, $h/r_* = 1$, and $\dot{M}_{\rm a} = 2.3 \cdot 10^{-9}\ M_\odot\ {\rm yr}^{-1}$.

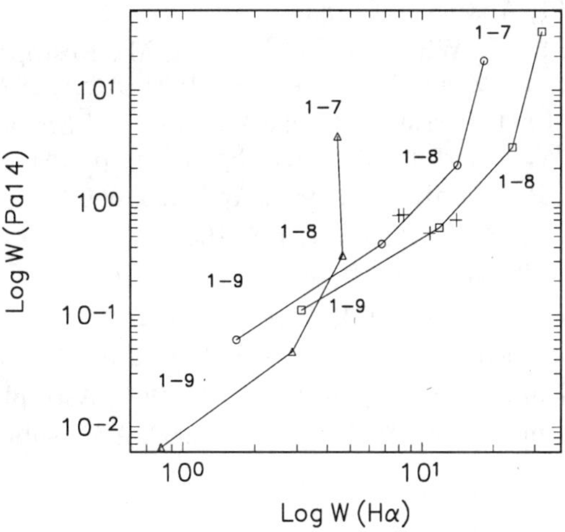

Figure 3. Equivalent width of the Hα and Pa_{14} lines depending on the accretion rate, $\dot{M}_{\rm a}$, changed from 10^{-9} to 10^{-7} $M_\odot$ per yr. $T(r_*)$ = 10 000 K (*triangles*), 15 000 K (*circles*) and 20 000 K (*squares*). The index of the temperature law is $n = 3$. *Crosses* mark equivalent widths evaluated from observations for UX Ori.

The values of $\dot{M}_a$ obtained from modeling are in agreement with the lack of veiling in the spectra of Herbig Ae/Be (Ghandour et al. 1994). Also it is clear that the energy output due to such low accretion rates are not able to contribute noticeably to the photometric activity of UXORs as suggested by Herbst & Shevchenko (1999). The detailed results of our modeling will be published in Tambovtseva, Grinin, & Rodgers (2000).

Acknowledgments. This work is partly supported by RFFI grants NN 99-02-18520 and 99-02-16336.

References

Camenzind, M. 1990, Rev. Mod. Astron., 6, 233

de Winter, D. 1995, PhD Thesis, Amsterdam

Graham J. A. 1992, PASP, 104, 479

Grady, C., et al. 1996, A&AS, 120, 157

Ghandour L., Strom, S., Edwards, S., & Hillenbrand, L. 1994, ASP Conf. Ser., Vol. 62, The Nature and Evolutionary Status of Herbig Ae/Be Stars, ed. P. S. Thé, M. R. Pérez, & E. P. J. van den Heuvel (San Francisco: ASP), p. 223

Grinin, V. P. 1988, Sov. Astron. Let., 14, 27

Grinin, V. P., & Mitskevich, A. S. 1990, Astrofizika, 35, 61

Grinin, V. P., Kiselev, N. N., Minikulov, N. Kh., Chernova, G. P., & Voshchinnikov, N. V. 1991, Ap&SS, 186, 283

Grinin, V. P., Thé, P. S., de Winter, D., Giampapa, M., Rostopchina, A. N., Tambovtseva, L. V., & van den Ancker, M. 1994, A&A, 392, 165

Grinin, V. P. 1998, in ESO Workshop, Cyclic Variability of Stellar Winds, ed. L. Kaper & A. W. Fullerton (München: Springer), p. 164

Hartmann, L., Hewett, R., & Calvet, N. 1994, ApJ, 426, 669

Herbst, W., & Shevchenko, V. S. 1999, AJ, 118, 1043

Königl, A. 1991, ApJ, 370, L39

Peréz, M., Grady, C., & Thé, P. S. 1993, A&A, 274, 381

Shu, F. H., Najita, J., Ruden, S. P., & Lizano, S. 1994, ApJ, 429, 797

Tambovtseva, L. V., Grinin, V. P., & Kozlova, O. V. 1999, Astrophysics, 42, 54

Tambovtseva, L. V., Grinin, V. P., & Rodgers, B. 2000, AZh, submitted

Disks, Planetesimals, and Planets
ASP Conference Series, Vol. 219, 2000
F. Garzón, C. Eiroa, D. de Winter, and T. J. Mahoney, eds.

Herbig Ae/Be and Vega-Type Stars as Objects at Different Stages in their Circumstellar Shell Evolution

R. V. Yudin

Central Astronomical Observatory of the Russian Academy of Sciences at Pulkovo, 196140 Saint-Petersburg, Russia

Abstract. We present the results of an investigation of correlations between the polarimetric and photometric characteristics of a sample of Herbig Ae/Be (138 objects) and Vega-type and young solar-type stars (144 objects). It is shown that for 85% of the sample stars there is a general relation between the degree of optical polarization and the infrared color index, $(V-L)_{\rm obs}$, and the color excess, $E(V-L)$, due to the contribution of a circumstellar dust shell.

1. The New $p/E(V-L)$ Diagram

The general aim of the proposed study is to compare all available polarimetric data for young early-type stars with their photometric behavior in the near and far IR, thus establishing real correlations between polarization and the stages of their circumstellar shell evolution.

First of all we plotted, in Figure 1, the observed polarization for the sample of selected young stars against their observed $(V-L)_{\rm obs}$ color index. We emphasize that a pronounced correlation between p and $(V-L)_{\rm obs}$ for a sample of 250 young stars is evident. It follows even from this relation that most Vega-type and young solar-type stars are well concentrated on the left of the diagram and represent a separate group. Next, we construct the $p/E(V-L)$ diagram for HAeBe and Vega-type stars (Figure 2), where $E(V-L)$ is the IR excess, which was corrected for the normal color index corresponding to the spectral type of the star and the amount of interstellar absorption. From Figure 2 we conclude that most HAeBe stars (as many as 85%) are well concentrated along a line with a pronounced positive correlation between p and $E(V-L)$. The main conclusions drawn from the diagram constructed here (see Figsure 1 and 2) are as follows:

First, most HAeBe stars show a common linear dependence between the $\log p$ and near IR excess. In the range of $E(V-L)$ from 1 to 8 mag the best fit for this dependence may be depicted by the following equation:

$$p \approx 0.22\ e^{0.58E(V-L)}. \tag{1}$$

In the range $E(V-L) < 1$ mag the relation is quite different. The behavior for this region is more complicated and probably may be depicted by an equation of the form:

$$p \approx k \times E(V-L). \tag{2}$$

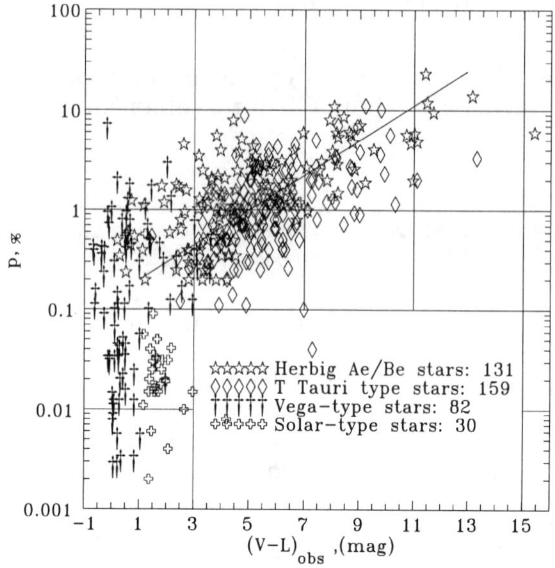

Figure 1. $p/E(V-L)_{\rm obs}$ diagram for HAeBe, Vega-type and young solar-type stars.

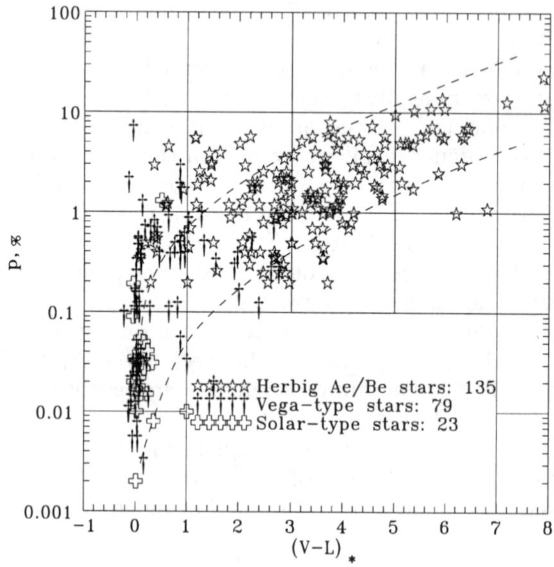

Figure 2. $p/E(V-L)$ diagram for HAeBe, Vega-type, and young solar-type stars.

Second, most of the young early-type stars that might be classified as Vega-type, β Pic-type or post-HAeBe stars (i.e., comparable young stars near the end of the pre-MS evolutionary phase or young MS stars) show statistically smaller near-IR excesses and polarization than HAeBe stars (see Figure 2). Note that about 65% of Vega-type and solar-type stars exhibit near-esIR excess less than $0\overset{m}{.}25$, and 85%—less than 1 mag. In the polarization distribution about 71% of Vega-type and solar-type stars show a polarization degree of less than 0.25%, and 90% of these have $p < 0.75\%$.

2. Young Vega-Type Stars

The prototypes for these stars (Vega, β Pic, and Fomalhaut) are classified as the young MS stars and are characterized by the existence of disk-like dust envelopes. In spite of the presence of cool dust in their CS disks, the polarization of these stars is very small, even for β Pic itself, for which the CS disk is viewed edge on. Two possibilities to explain this behavior exist:

1. The absence of a sufficient amount of hot dust (which is a primary source of polarization in most young stars) in their CS shells, i.e., the disks are optically thin and the unpolarized starlight dominates; and

2. The small geometrical thickness of the disks around them, which is unfavorable for producing polarization.

Note however, that the group of Vega-type stars is not homogeneous. Some objects might be considered as genuine MS stars and have small polarization ($p < 0.1\%$) and near-IR excess 0 mag $< E(V-L) <$ 1 mag, whereas others have 1 mag $< E(V-L) <$ 3 mag and a higher level of polarization ($0.1\% < p < 0.7\%$). The latest group includes the stars: HD 31648, HD 34282, HD 37411, HD 38120, 51 Oph, HD 139614, HD 142666, HD 144432, HD 163296, and HD 169142. Most of them have been studied early as the members of HAeBe group and there are no doubts as to their youth. Some stars have been investigated recently as Vega-type candidates by Coulson, Walther, & Dent (1998), who suggested, 'these stars are on an earlier evolutionary state than the archetypes'. It is felt that these objects are in the transition stage from HAeBe stars to young MS stars (see also Yudin, Clarke, & Smith 1999; Sitko et al. 1999).

The important conclusions from this section are as follows: most young MS stars and stars in the stage of evolution close to the MS exhibit very small levels of intrinsic polarization, as well as small near-IR excesses (or even an absence thereof). However, some differences between young MS stars and young stars close to the end of the PMS stage of evolution may be seen in the sense that the latter exhibit larger intrinsic polarization and near-IR excesses.

3. Polarization and Light Scattering

In the general case, polarization for the disk-shaped envelope can be described by the following formula:

$$P^0_{\rm int}(\%) \approx 300/16 \mid \bar{b} \mid \times f(d/R) \times \tau \sin^2\theta, \qquad (3)$$

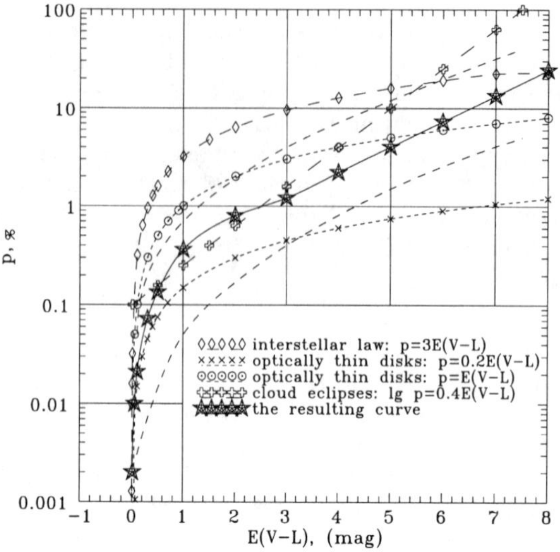

Figure 3. $p/E(V-L)$ relations for different sources of polarization.

where τ is the optical thickness of the envelope, $f(d/R)$ a function dependent on the ratio of envelope geometrical thickness to the radius, i the inclination angle of the disks to the line of sight, and $\bar{b} = 1/q = 1 - \sigma_a/\sigma_t$ (σ_a and σ_t are absorption and scattering cross-sections). Let us reconstruct this formula in terms of $E(V-L)$. Using the canonical equations $A_V = 1.086\tau_V$ and the relation between $E(B-V)$ and $E(V-L)$, the optical thickness $\tau_V \approx 0.98 \times E(V-L)$. Thus

$$P^0_{\rm int}(\%) \approx 300/16 \times f(d/R) \times E(V-L) \times \sin^2\theta. \tag{4}$$

For favorable geometry and $\sin^2\theta = 1$:

$$P^0_{\rm int}(\%) \approx E(V-L), \tag{5}$$

or, for the intermediate case,

$$P^0_{\rm int}(\%) \approx 0.2 \times E(V-L). \tag{6}$$

The relations (5) and (6) are plotted in Figure 3. Note that these relations cannot explain our dependence adequately. Incidentally, the relations (5) and (6) describe well the position of most Vega-type and post-HAeBe stars. This is in good agreement with the above suggestion concerning the presence of optically thin and homogeneous disks around them. Because of the assumption on non-variable IR fluxes we can rewrite $\Delta m \approx \Delta(V-L) = E(V-L)$ and

$$P_{\rm int}(\%) = 10^{0.4E(V-L)}. \tag{7}$$

This relation is also shown in Figure 3. Note that, as is evident from the present statistical study, the polarization is proportional to the near-IR excess, $E(V-L)$, in the form $P \sim e^{0.58E(V-L)}$. To make the coefficients consistent we must assume that the extinction curve for dust clouds/condensations is different from that of the standard curve. Put another way, the extinction in the near IR must be higher: $\tau_V \approx 0.6 \times E(V-L)$, which corresponds to the ratio $A_L/A_V \approx 0.2 - 0.4$; i.e., the extinction curve is flatter than the interstellar one. This partly non-selective extinction can occur only because of the presence of large dust grains in clouds from which stellar radiation is obscured. Rostopchina et al. (1997) noted recently that the modeling of the behavior in UXORs requires the minimum size of the dust particles to be about ten times larger than that of the interstellar matter (see also Chini, Krüger, & Kreyser 1990: the minimum size of the dust particles in CS shells around Vega, α PsA and ϵ Eri is $a_{\rm min} > 20$ μm). Simultaneous optical and IR photometry for some UXORs, which was carried out by Hutchinson et al. (1994), clearly indicates the decreasing of IR fluxes during a visual fade with a ratio that is in excellent correspondence with $A_L/A_V \approx 0.3$.

Taking into account the above discussion, the expression for the degree of intrinsic polarization for the majority of young stars has the form:

$$P_{\rm int}(\%) \approx 0.2E(V-L) \times e^{0.6E(V-L)} \tag{8}$$

for the region of 0 mag $< E(V-L) <$ 1 mag or $\leq$ 3 mag and

$$P_{\rm int}(\%) \approx 0.2 \times e^{0.6E(V-L)} \tag{9}$$

for the region of 3 mag $\leq E(V-L) <$ 8 mag.

4. Some Astrophysical Aspects of the $p/E(V-L)$ Diagram

It is interesting that some HAeBe stars with Algol-like brightness minima and low $v \sin i$ values show large photometric and polarimetric variations. If the value of $v \sin i$ in fact reflects the inclination of CS dust disks, the opening angle of these disks should not be small to explain the observed eclipses. For UXOR disks, the opening angle was estimated to be of the order 35° (see Natta, Grinin, & Mannings 1999). For relatively large inclination angles, the dust inhomogeneities responsible for the eclipse should have a size of the order of a stellar surface. However, these sizes of dust inhomogeneities are much less than those we can expect for younger objects (taking into account the timescale of their photometric and polarimetric variations). Conceivably, the formation of compact dust inhomogeneities in CS environments must reflect a definite stage of CS shell evolution. Van den Ancker, de Winter, & Tjin A Djie (1998) noted that the patchy dust clouds are only present during the pre-MS evolution of a star. They either vanish or become more homogeneous when a star has reached the ZAMS. On the other hand, for extreme young IR sources (such as GSS 30: $A_v > 20$ mag—Chrysostomou et al. 1997—or R Mon—Magnier et al. 1999) the opening angle of non-spherical dust envelopes was estimated to be of the order $> 50°$ whereas for β Pic the opening angle was estimated to be 10° (Artymowicz, Burrows, & Paresce 1989) or even 2° (Chini et al. 1991). The important conclusion which emerge from the above are as follows:

For the stars with geometrically thick dust disk-like shells, which screen most of the non-polarized stellar radiation, we observe mainly scattered radiation from their non-spherical CS envelopes (i.e., a high level of observed polarization).

These geometrically thick disks are homogeneous; i.e., not many clouds or holes exist in their circumstellar envelopes, or these clouds are very large (like that for R Mon, V 376 Cas, MWC 349, etc.; see, for example, Yudin 1996; Matsumura, Seki, & Kawabata 1999). Therefore, large polarimetric and photometric variations for these objects are apparent on a timescale of tens of days and are not obvious for shorter periods (for a similar discussion see Matsumura et.al. 1999 [for R Mon]). For UXORs the disks are still relatively geometrically thick but their width is less than those for younger objects. These disks (or non-spherical envelopes) are rather optically thin but contain numerous optically thick condensations. Besides, the dust cloud size is comparable to the stellar radius. For these objects significant variations in polarization and photometry occur over a few days, but over longer periods they show a low level of polarization and a small-amplitude photometric variability. At the later stage of evolution the young stars exhibit geometrically thin disks and dust inhomogeneities of size significantly less than the stellar surface. These disks and dust condensations cannot screen stellar radiation much, so the observed polarization is small (even if the small-amplitude Algol-like minima are observed; see Vieira, Pogodin, & Franco 1999 for HD 100546). On the other hand, the CS disks around Vega, Fomalhaut, and β Pic are still extended but geometrically and optically thin and homogeneous, therefore no significant polarization occurs in these objects. It can be said with confidence that possible evolutionary changes in the geometry of CS disks are in good agreement with observational features, at least in early-type young stars.

We conclude that there is clear evidence of changes in the polarimetric behavior of stars during their evolution to the MS.

References

Artymowicz, P., Burrows, C., & Paresce, F. 1989, ApJ, 337, 494

Chini, R., Krügel, E., & Kreysa, E. 1990, A&A, 227, L5

Chini, R., et al. 1991, A&A, 252, 220

Chrysostomou, A., et al. 1997, MNRAS, 285, 750

Coulson, I. M., Walther, D. M., & Dent, W. R. F. 1998, MNRAS, 296, 934

Hutchinson, M. G., et al. 1994, A&A, 285, 883

Magnier, E. A., et al. 1999, A&A, 346, 441

Matsumuta, M., Seki, M., & Kawabata, K. 1999, AJ, 117, 429

Natta, A., Grinin, V. P., & Mannings, V. 1999, in Protostars and Planets IV, ed. V. Mannings, A. P. Boss, & S. S. Russell (Tucson: University of Arizona Press), p. 559

Rostopchina, A. N., et al. 1997, A&A, 327, 145

Sitko, M. L., et al. 1999, ApJ, 510, 408

van den Ancker, M. E., de Winter, D., & Tjin A Djie H. R. E. 1998 A&A, 330, 145

Vieira, S. L. A., Pogodin, M. A., & Franco, G. A. P. 1999, A&A, 345, 559

Yudin, R. V. 1996, A&A, 312, 234

Yudin, R. V., Clarke, D., & Smith, R. A. 1999, A&A, 345, 547

Part 4
Exoplanets

Disks, Planetesimals, and Planets
ASP Conference Series, Vol. 219, 2000
F. Garzón, C. Eiroa, D. de Winter, and T. J. Mahoney, eds.

Mass Function and Distributions of the Orbital Elements of Substellar Companions

Michel Mayor and Stéphane Udry

Geneva Observatory, 51 ch. des Maillettes, CH-1290 Sauverny, Switzerland

Abstract. The orbital-element distributions of giant-planet candidates compared to the equivalent distributions for spectroscopic binaries strongly suggest different formation mechanisms for the two populations. In particular, the minimum mass function of secondaries of solar-type stars shows the paucity of companions in the brown-dwarf domain, whereas there is a large peak of giant-planet candidates. This result is further emphasized when the orbital-plane inclinations can be derived as, for example, from the *Hipparcos* astrometric data. Other characteristics such as the primary metalicity content or the eccentricity distributions in the $10 \leq P \leq 50$ day period range also present significant differences. Finally, the sharp low-end edge of the period distribution clearly indicates a cut-off in the separation range observed for giant planets, a trace of the formation or evolution processes.

1. Introduction

Systematic radial-velocity (RV) surveys carried out over the last 20 years have allowed the detection of numerous stellar companions for the stars in the solar vicinity. Distributions of orbital elements for these spectroscopic binaries and the mass function of secondaries are reasonably well determined, at least down to the substellar regime ($m_2 \sin i \geq 0.075\ M_\odot$). A rather complete census of spectroscopic binaries, followed with CORAVEL (Baranne, Mayor, & Poncet 1979) at moderate precision ($0.25\,\mathrm{km\ s^{-1}}$), exists for a volume-limited sample of G dwarfs in the solar neighborhood (Duquennoy & Mayor 1991, DM91) and for K dwarfs (Halbwachs et al. 1998, 2000a), comprising more than 500 stars. Thanks to the rich data collected for the nearby stars (visual and astrometric companions, common proper motion stars), these surveys are rather complete, at least for the detection of stellar companions.

Despite their moderate precision (as compared to the present RV surveys) these programs have already revealed a few companions with $m_2 \sin i$ small enough to be brown-dwarf candidates (DM91; Mayor et al. 1992, 1997). The most famous object discovered during that period is the companion of HD 114762 with a minimum mass $m_2 \sin i$ as small as 11 M_{Jup} (Latham et al. 1989). That star was not really part of our volume-limited sample; it had been followed up independently at CfA by Latham and collaborators and figured in our list of CORAVEL RV-standard stars.

The more recent developments of spectrographs with much higher precisions ($\epsilon_{\rm RV} \leq 10\,{\rm ms}^{-1}$) have brought much further insight into orbital element distributions and mass function of companions with minimum masses below 0.1 $M_\odot$. In particular, some 44 companions of stars of the solar vicinity have been detected with $m_2 \sin i$ less than 20 $M_{\rm Jup}$, an impressive series of discoveries, boosted by the announcement of the first Jovian-type exoplanet orbiting 51 Peg (Mayor & Queloz 1995). The new spectrographs used for these detections are either using the double optical fiber technique (Baranne et al. 1996) or the Iodine-cell approach (Marcy & Butler 1992; Butler et al. 1996).

The information collected so far on spectroscopic binaries, combined with the most recent detection of very low-mass companions (planets and brown-dwarf candidates), give us a synthetic view on companions to solar-type stars from $q = m_2/m_1 = 1$ to $q < 0.001$.

Recently, using different kinds of techniques, a fair number of orbital-plane inclinations have been determined allowing the real masses for a significant part of the companions to solar-type stars to be derived. The sources for the orbital elements and orbital-plane inclinations are presented in Section 2. The mass function for substellar companions is discussed in Section 3. As a result of the clear gap observed in the mass distribution between giant planets and spectroscopic binaries, it is suitable to analyze and compare the distributions of orbital elements of these two populations emphasizing their differences. This is presented in Section 4.

2. Orbital Elements and Orbital-Plane Inclinations for Substellar Companions with $m_2 \sin i$ less than 75 $M_{\rm Jup}$

Radial velocities give only access to the minimum mass $m_2 \sin i$ of low-mass companions. In the long run, most of the orbital planes of known brown-dwarf and planetary companions will be determined by precise astrometric measurements (often combined with spectroscopic data) obtained with either ground-based interferometers (VLTI, Keck I) or astrometric space missions (*Fame*, *SIM*, *Gaia*). A significant number of orbital planes of these very low-mass companions (VLMCs) have however already been determined thanks to several different sources of data, described in the following subsections. For these companions we have an estimate of their real mass, m_2. The histogram of minimum masses for VLMCs ($m_2 \sin i \leq 0.1$ $M_\odot$) is illustrated in Figures 3. The significant parameters for the VLMCs, namely their periods, eccentricities, inferred minimum masses of the companions as well as actual masses when $\sin i$ is available (method indicated in the last column), are summarized in Table 1. Except for the $\sin i$ derived by Halbwachs et al. (2000b) from the *Hipparcos* astrometric data (see Section 2.1), indicated by * indexes in the table, the given error intervals ε_{m_2} are the 1 σ errors.

2.1. The Orbital Planes Determined by *Hipparcos* Astrometry

Soon after the discovery of 51 Peg (Mayor & Queloz 1995), 70 Vir (Marcy & Butler 1996), and 47 UMa (Butler & Marcy 1996), the reflex motion induced by the planets have been searched for by analyzing the *Hipparcos* astrometric data (Perryman, Lindegren, & Arenou 1997). Unhappily, the precision of the

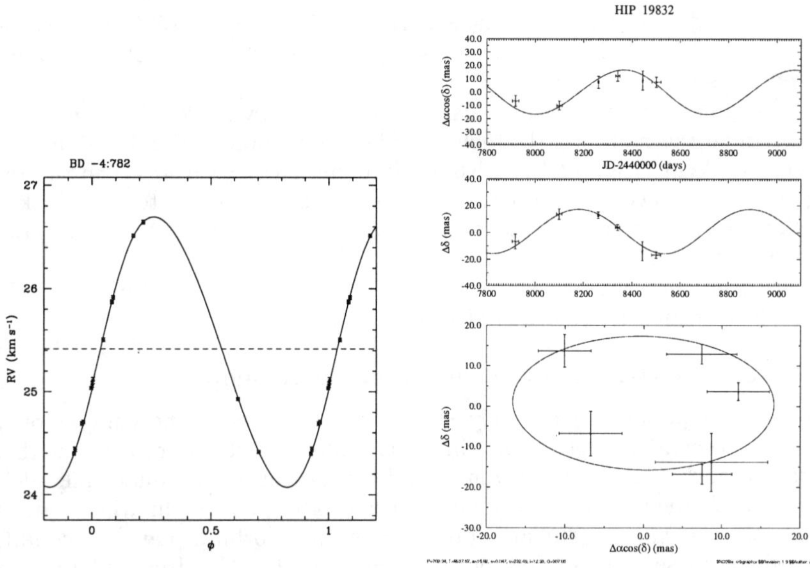

Figure 1. Orbital solution for BD −4:782. *Left:* Radial-velocity curve obtained with ELODIE. *Right:* Astrometric solution obtained from the *Hipparcos* data fixing the spectroscopic parameters.

astrometric data was not sufficient, and it was only possible to derive upper limits for the masses of 47 UMa and 70 Vir.

The CORAVEL and ELODIE surveys have revealed a few VLMCs with minimum masses in the domain of brown dwarfs (Mayor et al. 1997; Table 1), such as HD110833 ($m_2 \sin i = 19\ M_{\rm Jup}$), HD 112758 ($m_2 \sin i = 35\ M_{\rm Jup}$), BD $-4°782$ ($m_2 \sin i = 45\ M_{\rm Jup}$; Figure 1), etc. We were able to detect the astrometric reflex motion in the *Hipparcos* data (Halbwachs et al. 2000b) for most of them, i.e., when the orbital periods are compatible with the lifetime of the *Hipparcos* mission (short-period systems have too small reflex motions and longer-period orbits have their orbital and proper motions mixed). It is very impressive to see that most of the $\sin i$ of these brown-dwarf candidates are found to be small, leading to masses of secondaries above the hydrogen-burning limit. As in those cases, the astrometric motions amount only to a few milliarcsec, the orbital elements already known from radial-velocity orbits were imposed for the $\sin i$-value determination. An illustration of the results is given in Figure 1 for BD $-4°782$ (Halbwachs et al. 2000b).

Most of the determined $\sin i$ values are small enough to move a majority of the brown-dwarf candidates to the low-mass M dwarfs at the bottom of the main sequence. This is expected in the case of a mass cut-off in the secondary mass distribution (Halbwachs 1987). However, a few of the brown-dwarf candidates treated with the *Hipparcos* data have masses in the domain of brown dwarfs at a 97.7% confidence level[1]: HD 29587, HD 89707 and HD 127506 (Halbwachs et al. 2000b; Table 1). They are the first known brown dwarfs with dynamically derived masses.

The same astrometric approach has been used by Zucker & Mazeh (2000) to determine the mass of HD 10697b. The companion of HD 10697 has been detected by Vogt et al. (2000) with a minimum mass $\simeq 6.6\ M_{\rm Jup}$, in the range of giant-planet masses. The orbital plane inclination estimated by Zucker & Mazeh from the *Hipparcos* data gives a mass $m_2 = 38 \pm 13\ M_{\rm Jup}$ for this VLMC, thus an orbiting brown dwarf.

The technique was also successfully applied to most of the binaries with G- and K-dwarf primaries (not included here).

2.2. Very Short-Period Synchronized Companions

If the orbital period is extremely short and the mass of the companion not too small, the tidal interaction will synchronize the stellar rotation with the orbital motion. The observed $v \sin i$ and the synchronized equatorial velocity $V_{\rm eq} = 2\pi R/P$ then allow an estimate of the orbital-plane inclination under the hypothesis that $\sin i_{\rm rot}$ (stellar rotation) $= \sin i_{\rm orb}$ (orbital plane), a condition satisfied for these short-period synchronized objects. Good candidates in the substellar domain are HD 283750, HD 98230, and τ Boo (Fuhrmann, Pfeiffer, & Bernkopf 1998; Table 1).

If the star is not synchronized with the orbital motion and if there is an estimate of the rotational period (spots, chromospheric activity) and a precise $v \sin i$ determination from the broadening of stellar lines, the orbital inclination can also be evaluated. Apart from the three synchronized systems mentioned above, the constraints on $\sin i$ are not very strong for substellar companions. For our analysis of the mass function, $f(m)$, we prefer to include only masses derived for the three objects mentioned above.

2.3. Disks around Stars with Giant Planets

Disks have been detected around a few stars with giant planets by Trilling, Brown, & Rivkin (2000) (55 Cnc, ρ CrB, HD 210277). The ellipticity of these projected disks allows an estimation of the orbital plane inclination and thus of the mass of the giant planet (Table 1). Once again, the orbital plane is supposed to coincide with the detected large-scale disk. This is a valid hypothesis for most scenarios of planetary formation (including orbital migration), which could not , however, be satisfied in the case of strong gravitational scattering of giant planets.

2.4. Reflected Light of τ Bootis

The recent claim for the (possible) detection of the reflected light of τ Bootis (Cameron et al. 1999) also gives the mass of this VLMC. The detection of the reflected spectra provides the radial velocity of the planet and thus allows for the determination of $K_2/K_1 = m_1/m_2$ and hence the mass of the planet.

---

[1]Because of the asymmetry of the error on m_2 (Halbwachs et al. 2000b), the error interval $\varepsilon_{m_2} = \frac{1}{2}\{(m_2 + 2\sigma_+) - (m_2 - 2\sigma_+)\}$, given in Table 1, corresponds to a 97.7% confidence level.

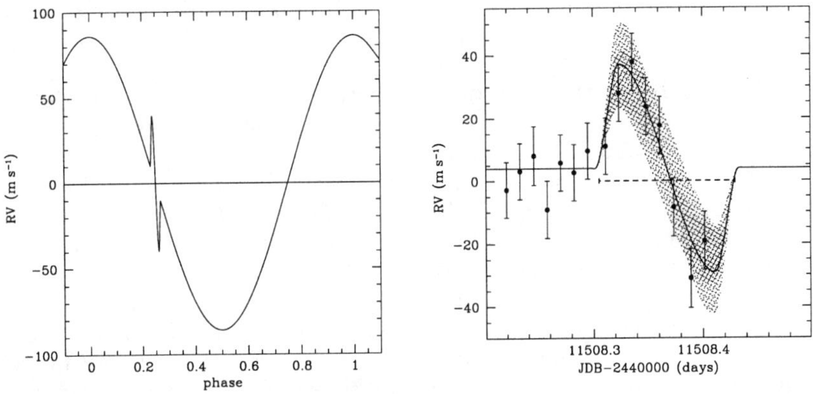

Figure 2. Spectroscopic transit of HD 209458. *Left:* Expected phased radial-velocity curve with the anomaly due to the spectroscopic transit. *Right:* Actual radial-velocity points obtained with ELODIE and best solution for the anomaly curve corrected from the orbital motion (Eggenberger 2000; Queloz et al. 2000). The 1 σ confidence level is indicated by the hatched area.

The mass of τ Bootis inferred from the reflected light ($m_2 \simeq 8\ M_{\rm Jup}$) is not in agreement with the mass determined from the synchronization (Section 2.2, Table 1). This difference, however, is of minor significance to our discussion of the mass function.

2.5. Planetary Transit of HD 209458b

Short-period giant planets ("hot jupiters"), close to their parent stars, have a significant probability of their orbital plane being sufficiently close to the line of sight to allow the observation of a transit of the planet in front of the stellar disk. That probability is about 10% for a 4 day period planetary system.

Two teams independently determined the orbit of the companion of the star HD 209458 (Mazeh et al. 2000; Henry et al. 2000). As for all other hot jupiters, these teams have applied photometric monitoring in the hope of detecting a transit at the precise time determined by the radial-velocity curve. Two full photometric transits were measured on 1999 September 9 and 16 by Charbonneau et al. (2000) and independently by Henry et al. (2000) on 1999 November 7. The luminosity dimming during the transit and the duration of the transit yield some precious information on the planetary radius and mass (via the $\sin i$):

$$\begin{aligned} m_2 &= 0.69 \pm 0.05 \quad (m_1/1.1\,M_\odot)^{2/3}\,M_{\rm Jup} \\ R_2 &= 1.40 \pm 0.17 \quad R_{\rm Jup} \\ \bar{\rho}_2 &= 0.31 \pm 0.07 \quad {\rm g\ cm^{-3}} \end{aligned}$$

where $\bar{\rho}_2$ is the mean density of the planet (Mazeh et al. 2000). For the first time, this transit provides direct proof that "hot jupiters" are gaseous giant planets.

When reanalyzing the series of photometric data from *Hipparcos* Robichon & Arenou (2000) were able to redefine the period and obtain an extremely precise value $P_{\rm Hipp} = 3.524739 \pm 0.000014$ days.

The transit of the planet in front of the disk of the rotating star also induces a small anomaly of the radial-velocity curve. On 1999 November 25, we were able to detect the spectroscopic transit for this object (Queloz et al. 2000). Figure 2 illustrates the RV anomaly during the transit. The positive sign of the anomaly during the ingress shows that the stellar rotational velocity is in the same direction as the orbital motion. The symmetry between the lobes of the anomaly curve is a strong indication that the angle between the apparent stellar equatorial plane and the orbital plane is rather small. The angle inferred from the data is 3.9^{+18}_{-21} [°] (Eggenberger 2000; Queloz et al. 2000).

3. The Mass Function of Substellar Companions

The histogram of minimum masses for the companions of solar-type stars with $m_2 \sin i$ less than 0.1 $M_\odot$ is drawn in Figure 3 (*left*). This distribution has been already discussed on several occasions (Butler & Marcy 1997; Marcy & Butler 1998; Mayor, Udry & Queloz 1998; Mazeh 1999), but the large number of recently detected VLMCs and the new information on orbital inclination justify revisiting the question. The distribution is remarkable with its impressive peak for $m_2 \sin i < 10\ M_{\rm Jup}$. Despite the huge observational bias against the detection of small-mass companions, we observe an increasing number of low-mass gaseous planets. At the same time, easier-to-detect brown-dwarf candidates are rare.

Taking into account the data on orbital inclinations (Section 2), a composite histogram has been built up for masses m_2 or $m_2 \sin i$ (if $\sin i$ not known) (Figures 3). The gap between spectroscopic binaries and planets is remarkable, emphasizing the paucity of brown-dwarf companions. Only very few secondaries of solar-type stars exist with masses in the range of 15 to 75 $M_{\rm Jup}$ (Table 1).

If we consider the still limited span of precise radial-velocity surveys, as well as the sometimes still limited number of measurements per star, the correction of observational biases is premature. In this sense the dotted line in Figure 3 (*right*) is only indicative of a potential estimate of the bias-corrected distribution.

What is the Maximum Mass for a Giant Planet? The sharp drop in the mass function observed below 8 $M_{\rm Jup}$ strongly suggests a maximum mass for giant planets close to 10 $M_{\rm Jup}$. In Figure 3 (*left*) the deuterium- and hydrogen-burning limits at 13 and 75 $M_{\rm Jup}$, respectively, are indicated by dotted lines. *The shape of the mass function does not suggest that the deuterium-burning limit is related to the maximum mass for giant planets.*

Four secondaries have been discovered with minimum masses in the range 10 to 20 $M_{\rm Jup}$ (HD 114762, 11 $M_{\rm Jup}$, Latham et al. 1989; HD 162020, 13.7 $M_{\rm Jup}$ and HD 202206, 14.7 $M_{\rm Jup}$, ESO 2000; Udry et al., in preparation; HD 110833, 19 $M_{\rm Jup}$, Mayor et al. 1997). The companion of HD 110833 was found to be a low-mass M dwarf of $m_2 = 0.12\ M_\odot$ (Halbwachs et al. 2000b). The period of

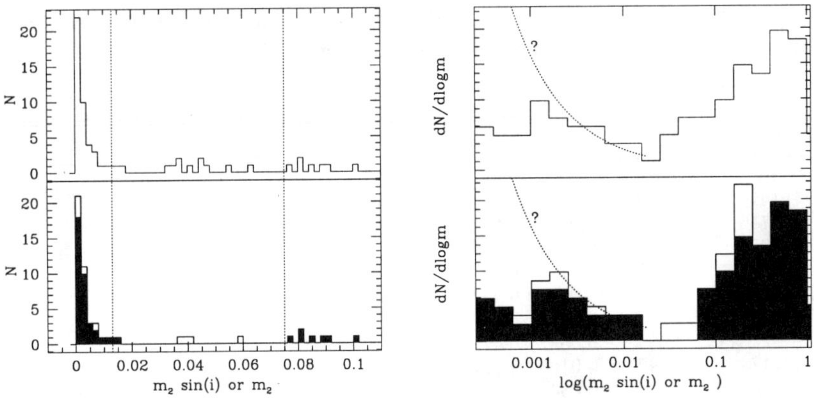

Figure 3. Mass function of companions to solar-type stars. *Left:* Linear scale. The *dotted lines* indicate the hydrogen- (75 $M_{\rm Jup}$) and the deuterium- (13 $M_{\rm Jup}$) burning limits. *Right:* Log scale. The *dotted line* is indicative of a potential estimate of a bias-corrected distribution. For both diagrams: *Top:* $m_2 \sin i$. *Bottom:* composite histogram of m_2 (*open part*) and $m_2 \sin i$ (if $\sin i$ is not known, *shaded part*).

HD 162020 (P = 8.4 d) is too small to expect a significant astrometric reflex motion. Owing to the distance and relatively short period of HD 114762, no real constraint on the orbital plane could be obtained either. However, the high metal deficiency [Fe/H] = −0.6 of that star is highly unusual when compared to the distribution of metalicities of stars with giant planets (Gonzales 1998; Santos, Israelian, & Mayor 2000).

Presently, we do not have any strong arguments to consider (or not) secondaries with $m_2 \sin i$ in the range of 10 to 20 $M_{\rm Jup}$ as the upper tail of the mass distribution of giant planets. If we consider the impressive peak of $f(m_2 \sin i)$ for $m_2 \sin i$ less than 8 $M_{\rm Jup}$, as well as the excess of metalicities of these stars compared to stars without giant planets, we are inclined to see secondaries with $m_2 \sin i$ less than 8 $M_{\rm Jup}$ as distinct from the more massive ones.

4. The Distribution of Orbital Elements of Giant Planets

The observed gap in the mass distribution between giant planets and stellar secondaries strongly suggests the existence of two distinct companion populations for solar-type stars. Giant planets and stellar binaries are believed to have different formation mechanisms whose fossil traces should be revealed by a comparison of their orbital properties.

4.1. The Distribution of Periods

Clearly, due to the still strong bias affecting the detection of long-period planets, a significant comparison of the period distributions of planetary candidates ($m_2 \sin i \leq 10$ $M_{\rm Jup}$) and spectroscopic binaries ($m_2 \sin i \geq 10$ $M_{\rm Jup}$) is only

Table 1. Orbital parameters of VLMCs. Periods are given in days, masses in 10^{-3} $M_\odot$. (For a ε_{m_2} description see Sections 2–2.1)

Star	P	e	$m_2 \sin i$	m_2	ε_{m_2}	$\sin i$ origin
HD 16141	75.82	0.28	0.215			
HD 168746	6.407	0.0	0.24			
HD 46375	3.024	0.0	0.249			
HD 108147	11.051	0.57	0.33			
HD 75289	3.51	0.0	0.42			
51 Peg	4.23	0.0	0.45			
BD −10°3166	3.487	0.05	0.47			
HD 187123	3.097	0.01	0.57			
υ And (b)	4.62	0.034	0.68			
HD 209458	3.52	0.00	0.685	0.69	0.05	Transit, Mazeh et al. 2000
HD 192263	24.13	0.0	0.72			
55 Cnc	14.7	0.03	0.93	1.9	$^{+1.1}_{-0.4}$	Disk, Trilling et al. 2000
ρ CrB	39.645	0.07	0.99	1.5	0.3	Disk, Trilling et al. 2000
HD 130322	10.72	0.044	1.0			
HD 37124	154.8	0.31	1.0			
HD 52265	119.16	0.35	1.03			
HD 177830	391.6	0.41	1.2			
HD 217107	7.11	0.14	1.28			
HD 210277	435.6	0.34	1.29	2.2	$^{+0.6}_{-0.2}$	Disk, Trilling et al. 2000
HD 134987	259.6	0.24	1.5			
16 Cyg B	804.41	0.67	1.67			
Gl 876	60.97	0.34	2.0			
υ And (c)	241.2	0.24	2.05			
HD 82943	443.5	0.59	2.2			
ι Hor	311.3	0.22	2.26			
47 UMa	1084.	0.13	2.45			
HD 12661	264.	0.33	2.8			
HD 169830	229.9	0.35	2.9			
HD 1237	133.7	0.51	3.35			
HD 195019	18.3	0.01	3.43			
Gl 86	15.84	0.042	3.6			
τ Boo	3.31	0.02	4.14	6.27	0.7	Synchronization
υ And (d)	1308.5	0.31	4.25			
HD 222582	575.9	0.71	5.2			
14 Her	1738.	0.37	5.4			
HD 10697	1072.3	0.12	6.1	38.	13.	*Hipp.* astr., Zucker & Mazeh 2000
HD 89744	265.	0.70	7.2			
70 Vir	116.7	0.40	7.4			
HD 168443	57.9	0.52	8.0			
HD 114762	84.	0.35	11.			
HD 162020	8.428	0.29	13.7			
HD 202206	256.45	0.43	14.3			
HD 110833	270.04	0.69	19.	137.	11.*	*Hipp.* astr., Halbwachs et al. 2000b
HD 112758	103.	0.16	32.	199.	41.*	*Hipp.* astr., Halbwachs et al. 2000b
HD 127506	2500.	0.63	35.	42.	20.*	*Hipp.* astr., Halbwachs et al. 2000b
HD 29587	1470.	0.37	37.	39.	25.*	*Hipp.* astr., Halbwachs et al. 2000b
HD 98230	3.980	0.0	37.	123.	40.	Synchronization
HD 18445	553.	0.56	40.	176.	19.*	*Hipp.* astr., Halbwachs et al. 2000b
HD 140913	147.96	0.61	44.	166	69.*	*Hipp.* astr., Halbwachs et al. 2000b
BD −04°782	710.95	0.07	45.	245.	14.*	*Hipp.* astr., Halbwachs et al. 2000b
HD 283750	1.788	0.0	47.	162	20.	Synchronization
HD 89707	298.	0.95	54.	60.	18.*	*Hipp.* astr., Halbwachs et al. 2000b
HD 217580	454.66	0.52	62.	161.	13.*	*Hipp.* astr., Halbwachs et al. 2000b

possible for relatively short-period systems ($P \leq 100$ days), for which the detection bias is vanishingly small. In the context of "hot jupiters", we will focus on the domain of very short periods ($P \leq 10$ days).

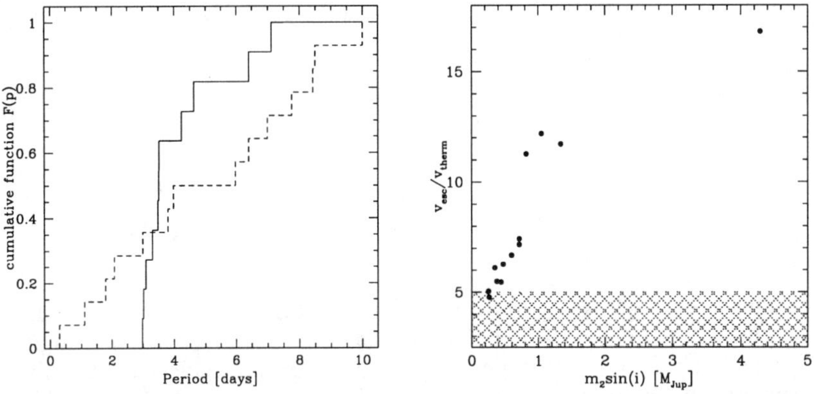

Figure 4. *Left:* Cumulative distributions of periods smaller than 10 days for planetary (solid line) and stellar companions (dashed line) to solar-type stars. The minimum period cut-off for planetary candidates just below 3 days is clearly emphasized. *Right:* Ratio $Q = V_{esc}/V_{th}$ as a function of separation for giant planets

The distribution of periods of giant planets is steeply rising for decreasing periods down to about 3 days. A cut-off is observed around that period. The shortest period observed to date is $P = 2.986$ days for HD 83443 (ESO 2000; Mayor et al., in preparation). It is clearly emphasized in Figure 4 (*left*), comparing the cumulative distributions of periods smaller than 10 days for giant planets and spectroscopic binaries. Several reasons may be invoked to set this limit, such as a central cavity of the accretion disk, tidal interaction (Lin, Bodenheimer, & Richardson 1996), or Roche-lobe overflow (Trilling et al. 1998).

Planet evaporation takes place when the thermal velocity of the gas (mainly hydrogen) becomes important with regard to the escape velocity from the planet, typically for $Q \equiv v_{\mathrm{esc}}/v_{\mathrm{therm}}$ smaller than about 5 (Lewis & Prinn 1984). From the dependency of Q on the stellar and planetary parameters, given below, it is apparent that one of the main quantities governing the evaporation is the planet mass:

$$Q = \frac{v_{\mathrm{esc}}}{v_{\mathrm{therm}}} \sim \left(\frac{m_{\mathrm{pl}}}{R_{\mathrm{pl}}}\right)^{1/2} T_\star^{-1/2} R_\star^{-1/4} a^{-1/4}.$$

Values of Q as a function of minimum masses for the short-period extrasolar planets ($P < 15$ days) are given in Figure 4 (*right*). The lighter candidates come close to the evaporation limit. If in the future some gaseous giant planets on short-period orbits are detected with masses well below $0.2\ M_{\mathrm{Jup}}$, they will be subjected to evaporation-related evolution. We can speculate that these objects will have hydrogen-depleted atmospheres.

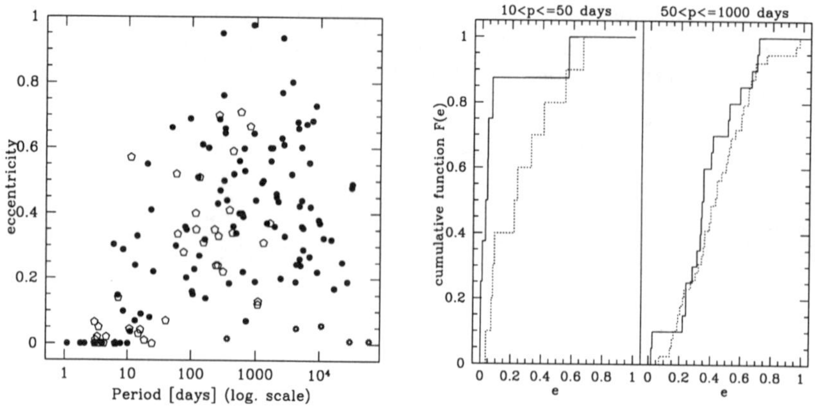

Figure 5. *Left:* $(e, \log p)$ diagram for planetary (*open pentagons*) and stellar companions (*filled circles*) to solar-type stars. The giant planets of the Solar System are indicated by star-like symbols. *Right:* Cumulative distributions of eccentricities for planetary (*solid line*) and stellar (*dotted line*) companions to solar-type stars, for two domains of periods: $10 \leq P \leq 50$ days and $50 \leq P \leq 1000$ days.

4.2. Distributions of Eccentricities

The comparison of orbital eccentricities of spectroscopic binaries for G and K primaries (DM91; Halbwachs et al. 2000b) with the equivalent distribution for giant planets is remarkable. In Figure 5 (*left*) we have plotted the eccentricities of orbits as a function of $\log P$, both for double stars and giant planets. At the first glance we do not observe significant differences in the orbital elements between planets and spectroscopic binaries. If the formation mechanisms for planets and double star are different, why do we observe such similar $(e, \log P)$ distributions? In fact some differences may be emphasized in particular domains of periods.

For planets with very short periods, most of the orbits are quasi-circular as for double stars. We know that orbits of short-period binaries are tidally circularized when the period is less than about 10 days. Planets having suffered a strong orbital migration are probably circularized by the tidal interaction with the accretion disk. We observe quite a number of quasi-circular planetary orbits with $P \leq 40$ days. It is however also true that two short-period planets present eccentric orbits: around HD 108147 ($P = 10.9$ d, $e = 0.56$) and HD 217107 ($P = 7.11$ d, $e = 0.14$). Such configurations could be explained by the presence of an additional companion revealed by a drift observed in the systematic velocity of those stars or by gravitational interaction between giant planets formed in the outer regions of the system (Weidenschilling & Marzari 1996; Lin & Ida 1997).

Most of orbits of double stars with long periods ($P \geq 50$ days) are fairly eccentric. Quasi-circular ($e < 0.01$) long-period orbits are very rare. A similar situation already exists for extrasolar giant planets (with $e = 0.13$, 47 UMa is just at the border), although they are still strongly observationally biased in that period range. We of course know one planetary system with long-period giant planets on quasi-circular orbits, namely our own Solar System (Figure 5, *left*).

The origin of the eccentricity of extrasolar giant planets in that range of periods ($P > 50$ days) has been searched in the gravitational interaction between multiple giant planets (Weidenschilling & Marzari 1996; Rasio & Ford 1996; Lin & Ida 1997), or between the planets and the planetesimals in the early stages of the system formation (Levison, Lissauer, & Duncan 1998). Among the giant-planet candidates several eccentric orbits show a drift in their mean velocity, indicating the presence of a long-period companion (probably stellar) whose gravitational perturbation can also be suspected to be responsible for the observed (high) planetary eccentricity, as for the planet orbiting 16 Cyg B (Mazeh, Krymolowski, & Rosenfeld 1997).

In summary, for periods larger than about 50 days, the comparison of double-star and planet eccentricities is really surprising. With the limited-size samples presently available, we cannot see any significant distinctions between both populations. This is shown in Figure 5 (*right*), which gives the cumulative function of eccentricities for periods between 50 and 1000 days for the two populations. If the orbital eccentricity of double stars finds its origin in the disruption of small N-body systems, maybe the generated distribution of eccentricities is close to the distribution generated by the gravitational interactions between giant planets.

Nevertheless, a clear difference may be noticed between the two populations for periods between 10 and 50 days. This is illustrated by the corresponding cumulative functions given in Figure 5 (*right*). In this range of periods, outside the circularization domain, planetary systems have smaller eccentricities than stellar binaries, indicating different formations or evolutions.

To conclude, the differences observed between planetary systems and double stars in the metalicity content of the primaries, in the shape of the secondary mass functions, and in the period and eccentricity distributions argue for the two populations being formed by distinct processes. More observations are, however, needed to put clear constraints on the possible formation and evolution scenarios.

References

Baranne, A., Mayor, M., & Poncet, J.-L. 1979, Vistas, 23, 279

Baranne, A., et al. 1996, A&AS, 119, 373

Butler, P., & Marcy, G. 1996, ApJ, 464, L153

Butler, P., Marcy, G., Williams, E., McCarthy, C., Dosanjh, P., & Vogt, S. 1996, PASP, 108, 500

Butler, P., & Marcy, G. 1997, in IAU Coll. 161, Astronomical and Biochemical Origins and the Search for Life in the Universe, ed. C. Cosmovici, S. Browyer, & D. Werthimer (Editrice Compositori), p. 331

Cameron, A., Horne, K., Penny, A., & James, D. 1999, Nat, 402, 751

Charbonneau, D., Brown, T., Latham, D., & Mayor, M. 2000, ApJ, 529, L45

Duquennoy, A., & Mayor, M. 1991, A&A, 248, 485

Eggenberger, A. 2000, Master's Thesis, Geneva Observatory

ESO 2000, May 4th Press Release, http://www.eso.org/outreach/press-rel/pr-2000/pr-13-00.html

Fuhrmann, K., Pfeiffer, M. J., & Bernkopf, J. 1998, A&A, 336, 942

Gonzales, G. 1998, A&A, 334, 221

Halbwachs, J.-L. 1987, A&A, 183, 234

Halbwachs, J.-L., Mayor, M., & Udr,y S. 1998, in ASP Conf. Ser., Vol. 134,Brown Dwarfs and Extrasolar Planets, ed. R. Rebolo, E. Martin, & M. R. Zapatero Osorio (San Francisco: ASP), p. 308

Halbwachs, J.-L., Arenou, F., Mayor, M., & Udry S. 2000a, in IAU Symp. 200, The Formation of Binary Stars, ed. H. Zinnecker & R. Mathieu R. (San Francisco: ASP), in press

Halbwachs, J.-L., Arenou, F., Mayor, M., Udry, S., & Queloz, D. 2000b, A&A, 355, 581

Henry, G., Marcy, G., Butler, P., & Vogt S. 2000, ApJ, 529, L41

Latham, D., Mazeh, T., Stefanik, R., Mayor, M., & Burki, G. 1989, Nat, 339, 38

Levison, H. F., Lissauer, J., & Duncan, M. J. 1998, AJ, 116, 1998

Lewis, J. S., & Prinn, R. G. 1984, Intern. Geophys. Ser., Planets and their Atmospheres. Origin and Evolution, 33

Lin, D., & Ida, S. 1997, ApJ, 477, 781

Lin, D., Bodenheimer, P., & Richardson, D. C. 1996, Nat, 380, 606

Marcy, G., & Butler, P. 1992, PASP, 104, 270

Marcy, G., & Butler, P. 1996, ApJ, 464, L147

Marcy G., & Butler, P. 1998, ARA&A, 36, 52

Mayor, M., & Queloz, D. 1995, Nat, 378, 355

Mayor, M., Duquennoy, A., Halbwachs, J.-L., & Mermilliod, J.-C. 1992, in ASP Conf. Ser., Vol. 32, IAU Coll 135, Complementary Approaches to Double and Multiple Star Research, ed. H. A. McAlister & W. Hartkopf (San Francisco: ASP), p. 73

Mayor, M., Queloz, D., Udry, S., & Halbwachs, J.-L. 1997, in IAU Coll. 161, Astronomical and Biochemical Origins and the Search for Life in the Universe, ed. C. Cosmovici, S. Browyer, & D. Werthimer (Editrice Compositori), p. 313

Mayor, M., Udry, S., & Queloz, D. 1998, in ASP Conf. Ser., Vol. 154, 10th Cambridge workshop, Cool Stars, Stellar Systems and the Sun", ed. R. Donahue & J. A. Bookbinder (San Francisco: ASP), p. 77

Mazeh, T., Krymolowski, Y., & Rosenfeld, G. 1997, ApJ, 477, 403

Mazeh, T. 1999, in ASP Conf. Ser., Vol. 185, IAU Coll. 170, Precise Stellar Radial Velocities, ed. J. Hearnshaw & C. Scarfe (San Francisco: ASP), p. 131

Mazeh, T., et al. 2000, ApJ, 532, L55

Queloz, D., et al. 2000, A&A, in press

Perryman, M. A. C., Lindegren, L., & Arenou, F. 1997, A&A, 310, L21

Rasio, F., & Ford, E. 1996, Sci, 274, 954
Robichon, N., & Arenou, F. 2000, A&A, 355, 295
Santos, N. C., Israelian, G., & Mayor, M. 2000, A&A, submitted
Trilling, D. E., et al. 1998, ApJ, 500, 428
Trilling, D. E., Brown, R. H., & Rivkin, A. S. 2000, ApJ, 529, 499
Vogt, S., Marcy, G., Butler, P., & Apps, K. 2000, ApJ, in press
Weidenschilling, S. J., & Marzari, F. 1996, Nat, 384, 619
Zucker, S., & Mazeh, T. 2000, ApJ, 531, L67

Disks, Planetesimals, and Planets
ASP Conference Series, Vol. 219, 2000
F. Garzón, C. Eiroa, D. de Winter, and T. J. Mahoney, eds.

The Search for Extended Atmospheres around the Planets Orbiting 51 Peg and τ Boo

Alan W. Harris and Heike Rauer

DLR Institute of Space Sensor Technology and Planetary Exploration, Rutherfordstrasse 2, 12489 Berlin, Germany

Andrew C. Cameron

School of Physics and Astronomy, University of St. Andrews, Fife KY16 9SS, Scotland, UK

EXPORT

Abstract. The stars 51 Peg and τ Boo were observed with the high-resolution Utrecht Echelle Spectrograph (UES) on the 4.2 m William Herschel Telescope (WHT) on La Palma during the times the planets were due to transit in front of them and at times outside of the predicted transit periods. The results of a detailed comparison between on-transit and off-transit spectra to search for signatures of extended planetary atmospheres are presented and discussed. No evidence of transient spectral features correlating with the predicted transit times was found. This project constitutes part of the 1998 La Palma International Time Programme on extrasolar planetary systems.

1. Introduction

The stars 51 Peg and τ Boo are each supposed to have a massive planet orbiting them (at 0.05 and 0.042 AU, respectively), as inferred from periodic variations in their radial velocities (Mayor & Queloz, 1995; Butler et al. 1997). Relevant stellar and planetary data are given in Table 1. We have attempted to confirm the existence of the planets via detection of atmospheric absorption features during transit of a planet in front of its star.

The observations were carried out in the visible region ($\sim$ 3600–6000 Å) at the WHT with resolutions of around 0.1–0.25 Å. Similar spectroscopic studies have been carried out in the infrared (e.g., Rauer et al. 2000) but so far no evidence for planetary exospheres has been found. The existence of extended atmospheres around hot giant planets, which can extend beyond a planet's Roche lobe to form a cometary tail, has been suggested by Schneider et al. (1998).

What might we expect to observe? The stars τ Boo and 51 Peg are supposed to be accompanied by massive planets of at least 3.9 and 0.47 Jupiter masses, respectively. We assume these are Jupiter-like gas-giant planets. The close proximity of these planets to their stars and the influence of the stellar winds and radiation fields may lead to extended planetary exospheres populated by

Table 1. Relevant stellar/planetary data

	51 Peg	τ Boo
The stars:		
Distance [pc]	15.4	15.0
Spectral type	G2 IV	F7 V
Apparent V mag	5.5	4.5
$T_{\rm eff}$ [K]	5770	6450
RA (2000)	$22^{\rm h}\ 57^{\rm m}\ 27^{\rm s}$	$13^{\rm h}\ 47^{\rm m}\ 17^{\rm s}$
Dec. (2000)	$+20°\ 46'\ 05''$	$+17°\ 27'\ 22''$
The planets:		
Period [days]	4.23	3.31
Semimajor axis [AU]	0.051	0.046
$M \sin i$ [$M_{\rm Jup}$]	0.47	3.87
Inclination of orbit, i	$< 85°$?	29°?

ions and dissociation products of atmospheric molecules. Among the possibly relevant species observable in the UES wavelength region are CH, CH^+, CO^+, CO_2^+, H_2O^+, and N_2^+, in addition to atomic species including H, He, and O II.

At the time the observations were carried out it was thought that $\sin i$ of both stars was sufficiently close to 1 that there was a reasonable chance that the extended exospheres of the planets would transit across the stellar disks as seen from the Earth. In the meantime, however, evidence has emerged (Cameron et al. 1999) that the inclination of the orbit of the τ Boo planet is only 29°, which would greatly weaken the chances of detecting exospheric absorption lines. In the case of 51 Peg, the results of Henry et al. (1997) suggest the inclination of the planet's orbit is less than 85°.

In both cases transits took place during the first of two nights of observations. We took spectra throughout and after the transit period (lasting an estimated $\sim$ 3 hr in both cases) to follow the development and disappearance of absorption lines as the planets passed in front of their stars. In each case similar sequences of reference spectra were taken on the second night.

2. Observations and Data Extraction

A total of 62 spectra of τ Boo, each with an integration time of around 100 s, were taken in the period 1998 May 14, 21:30–04:50 UT. Batches of five τ Boo spectra were interspersed with arc, sky, and occasional comparison star exposures. The predicted transit mid-point for τ Boo was at 23:01 $\pm$ 18 min, with an expected duration of about 3 hr. A similar reference sequence was taken on May 15.

A total of 48 spectra of 51 Peg, each with an integration time of around 300 s, were taken in the period 1998 October 23, 19:00–03:00 UT. The observation sequence was similar to that for τ Boo. The predicted transit mid-point for 51 Peg was at 20:43 $\pm$ 45 min, with an expected duration of about 3 hr. A similar reference sequence was taken on October 24.

The transit time predictions were kindly provided by G. W. Marcy.

The data were extracted at St. Andrews University using an automated pipeline reduction system built around the Starlink ECHOMOP and FIGARO packages. At the start of each night, median bias and flatfield frames were computed from 20 to 30 frames each, taken on the afternoon prior to the observations.

The initial tracing of the echelle orders on the CCD frames was performed manually on the spectrum of the B star η Hya, which was observed at the beginning or end of each night. The automated extraction procedure then subtracted the bias from each frame, cropped the frame, determined the form and location of the stellar profile on each image relative to the trace, subtracted a polynomial fit to the scattered-light background in the inter-order gaps, and performed a simple (unweighted) extraction of the orders, applying flatfield balance factors in the process.

Wavelength calibration was performed using a four-coefficient (cubic) polynomial fit to the wavelengths of lines identified in thorium–argon arc comparison spectra. In the case of τ Boo the full width at half maximum intensity of the thorium–argon lines in the arc calibration spectra was 2.1 pixels, giving an effective resolving power R = 48 900. In all, 62 orders were extracted from each frame, giving full spectral coverage from 3640 to 6120 Å. The corresponding figures in the case of 51 Peg are 2.3 pixels, R = 43 500 and 3617 to 6054 Å. There is considerable wavelength overlap between adjacent orders.

3. Data Analysis and Results

For each star all spectra taken within a 2 hr period centered on the predicted transit mid-point were corrected for small relative wavelength shifts and averaged (a total of $\sim$ 13 spectra). A corresponding mean spectrum from the same UT interval on the following night (off transit) was used as a reference. A detailed comparison of the on-transit and off-transit mean spectra was made for each of the 62 orders. The example shown in Figure 1 (*top frame*) is the H_β line of 51 Peg around 4861 Å. The plotted spectrum is the on-transit mean spectrum from 1998 October 23 superimposed on the mean reference (off-transit) spectrum from 1998 October 24. However, the two spectra are virtually indistinguishable, and no differences are discernible. Plotted above in the same frame, multiplied by an arbitrary amount, is the ratio of the on-transit mean spectrum to the off-transit mean spectrum, with 1 σ error bars. Small dips in the ratio plot correspond to strong narrow lines and are due to slight sampling differences between the on-transit and off-transit spectra. The maximum sensitivity of the ratio plot to transit-related absorption features can be estimated on the basis of the variance in the range 4860 to 4863 Å, in which there is little interference from strong narrow lines. The resulting signal/noise ratio is 428 which, given R = 43500, implies a minimum detectable equivalent width at the 3 σ level of 0.8 mÅ.

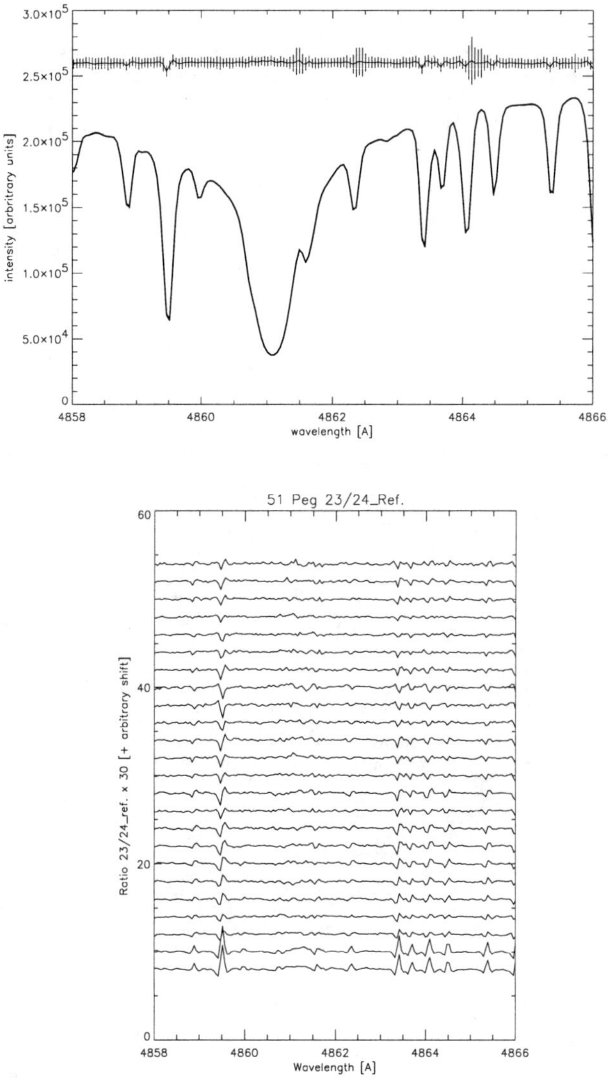

Figure 1. The H_β line of 51 Peg. *Top frame:* the on-transit mean spectrum from 1998 October 23 is plotted over the mean reference (off-transit) spectrum from 1998 October 24 (the two spectra are indistinguishable in this figure). The upper plot is the ratio of the on-transit mean spectrum to the off-transit mean spectrum, multiplied by an arbitrary amount, with 1 σ error bars. *Bottom frame:* successive individual on-transit spectra taken over a time interval of 3.5 hr, centered on the predicted time of mid-transit, divided by the corresponding order in the mean off-transit reference spectrum from 1998 October 24. Time increases downwards.

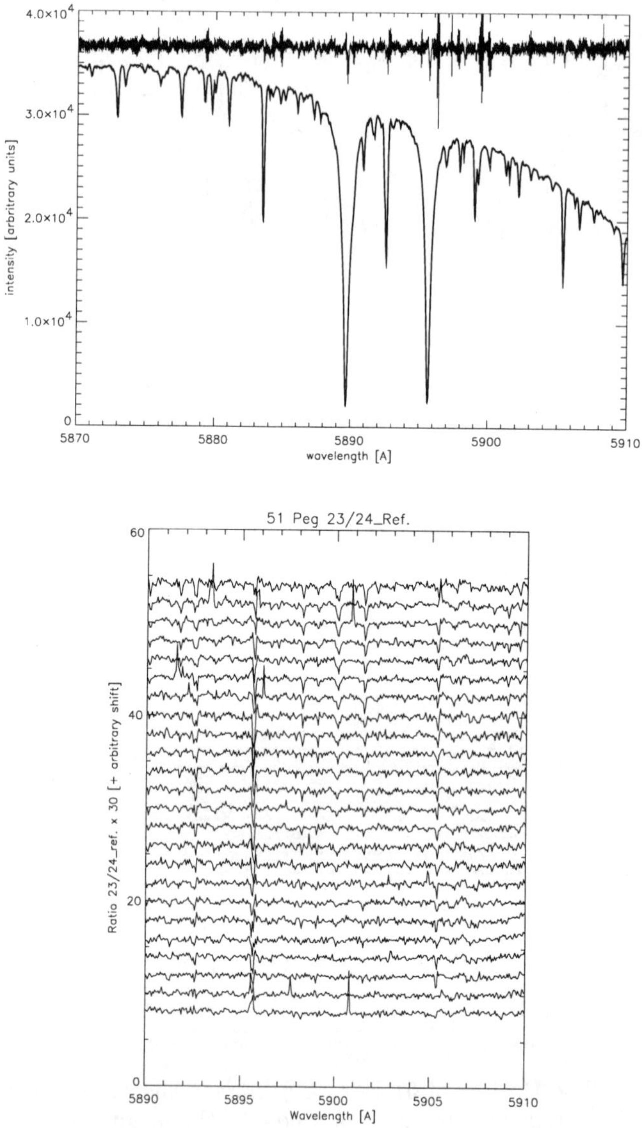

Figure 2. The Na D lines of 51 Peg. Details as in the caption to Figure 1. In this case the on-transit mean spectrum (*top frame, thin line*) is just distinguishable in parts from the mean reference (off-transit) spectrum (*thick line*).

In order to monitor the development of features during the course of the transit, the ratios of selected orders of successive individual on-transit spectra taken over a time interval of 3.5 hr, covering the predicted UT interval of the transit, to corresponding orders in the mean off-transit reference spectrum from 1998 October 24, were plotted in a sequence. As an example the sequence containing the H_β line of 51 Peg is shown in Figure 1 (*bottom frame*). Time increases downwards and the 12th plot corresponds to the time predicted for mid-transit. The time interval between successive plots is about 9 minutes. The dips in the ratio plot *bottom frame*, due to the aforementioned sampling differences, stand out clearly in the sequence frame but no time-dependent features attributable to the passage of the planetary exosphere across 51 Peg are evident. Nevertheless, it is clear that this type of sequence frame is a very useful and sensitive tool for the purposes of searching for weak time-dependent features in a large volume of spectral data.

The fact that there is considerable overlap in the wavelength coverage of adjacent echelle orders allowed many candidate features to be dismissed immediately as spurious by comparison of the same wavelength regions in different orders.

In the region around the Na D lines a number of apparently significant absorption enhancements are evident in the mean on-transit spectrum (Figure 2, *top frame*), in particular in the range 5898–5902 Å. Furthermore, the corresponding sequence frame (Figure 2, *bottom frame*) shows that these features appear to decrease in strength during the course of the transit. However, exactly the same features appear in the case of τ Boo, but enhanced in the *off-transit* reference spectrum, and also in the comparison star spectra, which demonstrates conclusively that they are telluric lines.

The maximum signal/noise ratio of the on-transit mean/off-transit mean order ratios is around 500. With a representative resolution interval of 0.15 Å the minimum (3 σ) detectable equivalent width is around 1 mÅ.

4. Conclusions

Detailed comparison of on-transit and off-transit spectra of the 51 Peg and τ Boo systems has revealed no evidence of transient spectral absorption features arising from the passage of extended planetary exospheres in front of the stars. The maximum sensitivity of this study is around 1 mÅ. The non-detection of transit-related absorption features in these cases may be due to the orbital inclinations of the planets being too small. Similar observations are now planned of the HD 209458 system, in which case transits have recently been observed via photometric monitoring (Henry et al. 2000; Charbonneau et al. 2000) and i has been determined by Charbonneau et al. to be 87.1°.

References

Butler, R. P., Marcy, G. W., Williams, E., Hauser, H., & Shirts, P. 1997, ApJ, 474, L115

Cameron, A. C., Horne, K., Penny, A., & James, D. 1999, Nat, 402, 751

Charbonneau, D., Brown, T. M., Latham, D. W., & Mayor, M. 2000, ApJ, 529, L45

Henry, G. W., Baliunas, S. L., Donahue, R. A., Soon, W. H., & Saar, S. H. 1997, ApJ, 474, 503

Henry, G. W., Marcy, G. W., Butler, R. P., & Vogt, S. S. 2000, ApJ, 529, L41

Mayor, M., & Queloz, D. 1995, Nat, 378, 355

Rauer, H., Bockelée-Morvan, D., Coustenis, A., Guillot, T., & Schneider, J. 2000, A&A, 335, 573

Schneider, J., Rauer, H., Lasota, J. P., Bonazzola, S., & Chassefiere, E. 1998, in ASP Conf. Ser., Vol. 134, Brown dwarfs and Extrasolar Planets, ed. R. Rebolo, E. L. Martin, & M. R. Zapatero Osorio (San Francisco: ASP), p. 241

Discussion

Jean-Loup Bertaux: Have you considered the effects of refraction, which bends the stellar rays away from the shadow of the planet?

Alan Harris: No. We were not aware that refraction presents a problem in the case of an extended planetary atmosphere transiting across a stellar disk.

Tristan Guillot: In answer to J.-L. Bertaux's comment, the star is not a point source when seen from the planetary atmosphere. Refraction is not a problem.

Alexander Krivov: Is it possible to give a crude theoretical estimate of the expected difference between the on-transit and off-transit depths of particular absorption lines? I.e., how strong an effect would you expect?

Alan Harris: Due to the very large uncertainties involved in predicting the characteristics of the atmosphere/exosphere of a large planet of unknown composition at a distance of only 0.05 AU from its star, I don't think it is possible at present to make useful estimates of the magnitude of the effect for particular lines. However, we are aware of some relevant theoretical work in progress in this field and we intend to follow this up and re-analyze the data in the future, should such estimates become available. At present our approach is to simply search the entire spectral range covered for any changes that correlate with the predicted transit period.

Disks, Planetesimals, and Planets
ASP Conference Series, Vol. 219, 2000
F. Garzón, C. Eiroa, D. de Winter, and T. J. Mahoney, eds.

Present and Near-Future Reflected-Light Searches for Close-In Planets

David Charbonneau[1,2] and Robert W. Noyes[1]

[1]*Harvard-Smithsonian Center for Astrophysics, 60 Garden St., Cambridge, MA 02138, USA*

[2]*High Altitude Observatory, National Center for Atmospheric Research, P. O. Box 3000, Boulder, CO 80307, USA*

Abstract. Close-in extrasolar giant planets may be directly detectable by their reflected light due to the proximity of the planet to the illuminating star. The spectrum of the system will contain a reflected-light component that varies in amplitude and Doppler shift as the planet orbits the star. Intensive searches for this effect have been carried out for only one extrasolar planet system, τ Boo. There exist several other attractive targets, including the transiting planetary system HD 209458.

1. Introduction and Motivation

Radial-velocity surveys of nearby Sun-like stars have uncovered a population of close-in orbiting companions of roughly Jupiter mass. The ten such objects with semi-major axes $a \lesssim 0.1$ AU are listed in Table 1, along with the values for the semi-major axes, (minimum) masses, spectral types of the stars, and equilibrium temperatures (calculated from the estimated values for the semi-major axes, stellar radii and effective temperatures, and assuming Bond albedo A).

A successful spectroscopic detection of an extrasolar planet in reflected light would yield the inclination, and hence the planetary mass, and would also measure a combination of the planetary radius and albedo. Furthermore, it would open the way to direct investigation of the spectrum of the planet itself. Conversely, a low enough upper limit would provide useful constraints on the radius and albedo of the companion.

The predicted albedo of a close-in extrasolar giant planet has been the focus of recent theoretical work (Marley et al. 1999; Seager, Whitney, & Sasselov 2000), and the possible values range by several orders of magnitude based on the atmospheric temperature, chemical composition (see, for example, Burrows & Sharp 1999), and the presence (or absence) of atmospheric condensates and their respective size distributions.

Table 1. Close-in planets ordered by increasing semi-major axis

Star	Spectral type	a [AU]	M_{p} [M_{Jup}]	$T_{\mathrm{eq}}(1-A)^{-\frac{1}{4}}$ [K]
HD 187123	G3 V	0.042	$0.52/\sin i$	1400
HD 75289	G0 V	0.046	$0.42/\sin i$	1600
τ Boo	F7 V	0.046	$3.87/\sin i$	1600
HD 209458	G0 V	0.047	0.69	1500
51 Peg	G2 V	0.051	$0.45/\sin i$	1300
υ And	F8 V	0.059	$0.71/\sin i$	1500
HD 217107	G7 V	0.072	$1.28/\sin i$	1000
HD 130322	K0 V	0.088	$1.08/\sin i$	800
55 Cnc	G8 V	0.11	$0.84/\sin i$	700
Gl 86	K0 V	0.11	$3.6/\sin i$	700

2. The Method of Reflected Light

The amplitude, relative to the star, of the observed flux reflected from a close-in planet viewed at a phase angle α (where α is the angle between the star and the observer as viewed from the planet) is given by

$$f_{\lambda}(\alpha) = \epsilon\ \phi_{\lambda}(\alpha) = p_{\lambda}\left(\frac{R_{\mathrm{p}}}{a}\right)^{2}\phi_{\lambda}(\alpha), \tag{1}$$

where R_{p} is the planetary radius, a is the semi-major axis, and p_{λ} is the wavelength-dependent geometric albedo. The quantity ϵ is the contrast ratio at opposition. The phase function, $\phi_{\lambda}(\alpha)$, is the brightness of the planet viewed at an angle α relative to its value at opposition, where by definition $\phi_{\lambda}(0) = 1$, and is a monotonically decreasing function of α for all physically reasonable atmospheres. For planets detected by the radial-velocity technique alone, a is determined, and α at a given time is prescribed by the known orbital phase, Φ, and the unknown orbital inclination, i, but R_{p}, p_{λ}, and $\phi_{\lambda}(\alpha)$ remain unknown.

A distant observer viewing the Solar System in the V band would measure a tiny reflected-light ratio of $f \simeq 4 \times 10^{-9}$ from Jupiter viewed at opposition. This ratio increases dramatically for close-in giant planets due to the small orbital separation. Scaling Equation (1), we find

$$f_{\lambda}(\alpha) \simeq 9.1 \times 10^{-5}\ p_{\lambda}\left(\frac{R_{\mathrm{p}}}{R_{\mathrm{Jup}}}\right)^{2}\left(\frac{0.05\mathrm{AU}}{a}\right)^{2}\phi_{\lambda}(\alpha). \tag{2}$$

This photometric modulation is most probably beyond the reach of ground-based observations, but will be accessible to upcoming photometric satellite missions such as *COROT* (see, for example, Rouan et al. 1999), provided such missions can achieve a precision of $\sim 20\ \mu$mag with stability over timescales of a few days.

Photometry alone will not yield the orbital inclination; hence, the planetary mass will be constrained only by the lower limit imposed by the radial-velocity observations. However, a spectrum of a close-in planet system will contain a secondary component that varies in brightness according to Equation (1) and in Doppler shift (relative to the star) according to

$$v_{\rm p}(\Phi) = -\, K_{\rm s}\, \frac{M_{\rm s} + M_{\rm p}}{M_{\rm p}}\, \cos 2\pi\Phi, \tag{3}$$

where $M_{\rm s}$ and $M_{\rm p}$ are the stellar and planetary masses. The stellar radial-velocity amplitude, $K_{\rm s}$, and orbital phase, Φ, are determined from the observed radial-velocity orbit of the star. For the close-in planets, $|v_{\rm p}| \simeq 100\ {\rm km\,s^{-1}}$, which is two orders of magnitude larger than the resolution of current spectrographs. If the reflected-light spectrum of the planet is detected, then the planetary mass is determined by Equation (3).

Furthermore, as first discussed in Charbonneau, Jha, & Noyes (1998), if the star has been tidally spun up so that its rotation period is equal to the planetary orbital period, then the planet would reflect a non-rotationally broadened spectrum, further distinguishing these features from the stellar lines.

In summary, the observational challenge is similar to that of transforming a single-lined binary system into a double-lined system, in the case of an exceptionally large contrast ratio between the two components. See Charbonneau et al. (1999, hereafter C99) for more details.

3. Results for τ Boo

Figure 1 summarizes our search for the reflected-light spectrum from the planet orbiting the star τ Boo, as described in C99. At the 99% confidence level, we find no evidence for a reflected-flux ratio in excess of 1×10^{-4}. For edge-on values of the inclination ($i \gtrsim 70°$), this ratio is further restricted to be less than 5×10^{-5}. Assuming a planetary radius of 1.2 $R_{\rm Jup}$ (Guillot et al. 1996), this limits the geometric albedo to $p_\lambda \leq 0.3$ for 466 nm $\leq \lambda \leq$ 498 nm.

These conclusions are marginally in conflict with the recent claim of a possible detection of reflected light for the same system by Cameron et al. (1999). They find a reflected-light ratio of $(1.9 \pm 0.4) \times 10^{-4}$ in the wavelength range from 456 nm to 524 nm, and do not detect the planet in wavelength bands on either side of this region (385 nm to 456 nm, and 524 nm to 611 nm). Figure 2 demonstrates the confidence intervals that we would have obtained on the two unknowns $\{\epsilon, i\}$ for inclinations near the one claimed by Cameron et al. (1999) and for a selection of contrast ratios.

Taking into account the different descriptions of the assumed phase function, $\phi_\lambda(\alpha)$, and the respective errors stated in the two papers, it is possible that the two results are consistent, i.e., the inclination claimed by Cameron et al. (1999) is correct, but the amplitude of the signal is less than their most probable value by 1.5 times their quoted error. However, Cameron et al. (1999) caution that there is a 5% possibility that their detection is spurious. This controversy should be resolved by observations in the near future that can now be tailored (in selection of wavelength region and orbital phase) to verify the claimed detection.

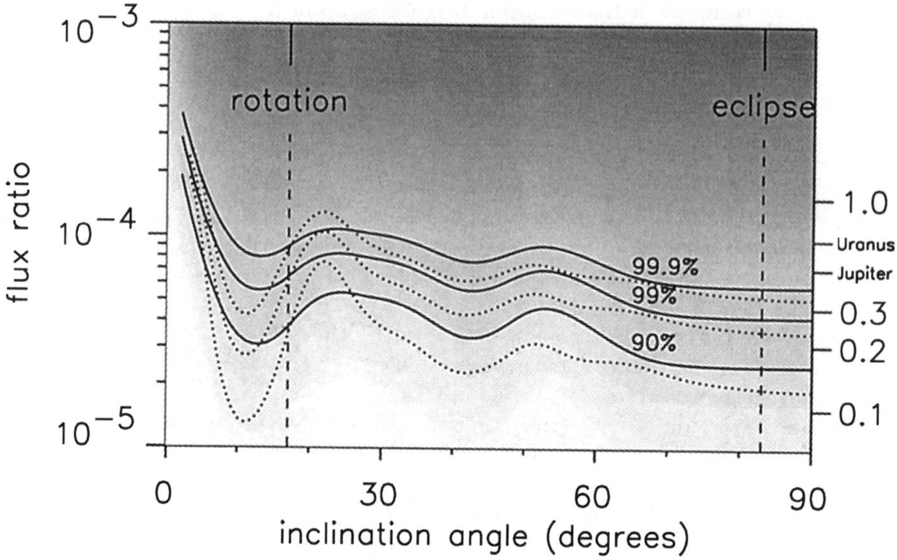

Figure 1. *Solid curves:* Confidence levels on the upper limit for the relative reflected flux, ϵ, in the τ Boo system as a function of orbital inclination, i, if the reflected light is a copy of the stellar spectrum. *Dashed curves:* The same confidence levels under the assumption that the system is tidally locked and thus the planet reflects a non-rotationally broadened copy of the stellar spectrum. Upper limits on the geometric albedo, p, under the assumption that $R_{\rm p} = 1.2\ R_{\rm Jup}$ are shown on the right-hand axis, and the values for Jupiter and Uranus are included for comparison. The lack of transits (Baliunas et al. 1997) excludes $i \gtrsim 83°$, and $i \lesssim 17°$ can be excluded under the assumption that the stellar rotation axis is co-aligned with that of orbital motion, as it would imply an abnormally large rotational velocity (Gray 1982).

4. Additional Targets

There are two principal considerations that enter into the choice of which planets may be most easily detected by reflected light. An ideal system is one with a small semi-major axis and a large stellar apparent brightness, and thus a large reflected-light signal relative to the photon noise of the star. There are two other systems that satisfy these criteria, and we compare below the prospects for detecting each of these with those for τ Boo ($a = 0.046$ AU, $V = 4.5$ mag).

The planet orbiting υ And is further from its star ($a = 0.059$ AU), but the star has a greater apparent brightness ($V = 4.1$ mag). Combining these two effects, the detection threshold that could be established (given an equivalent amount of observing time) would be 1.36 times higher relative to the primary star than that for τ Boo. For the planet orbiting 51 Peg, the larger semi-major axis

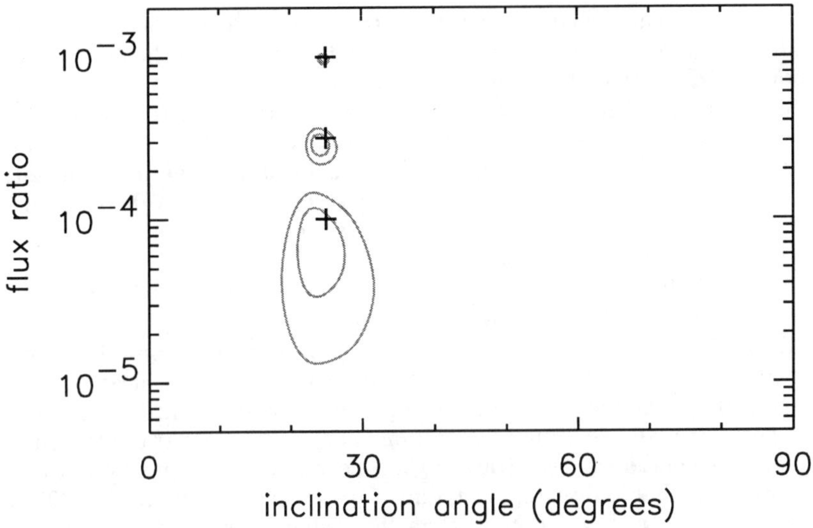

Figure 2. The detection thresholds shown in Figure 1 were tested by directly injecting a secondary into the data at a given amplitude and inclination. Three such tests are shown above, and the *black crosses* indicate the input parameters. The contours demarcate the 68% and 95% confidence regions for the derived parameters for the planet.

($a = 0.051$ AU) and fainter star ($V = 5.5$ mag) result in a detection threshold that is 1.93 times higher than that for τ Boo.

Near edge-on inclinations ($i \simeq 90°$) are desirable for reflected-light observations since they allow the planet to be viewed at the full range of phase angles. There are additional considerations that may constrain the inclination for these two systems. Assuming that the orbital planes of the three planets orbiting υ And (Butler et al. 1999) are co-aligned, stability arguments (Laughlin & Adams 1999) favor $\sin i \gtrsim 0.75$. However, *Hipparcos* astrometry favors (Mazeh et al. 1999) $\sin i \simeq 0.4$. The rotation periods of υ And and 51 Peg indicate that the stellar rotation has not become tidally locked to the orbital period, and this in turn might imply planetary masses which are near the minimum mass values (Drake et al. 1998).

We expect 51 Peg b and υ And b to have larger radii than the more massive τ Boo b (Guillot et al. 1996). If τ Boo b has been detected, it has a very large surface area. If 51 Peg b and υ And b have comparable albedos to that of τ Boo b, they should have a comparable or larger reflected-light ratio.

5. The Transiting System HD 209458

In the case of a planet that is observed to transit its star, the situation improves considerably. Since transit observations yield $R_{\rm p}$ and i, then the only remaining unknowns in Equation (1) are p_λ and $\phi_\lambda(\alpha)$.

The first transiting extrasolar planet to be discovered, HD 209458 b, was detected by Charbonneau et al. (2000) and Henry et al. (2000). The dominant uncertainty in deriving the planetary radius and orbital inclination is the estimation of the stellar radius, and not the photometric precision. A detailed analysis of the stellar modeling, and the resulting derivation of the planetary parameters and the systematic uncertainties is presented in Mazeh et al. (2000). They derive $R_{\rm p} = 1.40 \pm 0.17\ R_{\rm Jup}$, $a = 0.467$ AU, and $i = 86.^\circ1 \pm 1.^\circ6$. Substituting these values into Equation (3), we find $f_\lambda(\alpha) = 2.1 \times 10^{-4}\ p_\lambda\ \phi_\lambda(\alpha)$.

Transiting systems such as HD 209458 are particularly desirable targets for reflected-light measurements since by securing measurements near opposition, where $\phi_\lambda(0) \simeq 1$, an observed reflected-light amplitude is a direct measurement of the geometric albedo. Furthermore, one can use the spectra obtained while the planet is behind the star (rather than near inferior conjunction) to create the required stellar template spectrum. One difficulty is that HD 209458 is significantly fainter ($V = 7.6$ mag) than the systems described above.

We have simulated what could be achieved in a modest observing run using the Keck-1 telescope and HIRES spectrograph (Vogt et al. 1994), based on experience gained from our earlier observing run on τ Boo (C99). In three nights when the planet is near opposition, we would obtain a detection threshold of 4×10^{-5} for the reflected-flux ratio. Thus we would constrain the geometric albedo to be $p_\lambda \leq 0.2$, or directly detect the planet in reflected light.

Acknowledgments. We thank our collaborators in our reflected-light study of τ Boo, Sylvain Korzennik, Peter Nisenson, Saurabh Jha, Steven Vogt, and Robert Kibrick. We also thank Timothy Brown for many discussions regarding HD 209458.

References

Baliunas, S. A., Henry, G. W., Donahue, R. A., Fekel, F. C., & Soon, W. H. 1997, ApJ, 474, L119

Burrows, A., & Sharp, C. M. 1999, ApJ, 512, 843

Butler, R. P. Marcy, G. W., Fischer, D. A., Brown, T. M., Contos, A. R., Korzennik, S. G., Nisenson, P., & Noyes, R. W. 1999, ApJ, 526, 916

Cameron, A. C., Horne, K., Penny, A., & James, D. 1999, Nature, 402, 751

Charbonneau, D., Brown, T. M., Latham, D. W., & Mayor, M. 2000, ApJ, 529, L45

Charbonneau, D., Jha, S., & Noyes, R. W. 1998, ApJ, 507, L153

Charbonneau, D., Noyes, R. W., Korzennik, S. G., Nisenson, P., Jha, S., Vogt, S. S., & Kibrick, R. I. 1999, ApJ, 522, L145 (C99)

Drake, S. A., Pravdo, S. H., Angelini, L., & Stern, R. A. 1998, AJ, 115, 2122

Gray, D. F. 1982, ApJ, 261, 259

Guillot, T., Burrows, A., Hubbard, W. B., Lunine, J. I., & Saumon, D. 1996, ApJ, 459, L35

Henry, G. W., Marcy, G. W., Butler, R. P., & Vogt, S. S. 2000, ApJ, 529, L41

Laughlin, G., & Adams, F. C. 1999, ApJ, 526, 881

Marley, M. S., Gelino, C., Stephens, D., Lunine, J. I., & Freedman, R. 1999, ApJ, 513, 879

Mazeh, T., Zucker, S., Dalla Torre, A., & van Leeuwen, F. 1999, ApJ, 522, L149

Mazeh, T., et al. 2000, ApJ, 532, 55

Rouan, D., et al. 1999, Phys. Chem. Earth, 24, 567

Seager, S., Whitney, B. A., & Sasselov, D. D. 2000, ApJ, submitted

Vogt, S. S., et al. 1994, Proc. SPIE, 2198, 362

Discussion

Keith Horne: What fraction of the planet signal is removed when you form and subtract the stellar template spectrum (particularly near quadrature, where the planet velocity is not changing), and do your tests showing that you can correctly recover weak planet signals account for this effect?

David Charbonneau: We calculated the combined effect for our times of observation and found the suppression of the planetary signal to be less than 25%. This would not significantly alter our stated upper limits.

Disks, Planetesimals, and Planets
ASP Conference Series, Vol. 219, 2000
F. Garzón, C. Eiroa, D. de Winter, and T. J. Mahoney, eds.

Probable Detection of Reflected Starlight from Tau Boo b

Keith Horne[1], Andrew Collier-Cameron[1], Alan Penny[2], and David James[1,3]

[1] *Physics and Astronomy, University of St.Andrews, North Haugh, St.Andrews, KY16 9SS, Scotland, UK*
[2] *Rutherford Appleton Laboratories, Chilton, Didcot, OX11 0QX, UK*
[3] *Present whereabouts unknown*

Abstract. Hot jupiters in 4 day 0.05 AU orbits intercept and reflect $\sim 10^{-4}$ of the light from their host stars. The reflected starlight is Doppler-shifted by ~ 100 km s^{-1} due to the planet's orbital velocity. This provides a means of separating the direct starlight from the light reflected from the planetary atmosphere. We summarize our analysis of 1998 and 1999 observations of τ Boo, in which we find the signature of the reflected starlight, which gives new information about the orbital inclination, the planet's mass and radius, and the albedo spectrum of its atmosphere. Our detailed results are published in Collier-Cameron et al. (1999).

1. Direct Detection of Hot Jupiters

Direct detection of exoplanets is not easy. Planets are feeble sources, much fainter than their parent stars, and too close to be easily resolved. Light from a planet includes both thermal radiation in the infrared, and reflected starlight in the optical. In the case of Jupiter and the Sun, the planet-to-star flux ratio is $\epsilon \equiv f_{\rm p}/f_\star \sim 10^{-4}$ for infrared light and $\sim 10^{-8}$ for optical light. Because of the infrared's vastly more favorable contrast ratio, we have always assumed that infrared interferometry would enable the first direct detections of exoplanets.

Rather than directly detecting light from the planet, Doppler surveys infer the presence of orbiting planets from periodic wobbles in the motion of the host star. Some 30 extrasolar planets have now been identified by this indirect method. The four-day wobble of 51 Peg (Mayor & Queloz 1995) introduced a new class of planets—the "hot jupiters" (Table 1).

The discovery of hot jupiters has revived interest in photometric searches for planetary transits (Janes 1996; Charbonneau et al. 2000) and prompts also a re-evaluation of prospects for direct detection (Charbonneau et al. 1999; Collier-Cameron et al. 1999). Scaling from Jupiter using blackbody approximations, the contrast for infrared thermal emission is

$$\epsilon_{\rm IR} \sim 10^{-3} \left(\frac{R_{\rm p}}{R_{\rm Jup}}\right)^2 \left(\frac{a}{0.05\ {\rm AU}}\right)^{-1/2} \left(\frac{M_\star}{M_\odot}\right)^{-3/2},$$

Table 1. Hot-jupiter candidates for direct detection

Star	m_V [mag]	Sp.	$M_\star$ [$M_\odot$]	P [day]	a [AU]	$M_p \sin i$ [$M_{\rm Jup}$]	$f_{\rm p}/f_\star$ [10^{-5}]	V_p [km s^{-1}]
τ Boo	4.50	F7 V	1.2	3.31	0.046	3.9	7.9	152
υ And	4.63	F8 V	1.1	4.62	0.059	0.71	4.8	129
51 Peg	5.49	G2 IVa	1.0	4.23	0.05	0.47	6.7	134
HD 75289	6.35	G0 V	1.1	3.51	0.054	0.42	5.8	135
HD 209458	7.65	G0 V	1.1	3.52	0.045	0.69	8.3	148
HD 187123	7.79	G5	0.93	3.10	0.042	0.52	9.6	141
HD 46375	7.94	K1 IV	0.74	3.02	0.041	0.25	10	127

Assumed planet radius $R_{\rm p} = 1.2\ R_{\rm Jup}$ and albedo 0.55.

while that for optical reflected starlight is

$$\epsilon_{\rm opt} \sim 10^{-4} \left(\frac{R_{\rm p}}{R_{\rm Jup}}\right)^2 \left(\frac{a}{0.05\ {\rm AU}}\right)^{-2} .$$

On this basis, moving Jupiter 100 times closer to the Sun would boost its thermal emission by ~ 10 and its optical light by $\sim 10^4$. For hot jupiters, then, the infrared advantage may be only ~ 10.

The primary obstacle to direct detection is the difficulty of disentangling the planet's feeble light from the glare of its parent star. The angular separation,

$$\theta_{\rm p} \sim 5 \times 10^{-3} \left(\frac{a}{0.05\ {\rm AU}}\right) \left(\frac{D}{10\ {\rm pc}}\right)^{-1}\ {\rm arcsec},$$

is 100 times less for hot jupiters than for a planet at 5 AU, raising the hurdle for direct imaging and interferometry. However, a hot jupiter's orbital velocity,

$$V_{\rm p} \sim 150 \left(\frac{a}{0.05\ {\rm AU}}\right)^{-1/2} \left(\frac{M_\star}{M_\odot}\right)^{1/2}\ {\rm km\ s^{-1}},$$

is much larger than the rotationally broadened widths of stellar absorption lines. This makes it easier to use Doppler shifts rather than angular separations to resolve the star and planet light.

2. The Signature of Reflected Starlight

Given observed spectra of a star with an orbiting planet, how can we disentangle the planet's light, some 10^4 times fainter, from that of the star? The key insight is to realize that starlight reflected from the planet's atmosphere retains the same pattern of absorption lines that is present in the direct starlight, but with the strengths, wavelengths, and widths of the lines changing in characteristic ways that depend on the planet's atmosphere and its orbit around the star. These effects allow us to distinguish the planet's light from that of its star.

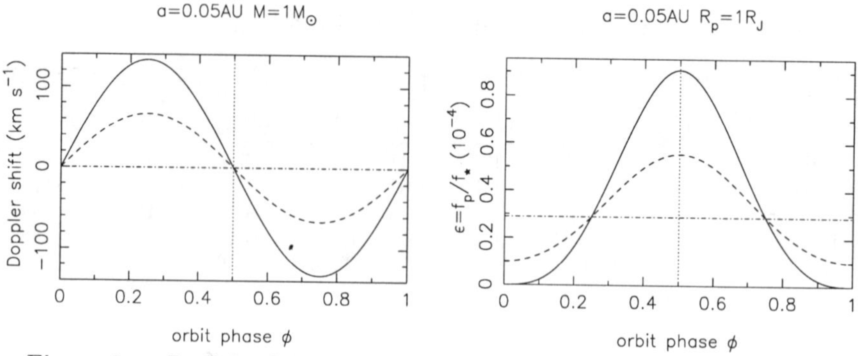

Figure 1. Radial-velocity curve (*left*) and lightcurve (*right*) for an albedo 1 Jupiter-radius planet in an 0.05 AU orbit around a 1 $M_\odot$ star. The adopted phase function is similar to that of Venus. Results are shown for three orbit inclinations, 90° (*solid*), 30° (*dashed*), and 0° (*dot-dashed*).

Figure 1 shows the expected radial-velocity curve and lightcurve for starlight reflected from an orbiting hot jupiter. We know from the star's Doppler wobble the phase, ϕ, of the planet's orbit. For a circular orbit, the planet's radial velocity curve is

$$V(\phi) = \left(\frac{GM_\star}{a}\right)^{1/2} \sin i \sin\phi = K_{\rm p} \sin\phi \; .$$

The velocity semi-amplitude $K_{\rm p} = V_{\rm p} \sin i$ vanishes for a face-on orbit and attains its full value $V_{\rm p} \sim 150$ km s^{-1} for an edge-on orbit. By observing $K_{\rm p}$, we determine $\sin i$ and hence the planet's mass, since $V_{\rm p}$ and $M_{\rm p} \sin i$ are known from the star's mass and the amplitude of its Doppler wobble.

The reflected light is strongest at phase 0.5, when the planet is on the far side of the star so that we view its illuminated face, and dimmest at phase 0, when we view its dark side. The modulation is complete for edge-on orbits and vanishes for face-on orbits. The planet-to-star flux ratio can be written as

$$\epsilon(\lambda,\phi) = p(\lambda) g(\lambda,\alpha) \left(\frac{R_{\rm p}}{a}\right)^2 .$$

By observing the albedo spectrum, $p(\lambda)$, and phase function, $g(\lambda,\alpha)$, where α is the star–planet–observer angle, we can compare with model predictions to learn about the composition and structure of hot-jupiter atmospheres.

The planet "sees" the star rotating at a different rate than we do. The lines we see in the light reflected from the planet can therefore be narrower or broader than those in the direct starlight. In the case of τ Boo, the star corotates with the planet's orbit, so that the planet receives starlight with no rotational broadening.

3. Echelle Spectroscopy of τ Boo

Table 1 summarizes parameters of hot jupiters that are candidates for direct detection. We selected the best candidate, τ Boo, for study with the Utrecht Echelle Spectrograph (UES) on the 4.2 m William Herschel Telescope (WHT) at La Palma. Our spectra cover 384–612 nm with $R \sim 49\,000$ (6 km s^{-1}). We had four clear nights in a six-night run in 1998 April. In 1999 we had five clear nights in April, May, and June, targeting orbit phases 0.4 to 0.6 when the planet is brightest.

4. Disentangling the Planet Signal from the Starlight

Our first challenge is to subtract the spectrum of the starlight without also subtracting the spectrum of the planet. Like Charbonneau et al. (1999), we construct a high-fidelity model of the starlight from the observed spectra and subtract this to remove the starlight. This unfortunately also removes part of the planet signal, but since the planet velocity changes appreciably during each night—except near quadrature phases—some of the planet signal remains behind.

Our starlight model is derived from high signal-to-noise ratio template spectra obtained by averaging the spectra obtained on each night. Before subtracting, we "tweak" the template spectra to account for small changes in the strength, shift, and width of the lines due to instrumental (flexure, focus, guiding) and atmospheric (seeing, differential refraction) effects. We fit each spectrum with a linear combination of the template and its first two derivatives, allowing the three coefficients to be smooth spline functions of wavelength and echelle order.

After starlight subtraction, the remaining planet signal, spread throughout the spectrum as faint Doppler-shifted copies of several thousand spectral lines, is completely buried in Poisson noise. We draw together this signal, using the known strengths and wavelengths of the stellar lines, to form a single velocity profile from each echelle spectrum. The planet signal at favorable orbital phases should now be just barely above the Poisson noise.

Examining the velocity profiles vs. orbit phase (Figure 2) reveals at low velocities a pattern of time-correlated residuals at the 10^{-4} level. We suspect artifacts of starspots crossing the face of the star, and to avoid these, we reject from our analysis all data inside $\pm$ 31 km s^{-1}. Our future observing strategy will target phases 0.3–0.4 and 0.6–0.7, rather than phase 0.5, to keep the planet signal away from these stellar artifacts.

Finally, we stack up the planet's signal over all the orbital phases covered by our spectra, taking account of the expected lightcurve and radial-velocity curve of the planet, and the partial suppression of the planet signal on each night. The orbit inclination is unknown, so we must do this for a range of possible inclinations (equivalently K_p). The evidence for a planet at each K_p and ϵ is assessed by noting the decrease in χ^2 compared with a no-planet assumption (Figure 3).

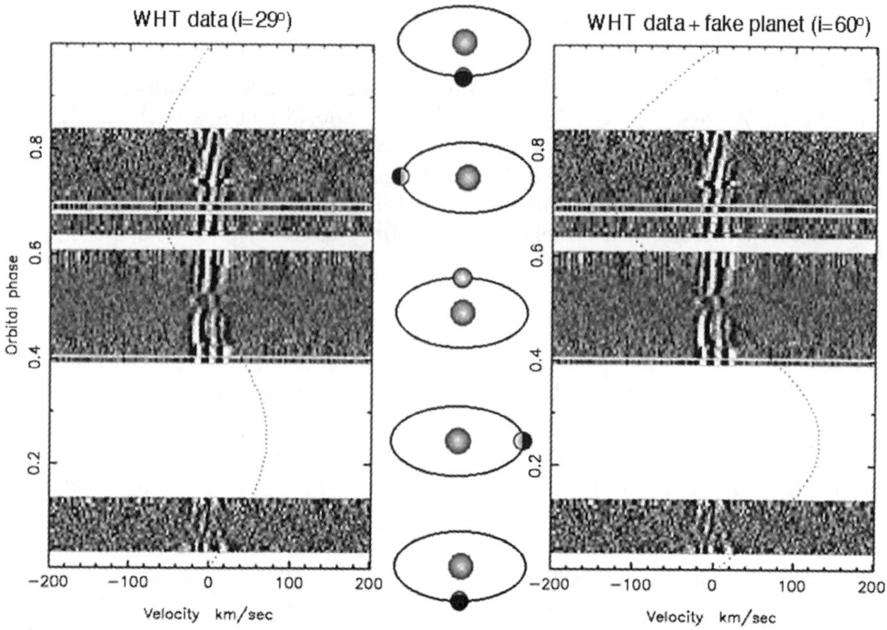

Figure 2. Residuals vs. velocity and orbit phase formed by subtracting the starlight spectrum and then combining several thousand absorption line profiles. The "barber pole" pattern inside ± 28 km s^{-1} may be an artifact of starspots. *Right:* Results of an identical analysis after adding a fake planet signal to our spectra. The fake planet's signature is visible here as a dark streak crossing from red to blue at phase 0.5. Velocity curves for $i = 29°$ and $60°$ are indicated by dotted curves in the *left* and *right panels* respectively.

5. Discussion of the Candidate Detection

We tested our method by injecting a variety of fake planet signals into our data, finding that our data set and analysis method are easily capable of recovering the correct $K_{\rm p}$ and ϵ for planets with Jupiter-like radii and albedos in moderate inclination orbits ($i > 60°$). Like Charbonneau et al. (1999), we see no such signals in our data (Figure 2) and can rule out such planets with high confidence (Figure 3).

Our strongest candidate planet signal (Figure 3) is a 2–3 σ peak at $K_{\rm p} = 74 \pm 3$ km s^{-1} and $\epsilon = (7 \pm 3) \times 10^{-5}$. This corresponds to $i = 29°$, $M_{\rm p} = 7.9\ M_{\rm Jup}$, and $R_{\rm p}\sqrt{p} = a\sqrt{\epsilon} = 0.8\ R_{\rm J}$.

By separate analysis of six subsets of the echelle orders, we construct a crude albedo spectrum. The signal is then stronger ($\epsilon \approx 2 \times 10^{-4}$, $R_{\rm p}\sqrt{p} \approx 1.4\ R_{\rm Jup}$) for blue–green light, but consistent with 0 for red and violet light. This pattern suggests that molecular absorption bands may be present.

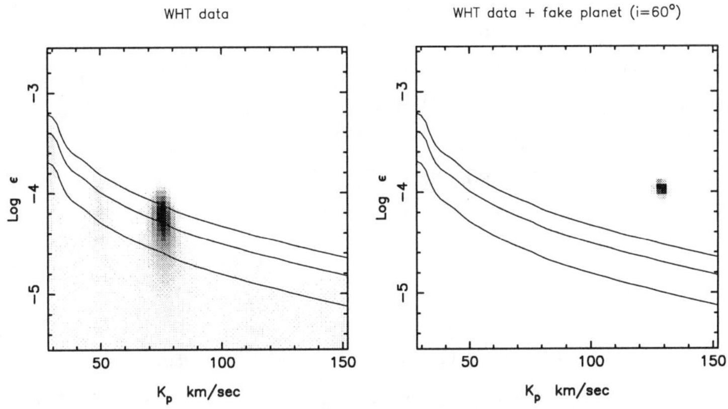

Figure 3. Probability map ($\exp(-\Delta\chi^2/2)$) derived from the residuals in Figure 2 for planets with orbit velocity K_p and planet-to-star flux ratio ϵ. Lines give 1, 2, 3 σ detection thresholds for K_p fixed. *Right:* A clean recovery of the fake planet signal. *Left:* A candidate planet signal at $K_p \sim 74$ km s^{-1} and $\epsilon \sim 7 \times 10^{-5}$.

The candidate planet signal would have a formal significance of 2–3 σ if K_p were known. Allowing K_p to be anywhere from 60 to 152 km s^{-1}, we estimate a false alarm probability of about 5% based on the bootstrap method (scrambling the orbit phases of the spectra) and on periodogram analysis (folding data on other planet periods).

Our candidate detection is a factor ~ 2 stronger than the upper limit set by Charbonneau et al. (1999) based on a similar analysis of three nights of Keck/HIRES spectra. Comparing the two analyses in detail, we attribute a 30% difference to adoption of different phase functions, and another 20% to different treatments of planet attenuation during starlight subtraction. The remaining 30% discrepancy is within the uncertainties, but might also reflect a Malmquist bias—the best-fit signal strength is too large because the fit optimizing K_p moves to whichever side of the true K_p happens to be enhanced by noise.

6. Future Prospects

We are observing τ Boo again with the WHT/UES in spring 2000. Weather permitting, these observations should increase our signal-to-noise ratio enough to eliminate the remaining 5% doubt about the detection, and hopefully measure the albedo spectrum with enough clarity to identify absorption features impressed by the planet's atmosphere. Other targets are amenable to similar study.

References

Charbonneau, D., et al. 1999. ApJ, 522, L145

Charbonneau, D., Brown, T., Latham, D., & Mayor, M. 2000. ApJ, 529, L45

Collier-Cameron, A., Horne, K., Penny, A., & James, D. 1999. Nat, 402, 751

Janes, K. 1996. JGR, E101, 14853

Mayor, M., & Queloz, D. 1995. Nat, 378, 355

Discussion

Tristan Guillot: What are the spectra that you have to throw away because the planetary signal is too close to the $\pm$ 28 km/s stellar signature?

Keith Horne: We don't throw away whole spectra, but we do omit all data inside $\pm$ 31 km s^{-1} because this range is affected by the "barber pole" residuals left over from subtraction of the stellar spectrum. We suspect this to be due to starspot activity at a level similar to that on the Sun. This gap has little effect for large K, but increases as K is reduced.

William Borucki: Did you calculate your false alarm rate by sweeping through possible orbital velocities?

Keith Horne: Yes. The false alarm probabilities are given for $K_{\rm min} < K < 152$ km s^{-1} and is about 5% for $K_{\rm min} \sim 60$ km s^{-1}.

David Charbonneau: You actively account for the suppression of the planetary signal by altering your matched template. In order to do this, you need to assume an orbital inclination to set the range of radial-velocity shifts for the times of observation. Are you concerned that this procedure injects the orbital information that you are searching for in the analysis?

Keith Horne: The compensation is made by averaging the expected planetary signal over time on each night and subtracting that average. This mimics the suppression of the true planetary signal that inevitably occurs when the starlight spectrum is subtracted using a template obtained by averaging spectra on each night. There is no "injection of orbital information".

Disks, Planetesimals, and Planets
ASP Conference Series, Vol. 219, 2000
F. Garzón, C. Eiroa, D. de Winter, and T. J. Mahoney, eds.

Late Planetesimal Delivery and the Composition of Giant Planets

Tristan Guillot and Brett Gladman

Observatoire de la Côte d'Azur, Lab. Cassini, CNRS UMR 6529, BP 4229, 06304 Nice Cedex 4, France

Abstract. The heavy elements in the giant planets (the total abundance, the fraction of the planet's mass, and the ratios of various chemical species) can tell us much about their formation. Solar and extrasolar giant planets give us different types of information: our four giants give us more detailed information than we can obtain about extrasolar giants, but will hopefully soon be dwarfed by the better statistical (but less precise) information on extrasolar planets. We discuss here the constraints available about giant-planet formation in both our Solar System and others. In particular, we propose that the heavy-element enrichment seen in the envelopes of our giants implies that they underwent an early phase with significantly enlarged effective capture radii.

1. Introduction

The last few years have shown that giant planets similar to our Jupiter and Saturn orbit stars other than the Sun, and even that they are relatively common companions of solar-type stars (Mayor & Queloz 1995; Marcy & Butler 1998). The definitive confirmation of their reality came from the detection of the slight dimming of the light of star HD 209458 due to the transit of a $\sim 0.6\,\mathrm{M_{Jup}}$ planet of inferred radius ≈ 1.2–$1.5\ R_{\mathrm{Jup}}$ (Charbonneau et al. 2000; Henry et al. 2000). The radius seems to agree with theoretical predictions for the evolution of solar-composition giant planets (Guillot et al. 1996; Guillot 1999b; Burrows et al. 2000) and thus confirms the expectation that these objects should be mainly made of hydrogen and helium. The possible detection of the stellar light reflected by the planetary companion of τ Boo (Cameron et al. 1999) serves as an example that other methods will also increase our knowledge of these systems in the near future.

Giant planets hold crucial information concerning the structures and compositions of the protostellar nebulae that bore stars and planets. In this article, we show how constraints on the composition of the giant planets in our Solar System, and soon-to-come characterization of extrasolar giant planets, can yield a better understanding of planet formation. In Section 2, recent observations and constraints from interior and evolution models are presented. In Section 3, we discuss current theories of formation of the giant planets. We then use dynamical simulations to understand the fate of planetesimals in the early Solar System. The results are discussed in the last section.

2. Relevant Observations and Constraints

The most important conclusion to be drawn from the discovery of extrasolar planets is that planet formation is not an extremely rare outcome of astrophysical evolution. Indeed, about 5% of solar-type stars in the solar neighborhood have been found to possess a giant planetary companion (e.g., Marcy & Butler 2000). Planets of smaller masses or orbiting further from their star are still waiting to be discovered, so the total fraction of planet-bearing systems may be significantly larger than 5%.

Observations thus confirmed theoretical expectations that planets should be found around other stars, but at the same time and somewhat mischievously Nature provided us with another puzzle: the unexpected presence of giant planets at only a few stellar radii of the primary. Although models of in situ formation exist, it seems difficult to avoid the conclusion that orbital migration of planetesimals or planets is the mechanism responsible for the existence of these "hot jupiters" (e.g., Lin, Bodenheimer, & Richardson 1996). Models of planetary formation have to either include migration processes or provide mechanisms to prevent them from occurring.

On the other hand, observations of star-forming regions indicate the presence of gas-disks around young stars, the lifetime of these being 1 to 10 Myr (Strom, Edwards, & Skrutskie 1993). Slightly older stars also have infrared excesses, even up to model ages of $\sim$ 400 Myr (Habing et al. 1999). This infrared emission is caused by the presence of dust. Since dust is generally efficiently destroyed (by spiraling in onto the star), collisions and the presence of more massive bodies are required to explain its presence.

However, most of the quantitatively useful information needed to constrain formation theories still lies inside our own Solar System. We will focus here on information obtained from observations of the outer Solar System, working from the outside in.

The roughly spherical Oort cloud, an ensemble of about 10^{8-10} of comets located at $\sim$ 1000–20 000 AU from the Sun, is witness to an early phase of the outer Solar System, when km-sized planetesimals formed near the giant planets were ejected to the this region by powerful gravitational slingshots (see Duncan, Quinn, & Tremaine 1987; Hahn & Malhotra 1999).

Extending outwards from about 30 AU we find the Kuiper Belt, representing the inner edge of the "disk" of planetesimals which have resisted the erosive effects of the giant planets over the age of the Solar System. Far from being a dynamically cold vestigial disk, the Kuiper Belt is beginning to show, as more of its objects have their orbits better determined, extensive dynamical structure. This structure is characteristic of the processes that occurred during the final stages of giant-planet formation, a topic currently of intense research interest from both the theoretical and observational side.

Still closer in, the Sun possesses two giant planets as companions, Jupiter and Saturn, and two so-called *ice giants*, Uranus and Neptune. It is not clear at the moment if this is very different from "standard" extrasolar planetary systems, given the fact that the sensitivities of the radial-velocity techniques are still too low to detect companions with masses significantly smaller than that of Saturn, and that objects with orbital periods larger than a few years are not detected.

Jupiter and Saturn are mainly made of hydrogen and helium, and therefore have chemical compositions that are reminiscent of that of the Sun. However, they are enriched in heavy elements (elements other than hydrogen and helium) compared to our star: Jupiter's atmosphere contains about three times more carbon, nitrogen, sulfur, argon, krypton, and xenon than our Sun (Owen et al. 1999). The composition of the atmosphere of Saturn is more poorly determined, but it clearly contains an even larger fraction of carbon and probably nitrogen (e.g., Gautier & Owen 1989). Uranus and Neptune are globally mainly made of ices (a mixture of oxygen, carbon, and nitrogen compounds); their hydrogen–helium atmospheres are enriched in carbon and probably nitrogen by a factor of about 30.

The composition of the visible part of the atmosphere is not necessarily relevant to the composition of the entire planet. Indeed, constraints from gravitational measurements shown in Figure 1 reveal the presence of a dense central core of less than about 15 $M_\oplus$ for all four planets, but with an ambiguity in the case of Jupiter (interior models of this planet are also consistent with no core). The presence of such a core has traditionally been used as a constraint for formation models, with the idea that it was formed early and remained untouched throughout the evolution of the planet (Stevenson 1982). However, formation models should also reproduce the *total* amount of heavy elements found to be present in the giant planets (Figure 1).

3. The Formation of Planets

Although far from being well understood, the formation of stars and brown dwarfs is believed to be the natural outcome of the collapse and fragmentation of a giant molecular cloud (e.g., Bodenheimer, Ruzmaikina, & Mathieu 1993). This process probably cannot form objects less massive than about 10 $M_{\rm Jup}$, a prediction that will be directly tested by observations of very young clusters in which low-mass objects can be detected because of their high initial heat (Zapatero-Osorio et al. 1999). Planets subsequently form within the circumstellar accretion disk (e.g., Lissauer 1993), which is both accreted onto the central star(s) and dissipated by stellar wind, photoevaporation or other mechanisms. Observations of young stars prove these disks to be common, generally not as massive as that inferred for the Solar System, and with lifetimes typically between 1 and 10 Myr (e.g., Strom et al. 1993).

A crucial step in the process of planetary formation is the creation of micron-sized grains in the nebula, with their composition depending on the local temperature and pressure, which must have varied with location and time. Conventional wisdom says that rocks, more refractory materials such as silicates and iron, will form almost everywhere in the nebula whereas ices, a combination of various materials among which water is the most abundant species, will form further from the star, beyond the so-called ice line. The different condensation temperatures for given chemical species qualitatively explain the decreasing mean densities of terrestrial planets with increasing distance from the Sun. The

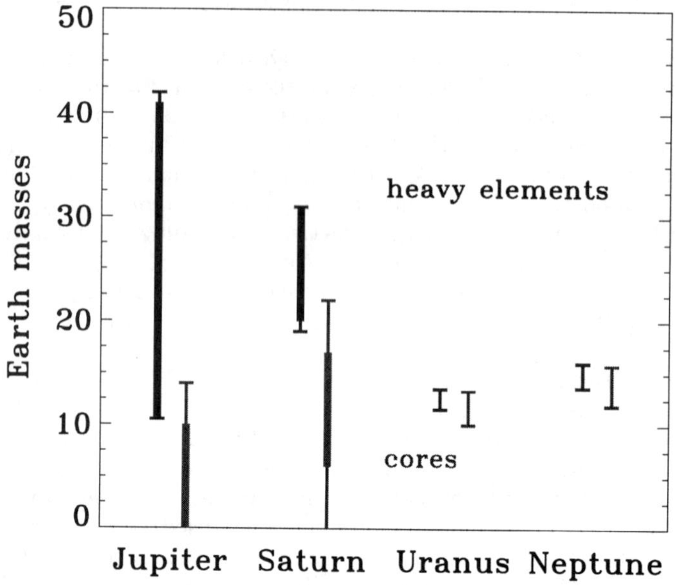

Figure 1. Limits on the abundances of heavy elements in the four Jovian planets in our Solar System, according to interior models matching their measured radius and gravitational moments. For each planet, the error bar on the left represents the total amount of high-Z material (heavier than He), whereas the (lower) error bar on the right shows the mass of heavy elements that are segregated into the planet's core. For Jupiter and Saturn, the thicker ranges take into account additional constraints obtained from planetary evolution models. Note the high level of uncertainty, especially regarding the core masses of Jupiter and Saturn. Models of Jupiter with small cores (i.e., less than 2 $M_\oplus$) require significant enrichments in total heavy-element abundance (i.e., more than 20 $M_\oplus$). (From Wuchterl, Guillot, & Lissauer 2000).

same feature is also observed in the case of the Galilean satellites, with close-in Io being twice as dense as distant Callisto.

One of the great puzzles for the next decade will certainly be understanding why this simple theory works at all, with its implicit assumption that planets are assembled from material that formed at their present distances! Many mechanisms are expected to produce significant orbital migration of solid material in the protosolar nebula, from micron-sized grains to giant planets. But how could rather sharp gradients of chemical composition, such as that observed in spectral taxonomic types in the asteroid belt, persist in the presence of such large and seemingly unpredictable orbital displacements of solid material? Another puzzle is the recent observation by the *Galileo* probe of a 3 × solar enrichment in all of C, N, S, Ar, Xe, and Kr in the Jovian upper atmosphere (Owen et al.

1999). Because Ar, Xe, and Kr are only trapped in solid condensates at very low temperature, this seems to imply that the planetesimals that bombarded the atmospheres of the giant planets were formed in a very cold nebula, either very far from the Sun or very late in the history of the disk; both possibilities seem difficult to reconcile with the fact that many earth masses of these objects must have then successfully reached the giant planets!

A useful first indication of the structure of the early protosolar nebula comes from the estimation of the *minimum* amount of gas that initially had to be present in the disk in order to form the planets that we see today. The result is commonly called the *minimum mass nebula* (see Weidenschilling 1977; Hayashi 1980). We present an update of the Weidenschilling work, accounting for the better-known internal structures of the giant planets. Figure 2 shows the minimum surface density (g cm^{-2} projected onto the plane of the Solar System) of hydrogen and helium, as a function of distance to the Sun, *assuming that the planets were formed at their present locations.* The gas-to-solid ratio was assumed to lie between 55 and 90; these two extremes correspond to 1) the condensation of all species except H, He, and noble gases, and 2) the condensation of only water and metals and only small amounts of condensed C (which is then assumed mostly to remain in the form of gaseous CO) and N (remaining mostly bound up as N_2), respectively.

In Figure 2, the surface density required by two models of formation of the giant planets are indicated. The most "standard" model of formation of the giant planets is based on the formation of a solid protoplanetary embryo followed by the capture of the surrounding hydrogen and helium, on a few-million-year timescale (Pollack et al. 1996). The density required to form Jupiter in less than 10 million years is just slightly over the the new median estimate for the minimum-mass nebula. Neglecting any migration process, this implies that between 20 and 75% of the solids in that region have been lost, probably due to dynamical evolution. (These numbers account for the fact that the gas-to-solid ratio has to be the same when comparing different results). The density increases over the minimum one required to form Saturn, Uranus, and Neptune is higher, implying an even larger ejection efficiency (or other loss mechanism) for solids in those regions.

An alternative model that might explain the formation of Jupiter is by direct gravitational instability of the gas itself, on much shorter timescales (Boss 1998, 2000). This requires even larger densities than in the core-accretion scenario, but it could be advocated that part of the gas present in the disk at that early times has been accreted onto the star. The subsequent formation of central cores in these models would require an early settling of heavy elements.

In any scenario, it appears that a significant fraction of the heavy elements that are present in the *envelopes* of Jupiter, Saturn, Uranus, and Neptune had to be brought there during or after the planets had acquired a significant fraction of their mass because it is extremely difficult for convection to overcome the strongly stabilizing compositional gradient due to the presence of the central core. Furthermore, the atmospheres of the giant planets are observed to be enriched in heavy elements compared to the Sun. Interior models also point to significant amounts of heavy elements that are not embedded in the central core (see Figure 1).

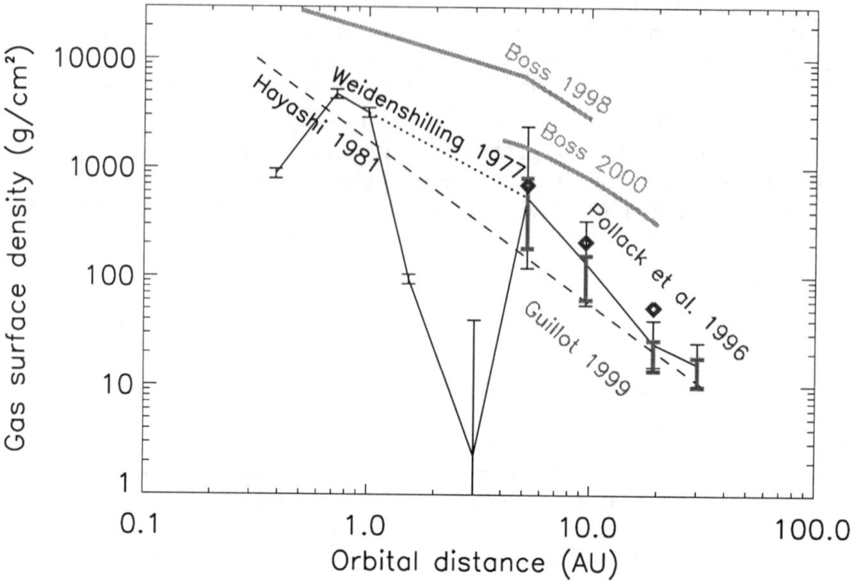

Figure 2. Surface density of hydrogen and helium as a function of distance to the Sun, as estimated by various workers. The *dashed* and *plain black lines* correspond to the minimum-mass protosolar nebula as derived by Hayashi (1980) and Weidenshilling (1977) respectively. The extreme dip in the 2–4 AU asteroid-belt region simply indicates that the current asteroid population represents only a small fraction of the primordial one (see Petit et al. 2000). The thicker vertical error bars outside 5 AU are updates of the Weidenshilling values using interior models of the outer planets from Guillot (1999a). The *diamonds* are the optimal surface densities for giant-planet formation in a core-accretion scenario, assuming a gas-to-solid ratio of 70 (Pollack et al. 1996). The two *upper lines* (Boss 1998, 2000) correspond to a scenario of formation of Jupiter and possibly Saturn by direct gravitational instability in the gas.

4. Delivering Planetesimals to the Giant Planets

To study the delivery of heavy-element-rich planetesimals to the forming giant planets, we have performed extensive numerical dynamical simulations. We used a regularized mixed-variable symplectic integrator (rmvs3) written by Levison & Duncan (1994). For this work, we assumed that the giant planets begin on their current orbits, although they then interact gravitationally in a realistic manner. Massless test particles (that is, planetesimals with masses negligible compared to the planets) were initially linearly distributed between 4 and 35 AU, corresponding to an r^{-1} heliocentric distance surface-density profile.

As a baseline case, we set the masses and radii of the giant planets to their present-day values, exploring a scenario in which the planets "suddenly" reach nearly their current masses by a rapid gas accretion onto a much smaller core. Before such a rapid gas collapse the gravitational effects of the cores would be comparatively small over most of the outer Solar System. We integrated 10 000 planetesimals for 100 Myr. After that time, 61% of the initial particles had been ejected out of the system, 23% had been sent to the Oort cloud (aphelia larger than 10 000 AU), with only 13% remaining in the system. Only 4% of the particles impacted one of the four giant planets. In this physical scenario, the probability of impact is low compared to that of ejection, mainly due to the presence of Jupiter, to which the other giant planets efficiently "pass" their planetesimals. However, this inefficiency of planetesimal accretion poses grave problems when we consider the known mass of heavy elements in the giant planets.

These results can be compared (Figure 2) to the accretion efficiencies needed to provide the giant planets with their present-day heavy-element inventories. In this article, we choose to focus on the core-accretion scenario and leave the gas-instability model for further studies. The ratio between the inferred amount of heavy elements in the giant planets (Guillot 1999a) and the amount of solids required for their formation (Pollack et al. 1996) yields the *accretion efficiency*. That is, the fraction of planetesimals that would need to be accreted to give their current heavy-element contents. This is shown by thin error bars in Figure 3. (Note that in the case of Neptune, which is not included in Pollack et al. 1996, we assumed the optimal surface density to be 0.30 g cm^{-2}.) The accretion efficiencies thus derived are quite large, i.e., over 30% for the three planets.

These kinds of efficiencies are impossible to reach by gravitational accretion of planetesimals by planets with physical properties (masses and radii) similar to the current giant planets. The planets are too efficient at ejecting material from the system compared to accreting it. However, the first phase of runaway growth is expected to proceed with a high accretion efficiency and lead to the formation of the cores of the giant planets. A better estimate of the required efficiency after capture of the hydrogen–helium envelope hence stems from the ratio between the amount of heavy elements that are *in the gaseous envelope only* (thick vertical error bars, as opposed to the thin bars representing the entire planet).

The results of our work are shown as diamonds in the case of the run with the present-day planets. Except for Jupiter, the accretion efficiencies are *still* too low, and cannot explain the enrichment of the envelopes of these planets.

However, we observed that most of the planetesimal accretion probably took place in the first 1 to 10 Myr, an epoch in which gas was still present in the nebula and planets had probably not yet grown to their present masses. The work of Pollack et al. (1996) predicts a relatively extended phase in which planetary embryos have masses between 10 and 20 $M_\oplus$. Furthermore, due to the filling of their Hill spheres by gas, the protoplanets are expected to have effective capture radii that are larger than their present-day radii (Podolak, Pollack, & Reynolds 1988; Pollack et al. 1996). This motivated an alternative run of 1000 particles between 4 and 35 AU as before, in which Jupiter and Saturn were assumed to have 20 $M_\oplus$ and Uranus and Neptune 10 $M_\oplus$. The radius of all four

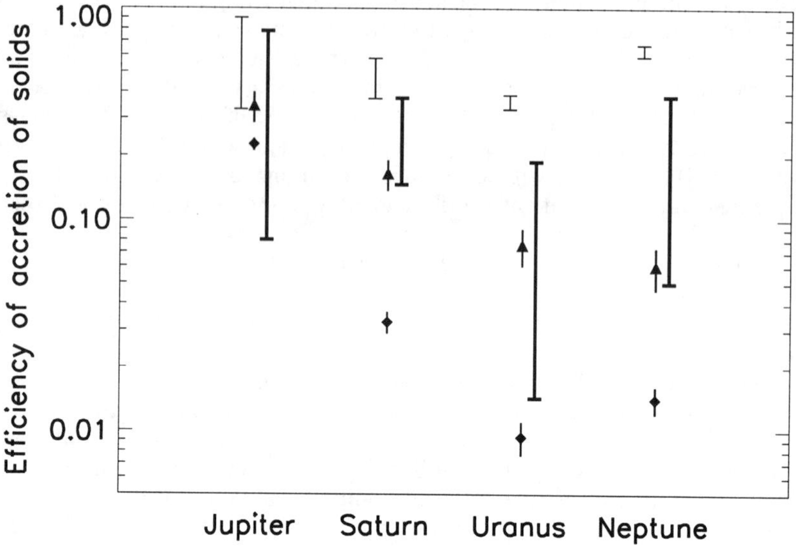

Figure 3. Accretion efficiences required to form the giant planets in the core-accretion scenario. The accretion efficiency for a given planet is defined as the quantity of heavy elements that had to impact the planet divided by the total amount of heavy elements within the annulus that extends halfway to the next planet. The *upper left error bar* for each planet corresponds to the ratio between the total mass of heavy elements and the quantity of solids required by Pollack et al. (1996). The *lower thick error bars* account for core formation by runaway growth and include only heavy elements in the envelopes. *Diamonds* are accretion probabilities in the Jupiter, Saturn, Uranus, and Neptune regions, respectively, as calculated in the standard case. *Triangles* correspond to the reduced-mass case (see text). The *vertical lines* represent the statistical error due to the finite number of events.

planets was set to 3 $R_{\rm Jup}$ (210 000 km). We let the integration proceed for only 10 Myr.

In this second case, accretion efficiencies (shown as diamonds in Figure 3) were found to be consistent with the constraints. Thus, the enriched envelopes of the giant planets may *prove* that those planets passed through a phase of $\sim$ 1–10 Myr, during which they had greatly increased planetesimal cross-sections. Further work is needed to check whether such large radii are effectively possible, and whether a more likely scenario with a shorter embryonic phase may yield similar results. In any case, it appears that the contraction of giant planets after they have reached their present masses (Guillot et al. 1995) is too fast to explain the large enrichments of their envelopes.

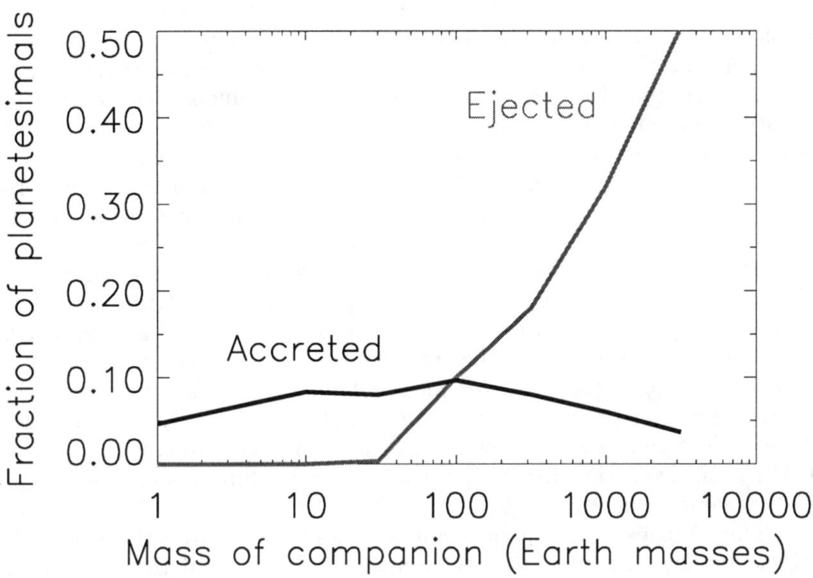

Figure 4. Probability of ejection and of accretion as a function of the mass of the planet/brown dwarf, assuming an $a_p = 5.2$ AU orbit around a 1 $M_\odot$ star. The planet is initially embedded in a disk of planetesimals that are linearly distributed between 4 and 13 AU. The planet/brown dwarf was assumed to have the same radius at that of Jupiter, independently of its mass.

5. Extrasolar Giant Planets: What Can Be Learned from Radius Measurements

Crucial constraints for planet formation will be derived from the determination of the radii of giant planets around other stars. This is certainly a challenging task, both observationally, because a large number of stars have to be monitored, and theoretically, because accurate atmospheric models must be developed. To first order, the radius of a planet is inversely proportional to the amount of heavy elements it contains. Theoretical uncertainties in radius determinations can be quantified and turn out to be small enough to reach interesting conclusions on planetary composition (Guillot 1999b).

The case of HD 209458b is enlightening, although it is probably more complex than previously thought. From observations of the planetary transit, the large measured radius of $\sim 1.4\ R_{\rm Jup}$ (Charbonneau et al. 2000; Henry et al. 2000) shows that this 0.69 $M_{\rm Jup}$ planet is mostly made of hydrogen and helium (Burrows et al. 2000). However, uncertainties in the atmospheric models used to calculate the evolution of "hot jupiters" forbid a more quantitative determination of its composition for the time being.

A dynamical model is useful to predict the amount of heavy elements expected for extrasolar planets, depending on their mass and orbital distance. Here, we show an example of a calculation of the resulting accretion efficiency for a giant planet at 5.2 AU from a solar-type star, as a function of the planet's mass, m_p (Figure 4). The gross features are easily understood from elementary celestial mechanics. Only a planet of $\sim 30\ M_\oplus$ is able to eject planetesimals out of the system, because it is at this mass that the escape velocity of the planet, $(2Gm_\mathrm{p}/R_\mathrm{p})^{1/2}$, reaches the local orbital speed, $(GM_\odot/a_\mathrm{p})^{1/2}$; at smaller masses a planet on a low-e orbit is effectively never able to scatter objects out of the system, but can easily accrete them. The planetesimal fraction accreted initially steadily grows with mass, because the gravitational sphere of influence of the planet increases and thus the number of planetesimals "available" to cross the orbit of the planet increases as the cube root of the planetary mass (Gladman 1993). On the other hand, the fraction accreted decreases when the ejection process becomes significant (for masses larger than that of Saturn) since the planet becomes more efficient at disposing of the small bodies than at accreting them. (Note that a slight increase in the planetary radius with mass up to $\sim 4\ M_\mathrm{Jup}$ will lead to a displacement of the maximum of accretion efficiency to slightly larger masses.)

Therefore, extrasolar giant planets of large masses (over the mass of Jupiter) might expected to be of similar composition to that of the star. Planets of smaller masses should be substantially enriched in heavy elements. However, one not must overinterpret this simple model: for example, if a second planet is present, it can "pass" planetesimals to a larger neighbor, thus increasing the latter's supply of planetesimals. The observation of extrasolar planets and brown dwarfs will directly address this question.

References

Bodenheimer, P., Ruzmaikina, T., & Mathieu, R. D., 1993, in Protostars & Planets III, ed. E. H. Levy & J. I. Lunine (Tucson: Univ. of Arizona Press), p. 367

Boss, A. P. 1998, ApJ, 503, 923

Boss, A. P. 2000, submitted to ApJ

Burrows, A., et al. 2000, ApJ, in press

Cameron, A. C., Horne, K., Penny, A., & James, D. 1999, Nat, 402, 751

Charbonneau, D., Brown, T. M., Latham, D. W., & Mayor, M. 2000, ApJ, 529, L45

Duncan, M., Quinn, T., & Tremaine, S. 1987, AJ, 94, 1330

Gautier, D., & Owen, T. 1989, in Origin and Evolution of Planetary and Satellite Atmospheres (Tucson: University of Arizona Press), p. 487

Gladman, B. 1993, Icarus, 106, 247.

Guillot, T., Chabrier, G., Gautier, D., & Morel, P. 1995, ApJ, 450, 463

Guillot, T., et al. 1996, ApJ, 459, L35

Guillot, T. 1999a, Plan. Space Sci., 47, 1183

Guillot, T. 1999b, Sci, 286, 72

Habing, H. J., et al. 1999, Nat, 401, 456
Hahn, J., & Malhotra, R. 1999, AJ, 117, 3041
Hayashi, C. 1980, Prog. Theor. Phys. Suppl., 70, 35
Henry, G. W., Marcy, G. W., Butler, R. P., & Vogt, S. S. 2000, ApJ, 529, L41
Levison, H., & Duncan, M. 1994, Icarus, 108, 18.
Lin, D. N. C., Bodenheimer, P., & Richardson, D. 1996, Nat, 380, 606
Lissauer, J. J. 1993, ARA&A, 31, 129
Marcy, G. W., & Butler, R. P. 1998, ARA&A, 36, 57
Marcy, G. W., & Butler, R. P. 2000, PASP, 112, 137
Mayor, M., & Queloz, D. 1995, Nat, 378, 355
Owen T., et al. 1999, Nat, 402, 269
Petit, J-M., et al. 2000, Icarus, in press
Podolak, M., Pollack, J. B., & Reynolds, R. T. 1988, Icarus, 73, 163
Pollack, J. B., et al. 1996, Icarus, 124, 62
Stevenson, D. J. 1982, Plan. Space Sci., 30, 755
Strom, S. E., Edwards, S., & Skrutskie, M. F. 1993, in Protostars & Planets III, ed. E. H. Levy & J. I. Lunine (Tucson: Univ. of Arizona Press), p. 837
Weidenschilling, S. J. 1977, MNRAS, 180, 57
Wuchterl, G., Guillot, T., & Lissauer, J. J. 2000, in Protostars & Planets IV, eds. V. Mannings, A. P. Boss, & S. S. Russell (Tucson: University of Arizona Press), p. 1081 in press
Zapatero Osorio, M. R., et al. 1999, ApJ, 524, L115

Discussion

William Borucki: Are the more massive planets being found around stars with high metalicity expected to have a significantly lower proportion of metals because of their ability to scatter planetesimals efficiently?

Tristan Guillot: The more massive objects are expected to have a similar composition to that of their parent star. We don't expect these to have a lower proportion of heavy elements than their star.

Tsevi Mazeh: The classification of a 10 $M_{\rm Jup}$ as a brown dwarf is not well established. The mass of Upsilon And d has recently been estimated to be 10 $\pm$ 5 $M_{\rm Jup}$, and this object is very probably a planet.

Tristan Guillot: I agree. It's sometimes convenient to call a massive giant planet a brown dwarf, but it's unclear where the limit between the two is.

Alan Boss: In the models of HD 209458 and Tau Boo where you try to match the observed planetary radii, what were the assumptions regarding the planetary compositions?

Tristan Guillot: I assumed solar composition for all models presented here. Of course, these yield maximum values of the radii since any enrichment in heavy elements will tend to produce smaller objects.

Disks, Planetesimals, and Planets
ASP Conference Series, Vol. 219, 2000
F. Garzón, C. Eiroa, D. de Winter, and T. J. Mahoney, eds.

Theoretical Models for Substellar Objects

Jean-François Gonzalez, France Allard, Isabelle Baraffe, and Gilles Chabrier

Centre de Recherche Astronomique de Lyon (CNRS-UMR 5574), Ecole Normale Supérieure, 46 allée d'Italie, F-69364 Lyon Cedex 07, France

Peter H. Hauschildt

University of Georgia, Department of Physics and Astronomy, Athens, Georgia 30602-2451, USA

Abstract. We present theoretical models for substellar objects (SSOs) covering a mass range from 1 $M_{\rm Jup} = 10^{-3} M_{\odot}$ to $\sim 0.1 M_{\odot}$, characteristic of giant planets and brown dwarfs. Evolutionary models are based on the most recent theoretical developments in the field: equation of state (EOS), non-gray atmosphere models, and synthetic spectra including grain formation. The influence of irradiation by a close stellar companion is explored.

1. Input Physics for the Models

1.1. Interior

SSOs are dense objects, with typical central conditions $T_{\rm c} \lesssim 10^5$ K and $\rho_{\rm c} \sim 10^2$–10^3 g cm^{-3}. One therefore needs to apply corrections to the ideal gas, in order to take into account particle interactions (ion–ion, ion–electron, electron–electron) and pressure ionization. Recent laser-driven shock-wave experiments at Livermore (Collins et al. 1998) explored conditions typical of SSO interiors, reaching pressures of 5 Mbar on liquid D_2 at high temperature, and showed a very good agreement with the Saumon–Chabrier EOS (described in Saumon, Chabrier, & VanHorn 1995, and references therein). Our models for SSOs use this EOS for a hydrogen and helium mixture. The contribution of metals does not exceed 1%, and they are therefore neglected (Chabrier & Baraffe 1997).

1.2. Atmosphere and Spectrum Synthesis

The atmospheres of low-mass stars (LMSs) and SSOs ($T_{\rm eff} \lesssim 5000$ K), contain numerous molecules, such as TiO and VO, main absorbers in the optical, or H_2 and H_2O which contribute mostly in the near-infrared (a review is given in Allard et al. 1997). They are well reproduced by the NextGen grid of non-gray model atmospheres (Hauschildt, Allard, & Baron 1999), which includes hundreds of millions of molecular transitions. As one goes down the effective temperature scale towards Jovian conditions ($T_{\rm eff} \sim 170$ K), the chemistry becomes more and more complicated as many species condense into grains, in particular Al_2O_3, $CaTiO_3$, or $MgSiO_3$, below $T_{\rm eff} \sim 2600$ K. At temperatures cooler than ~ 1400 K, for the pressures considered, carbon is mostly found in CH_4. Our latest model atmospheres, the so-called DUSTY models (Allard et al. 2000) include grain formation, taking into account their effect on the EOS, as well as on the radiative transfer. We also considered the case, in the so-called COND models, where the grains have settled below the photosphere: they still modify the atmosphere EOS but no longer affect the opacity. The latter models are similar to what was done by the Tucson group (Burrows et al. 1997).

2. Evolution

Consistent evolutionary calculations have been performed with our model interiors and three different grids of atmospheres. For objects hot enough to prevent the formation of grains ($T_{\rm eff} \gtrsim 2600$ K), we have used the NextGen models (Baraffe et al. 1997, 1998), for cooler objects, we have used both the DUSTY and the COND models (Chabrier et al. 2000, see Chabrier & Baraffe 2000 for a review).

We find that the inclusion of grains in the atmosphere affects the evolution only moderately. Their opacity produces a blanketing effect that lowers the effective temperature by 10% at most at a given age. The hydrogen-burning minimum mass changes very little as well, it amounts to 0.072 $M_\odot$ ($T_{\rm eff}$ = 1700 K, $L = 5 \times 10^{-5}$ $L_\odot$) for grainless models, whereas DUSTY models give a slightly lower value of 0.07 $M_\odot$ ($T_{\rm eff} = 1550$ K, $L = 4 \times 10^{-5}$ $L_\odot$).

Evolutionary tracks for LMSs and SSOs from 1 to 0.001 $M_\odot$ are displayed in Figure 1 on a $T_{\rm eff}$–$\log g$ diagram, together with isochrones from 1 Myr to 5 Gyr. They are produced with DUSTY models, but differ only slightly from NextGen tracks since the treatment of grains in the atmosphere scarcely affects the star's evolution. This diagram is of particular interest in deriving the approximate intrinsic parameters of a star, its mass and age, directly from two of its observed properties, its effective temperature and gravity.

3. Color–Magnitude Diagram

The most recent models of LMSs and SSOs are compared to the photometric observations of field stars and brown dwarfs of known distance in the M_K vs. J–K color–magnitude diagram (CMD) shown in Figure 2. The NextGen models follow the observations closely down to the bottom of the main sequence but fail in the brown-dwarf regime, where the DUSTY models show the onset of grain formation and bring an improved agreement with the observed sequence. However, the DUSTY models predict redder colors as mass decreases, due to the presence of dust in the atmosphere. On the other hand, the COND models, where all the dust has gravitationally settled, predict bluer colors, mainly because of the strong CH_4 absorption in the K band, and give a good fit to the position of Gl 229B at the lower end of the mass scale. Actually, the dust is expected to settle as mass decreases, and the DUSTY tracks should turn back (in near-IR colors) towards the blue and meet the COND tracks. The redder brown dwarfs observed give a hint of this deviation from the DUSTY tracks. A full dynamical model

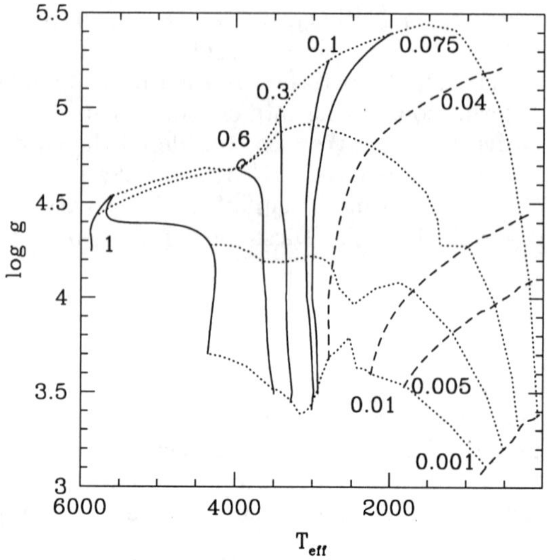

Figure 1. Evolutionary tracks in a $\log g$ (cm s^{-2}) vs. $T_{\rm eff}$ (K) diagram for LMSs (*solid curves*) and SSOs (*dashes*) from 1 to 0.001 $M_\odot$ (masses in $M_\odot$ are indicated on the curves). Isochrones for 1 Myr, 10 Myr, 100 Myr, and 5 Gyr (*bottom* to *top*) are shown in *dotted lines*.

for the grains would be required to tell whether there is a complete sequence between the L dwarfs and the methane dwarfs or if the transition is sharp and a gap is to be expected.

4. Spectral Sequence

Figure 3 shows a spectral sequence of model atmospheres of brown dwarfs to giant planets at 20 pc. They are calibrated to absolute flux using radii from consistent evolutionary models (Chabrier et al. 2000) and matched to a 5 Gyr isochrone. The dust grains in the atmosphere are allowed to settle when they exceed the local gas mass density. The spectra shown here change rather continuously from the warmer objects to the cooler ones as some molecular bands appear and others vanish. No clear boundary can therefore be defined between brown dwarfs and giant planets from their atmospheric properties.

The effect of dust opacity can be seen by comparison with a model where the dust is not allowed to settle and stays in the atmosphere: the larger opacity lowers the flux at the blue end of this wavelength range whereas the flux at the redder end is larger due to a more important backwarming effect. The opposite can be seen with a model with no dust opacity at all.

The detection limits of ground- and space-based instruments displayed in Figure 3 show that it is possible to detect isolated SSOs down to fairly low effective temperatures.

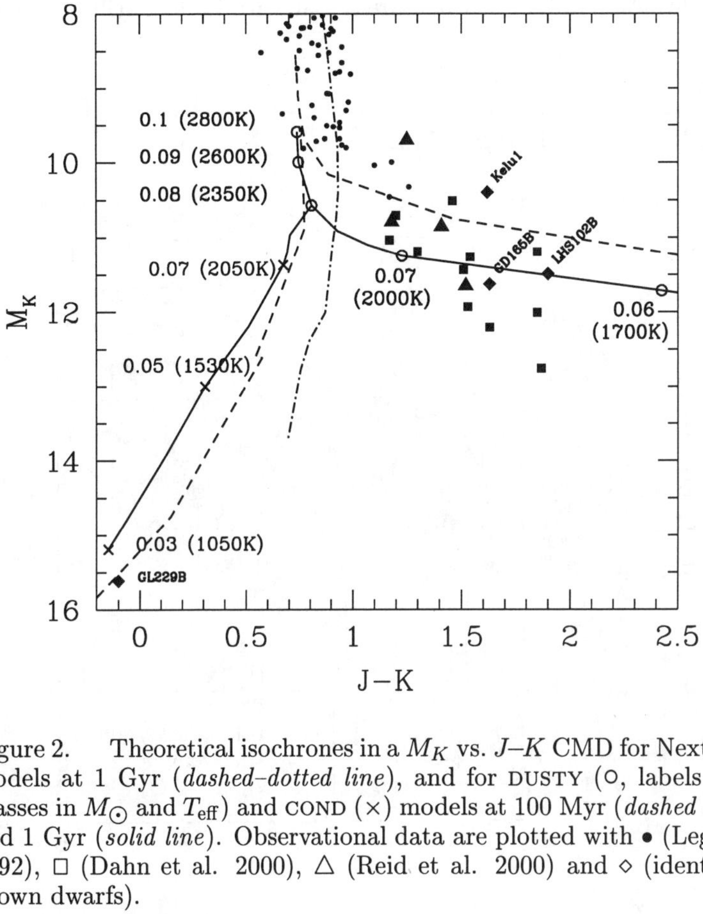

Figure 2. Theoretical isochrones in a M_K vs. J–K CMD for NextGen models at 1 Gyr (*dashed-dotted line*), and for DUSTY (○, labels give masses in $M_\odot$ and $T_{\rm eff}$) and COND (×) models at 100 Myr (*dashed line*) and 1 Gyr (*solid line*). Observational data are plotted with • (Leggett 1992), □ (Dahn et al. 2000), △ (Reid et al. 2000) and ◇ (identified brown dwarfs).

5. Irradiation by a Parent Star

It is a different story altogether for brown dwarfs or giant planets orbiting a star: they can easily be hidden in the flux of the much brighter star. However, dust clouds in the SSO's atmosphere may increase its albedo sufficiently to make it detectable in the optical, where the clouds are brightest.

We have computed models of a $T_{\rm eff} = 1000$ K planet irradiated by a solar-type (G2 V) star at orbital distances of 0.05, 0.1, and 0.2 AU. The resulting spectra are shown in Figure 4 together with the spectrum of the free-floating planet. In this simple test case, the irradiation is assumed to be isotropic. The planet model includes dust formation and opacities and is iterated to conserve a constant intrinsic flux in the radiative layers.

As can be seen, the parent star can substantially heat the planet's atmosphere, causing the dust clouds to vanish and the molecular bands even to disappear from its spectrum. However, the expected effect is small, and the planet stays significantly fainter than its parent star with its flux still peaking in the near infrared.

6. Conclusion and Perspectives

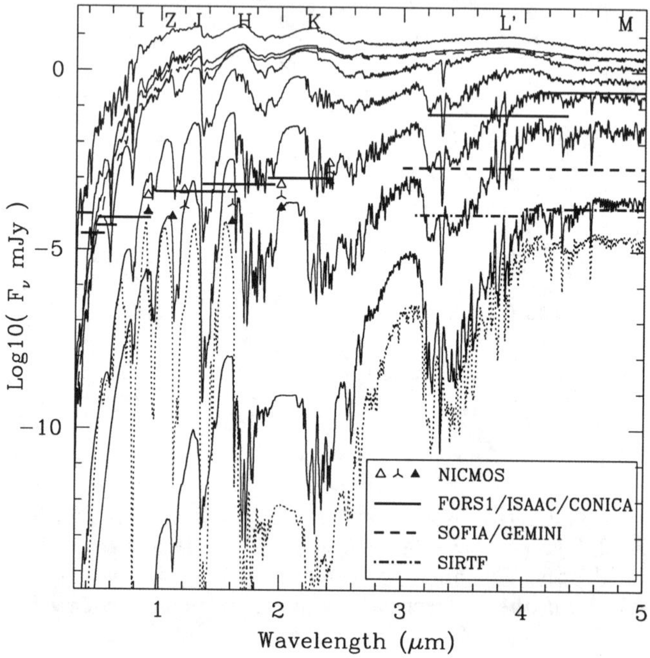

Figure 3. Spectral sequence of dusty SSO atmospheres at 20 pc. *From top to bottom:* $T_{\rm eff}$ = 2500, 1900, 1800, 1300, 700, 400, and 200 K (*solid curves*). Models at 1800 K with no dust settling (*dashed curve*), and at 400 K without dust opacity (*dotted curve*) are also shown. Detection limits of a few instruments are indicated.

The CMD of Figure 2 and others not shown here (see Chabrier et al. 2000) demonstrate the good agreement between the present theoretical models of SSOs including dust formation and opacity and the observations. They can be used to calibrate the intrinsic properties of late M to L dwarfs.

However, the models do not reproduce a full sequence down to the methane dwarfs and the gap between the DUSTY and COND tracks needs to be investigated. The next step of our theoretical work will be to include the dynamical processes of grain formation, diffusion, and settling.

We also intend to improve the reflected-light calculations with a consistent treatment of the evolution of irradiated planets for different values of the orbital distance and of the masses of the planet and parent star.

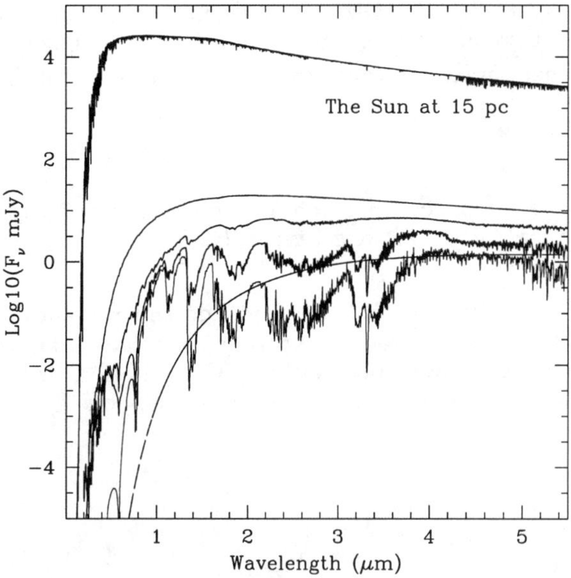

Figure 4. Spectrum of a young extrasolar giant planet ($T_{\rm eff}$ = 1000 K) irradiated by a G2 V primary at 15 pc (*topmost spectrum*) at orbital distances of 0.05, 0.1, 0.2 AU and ∞ (*from top to bottom*). The *long-dashed curve* is the spectrum of a 1000 K blackbody.

References

Allard, F., Hauschildt, P. H., Alexander, D. R., & Starrfield, S. 1997, ARA&A, 35, 137

Allard, F., Hauschildt, P. H., Alexander, D. R., Tamanai, A., & Ferguson, J. 2000, submitted to ApJ

Baraffe, I., Chabrier, G., Allard, F., & Hauschildt, P. H. 1997, A&A, 327, 1054

Baraffe, I., Chabrier, G., Allard, F., & Hauschildt, P. H. 1998, A&A, 337, 403

Burrows, A., Marley, M., Hubbard W. B., Lunine, J. I., Guillot, T., Saumon, D., Freedman, R., Sudarsky, D., & Sharp, C. 1997, ApJ, 491, 856

Chabrier, G., & Baraffe, I. 1997, A&A, 327, 1039

Chabrier, G., & Baraffe, I. 2000, ARA&A, in press

Chabrier, G., Baraffe, I., Allard, F., & Hauschildt, P. H. 2000, submitted to ApJ

Collins, G. W., et al. 1998, Sci, 281, 1178

Dahn, C. C., et al. 2000, in APS Conf. Ser., From Giant Planets to Cool Stars, ed. C. A. Griffith & M. S. Marley, in press

Hauschildt, P. H., Allard, F., & Baron, E. 1999, ApJ, 512, 377

Leggett, S. K. 1992, ApJS, 82, 351

Reid, I. N., Kirkpatrick, J. D., Gizis, J. E., Dahn, C. C., Monet, D. G., Williams, R. J., Liebert, J., & Burgasser, A. J. 2000, AJ, 119, 369

Saumon, D., Chabrier, G., & VanHorn, H. M. 1995, ApJS, 99, 713

Discussion

Tristan Guillot: A comment: your 0.05 AU models assume $T_{\mathrm{eff,intrinsic}} = 1000$ K. In the case of a 1 M_{Jup} planet, we get typically after a few Myr $T_{\mathrm{eff,intrinsic}} \sim 100$ K.

Jean-François Gonzalez: Yes, but this model doesn't necessarily correspond to a real exoplanet. It is a test case, done with models that we know to be reliable, as supported by brown-dwarf observations. Computations with colder models are being currently being done and will be available soon.

Disks, Planetesimals, and Planets
ASP Conference Series, Vol. 219, 2000
F. Garzón, C. Eiroa, D. de Winter, and T. J. Mahoney, eds.

Towards the Detection of Planets around Brown Dwarfs

Eduardo L. Martín
Division of Geology and Planetary Sciences, Caltech, MC 150-21, Pasadena, CA 91125, USA

Abstract. I summarize some pioneering efforts for the detection of companions around brown dwarfs. Brown-dwarf binaries appear to be frequent and they prefer smaller orbital separations than stellar binaries. I discuss several scenarios of brown-dwarf formation and speculate about the implications for the presence of planets around brown dwarfs. I argue that the best place to detect a planet could be around a very-low-mass star or brown dwarf. I present estimates of the discovery space for planets around a very-low-mass star using instrumentation planned for the near future on the ground and in space.

1. Introduction

The formation of binary and higher-order multiple systems has been recognized as a common product of star formation. We must accept that the formation of a single star with a planetary system like our own may more the exception than the rule. The majority (57%) of G- and K-type dwarfs in the solar vicinity have H-burning companions (Abt & Levy 1976; Duquennoy & Mayor 1991, DM91 hereafter). M-dwarf primaries seem to have a lower occurrence of companions (38%; Henry & McCarthy 1990; Fisher & Marcy 1992; Reid & Gizis 1997). What is the frequency of companions in brown dwarf primaries? Is there a trend of lower binary frequency for lower primary mass?

One motivation to search for companions to brown dwarfs is to learn about the origin of these objects. Free-floating brown dwarfs can form in several ways: a) ejection from an unstable multiple system, which would disrupt most brown-dwarf binaries (except the very hard ones), leading to a very low brown-dwarf binary fraction; b) fragmentation and collapse of a small molecular core, which would be a lower-mass analog to the formation of a star and would probably yield a significant binary fraction among brown dwarfs, in analogy to stars; and c) condensation from a circumstellar disk, which would be a higher-mass analog to the formation of a planet and would probably yield a low binary fraction among brown dwarfs, because planets do not seem to be double (at least in our Solar System, with the possible exception of the Pluto–Charon system).

Another motivation for searching for companions around brown dwarfs is that they can provide dynamical masses, which are powerful tests of evolutionary models of substellar objects, and can help to derive an observational mass–luminosity relationship below the substellar limit.

Last but not least, any faint physical companion to a brown dwarf must be another brown dwarf, a planet or a cool compact object (white dwarf, neutron star, black hole). Brown dwarfs offer the opportunity of finding the coolest objects outside the Solar System, which can be used as benchmarks for further studies of extrasolar planets.

2. Searches for Companions to Brown Dwarfs

The first free-floating brown dwarfs were discovered in the Pleiades cluster five years ago (Rebolo, Zapatero Osorio, & Martín 1995), and several deep surveys have ensued (Zapatero Osorio et al. 1999; Pinfield et al. 2000). G- and K-type Pleiades members have a distribution of both orbital periods and mass-ratios similar to those of field G-dwarf binaries (Bouvier, Rigaut, & Nadeau 1997). Thus, it seems that the binary properties of the Pleiades stars might not be very different from those of the disk population. Martín et al. (2000) recently presented a search for companions among 34 very-low-mass Pleiades members ($M \leq 0.09\ M_{\odot}$) using high-spatial-resolution images obtained with the *Hubble Space Telescope* and the adaptive optics system of the Canada–France–Hawaii Telescope. They did not find any resolved binaries with separations larger than 0.2 arcsec ($a \sim 27$ AU). Nevertheless, they found evidence for a binary sequence in the color–magnitude diagrams, in agreement with the results of Steele & Jameson (1995) for higher-mass stars.

Martín et al. (2000) applied the lithium test to two objects: CFHT-Pl-16, which lies in the cluster binary sequence but is unresolved in images obtained with the *Hubble Space Telescope*, and CFHT-Pl-18, which is binary with 0.33″ separation (Martín et al. 1998). The first object passes the test, but the second object does not. They concluded that CFHT-Pl-16 is a Pleiades brown-dwarf binary with separation < 11 AU, and that CFHT-Pl-18 is a foreground system.

A comparison of the multiplicity statistics of the Pleiades very-low-mass stars and brown dwarfs with that of G- and K-type main-sequence stars in the solar neighborhood (DM91), indicates that 3.4 binaries were expected with separation > 27 AU, but none was found. Martín et al. (2000) concluded that there is some evidence for a deficiency of wide binary systems among the Pleiades very-low-mass members.

On the other hand, the discovery that PPl 15 is a double-lined brown-dwarf binary with a period of only 5.8 days (Basri & Martín 1999) is surprising because the chance of finding such a hard binary in a DM91-type population is only 10% (Heacox 1998). Among the first brown dwarfs discovered in the field, there are already three brown-dwarf binaries with separations in the range 5–10 AU (Martín, Brandner, & Basri 1999; Koerner et al. 1999). It is noteworthy that all three binaries were found near the resolution limit of the images. Wider binaries could have been easily detected, but closer binaries could not.

The Pleiades and field results are consistent and suggest that brown-dwarf binaries are common ($>$ 10% binary frequency) and prefer smaller separations than stellar binaries.

3. Origins of Brown Dwarfs

I briefly consider how possible scenarios for the formation of brown dwarfs could explain the tendency of brown-dwarf binaries to have smaller separations and shorter orbital periods than stellar binaries.

- *Capture:* Brown dwarfs are more likely to be captured by a star than to capture another brown dwarf. This scenario is not supported by observations because brown dwarfs are rare as companions to stars (brown-dwarf desert: Marcy & Butler 1998; Halbwachs et al. 2000), but common as companions to brown dwarfs. Moreover it fails to explain why wide brown-dwarf binaries are rare.

- *Disk Fragmentation:* This tends to produce high-mass-ratio systems. Not supported: brown-dwarf companions to stars are rare.

- *Fragmentation:* This is the favored mechanism for producing most binary stellar systems (Bate 2000) because it appears possible to form binaries with similar properties to those that are observed. Supported: the fraction of substellar systems which are binary may not be much lower than the fraction of stellar binaries. Numerical simulations of molecular-cloud fragmentation have shown that more slowly rotating cloud cores form closer binaries than when the clouds are rapidly rotating (Boss 1993). Small cloud cores that start collapse from high initial densities form even closer binaries. This scenario seems provide a natural way of explaining the absence of wide brown-dwarf binaries and an excess of close systems.

In addition to formation scenarios, the brown-dwarf binaries in the Pleiades and the field may have suffered dynamical evolution. Wide binaries can be disrupted due to dynamical interactions with stars. Such process has been advocated by Kroupa, Petr, & McCaughrean (1999) to explain the binary frequency of stars in the Trapezium Cluster. An important test of this scenario is the binary frequency of hard binaries, which should be independent of primary mass. Another test is the binary frequency of brown dwarfs in very young associations and clusters. For example, Najita, Tiede, & Carr (2000) have identified a candidate wide brown-dwarf binary in IC 348. Spectroscopic follow-up is necessary to confirm it.

4. Disks and Planets around Brown Dwarfs

The lack of brown-dwarf binaries with separations larger than 30 AU makes one wonder if brown dwarfs typically have smaller planetary systems than stars. Jupiter's moon system would be an extreme example. Some speculations and prospects about disks and planets around brown dwarfs follow:

- Brown dwarfs may form as stars do but from smaller clouds. If so, a prediction is that brown dwarfs have smaller protoplanetary disks than stars. Disks around brown dwarfs may take longer to dissipate than disks around stars because of weaker brown-dwarf radiation and wind. Planets

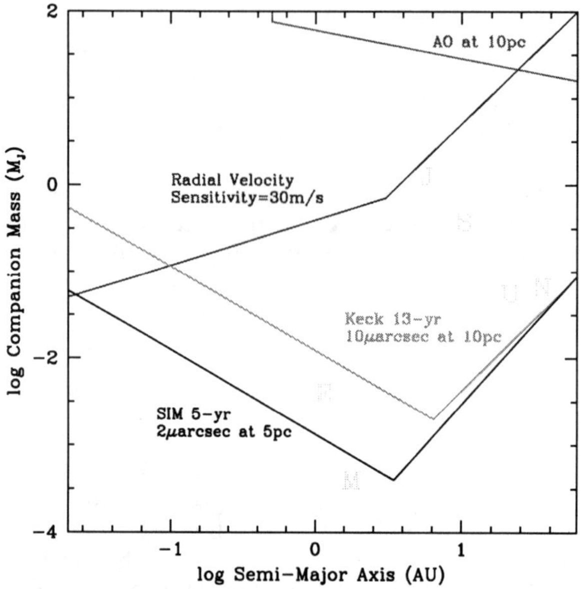

Figure 1. Parameter space for planet searches around a 0.1 $M_\odot$ primary using a variety of techniques from the ground (adaptive optics, Keck interferometer, high-precision radial velocity) and space (*Space Interferometry Mission*).

around brown dwarfs could be systematically closer to the central object than planets around stars.

- Jovian planets around brown dwarfs can be detected with radial-velocity measurements and direct imaging. Telluric planets around brown dwarfs can be detected with interferometry and occultations.
- The best place to find a brown dwarf companion is around a very-low-mass primary. Planets around VLM stars and brown dwarfs are detectable with planned future instrumentation.

The last statement is summarized in Figure 1, where I show the discovery space for companions around a very-low-mass star (0.1 $M_\odot$) using different techniques. For a brown-dwarf primary (mass $\leq$ 0.075 $M_\odot$) the situation is even more favorable. I have marked in the figure the locations of some of the planets of the Solar System. It is noteworthy that SIM may be able to detect an Earth-mass planet at 1 AU around a VLM star.

In addition to the techniques illustrated in Figure 1, planets around brown dwarfs can also be detected by occultations and direct imaging in the mid infrared.

- *Occultations:* The ratio of the areas between a brown dwarf and a telluric planet is about 1%, a precision which is customarily reached with

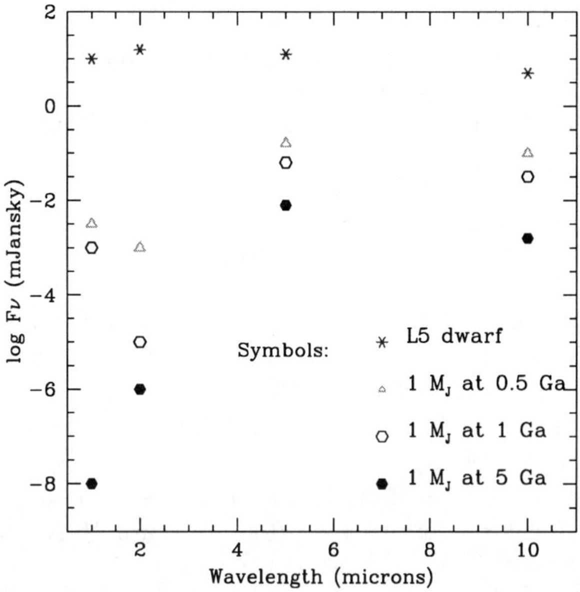

Figure 2. Flux contrast between an L5 BD (DENIS-P J0205-1159) at 10 pc and theoretical models for different ages for a 1 Jupiter-mass planet (Burrows et al. 1997).

differential CCD photometry from the ground. A Jovian planet around a brown dwarf could have a radius larger than its primary and produce a total eclipse.

- *Direct imaging:* The brightness contrast between a central object and a planet is minimized for the case of a brown-dwarf primary and a planet at wavelengths longer than 4 microns (Figure 2). *SIRTF* can potentially detect planets around the nearest brown dwarfs. The NICI coronagraph, which is being designed for Gemini South, may also achieve the desired sensitivity.

Acknowledgments. I thank the SOC for accepting my contributed talk. I acknowledge travel support from a UE grant. I thank I. Baraffe, G. Chabrier, X. Delfosse, and D. Lin for useful discussions.

References

Abt, H. A., & Levitt, S. G. 1976, ApJS, 30, 273

Basri, G., & Martín, E. L. 1999, AJ, 118, 2460

Bate, M. 2000, MNRAS, 314, 33

Boss, A. 1993, in "The Realm of Interacting Binary Stars", ed. J. Sahade (Dordrecht: Kluwer), p. 355

Bouvier, J., Rigaut, F., & Nadeau, D. 1997, A&A, 323, 139

Burrows, A., et al. 1997, ApJ, 491, 856
Duquennoy, A., & Mayor, M. 1991, A&A, 248, 485 (DM91)
Fisher, D. A., & Marcy, G. W. 1992, ApJ, 396, 178
Halbwachs, J. L., Arenou, F., Mayor, M. Udry, S., & Queloz, D. 2000, A&A, 355, 581
Heacox, W. D. 1998, AJ, 115, 325
Henry, T. J., & McCarthy, D. D. Jr. 1990, ApJ, 350, 334
Koerner, D. W., Kirkpatrick, J. D., McElwain, M. W., & Bonaventura, N. R. 1999, ApJ, 526, L25
Kroupa, P., Petr, M. G., & McCaughrean, M. J. 1999, New Astron., 4, 495
Marcy, G., & Butler, R. P. 1998, ARA&A, 36, 57
Martín, E. L., et al. 1998, ApJ, 509, L113
Martín, E. L., Brandner, W., & Basri, G. 1999, Sci, 283, 1718
Martín, E. L., et al. 2000, ApJ, in press
Najita, J. R., Tiede, G. P., & Carr, J. S. 2000, ApJ, submitted
Pinfield, D.J ., Hodgkin, S. T., Jameson, R F., Cossburn, M. R., Hambly, N. C., & Devereux, N. 2000, MNRAS, 313, 347
Rebolo, R., Zapatero Osorio, M. R., & Martín, E. L. 1995, Nat, 377, 129
Reid, I. N., & Gizis, J. E. 1997, AJ, 114, 1992
Steele, I. A., & Jameson, R. F. 1995, MNRAS, 272, 630
Zapatero Osorio, M. R., et al. 1999, A&AS, 134, 537

Disks, Planetesimals, and Planets
ASP Conference Series, Vol. 219, 2000
F. Garzón, C. Eiroa, D. de Winter, and T. J. Mahoney, eds.

Substellar Companions to Nearby Stars from NICMOS Surveys

Glenn Schneider

Steward Observatory, University of Arizona, 933 N. Cherry Ave., Tucson, AZ 85721, USA

Eric E. Becklin and Patrick J. Lowrance

UCLA, 405 Hilgard Ave., Los Angeles, CA 90095, USA

Bradford A. Smith

University of Hawaii, 82-6012 Puuhonua Rd. Naoopoo, HI 96704, USA

NICMOS IDT/EONS and *HST* GO/8176 Teams

Abstract. A coronagraphic imaging survey of 65 nearby stars was conducted using the Near Infrared Camera and Multi-Object Spectrometer (NICMOS) on the *Hubble Space Telescope* by the Instrument Definition Team. Using these guaranteed-time observations we searched for very-low-mass stellar and substellar (extrasolar giant-planet and brown-dwarf) companions to young main-sequence and late M-dwarf stars. Nineteen additional stars, similarly observed in our circumstellar disk program, were also examined for evidence of companionship. We discuss the large depth and dynamic range of these surveys, which have given rise to several putative and confirmed detections. Using follow-up spectrographic observations obtained with the Space Telescope Imaging Spectrograph (STIS) we have begun to characterize the physical nature of the objects found in the companion search program.

1. Introduction

Is there a continuity of companion objects across the substellar mass spectrum bridging the stellar main sequence into the planetary domain? If so, in what sort of local environments will these objects form? At what distances will they be found from their primaries, and how might this be affected by the characteristics of the primary and companion objects, and of the circumstellar regions? What implications will the discovery and characterization of such objects have for our understanding of the formation mechanisms of extrasolar planetary systems?

To begin answering these fundamental questions we have searched for candidate brown-dwarf and extrasolar giant-planet (EGP) companions and circumstellar debris disks in a direct-imaging survey using the *HST*/NICMOS coronagraph. Our survey of 84 stars is sensitive to transitional and substellar companions over a broad range of masses and separations. Our disk imaging program is discussed elsewhere in these proceedings (Weinberger et al., p. 329); here we concentrate on our companion survey. With the NICMOS coronagraph, and exploiting the highly stable characteristics of the *HST* + NICMOS point spread function (PSF) on sub-orbit timescales, we are able to image unresolved objects 13 to 15 magnitudes fainter than the occulted stars at distances of only a few arc seconds (Schneider 1998), thus identifying companion candidates for follow-up observations. To unequivocally establish the physical association of a companion candidate with its suspected primary, common proper motions must be demonstrated through astrometric measurements. Spectroscopy of a putative companion is necessary to ascertain its physical nature.

2. The NICMOS Coronagraphic Survey

The first phase of this program, designed to identify candidate companion objects, was carried out as a single-color survey using the F160W filter (a very close analog to the H band) where the instrumental performance for low-mass point-source detection is optimized (Lowrance et al. 1998). Each star was imaged coronagraphically at two field orientations (spacecraft roll angles) in the same orbit. The coronagraph reduces the background scattered light from the primary target by factors of several to more than an order of magnitude, varying with radial distance. The PSF of the primary and residual scattered-light artifacts remain fixed as the detector corotates with the spacecraft. Subtraction of the two differentially rolled frames further reduces the background light by factors of several hundred in a spatially dependent fashion. This leaves "conjugate" (positive and negative) images of any astronomical sources rolled about the target axis which may be later recombined. By combining all frames in the one orbit per target allocated for the NICMOS coronagraphic survey, companion objects as faint as $H \sim 23$ mag can be detected a few arc seconds from many of our primary targets (Schneider et al. 1998). For the youngest stars in our census, evolutionary (cooling) models by Burrows et al. (1997) and others indicate that this sensitivity limit corresponds to companion masses extending into the Jovian domain. This differential imaging strategy was generally adopted in our coordinated observing programs under the umbrella of the NICMOS IDT's Environments of Nearby Stars investigations. While different in detail, those programs may be summarized as follows:

- *A Search For Massive Jupiters:* Thirty-eight very young main-sequence stars with a mean distance of $\sim$ 30 pc were observed. The median age for the selected targets with well established ages is $\sim$ 90 Myr, with eight targets as young as $\leq$ 10 Myr. Substellar companions to young stars are still in early cooling phases and are more readily detectable as a result of their elevated luminosities ($L \sim t^{-1.3} M^{2.24}$). Among the youngest targets are several members of the TW Hydrae Association (TWA), a loose stellar association of (perhaps) two dozen identified members (Webb & Zuckerman 1999), all $\sim$ 10 Myr of age, comprising the nearest site of recent star formation to the Earth ($d \sim 50$ pc).

- *A Search for Low-Mass Companions to Nearby M-Stars:* Twenty-seven M-dwarfs which are a) very nearby ($d \leq 6$ pc) with spectral types later than M3.5; b) young (age ≤ 10 Myr) with $d \leq 25$ pc; and c) spectrally the latest known (i.e., "ultra-cool" dwarfs later than $\sim$ M8.5) were observed. For the nearest stars in this sample, the minimum separations for companion detections are as small as 1.2 AU.

- *Dust Disks around Main-Sequence Stars:* Stars with large *IRAS* excesses, and other indicators of the likely presence of circumstellar dust, were observed at 1.1 and/or 1.6 microns. Shorter-wavelength observations provide better coronagraphic sensitivity for circumstellar scattered light which will closely follow the spectral energy distributions of the central stars. While observations were tailored for disk detection, all images were analyzed for the possible presence of close low-mass companions.

In a limited number of cases, we obtained follow-up observations of newly found candidates in other NICMOS filters before the cryogenic coolant in NICMOS was exhausted. The color information was used to discriminate ill-placed unresolved background objects from likely physical companions.

3. Extrasolar Giant-Planet Detection

NICMOS coronagraphy enables the direct imaging of young EGPs. This is demonstrated in the case of a candidate EGP companion to TWA 6 (ROSAT 116, $H = 6.9$, K7 V). This X-ray source is a TWA member and has an estimated age of $\sim$ 10 Myr, concurrent with the ensemble age of other association members. As shown in Figure 1 (*left*), the $H = 20.1$ putative companion is seen with an S/N of 50 at an angular separation of $2''.5$ (125 AU projected distance) from its primary, which is brighter by 13.2 magnitudes. In a follow-up NICMOS camera 1 observation at 0.9 μm (F090M filter) the object was unseen to a (less sensitive) 3 σ limiting magnitude of > 22, indicative of a very red source (if not a highly reddened background object). If the object is physically associated and coeval with ROSAT 116, its absolute H magnitude suggests an effective surface temperature of $\sim$ 800 K (for a surface gravity of 7.5×10^4) and a mass of $\sim$ 1 Jupiter based upon evolutionary models by Burrows et al. (1997).

In the case of TWA 6 "B", as well as a number of other substellar companion candidates, we are obtaining *HST*/STIS 0.75–1.0 μm spectra of the objects to help elucidate their physical natures. We are also attempting to measure differential proper motions to establish (or reject) physical association with ground-based adaptive-optics facilities. Such astrometric measures may be difficult with the current temporal baseline of less than two years. Re-observations with NICMOS, when it is returned to service in *HST* Cycle 10, should unequivocally establish whether or not TWA 6"A" and "B" are gravitationally bound, and hence if TWA 6"B" is a young Jovian planet.

While the unambiguous determination of the planetary nature of a suspected companion demands dynamical and spectroscopic confirmation, the rejection of putative young EGPs may be done with high confidence based on their color indices. For example, a more luminous candidate companion object with an apparent H magnitude of 16.8 was identified at a similar distance from

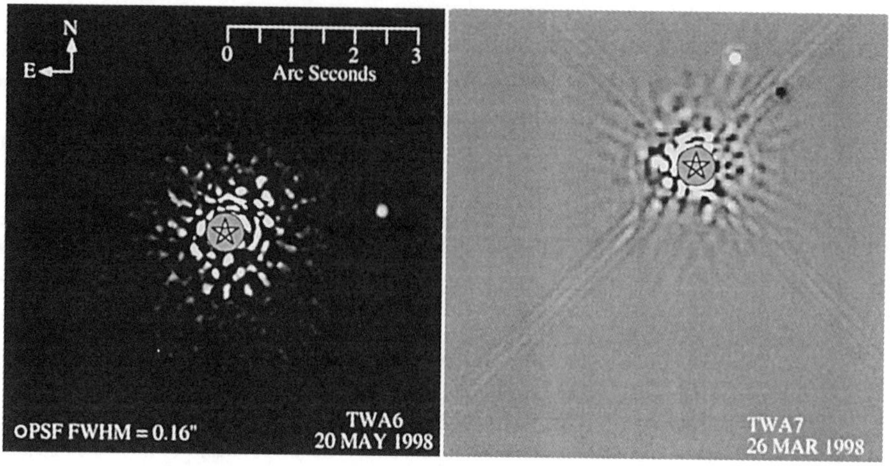

Figure 1. NICMOS F160W coronagraphic images of TW Hydrae association members TWA 6 (*left*, reconstructed from rotated difference image conjugates), and TWA 7 (*right*, difference image).

another TW Hydrae association member, TWA 7 (M1 V, $H = 7.4$) as shown in Figure 1 (*right*). A later NICMOS observation found its $I'[F090M(0.9\mu\text{m})] - H$ color to be +0.88, too blue to be a substellar object at the age of the primary and most probably a background K-star. One might conjecture extrinsic temporal variations in the color indices arising, for example, from cometary impacts. This is an unlikely scenario as a sustained blueing of more than an order of magnitude would be required for a 10 Myr EGP of this H-band luminosity to mimic a background star of the same color.

4. Brown Dwarfs

Recently, many old isolated "field" brown dwarfs have been discovered from large ground-based surveys such as the 2-Micron All-Sky and Sloan Digital Sky surveys and were subsequently confirmed spectroscopically. Indeed, it has been suggested that brown dwarfs, once considered somewhat exotic objects, may be as common as stars. Yet, the companion mass fraction of these substellar objects, which occupy a niche in the mass function between planets and stars, remains largely unknown. How common are they? Do companion brown dwarfs form in a process more like planets than stars? How does the presence of a brown-dwarf companion affect the evolution of a newly forming solar system? Our small NICMOS survey begins to address these questions but many more objects will have to be studied before they are fully answered.

The first brown-dwarf companion candidate identified from the NICMOS surveys, TWA 5B (CD −33°7796B, M1.5 V, $H = 7.2$), as reported by Lowrance et al. (1999), is also a member of the TW Hydrae association. The $H = 12.14$ magnitude (Vega system) companion is readily seen in the NICMOS coronagraphic difference image shown in Figure 2 (*left*) with a separation of 1.″96 (98 AU

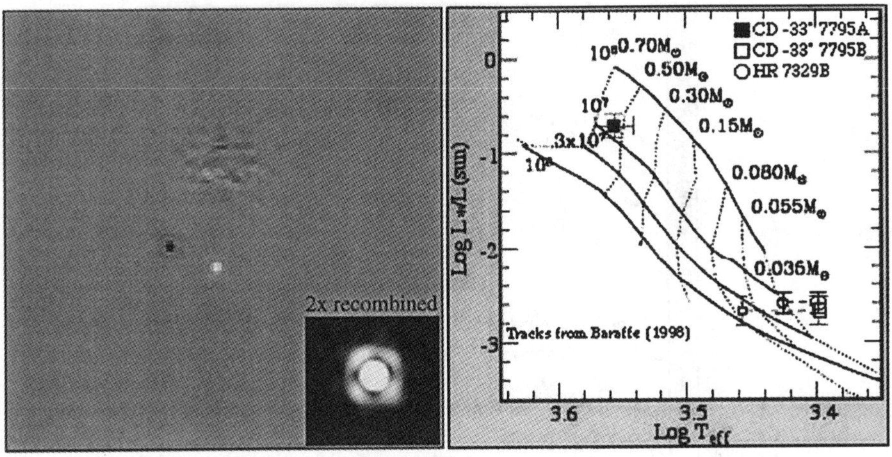

Figure 2. *Left:* F160W coronagraphic difference image of TWA 5B (recombined image inset). *Right:* Placement of TWA 5A/B and HR 7329B on evolutionary tracks.

projected distance) at a PA of 358°.9. Direct imaging with the KeckI/NIRC and KeckII/LRIS acquisition cameras (possible given the modest primary/secondary contrast ratio and sufficiently large separation) show color indices of $I-J=3.2$, $I-H=3.7$, and $I-K=4.4$. With an absolute H magnitude of 8.41 (assuming physical association: $d=50$ pc, and a bolometric correction of $BC(H)=2.8$) we ascertained its luminosity as 0.0021 $L_\odot$ and estimate an effective temperature $\sim$ 2600 K.

Young, and therefore hot, brown dwarfs have spectra resembling M-dwarf stars of the same photospheric temperatures. Thus we assigned a photometric MKL classification for TWA 5B of M8.5 V. Using evolutionary tracks from Baraffe et al. (1998, see Figure 2 [*right*]) and others we estimate the mass of the object to be $\sim$ 0.02 $M_\odot$ (20 Jupiter masses). We recently obtained a short-wavelength NIR spectrum using the G750M grating in STIS. We compare the 8000–9000 Å spectrum of TWA 5B to that of two late M-dwarf standards, J2309+1594 (M8.5 V) and J1239+2029 (M9 V)—see Figure 3—for which we find a very good fit particularly in the Na I doublet, TiO absorption bandheads, and slope of the continuum longward of 8450 Å confirming our photometrically determined photospheric temperature and spectral classification.

We compare TWA 5B with a second brown-dwarf companion identified with NICMOS and confirmed spectroscopically with STIS, HR 7329B (Lowrance, et al. 2000). The older ($\sim$ 30 Myr) A0 V star ($H=5.05$) is at a comparable distance of 47 pc. Assuming companionship for the 4″.17 distant unresolved object (196 AU projected distance), the $H=11.90$ companion candidate imaged by NICMOS at a PA of 166°.8 was estimated to have a luminosity of 0.0026 $L_\odot$ ($H_{\rm abs}=8.54$, assuming a $BC(H)=2.67$). As with CD $-33°7795$ we place HR 7329B on evolutionary models and find a mass of $\sim$ 0.04 $M_\odot$ (40 Jupiter masses). Its spectrum, from which we deduce a spectral class of M7.5 V, is shown

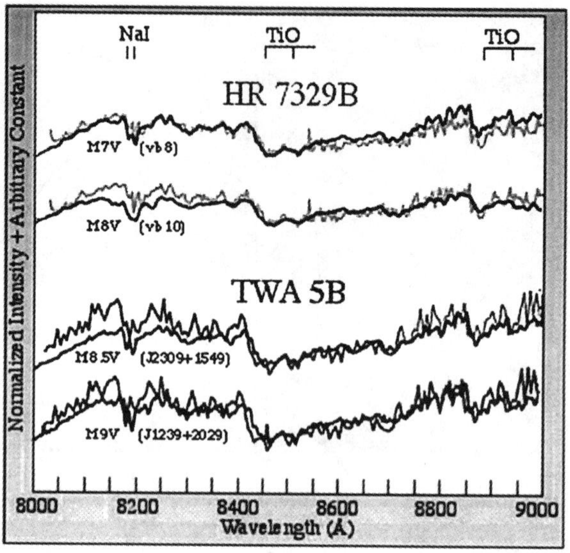

Figure 3. STIS spectra of TWA 5B, HR 7329B, and late M-dwarfs.

in Figure 3 along with TWA 5B and the M7 V and M8 V spectral standards VB8 and VB10, respectively.

Acknowledgments. This work is based on observations with the NASA-ESA *Hubble Space Telescope*, obtained at STScI, which is operated by AURA, Inc., under NASA contract NAS2-6555 and supported by NASA grants NAG5-3042 and GO-98.8176A to the NICMOS IDT and EONS teams.

References

Baraffe, I., Chabrier, G., Allard, F., & Hauschildt, P. H. 1998, A&A, 337, 403

Burrows, A., et al. 1997, ApJ, 491, 856

Lowrance, P. J., et al. 1998, NICMOS and the VLT, ed. W. Freufling & R. Hook (Garching: ESO), p. 96

Lowrance, P., Becklin, E., Schneider, G., Kirkpatrick, D., Zuckerman, B., Plait, P., Malimuth, E., Heap, S., Weinberger, A., Smith, B., Terrile, B., Schultz, A., and Hines, D. 2000, ApJ, in press

Lowrance P. J., et al. 1999, ApJ, 512, L69

Schneider, G. 1998, NICMOS and the VLT, ed. W. Freufling & R. Hook (Garching: ESO), p. 88

Schneider, G., Thompson, R. I., Smith, B. A., & Terrile, R. J. 1998, SPIE, 3356, 222

Webb, R. A. & Zuckerman, E. 1999, BAAS, 195, 3202W

Discussion

Alan Boss: what is the projected separation from the star of the giant-planet candidate around ROSAT 116?

Glenn Schneider: Assuming a distance of 50 pc, which is a reasonable canonical value for a member of the TW Hydrae association (ROSAT 116 $\equiv$ TWA 6), the projected distance is 125 AU ($\pm \sim 5$ AU, internal error). Note that, for CD −33.7795B also a member, we find $p = 100$ AU ($\pm$ 5 AU), and for HR 7329B 196 AU ($\pm$ 9 AU). So we are finding brown dwarfs (CD −33.7795, $\sim 20\ M_{\rm Jup}$; HR 7329B, $\sim 40\ M_{\rm Jup}$) in the region where there is a dearth of radial-velocity observations.

Disks, Planetesimals, and Planets
ASP Conference Series, Vol. 219, 2000
F. Garzón, C. Eiroa, D. de Winter, and T. J. Mahoney, eds.

Ground-Based Direct-Imaging Search for Substellar Companions Next to Young Nearby Stars

João Alves

European Southern Observatory, D-85748 Garching, Germany

Ralph Neuhäuser

MPE, D-87540 Garching, Germany

Abstract. We report the first results from our ground-based direct-imaging search for substellar companions of young (up to 100 Myr) nearby (up to 75 pc) stars, where companions should be well separated from the central stars and still be relatively bright due to ongoing accretion and/or contraction. Our target list includes known young stars, such as the entire young population of the TW Hya association, as well as recently discovered young stars among *ROSAT* sources (some of which are as yet unpublished). Our observations are performed mainly with SOFI and SHARP at the ESO 3.5 m NTT on La Silla and with ISAAC at the ESO 8.2 m ANTU (VLT-UT1) on Paranal, all in the H and K bands. In this contribution we present direct-imaging data of a faint object detected next to TWA 7, which if at the same age and distance as TWA 7 could be an object with only a few Jupiter masses. More recently, we took an H-band spectrum of this object using ISAAC at the VLT. The object turned out to be an unrelated background K-dwarf. We show, nevertheless, that *direct-imaging detection and spectroscopic follow-up of faint substellar companions candidates very close (~ 2 arcsec) to young nearby stars (contrast ~ 10^4), is possible from ground-based telescopes.*

1. Introduction

Several extrasolar planet candidates have been detected indirectly by radial-velocity variations of stars (review by Marcy & Butler 1998), one such candidate is confirmed by a transit event (Charbonneau et al. 2000). There is no direct-imaging detection of an extrasolar planet, yet. Direct imaging detection of planets like those in our Solar System but orbiting other stars is difficult due to the limited dynamical range: extrasolar planets are simply too faint and too close to their bright host stars. One can try to avoid the problem of spatial resolution by searching for planetary companions around nearby stars, where the orbit of the outermost solar system planet corresponds to several arcsec, sufficient to resolve a faint object next to a bright star. However, very nearby stars usually are too old, so that their hypothetical planets are correspondly too faint (Burrows et al. 1997) for direct detection with current technology.

Young planets, on the other hand, are still self-luminous due to ongoing accretion and/or contraction (Burrows et al. 1997; Brandner et al. 1997; Malkov, Piskunov, & Zinnecker 1998) and, if also nearby, they would be sufficiently bright and resolved for direct detection. Direct imaging of these young planets is optimal in the near-infrared bands H and K, where the brightness difference between young stars and young planets is expected to be the lowest (Burrows et al. 1997). In addition, nearby stars usually have large proper motion, so that one can decide after only a few years whether a companion candidate is co-moving. Finally, there is a crucial advantage in studying companion planet candidates instead of free-floating planet candidates: the mass can be better constrained for companions than for free-floating planets, because the age and distance of the primary is usually well-known.

A few substellar companions to normal stars have already been detected by direct imaging, namely Gl 229B (Nakajima et al. 1995) and G1 96-3B (Rebolo et al. 1998), brown-dwarf companions confirmed by spectroscopy and proper motion. The extrasolar planet candidate directly detected by the L*HST* near a T Tauri star in Taurus (Terebey et al. 1998) has not yet been confirmed by spectroscopy or proper motion.

2. Our Sample: Young Nearby Stars

We selected young ($\leq$ 100 Myr) nearby ($\leq$ 75 pc) stars, some of them from the literature, others discovered recently among *ROSAT* sources by ourselves (some of these as yet unpublished). We are confident of the young age of our sample of stars because of lithium 6708 Å absorption lines (and/or Hα emission, IR excess, and/or kinematic membership to a young cluster). We know the distances of most target stars from *Hipparcos* (and for some of them, from kinematic membership to a cluster of known distance). Our target list includes all members of the TW Hya association (TWA) and the T Tauri stars in the nearby and even younger MBM 12 cloud (Hearty et al. 2000a,b).

Since we know the distances and ages of the stars in our sample we can predict the H- and K-band magnitudes of possible substellar companions for different masses, e.g., from 1 to 80 M_{Jup}, using the non-gray theory by Burrows et al. (1997). In Figure 1, we present the K-band magnitudes of the program stars (either known from infrared photometry, or estimated from their known spectral types and optical magnitudes), their distance distribution, the corresponding angular separation between those stars and possible companions at assumed 50 AU physical separations, and the K-band magnitude distribution of assumed $\sim$ 10 M_{Jup} companions (estimated from Burrows et al. (1997) for the known ages and distances of the stars). These distributions clearly show that such companions can be detectable and resolvable with current technology.

Hence, we have started an observational program using mainly the SOFI and SHARP infrared cameras at the ESO 3.5 m NTT at La Silla, and the ISAAC imaging camera at the ESO 8.2 m Antu (VLT-UT1), but also other state-of-the-art instruments.

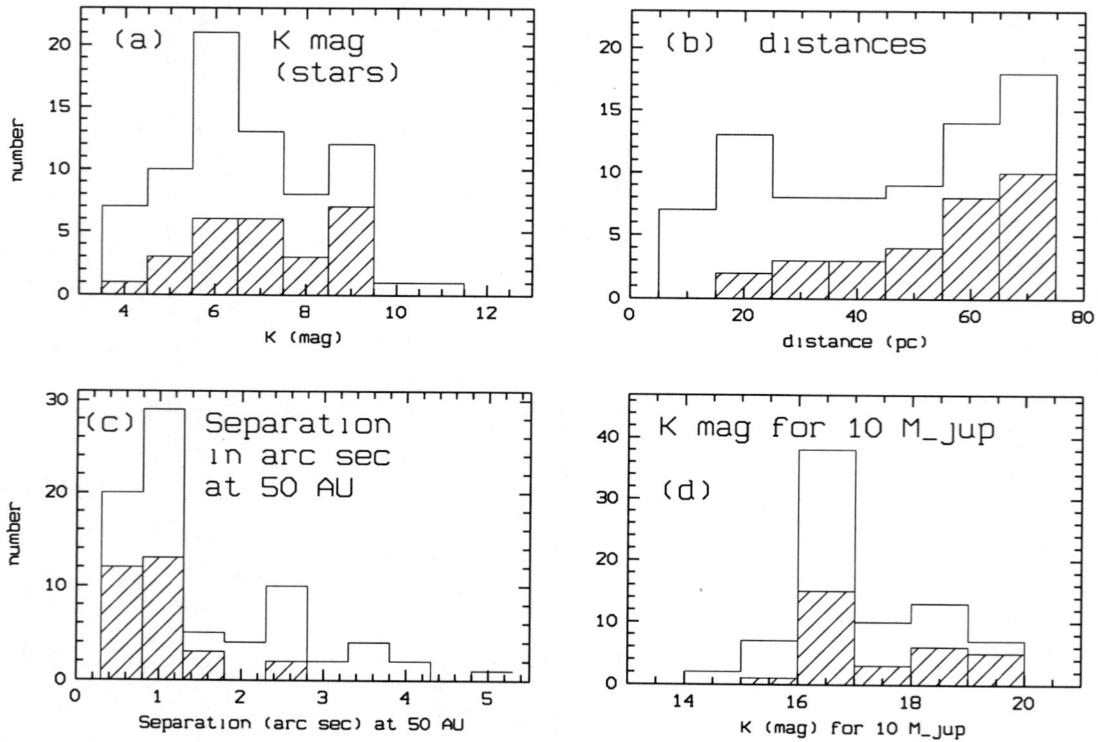

Figure 1. Properties of our sample of young nearby stars and possible companions: primary star K-band magnitude distribution (*upper left*), distance distribution (*upper right*), distribution of angular separations at 50 AU (*lower left*), and distribution of companion K-band magnitudes (*lower right*), as expected from Burrows et al. (1997) given the ages and distances of our target stars. Hatched are those stars recently discovered by *ROSAT*.

3. First Results: Case Study of TWA 7

Using the MPE-built SHARP speckle camera, we have detected a very faint object 2.5 arcsec south-east of the young nearby star T Tauri TWA 7, a member of the TW Hya association (called TWA, see Webb et al. 1999). Four stars of the TWA association have been observed by *Hipparcos*, their mean distance being 55 pc. From JHK photometry of the M1-type star TWA 7, our case study, we can estimate its bolometric luminosity to place it into the H-R diagram. By comparison with isochrones, we obtain an age of 1 to 6 Myr.

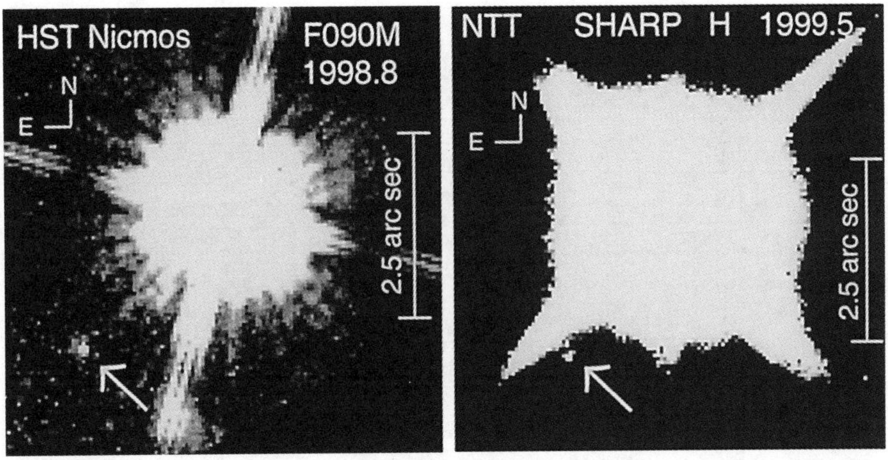

Figure 2. Images of TWA 7A and B: *HST* NICMOS F090M filter (*left*) and NTT SHARP H band (*right*).

The faint object (TWA 7B) detected by SHARP, SE of TWA 7A has $H =$ 16.4 and $K = 16.3$ mag, which is more than 9 mag fainter than the primary star at these wavelengths. If TWA 7B were to be a companion to TWA 7A, i.e., if it were to be at the same distance and age, then its apparent H- and K-band magnitudes would correspond to absolute magnitudes ($d \sim 55$ pc) consistent with an effective temperature $T_{\rm eff} \simeq 1050$ K and a surface gravity $\simeq 3000$ g s^2 (see Table 5 in Burrows et al. 1997). These values are then consistent with an object with a mass of $\sim 3\ M_{jup}$ and an age of $\sim 10^{6.5}$ yr (see Figure 9 in Burrows et al. 1997). This derived age is in agreement with the ages of TWA 7A and the other TWA members. The angular separation of 2.5 arcsec (at a distance of 55 pc) corresponds to a physical separation of 138 AU, well within typical T Tauri disk sizes.

TWA 7B has previously been detected by *HST* NICMOS observations in the F160W, F090M, F165M, and F180M filters. The *HST* F160W image has been shown in Neuhäuser et al. (2000), where a coronagraph has been used as in the F165M and F180M images. Only in the F090M image with NIC1 was no coronagraph used. In Figure 2, we show the NIC1 F090M image together with our SHARP H-band image. The NIC2 F160W image and our SHARP K-band image have been shown before in Neuhäuser et al. (2000). See Neuhäuser et al.

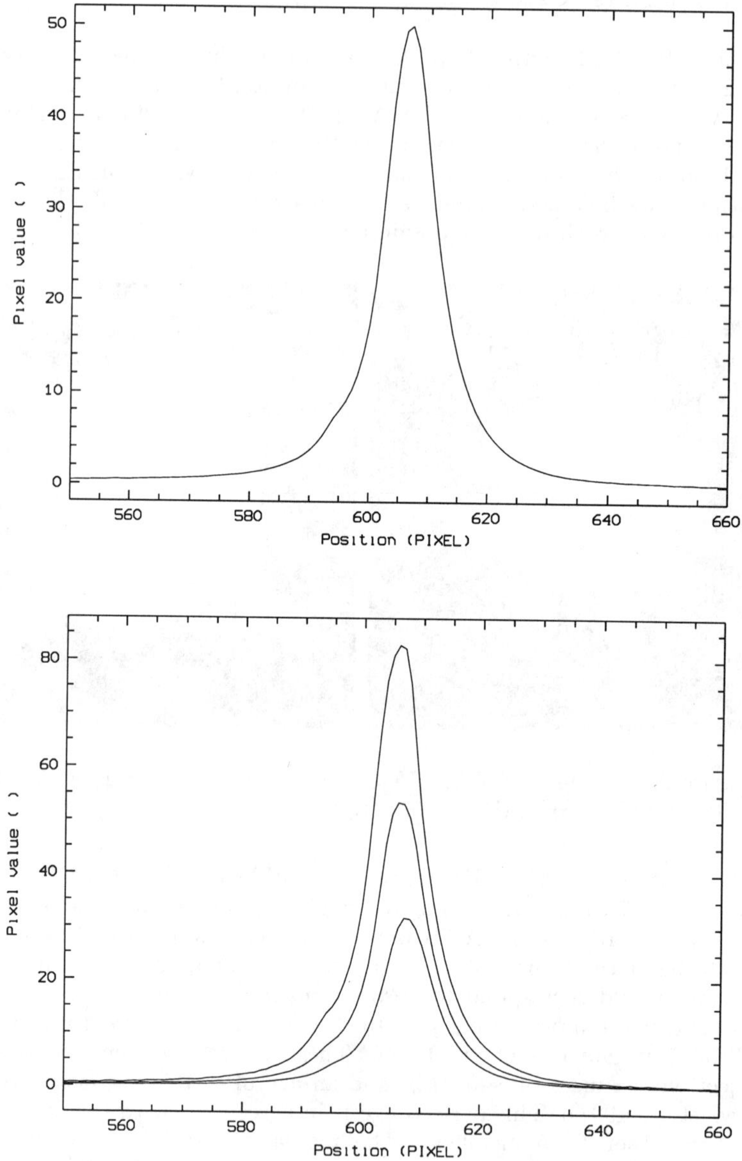

Figure 3. Cut through the spectrum of TWA 7B, for full wavelength range (*left*) and three bins (*right*), namely 1.35 μm to 1.50 μm (where there would be water absorption in a T-dwarf), 1.50 μm to 1.65 μm (outside water and methane absorption), and 1.65 μm to 1.80 μm (where there would be methane absorption in a T-dwarf). See text for discussion.

(2000) for the *HST* magnitudes, the position angles between TWA 7A and B, and more details on data reduction.

Although all the indications from photometry and astrometry point to a substellar nature for TWA 7B (Neuhäuser et al. 2000), a confirmation from spectroscopy is nevertheless required. To check for a possible spectral signature of a very cool object we took an H-band spectrum of TWA 7B using ISAAC at the ESO 8.2 m Antu (VLT-UT1). In fact, the derived temperature of $\sim$ 1050 K for TWA 7B (if indeed a companion) is similar to those of known old T-dwarfs, so that one might, under some conditions, expect to see methane absorption features in the spectrum of TWA 7B. The technical problem with taking and analyzing a spectrum of such a faint object so very close ($\sim$ 2.5 arcsec) to a much brighter star (contrast $\sim 10^4$) is again one of dynamical range. To maximize both the separation between TWA 7B and A and also the fraction of the light coming from TWA 7B, we did not place the slit in alignment with the position angle (PA) of the stars, or perpendicular to this PA (with only B in the slit). In the latter case, the signal from B would be above the dominant signal from TWA 7A scattered light in the collapsed spectrum (zero separation). In the former case (maximum separation), the light from TWA 7A would have "swamped" the signal from TWA 7B. We used, instead, a slit orientation between these two extremes.

In Figure 3, we show the profile of the spectrum, i.e., cuts through the spatial direction for both the full spectrum as well as in three different wavelength bins; namely, i) from 1.35 μm to 1.50 μm, i.e., in the water absorption band, ii) from 1.50 μm to 1.65 μm, i.e., outside both the water and the methane absorption bands, and iii) from 1.65 μm to 1.80 μm, i.e., in the methane absorption band. Then, if TWA 7B were a methane T-dwarf, as expected for a young hot planet, the small bump clearly seen in Figure 3 (*left*) would be most pronounced in the 1.50 μm to 1.65 μm region, i.e., outside the water and methane absorption bands. This, however, is not the case. Hence, the object is not likely to be a methane T-dwarf.

The full extraction of the spectrum has been finished in the meantime and will be published elsewhere. We can already report here that TWA 7B turned out to be of spectral type K. This means that, given its brightness, TWA 7B cannot be at the same small distance as TWA 7A (55 pc) and it is most probably a background K-type main-sequence star. Given that it is a faint source ($H = 16.4$ mag), this K dwarf should be at a distance between $\sim$ 2 and 4 kpc, in the halo of our Galaxy ($b = 21°$). Curiously, the probability of finding a source as faint as TWA 7B within a 2.5 arsec radius circle in the sky, towards this direction of the Galaxy, is just about 1%.

4. Summary

We show that ground-based direct-imaging detection of extrasolar planets is in principle possible. As an example, we present the detection of a very faint object ($H = 16.4$ mag, $K = 16.3$ mag, i.e. $\sim$ 9.5 mag fainter than a nearby young star), which could have been a few Jupiter-mass giant extrasolar planet, given its brightness and color, its astrometry suggesting that its co-moving with its hypothetical host star, and given the fact that this star is a *young* star.

At the age and distance for this star, this faint object could have been the first direct-imaging detection of an extrasolar planet. However, our H-band spectrum shows that it is likely to be a background K-dwarf. We demonstrate, nevertheless, that spectroscopic observation of such a faint object, so close to a brighter star (contrast $\sim 10^4$) is possible with current technology from ground-based telescopes.

We are living in very special times in the history of humankind where we are starting to be able to *directly* image extrasolar planets and obtain spectra of their atmospheres, using ground-based astronomical facilities. We believe that the first direct-imaging detection of an extrasolar planet is imminent.

Acknowledgments. We would like to thank our collaborators: Eike Guenther, an expert on spectroscopy, Wolfgang Brandner, an expert on *HST* NICMOS observations, Nuria Huélamo, an expert on young nearby stars, Andreas Eckart and Thomas Ott, experts on SHARP; Jean-Gabriel Cuby, an expert on ISAAC, and Matilde Fernández, an expert on TWA stars.

References

Baraffe, I., Chabrier, G., Allard, F., & Hauschildt, P. 1998, A&A, 337, 403

Brandner, W., Alcalá, J. M., Frink, S., & Kunkel, M. 1997, Messenger, 89, 37

Burrows, A., et al. 1997, ApJ, 491, 856

Charbonneau, D., Brown, T. M., Latham, D. W., & Mayor, M. 2000, ApJ, 529, L45

D'Antona, F., Mazitelli, I. 1994, ApJS, 90, 467

Hearty, T., Neuhäuser, R., Stelzer, B., Fernández, M., Alcalá, J. M., Covino, E., & Hambaryan, V. 2000a, A&A, 353, 1044

Hearty, T., Fernández, M., Alcalá, J. M., Covino, E., & Neuhäuser, R. 2000b, A&A, in press

Malkov, O., Piskunov, A., & Zinnecker, H. 1998, A&A, 338, 452

Marcy, G. W., & Butler, R. P. 1998, ARA&A, 36, 57

Nakajima, T., et al. 1995, Nat, 378, 463

Neuhäuser, R., Brandner, W., Eckart, A., Guenther, E. W., Alves, J., Ott, T., Huélamo, N., & Fernández, M. 2000, A&A, 354, L9

Rebolo, R., et al. 1998, Sci, 282, 1309

Terebey, S., Van Buren, D., Padgett, D. L., Hancock, T., & Brundage, M. 1998, ApJ, 507, L71

Webb, R. A., et al. 1999, ApJ, 512, L63

Discussion

Glenn Schneider: Assuming an age indicated for TWA ($\sim$ 10 Myr), TWA 7"B" is far too luminous to be an EGP. We deduce a comparable mass for TWA 6B (which is $H = 20.1$) as opposed to TWA 7"B" (which is $H = 16.8$). Assuming coevality for the association members, how can you explain the disparity? Also, your I'(F090M) magnitude is apparently too high by ~ 0.2 mag. The photometric calibration should be good to ~ 0.05 mag. Correcting this would bring the I' magnitude down as expected. Finally, the NICMOS astrometric error should be much less (~ 10 mas) in camera 2. You should use the target acquisition image and offest slew information to get the position in the coronagraph. Was this done?

João Alves: The mean distance of the four TWA members observed by *Hipparcos* is 55 pc. The JHK magnitudes of TWA 7 (from both Webb et al. 1999 and our SHARP H and K data) are consistent with an unabsorbed M1 star. Given these magnitudes and 55 pc as the distance, we derive $\log L_{\rm bol}/L_{\odot} = -0.41$. Together with spectral type M1, i.e. $T_{\rm eff} = 3700$ K, this corresponds to an age of 1 to 6 Myr, according to D'Antona & Mazitelli (1994) and Baraffe et al. (1998). Then, at 55 pc, our SHARP H and K magnitudes for TWA 7B correspond to absolute magnitude $M_H = 12.72$ mag and $M_K = 12.64$ mag. According to Table 5 in Burrows et al. (1997), this is consistent with an object of gravity 3×10^3 cm s^{-2} and temperature 1050 K. Then, according to Figure 9 in Burrows et al. (1997), this is consistent with an object with an age of $10^{6.5}$ yr (like TWA 7A) and a mass of just about 3 $M_{\rm Jup}$. I suspect that the apparent inconsistency between our estimates is due to a smaller age and a larger distance used by us. If the value we assumed for TWA 7 is the correct one, it would imply that the TWA association members are not that coeval. In his talk at AAS (January 2000), R. Webb has shown that all TWA members are coeval and co-moving, if on the 7 Myr isochrone, implying a distance of 32 pc towards TWA 7.

According to the NICMOS home page[1], there is a calibration inconsistency between the *HST* Vega system and ground-based systems of up to 0.3 mag, especially for red objects. Furthermore, the F090M filter has very low quantum efficiency (15%) and there may be spectral features just inside or outside this filter (for red objects). The F090M magnitude for TWA 7A is inconsistent with its JHK data and its M1 spectral type by 4 σ. One has to be open to the possibility of strong variability even in the red and infrared in young planets due to accretion and possible comet impacts (the *HST* F090M and F0160W observations were done about half a year apart, i.e., not simultaneously).

In the NIC2 images (with coronagraph), we estimated the centroid of TWA 7A both by fitting elliptical isophots to the reflected light and by fitting and extrapolating the diffraction spikes to their crossing. The uncertainty in the north–south alignment of NICMOS during this observation was determined by guide stars.

Glenn Schneider: The performance for NNT direct imaging was shown without subtraction of a reference PSF. If you do such a subtraction, in general, how much better would you do in terms of background light rejection and sensitivity?

João Alves: According to our broad experience with SHARP (our coauthors/Co-Is T. Ott and E. Eckart are on the MPE-SHARP team), we did not attempt PSF subtraction (it would not give better results). The seeing and atmospheric conditions vary too fast compared to the relatively long total exposure times, 1000 times 0.5 seconds.

[1] www.stsci.edu/instruments/nicmos/nicmos_doc_phot.html

Alan Boss: If the candidate giant planets TWA 6 and TWA 7 are true giant planets, because it is very hard to imagine them forming at $\sim$ 120–40 AU, they would seem to be likely to have been kicked out there by a close encounter with other giant planets in the system (closer in) and so should be in eccentric orbits.

João Alves: You raise a good point. Kicking these possible giant planets out by a close encounter with another giant planet is probably the most likely way of having them at 120–140 AU. For the TWA 7B candidate we estimate a period of $\sim$ 2000 years and so the eccentric orbits might be checked in the near future with VLTI, for example.

Nevertheless, I think we should have an open mind when it comes to forming planets at these distances from the star. During this conference Jane Greaves [this volume, p. 296—*editors*] has shown the detection of a possible condensation of dust at a distance of 160 AU from β Pictoris. Also, Alycia Weinberger [this volume, p. 329—*editors*] has presented NICMOS data for HD 141569, which has a disk with a gap at 250 AU that could have been carved by a young planet.

Alycia Weinberger: What models predict that an $H = 16$ mag object at $d = 60$ pc (i.e., ROSAT 7) would have a mass of 3 $M_{\rm Jup}$? It seems that an object with such a large absolute H-band magnitude would have a larger mass from the Burrows models.

João Alves: [See reply to Glenn Schneider's first question above.—*Editors* .]

Disks, Planetesimals, and Planets
ASP Conference Series, Vol. 219, 2000
F. Garzón, C. Eiroa, D. de Winter, and T. J. Mahoney, eds.

Very Young Free-Floating Planets in the σ Orionis Star Cluster

R. Rebolo

Instituto de Astrofísica de Canarias and Consejo Superior de Investigaciones Científicas, E-38205 La Laguna, Tenerife, Spain

M. R. Zapatero-Osorio and V. J. S. Béjar

Instituto de Astrofísica de Canarias, E-38205, La Laguna, Tenerife, Spain

D. Barrado-Navascués, C. Bailer-Jones, and R. Mundt

Max Planck Institut für Astronomy, Königstuhl 17, D-69117, Heidelberg, Germany

E. L. Martín

Division of Geological and Planetary Sciences, California Institute of Technology, MS 150-21, Pasadena, CA 91125, USA

Abstract. We present a deep optical and infrared survey of the stellar cluster around the very young (1–5 Myr), multiple massive star σ Orionis. We have discovered a large number of very red ($I-K \geq 4$), low-luminosity objects with masses estimated in the range 5–15 $M_{\rm Jup}$ according to evolutionary models. These objects are in most cases well below the mass necessary to burn deuterium and cannot sustain any stable nuclear burning in their interiors. Contrary to similar-mass extrasolar planets discovered by Doppler techniques, our objects are not bound to stars. They may have resulted from the fragmentation of low-mass molecular clouds during collapse, or, for example, may have been formed in massive disks around protostars and eventually ejected due to perturbations. This new kind of isolated giant planet offers a challenge to our understanding of the formation processes of planetary-mass objects. Our statistics indicate that these free-floating planets are as common as brown dwarfs; both kinds of objects seem to be more numerous than stars in the σ Orionis cluster. We predict that older and much lower-luminosity free-floating planets with spectral characteristics similar to Jupiter are likely to populate the disk of the Galaxy and in particular the solar neighborhood.

1. Introduction

Radial-velocity searches have shown that giant planets from $\sim$ 10 times to one quarter of the mass of Jupiter exist in orbit around solar-type and low-mass stars (Mayor & Queloz 1995; Marcy & Butler 2000; Cochran et al. 1997).

Whether objects with such low masses can form isolated from stars is completely unknown. Theoretical models describing the fragmentation of collapsing clouds are uncertain about the lower mass limit for this process. The extrapolation of recent observations on the number of brown dwarfs per mass interval in young star-forming regions (Tamura et al. 1998; Béjar, Zapatero-Osorio, & Rebolo 1999; Zapatero-Osorio et al. 1999; Hillenbrand & Carpenter 2000; Najita, Tiede, & Carr 2000) to lower masses suggest that isolated planetary-mass objects could exist in large numbers. We report here on the detection of 18 free-floating objects in the σ Orionis cluster whose photometry fulfills the expectations for very young giant planets.

2. Observations

We have carried out a deep photometric survey both in the optical and near-infrared wavelengths over a region of 847 arcmin^2 around the multiple O9.5-type star σ Orionis. This is a very low-extinction region in Orion located at a distance of 352 pc from the Sun (see, for example, Béjar et al. 1999, and references therein). Photometry in the J band has been obtained with the Omega Prime near-infrared camera of the 3.5 m Telescope at the Calar Alto Observatory (Spain) during 1998 October 27–31. Optical I- and Z-band images were acquired with the Wide Field Camera (1998 November) of the 2.5 m Isaac Newton Telescope at the Roque de los Muchachos Observatory (Spain). A cross-correlation of the list of objects identified in the three bands was then performed. The limiting I (Cousins photometric system) and J (UKIRT photometric system) magnitudes of our survey are 23.8 and 21.2, respectively, while completeness magnitudes are estimated at 21.5 and 19.5, respectively. Uncertainties in the photometry range from about 0.05 mag at completeness to 0.2 mag close to the limit of our search. Further observations in the K band (2.2 micron) with an accuracy of 0.15 mag or better have been undertaken on selected individual targets using the 1.5 m Carlos Sánchez Telescope at the Teide Observatory (Spain), the 2.2 m at Calar Alto Observatory, and the 3.8 m United Kingdom Infrared Telescope at the Mauna Kea Observatory (USA).

3. Results

In Figure 1 (*left panel*), we show the brown-dwarf sequence in σ Orionis in the I versus $I - J$ color for the magnitude interval I = 16–20.5. Membership for objects plotted here has been confirmed via optical spectroscopy (Béjar et al. 1999; Zapatero-Osorio et al. 1999). The coolest of these members is the L1.5-type substellar object S Ori 47; with a mass around 15 $M_{\rm Jup}$, it lies very close to the deuterium-burning mass limit. In the same panel we plot those candidates found in our survey which delineate a smooth extension of the previous sequence to lower luminosity and redder colors as expected for even less massive objects. We concentrate here on 18 new candidates fainter and redder than S Ori 47, marked with filled circles in the figure. The *right panel* of Figure 1 illustrates the I vs. $I - K$ color–magnitude diagram whenever K photometry is available.

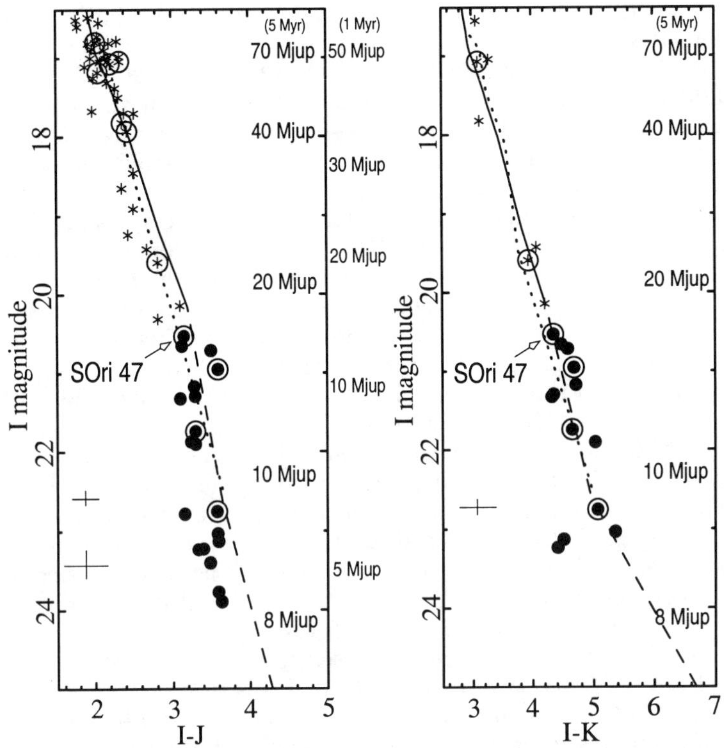

Figure 1. Optical–infrared color–magnitude diagrams for σ Orionis cluster candidate members found in our *IZJ* survey. *Filled circles* denote substellar objects fainter and redder than S Ori 47. *Asterisks* indicate brown dwarfs. Cluster members spectroscopically confirmed (Béjar et al. 1999; Zapatero-Osorio et al. 1999, 2000) are indicated with an *open circle* around their corresponding symbols. Error bars are given to the left side of the diagrams when they are larger than the symbol size. Overplotted to the data are the theoretical, solar metallicity, 5 Myr isochrones provided by the Lyon group and by the Arizona group. Masses for two different ages of 5 and 1 Myr are provided to the *right*.

Visual inspection of the candidates guarantees that there are no resolved objects. Galaxies can be distinguished from stellar-like sources down to $K = 19$ mag in images with typical FWHMs of 1.1″ (Thompson et al. 1999). Unresolved galaxies, quasars and other extragalactic objects exhibiting colors as red as those of our candidates could be contributing to the list with less than 1–2 objects, according to the statistics inferred from the Hubble Deep Field (Fergurson, Dickinson, & Williams 2000). Extinguished red-giant stars are a negligible source of contamination in our search because of the high Galactic latitude ($b = -17°$) of the cluster (Kirkpatrick et al. 1994). The main source of contamination in our survey could be due to L-type (Kikpatrick et al. 1999; Martín et al. 1999) very-low-mass stars and brown dwarfs located in the foreground. The DEep Near-Infrared Survey (DENIS, Delfosse et al. 1999) and the 2 Micron All Sky Survey (2MASS, Kirkpatrick et al. 2000) have found that the sky density of objects with early- to mid-L spectral types in the solar vicinity is about 0.0053 objects in a volume of 1 pc^3, and about 0.0073 objects/pc^3 for the latest L classes. Properly correcting for the depth and area coverage of our search in the σ Orionis cluster, we would expect five foreground objects with spectral type L0–L5 in our list of candidates, while late-L field dwarf contaminants would be rare. Recent deep optical–infrared searches for EROs (Thompsom et al. 1999) show a contamination level of Galactic L dwarfs consistent with our estimates.

We are confident that a large fraction of our candidates are true members of the σ Orionis cluster. For three of these candidates we have obtained follow-up optical and near-infrared (1.4 up to 2.5 μm) spectroscopy with LRIS and NIRC, respectively (Keck I and Keck II Telescopes at the Mauna Kea Observatory, see Zapatero-Osorio et al. 2000). They show energy distributions within the L-class (L0–L4) as we expected for young substellar objects members of the σOrionis cluster. We estimate their masses comparing observed magnitudes with those predicted by evolutionary tracks. Mass determinations are then model-dependent and are subject to possible systematic effects arising from the description of atmospheres and interiors. The age of σ Orionis stars and brown dwarfs has been estimated in the range 1–5 Myr (Wolk 1996; Brown, de Geus, & de Zeeuw 1994). Overplotted on the observed photometry of our candidates in Figure 1 are the 5 Myr isochrones from the Lyon group (Baraffe et al. 1998) and from the Arizona group (Burrows et al. 1997); all models shifted to the distance of σ Orionis using the *Hipparcos* distance modulus ($m - M = 7.73$). The "dust-free" NextGen Lyon's models are used down to $I = 20$, since we do not expect dusty effects in the atmospheres of M-type cluster brown dwarfs; for fainter magnitudes, we use recent computations of dusty models. The isochrones in Figure 1 nicely reproduce the trend of our near-infrared observations. A direct comparison with the 5 Myr tracks yields masses for our candidates in the interval 15–8 $M_{\rm Jup}$. If the cluster is younger, the masses of our fainter candidates could be as low as 5 $M_{\rm Jup}$. This is well below the minimum mass for deuterium burning $\sim$ 13 $M_{\rm Jup}$ (the proposed boundary between brown dwarfs and giant planets, Saumon et al. 1996; Basri & Marcy 1997). These very young isolated giant planets will be unable to stably sustain thermonuclear reactions in their interiors. Remarkably, our findings suggest that the formation of unbound objects a few times more massive than Jupiter is almost over at the short age of a few million years.

4. Mass Function

Our observations allow the mass function (number of objects per interval of mass) for the σ Orionis cluster to be derived from the substellar limit (the star–brown dwarf boundary) down to the deuterium-burning threshold with completeness. Due to the very young age of the cluster, no strong dynamical evolution is expected; therefore, the MF of σ Orionis is very probably close to the initial MF. The MF is illustrated in Figure 2 where the mass spectrum

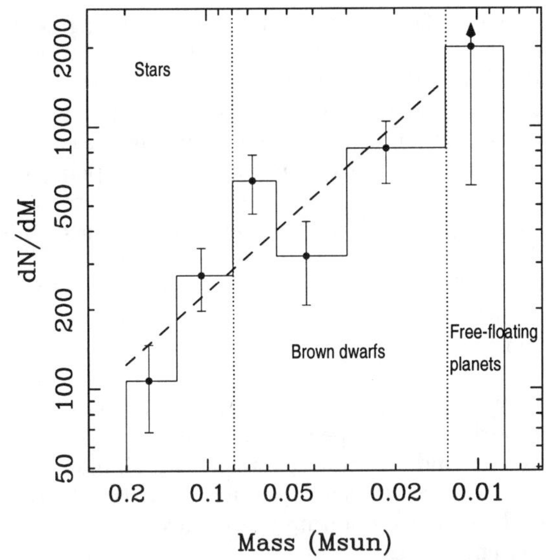

Figure 2. The mass function of the σ Orionis cluster in the substellar domain (adopting an age of 5 Myr). The power-law fitting down to the brown dwarf–planet boundary ($\sim 13\ M_{\rm Jup}$) provides $\alpha = 0.8 \pm 0.4$ Error bars correspond to Poissonian uncertainties, except for the planetary mass interval where the upper limit denotes the incompleteness of the photometric and spectroscopic searches and the lower error bar accounts for the possible contamination of cluster non-members.

representation ($dN/dM\ M^{-\alpha}$) on a logarithmic scale is provided (Salpeter's value is 2.35 for the regime of solar-type stars). We have adopted a plausible age of 5 Myr for the cluster and the mass–luminosity relationship given by the evolutionary models from the Lyon group. A single power-law with index $\alpha = 0.8 \pm 0.4$ fits the points reasonably in the mass range which goes from very-low-mass stars (0.2 solar masses) through the whole brown-dwarf domain to 13 $M_{\rm Jup}$. The uncertainty of $\pm$ 0.4 in the α parameter results from the possible age range for the cluster ($\pm$ 0.2) and the use of other evolutionary models ($\pm$ 0.2). The lowest-mass bin in Figure 2 is well populated in spite of the incompleteness of our survey at these very low masses. Free-floating planets with masses 13–5 $M_{\rm Jup}$ appear to be abundant in σ Orionis. We do not find any evidence for an "end" of the MF down to several times the mass of Jupiter.

Such a numerous population of free-floating giant planets in this cluster challenges our knowledge of how objects with planetary masses form in Nature. The smoothness of the cluster MF appears to indicate that they originate similarly to stars and brown dwarfs, i.e., from the fragmentation of collapsing clouds. However, models describing the collapse of low-mass molecular clouds are unable to explain the formation of single planetary bodies down to a few Jupiter masses, unless external mechanisms prevent the mass accretion onto the central object. Other alternative processes have been proposed (Bodenheimer 1998; Burkert,

Bate, & Bodenheimer 1997), such as multiple cloud fragmentation during collapse, disintegration of massive, gravitationally unstable protoplanetary disks, and core accretion followed by capture of gas in disks around protostars. These processes typically produce giant planets and brown dwarfs around similar-mass objects or more massive objects, forming multiple systems. These scenarios would require ejection mechanisms to separate giant planets from their orbits around stars. Searches for free-floating planets in other star-forming regions will be extremely important to shed light on the formation mechanisms of these new kinds of giant planets.

The extrapolation of the σ Orionis cluster MF to the Galactic disK reveals that isolated planets in the mass range 13–1 $M_{\rm Jup}$ could be as numerous as brown dwarfs; and the number of brown dwarfs and free-floating planets together may be comparable to that of stars in the Galaxy. However, their contribution to the total mass of the disk would be less than 10%. Most free-floating planets in the solar neighborhood are expected to be old and consequently will display effective temperatures and spectral characteristics similar to those of the giant planets in the Solar System. If current models are correct predicting their fluxes in the mid infrared, their detection up to distances of several parsecs is feasible with present-day technologies.

References

Baraffe, I., Chabrier, G., Allard, F., & Hauschildt, P. 1998, A&A, 337, 403

Basri, G., & Marcy, G. 1997 in Star Formation Near and Far, ed. S. S. Holt & L. G. Mundy (New York, AIP Press), 393, 228

Béjar, V. J. S., Zapatero Osorio, M. R., & Rebolo, R. 1999, ApJ, 521, 671

Bodenheimer, P. F. 1998, in ASP Conf. Ser., Vol. 134, Brown Dwarfs and Extrasolar Planets, ed. R. Rebolo, E. L. Martín, & M. R. Zapatero Osorio (San Francisco: ASP), p. 115

Brown, A. G. A., de Geus, E. J., & de Zeeuw, P. T. 1994, A&A, 289, 101

Burkert, A., Bate, M. R., & Bodenheimer, P. 1997, MNRAS, 289, 497

Burrows, A., et al. 1997, ApJ, 491, 856

Cochran, W. D., Hatzes, A. P. Butler, R. P., & Marcy, G. W. 1997, ApJ, 483, 457

Delfosse, X., et al. 1999, A&A, 135, 41

Fergurson, H. C., Dickinson, M., & Williams, R. 2000, ARA&A, 38

Hillenbrand, L. A., & Carpenter, J. M., 2000, ApJ, in press

Kirkpatrick, J. D., McGraw, J. T., Hess, T. R., Liebert, J., & McCarthy, D. W. Jr. 1994, ApJS, 94, 749

Kirkpatrick, J. D., et al. 1999, ApJ, 519, 802

Kirkpatrick, J. D., et al. 2000, AJ, in press

Marcy, G. W., & Butler, R. P. 2000, PASP, 112, 137

Martín, E. L., et al. 1999, AJ, 118, 2466

Mayor, M., & Queloz, D. A. 1995, Nat, 378, 355

Najita, J. R., Tiede, G. P., & Carr, J. S. 2000, ApJ, in press

Reid, I. N., et al. 1999, ApJ, 521, 613

Saumon, D., et al. 1996 ApJ, 460, 993

Tamura, M., Itoh, Y., Oasa, Y., & Nakajima, T. 1998, Sci, 282, 1095

Thompson, D., et al. 1999, ApJ, 523, 100

Wolk, S. J. 1996, PhD Thesis, Univ. New York at Stony Brook

Zapatero Osorio, M. R., Béjar, V. J. S., & Rebolo, R., Martín, E. L., & Basri, G. 1999, ApJ, 525, 115

Zapatero-Osorio, M. R., et al. 2000, Sci, submitted

Discussion

João Alves: From the radial profile of the distribution of sources that you showed, it looks like you have reached some kind of uniform background that extends to the edges of your survey. Aren't you worried about contamination? Did you take control fields?

Rafael Rebolo: We have considered the possible contribution of contaminants in our list of candidates. While we discard extreme red galaxies or extinguished red giants as possible contaminants, about 1–5 foreground L-dwarfs could indeed be contaminating the sample. This can be estimated with proper corrections from other surveys comparable in depth and size or from extended surveys, albeit much limited in sensitivity. The foreground objects could be identified by proper-motion and/or spectroscopic studies. Since they are nearer than cluster members and most probably have higher gravities.

Lee Hartmann: You are plotting your IMF as dN/dU vs. $\log M$. To compare with other researchers, you could plot $dN/d\log M$, which would then show the "flattening" of the IMF at low mass as well in other regions.

Rafael Rebolo: Yes, of course. In the $dN/d\log M$ representation we would have a "flattening" of the IMF at low masses.

Vladimir Grinin: Could you show the direction of interstellar reddening in the I vs. $I - J$ and I vs. $I - K$ diagrams?

Rafael Rebolo: The direction of interstellar reddening is along the sequence of substellar candidates in the cluster. We are dealing with a region with very little extinction. The reddening correction is estimated to be smaller than the size of the symbols in Figure 1.

Thomas Henning: Could the presence of the luminous star influence the formation of the lower-mass stars and shift the lower-mass limit to rather small values?

Rafael Rebolo: Yes, the luminous star may have influenced the formation of very-low-mass brown dwarfs and free-floating planets. The standard scenarios for the collapse and fragmentation of clouds do not predict the formation of such small-mass bodies. Accretion after fragmentation is expected to produce more massive ones, but winds or other phenomena originating in the luminous stars of the cluster may have prevented this accretion.

Günter Wuchterl: As someone who calculates evolution models of brown-dwarf planets, I'm worried that the uncertainty in masses that you infer may be quite a bit larger; the ages and distances are uncertain. Furthermore, we're getting close to early ages for which theoretical evolution models depend on initial conditions.

Rafael Rebolo: Theoretical models are uncertain at these young ages and we caution that our absolute masses may have errors as high as 50%. Relative masses along the substellar sequence of the cluster are more reliable. As can be seen in Figure 1, models can predict the observed sequence reasonably well. It will be very interesting to test these models through observations of deuterium and empirical determinations of the deuterium depletion boundary.

Tsevi Mazeh: I would like to congratulate you for a wonderful and important result. This has implications for the stellar mass in the Galaxy. Can you comment on the comparison between your result and the field frequency of BDs?

Rafael Rebolo: Field searches for brown dwarfs have been able to measure the upper mass range of the substellar domain. There is very good agreement between the mass function for field massive brown dwarfs (see Reid et al. 1999) and what we have found in Sigma Orionis.

Searches for MACHOS via microlensing have concentrated on the halo and bulge of the Galaxy. There are no searches through the disk which would be adequate for comparison. Our claim that brown dwarfs and free-floating planets more massive than Jupiter will not contribute more than 10% of the total mass of the disk of the Galaxy is consistent with the upper limits established by microlensing searches in the halo and bulge.

Disks, Planetesimals, and Planets
ASP Conference Series, Vol. 219, 2000
F. Garzón, C. Eiroa, D. de Winter, and T. J. Mahoney, eds.

Chemical-Abundance Trends among Stars with Planets

Guillermo Gonzalez
University of Washington, Astronomy Department, Box 351580, Seattle, WA 98195, USA

Abstract. The metallicities and relative abundance trends among a subset of the known solar-type stars with planets are reviewed. Seventeen stars with planets are shown to have a significantly higher mean metallicity compared to the general field G-dwarf population. In addition, they have smaller [C/Fe] values at a given [Fe/H].

1. Introduction

Since 1995 December we have been obtaining high-resolution spectra of the solar-type parent stars of extrasolar planets. Our motivations are simple—to derive accurate stellar atmospheric parameters in the hope of better constraining planetary formation models. Of particular importance is the distribution of the metallicities of the stars with planets. If the formation of giant planets is dependent on the concentration of planetesimals in the protoplanetary disk, then one would expect a significant difference between the metallicity distribution of stars with planets and that of stars without planets. The unexpected discovery of "hot jupiter" systems presents us with a challenge to account for their formation and evolution. The metallicity distribution of these systems may also give us insight into their history. For example, if giant-planet migration occurs via scattering of planetesimals (some of which fall into the parent star), then one would expect a correlation between metallicity and the presence of hot jupiters.

In the following we discuss a sample of 17 solar-type stars with planets with spectroscopic analyses. We compare their metallicity distribution to the general field-star population and present evidence of suggestive trends.

2. The Sample

We present the latest estimates of the physical characteristics of 17 stars with planets in Table 1. The data for the 13 stars not in italics are from Gonzalez (1997, 1998), Gonzalez & Vanture (1998), Gonzalez, Wallerstein, & Saar (1999), and Gonzalez & Laws (2000). Several of these stars have also been studied by Fuhrmann, Pfeiffer, & Bernkopf (1997, 1998), who generally confirm our results. The fundamental atmospheric parameters have been determined using standard fine analysis methods, which make use of equivalent widths of Fe I and Fe II lines in the optical. Most of our spectra have a resolving power near 60 000 and an S/N ratio between 400 and 600. The age and mass estimates were obtained from the theoretical evolutionary tracks of Schaller et al. (1992) and Schaerer et al. (1993). For additional details on the observations and analysis methods, see the papers listed above.

The locations of the stars with planets on the Hertszprung–Russell (HR) diagram are shown in Figure 1, along with the stellar isochrones from Schaller et al., and Schaerer et al. The luminosities are derived from the *Hipparcos* parallaxes. Most of the sample stars are located in regions of the HR diagram where the isochrones are well separated and hence useful for deriving ages and masses; due to the convergence of the tracks at low luminosities, age and mass estimates are not possible for stars later than about spectral type G6. The very high metallicities of 14 Her and ρ^1 Cnc are apparent from their large separations from the lower ZAMS. Of course, the application of these theoretical isochrones to a planet-bearing star assumes it has a homogeneous initial composition, an assumption that may be violated if it experienced self-pollution.

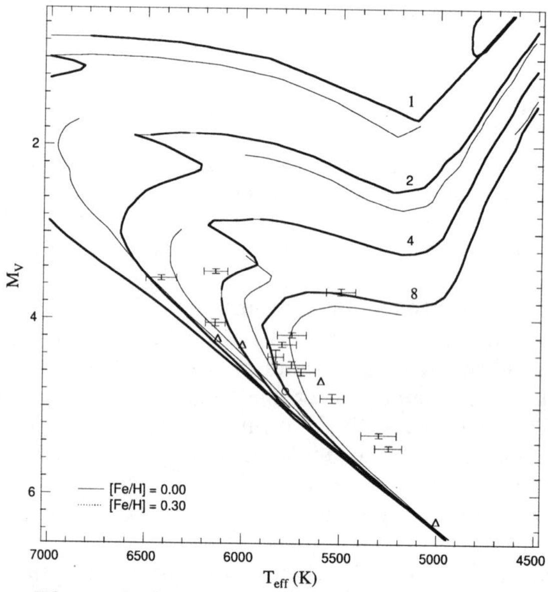

Figure 1. Theoretical stellar evolutionary isochrones of Schaller et al. (1992) and Schaerer et al. (1993) are shown for 1, 2, 4, and 8 Gyr for two [Fe/H] values. The stars from Table 1 are shown as *dots* with error bars; *open triangles* correspond to HR 810, HD 192263, HD 209458, and HD 217107; the Sun is shown as an *open circle*.

3. Comparisons with Field Stars

As a first-order analysis, we compare the metallicity distribution of our stars-with-planets sample to a sample of 62 nearby single G-dwarfs without known planets from Favata, Micela, & Sciortino (1997) in Figure 2. The mean of the Favata et al. sample is -0.12 ± 0.25 (std. dev.) and the mean of our sample is $+0.11 \pm 0.27$. The 99% confidence interval for the difference between these two means is $0.03 < \mu_1 - \mu_2 < 0.43$. It is not clear if HD 114762 should be included in the stars-with-planets sample, as its minimum mass (11.02 $M_{\rm Jup}$; Marcy et al. 1999) is near the deuterium-burning limit near 13 $M_{\rm Jup}$ (Burrows et al. 1998). In addition, HD 114762 was discovered as part of a large low-precision radial-velocity survey, in which one would expect to find one star viewed nearly pole on from simple probabilities. For these reasons, it is useful to calculate the 99% confidence interval without HD 114762: $0.11 < \mu_1 - \mu_2 < 0.43$. Finally, for the fairest comparison with field G-dwarfs, we should remove τ Boo, a mid-F dwarf, from the stars-with-planets sample. Leaving out both HD 114762 and τ Boo, the 99% confidence interval becomes $0.09 < \mu_1 - \mu_2 < 0.43$. Two of the stars with planets, 14 Her and ρ^1 Cnc, are the most metal-rich solar-type stars known in the solar neighborhood that have modern spectroscopic abundance analyses.

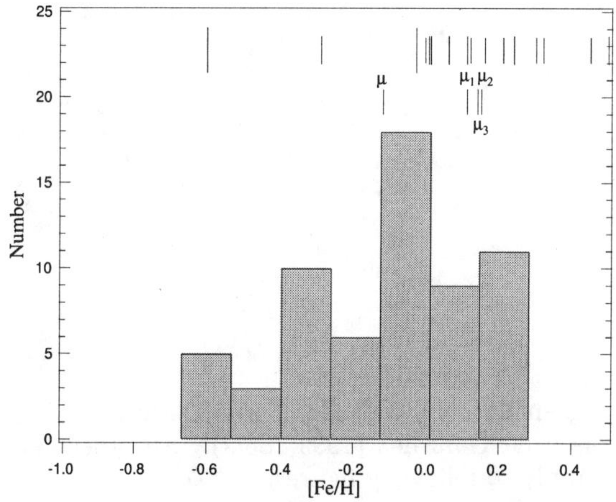

Figure 2. Histogram of [Fe/H] for nearby single G-dwarfs with $T_{\rm eff} > 5200$ K from Favata et al. (1997). The mean of the nearby star distribution is indicated with a μ. The stars with planets with P > 15 days are shown with *solid bars*; those with P < 15 days are shown as *dashed bars*. The *long bars* correspond to minimum masses greater than 6 $M_{\rm Jup}$. μ_1 is the mean of all the planet-bearing stars; μ_2 excludes HD 114762; μ_3 excludes both HD 114762 and τ Boo.

We can perform a second-order comparison between stars with planets and field stars by including stellar age as a parameter (Figure 3). The large scatter in [Fe/H] at a given age is still a matter of some controversy, but the stellar orbital diffusion model of Wielen, Fuchs, & Dettbarn (1996) is able to account for it well. According to the diffusion model, stars that we see near the Sun today originated at different galactocentric radii, and since there is a significant metallicity gradient in the disk, they will exhibit an apparent scatter in [Fe/H] at a given age and location. As can be seen in Figure 3, stars with planets occupy the upper half of the distribution. One could explain this pattern by having the stars with planets originate from the inner, more metal rich, regions of the disk. However, we cannot determine from its present kinematics alone where a particular star originated.

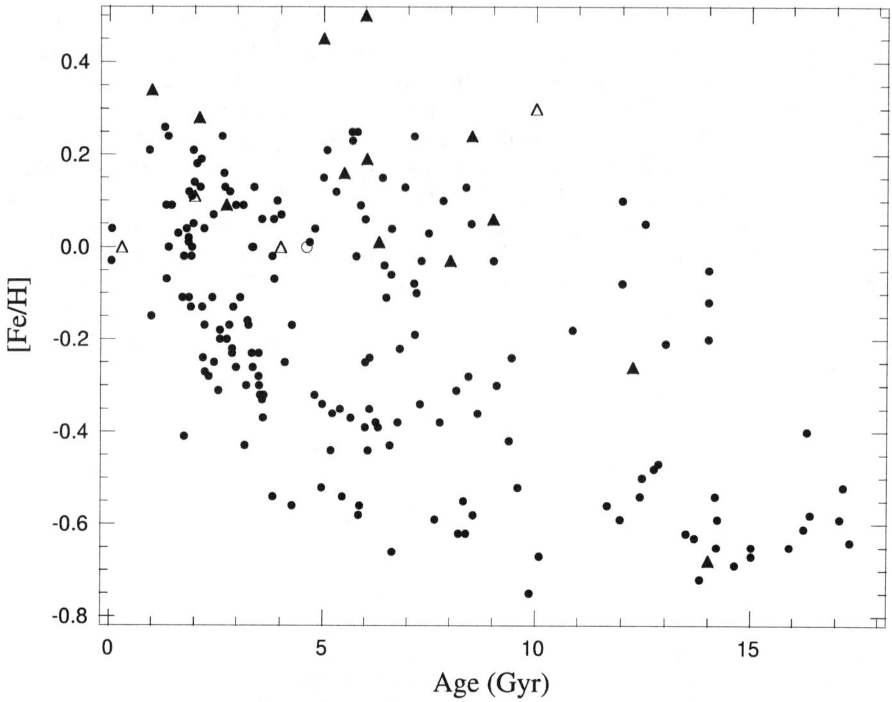

Figure 3. Age–metallicity plot for nearly 200 nearby single stars using the sample from Gonzalez (1999; *dots*). All other symbols as in Figure 1 and apply to following figures.

To circumvent the uncertainties resulting from stellar migration, we can look at only young stars. Restricting our attention to stars less than about 2 Gyr old, we find that the four youngest stars with planets in our sample are metal rich relative to the general field population at the same galactocentric distance (Figure 4). It is still too early in the game to make too much of this pattern given the small-number statistics, but it is worth giving extra attention to young nearby stars in future studies.

4. Abundance Trends

So far, orbital period is the only measured parameter for extrasolar planets that correlates with the compositions of their parent stars. As can be seen in Figure 2, hot jupiters have, on average, the most metal-rich parent stars. If one removes the hot-jupiter systems from the sample, then it is not clear that the remaining stars have a significantly different mean metallicity compared to the field population given the small-number statistics. It should be possible to answer this question within the next year.

In addition to the high mean metallicity of the stars-with-planets sample compared to the general field population, stars with planets also display sugges-

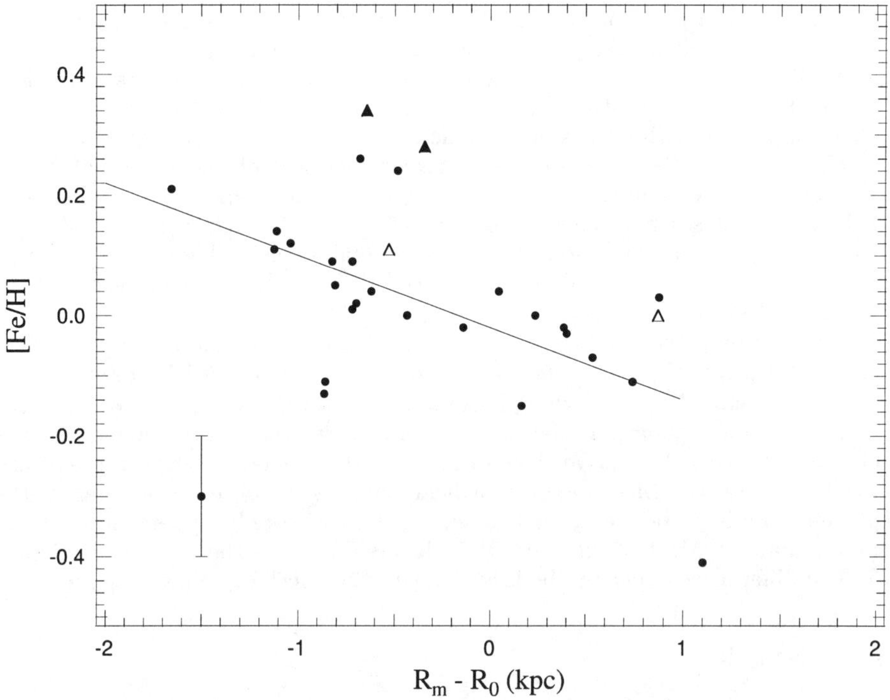

Figure 4. Gradient in [Fe/H] among nearby young stars from Edvardsson et al. (1993; *dots*); the error bar shows the typical uncertainty for the field stars; $R_{\rm m}$ is the mean galactocentric radius. The *dotted line* is a least-squares fit to the young stars. The four young planet-bearing stars are τ Boo, HD 75289, HR 810, and HD 192263.

tive relative abundance (such as [Me/Fe]) anomalies. Among the elements we measured, C is the best case for anomalous behavior in our sample. We show [C/Fe] for a nearby field-star sample and the stars-with-planets sample in Figure 5. Apart from the mean trend of decreasing [C/Fe] with increasing [Fe/H], the nearby stars sample displays a scatter consistent with measurement uncertainty only. The stars-with-planets cluster at high [Fe/H] and low [C/Fe]. When we (the author and Chris Laws) first noticed this pattern 6 months ago, we searched the literature for additional stars with published C and Fe abundances that fall in the lower right-hand corner of Figure 5. The best such candidate was HD 89744, which we show in the figure with a plus sign. We suggested to Geoff Marcy that he place this star on his Doppler planet survey program. Unbeknownst to us, Robert Noyes had already been monitoring this star for over a year. His group announced the presence of a planet around this star at this meeting! While the confirmation of our prediction lends support to the reality of the C/Fe anomaly among stars-with-planets, it is still too early to declare this as a certain trend.

A potentially important quantity is [Li/H], given that the abundance of Li in the atmosphere of a star can be altered by: 1) mixing into deeper layers where it is destroyed, or 2) accretion of rocky material (i.e., planets, asteroids, dust). The most interesting case regarding Li at present is the 16 Cyg system. The component without a known planet, 16 Cyg A, has a significantly larger Li abundance. Otherwise, they have nearly identical physical characteristics. Within the uncertainties, $\sim$ 0.1 dex, they have the same [Fe/H] value, but published analyses have consistently quoted a slightly higher [Fe/H] value for 16 Cyg A. A very careful modern abundance analysis using high-quality spectra should be able to detect a difference in [Fe/H] as small as $\sim$ 0.05 dex between this pair of stars.

To compare the Li abundances in our sample to field stars, we must first correct for physical parameters that are known to correlate with Li abundance: $T_{\rm eff}$, [Fe/H], and $R'_{\rm HK}$. Following Pasquini, Liu, and Pallavicini (1994), we have derived a simple equation relating Li abundance to these three quantities (see Gonzalez & Laws 2000). We show in Figure 6 the difference between the calculated and observed Li abundance for field stars and a subset of our stars-with-planets sample; note, we are not including stars with only upper limits on the Li abundances. Most of the stars with planets fall below the zero line in Figure 6. The differences appear to be largest for the smallest Li abundances.

5. Discussion

Given the results presented above, we are not yet in a position to choose unambiguously between the two main planet-formation mechanisms discussed in the literature: core instability accretion (CIA; reviewed by Podolak, Hubbard, & Pollack 1993) or gravitational instability (GI; Boss 1997). The CIA process should be highly dependent on metallicity, since a rocky core must be in place before runaway growth via gas accretion can occur; GI does not require a rocky core to be in place, because the initial growth is triggered by self gravity in the disk (which could be purely gaseous). This does not imply, though, that a giant planet formed via the GI process cannot be metal rich; such a planet can acquire high-metallicity material after the initial growth begins. It does mean that the GI process should have a much weaker dependence on metallicity.

Also, we cannot yet determine whether planetary migration is metal-dependent or whether pollution of stars with planets has occurred. However, the high mean [Fe/H] value of the hot-jupiter systems (especially the youngest ones) does suggest that either planet migration is metallicity-dependent and/or pollution did occur. If planet migration is due to scattering of planetesimals, as Quillen & Holman (2000) argue, then one would expect there to be a strong dependence on metallicity. Migration via planetesimal scattering would also naturally lead to pollution of the parent star's atmosphere.

The low [C/Fe] values among stars with planets is consistent with pollution by bodies formed in the inner protoplanetary disk. Kargel & Lewis (1993) have published their best estimates for the bulk composition of the Earth, finding that C is a trace element in it. Therefore, accretion of bodies similar in composition to the Earth would lead to a reduced [C/Fe] value in the atmosphere of a solar-

Table 1. Spectroscopically determined parameters of the stars-with-planets subsample

Name[a]	$T_{\rm eff}$ [K]	$\log g$	$\xi_{\rm t}$ [km s^{-1}]	[Fe/H]	mass [$M_\odot$]	age [Gyr]
HD 75289	6140 ± 50	4.47 ± 0.24	1.48 ± 0.10	$+0.28 \pm 0.05$	1.22 ± 0.02	2.1
51 Peg	5750 ± 75	4.40 ± 0.10	1.00 ± 0.10	$+0.21 \pm 0.06$	1.05 ± 0.03	6
HD 187123	5830 ± 40	4.40 ± 0.07	1.00 ± 0.08	$+0.16 \pm 0.05$	1.08 ± 0.04	4
HD 209458	*6000*	*4.25*	*1.15*	*+0.00*	*1.1*	*4*
υ And	6140 ± 60	4.12 ± 0.11	1.35 ± 0.10	$+0.12 \pm 0.05$	1.28 ± 0.02	3.3
HD 192263	*5000*	...	...	*+0.00*	...	*$\sim$ 0.3*
ρ^1 Cnc	5250 ± 70	4.40 ± 0.15	0.80 ± 0.09	$+0.45 \pm 0.05$	~ 1	~ 5
ρ CrB	5750 ± 75	4.10 ± 0.05	1.20 ± 0.10	-0.29 ± 0.06	0.96 ± 0.03	11
HD 217107	*5597*	*4.3*	...	*+0.30*	*1.0*	*10*
HD 210277	5540 ± 60	4.35 ± 0.10	0.85 ± 0.08	$+0.24 \pm 0.05$	0.92 ± 0.04	12
16 Cyg B	5700 ± 75	4.35 ± 0.05	1.00 ± 0.10	$+0.06 \pm 0.05$	0.97 ± 0.03	9
HR 810	*6125*	*4.43*	*1.64*	*+0.11*	*1.2*	*2*
47 UMa	5800 ± 75	4.25 ± 0.05	1.00 ± 0.10	$+0.01 \pm 0.06$	1.03 ± 0.03	8
14 Her	5300 ± 90	4.27 ± 0.16	0.80 ± 0.12	$+0.50 \pm 0.05$	~ 1.05	~ 6
τ Boo	6420 ± 80	4.18 ± 0.08	1.25 ± 0.11	$+0.32 \pm 0.06$	1.34 ± 0.02	1.5
70 Vir	5500 ± 75	3.90 ± 0.10	1.00 ± 0.10	-0.03 ± 0.06	1.10 ± 0.02	8.5
HD 114762	5950 ± 75	4.45 ± 0.05	1.00 ± 0.10	-0.60 ± 0.06	0.82 ± 0.03	14

[a]The data given in the rows shown in italics are preliminary estimates from a variety of sources.

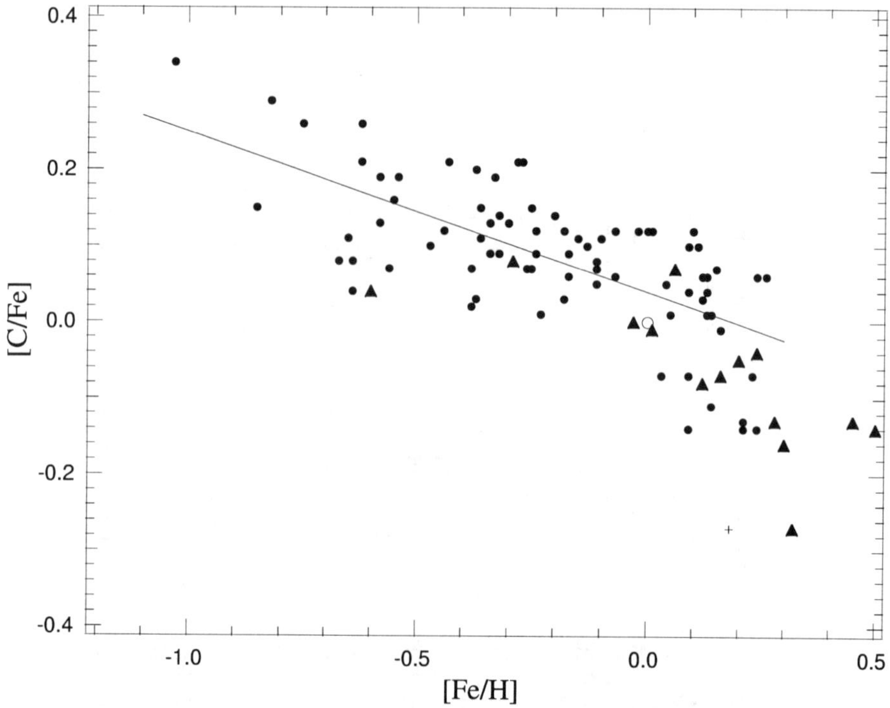

Figure 5. [C/Fe] for 73 nearby F- and G-dwarfs from Gustafsson et al. (1999) and Tomkin et al. (1997). The *plus sign* is HD 89744. A least-squares fit to the nearby field star sample is shown as a *dashed line*.

type star. The relative abundances of both O and Fe are very high in the Earth, so accretion of such bodies by a star would not be expected to alter its [O/Fe] value significantly. If this hypothesis is correct, then it may be possible to reconstruct the initial stellar metallicity of a polluted star by assuming the deviation between [C/Fe] and the mean trend line in Figure 5 is due entirely to accretion of C-poor material.

The alternative explanation for the high mean metallicity of the stars-with-planets sample is that they were born in metal-rich regions of the Milky Way and that the CIA process operates. Note that this explanation can be simultaneously true with the alternate hypothesis given above. The lack of giant planets around metal-poor stars is consistent with the CIA process, but the minimum metallicity required to build a giant planet is still not clear. ρ CrB is an important case in this regard, since it is the most metal-poor star with a planet significantly below the deuterium-burning mass limit. However, the inclination of its orbit is poorly constrained, and, hence, so is the true mass of its planet. The high relative metallicities of the youngest stars with planets shown in Figure 4 tend to argue against high primordial abundances as the source of their present compositions.

It might be possible to test the self-pollution hypothesis with observations of dwarfs in a cluster over a wide range in spectral type. Assuming the same amount

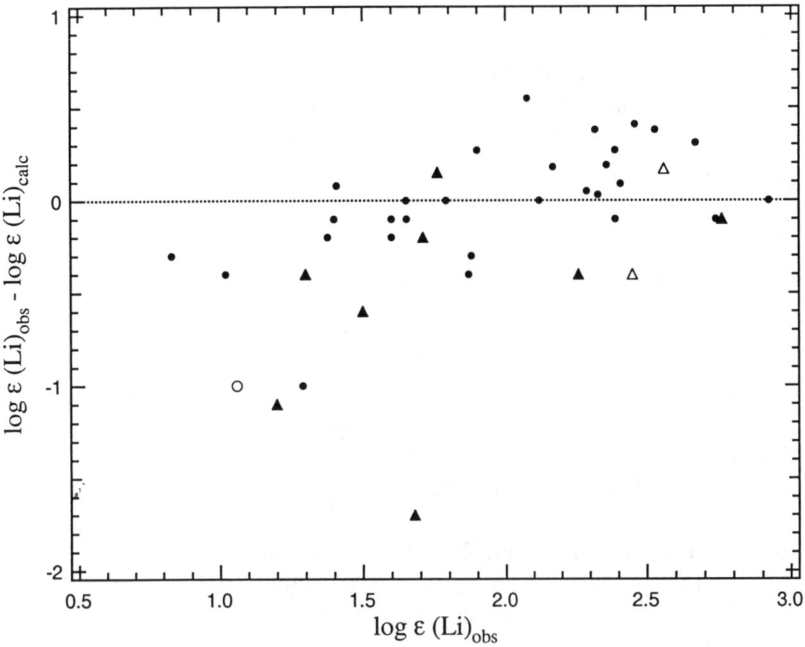

Figure 6. Difference between the observed and "calculated" Li abundances for field stars from Pasquini et al. (1994), Favata et al. (1997), and Randich et al. (1999).

of rocky material is accreted in each case, the resulting increase in [Fe/H] at the surface of a K0 dwarf will be less than that at the surface of a G0 dwarf. This is due to the greater mass of the outer convection zone in dwarfs of later spectral type. An observed correlation between metallicity and spectral type among dwarfs in an open cluster which is known not to have a spread in metallicity among its giants would be evidence for pollution. Giants, having a much deeper convection zone, would have erased any previous abundance inhomogeneities as they ascended the subgiant branch. Along these lines, another possible test would be a search for a variation in metallicity along the subgiant branch; as a star ascends the subgiant branch, not only will the surface layers be mixed with deeper ones, but any close-in planets will be accreted. The current search for planets in the Hyades by William Cochran may soon allow us to apply some of these tests.

References

Boss, A. P. 1997, Sci, 276, 1836

Burrows, A., et al. 1998, in ASP Conf. Ser., Vol. Brown Dwarfs and Extrasolar Planets, ed. R. Rebolo, E. L. Martin, & M. R. Zapatero (San Francisco: ASP), p. 354

Favata, F., Micela, G., & Sciortino, S. 1997, A&A, 323, 809

Fuhrmann, K., Pfeiffer, M. J., & Bernkopf, J. 1997, A&A, 326, 1081

Fuhrmann, K., Pfeiffer, M. J., & Bernkopf, J. 1998, A&A, 336, 942

Gonzalez, G. 1997, MNRAS, 285, 403

Gonzalez, G. 1998, A&A, 334, 221

Gonzalez, G. 1999, MNRAS, 308, 447

Gonzalez, G., & Vanture, A. D. 1998, A&A, 339, L29

Gonzalez, G., Wallerstein, G., & Saar, S. H. 1999, ApJ, 511, L111

Gonzalez, G., & Laws, C. 2000, AJ, 119, 390

Gustafsson, B., Karlsson, T., Olsson, E., Edvardsson, B., & Ryde, N. 1999, A&A, 342, 426

Kargel, J. S., & Lewis, J. S. 1993, Icarus, 105, 1

Marcy, G. W., Butler, R. P., Vogt, S. S., Fischer, D., & Liu, M. C. 1999, ApJ, 520, 239

Pasquini, L., Liu, Q., & Pasquini, R. 1994, A&A, 287, 191

Podolak, M., Hubbard, W. B., & Pollack, J. B. 1993, in Protostars and Planets III, ed. E. Levy & J. Lunine (Tucson: University of Arizona Press), p. 1109

Quillen, A. C., & Holman, M. 2000, AJ, 119, 397

Randich, S., Gratton, R., Pallavicini, R., Pasquini, L., & Carretta, E. 1999, A&A, 348, 487

Schaerer, D., Charbonnel, C., Meynet, G., Maeder, A., & Schaller, G. 1993, A&AS, 102, 339

Schaller, G., Schaerer, D., Meynet, G., & Maeder, A. 1992, A&AS, 96, 269

Tomkin, J., Edvardsson, B., Lambert, D. L., & Gustafsson, B. 1997, A&A, 327, 587

Wielen, R., Fuchs, B., & Dettbarn, C. 1996, A&A, 314, 438

Discussion

Caroline Terquem: How many planets would you need to swallow to account for the abundances of 14 Her and ρ^1 55 Cnc? It seems as if the whole dust content of the disk would have to be dumped onto the star!

Guillermo Gonzalez: If you dump 20 $M_\oplus$ into the present Sun, then [Fe/H] goes up by ~ 0.1 dex, because the Fe gets mixed only in the outer convection zone (2% of the Sun's mass). The convection zones of 14 Her and ρ^1 55 Cnc are much deeper, so you need to dump much more material to get a similar increase in [Fe/H]. I haven't done the calculation for these two stars yet, so I don't know if there was enough dust in its disk to enhance the atmospheric [Fe/H] to the present value (and we don't know the original [Fe/H]).

Eduardo Martın: If the high metal abundances of stars with planets like 55 CnC and 14 Her are primordial, do you know where they formed in the Galaxy? Are they kinematically peculiar?

Guillermo Gonzalez: It is difficult to understand the high metallicities of 14 Her and 55 Cnc as being primordial, even if they were born in the more metal-rich inner region of the Milky Way. Their orbits are not sufficiently eccentric to have come from very far in the disk, but the possibility of orbital diffusion in the disk cannot be ruled out, especially if these two stars are $\sim$ 10 Gyr old.

Robert Noyes: Here is one possible test for the hypothesis that stellar "pollution" due to inward migration of planets or planetesimals occurs preferentially for stars with thin convection zones (earlier spectral types): If almost none of the stars with close-in giant planets and convection-free zones have low [Fe/H], then this is an argument for the hypothesis. Conversely, if a significant fraction of stars with close-in giant planets and thin convection zones have low [Fe/H], this would argue against the hypothesis.

Keith Horne: Can we be quite certain that the correlations of abundances with planets is not a selection effect due to the stars chosen for radial-velocity study, or the stronger lines giving better velocities and hence better sensitivity to planets?

Guillermo Gonzalez: Geoff Marcy has told me that, at least down to [Fe/H] $\simeq -1$, he would be able to detect the planets announced so far. A high $v \sin i$ does affect sensitivity to planets, so high $v \sin i$ stars are avoided in the searches. That biases the searches to older stars, which are likely to be more metal poor.

Disks, Planetesimals, and Planets
ASP Conference Series, Vol. 219, 2000
F. Garzón, C. Eiroa, D. de Winter, and T. J. Mahoney, eds.

The stability of the Orbits of Terrestrial Planets in the Habitable Zones of Known Exoplanetary Systems

Barrie W. Jones and P. Nick Sleep

Astronomy Group, The Open University, Milton Keynes, MK7 6AA, UK

Abstract. We have investigated whether terrestrial planets can exist in orbits in known exoplanetary systems such that life could have emerged on those planets. Four contrasting systems have been examined in which giant planets have been detected. Mixed-variable symplectic numerical integration has been used to investigate the orbits of simulated terrestrial planets within the habitable zone of each system. We have shown that Rho CrB and 47 UMa could have terrestrial planets that remain confined to their habitable zones for biologically significant lengths of time. We have also shown that the Gliese 876 and Ups And systems are very unlikely to have such orbits.

1. Introduction

The question we have addressed is: 'can terrestrial planets exist in orbits in known exoplanetary systems such that life could have emerged on those planets?' The planets so far discovered around other stars are much more massive than the Earth. It is believed (Lissauer 1987; Pollack et al. 1996) that they are rich in hydrogen and helium, and therefore resemble the giant planet Jupiter rather than a supermassive terrestrial planet. This belief is supported by the mass and radius of HD 209458b (Charbonneau et al. 2000), about 0.69 $M_{\rm Jup}$ and 1.5 $R_{\rm Jup}$ respectively, where $M_{\rm Jup}$ and $R_{\rm Jup}$ are the mass and radius of Jupiter. Therefore, all of the known exoplanets must have formed much further out than their present positions and then migrated inwards (Boss 1995). This probably left the zone traversed by the giant, and much of the space interior to this zone, devoid of terrestrial bodies. However, given that migration is caused either by planetesimal scattering by the giant (Murray et al. 1998) or by the interaction of the giant with the nebular disk (Goldreich & Tremaine 1980; Ward 1996; Lin et al. 1998), it is possible that there was enough material left after giant migration to create terrestrial planets.

The detection of terrestrial planets is beyond present capabilities, and so one can only theorize about their existence. We assume that terrestrial planets could be present in at least some of the exosystems, and we have concentrated on four contrasting systems—Gliese 876, Rho CrB, Ups And, and 47 UMa. Whether life could have emerged on terrestrial planets in any of these systems depends on whether terrestrial orbits remain confined to the habitable zone of each system.

2. The Habitable Zone

All life on Earth requires liquid water during at least part of its life cycle. Consequently, it is usual to define the habitable zone (HZ) as the range of distances from a star within which any water at the surface of a terrestrial planet would be in the liquid phase (Kasting, Whitmire, & Reynolds 1993). A variety of criteria have been used to define the boundaries of the HZ. For the inner boundary we use the maximum distance from the star where a runaway greenhouse effect occurs leading to the evaporation of all surface water, and for the outer boundary we use the maximum distance at which a cloud-free CO_2 atmosphere can maintain a surface temperature of 273 K. Kasting et al. (1993) have used these criteria in conjunction with climate models to obtain the boundary distances for various stars, and we have used these values. The values are conservative because of simplifying features in the models. Thus, at the inner boundary H_2O cloud formation is neglected; such clouds would cause this boundary to move inwards. At the outer boundary CO_2 cloud formation is neglected. Forget & Pierrehumbert (1997) have shown that the net effect of CO_2 clouds is to warm the surface, through the greenhouse effect of scattering from cloud particles, and this moves the outer boundary of the HZ outwards.

For zero-age main-sequence (ZAMS) stars the boundaries of the HZ lie closer to the star the later its spectral type. This is because of the combined effects of the star's lower luminosity and the shift in its spectrum to longer wavelengths. As a star ages its luminosity increases and the inner boundary moves outwards. The outer boundary also moves outwards unless 'cold starts' are prohibited, in which case it remains fixed at the ZAMS value. The estimated ages of Rho CrB, Ups And, and 47 UMa are 6 Gyr, 3.3 Gyr, and 7 Gyr respectively (Gonzalez 1998; Gonzalez & Laws 2000), and so outward movement of the boundaries will have been significant. For Gliese 876 the main-sequence lifetime is of the order of 10^3 Gyr, and so boundary movement has been negligible.

3. The Method of Orbital Investigation

To establish whether terrestrial orbits could remain confined to the habitable zone of a system we must first investigate orbital stability. Analytical integration is restricted to three bodies in (near) coplanar, (near) circular orbits. By contrast, the giants are in eccentric orbits, the terrestrial bodies would acquire eccentric orbits, and we cannot be restricted to three bodies nor to coplanar orbits. We therefore use numerical integration. In particular we use the extensively tested second-order mixed-variable symplectic (MVS) integrator contained within the Mercury integrator package (Chambers 1999). This integrator originated with Levison & Duncan (1994). MVS integrators cannot handle close encounters between planets accurately. The smallest distance at which it is safe to use the integrator is about three times the Hill radius of the planet in the encounter with the larger Hill radius. The Hill radius, R_H, is defined as

$$R_\mathrm{H} = \left(\frac{m}{3M_*}\right)^{1/3} a, \tag{1}$$

where m is the mass of the planet, a is its orbital semi-major axes, and M_* is the mass of the star. When the two planets are separated by R_H their gravitational interaction is of the same order as the gravitational interaction of each planet with the star, and so considerable orbital modification will occur, particularly for the terrestrial planet in a giant-terrestrial encounter. We halt integration at $3R_H$. This not only avoids using the MVS integrator in an inaccurate domain, it is also a conservative definition of the point at which orbits become unstable.

For there to be a stable orbit we require that the $> 3R_H$ criterion be met, but this alone is not sufficient. We also require that the orbit remains confined to the HZ, otherwise it is unlikely that life could evolve. We take confinement to mean that the semi-major axis remains in the HZ. To perform an integration we set up one or two terrestrial planets in orbits with zero eccentricity and, in most cases, with zero inclinations with respect to the plane of the giant's orbit. The planets are started on the other side of the star from the giant, and at least one of the terrestrial planets is placed in the HZ. If there is just one terrestrial planet then it is called EM and given a multiple of the mass of the Earth plus the Moon, up to eight times the combined Earth–Moon mass. If there is a second terrestrial planet it is called V, is given up to eight times the mass of Venus, and is placed interior to EM. The mass, m, of the giant planet in each of the four systems, being detected by the radial-velocity technique, is only known as $m \sin(i)$, where $i = 0$ corresponds to a face-on view of the orbit. The mass is set at multiples of its minimum mass, again up to eight times, which corresponds to a view of its orbit at $i = 7.2°$, or a smaller multiple if this is necessary to keep the giant from exceeding the 13 M_{Jup} threshold for becoming a brown dwarf. We then integrate for simulated times up to 10^9 years, unless the integration is halted automatically by a close encounter ($3R_H$).

4. Results

Details of the four contrasting exosystems selected for study are given in Table 1, where m is the mass of the giant (in Jupiter masses, M_{Jup}), a is the semi-major axis (sma) of the orbit, and e is its orbital eccentricity. Figure 1 displays each of the systems in Table 1, showing the ZAMS habitable zone (*shaded*), the boundaries of the habitable zone today (*vertical dashed lines*), and the giant(s) (*black disks*). The *solid line* shows the *total* excursion, $2\Delta r$, of the giant planet due to its eccentricity, and the L*dashed line* extends to $(3R_H + \Delta r)$ each side of the giant for the case when the giant has its minimum mass.

4.1. Gliese 876 and Ups And

From Figure 1 it can be seen that in the Gliese 876 system, even if the giant has its minimum mass, the distance $(3R_H + \Delta r)$ from the giant spans the entire HZ. It might therefore be expected that there are no stable orbits in the HZ. Numerical integration confirms this expectation; the $3R_H$ criterion is violated after no more than a few thousand years even with the minimum giant mass and EM launched near the inner boundary of the HZ.

Figure 1 shows a similar expectation for Ups And. Again this is borne out by numerical integration. Close encounters with a giant occur no later that 12.5

Table 1. The exosystems selected for study

Star			Planet(s)			Comment
Name	Type	Mass $[M_\odot]$	$m \sin(i)$ $[M_{\rm Jup}]$	a [AU]	e	
Gl 876	M4 V	0.336	2.1 ± 0.2	0.21 ± 0.1	0.27 ± 0.03	Low mass star
Rho CrB	G0 Va	1.00	1.1	0.23	0.028 ± 0.04	Small a and e
Ups And	F8 V	1.3	0.71	0.059	0.034 ± 0.15	A family of giants
			2.11	0.83	0.18 ± 0.11	
			4.61	2.50	0.41 ± 0.11	
47 UMa	G1 V	1.03	2.41	2.10	0.096 ± 0.03	Most like the Solar System

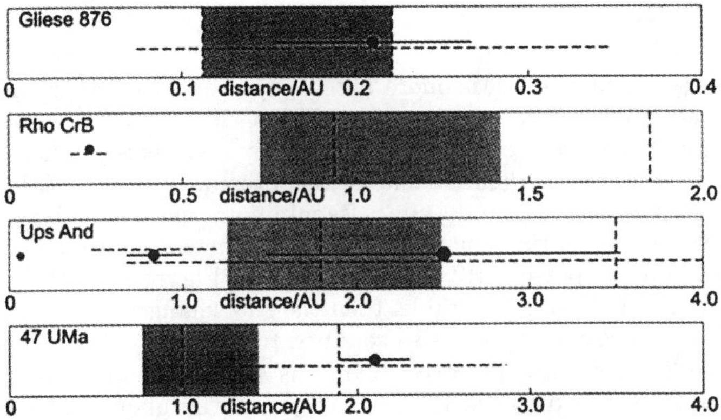

Figure 1. The four exosystems studied here. See the text for details.

million years into the integration. Lissauer & Rivera (2000) have studied small objects in Ups And, and find no stable orbits within about 8 AU of the star.

The mass of EM, up to eight times the Earth–Moon mass, and the inclination of the EM orbit, up to 30°, makes little difference to the outcome, and this is also the case in the remaining two systems. Another common feature is that the main effect of the giant is to pump up the eccentricity of the terrestrial planet's orbit. The change in sma is smaller, and usually a lot smaller, than the excursion associated with the maximum eccentricity. This is in accord with analytical studies of systems of two planets of low mass in near-circular, low-inclination orbits (Hénon & Petit 1986).

4.2. Rho CrB

Inspection of Figure 1 shows that Rho CrB could well have stable orbits confined to the HZ. We have performed integrations for simulated times up to 3×10^8 years. We have EM launched from several places in the ZAMS HZ and never found an unstable orbit. In addition, even if the giant has eight times its minimum mass, then only when EM is launched beyond about 1.3 AU does it move outside the ZAMS HZ. The sma quickly settles into a range that extends outwards from its starting value, and so EM can be launched at the inner boundary of the ZAMS HZ and thereafter remain confined.

If V is added, the main change is that EM and V need to be launched in sufficiently separated orbits to avoid close encounters between them. With eight times the minimum giant mass, initial separations greater than about 0.1 AU suffice in the ZAMS HZ.

The present HZ extends further from the star (Figure 1), and so stability and confinement are no less prevalent than in the ZAMS HZ.

4.3. 47 UMa

We have investigated 47 UMa more extensively than the other three systems, and have adopted 10^9 years as the integration time. This is a solar-type star with the giant outside the HZ, and so of the four systems studied it resembles the Solar System most. Figure 2 shows the outcome with just EM present, and with the giant at 1, 1.5, 2, and 4 times its minimum mass. The bar at each mass shows the range of initial sma within which we have found stable orbits. The ZAMS HZ and the present HZ are denoted as in Figure 1. If the initial sma is in the HZ, and if the orbit is stable, then the sma changes very little during the integration, and so the orbit is also confined to the HZ.

Figure 2 shows that the orbit of EM is stable provided its initial sma is (nearly) outside the distance $(3R_{\rm H} + \Delta r)$ from the giant. This is a reasonable outcome, but it is not obvious because the orbit of EM evolves, requiring numerical integration to explore accurately.

The gaps in the bars in Figure 2 represent orbital instabilities that result from mean motion resonances with giant:EM orbital period ratios of 4:1 ($a_{\rm EM}$ = 0.83339 AU) and 3:1 ($a_{\rm EM}$ = 1.00957 AU). In addition, at the minimum giant mass the bar might have extended to larger sma but for the 2:1 mean motion resonance at $a_{\rm EM}$ = 1.32292 AU.

When V is added to EM there is further instability due to collisions between EM and V. The results are shown in Figure 3 for the case of a giant planet with twice the minimum mass ($i = 30°$). The range of distances along the x-axis corresponds to the ZAMS HZ (Figure 2). The initial sma of EM is plotted along the x-axis and the initial distance between EM and V along the y-axis. Remember that V is always interior to EM. The boundary line is shown that has unstable orbits for EM to the right, and stable, confined orbits to the left except at the giant:EM 3 : 1 resonance. When $(a_{\rm EM} - a_{\rm V})$ is greater than 0.275 AU the instability lies in close encounters between EM and the giant, except at the 3 : 1 resonance where the encounter is between EM and V. When $(a_{\rm EM} - a_{\rm V}) \leq 0.275$ AU the instability on the boundary line and for some way to its right lies in close encounters between EM and V; it is these that turn the boundary line to the left in Figure 3. The position with the present day HZ can be seen by cutting Figure 3 to the left of 1 AU.

For all close encounter that we observed in 47 UMa, the time to the encounter was less than 10^8 years, and usually a *lot* less.

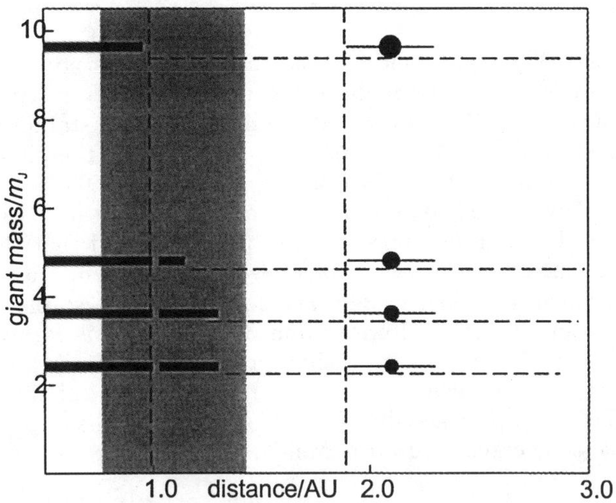

Figure 2. The 47 UMa exosystem, with the giant at various masses, and the stable regimes for EM. See the text for details.

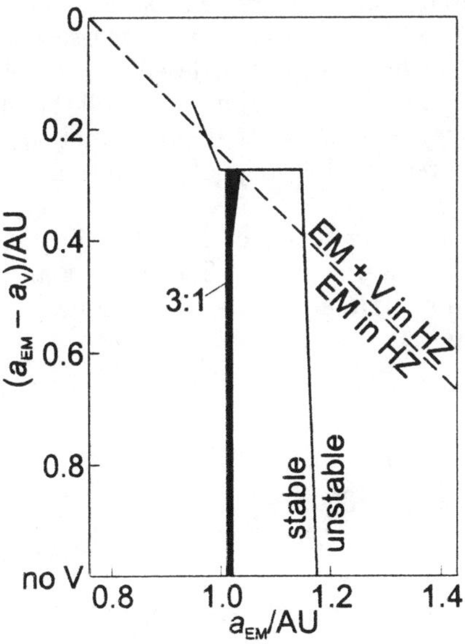

Figure 3. The ZAMS HZ of the 47 UMa exosystem, with V and EM present, showing the stable regimes. See the text for details.

5. Comparison with Other Work

Gehman, Adams, & Laughlin (1996) made an analytical study of systems consisting of a giant planet in a circular orbit, plus a single terrestrial planet in a circular coplanar orbit. They obtained conditions for Hill stability. One of the systems studied was 47 UMa. At the time of their work the giant's mass was estimated as 2.4 $M_{\rm Jup}\sin(i)$ and its orbital eccentricity as 0.03, not very different from the zero they needed to assume. The current values are 2.41 $M_{\rm Jup}\sin(i)$, and 0.096. For the minimum mass (2.4 $M_{\rm Jup}$) they found that the Hill stable zone interior to the giant's orbit extends inwards from about 1.6 AU.

We have integrated their system and found that the stable zone extends inwards from about 1.4 AU. However, our $3R_{\rm H}$ criterion is rather stiffer than their Hill stability criterion and so we have used a hybrid integrator (Chambers 1999) that is stable and accurate well inside the $3R_{\rm H}$ limit of the "pure" MVS integrator. When we apply a stability criterion similar to theirs then we too find that the stable zone extends inwards from about 1.6 AU.

6. Conclusions

Rho CrB and 47 UMa could have terrestrial planets in orbits that remain confined to their habitable zones (HZs) for biologically significant lengths of time—for at least a few hundred million years. Gliese 876 and Ups And are very unlikely to have such orbits. To a first order the masses and initial orbital inclinations of the terrestrial planets have little effect on the outcome; it is the mass of the giant, and the initial semi-major axes of all the planetary orbits that matter. When a second terrestrial planet is present, close encounters between the two terrestrial bodies reduce the range of initial semi-major axes for which the orbits are stable.

In 47 UMa there is a strong tendency for instability in terms of the $3R_{\rm H}$ criterion to show itself within 10^8 years. The 3 : 1 giant:EM orbital period resonance lays an instability strip across the stable zone.

References

Boss, A .P. 1995, Sci, 267, 360

Chambers, J. E. 1999, MNRAS, 304, 793

Charbonneau, D., Brown T. M., Latham D. W., & Mayor, M. 2000, ApJ, 529, L45

Forget, F., & Pierrehumbert, R. T. 1997, Sci, 278, 1273

Gehman, C. S., Adams F. C., & Laughlin, G. 1996, PASP, 108, 1018

Goldreich, P., & Tremaine, S. 1980, ApJ, 241, 425

Gonzalez, G. 1998, A&A, 334, 221

Gonzalez, G., & Laws, C. 2000, AJ, 119, 390

Hénon, M., & Petit, J-M. 1986, Mech. Dynam. Astron., 38, 67

Kasting J. F., Whitmire, D. P., & Reynolds, R. T. 1993, Icarus, 101, 108

Levison, H. F., & Duncan, M. J. 1994, Icarus, 108, 18

Lin, D. N. C., Papaloizou, J. C. B, Terquem, C., Bryden, G., & Ida, S. 2000, in V. Protostars and Planets IV, ed. V. Mannings, A. P. Boss, & S. S. Russell (Tucson: University of Arizona Press), p. 1111

Lissauer, J. J. 1987, Icarus, 69, 249

Lissauer, J. J., & Rivera, E. J. 2000, ApJ, 530, 454

Murray, N., Hansen, B., Holman, M., & Tremaine, S. 1998, Sci, 279, 69

Pollack, J. B., Hubickyj, O., Bodenheimer, P., Lissauer, J. J., Podolak, M., & Greenzweig Y., 1996, Icarus, 124, 62

Ward, W. R. 1996, Icarus, 126, 261

Disks, Planetesimals, and Planets
ASP Conference Series, Vol. 219, 2000
F. Garzón, C. Eiroa, D. de Winter, and T. J. Mahoney, eds.

The Galactic Exoplanet Survey Telescope: A Proposed Space-Based Microlensing Survey for Terrestrial Extrasolar Planets

David P. Bennett and Sun Hong Rhie

Department of Physics, University of Notre Dame, Notre Dame, IN 46556, USA

Abstract. We present a conceptual design for a space-based Galactic Exoplanet Survey Telescope (GEST) which will use the gravitational microlensing technique to detect extrasolar planets with masses as low as that of Mars at all separations $\gtrsim 1$ AU. The microlensing data would be collected by a diffraction-limited, wide-field imaging telescope of ~ 1.5 m aperture equipped with a large array of red-optimized CCD detectors. Such a system would be able to monitor $\sim 2 \times 10^8$ stars in ~ 6 square degrees of the Galactic bulge at intervals of 20–30 minutes, and it would observe ~ 12000 microlensing events in three bulge seasons. If planetary systems like our own are common, GEST should be able to detect ~ 5000 planets over a 2.5 year lifetime. If gas giants like Jupiter and Saturn are rare, then GEST would detect ~ 1300 planets in a 2.5 year mission if we assume that most planetary systems are dominated by planets of about Neptune's' mass. Such a mission would also discover ~ 100 planets of an Earth mass or smaller if such planets are common. This is a factor of ~ 50 better than the most ambitious ground-based programs that have been proposed. GEST will also be sensitive to planets which have been separated from their parent stars.

1. Microlensing Planet Detection: GEST's Main Goal

The main strength of the gravitational microlensing planet search technique is that it is sensitive to lower-mass planets than other techniques. Observed from space, signals for planets down to the mass of Mars are detectable, but they are much rarer and have a shorter duration than higher-mass planetary signals. Thus, a large number of stars must be followed with a high sampling frequency in order to detect low-mass planets. With GEST, we are able to monitor $\sim 2 \times 10^8$ stars once every 30 minutes or so with a photometric accuracy of $\sim 1\%$. The microlensing event will not repeat, so high-quality photometric data must be obtained while the event is in progress. Our proposed GEST mission will accomplish this.

The planetary systems studied by the microlensing technique are located 1–8 kpc away towards the Galactic Center rather than in the local neighborhood. The planetary signals will usually be detected as a modifications of the single-lens light curve due to the gravitational effect of the planet. Microlensing is most sensitive to planets near the Einstein-ring radius which corresponds to a distance of 1–10 AU from the lens star, but because GEST does not require the discovery of stellar microlensing event to begin intensive monitoring for planets, GEST will be able to detect planets at arbitrarily large separations from their host stars. One such isolated planet may have already been observed (Bennett et al. 1997) by the MACHO Collaboration.

2. Required Features of the Satellite Design

The main challenge for a microlensing search for low-mass planets is that, although the planetary signals can be strong, they are both very rare and and have durations as short as ~ 2 hours (Mao & Paczynski 1991; Gould & Loeb 1992). Furthermore, the relatively large angular size of giant source stars makes them poor targets for a low-mass microlensing planet search project as finite source effects (Bennett & Rhie 1996) tend to wash out the microlensing signal. Thus, GEST must be able to monitor large areas of sky rapidly while attaining $\sim 1\%$ photometry on main-sequence source stars. In the central Galactic bulge fields where the microlensing rate is highest, ground-based images are seriously incomplete at or above the bulge main-sequence turn-off, so there is a great advantage to be gained from higher-resolution imaging from space which will allow many of the main-sequence stars to be separately resolved. Since areas of high star density must be observed, we need high angular resolution to maintain photometric accuracy. So, we are led to a requirement of a large field of view and a high angular resolution—for a very large number of CCD pixels. Our baseline design calls for a ~ 1 square degree field of view with pixels of $0.1''$ or smaller, for a total of $\gtrsim 1.3 \times 10^9$ pixels. Our suggested observing strategy will be to cycle over 4–6 selected fields near the Galactic Center every 20–30 minutes. Given the high data rate implied by this strategy, a geosynchronous orbit seems sensible.

Our selected Galactic bulge fields fields have high reddening, so it is advantageous to observe at long wavelengths. Fortunately, recent developments in CCD technology have produced devices with a quantum efficiency of $\gtrsim 80\%$ at 750–950 nm (Groom et al. 2000). This is ~ 4 times better than the CCDs used by *HST*'s WFPC2 camera.

3. A Simulation of the GEST Mission

We have carried out a detailed simulation of the expected performance of the GEST mission. Here are the relevant features of our simulation: We assume that 6 square degrees of the Galactic bulge are observed once every 27 minutes. Photometry is assumed to be done using a difference-imaging technique which is photon-noise limited with a 0.3% systematic error added in quadrature. The photon noise is due to the target star and also any neighboring stars which have a PSF that overlaps with the target star. The effect of the neighboring stars was taken into account by constructing an artificial image with a resolution of $0.16''$ (the diffraction limit for a 1.5 m telescope at $\lambda = 900$ nm). We assume a signal to noise of 60 for a source with $I = 22$, or about 16 detected photo-electrons per

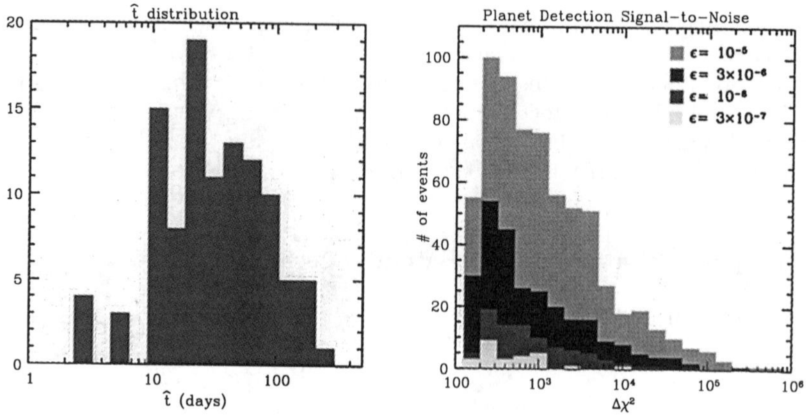

Figure 1. The distribution of Einstein-diameter crossing times ($\hat{t}$), corrected for detection efficiencies (*left-hand panel*) and the planetary detection signal-to-noise distribution (*right*) as a function of the planetary mass fraction, ϵ.

second for a 225 s exposure. This estimate assumes the use of high-sensitivity CCDs and a broad passband of 750–950 nm.

We assume that the luminosity function found by Holtzman et al. (1998) for Baade's Window applies to the entire 6 square degrees. This is conservative in that most of our fields will be closer to the Galactic Center with a higher star density, but this will be partially compensated for by the higher reddening in these fields. We conservatively assume a microlensing optical depth of $\tau = 3\times10^{-6}$, which is slightly below the measured values (Udalski et al. 1994; Alcock et al. 1997), and we assume the event timescale distribution given by Alcock et al. (2000) as shown in Figure 1.

4. GEST Simulation Results

Our simulated GEST mission observes a total of 18 000 microlensing events with peak magnification > 1.34 for source stars down to $I = 24.5$ over an assumed 3 year mission which observes the Galactic bulge for 8 months per year. Our detection threshold is a cut on $\Delta\chi^2$, which is the χ^2 difference between a single lens and planetary binary-lens fit. We have simulated planetary microlensing lightcurves for planets at a range of separations ranging from 0.7–30 $(M_{\rm star}/M_\odot)$ AU, and the *left-hand panel* of Figure 2 shows the number of detected planets if each lens system has a planet of the assumed mass fraction (ϵ) at the assumed separation. The *right-hand panel* of Figure 2 shows the number of detected "isolated" planets as a function of the planetary mass fraction, ϵ, which is now compared to the average stellar mass. Figure 2 indicates that we will be able to detect 50–100 Earth-mass planets if they are common at distances of a few AU or more, and 10–20 Earth-mass planets if they are common at 1 AU. We also expect to detect ~ 10 Mars-mass planets if they are common at a few AU. The *right panel* of Figure 1 shows the distribution of low-mass planet detections

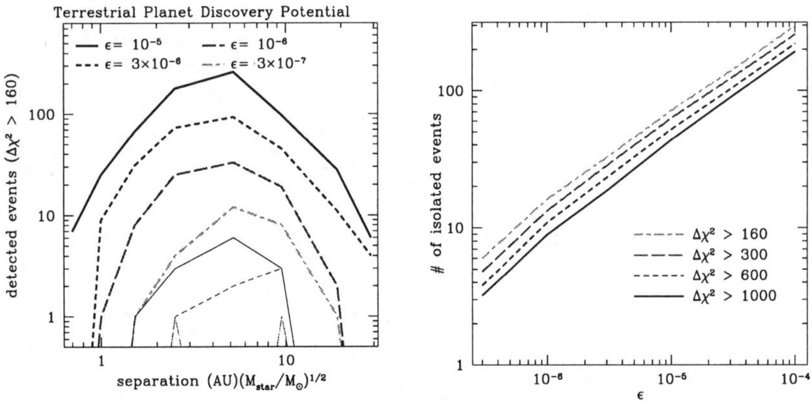

Figure 2. *Left:* The number of potential planetary detections plotted as a function of planetary mass fraction and separation for a selection criteria of $\Delta\chi^2 \geq 160$ for the GEST (*upper thicker lines*), and the VST (*lower lines*) surveys. This plot is normalized to the expected number of GEST planetary discoveries assuming every planetary system has a planet of the given mass fraction at the given separation. *Right:* The number of isolated planet detections as a function of the planetary mass fraction (ϵ) under the assumption that there is one isolated planet for every star in the Galaxy.

as a function of our signal-to-noise parameter $\Delta\chi^2$, and it indicates that most of the detections are well above our nominal detection threshold of $\Delta\chi^2 = 160$. Examples of simulated GEST low-mass planetary lightcurves are shown in Figures 3–7.

For more massive planets, our sensitivity is much better. For Solar System analogs, our simulations indicate that GEST would detect $\sim$ 5000 gas-giant planets if our Solar System is typical. If the typical planetary system resembles the Solar System with Saturn and Jupiter replaced by Neptune-mass planets, then GEST would detect $\sim$ 1300 of these Neptune-like planets.

5. Comparison to Ground-Based Surveys and Conclusions

In addition to detecting the planetary perturbation to the microlensing lightcurve, it is also important to determine the characteristics of the planet that has been detected. Microlensing generally allows the determination of the planetary mass fraction, ϵ, and the transverse separation of the planet from the lens star in units of the Einstein radius, which is typically about $3(M_{\rm star}/M_{\odot})$ AU. Gaudi & Gould (1997) have shown that these parameters can be accurately determined if the lightcurve deviations are well sampled. For caustic-crossing planetary microlensing events, which comprise a large fraction of the low mass planet detections, it is also possible to determine the planetary mass to about a factor of 2 or 3.

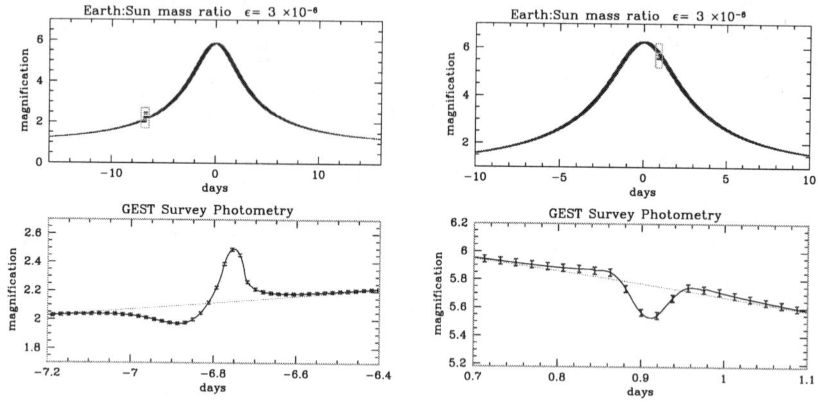

Figure 3. The simulated data for two Earth mass ratio (planetary mass fraction $\epsilon = 3 \times 10^{-6}$) planets with planet–star separations of $a = 0.77$ and $a = 0.93$ Einstein-ring radii, respectively. *Top panels:* The simulated data for each event; the region of the planetary deviation is indicated with the *dotted box*. *Lower panels:* Blow-ups of the planetary deviation region along with the planetary lensing lightcurve and the best-fit standard single-lens lightcurve (*dotted line*). The lightcurve on the *left* is a high-signal-to-noise detection with $\Delta\chi^2 = 4100$ while the lightcurve on the *right* is a detection just above our signal-to-noise cut at $\Delta\chi^2 > 160$.

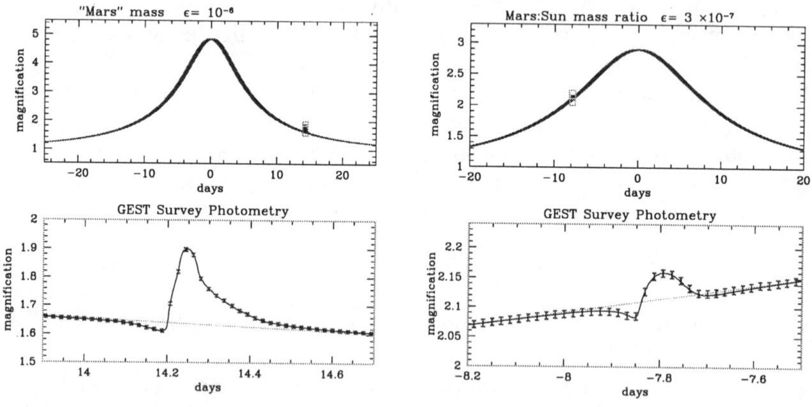

Figure 4. Simulated data for two planetary microlensing events. The event on the *left* has a mass fraction of $\epsilon = 10^{-6}$ and a separation of $a = 1.43$, and the event on the *right* has $\epsilon = 3 \times 10^{-7}$ and $a = 1.30$. These represent a Mars-mass planet (for a typical $M_{\rm star} \sim 0.3\ M_\odot$) and a Mars mass ratio planet, respectively. These events have $\Delta\chi^2 = 7800$ and $\Delta\chi^2 = 200$, respectively.

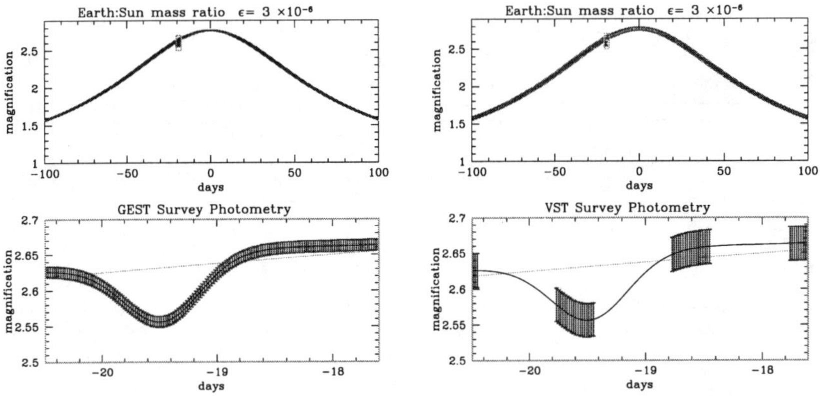

Figure 5. A comparison of the same planetary microlensing event as seen by the GEST and VST simulations. The planet has the same mass fraction as the Earth, $\epsilon = 3 \times 10^{-6}$ with a separation of $a = 0.97$ Einstein-ring radii. GEST detects the planet with a signal of $\Delta\chi^2 = 3100$, while The VST survey signal is $\Delta\chi^2 = 160$. The VST survey data are probably insufficient to determine the planetary parameters.

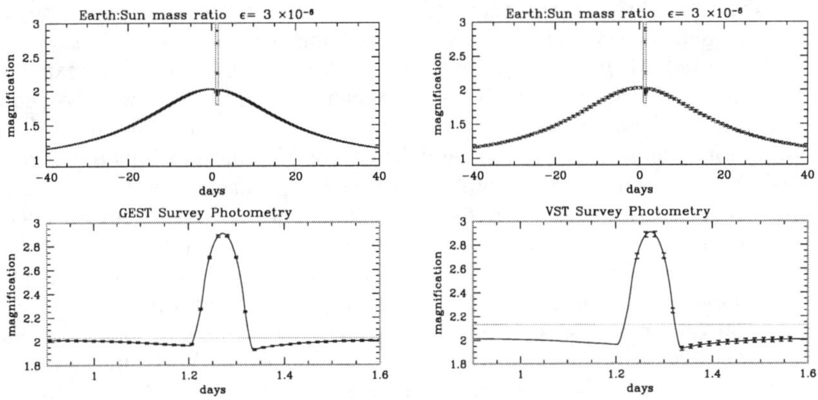

Figure 6. The most significant detection "fully sampled" Earth mass fraction planet detection from the simulated VST survey is compared with the same event as seen in the GEST simulation. The separation is $a = 1.31$ and the detection signal strengths are $\Delta\chi^2 = 39\,000$ and $\Delta\chi^2 = 5400$ for the GEST and VST simulations, respectively. Note that, despite passing our criteria, the VST data do not fully sample the planetary deviation.

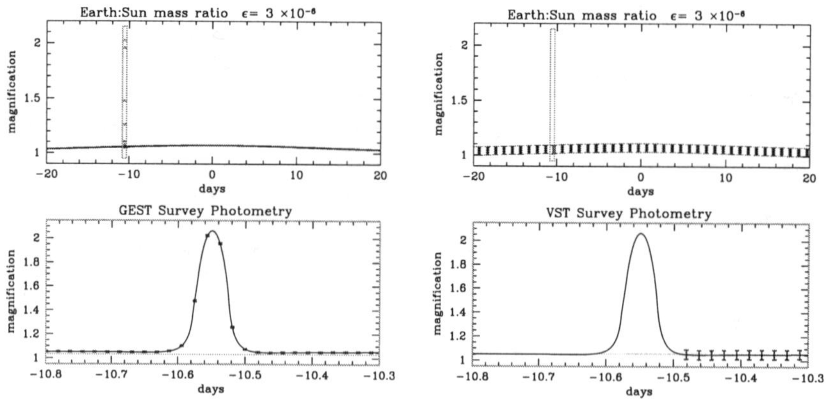

Figure 7. A large-amplitude planetary signal due to an $\epsilon = 3 \times 10^{-6}$, $a = 2.47$ planet is seen in a low-magnification stellar microlensing event for the GEST simulation, but the lensing event is entirely missed in the VST simulation.

The gravitational microlensing planet search technique has previously been been considered for ground-based observations (Peale 1997; Sackett 1997; Rhie et al. 2000; Albrow et al. 2000), but ground-based surveys face some difficulties due to the requirement of continuous lightcurve monitoring. This can be accomplished with a network of microlensing follow-up telescopes spanning the globe at southern latitudes, but this requires that the survey rely upon observing sites that often have poor weather or seeing conditions. Sackett (1997) has argued that a dedicated 2.5 m telescope such as the VST with a wide-field camera at an excellent site such as Paranal could efficiently search for low-mass planets where the deviations are expected to last only a few hours. We have done a simulation of such a survey (optimistically) assuming observations in consistent $0.7''$ seeing for 8 hours every night for three bulge seasons, and we find that low-mass planets are not easily detected by such a ground-based survey because the highest-signal-to-noise events generally last longer than a few hours. If we demand that more than 90% of the $\Delta\chi^2$ signal occur in the 8 hour observing window for an event to have measurable parameters, then we find that this "VST" survey is 30–50 times less sensitive to low-mass planets than GEST as shown in Figure 2. Also, even the "best" planet detection in the VST survey misses a significant part of the planetary deviation as seen in Figure 6. A study by Peale (1997) of a ground-based microlensing planet search program with follow-up telescopes in Chile, Australia, and South Africa also only manages a handful of low-mass planet detections after eight years of observations and is not sensitive to isolated or Mars-mass planets. So, the proposed GEST mission is $\gtrsim 50$ times more sensitive to low-mass planets than both types of proposed ground-based surveys.

References

Albrow, M. D., et al. 2000, ApJ, in press

Alcock, C. A., et al. 1997, ApJ, 479, 119

Alcock, C. A., et al. 2000, ApJ, submitted (astro-ph/0002510)

Bennett, D. P. et al. 1997, in ASP Conf. Ser., Vol. 119, Planets Beyond the Solar System and the Next Generation of Space Missions, ed. D. R. Soderblom (San Francisco: ASP), p. 95

Bennett, D. P., & Rhie, S. H. 1996, ApJ, 472, 660

Gaudi, B. S., & Gould, A. 1997, ApJ, 486, 85

Gould, A., & Loeb, A. 1992, ApJ, 396, 104

Groom, D. E., et al. 2000, Nucl. Instrum. Methods A, in press

Holtzman, J. A., et al. 1998, AJ, 115, 1946

Mao, S., & Paczynski, B. 1991, ApJ, 374, L37

Peale, S. J. 1997, Icarus, 127, 269

Rhie, S. H., et al. 2000, ApJ, in press

Sackett, P. 1997, in Appendix C of the Final Report of the ESO Working Group on the Detection of Extrasolar Planets (astro-ph/9709269)

Udalski, A., et al. 1994, Acta Astron., 44, 165

Discussion

Alan Boss: What is the limiting mass ratio (of planet to star) set by the source-star characteristics for the GEST project?

David Bennett: Our target stars are generally main-sequence stars, and this sets a limiting mass ratio of about $\epsilon \geq 3 \times 10^{-7}$, or the ratio of Mars' mass to that of the Sun.

Disks, Planetesimals, and Planets
ASP Conference Series, Vol. 219, 2000
F. Garzón, C. Eiroa, D. de Winter, and T. J. Mahoney, eds.

The PLANET Microlensing Collaboration Search for Extrasolar Planets: Status Report

J. P. Beaulieu[1], M. Albrow[2], J. An[3], J. A. R. Caldwell[4], D. L. DePoy[3], M. Dominik[5], B. S. Gaudi[3], A. Gould[3], J. Greenhill[6], K. Hill[6], S. Kane[6], R. Martin[7], J. Menzies[4], R. Naber[5], R. Pogge[3], K. R. Pollard[8], P. D. Sackett[5], K. C. Sahu[2], P. Vermaak[4], R. Watson[6], and A. Williams[7] (The PLANET Collaboration)

[1]*IAP, 98Bis Boulevard Arago, 75014 Paris, France*
[2]*STSCI, 3700 San Martin Drive, Baltimore, MD 21218, USA*
[3]*Ohio State University, Colombus OH 43210, USA*
[4]*SAAO, P.O. Box 9 Observatory 7935, South Africa*
[5]*Kapteyn Institute, PO Box 800, 9700 AV Groningen, The Netherlands*
[6]*University of Tasmania, GPO 252C, Hobart, TAS 7001, Australia*
[7]*Perth Observatory, Walnut road, Bickley, Perth 6076, Australia*
[8]*University of Canterbury, P. Bag 4800, Christchurch, New Zealand*

Abstract. PLANET, the Probing Lensing Anomaly NETwork, is an international collaboration that makes coordinated observations of ongoing gravitational microlensing events from four sites in the southern hemisphere. Our primary goal is to detect or to put constraints on substellar companions.

We will show that the lightcurve of MACHO 97-BLG-41 can be modeled as a source crossing two caustics of a physically realistic rotating binary, so that, contrary to a recent suggestion of the MPS team, the additional effects of a postulated planetary companion to this binary lens is neither required nor supported.

We present the preliminary results on the constraints of the frequency of Jupiter-mass planets via microlensing. Over the last five years, PLANET has monitored 100 events of which 23 have sensitivity to the perturbations that would be caused by a Jupiter-mass companion to the primary lens. Since no such perturbation has been observed, an upper limit can be put: less than 1/3 of stellar system towards the Galactic bulge would have a Jupiter companion over orbits of 1.5–3 AU, or a three Jupiter companions over orbits of 1–4 AU.

1. PLANET: The Probing Lensing Anomalies NETwork

The Herculean microlensing surveys launched at the beginning of the 1990s have borne fruit (see Paczyǹski 1996; Ferlet & Maillard 1997; Menzies & Sackett 2000, and references therein). Hundreds of microlensing events have been detected, the overwhelming majority with bulge stars as sources. Because these events are detected and alerted in real time by OGLE and EROS (and previously by MACHO until 1999 December 31), they give birth to new collaborations specialized in follow-up observations such as PLANET and MPS (see Bennett & Rhie, this volume, p. 542).

The aim of the PLANET collaboration (Albrow et al. 1998) is to perform precise (1–2% relative photometry in crowded fields), frequent (median sampling rate of 1–2 hour typically), multiband (V amd I), round-the-clock photometric observations of ongoing microlensing events in order to search for departures from a lightcurve due to lensing of a point source by a single point-like lens. These perturbations can be due to the blending of the source star by other stars close to the line of sight, the parallax effect due to the motion of the Earth around the Sun, the finite size of the source star in high-amplification events, and lensing binaries (binary stellar system, or a system made of star and planet)

The ultimate goal of PLANET is the detection of or constraint on the abundance and distribution of extrasolar planets in our Galaxy with masses down to that of Jupiter and orbits between 1 and 10 AU. PLANET has been operating since 1995 with a network of telescopes in South Africa, Chile, and Australia. It will be using for the seasons 2000 and 2001 a network of four 1 m telescopes equipped with CCD cameras (the Canopus 1 m, the Hobart and Perth 0.6 m Bickley, Australia), or Dual Optical-Infrared Camera (the Yale 1 m, CTIO, Chile, and SAAO 1 m, Sutherland, South Africa). Dedicated observations of ongoing microlensing in the Galactic bulge will be made from April to September, with most of the observing done in the V, I and H bands. Each site reduces the data online using an automatic pipeline built around DoPhot, and then send the data to the "home base". The homebase combines the data, and decides the observing strategy for the next 24 hours. It is critical to be able to adjust the strategy, sampling rate, and proportion of observations in different bands on a daily basis since anomalies (a caustic crossing in a binary star system or a planetary perturbation) will last only for a few hours. When an anomaly is in progress, or is predicted to happen in the near future, an alert is made to the astronomical community (you can register at http://www.astro.rug.nl/ planet).

We present here the controversial binary MACHO 97-BLG-41 (Bennett et al. 1999; Albrow et al. 2000a) and summarize the preliminary constraints put by PLANET on the distribution of Jupiter-mass planets (Gaudi et al. 2000).

2. MACHO-1997-BLG-41: A Rotating Binary System, or a Jovian Planet Orbiting a Binary Star?

The microlensing event MACHO-1997-BLG-41 was alerted by the MACHO collaboration in 1997 June. It was observed extensively by PLANET and MACHO/GMAN/MPS. This anomalous microlensing event turned out to be very difficult to model as we will see below. Detailed analysis, on two independent datasets have been published by MPS (Bennett et al. 1999) and PLANET (Albrow et al. 2000a).

PLANET obtained 146 V-band and 467 I-band observations from five observatories temporally spanning from the initial alert until the end of the event. The data have been processed by the normal pipeline, and then reprocessed using ISIS, an image-subtraction package with non-constant kernel (Alard 2000).

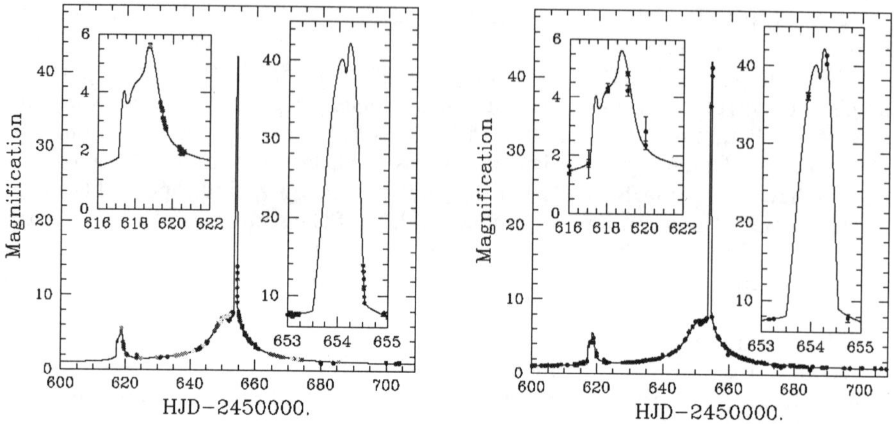

Figure 1. The PLANET (*left*) and the MACHO/GMAN (*right*) data are presented for MACHO 97-BLG-41. Insets show zooms over the two caustic-crossing regions. The plotted line is the best rotating binary model fitted to PLANET data. Note that, although the MACHO/GMAN data were not used in any way in the fitting, the PLANET model provides a good fit to their data too (Albrow et al. 2000a).

The PLANET and the MACHO/GMAN lightcurves are shown in the *left* and *right panels* of Figure 1, respectively.

The two peaks near days 619 and 654 are due to the source-transiting caustics created by a multiple lens. No static double-lens model can reproduce the observed light‘curve. However, all physical binaries must orbit their common center of mass. We include a first-order term to take into account the rotation, the relative motion of the lenses around their common center of mass. The plotted line in Figure 1 is a successful rotating-binary solution. The binary separation changes by $\delta d = -0.07 \pm 0.009 R_{\rm E}$ (where $R_{\rm E}$ is the Einstein radius) and rotates by $\theta = 5.61 \pm 0.36$ degrees during the 35.17 days between the separate caustic events. A Monte Carlo integration shows that the statistical distributions of the lens distance, mass, and orbital period peak at 5.5 kpc, 0.35 $M_\odot$, and 1.6 yr, respectively. We emphasize that, despite the fact that the exact mass, distance, and parameters of the binary are not known, the most likely combinations for the PLANET rotating-binary solution yields orbital velocities small enough for the system to be quite comfortably bound. Other data that did not enter into the fitting process in any way (including the choice of initial parameters) are available from the MACHO/GMAN collaboration. We show them in the *right-hand panel* of Figure 1 with the best rotating-binary model from PLANET. The PLANET model is in good agreement with this independent dataset.

Bennett et al. (1999) analyzed the MACHO/GMAN (Alcock et al. 2000) and MPS data for MACHO BLG 97-41. They showed that the static-binary

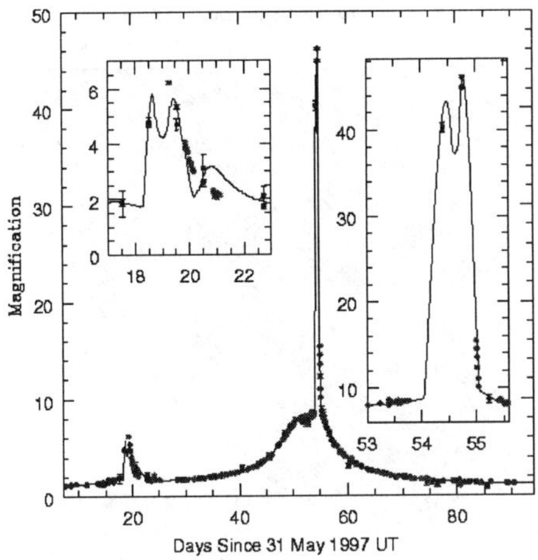

Figure 2. The *solid curve* is the static binary star with a Jovian-planet model from MPS. The model fits the MACHO/GMAN data, but deviates significantly from the PLANET data over the first peak. This model is therefore ruled out.

solution failed to provide an acceptable fit to their data. They investigated rotating-binary solutions and also failed to generate an acceptable fit to the data. Moreover, they found that their best rotating solution would lead to an unbound binary system (the parameters of this rotating binary system are not published). Turning to triple-lens systems, they found that a binary-star system (mass ratio 3.8 separated by 1.8 AU) with a Jupiter-mass planet orbiting the system at 7 AU provides a good fit to their data. However, when the data from PLANET are included, This static binary and Jupiter-planet model are ruled out, as illustrated by Figure 2.

A combined fit to both data set PLANET/MACHO GMAN has not been made, but it is clear that the PLANET rotating-binary solution provides an excellent fit to the currently available data. A triple-lens model is neither required nor supported. The main argument of Bennett et al. (1999) in favor of their triple-lens model was the claim that rotating-binary solutions are unbound. As far as the PLANET rotating-binary model is concerned, this statement is not supported.

3. Limits on Jupiter-Mass Planets

So far, no extrasolar planets have been detected by microlensing (Rhie et al. 2000 show that a low-mass companion to the lens of MACHO BLG 98-35 provides a better fit to their data, but is below their detection limit). However the large number of microlensing events observed by PLANET provide a very good

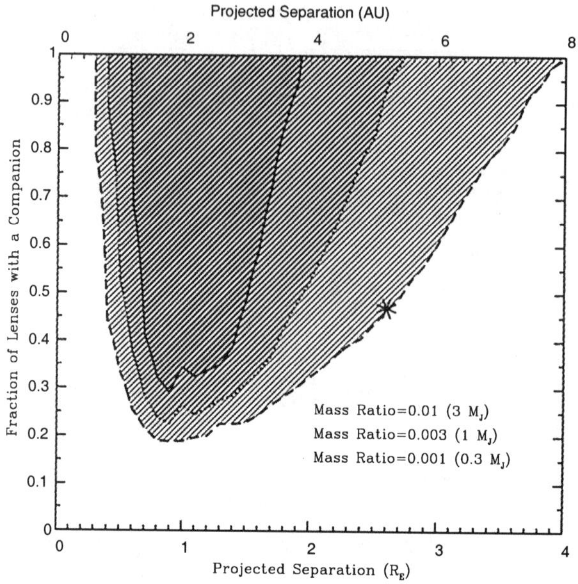

Figure 3. Upper limit to the fraction of primary lenses with a companion as a function of the projected separation for three different mass ratios, $q = 10^{-2}$ (dashed), $q = 10^{-2.5}$ (*dotted*), and $q = 10^{-3}$ (*solid*). The projected separation in AU, assuming all primary lenses are at 6 kpc and have masses of 0.3 $M_\odot$, is shown along the *top axis*. Masses of the secondary are represented by the different shaded regions. These upper limits, based on a preliminary analysis of 23 events observed by planet in 1998–1999, are at the 95% confidence level (Gaudi et al. 2000).

database with which to constrain the potential companions of the lenses. For a high-amplification event for which dense sampling and excellent photometry have been obtained, the efficiency in detecting companions with a mass ratio lower than 10^{-2} is quite high. PLANET adopted the method of Gaudi & Sackett (2000; for an example of implementation see Albrow et al. 2000b on OGLE-1998-BUL-14). This is to search for a binary-lens fit to calculate the detection efficiency as a function of mass ratio, q, and separation, d, for the different microlensing events. A large parameter space in q and d is explored, typically $10^{-5} < q < 1$ and $0 < d < 4$. The detection efficiency is high enough in many single microlensing events of high amplification, that non-detection in several microlensing events will be sufficient to place an upper limit on the abundances of stellar companions.

The 23 high-quality microlensing events of 1998–1999 have been analyzed in the same manner as OGLE-98-BUL-14. All of them are consistent with a single lens within our detection threshold. It is possible to combine them and to give an upper limit on the fraction of lensed stars with a substellar companion over a range of projected separations, as shown in Figure 3 (see Gaudi et al. 2000). We can draw the following upper limits:

- Less than 1/2 of the lenses have a three Jovian-mass planet between 0.8–5.4 AU
- Less than 1/3 of the lenses have a Jovian-mass planet between 1.5–3 AU
- Less than 1/2 of the lenses have a 0.3 Jovian-mass planet between 1.5–3 AU

Acknowledgments. We thank the MACHO, OGLE, and EROS collaborations for providing real-time alerts. We are especially grateful to the observatories that support our science (Canopus, CTIO, Perth, and SAAO) via the generous allocations of time that make this work possible. This work was supported by the grants AST 97-27520 and AST 95-30619 from the NSF, by grants NAG5-7589 from NASA, by a grant from the Dutch ASTRON foundation through GBE 614-21-009, by a Marie Curie Fellowship from the European Union, by a *coup de pouce 1999* award from the Ministère de l'Education Nationale, de la Recherche et de la Technologie and by a presidential Fellowship from the Ohio State University.

References

Alard, C. 2000, A&A, in press (astro-ph 9903111)

Albrow, M., et al. 1998, ApJ, 509, 687

Albrow, M., et al. 2000a, ApJ, 534, 894

Albrow, M., et al. 2000b, ApJ, 535, in press (astro-ph 9909325)

Alcock, C., et al. 2000, ApJ, submitted (astro-ph/9907369)

Bennett, D., et al. 1999, Nat, 402, 57

Ferlet, R., & Maillard, J. P. 1997, in Variable Stars and the Astrophysical Return of Microlensing Surveys (Gif sur Yvette:Éditions Frontières)

Gaudi, B. S., & Sackett, P. D. 2000, ApJ, 528, 56

Gaudi, B. S., et al. 2000, in ASP Conf. Ser., Microlensing 2000: A New Era of Microlensing in Astrophysics, ed. J. W. Menzies & P. D. Sackett, in press (astro-ph0004269)

Paczyński, B. D. 1996, ARA&A, 34, 419

Menzies, J. W., & Sackett, P. D. 2000, in ASP Conf. Ser., Microlensing 2000: A New Era of Microlensing in Astrophysics, ed. J. W. Menzies & P. D. Sackett, in press

Rhie, S., et al. 2000, ApJ, 533, 378

Discussion

David Bennett: I just wanted to comment on your comparison to the MACHO/GMAN data. It is not visible on the scale of your plot, but the PLANET model actually fails to predict the increase in brightness seen in the two Wise observations near the last caustic crossing. It may be that a fit which includes these data will be able to fit them, but this fit will probably have a shorter caustic crossing time and may imply a velocity larger than the escape velocity.

Disks, Planetesimals, and Planets
ASP Conference Series, Vol. 219, 2000
F. Garzón, C. Eiroa, D. de Winter, and T. J. Mahoney, eds.

Detecting Planets Using Transit Observations

Timothy M. Brown

High Altitude Observatory/National Center for Atmospheric Research, 3450 Mitchell Lane, Boulder, CO 80307, USA

Abstract. Extrasolar planets that transit the disks of their parent stars will play a special role in the study of planetary systems for the same reasons that eclipsing binaries play a special role in the study of stars. Such systems will be amenable to study in a variety of ways, and the accessibility of diverse kinds of information will allow fairly easy interpretation in terms of physical and orbital parameters of the planets. For this reason, transit searches for planets have become an active research topic. Here I describe the motivations and the methods for these searches, including one ground-based search for giant planets (STARE), and recent observations of the globular cluster 47 Tucanae with the *HST*. Finally, I describe the kinds of follow-up observations that are desirable for transiting extrasolar planets.

1. Basic Concepts

The idea of measuring the dip in stellar brightness as an extrasolar planet passes between its star and the Earth has been studied for many years (e.g., Rosenblatt 1971; Borucki et al. 1985). The important characteristics of a search for such events can be derived from elementary geometry and Kepler's laws; here I merely summarize the main points. 1) The signal amplitude is about $(R_{\rm p}/R_*)^2$, where $R_{\rm p}$ and R_* are the radii of the planet and star, respectively. For a Sun-sized star, Jupiter causes a dip of 1%, Earth 0.01%. 2) The likelihood that Earth lies near enough to the planet's orbital plane for transits to occur is $R_*/R_{\rm orb}$, where $R_{\rm orb}$ is the radius of the planet's orbit. For 51 Peg-like close-in planets, this likelihood is about 10%; for the Earth orbiting the Sun it is about 0.5%. To find a few planets, one must therefore look at a lot of stars. 3) Transits happen once per orbital period and have durations that (for Sun-sized stars) range from about 2 hours for planets at 0.05 AU up to 25 hours at 5 AU. This fact and 2) combine to make transit discoveries strongly biased in favor of planets in small orbits.

The above estimates, together with knowledge from radial-velocity searches concerning the frequency of occurrence of close-in giant planets, allow one to understand why searches for transiting planets are suddenly popular. If every solar system were like our own, then ground-based transit searches would be extravagantly difficult. The terrestrial planets would create transit signals too small for plausible ground-based detection, while jupiters would generate transits in only 0.1% of the stars where they exist, and then the events would last only for 1 day every 12 years. But the existence of even a modest fraction of stars accompanied by giant planets at distances of less than 0.1 AU changes the situation completely. The expected transit amplitude (1% or so) is small but detectable, and the frequency of transits is one every few days, rather than one per decade. The sampling rate must be short enough that one can resolve the transits that do occur; once per 15 minutes is sufficient (though when interpreting the lightcurves, shorter sample times are better). If 2% of Sun-like stars harbor close-in giant planets ("hot jupiters"), and the probability of favorable orbital alignment is about 10%, then one must search about 500 randomly chosen Sun-like stars in order to find one with a transiting planet. In many circumstances one cannot easily discriminate against background giant or subgiant stars, in which case the increased stellar radius causes the transit signal to become unusably small. For bright-star searches near the Galactic plane, this problem increases the number of stars to be searched to something like 1000 stars per transiting planet. All of this may be summarized by saying that, to detect one hot-jupiter transit, one must obtain something like 5×10^5 photometric measurements of a random set of stars, each measurement lasting 15 minutes and having relative precision better than 1%. With modern detectors and computer analysis, a program with these goals is feasible, if arduous. The following sections describe two such searches, one with a (very) small ground-based telescope, the other with a large space-borne one.

2. Desired Telescope Configuration

The figure of merit for transit planet searches is simply the number of stars that one can observe at once with the necessary photometric precision. Thus, if F_m is the detectable photon flux from a star of magnitude m, and the telescope's effective collecting area is A cm^2, then one requires simply

$$F_m A = C, \tag{1}$$

where C is an appropriate threshold count rate. This may usefully be rewritten as

$$C' = F_m L^2/f^2, \tag{2}$$

where L is the focal length, f is the f-ratio, and C' is connected to C by geometrical and transmission-related factors. The number of stars visible to the optical system that have fluxes greater than the threshold flux is evidently

$$N_m = D_m l^2/L^2, \tag{3}$$

where l is the edge length of the detector (which is assumed square), and D_m is the number of stars per steradian with fluxes at least as large as F_m. For most viewing directions and for $m \leq 12$ (approximately), D_m is fairly well represented by

$$D_m = D_0 F_m^{-q}, \tag{4}$$

where D_0 depends mostly on Galactic latitude, and where the exponent $q \simeq 1.2$ (e.g., Allen 1973, p. 244). Combining equations (1)–(4), one obtains

$$N_m \simeq \frac{D_0 l^2 L^{0.4}}{C'^{1.2} f^{2.4}}. \tag{5}$$

The principal desiderata are therefore that the detector be large, and that the focal ratio be small. The absolute scale of the system (parameterized by the focal length L) is less important. For a given detector size and f-number, a small focal length shows bright stars spread over a large field, a long focal length shows fainter stars over a smaller field, but the total number of stars for which one can achieve specified precision does not change very much. This means the overall scale of the system must be decided on other grounds, such as instrumental convenience or ease of performing follow-up observations. Both of these considerations drive one to small systems looking at relatively bright stars. The principal limitation to this trend is one's ability to correct for atmospheric extinction over very wide fields; combined with the ready availability of fast moderate-sized optics, these considerations lead to a focal length near 30 cm at f-ratios of $f/2$ to $f/3$.

3. The STARE Project

The STARE telescope was designed to search for giant extrasolar planets circling Sun-like stars (Brown & Kolinski 1999), according to the principles just described. In collaboration with similar systems run by W. Borucki at Lick Observatory (Vulcan, Borucki et al. 2000) and by E. Dunham at Lowell Observatory, it monitors a single star field all night long for several months at a time, looking for transits and other stellar variability. The STARE optical system is an $f/2.9$ Schmidt with a focal length of 28.6 cm. It images a field 6° square onto a 2k × 2k frontside-illuminated CCD. Most images are obtained in a red bandpass approximating Johnson R, though B and V filters are also available, for characterizing the target stars before beginning time-series observations. The usual exposure time is 107 s; including chip readout and other overhead, this leads to a sampling time of 2 minutes. With this exposure time, stars with R magnitudes between approximately 9.0 and 12.0 are suitable for photometry at better than 1% precision; in a typical field in the Milky Way, there may be 3000–5000 such stars, with ten times as many easily detectable above the sky background. Except for the brightest of unsaturated stars, the precision of derived time series is as expected from a physical noise model, with the largest term coming from sky background photon noise (our present site in Boulder, Colorado is convenient, but very bright). A useful by-product of the transit searches is an inventory of variable stars in the chosen fields that is apparently much more complete than any hitherto available. For example, a STARE field in Auriga contains 40 variables listed in the *Combined General Catalog of Variable Stars* (Kholopov et al. 1998); a search of the brightest 6000 stars in the STARE star list (reaching to approximately $R = 12.1$) yields about six times this many variables, including large numbers of eclipsing binaries and δ Scuti stars. Figure 1 illustrates two low-amplitude eclipsing binaries found with the STARE system. Both have eclipse depths of only about 5%, only a few times larger than those expected for planetary transits. The clarity with which the data define these lightcurves is encouraging for future success in finding Jupiter-sized planets. Information on variables in the STARE database (including positions and light curves) may be retrieved directly from the STARE web site (Brown & Kolinski 1999).

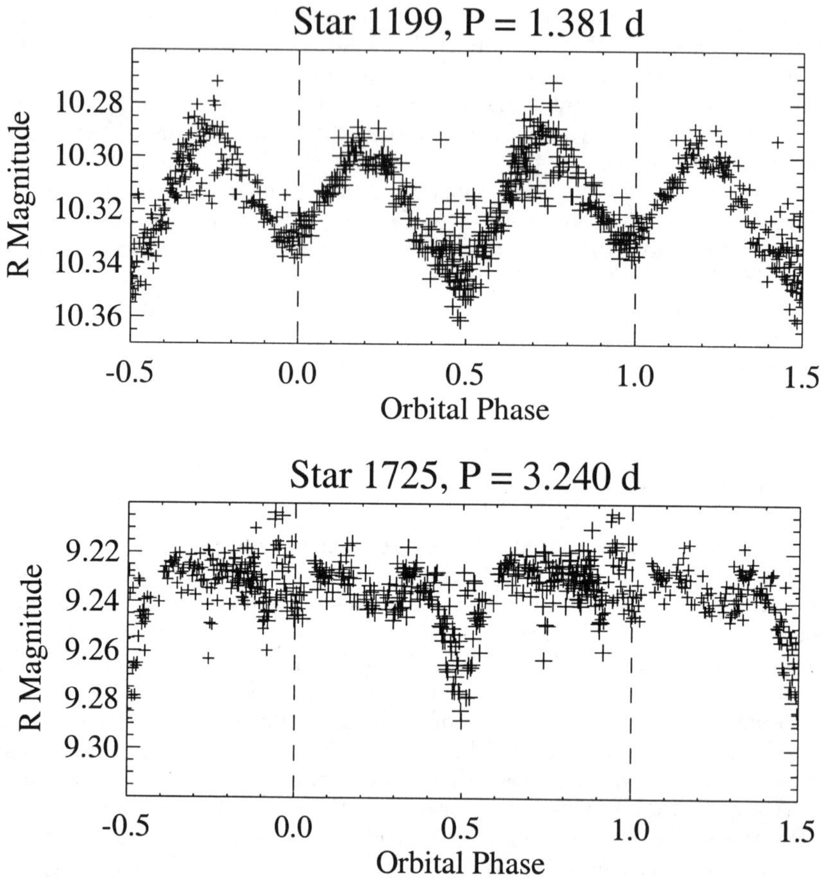

Figure 1. Two low-amplitude eclipsing binaries detected by the STARE telescope. Star 1199 is evidently a nearly face-on contact binary with a period of 1.38 days. Star 1725 is a grazing detached system with a period of 3.24 days; it shows a weak secondary eclipse. Both of these stars have eclipse amplitudes that are only slightly larger than those seen in the transiting giant planet of HD 209458.

4. Transits in HD 209458

In 1999 September, D. Charbonneau and I used the STARE telescope to observe the G0 main-sequence star HD 209458. Based on radial-velocity observations from the Keck Observatory and from the ELODIE spectrograph, this star was known to have a close companion in a 3.52 day orbit, with a minimum mass of 0.69 $M_{\rm Jup}$; information regarding this planet and its ephemeris was kindly made available to us by D. Latham and M. Mayor before publication. Analysis of the September observations revealed transits consistent with a planet of about

1.4 $R_{\rm Jup}$ (Charbonneau et al. 2000; Henry et al. 2000; Mazeh et al. 2000; see also Brown & Charbonneau, this volume, p. 584). The chief importance of observing such transits is precisely that they allow accurate estimates of the planet's size (the principal uncertainty is in the radius of the parent star), which in turn allows inferences about the planet's composition. The observed radius is consistent with models of a planet composed mostly of hydrogen and helium, inflated in size because the irradiation from its star holds the surface temperature at about 1400 K (Guillot 1999; Burrows et al. 2000). A further opportunity offered by this transiting planet is that a variety of follow-up observations are feasible, each probing a different aspect of the planet's structure. These are discussed in some detail in Charbonneau et al. (2000), and reviewed briefly below.

5. *HST* Observations of 47 Tuc

The preference for small f-number optical systems would seem to disqualify the *Hubble Space Telescope* from success as a transit finder. This disadvantage can be overcome, however, by observing a globular cluster, in which the star density is very high. In this way one may exploit the extreme stability of space-based photometry, and in addition can hope to learn about planet formation in circumstances of comparatively low metallicity and high crowding. This opportunity was first noticed by R. Gilliland, who successfully proposed for 120 orbits (8.3 continuous days) of *HST* time to observe the cluster 47 Tucanae. This is among the richest, nearest, and least metal-depleted of the globular clusters. Observations were successfully obtained in 1999 July, with about 40 000 stars available for photometry with $17 \leq m_V \leq 21$.

The noise limits for these observations are set by the Poisson noise from the combination of the stars observed and the sky background. It is illustrated as a function of stellar magnitude in Figure 2, along with the realized time-series rms for a sample of stars. A convenient feature of observing a well studied globular cluster is that the properties of the observed stars (in particular their mass and radius) are well known and, on the main sequence, depend only on their apparent brightness. It is therefore possible to estimate for each star the depth and duration of a transit by a planet of known radius and orbital period. From this information, one can calculate the maximum noise level that would allow detection of the planet. This relation is also shown in Figure 2. Bright stars require high precision because of their large radii, especially near and above the cluster's main-sequence turn-off. Fainter stars are smaller, giving a bigger relative signal for a planet of given size, and allowing lower precision for their detection. Evidently most of the stars observed have realized noise levels small enough to permit clear detection of Jupiter-sized planets. This conclusion is reinforced by Figure 3, which shows the time series for a star of magnitude 18.5 as observed by *HST*, along with the same time series with imposed transits from a 1.3 $R_{\rm Jup}$ planet. For this star, such transits would be easily detectable. Analysis of the time series for all of the stars observed in 47 Tucanae is now proceeding, and we expect to announce results soon (Gilliland et al. 2000; Brown et al. 2000).

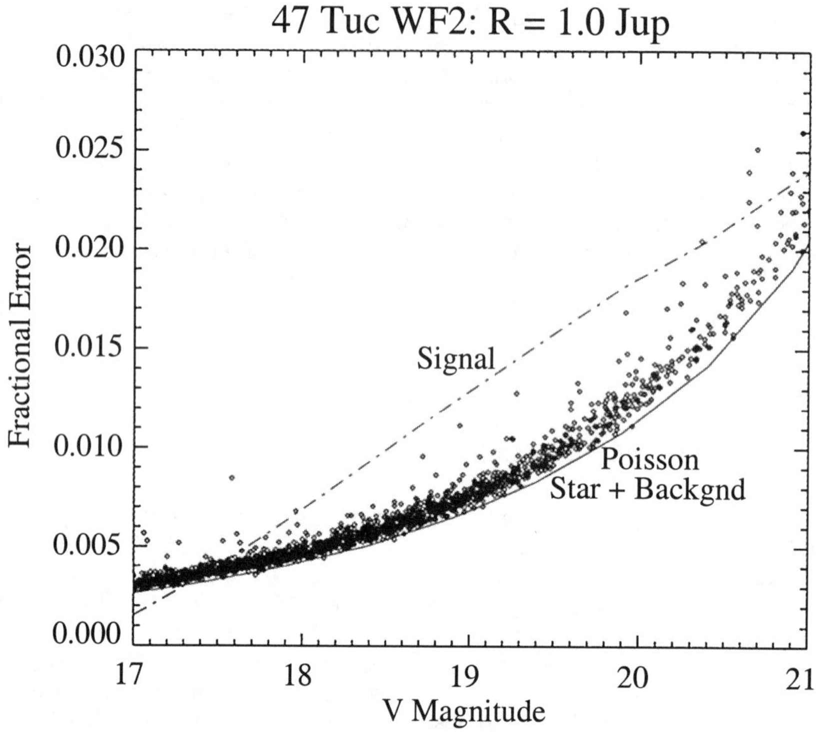

Figure 2. Time-series rms (*points*) plotted against V magnitude for V-band *HST* data of 2000 stars in 47 Tucanae. The *solid line* shows the expected noise limit resulting from Poisson noise from the star and the background (which is mostly faint cluster stars). The *dot-dashed line* shows the noise level that must be achieved to detect transiting planets of 1-Jupiter radius with 7 σ confidence in two transits.

6. Physical Observations of Transiting Planets

Transit photometry by itself yields an estimate of the planetary diameter. By comparison with models, this allows inferences about planet's compositions and even migration histories (Burrows et al. 2000). But other kinds of observations of transiting planets are possible, and these can give even more detailed information about the physical nature of the planets and their atmospheres. Two of the most promising approaches are transit spectroscopy and infrared photometry.

Because of its high surface temperature and relatively small mass, and because it evidently consists mostly of light elements, the pressure scale height in the atmosphere of HD 209458b is very large—about 500 km. As a result, the $\approx$ 10 scale heights above the 0.1 bar pressure level obstruct a cross-sectional area that is $\approx$ 10% of that of the totally opaque part of the planet; the difference is that the outer layers probably are opaque only at the wavelengths absorbed

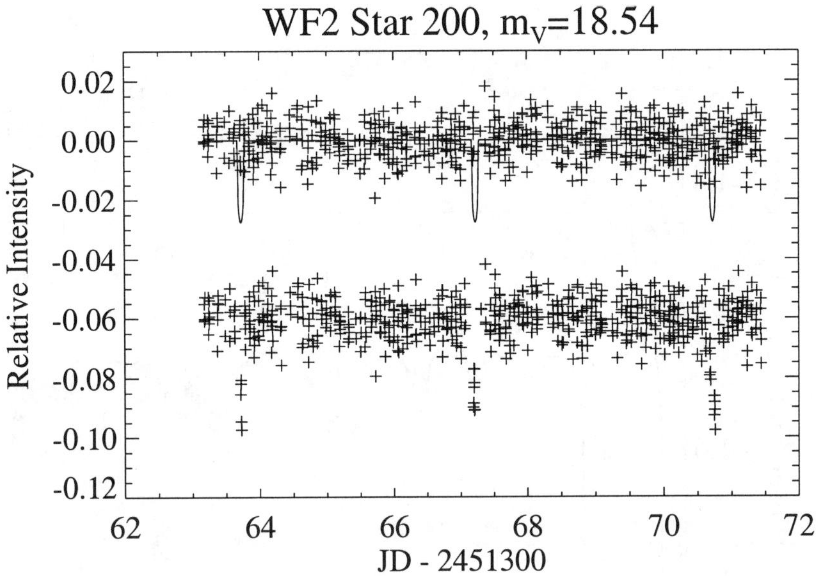

Figure 3. *Top trace* shows an I-band photometric time series of the relative brightness variation of a typical star in 47 Tucanae from the WF2 detector; rms variation is about 5 mmag. Overplotted is the expected lightcurve from transits by a 1.3 $R_{\rm Jup}$ planet with a 3.5 day period. The *lower trace* is the same as the upper one, but offset and with the planetary transit signal added, illustrating that even with limited sampling, transit signals, if present, should be clearly visible.

by abundant constituents. These might be condensed particles (silicates, iron, or photochemical products) or atoms and molecules (Na, K, CO, H_2O, and in smaller quantity CH_4; Sudarsky, Burrows & Pinto, 2000). If the main opacity source in the outer layers is particulates, then the wavelength dependence will be weak and the diagnostic potential of spectroscopy may be low. But if atomic or molecular absorption dominates, then it should be possible to make observations of considerable diagnostic power. As illustrated in Figure 4, the depths of the strongest lines imposed by the planetary atmosphere should be about 10^{-3} relative to the continuum; since HD 209458 is fairly bright, and since transits last for about three hours, it should be fairly straightforward to accumulate spectra with the necessary S/N, using existing large telescopes. Because of the diagnostic importance of molecular features, the recent advent of efficient near-IR spectrographs such as NIRSPEC (McLean et al. 1998) is particularly advantageous for such studies.

The reflected-light signature of even close-in giant planets is quite small in visible light (of order 10^{-5} of the stellar intensity), because only a small fraction of the star's light is intercepted by the planet. The relative flux levels are much more favorable at infrared wavelengths, however. The planets' high surface

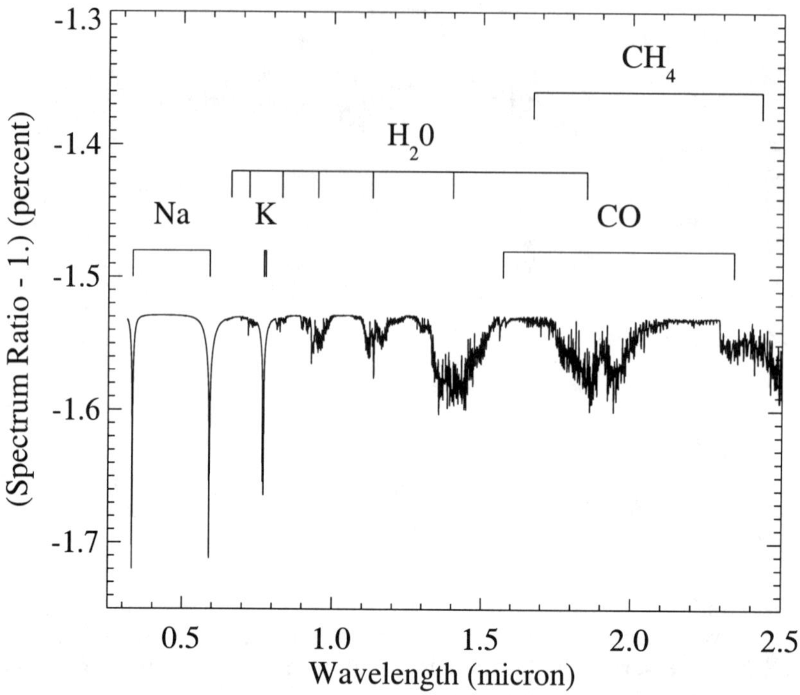

Figure 4. Synthetic transmission spectrum of the atmosphere of HD 209458b, shown as the difference between unity and the ratio of stellar spectra taken in and out of transit. In the visible, this spectrum is dominated by pressure-broadened lines of the alkali metals; in the near infrared, the important opacity comes from vibration–rotation bands of H_2O and CO. At the relatively high temperature (1400 K) chosen for this model, CH_4 absorption is unimportant.

temperatures imply that the peak of their thermal emission occurs in the near infrared, around 2 micron wavelengths. For significantly longer wavelengths, the relative fluxes are proportional to the ratio of effective temperatures and surface areas. For HD 209458, the flux ratio should be about 400:1, depending on the actual planetary temperature. The easiest observation would be of the system's secondary eclipse; this would be a short-lived event with a depth of a few millimagnitudes. This would provide an estimate of the planet's dayside temperature. Because the temperature is determined by the amount of absorbed stellar radiation, this would in turn allow an estimate of the visible-light albedo, with implications about the constituents of the atmosphere. A more difficult but more rewarding measurement would be of the entire infrared phase function of the planet. Such a measurement would say something about the temperature distribution with planetary longitude, and hence about mechanisms of energy distribution in the planet's atmosphere.

7. Summary

The study of extrasolar planets is rapidly moving from the discovery phase into a characterization phase, in which it is possible to say something about the physical nature of the planets. For the near future, a prominent role in this characterization will be played by planets that happen to transit their parent stars, since such transits permit several kinds of measurements that can only be performed in such a favorable geometry. The first such transiting extrasolar planet (HD 209458b) has now been found, and several searches for others are under way. Since the technology necessary to make at least the first steps seems well in hand, one can anticipate rapid progress in this field.

References

Allen, C. W. 1973, Astrophysical Quantities, edn. 3 (London: Athlone Press)

Borucki, W. J., Scargle, J. D., & Hudson, H. S. 1985, ApJ, 291, 852

Borucki, W. J., Caldwell, D., Koch, D., Webster, L., Jenkins, J., Ninkov, Z., & Showen, R. 2000, SPIE, 4013, in preparation

Brown, T. M., & Kolinski, D. 1999,http://www.hao.ucar.edu/public/research /stare/stare.html

Brown, T. M., et al. 2000, AAS, 196.2.03

Burrows, A., Guillot, T., Hubbard, W. B., Marley, M. S., Saumon, D., Lunine, J. I. & Sudarsky, D. 2000, ApJ, 534, L97

Charbonneau, D., Brown, T. M., Latham, D. W., & Mayor, M. 2000, ApJ, 529, L45

Gilliland, R. L., et al. 2000, AAS, 196.2.02

Guillot, T. 1999, Sci, 286, 72

Henry, G. W., Marcy, G. W., Butler, P., & Vogt, S. S. 2000, ApJ, 529, L41

Kholopov, P. N., et al. 1998, Combined General Catalog of Variable Stars, edn. 4.1, http://cdsweb.u-strasbg.fr/Cats.html

Mazeh, T., et al. 2000, ApJ, 532, L55

McLean, I. S., et al. 1998, SPIE, 3354, 556

Rosenblatt, F. 1971, Icarus, 14, 71

Sudarsky, D., Burrows, A., & Pinto, P. 2000, ApJ, 538, 885

Disks, Planetesimals, and Planets
ASP Conference Series, Vol. 219, 2000
F. Garzón, C. Eiroa, D. de Winter, and T. J. Mahoney, eds.

EXPORT: Search for Transits in Open Clusters with the Jakobus Kapteyn and Lick 1 m Telescopes

Andreas Quirrenbach, Jeff Cooke, David Mitchell, and Neda Safizadeh

University of California, San Diego, Physics Department and Center for Astrophysics and Space Sciences, Mail Code 0424, La Jolla, CA 92093-0424, USA

Hans Deeg[1]

Instituto de Astrofísica de Andalucía, 18080 Granada, Spain

EXPORT

Abstract. A photometric search for transits by extrasolar planets in three open clusters has been carried out as part of the EXPORT project. We discuss the rationale for this program, describe the observations carried out with the 1 m JKT (La Palma) and Nickel Telescope (Lick Observatory), and give a status report on the data reduction.

1. Introduction

After many years of trying to find extrasolar planets, the first discovery of a planet orbiting the solar-type star 51 Peg was made in 1995 by Mayor and Queloz; soon thereafter other planets outside our Solar System were found (Butler & Marcy 1996). Their search method, based on variations in the radial velocity of the parent star, is strongly biased towards finding giant planets in very close orbits. The—unanticipated—result of these programs is that planets with several Jupiter masses and orbital periods of just a few days are fairly common. As part of the EXPORT program, we have embarked on a complementary search for similar planets, by looking for transits of the planet in front of the stellar disk.

2. Why Transit Searches in Open Clusters?

The technique of planet transits is based on a photometric method, measuring the brightness drop in the lightcurve of the star produced by the eclipsing planet. For giant planets in close orbits, such as the 51 Peg planet, through transits we have a good chance of assessing the number of such objects and determining their physical size. The latter information cannot be obtained with the radial-velocity method. Furthermore, transiting planets lend themselves to

[1] Also at Instituto de Astrofísica de Canarias, 38200 La Laguna, Tenerife, Spain.

Table 1. Summary of the observations

Telescope	Object	# Nights	# Images	Total hours	Seeing
JKT	NGC 6819	7	302	56	0″.8–2″.2
	NGC 7789	9	202	38	1″.2–2″.0
	NGC 2682	9	57	11	0″.9–1″.7
Nickel	NGC 6819	43	956	212	1″.0–2″.5
	NGC 7789	44	909	180	1″.0–2″.6
	NGC 2682	9	104	24	1″.0–2″.7

spectroscopic follow-up studies, which may reveal material that evaporates from the planet. According to the recent work carried out by Mayor, Queloz, Marcy, Butler and others (see, for example, Marcy & Butler 2000), the probability of a solar-type star having such a Jupiter-like companion is a few per cent and the probability of having a suitable inclination of the system is around 0.05 (for a 0.1 AU orbit). Therefore the probability of success in detecting a transit is about 0.1% or slightly better, for any randomly chosen main-sequence star. The time of transit of the planet in front of the star is one to a few hours (considering 3–10 day orbits). The expected brightness drop for a Jupiter-like planet transiting a star is around 1%. These numbers mean that we should look at a few thousand stars, which immediately suggests to observe old open clusters with a photometric precision of around 0.5% or better.

A similar strategy has been suggested by Janes (1996) for searches for Earth-like planets. For these objects, eclipsing systems are rare (because of the comparatively wide orbits); eclipses repeat only about once per year, and are quite shallow (because of the small diameter of the Earth). Janes suggested that one or several dedicated 4 m-class telescopes would be needed for such a program. A similar strategy will be used in searches for Earth-like planets in future space missions such as *COROT*, to be launched in 2004, and in the more ambitious proposals for the *Kepler* (NASA) and *Eddington* (ESA) missions. In the case of 51 Peg B-like objects, the parameters are much more favorable, and a search with a realistic chance of success can be carried out with 1 m-class telescopes. In order to see planet transits which last one hour or more, the observations need to be repeated at least once every 30 minutes. Furthermore, to successfully detect a 1% drop in brightness a photometric accuracy of about 0.3% is needed. This translates to an integration time of 12–15 minutes for 17th magnitude stars with a 1 m telescope, if a photometric accuracy close to the photon noise limit can be achieved. Transits repeat every few nights, so that multiple eclipses of the same planet can be observed with a reasonable number of observing nights (about 10 to 15); this gives a robust criterion for successful detections.

The choice of open clusters as target fields has the added benefit of providing a well defined, chemically homogeneous parent population. It has been suggested that most of the stars with planetary companions found in radial-velocity surveys are metal rich with respect to the Sun and the general field (e.g., Gonzalez, Wallerstein, & Saar 1999). Two possible explanations have been advanced for

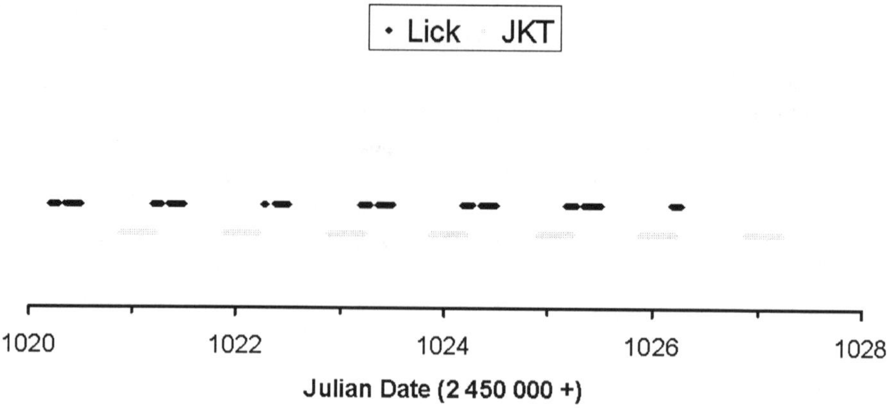

Figure 1. Combined time coverage of the Lick and JKT data on the cluster NGC 6819 for 1998 July.

this phenomenon. High metallicity may favor the formation of rocky planets (or rocky cores that form the nuclei of giant gaseous planets); the existence of a planet may thus be the consequence of high metallicity. Conversely, the metallicity may get enhanced through the capture of rocky material by the star; in this scenario the high metallicity is a consequence of the existence of a planetary system. Planet searches in open clusters can provide the means to distinguish between these two hypotheses. In the first case, the detection rate of planets is expected to depend strongly on the cluster metallicity, whereas in the second case it should be nearly independent of it, but individual parent stars of planets should have metallicities significantly above the cluster mean.

3. Observing Strategy

The target clusters should be relatively old, metal rich, and compact, and provide a large number of target stars within the field-of-view of the CCD. Based on these criteria, and a suitable distribution over the sky, three target clusters were selected for the EXPORT planet search program: NGC 6819, NGC 7789, and NGC 2682. We concentrated on NGC 6819 and NGC 7789, and observed NGC 2682 only at times when the two other clusters were not accessible. We decided to observe only one field in the center of each cluster; mosaicking would have resulted in a larger total number of observed stars, but poorer time coverage.

The first observations were carried out with the 1 m Nickel Telescope at Lick Observatory. Two intensive campaigns were scheduled simultaneously at this telescope and at the 1 m Jacobus Kapteyn Telescope (JKT) on La Palma. Both telescopes are equipped with cameras that provide a field of view of about $6' \times 6'$. All observations were carried out in the R band with an exposure time of 10 minutes for each frame. (A few additional exposures in V, B, and I were taken at Lick for each field to determine the colors of the stars.) We alternated between NGC 6819 and NGC 7789 whenever both clusters were visible, and took repeated exposures on a single position when only one of these clusters was accessible.

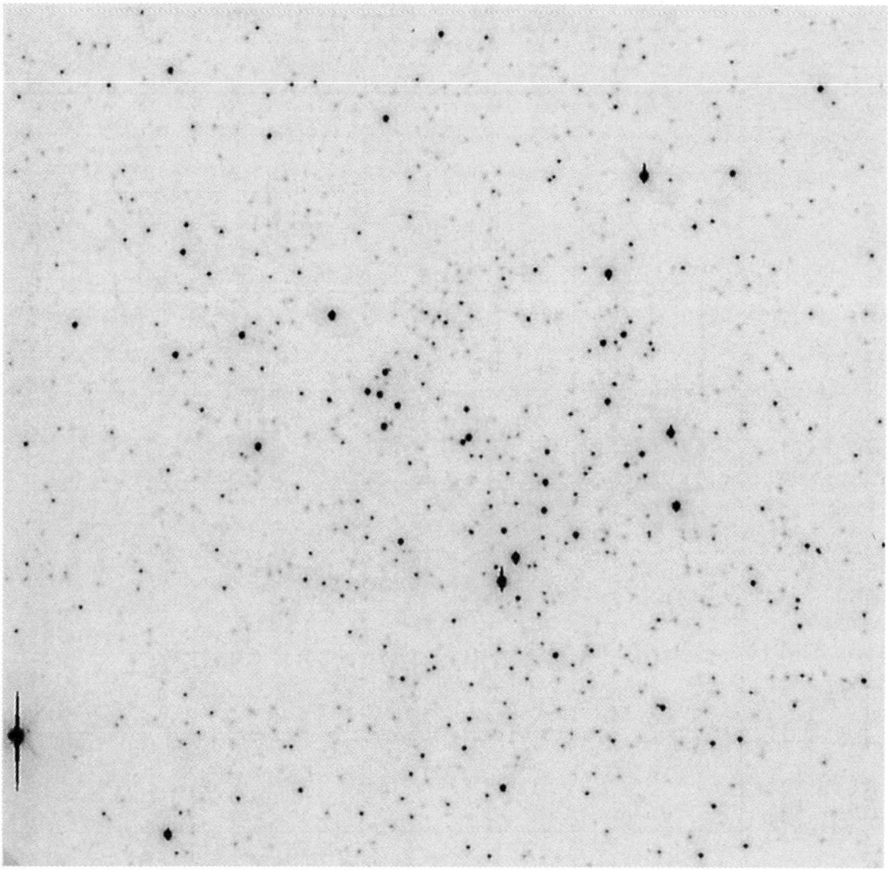

Figure 2. 6′ × 6′ field towards NGC 6819 in the *R* band.

Table 1 gives an overview of the amount of images obtained in this way. Most of the data were taken in photometric conditions. The longitude difference between La Palma and Lick Observatory resulted in fairly good time coverage during the two simultaneous campaigns; an example is given in Figure 1.

4. Status of the Data Reduction

The initial steps of the data reduction follow standard procedures: bad pixel removal, bias correction, and flatfielding. Figure 2 gives an example of an image of NGC 6819 processed in this way. It is obvious that a few bright stars are saturated and must be discarded in the following photometric reduction. The crowding is sufficiently strong that it has to be taken into account for most of the stars. However, we selected a few isolated stars for tests with a simple simulated aperture-photometry algorithm. These tests show that the systematic errors are sufficiently small so that lightcurves with rms variations of $\lesssim 0.3\%$ for

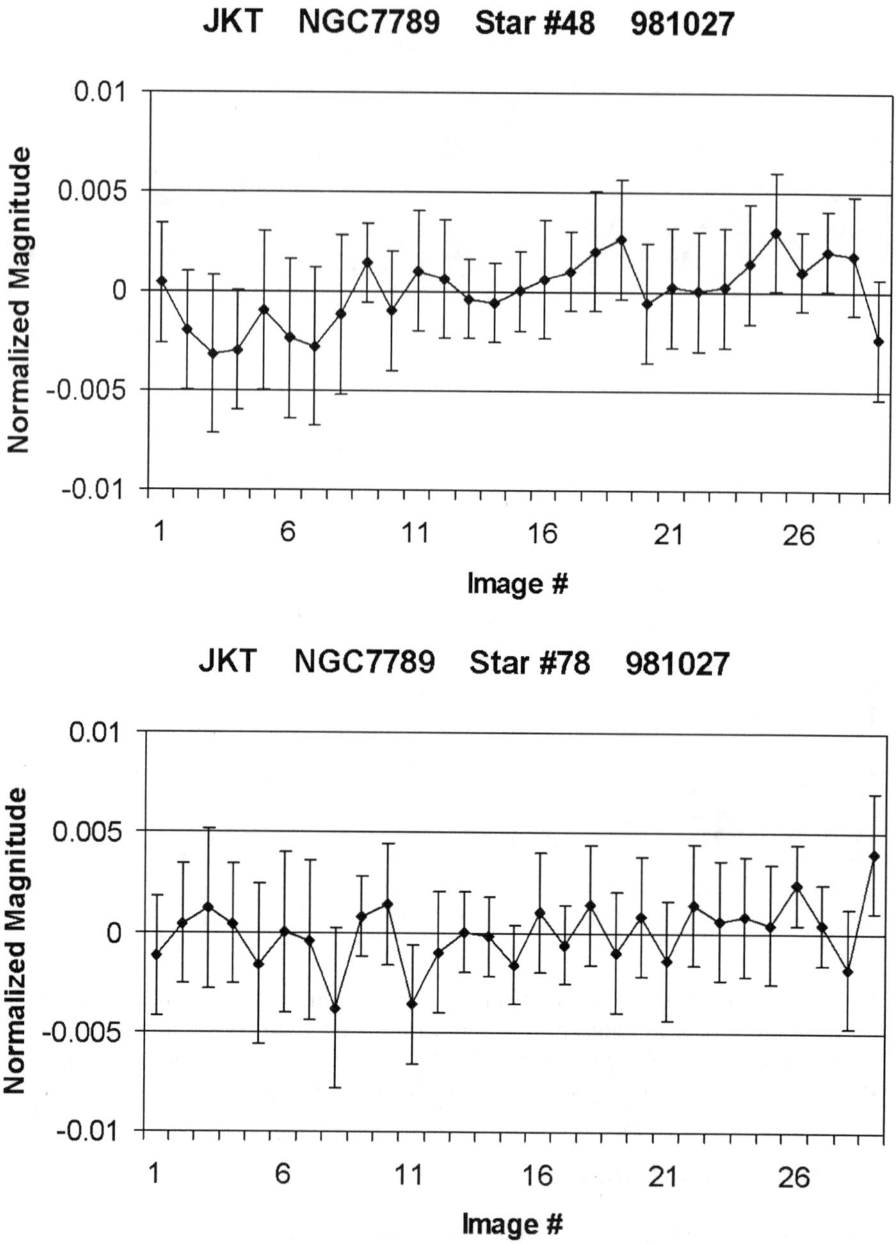

Figure 3. Lightcurves of two stars in NGC 7789 for a good night.

the individual frames can be obtained. We have developed a photometric data-reduction pipeline for crowded fields, based on the psf-fitting package DAOPHOT in IRAF. Good results have been obtained with this pipeline in tests on partial data sets (see Figure 3), again with frame-to-frame variations of $\lesssim$ 0.3% for many of the brighter stars. We are currently working on getting consistent photometry also for the nights that had less than optimum conditions (seeing variations, cirrus, focus drifts). It appears that psf fitting is not the optimum method for these data. A hybrid approach using aperture photometry on images in which models for all stars but one have been subtracted will be tested soon.

5. Outlook

We expect that we can extract lightcurves with rms variations of 0.3% to 0.5% for several hundred stars in each of the target clusters. These lightcurves will be searched for the characteristic dips due to planetary transits. Because of the dense sampling (about 15 to 30 minutes between exposures of the same cluster), each transit must show up in a number of successive frames. Stars showing at least two dips will be considered candidate planetary systems. Further observations would then be needed to establish periodic transits and to rule out grazing eclipses by stellar companions or contamination by background eclipsing binaries.

A follow-up program to the EXPORT planet search project has been initiated at the Isaac Newton Telescope on La Palma. This program takes advantage of the large area covered by the Wide Field Camera, and of the larger aperture size, to monitor a much larger number of fainter stars in the same clusters. A status report on the INT project is given by Street et al., this volume, p. 572).

References

Butler, R. P., & Marcy, G. W. 1996, ApJ, 464, L153

Gonzalez, G., Wallerstein, G., & Saar, S. H. 1999, ApJ, 511, L111

Janes, K. 1996, JGR, E101, 14853

Marcy, G. W., & Butler, R. P. 2000, PASP, 112, 137

Mayor, M., & Queloz, D. 1995, Nat, 378, 265

Disks, Planetesimals, and Planets
ASP Conference Series, Vol. 219, 2000
F. Garzón, C. Eiroa, D. de Winter, and T. J. Mahoney, eds.

A Search for Planetary Transits in Open Clusters

R. A. Street[1], K. Horne[1], A. Penny[2], Y. Tsapras[1], A. Quirrenbach[3], N. Safizadeh[3], J. Cooke[3], D. Mitchell[3], and A. C.Cameron[1]

[1] *Dept. of Physics and Astronomy, University of St. Andrews, North Haugh, St. Andrews, Fife, KY16 9SS, Scotland*
[2] *Rutherford Appleton Laboratories, Chilton, Didcot, Oxon, OX11 0QX, UK*
[3] *Center for Astrophysics and Space Sciences (CASS), University of California, San Diego, CA 92093, USA*

Abstract. We report on the progress of a photometric search for planetary transits of stars in the open clusters NGC 6819, NGC 6940, and NGC 7789. The data were obtained during two allocations of ten nights with the Wide Field Camera on the Isaac Newton Telescope, La Palma, in mid-1999. These old, rich open clusters give numerous bright stars within the WFC field of view, allowing roughly 40 000 lower-main-sequence cluster stars to be monitored with comparable numbers of field stars. We have developed a semi-automated data-reduction pipeline to process the data and highlight potential transit candidates. We report the accuracy achieved in the photometry using a subset of the data and on factors affecting this accuracy, in addition to the variables found and the progress made in searching for transits.

1. Introduction

The regular photometric monitoring of large numbers of stars is the key to any search for planetary transits, given the small probability that the planetary orbit will be suitably inclined (Sackett 1999). With this in mind, Janes (1996) suggested that imaging open clusters would allow sufficient numbers of suitable stars to be monitored frequently enough to detect transit events. From the ground, this technique lends itself particularly well to the detection of "hot jupiter"-type planets, and the Wide Field Camera (WFC) on the 2.5 m Isaac Newton Telescope (INT) gives the large area coverage ideal for this type of study.

Of the clusters visible from La Palma, three were chosen in order to include as many stars as possible, while maintaining a good sampling rate. Bright, rich clusters, not widely dispersed were obviously important potential targets, while old age and high metallicity were the deciding factors. We settled on NGC 6819, NGC 6940, and NGC 7789, which were among those recommended by Janes (1996). Their properties are summarized in Table 1.

Table 1. Selected target clusters

Name	RA(J2000.0)	Dec(J2000.0)	Age [Gyr]	Dist [pc]	$E(B-V)$	[Fe/H]
NGC 6819	19 41 18.70	40 11 05.27	2.51	1689	0.21	+0.07
NGC 6940	20 34 35.98	28 18 00.00	0.87	1042	0.23	+0.01
NGC 7789	23 57 30.81	56 43 41.84	1.78	1841	0.25	−0.08

2. The Dataset

2.1. Observations

We were awarded two allocations of ten bright nights each in the 1999A semester on the 2.5 m INT, running June 21–30 and July 22–31, and were fortunate to have very good conditions almost all of the time. The instrument used was the WFC, which consists of four EEV CCDs, each measuring 2k ×4k pixels, allowing us to cover a $34' \times 34'$ field of view (Ives, Tullock, & Churchill, 1996). Our strategy each night was to cycle from cluster to cluster as they became visible, taking two 300 s exposures on each visit, with a readout time of 165 s. This enabled us to monitor the stars in each cluster at least once an hour, sufficient to detect the transits of hot jupiters, which have a typical duration of $\sim$ 2–4 hours. The majority of our observations were made using the Sloan r filter, though a few images per cluster were taken using the Sloan g, Sloan i, and RGO U filters in order to obtain color data for the stars.

2.2. Data Reduction

We have written a semi-automated data-reduction pipeline based around PSF fitting by the STARMAN photometry package (Penny 1995). This pipeline was developed around the reduction of approximately 2 per cent of the dataset; about half of the central CCD of the first run Sloan-r data on NGC 6819, centered slightly west of the cluster center. From this alone, we find we are covering 5820 stars. To roughly calibrate the photometry, we used the zero points quoted on the INT Wide Field Survey webpage to convert our instrumental magnitudes to magnitudes based in the Landolt (1992) system.

3. First Reduction Results

3.1. Color–Magnitude Diagram for NGC 6819

Figure 1 shows the color–magnitude diagram for the region of NGC 6819 covered by our data subset. This clearly shows the cluster main sequence, as well as a substantial "tail" of background stars. Our exposure times were tailored towards lower-main-sequence cluster stars of approximately 1 $M_{\odot}$; therefore, any stars below the main sequence on Figure 1 will be underexposed and subject to levels of noise that make them unsuitable for transit detection. The unsaturated stars

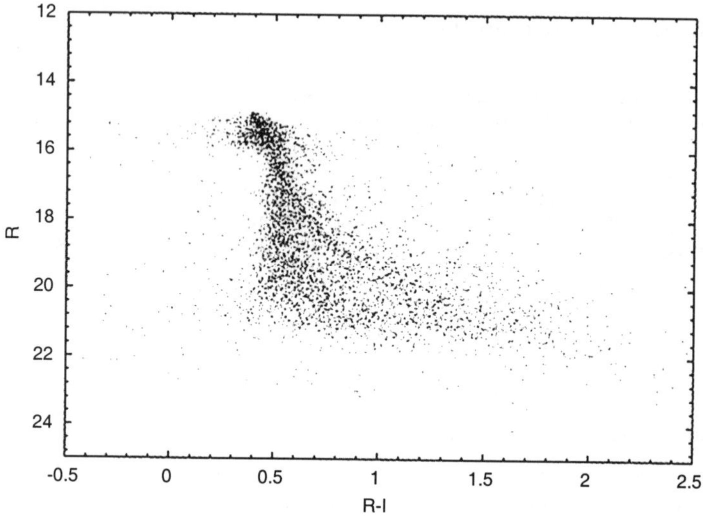

Figure 1. The color–magnitude diagram for NGC 6819, showing the cluster main sequence and background stars.

in our dataset comprise roughly equal numbers of cluster and background stars. Unfortunately, the cluster main-sequence stars saturated prior to the turn-off point, accounting for the bulge in the figure.

3.2. Achieved Photometric Precision

Figure 2 illustrates the photometric precision achieved so far on a typical night of the run. The effects of saturation are clearly visible for stars brighter than about 18th magnitude, and the accuracy of the fainter stars is dominated by the sky brightness, owing to the observations having been made in bright time. For comparison, we also plot the expected transit depths for a planet with Jupiter's radius transiting cluster main-sequence stars of masses ranging from 0.4 to 1.0 $M_{\odot}$. Any star lying below this line therefore has the measurement accuracy required to detect planetary transits. This figure clearly shows that this accuracy has been reached for some of the brighter stars and illustrates the levels of the various sources of noise in the data. Given the importance of sky brightness as a source of noise, future surveys should consider dark time as being more appropriate.

3.3. Systematic Effects

One reason for the larger-than-expected scatter in magnitude is a number of systematic effects we have discovered in the data. To illustrate these problems, we include the lightcurves of two example stars (see Figure 3). The first has one of the smallest rms scatters in the sample, while second has one of the largest. Both lightcurves display two noticeable systematic effects, seen in every star: an apparent steady decrease in brightness during each night (subsequently

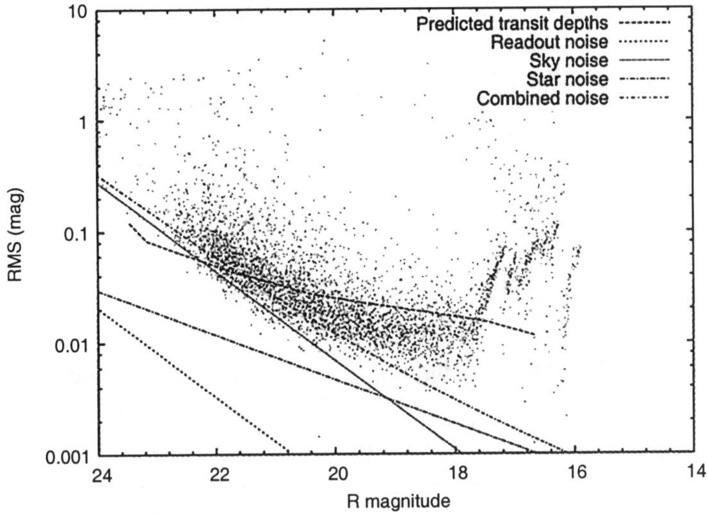

Figure 2. Magnitude scatter versus magnitude for the 17 images from the first night. Various sources of noise have been plotted, together with the predicted transit depth for a Jupiter-radius planet transiting a main-sequence cluster star.

attributed to secondary extinction effects), and a sinusoidal-like trend across the ten days.

Identifying and removing these and other systematics is our priority for the next stage of the project. We are currently applying a linear regression to the magnitudes and spatial positioning of the stars in order to identify and correct for these systematic effects.

3.4. Discovered Variable Stars

In spite of the improvements that remain to be made in the measurement precision our observational strategy lends itself quite naturally to the discovery of variable stars, and several large-amplitude candidates have already presented themselves. Of these, three stars have been found to be unequivocally variable; their respective lightcurves are shown in Figure 4. Star 3801 may be a dwarf nova, while star 3563 is clearly an eclipsing binary with a period of about 1.462 days. Star 4512 remains unclassified, although it may be a dwarf Cepheid. Further detailed study of these variables is planned and will be published in due course.

4. Conclusions

We have acquired a large dataset and are approaching the photometric accuracy required to detect transits for some of the stars in our sample. Ongoing investigation into the various systematic effects should enable us to considerably

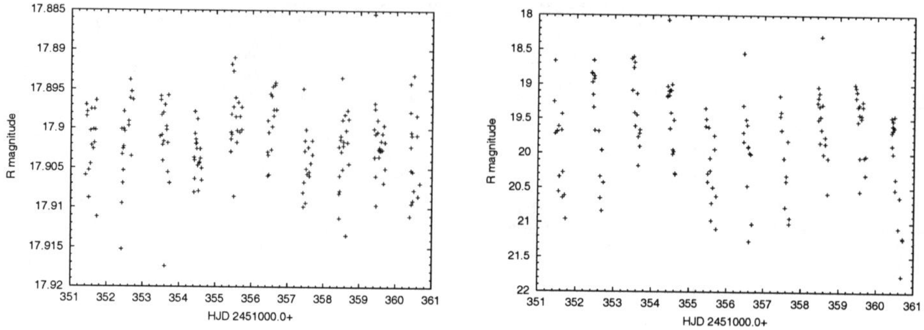

Figure 3. Two example lightcurves, showing the systematic variations in magnitude.

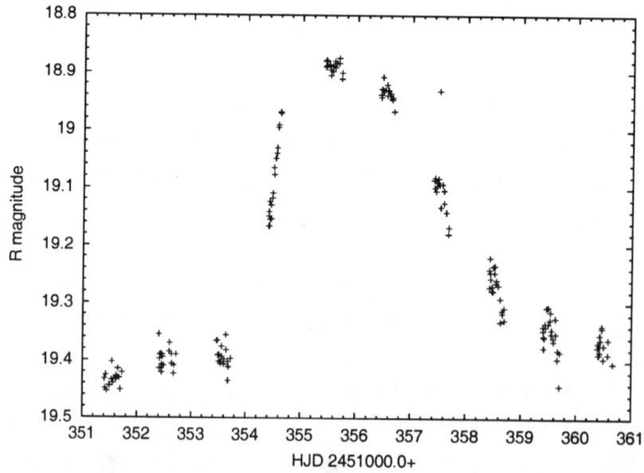

Figure 4. The lightcurve of star 3801, a suspected dwarf nova.

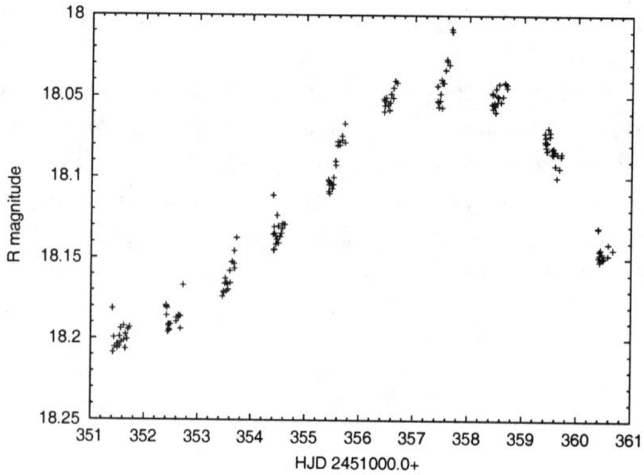

Figure 5. The lightcurve of star 4512, a suspected Cepheid.

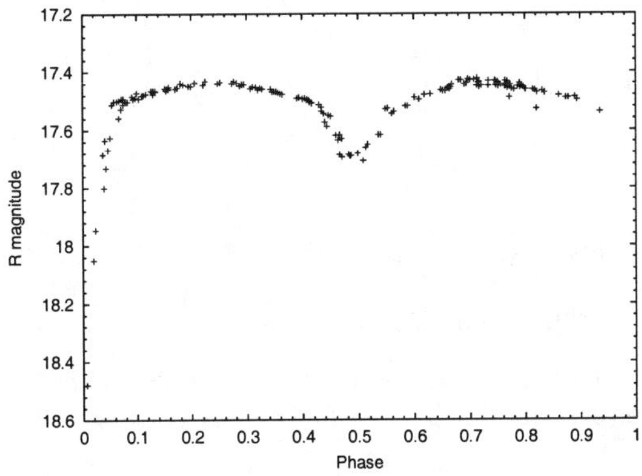

Figure 6. The phase diagram of star 3563, an eclipsing binary with a period of about 1.462 days.

increase the number of stars measured with an accuracy below the detection threshold. Some of these improvements are already in place. We have identified a number of variable and binary stars within the small-sample dataset that has been reduced as well as a number of potential candidate variables, and plan further investigation of these as soon as is feasible. Naturally, we expect many more to be discovered when the full dataset is reduced.

5. Acknowledgements

RAS would like to thank Tim Lister for his help and advice. RAS was sponsored by a PPARC studentship during the course of this work, and would like to thank the conference organizers for their financial support during the conference.

References

Ives, D., Tullock, S., & Churchill, J. 1996, SPIE, 2654, 226

Janes, K. 1996, JGR, 101, Iss. E6, p. 14853

Landolt, A. U. 1992, AJ, 104, 340

Penny, A. 1995, Starlink User Note 141.2 (Rutherford Appleton Laboratory)

Sackett, P. D. 1999, in NATO-ASI series, Planets outside the Solar System: Theory and Observations, ed. J.-M. Mariotti & D. Allion (Dordrecht: Kluwer), p. 189

Discussion

William Borucki: How do you correct for extinction?

Rachel Street: As yet we have included no correction for extinction in our pipeline data reduction—this issue is a priority for us to address in the very near future.

Disks, Planetesimals, and Planets
ASP Conference Series, Vol. 219, 2000
F. Garzón, C. Eiroa, D. de Winter, and T. J. Mahoney, eds.

Detection Limits in Space-Based Transit Observations

H. J. Deeg[1]

Instituto de Astrofísica de Andalucía, E-18080 Granada, Spain

F. Favata and The *Eddington* Science Team

Astrophysics Division—Space Science Department of ESA, ESTEC, Postbus 299, NL-2200 AG Noordwijk, The Netherlands

Abstract. From simulations of transit observations, it is found that the detectability of extrasolar planets depends only on two parameters: the signal-to-noise ratio during a transit and the number of data points observed during transits. All other physical parameters describing transit configurations (planet and star size, orbital period, orbital half-axis, latitude of the transit across star) can be reduced to these two parameters. In turn, once the dependency between transit detectability and these two parameters has been determined, predictions for an instrument's ability to detect transits by any combination of physical parameters can be derived with ease. These predictions are applied to the *Eddington* proposal of a combined asteroseismology/transit-detection space mission currently under study by ESA, which is described briefly.

1. Introduction

To evaluate the performance for the detection of extrasolar planets of observations with the transit method, Jenkins, Doyle, & Culler (1996) introduced the method of the calculation of two distributions, one of the detection statistics for transitless noise-only data (which may be observed or simulated), and one detection statistics for data with known transits in them. This method allows the determination of detection probabilities while at the same time setting limits for the false alarm rate—important in the case of transit detections, as one knows neither if a transit signal is in the data, nor the signal's duration, amplitude, or period. Doyle et al. (2000) used this method to set detection limits on ground-based observational data, taken with the goal of detecting transits across the eclipsing binary CM Draconis. The evaluations shown here were performed during the course of specifying the *Eddington* proposal currently pending at ESA for a combined asteroseismology/planet-finding mission with a 1 m-class space telescope, which is described at the end of this paper. In the following, the dependence of the parameters describing a transit configuration (planet's size, star's size, planet's period, for example) on detectability is shown.

[1] Also at Instituto de Astrofísica de Canarias, E-38205 La Laguna, Tenerife, Spain.

2. Extrasolar Planet Detectability by Transits

Planet-detection performance may be expressed in terms of *detectability* or the *detection probability* p_d, which is the probability that a physically present *transiting* planet is going to be detected by the experiment. The detectability of a planet in an observed transit sequence depends on a number of physical parameters of the planet–star system and of the observing instrument. The most obvious one is the size of the planet, with the amount of brightness loss during a transit being proportional to the square of the planet radius, R_pl. Similarly, detectability will remain roughly constant, if the ratio R_pl/R_* is kept constant.

To asses the performance in the detection of planetary transits, simulated lightcurves with the instrument's noise characteristics were submitted to a *detection test*. This detection test gives some "detection value", which may be calculated by any kind of procedure that performs some evaluation between a given model lightcurve and the data. The most suitable one for data dominated by "white" noises and without (or with correctable) zero-point drifts is a scalar multiplication between the data vector and the model vector:

$$C = \sum d_i * m_i,$$

where C is the detection value, and d_i, m_i are data and model values. The detection test was performed in large numbers on lightcurves with—and without—simulated transits added to the noise The highest "detection values" from a large number (between $10^5 - 10^7$) of lightcurves *without* transits (that is, with noise only) give a threshold for detection values that constitute false alarms. The fraction of lightcurves *with* transits (again using a large number of randomly generated curves, but with known transit parameters) that results in detection values *higher* than the false alarm threshold gives the detection probability. This procedure is similar to one used for an assessment of detection limits in the ground-based transit detection project "TEP" (Doyle et al. 2000).

During the course of evaluating a variety of transit configurations, it became obvious that the physical parameters governing transit lightcurves may be reduced to two parameters which solely govern the detectability in any observed sequence of transits around single stars (the case for binaries would be more complicated): the data's S/N during a transit, and the total number of data points within transits, N_tp. Thus, if the function $p_\mathrm{d}(S/N, N_\mathrm{tp})$ can be established, the detectability can be calculated for any combination of physical parameters. The physical parameters influence S/N and N_tp as follows:

- The Signal, S, depends on the planet and star sizes, the latitude of transit, and on the star's limb-darkening (mostly for high-latitude transits). For a precise calculation of model transit lightcurves (and hence S), a code dealing correctly with the planet's ingress and egress and with differential limb-darkening under the planet's disk should be used, such as "utm" (Deeg 1999).

- The noise, N, depends on the noises affecting the observing instrument's performance (which has always a dependency on a star's apparent brightness) and on intrinsic brightness fluctuations (stellar activity, oscillations, variability)

- The number of on-transit data points, N_{tp}, depends on exposure time, the number of transits observed (= observational coverage) and on the duration of a transit. The duration in turn is governed by the planet's orbital period, by the star's mass and size, by the latitude of the transit, and to a small amount by the size of the planet. It can be calculated by straightforward application of basic geometry and Kepler's third law.

For detectability evaluations, a "standard case" was chosen, which has a detection probability (p_d) of 0.99. In Figure 1, the sequence of six transits of the standard case is shown. That case comprised the following setup:

- Physical parameters: K0 star (0.85 $R_\odot$, 0.78 $M_\odot$), orbited by a planet with 1.4 $R_\oplus$, 200 d orbital period, transiting at a latitude of 45°across the star.

- Observational parameters: 3 years (1095 days) observing duration; lightcurve with a noise of 450 ppm, or approximately 0.5 mmag from 600 s exposure time. In the case of the *Eddington* proposal, this is expected for a star with 16 mag.

As an example of parameter variations around the standard case, the dependency of p_d on planet-size is shown:

$R_{pl}/R_\oplus$	S/N	p_d
1.0	0.25	0.16
1.3	0.42	0.91
1.4	0.48	0.99
$\gtrsim$1.5	$\gtrsim$ 0.55	1.00

More useful than detection probability, however, is the determination of the *minimum planet size that can be detected reliably*. A "reliable detection" is considered to have $p_d \geq 0.99$. From the table above, the detection limit is therefore a planet of 1.4 $R_\oplus$. Similar tables where also generated for the "standard case" but with orbital periods of 50, 100, and 365 days, thus performing an empirical determination of the planet size, $R_{pl,min}$, where $p_d \approx 0.99$. These results are shown in the next table, where besides $R_{pl,min}$ the following quantities are given: n_{tr}, the number of transits in the three-year-long lightcurve; t_{tr}, the duration of one transit; N_{tp}, the total number of on-transit data points at 600 s exposure time; and S/N, the signal-to-noise ratio, where the signal is taken as the maximum relative brightness variation during a transit.

P_{orb}/day	50	100	200	365
$R_{pl,min}$	1.16	1.30	1.40	1.62
n_{tr}	22	11	6	3
t_{tr}/day	0.186	0.234	0.295	0.361
N_{tp}	589	371	255	156
S/N	0.33	0.42	0.48	0.64

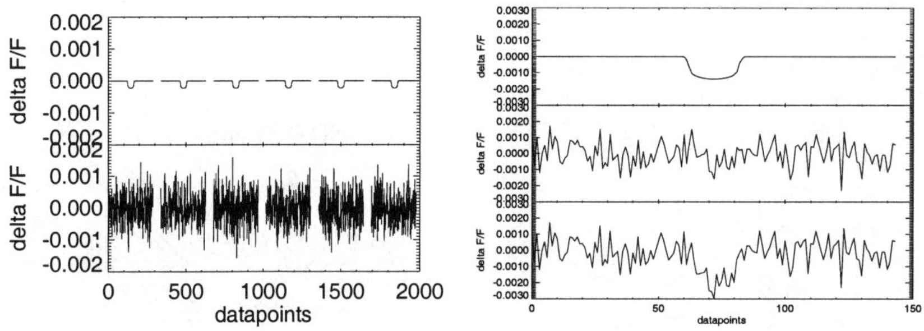

Figure 1. *Left:* Sequence of six transits corresponding to observations of the "standard case". The *upper panel* gives noiseless model transits, the *lower panel* is the simulated noisy lightcurve with the model subtracted. Of the whole lightcurve, only segments near the six transits are shown. As the detection limit of $p_{\rm d} = 0.99$ corresponds here to an S/N of 0.5, *individual* transits are not detectable—either to the eye or to any algorithm. *Right:* The same transit sequence, but data are rebinned by 2 and all six transits are coadded (that means that the model-transit is six times deeper, and the noise is $\sqrt{(6/2)}$ times larger). The *center panel* is pure noise, the *lower panel* is noise + transit. Planet detectability is practically independent of exposure time or binning of the data. This holds, as long as noises dominate that are proportional to $(\text{exp.time})^{-1/2}$, such as photon and background noise.

From above table, a linear relationship between $\log S/N$ and $\log N_{\rm tp}$ was established, which can be expressed as:

$$\log S/N = 0.90 - 0.50 \log N_{\rm tp},$$

or more generally:

$$S/N * \sqrt{(N_{\rm tp})} = \text{const. for } \mathrm{p_d} = \text{const.}$$

It should be noted, that this expression is valid only when using $C = \sum d_i * m_i$ for the determination of the "detection value". Other possibilities to derive detection values, such as the calculation of χ^2 values between model and data, are expected to lead to other dependencies, and may not allow the reduction of the physical transit parameters to the two parameters S/N and $N_{\rm tp}$. The above expression between the fundamental parameters governing transit detectability (using the level $p_{\rm d} = 0.99$) allows the determination of $R_{\rm pl,min}$ for any combination of physical parameters constituting a transit configuration.

In Figure 2, $R_{\rm pl,min}$ is given for a variety of stellar types and orbital periods and apparent magnitudes as expected by *Eddington*. As the calculations were performed for transits at latitudes of 45°, the given planet sizes are valid for $\approx$ 70% of theoretically possible transit configurations. For the remaining 30% of polar transits, detection probabilities may be significantly lower, and planets with $R_{\rm pl,min}$ cannot be detected reliably. Indicated as hatched regions in Figure 2 is the extent of the circumstellar habitable zone based on values from Whitmire

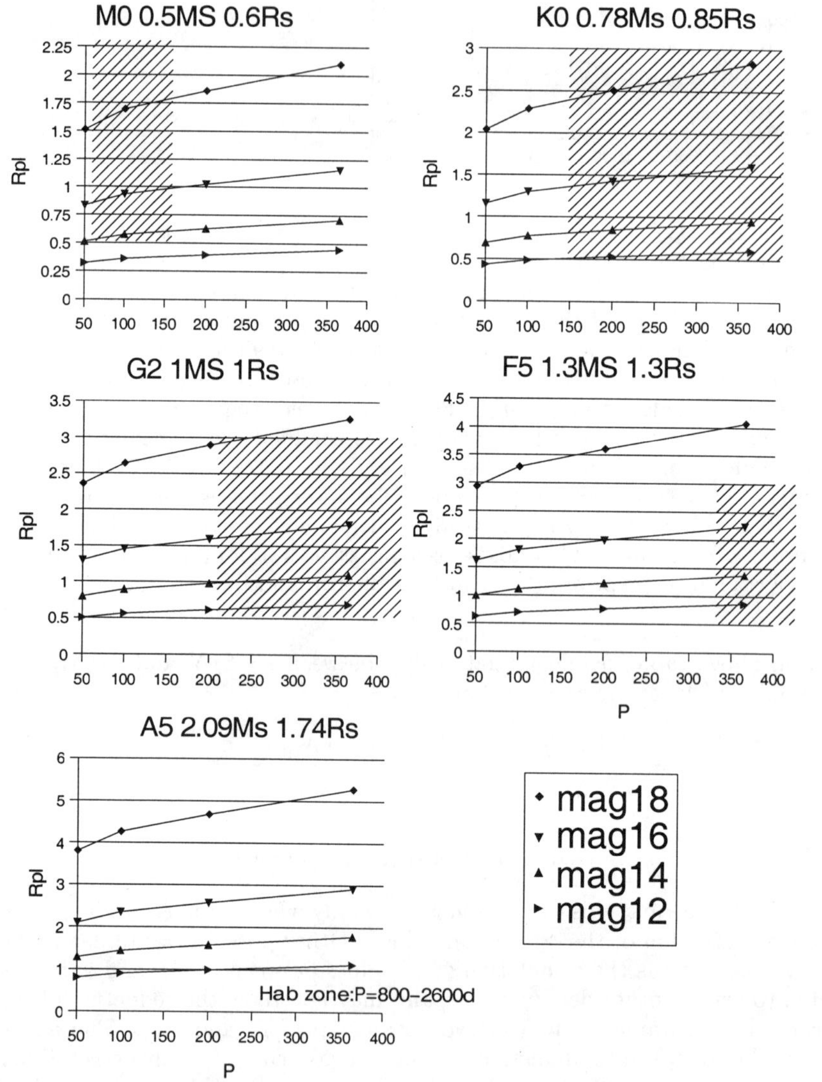

Figure 2. Minimum detectable planet size (in $R_\oplus$) for a range of stars in dependence of orbital period (in days) and apparent magnitude for the *Eddington* proposal. A transit latitude of 45° was used. L*Hatched regions* indicate the circumstellar habitable zone.

& Reynolds (1996). It can be seen that Earth-sized planets around a solar-type star of $m_V = 14$ may be detected, and even stars with $m_V = 18$ may still allow for the detection of "large terrestrial" planets.

3. The *Eddington* Proposal

This work was performed in the context of the *Eddington* study. *Eddington* is a a combined asteroseismology and planet-detection mission, based on a a 1 m-class space telescope, initiated by a European consortium of scientists (Favata, Roxburgh, & Christensen-Dalsgaard 2000) in response to ESA's F2/F3 mission call. *Eddington* was successfully selected for the initial currently ongoing study phase, and will undergo a final selection in 2000 September. With foreseen photometric precision of $\simeq 1$ part in 100 000, the major science goal for the planet-detection part will be the search for terrestrial-size planets in the habitable zone. With its large mirror size (1.2 m diameter) and field of view (3°diameter), it will monitor several hundred thousand stars during the three years of its mission dedicated to the transit search. To enable long, uninterrupted observations within a stable thermal environment, an L2 halo orbit is being baselined. These parameters will allow *Eddington* to achieve qualitatively different science goals from those of the small missions currently in preparation (*COROT*, *MONS*, *MOST*). The discovery of extrasolar terrestrial planets, in addition to being a major achievement on its own, will serve as a fundamental stepping stone for future missions specifically designed for the detection of extraterrestrial habitats, such as ESA and NASA missions *IRSI-Darwin* and *Terrestrial Planet Finder* (*TPF*). *Eddington* will also contribute in a major way to the field of asteroseismology, enabling an accurate characterization of the stellar structure for stars spanning a wide range of masses, ages, and chemical compositions.

References

Deeg, H. J. 1999, utm—Universal Transit Modeller program (for IDL), http://www.iac.es/galeria/hdeeg/idl_hans_lib/utm/

Doyle, L. R., et al., 2000, ApJ, 535, 338

Favata, F., Roxburgh, I., & Christensen-Dalsgaard, J. 2000, in Third MONS Workshop, Science Preparation and Target Selection, ed. T. C. Teixeira & T. R. Bedding (Aarhus: Aarhus Universitet), in press

Jenkins, J. M., Doyle, L. R., & Cullers, D. K., 1996, Icarus, 119, 24

Whitmire, D. P, & Reynolds, R. T. 1996, in Circumstellar Habitable Zones, ed. L. R. Doyle (Menlo Park: Travis House), p. 127

Discussion

William Borucki: Are you making available your predictions for future transits of CM Draconis?

Hans Deeg: For the next observing season (spring–summer 2000) we have scheduled verification observations. Afterwards, we will make the lightcurve publicly available for further analysis.

Disks, Planetesimals, and Planets
ASP Conference Series, Vol. 219, 2000
F. Garzón, C. Eiroa, D. de Winter, and T. J. Mahoney, eds.

The STARE Project: A Transit Search for Hot Jupiters

Timothy M. Brown[1] and David Charbonneau[2,1]

[1]*High Altitude Observatory, National Center for Atmospheric Research, PO Box 3000, Boulder, CO 80307, USA*

[2]*Harvard-Smithsonian Center for Astrophysics, 60 Garden St., Cambridge, MA 02138, USA*

Abstract. The STARE instrument is a small-aperture, wide-field, CCD-based telescope that delivers high-cadence time-series photometry on approximately 40 000 stars in a typical field centered on the Galactic plane. In a two-month observing run on a field, we obtain sufficient precision on roughly 4000 stars to detect a close-in Jupiter-sized companion in an edge-on orbit. We also used this instrument to detect the planetary transits across the Sun-like star HD 209458. The project is now in its third season, and we have acquired a large dataset on several fields. Given the frequency of close-in extrasolar planets found by the radial-velocity surveys, and the recent confirmation that at least some of these are indeed gas giants, the STARE project should be able to detect roughly a dozen Jupiter-sized planets in its existing dataset.

1. Introduction and Motivation

Radial-velocity surveys of nearby F-, G-, K- and M-dwarf stars have revealed a class of close-in extrasolar massive planets that orbit their stars with an orbital separation of $a \lesssim 0.1$ AU. Prior to the transit results for HD 209458, the radial-velocity method has been the only method by which we have learned anything about these planets. The radial-velocity technique measures the period, semi-amplitude, and eccentricity of the orbit, and by inference the semi-major axis. It also yields a value for the minimum mass, dependent upon the assumed value for the stellar mass, but aside from this it gives no direct information on the structure of the planet itself. The search to measure the transit photometrically is motivated by fact that, for a star for which both the radial velocity and transits are observed, one can estimate both the mass (with negligible error due to $\sin i$) and radius of the planet. These can then be combined to calculate such critically interesting quantities as the surface gravity and average density of the planet, and thus provide constraints on structural models for these low-mass companions. Assuming random orbital alignment for systems with $a = 0.05$ AU and Sun-like primaries, the chance of a transiting configuration is roughly 10 per cent.

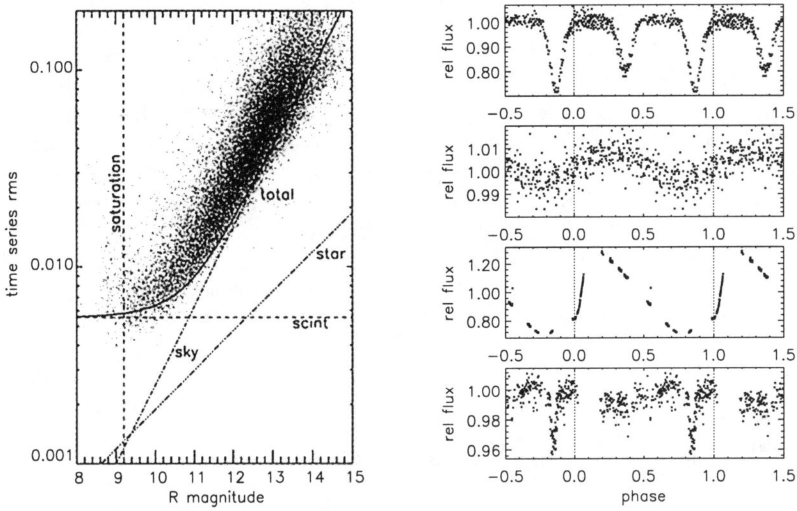

Figure 1. *Left panel:* The achieved time series rms variation for roughly 24 000 stars in a field in Auriga, as a function of R magnitude, for one night. The dominant noise source is sky background for all but the brightest stars in the field. *Right:* A selection of variable stars as observed by STARE. From *top* to *bottom* these are: 1) an eclipsing binary with a period of 0.876 days; 2) a δ Scuti with a period of 0.078 days; 3) a δ Cepheid with a period of 18.2 days, and 4) an eclipsing binary with a 4% primary eclipse and a period of 2.55 days. This small eclipse depth is approaching the level that would be predicted for a transit of a Jupiter-sized companion across a Sun-like star.

2. The STARE Instrument

The telescope is a Schmidt camera of focal length 286 mm and $f/2.9$. The modest 10 cm aperture images a 6 degree square field of view onto a 2034 $\times$ 2034 pixel CCD with 15 μm (11 arcsecond) pixels. The majority of observations are taken through a red (approximately Johnson R) filter, with supporting observations in B and V to determine stellar colors. The data are analyzed via a pixel-weighted photometry scheme, with post-photometry linear regression to account for gray and color-dependent extinction. Prior to 1999 July, the main instrument was an Aero-Ektar lens. The new system has a much more Gaussian-shaped point spread function, which increased the precision and the throughput by a factor of two.

The STARE instrument delivers high-cadence (2 minute) times-series photometry on roughly 40 000 stars ($9 < V < 14$) in a typical field centered on the Galactic plane. As shown in Figure 1, sky background is the dominant source of noise for the majority of stars in the field, while the brightest stars are scintillation limited. In each field, we obtain sufficient precision on roughly

4 000 stars to detect a close-in Jupiter-sized companion. Since fields centered on the Galactic plane are not always available, we have also stared at some high-Galactic-latitude fields. These fields are significantly less crowded and will allow us to measure the effects of crowding on our photometry pipeline.

The telescope is currently located in Boulder, Colorado, but is portable. In the future, it may be relocated in longitude so as to provide much greater time-series coverage when operated as part of a network with similar instruments, such as those run by W. Borucki at Lick Observatory (Vulcan, Borucki et al. 2000) and by E. Dunham at Lowell Observatory.

A by-product of the STARE observations is the high-cadence monitoring of numerous variable stars, the majority of which are new detections. We have made many of the lightcurves available electronically (Brown & Kolinski 1999), and plan to continue to do so in the future.

3. The First Transiting Extrasolar Giant Planet

We have detected the first planetary transits across a Sun-like star, HD 209458, as described in Charbonneau et al. (2000). These have also been reported by Henry et al. (2000).

Motivation for observing HD 209458 came from D. W. Latham and M. Mayor (personal communication) in 1999 August. The times at which a potential transit could occur were calculated from the preliminary orbital period and ephemeris from the radial-velocity observations. We observed HD 209458 for ten nights in August and September. Most of these nights occurred when no transit was predicted, and the residuals are consistent with no variation. On the other two nights (UT Sep 9 & Sep 16), we see a conspicuous dimming of the star for a time of several hours, at a time consistent with the prediction from the observed radial-velocity orbit. We attribute this dimming to the passage of the planet across the stellar disk. Our precise measurement of two complete transits allows us to accurately determine the planetary radius, orbital inclination, and mass, and hence derive quantities such as the surface gravity and average density.

Assuming a value for the stellar radius and mass, and a description for the stellar limb-darkening, we can determine the planetary radius with an accuracy of several per cent (see Figure 2). However, the dominant uncertainties in determining the planetary radius and orbital inclination are the uncertainty in the value of the stellar radius, and to a lesser extent the stellar mass and limb-darkening. Given time series observations of the transit in a single photometric band, it is possible to fit the data with a family of models, since a larger planetary radius can be accommodated by increasing the stellar radius and reducing the orbital inclination. In Mazeh et al. (2000), we undertook a detailed study of the stellar parameters, as well as the respective uncertainties. These allowed us to measure the planetary radius to be $R_{\rm p} = 1.40 \pm 0.17\ R_{\rm Jup}$. This value for the radius is consistent with early predictions (Guillot et al. 1996). The inflated value of the radius relative to that of Jupiter is a result of the slower rate of planetary contraction due to exposure to high stellar insolation soon after the formation of the planet (Burrows et al. 2000). We also calculated several derived quantities, in particular the average density $\rho = 0.31 \pm 0.07\ {\rm g\,cm^{-3}}$, surface gravity $g = 870 \pm 160\ \mathrm{cm\,s^{-2}}$, and escape velocity $v_e = 42 \pm 4\ \mathrm{km\,s^{-1}}$ of the planet.

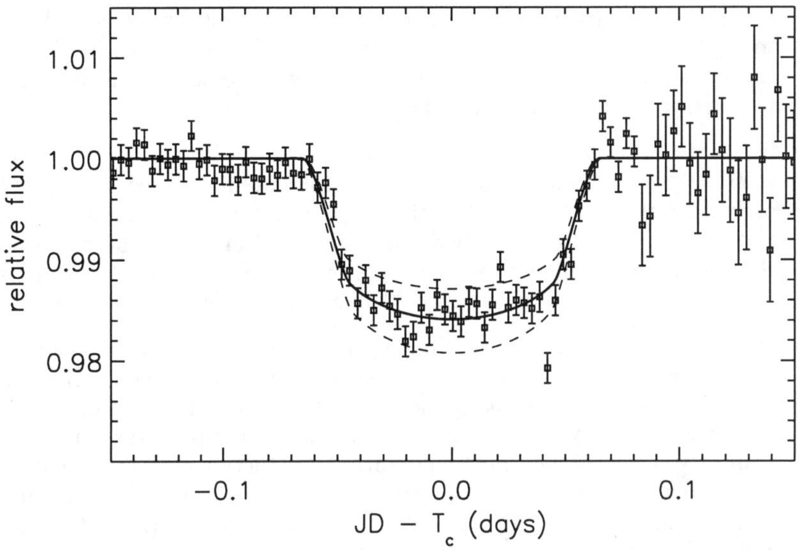

Figure 2. STARE observations of the planetary transit of HD 209458, binned into 5 m averages. The rms variation at the beginning of the time series is roughly 1.5 millimag, and this precision is maintained throughout the duration of the transit. The increased scatter at the end of the time series is due to increasing airmass. The *solid line* is the transit shape that would occur for our best-fit model. The *lower* and *upper dashed line*are the transit curves that would occur for a planet 10% larger and smaller in radius, respectively. The rapid initial fall and final rise of the transit curve correspond to the times when the planet is crossing the edge of the stellar disk, and the curvature across the center of the transit is due to the stellar limb-darkening. These data are available electronically from the STARE project website (Brown & Kolinski 1999).

4. Follow-Up Observations

The existence of a transiting planet suggests many fruitful follow-up observations, some of which may be accomplished within the next six months:

- High-cadence, high-precision photometry in other bandpasses would break the degeneracy shared between the stellar and planetary parameters by exploiting the color-dependent limb-darkening of the star. Furthermore, it may be possible to observe color-dependent variations in the observed planetary radius, since the planet would appear slightly larger when observed at wavelengths where the atmosphere contains strong opacity sources (Brown, this volume, p. 557; Burrows et al. 2000).

- If there are other planets in approximately coplanar orbits, then the likelihood that they too will generate transits is substantially enhanced relative to that for a randomly oriented system. A central transit by a Uranus-sized planet at 0.2 AU would yield a dimming some 6 hours in duration, with a depth of about 1 millimag.
- Reflected-light observations such as those for the τ Boo system by Cameron et al. (1999) and Charbonneau et al. (1999) would, if successful, yield the planetary albedo directly (Charbonneau & Noyes, this volume, p. 461). Predicted values for the albedo are highly sensitive to the atmospheric chemistry and condensates (Marley et al. 1999; Seager, this volume, p. 615; Seager, Whitney & Sasselov 2000; Sudarsky, Burrows, & Pinto 2000).
- Observations at wavelengths longer than a few microns may detect the secondary eclipse (perhaps 4 millimag) as the planet passes behind the star. This would allow the planet's dayside temperature to be estimated and hence quantify the net energy deposition in the planetary atmosphere. Similarly, it may be feasible to measure the reduction of the primary eclipse depth in the IR relative to shorter wavelengths and hence measure the planet's nightside temperature. If one was able to determine the dayside and nightside temperatures, then one would learn if the planetary atmosphere is effective at redistributing heat due to stellar insolation with a timescale less than the rotational period.
- If high-cadence photometry with a signal-to-noise ratio of 0.1 millimag can be achieved, it would be possible to detect planetary rings and/or large rocky satellites (Sartoretti & Schneider 1999).
- By taking the ratio of high-precision spectra in and out of transit, it may be possible to see additional absorption features during transit due to the absorption of light passing through the limb of the planetary atmosphere (Brown, this volume, p. 557; Seager & Sasselov 2000). The amplitude of the features may be as large as 0.2% in the case of the alkali-metal lines, and will be very sensitive to the height of the cloud layer in the planetary atmosphere.

Acknowledgments. We thank G. Card, C. Chambellan, D. Kolinski, A. Lecinski, R. Lull, T. Russ, and K. Streander for their assistance in the fabrication, maintenance, and operation of the STARE photometric camera, and we thank R. Noyes and S. Jha for many valuable discussions. D. C. is supported in part by a Newkirk Fellowship of the High Altitude Observatory. This work was supported in part by NASA grant W-19560.

References

Borucki, W. J., Caldwell, D., Koch, D., Webster, L., Jenkins, J., Ninkov, Z., & Showen, R. 2000, in preparation

Brown, T. M., & Kolinski, D. 1999, http://www.hao.ucar.edu/public/research/stare/stare.html

Burrows, A., Guillot, T., Hubbard, W. B., Marley, M. S., Saumon, D., Lunine, J. I., & Sudarsky, D. 2000, ApJ, 534, 97L

Cameron, A. C., Horne, K., Penny, A., & James, D. 1999, Nat, 402, 751

Charbonneau, D., Brown, T. M., Latham, D. W., & Mayor, M. 2000, ApJ, 529, L45

Charbonneau, D., Noyes, R. W., Korzennik, S. G., Nisenson, P., Jha, S., Vogt, S. S., & Kibrick, R. I. 1999, ApJ, 522, L145

Guillot, T., Burrows, A., Hubbard, W. B., Lunine, J. I., & Saumon, D. 1996, ApJ, 459, L35

Henry, G. W., Marcy, G. W., Butler, R. P., & Vogt, S. S. 2000, ApJ, 529, L41

Marley, M. S., Gelino, C., Stephens, D., Lunine, J. I., & Freedman, R. 1999, ApJ, 513, 879

Mazeh, T., et al. 2000, ApJ, 532, 55

Sartoretti, P., & Schneider, J. 1999, A&AS, 134, 553

Seager, S., & Sasselov, D. D. 2000, ApJ, 537, 916

Seager, S., Whitney, B. A., & Sasselov, D. D. 2000, ApJ, 540, 504

Sudarsky, D., Burrows, A., & Pinto, P. 2000, ApJ, 538, 885

Disks, Planetesimals, and Planets
ASP Conference Series, Vol. 219, 2000
F. Garzón, C. Eiroa, D. de Winter, and T. J. Mahoney, eds.

The MELODIE M-Dwarf Planet Survey

Xavier Delfosse

Instituto de Astrofisica de Canarias, E-38205 La Laguna, Tenerife, Spain

Thierry Forveille

Canada–France–Hawaii Telescope Corporation, PO Box 1597, Kamuela, HI 96743, USA

Damien Ségransan, Christian Perrier, and Jean-Luc Beuzit

Observatoire de Grenoble, Domaine Universitaire de St Martin d'Hères, F-38041 Grenoble, France

Michel Mayor and Stéphane Udry

Observatoire de Genève, CH-1290 Sauverny, Switzerland

Abstract. We present the current status of a survey for planets around M-dwarfs, using the ELODIE spectrograph at OHP. This observing program has detected the only planet known to date around an M-dwarf (Gl 876).

1. Introduction

Nearly all planet searches concentrate on solar-type stars (F–K dwarfs), even though the compared planetary statistics for different stellar types contain vital information on the planetary formation mechanism. M-dwarf planets are important in this respect, as these stars are the most common (over 70%) in the galaxy, and as their masses differ from that of the Sun by a significant factor.

These observations are challenging, though, because these very low-mass stars have low luminosities, and also because they commonly have strong chromospheric activity which affects radial-velocity measurements. For the last four years we have been monitoring the radial velocities of 120 M-dwarfs with the ELODIE spectrograph at Observatoire de Haute Provence. The main goals of this program are *not* planet search but rather:

- *Multiplicity statistics* of systems with an M-dwarf primary. This is an important constraint for stellar formation models, and it is also needed to correct the luminosity function for unresolved binaries (Delfosse et al. 1999a; Beuzit et al. 2000).
- Accurate *masses* for stars less massive than 0.5 $M_{\odot}$ (Forveille et al. 1999; Delfosse et al. 1998a, 1999b, 2000; Ségransan et al. 2000)

- *Rotational velocities* of M-dwarfs (Delfosse et al. 1998b)

For $\sim$ 50 stars, however, our measurement accuracy is sufficient to detect giant planets, as our detection of the Gl 876 companion illustrated. In the present paper we discuss the current status of this program and present its ongoing extension to a larger southern sample.

2. Sample and Observations

The observed radial-velocity sample is presented in detail by Delfosse et al. (1998b), who discuss the rotational properties of those stars. Briefly, it contains all M-dwarfs listed as closer than 9 pc in the third edition of the nearby star catalog, above a B1950.0 declination of $-16°$. 136 stars fulfill these criteria, a few of which had to be excluded because of the proximity of a much brighter primary.

Radial-velocity observations have been obtained at the Observatoire de Haute-Provence (OHP) with the ELODIE spectrograph (Baranne et al. 1996) on the 1.93 m Telescope, since 1995 September. The observational and data processing procedures are detailed in Delfosse et al. (1999a). The extracted spectra are analyzed for velocity by digital cross-correlation with a one-bit (0/1) template, derived from an observed spectrum of Barnard's Star and optimized for very low-mass stars. The instrumental velocity accuracy range between 12 m s^{-1} for $V < 8$ and 70 m s^{-1} for $V = 15$. Rotational and magnetic activity however together induce radial-velocity noise, as starspots and convective cells rotate over the stellar surface (Saar, Butler, & Marcy 1998). M-dwarfs are well known for their frequently large magnetic activity level. The dispersion of the measured radial velocitiesvaries greatly from star to star (see Figure 1), and our measurement accuracy is often degraded by stellar activity.

For the brighter and less active M-dwarfs in our sample, the radial-velocity observations are sensitive to all stellar and brown-dwarf companions with periods up to a few dozen years, and to Jupiter-mass planets with periods shorter than a few years (Figure 2).

3. Results and Prospects

The main result of this program is the discovery of the planet around the M4 dwarf Gl 876 (Figure 3, Delfosse et al., 1998c), independently discovered by Marcy et al. (1998). Gl 876b is the only example to date of a planet around an M-dwarf. It proves than planets exist around the most common stars of the Galaxy and that they can form around stars that greatly differ from our Sun. This has important implications for the total number of planets to be found within a given Galactic volume. It also constrains planetary formation theories: the protoplanetary disk around an M4 dwarf such as Gl 876 ($\sim 0.3\ M_{\odot}$) must be sufficiently massive to generate a 2 $M_{\rm Jup}$ planet like Gl 876b, unless one assumes that this system formed through capture.

Gl 876b is also the closest extrasolar planet (d = 4.7 pc), and orbits our fortieth-closest neighbor.

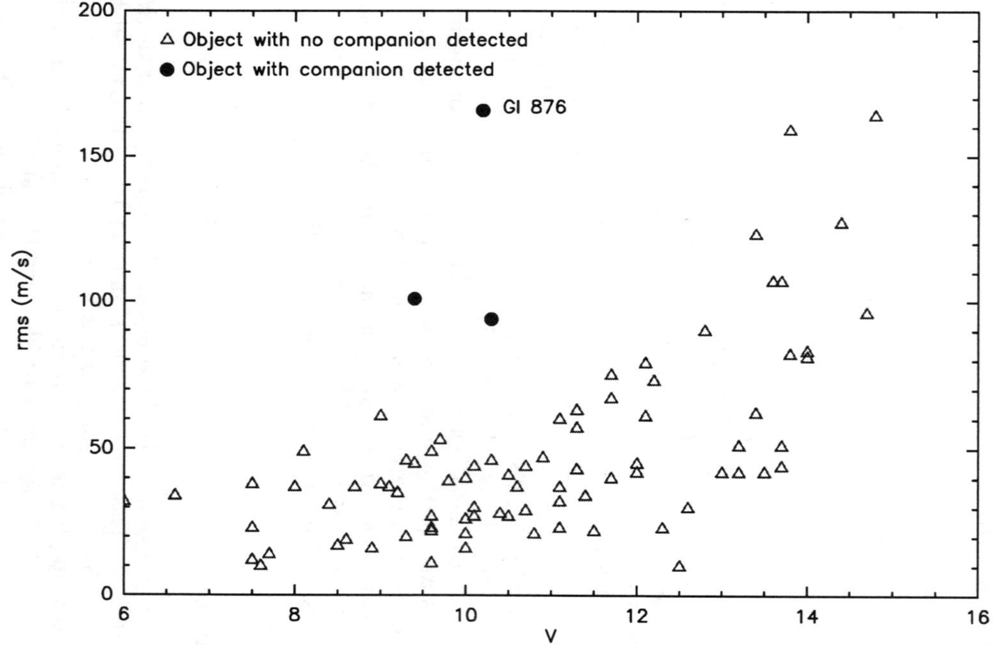

Figure 1. Rms dispersion of the radial-velocity measurements as a function of visual magnitude for the whole sample except identified binaries (with rms > 200 m s^{-1}). The dispersions range from 15 to 70 m s^{-1} for M0- to M6-dwarfs; stars with larger dispersions are either binaries or (easily identified) rapidly rotating stars. Gl 876 has a very large dispersion.

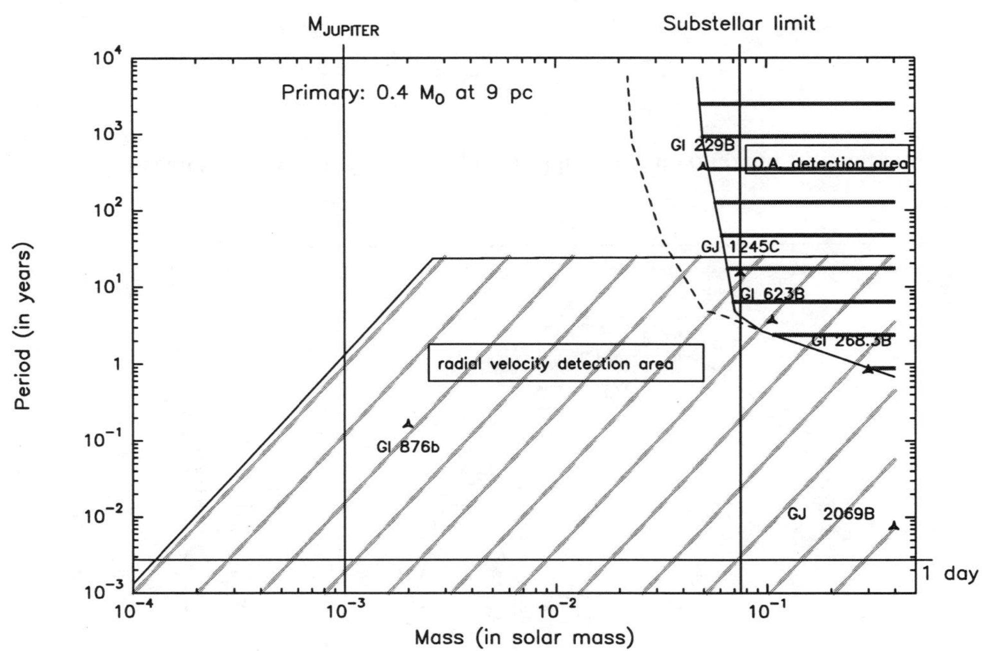

Figure 2. Companion detectability in the $\log(P)$ versus $\log(M)$ plane for our radial-velocity and adaptive-optics programs (the adaptive-optics program is only sensitive to brown dwarfs and stellar companions, not to planets). This diagram is scaled for a 0.4 $M_\odot$ primary at 9 pc. The radial-velocity detectability limits correspond to 4 σ (50 m s^{-1}) minimum amplitudes and a 25 year maximum period. Some longer-period binaries will be detected, but the observing program is unlikely to last long enough to determine their orbital elements.

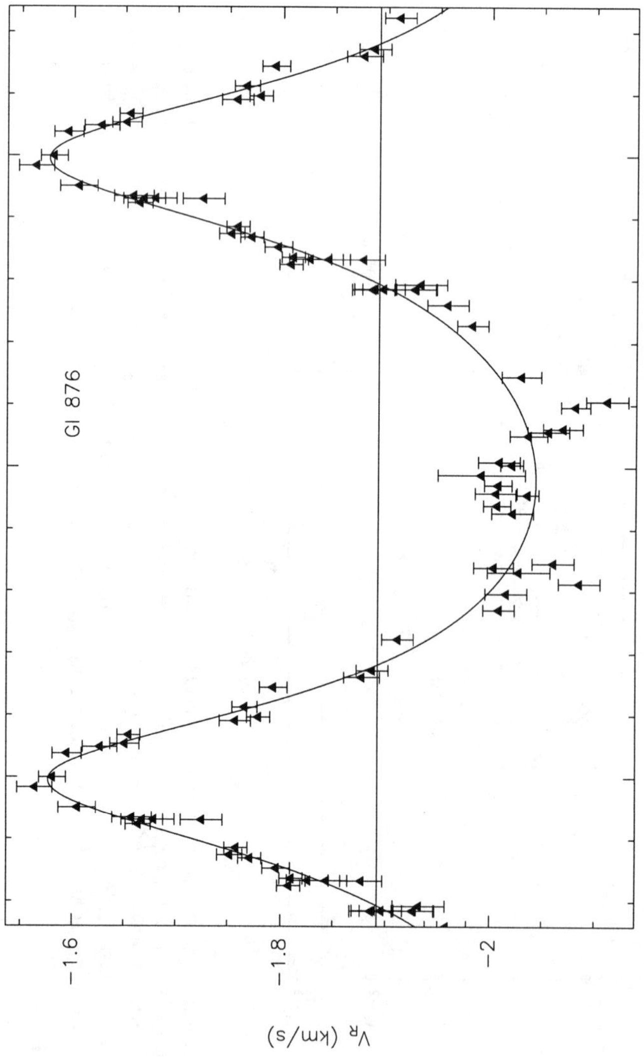

Figure 3. Our radial-velocity curve for Gl 876 (M4 V). The Gl 876 companion remains to date the only example of a planet around a very low-mass star.

To constrain planetary formation models, one needs *statistics* on very low-mass star companions. For this purposes we continue our observations at Observatoire de Haute Provence: a few objects have possible radial-velocity variations (Figure 1), but need more measurements and a longer timebase for confirmation and orbit determination.

Most importantly, since 1999 November we have started observing a larger sample with better accuracy, using the FEROS spectrograph on the 1.5 m ESO telescope. Improvements relative to the MELODIE program are:

- Exposure times are defined for planet detection, while the MELODIE program was optimized for stellar and brown-dwarf companion detection.
- FEROS has high throughput and its spectral coverage extends to 920 nm, providing an excellent match for the faint, red, M-dwarfs.
- We observe 200 M-dwarfs for planet detection, instead of $\sim$ 40 for the MELODIE program (one third of the total sample).

References

Baranne A., et al. 1996, A&AS, 119, 373

Beuzit J.-L., et al. 2000, in preparation for A&A

Delfosse, X., Forveille, T., Mayor, M., Burnet, M., & Perrier, C. 1998a, A&A, 341, L63

Delfosse, X., Forveille, T., Perrier, C., & Mayor M. 1998b, A&A, 331, 581

Delfosse, X., Forveille, T., Mayor, M., Perrier, C., Naef, D., & Queloz, D. 1998c, A&A, 338, L67

Delfosse, X., Forveille, T., Beuzit, J.-L., Udry, S., Mayor, M., & Perrier, C. 1999a, A&A, 344, 897

Delfosse, X., Forveille, T., Udry, S., Beuzit, J.-L., Mayor, M., & Perrier C. 1999b, A&A, 350, L39

Delfosse, X., Forveille, T., Ségransan, D., Beuzit, J.-L., Udry, S., Perrier, C., & Mayor, M. 2000, in preparation for A&A

Forveille, T., et al. 1999, A&A, 351, 619

Marcy, G. W., Butler, R. P., Vogt, S. S., Fischer, D., & Lissauer, J. J. 1998, ApJ, 505, L147

Ségransan, D., et al., 2000, in preparation for A&A

Saar, S. H., Butler, R. P., & Marcy, G. W. 1998, ApJ, 498, L153

Disks, Planetesimals, and Planets
ASP Conference Series, Vol. 219, 2000
F. Garzón, C. Eiroa, D. de Winter, and T. J. Mahoney, eds.

G Dwarf Planet Search

David W. Latham
Harvard-Smithsonian Center for Astrophysics, 60 Garden Street, Cambridge, Massachusetts, 02138, USA

Abstract. In 1997 August we made the first observations for our G Dwarf Planet Search using the High Resolution Echelle Spectrograph (HIRES) with its iodine gas absorption cell on Keck 1 under NASA time. As of 2000 February we had been assigned 21 nights and had accumulated 4180 stellar spectra of more than 900 targets chosen from a sample of nearby dwarfs with spectral types between late F and early K. Our strategy is to carry out an initial reconnaissance of approximately 1000 stars to identify candidate systems for follow-up studies. An innovative approach to the analysis of the HIRES spectra by team members in Tel Aviv, led by T. Mazeh, has achieved a precision of about 10 m s^{-1} per velocity observation as judged by the median rms for the stars with multiple observations. Follow-up observations of interesting candidates are under way.

1. Introduction

Much of the progress in the search for extrasolar planets over the past ten years has resulted from radial-velocity observations of solar-type stars. Perhaps the best way to keep up to date in this rapidly developing field is to consult the website maintained by Jean Schneider (2000).

Radial-velocity searches for extrasolar planets are now in a reconnaissance phase, where the velocities of more than two thousand solar-type stars are being monitored with a precision of 10 m s^{-1} or better by several different teams. Over the next few years we can expect these surveys to uncover several dozen new giant planets. This should help clarify the overall characteristics of the population of giant planets orbiting solar-type stars. These characteristics are now just beginning to emerge and are poorly defined because of the small number of orbits determined so far and because of selection effects that are difficult to model.

The planet candidates discovered so far have produced a number of surprises. Many are more massive than Jupiter, by factors as large as 5 or 10, and their orbits can be as small as 0.04 AU. The wider orbits are mostly non-circular, with eccentricities up to even 0.7. Upsilon Andromedae has a system of three giant planets, the outer two with eccentric orbits that are barely stable. The fact that none of these features were predicted underscores the importance of studying large populations.

In one case, HD 209458, the discovery of photometric transits (Charbonneau et al. 2000; Henry et al. 2000) by the unseen companion orbiting the star has provided dramatic confirmation that the companion really is a giant planet with properties similar to the gas giants in our own Solar System (Mazeh et al. 2000). This discovery has opened a new era of physical studies of extrasolar planets. The transits allow us to determine the exact mass, size, bulk density, and surface gravity of the planet. As these numbers get refined, it will be possible to determine whether the planet has a rocky high-density core or not, a critical issue for deciding whether core accretion or disk instability was the formation mechanism. Future observations may even be able to tell us something about the composition of the planet's atmosphere. These are remarkable developments indeed and they underscore the need for finding more examples of transiting planets.

In some cases it is possible to derive the orbital inclination from other information. An example is the astrometric orbit derived from the *Hipparcos* intermediate data for HD 10697 (Zucker & Mazeh 2000), a radial-velocity planet candidate discovered by Vogt et al. (2000). Although the minimum mass reported for the companion of HD 10697 is 6.4 $M_{\rm Jup}$, the orbit proves to be nearly face on, and the actual mass is about 38 ± 13 $M_{\rm Jup}$ (Zucker & Mazeh 2000), which moves this planet candidate into the brown-dwarf regime. However, the inclinations of most of the spectroscopic orbits for planetary candidates will remain unknown until astrometric space missions such as *FAME* and *SIM* can be brought to bear on the problem.

2. G Dwarf Planet Search

Since 1997 August we have been using HIRES with its iodine gas absorption cell on Keck 1 under NASA time to monitor the radial velocities of approximately 1000 G-dwarfs with a precision of 10 m s^{-1} or better. Our target list was drawn from a sample of 1840 northern stars identified by E. Olsen based on his extensive Strömgren photometry (Olsen 1983, 1994) of stars in the Henry Draper Catalogue. The $b - y$ color was limited to the range 0.340 to 0.505, the δc_1 index to 0.120, and the photometric distance to 60 pc. Subsequently, when the *Hipparcos* Catalogue (ESA 1997) became available, we eliminated stars with parallaxes smaller than 10 mas. We also eliminated stars with known visual companions with separations smaller than about 4 as.

The radial velocities of the stars in the parent sample have been monitored for more than ten years with the CORAVELs (Mayor 1985) and CfA Digital Speedometers (Latham 1985, 1992), and rotational velocities have also been determined. Based on the data accumulated so far for our G Dwarf Binary Survey (more than 50 000 velocities with precisions mostly in the range 200 to 600 m s^{-1}), we have already derived orbital solutions for more than 320 spectroscopic binaries and have identified a similar number of additional binary candidates (Udry et al. 1997). Thus, we have identified a large fraction of those stars in our sample where the presence of a stellar companion would have made it difficult for planets to form or survive in orbits with periods shorter than 10 or 20 years, and these stars have been eliminated from the target list for our Keck observing.

Our strategy has been to carry out an initial reconnaissance with HIRES of a sample of more than 1000 stars, with the immediate goal of identifying those systems which are orbited by substellar companions with periods shorter than a few years. For this reconnaissance we have increased the number of target visits by conceding velocity precision, and we have chosen to expose only until the signal-to-noise ratio exceeds the level needed to achieve a precision of 10 m s^{-1}. After we have identified a target as an interesting candidate, we take longer exposures for subsequent observations of that target, with the goal of deriving better orbits, and ultimately to look for the orbits of additional planets in the same system, but with lower amplitudes and/or longer periods. We also supplement our Keck observations of interesting candidates with precise radial velocities obtained with the AFOE (Brown et al. 1994), ELODIE (Baranne et al. 1996), and CORALIE (Queloz et al. 2000). This allows us to get better phase coverage and to derive orbital solutions in a timely manner for systems with short periods.

2.1. Status Report

This project has been scheduled for 21 nights, starting in 1997 August. As of 2000 February we had obtained a total of 4180 stellar spectra (this includes template spectra taken without the iodine cell and nightly spectra of rapidly rotating B-stars, which we use in our reductions). The team led by Prof. Tsevi Mazeh at Tel Aviv University has taken the lead role in the data reductions. Gordon Drukier, a postdoc, and Shay Zucker, a PhD student, have played key roles in this effort. The Tel Aviv team has developed an innovative approach to the derivation of precise velocities based on the two-dimensional correlation technique TODCOR (Mazeh & Zucker 1994).

The results from a global analysis of 803 stars with two or more iodine observations is shown in Figure 1, where we plot a histogram for the number of stars as a function of the standard deviation of the velocity residuals on a star-by-star basis. Notice that there is a tail of targets with larger velocity residuals than expected, assuming that the residuals are characterized by a Gaussian that fits the main body of residuals near 10 m s^{-1}. These are the interesting candidates that we are now following up.

2.2. HD 209458

An especially interesting example of a planetary companion that we discovered with our HIRES data is HD 209458b. Even after our first observing run in 1997 August we could see that the velocity of HD 209458 was variable, and by 1999 June we had enough observations to solve for an orbit. Additional observations were obtained by our Swiss collaborators, who had also been observing this same star. In 1999 August the identity of HD 209458 and its orbital elements were provided to D. Charbonneau and T. Brown so that they could look for transits with the STARE photometric instrument (Brown & Kolinski 2000). Two transits were successfully observed in 1999 September (Charbonneau et al. 2000). These transit observations, together with our orbital solution based on the combined HIRES, ELODIE, and CORALIE data, allowed us to determine directly the mass, radius, and density of the planet (Mazeh et al. 2000). This analysis depends critically on estimates for the mass, radius, and limb-darkening of the

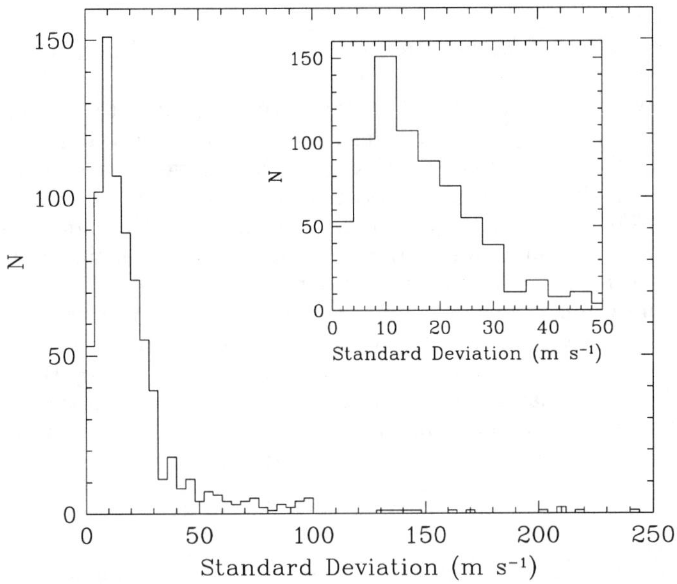

Figure 1. Histogram of the velocity residuals for 803 stars with multiple HIRES observations.

parent star, which in turn depend critically on the metallicity of the star. We used our HIRES template spectrum of HD 209458 to derive the iron abundance and effective temperature, and found good agreement between the stellar mass and radius derived from comparisons with four different sets of evolutionary models from the literature. The independent discovery in 1999 November of the planetary orbit and the detection of a planetary ingress are reported by Henry et al. (2000).

The case of HD 209458 is especially significant because the mass ($0.69 \pm 0.05 M_{\rm Jup}$), radius ($1.4 \pm 0.17 R_{\rm Jup}$), and density ($0.31 \pm 0.07$ g cm^{-1}) that we derive prove that this planet really is a gas giant, with just the properties expected for a planet that is heated by the parent star (Guillot 1999). Furthermore, this discovery has ignited a lot of excitement about follow-up studies that can now be contemplated, such as observations of the absorption lines from the planetary atmosphere as it is illuminated by the parent star during transit.

Acknowledgments. Our Keck G Dwarf Planet Search is a large collaboration involving many team members contributing in many different ways. The selection of the initial sample of G dwarfs with photometric distances less than 60 pc was based on E. H. Olsen's heroic effort to measure Strömgren photometry of tens of thousands of F- and G-stars in the Henry Draper Catalogue. Then the radial velocities were monitored to identify stars with stellar companions in orbits with periods shorter than about 10 or 20 years. Two teams contributed to this massive effort: M. Mayor, S. Udry, J. Andersen, B. Nordström, and others with the CORAVELS, and DWL, R. P. Stefanik, G. Torres, R. J. Davis, and others with the CfA Digital Speedometers. All the observing with HIRES for this project has been planned and carried out by DWL and G. Torres. A team led by T. Mazeh at Tel Aviv University has taken the lead role in the data reductions. G. A. Drukier and S. Zucker have played key roles in this effort. The Tel Aviv team has developed an innovative approach to the derivation of precise velocities based on the two-dimensional correlation technique TODCOR (Mazeh & Zucker 1994). Follow-up observations have been carried out with ELODIE and CORALIE by M. Mayor, S. Udry, D. Queloz, D. Naef, and others, and with AFOE by R. W. Noyes, S. G. Korzennik, and others. Photometric monitoring has been undertaken by S. L. Baliunas, R. Donahue, and W. Soon, and others, to look for evidence of rotation and chromospheric activity. Photometric monitoring to look for transits has been undertaken by T. M. Brown and D. Charbonneau and others. The analysis of the stellar metallicities, effective temperatures, and surface gravities has been undertaken by B. W. Carney, J. B. Laird, L. Buchhave, and others, both with spectra from the CfA Digital Speedometers and with the template spectra from HIRES. R. A. Brown has been a continuing source of advice and encouragement.

References

Baranne, A., et al. 1996, A&AS, 119, 373

Brown, T. M., & Kolinski, D. 2000, www.hao.ucar.edu/public/research/stare/stare.html

Brown, T. M., Noyes, R. W., Nisenson, P., Korzennik, S., & Horner, S. 1994, PASP, 106, 1285

Charbonneau, D., Brown, T. M., Latham, D. W., & Mayor, M. 2000, ApJ, 529, 45L

ESA 1997, The Hipparcos and Tycho Catalogues, European Space Agency, SP-1200 (Noordwijk: ESA Publications Division: ESTEC)

Guillot, T. 1999, Sci, 286, 72

Henry, G. W., Marcy, G. W., Butler, R. P., & Vogt, S. S., 2000, ApJ, 529, L41

Latham, D. W., 1985, in IAU Colloq. 88, Stellar Radial Velocities, ed. A. G. D. Philip & D. W. Latham (Schenectady: L. Davis Press), p. 21

Latham, D. W. 1992, in ASP Conf. Ser., Vol. 32, IAU Colloq. 135, Complementary Approaches to Double and Multiple Star Research, ed. H. A. McAlister & W. I. Hartkopf (San Francisco: ASP), p. 158

Mayor, M. 1985, in IAU Colloq. 88, Stellar Radial Velocities, ed. A. G. D. Philip & D. W. Latham (Schenectady: L. Davis Press), p. 35

Mazeh, T., et al. 2000, ApJ, 532, L55

Mazeh, T., & Zucker, S. 1994, ApJ, 420, 806

Olsen, E. H. 1983, A&AS, 54, 55

Olsen, E. H. 1994, A&AS, 106, 257

Schneider, J. 2000, the original site is http://www.obspm.fr/encycl/encycl.html; a mirror site can be found at http://www-cfa.harvard.edu/planets/

Queloz, D., et al. 2000, A&A, 354, 99

Udry, S., et al. 1997, in ASP Conf. Ser., Vol. 154, Tenth Cambridge Workshop, Cool Stars, Stellar Systems, & the Sun, ed. R. A. Donahue & J. A. Bookbinder (San Francisco: ASP), p. 2148

Vogt, S. S., Marcy, G. W., & Apps, K. 2000, ApJ, submitted

Zucker, S., & Mazeh, T. 2000, ApJ, 531, L67

Disks, Planetesimals, and Planets
ASP Conference Series, Vol. 219, 2000
F. Garzón, C. Eiroa, D. de Winter, and T. J. Mahoney, eds.

A Planetary Companion around the Young G6 Dwarf Gliese–Jahreiss 3021

D. Naef, M. Mayor, F. Pepe, D. Queloz, N. C. Santos, S. Udry, and M. Burnet

Observatoire de Genève, CH-1290 Sauverny, Switzerland

Abstract. We present here a new planetary companion detected with the CORALIE spectrograph mounted on the 1.2 m Leonard Euler Swiss Telescope at ESO La Silla Observatory (Chile). It is the fifth detection of the Geneva Southern Planet Survey (Udry et al. 2000). From our radial-velocity measurements, we infer an orbital period of 133.8 days and a semi-amplitude of 164 m s^{-1}. The orbit is eccentric ($e = 0.5$). The inferred minimum mass of the companion is $m_2 \sin i = 3.32\ M_{\rm Jup}$.

1. Gliese–Jahreiss 3021: Basic Properties

Gliese–Jahreiss 3021 (GJ 3021, HD 1237, HIP 1292) is a nearby dwarf (17.62 pc away) in the constellation of Hydrus. The basic stellar properties of GJ 3021 are summarized in Table 1. More details about this star and its planet will be presented in Naef et al. (2000).

Table 1. Basic data about GJ 3021

Parameter	Value	Reference
Spectral type	G6 V	*Hipparcos* (ESA 1997)
Parallax (mas)	56.76	*Hipparcos* (ESA 1997)
Distance (pc)	17.62	–
m_V	6.59	*Hipparcos* (ESA 1997)
M_V	5.36	–
$(B - V)$	0.749	*Hipparcos* (ESA 1997)
Mass ($M_\odot$)	0.9	–
$V \sin i$ (km s^{-1})	5.5	CORAVEL[a]
[Fe/H]	$\simeq 0.2$	CORAVEL[b]
$\log(R'_{HK})$	-4.44	Henry et al. (1996)
L_X (erg s^{-1})	$103.6 \cdot 10^{27}$	Hünsch et al. (1998)

[a] Using CORAVEL cross-correlation dip width and the calibration by Benz & Mayor (1984).
[b] Using CORAVEL cross-correlation dip surface and the calibration by Pont (1997).

Using Flower's bolometric correction calibration (Flower 1996) and *Hipparcos* data, we can compute the effective temperature and the luminosity of GJ 3021:

$$T_{\rm eff} = 5417 \text{ K}$$
$$L = 0.66\ L_{\odot}$$

With the chromospheric activity indicator, $\log(R'_{HK})$, we can infer estimations for the rotational period (using the calibration by Noyes et al. 1984) and the stellar age (with the calibration by Donahue 1993):

$$P_{\rm rot} = 10.4 \text{ days}$$
$$\text{age} = 0.6 \text{ Myr}$$

The high value of the activity indicator, the high value of the X-ray luminosity, and the rather high value of the rotational velocity suggest that GJ 3021 is a very young star. This interpretation is corroborated by the detection of a strong lithium feature in the stellar spectrum (see Figure 1).

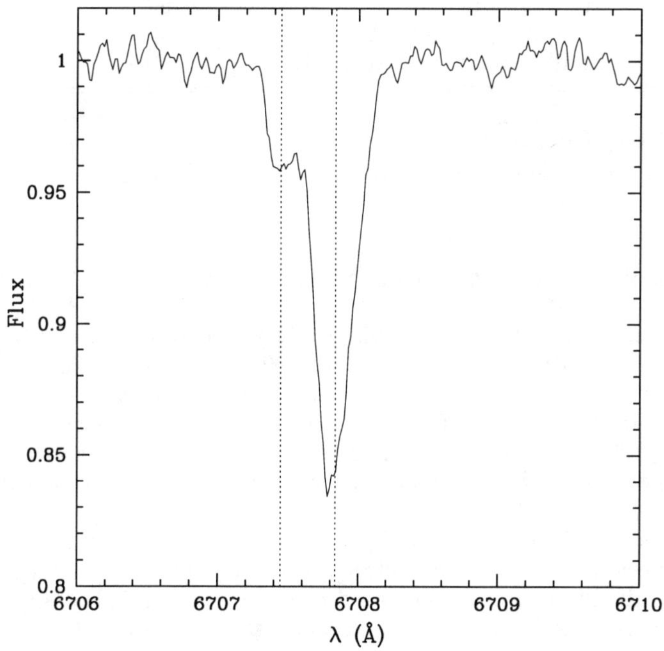

Figure 1. 6707.8 Å lithium absorption feature of GJ 3021. This spectrum is the result of the addition of 39 CORALIE spectra. The signal-to-noise ratio in the continuum is more than 300. The 6707.4 Å iron absorption feature is also clearly visible.

2. Radial-Velocity Data and Orbital Solution

Fifty radial-velocity measurements of GJ 3021 were obtained with the CORALIE echelle spectrograph between HJD = 51128 and HJD = 51564. The mean photon-noise error of these measurements is 8 m s^{-1}. Table 2 shows the results of the Keplerian orbit fitted to the data. Figure 2 shows the phase-folded curve and the velocities as a function of time for GJ 3021.

Table 2. Orbital solution for GJ 3021

Period (days)	133.82 ± 0.20
T (Julian Date)	51412.48 ± 0.62
Eccentricity	0.505 ± 0.018
γ velocity (km s^{-1})	-5.808 ± 0.003
ω (°)	291 ± 3
K (m s^{-1})	164 ± 4
$a_1 \sin i$ (Gm)	0.260 ± 0.007
$f_1(m)$ (10^{-8} $M_\odot$)	3.92 ± 0.31
m_2 sini ($M_{\rm Jup}$)	$3.32^\dagger \pm 0.1^\ddagger$
$N_{\rm meas}$	50
$\sigma(O-C)$ (m s^{-1})	18.5
Orbital separation (AU)	0.49
Equilibrium temperature (K)	$\simeq 315$

† Assuming a primary mass $m1 = 0.9\ M_\odot$.
‡ Error on $m1$ not taken into account.

3. Extrasolar Planets Orbiting Active Stars

It is now well known that chromospherically active stars can exhibit intrinsic radial-velocity variations (Saar & Donahue 1997; Santos et al. 2000b). Intrinsic processes correlated with the stellar rotational period such as spots can produce short-term variations. Long-life spots locked on the stellar surface can produce an almost perfectly periodic radial-velocity signal that can mimic the signal induced by a planetary companion as, for example, for HD 166435 (Queloz et al. 2000).

In the case of GJ 3021, the obtained period is much longer than the expected rotational period for such a high activity level. The planetary interpretation is thus beyond doubt. Intrinsic processes are nevertheless probably responsible for the abnormally high scatter observed around the fitted orbit ($\sigma(O-C) = 19$ m s^{-1} to be compared with the typical CORALIE precision of 7–8 m s^{-1}). It has recently been shown (Saar et al. 1998; Santos et al. 2000b) that the extra radial-velocity scatter depended on the spectral type of the target. A smaller scatter is expected for active late G dwarfs than for early active G dwarfs. For GJ 3021, the extra scatter is of the order of 15 m s^{-1}. For cooler active stars, the observed extra scatter is even smaller. For example, HD 130322 and HD

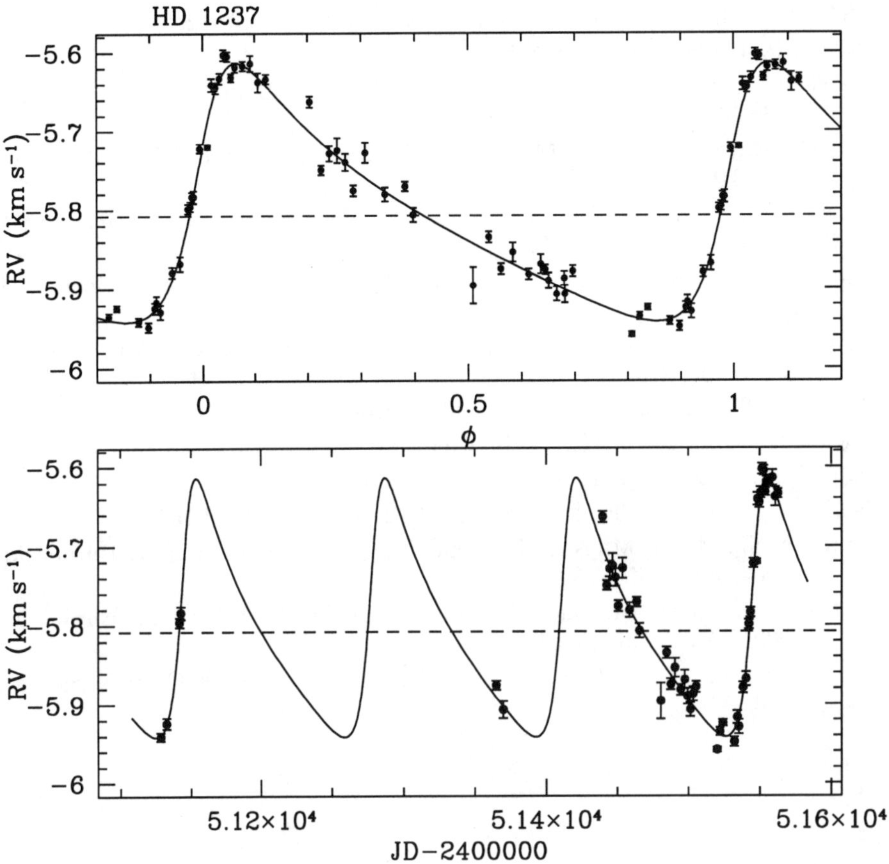

Figure 2. CORALIE radial-velocity data: phase-folded curve (*top*) and velocities versus time (*bottom*).

192263 (both stars are active K0 dwarfs with planetary companions) exhibit an extra jitter below 10 m s^{-1} (Udry et al. 2000; Santos et al. 2000a).

Intrinsic processes have very short timescales so it remains possible to detect long-period massive planets orbiting active stars (see, for example, the planet found around the active G0 dwarf ι Hor; Kürster et al. 2000). The small observed scatter around late-type active stars makes it possible to detect even short-period planets around them (the planet around HD 130322 has a period of 10.72 days). Active stars remain good candidates for planet searches (at least the coolest ones).

References

Benz, W., & Mayor, M. 1984, A&A, 138, 183

Donahue, R. A. 1993, PhD Thesis, New Mexico State University

ESA 1997, The Hipparcos and Tycho Catalogues, ESA SP-1200

Flower, P. J. 1996, ApJ, 469, 355

Henry, T. J., Soderblom, D. R., Donahue, D. A., & Baliunas, S. L. 1996, AJ, 111, 439

Hünsch, M., Schmitt, J. H. M. M., Sterzik, M. F., & Voges, W. 1998, A&AS, 135, 319

Kürster, M., Endl, M., Els, S., Hatzes, A., Cochran, W., Dobereiner, S., & Dennerl, K. 2000, A&A, 53, L33

Naef, D., Mayor, M., Pepe, F., Queloz, D., Santos, N. C., Udry, S. & Burnet, M. 2000, A&A, in preparation

Noyes, R. W., Hartmann, L. W., Baliunas, S. L., Duncan, D. K., & Vaughan, A. H. 1984, ApJ, 279, 763

Pont, F. 1997, PhD Thesis, Geneva Observatory

Queloz, D., et al. 2000, A&A, submitted

Saar, S. H., & Donahue, R. A. 1997, ApJ, 485, 319

Saar, S. H., Butler, R. P., & Marcy, G. W. 1998, ApJ, 498, L153

Santos, N. C., Mayor, M., Naef, D., Pepe, F., Queloz, D., Udry, S., & Burnet, M. 2000a, A&A, 356, 599

Santos, N. C., Mayor, M., Naef, D., Pepe, F., Queloz, D., Udry, S., & Burnet, M. 2000b, A&A, submitted

Udry, S., Mayor, M., Naef, D., Pepe, F., Queloz, D., Santos, N. C., & Burnet, M. 2000, A&A, 356, 590

Disks, Planetesimals, and Planets
ASP Conference Series, Vol. 219, 2000
F. Garzón, C. Eiroa, D. de Winter, and T. J. Mahoney, eds.

First Results of the EMILIE Spectrograph with the AAA System on Extrasolar Planets

J. Schmitt, F. Bouchy, and J. L. Bertaux

Service d'Aéronomie du CNRS, BP No 3, 91371 Verrières le Buisson, France

Abstract. The method of stellar radial-velocity variation measurements has recently shown its capability by the discovery of tens of extrasolar planets. Accuracies achieved today are in the range 3–10 m s^{-1}. The Absolute Astronomical Accelerometer (AAA) is an instrument which aims at reaching the photon noise for the variation velocity measurement, with a target accuracy of less than 1 m s^{-1}, long term. The principle is to use a servo-controlled CCD spectrograph used in a zero-mode detector, in order to register the spectrum of the star always on the same CCD pixels.Thus, systematic errors linked to large Earth-induced variations are cancelled out. A tunable Fabry–Perot also follows the star spectrum, and its thickness is measured with heterodyne detection of the beat of a tunable laser diode with a stabilized laser diode. The instrument is now implemented at the 152 Coudé Telescope of Observatoire de Haute Provence.The first results on the Sun (where 5 min solar oscillations are readily observed) and one star are presented. The presence of a planet around Upsilon Andromedae is confirmed, with a period similar to that derived from Marcy and Butler, and also a similar phase. The semi-amplitude is slightly larger (in 1999 November) than the previously reported 74 m s^{-1}.

1. The Principle of the AAA

The AAA method aims at achieving a high accuracy and long-term stability in the measurements of stellar v_r (radial velocity) in order to detect an orbiting planet around the star by the reflex motion.

While the claimed accuracies of the best spectrometer extrasolar planets survey are in the range 3 to 10 ms^{-1} (i.e., Butler & Marcy 1996), no discovery was announced so far below an amplitude of about 55 m s^{-1}. The targeted accuracy of the method of absolute accelerometry is 1 m s^{-1}. Speculating that the planets discovered so far are only the tip of the iceberg, it would mean that maintaining such an accuracy over a long period of time would yield the discovery of many more extrasolar planets, with masses significantly lower than that of Jupiter. The purpose of this paper is to describe briefly the principle of the Astronomical Absolute Accelerometer and to present the first astronomical results on stars and on the Sun.

Described earlier by Connes (1984), the AAA method is intended to make the best possible use of stellar photons by attempting to eliminate the systematic errors usually inherent in standard spectrometers. The principle is outlined in Figure 1.

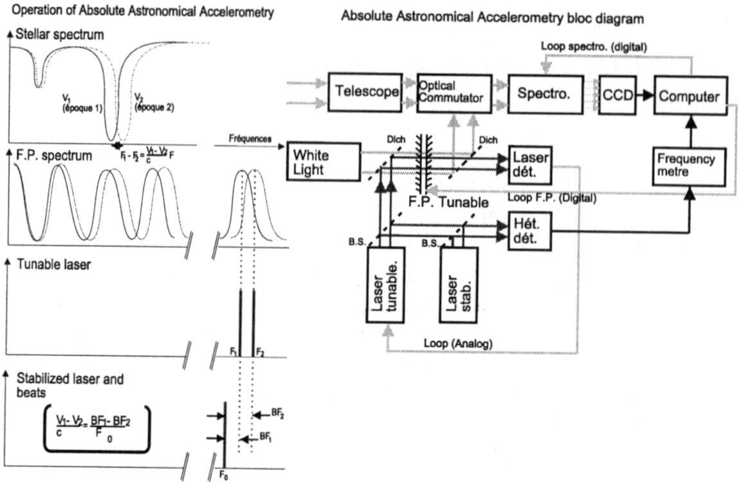

Figure 1. Principle of the AAA system associated with the EMILIE spectrograph. *Right:* The various elements of the AAA are schematized, with servo-control loops indicated. The servo-loop which controls the laser frequency is of analog type, while the others are digitally controlled through the main computer. *Left:* Schematized the stellar spectrum , the channeled spectrum and the beating of the two laser diodes (see text).

AAA makes use of a high-resolution fiber-fed spectrometer named EMILIE, which records a large portion of the stellar spectrum on one CCD, thanks to a cross-disperser. This spectrometer is used as a detector of the -changes in v_r, induced both by a potential planetary companion, and by the much larger Earth velocity changes (up to $\pm$ 30 km s^{-1}). A tilting glass plate inside the spectrometer is moved until the spectrum is recorded on the same pixels of the CCD detector at all epochs. Therefore, the spectrometer is servo-controlled on the radial velocity. v_r, of the star. Clearly, a very small displacement (typically 1/1000 of a pixel = 1.4 m s^{-1}) may be measured much more accurately than if the stellar spectrum were left to drift several tens of CCD pixels of uncontrolled widths and spacings.

Similary, a reference channeled spectrum, obtained with white light through a tunable Fabry–Perot, is recorded through the same optical path as the stellar spectrum, and in alternation with the stellar spectrum. The finesse of the FP must be small, to cover the whole CCD with a signal oscillating with wavelength (with a typically 4 to 6 pixel periodicity). The thickness of the FP is adjusted with piezo-electric spacers in such a way that the channeled spectrum will also always remain on the same pixels as during the reference epoch. Changing the thickness of the FP results in a stretching of the white-light channeled spectrum, similar to the wavelength changing of the stellar spectrum when v_r is changing (indeed, a Doppler stretch, $\Delta\lambda/\lambda = \Delta v_r/c$, rather than the improperly worded Doppler "shift"). The changes in thickness of the tunable Fabry–Perot are measured with a tunable laser diode, whose frequency is servo-controlled on the FP thickness. This frequency, F, is in turn measured by heterodyne mixing with another tunable laser diode, whose frequency, F_0, is stabilized on an atomic caesium line, providing both the short-term and long-term absolute reference. The beating frequency, $BF = F - F_0$, is measured directly with an electronic frequency meter. The change in v_r between two epochs t_1 and t_2 when $v_r = v_1$ and $v_r = v_2$ is $\mathrm{V}_1 - v_2 = c(BF_1 - BF_2)/F_0$. The barycentric velocity of the Earth must then be subtracted to get the barycentric variations in v_r.

The chain of servo-control loops may be summarized below:

- The spectrometer is servo-controlled on the radial velocity of the star, v_r, with the tilting glass
- The thickness of the FP is servo-controlled on the spectrometer, and thus on v_r
- The frequency, F, of the tunable laser is servo-controlled on the thickness of the FP and thus on v_r
- The frequency F is measured with heterodyne mixing with the stabilized laser of frequency F_0, and therefore the beating frequency $BF = F - F_0$ is servo-controlled on v_r

No attempt is made to measure accurately the absolute value of the thickness of the Fabry–Perot (typically 6 mm), which would be necessary to provide an absolute wavelength reference equivalent to a set of atomic lines, allowing the absolute value of v_r to be measured. Rather, only changes in v_r with time are accurately tracked and measured, giving the name of accelerometry to the method.

The numerical algorithm which computes the small spectrum shift (used both for the FP and the star) makes use of the slopes of spectral features and associated variations of intensity on CCD pixels. It is much faster than a correlation algorithm, but can cope only with small displacements (a fraction of a pixel), which is enough since the spectrometer is used as a null detector. Actually, the small displacements are made with the tilting glass plate, while larger displacements needed to compare two measurements further away in time (larger v_r difference due to the Earth's orbital motion) are performed by moving the fiber output mechanically inside the spectrometer.

2. Observations of the Sun

Observations of the Sun were organized in 1999 July in order to test the combination spectrometer plus AAA. This was not an end-to-end test, since the 152

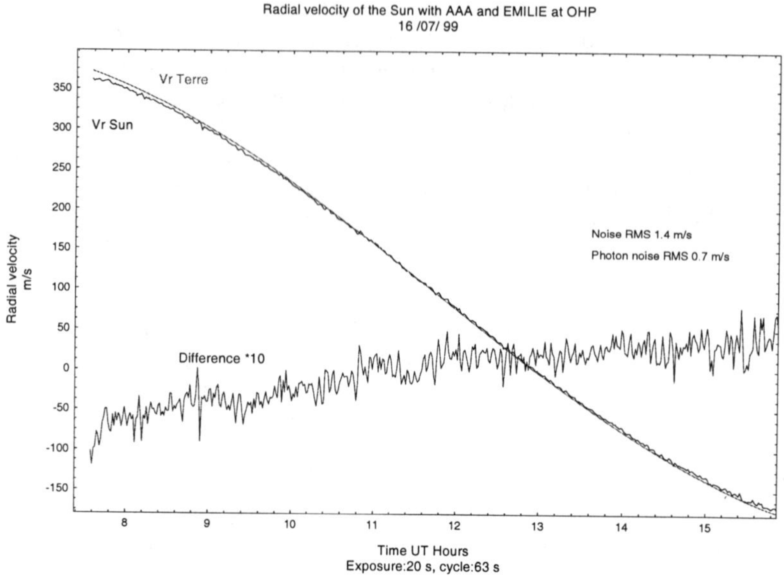

Figure 2. Variations in radial velocity of the Sun measured during one day. The continuous curve is the celestial-mechanics model prediction. The wiggly curve are the measurements, following the model. The difference is plotted after amplification by a factor of ten. Solar oscillations are clearly visible in the time series, with a period around 5 min.

cm telescope was not used. Instead, the instrumental set-up to collect the solar light was composed of two lenses increasing the solar flux by a factor 25, impinging on a diffuser plate. Behind the diffuser was placed the entrance of an optical fiber, 1 mm in diameter, the whole system being placed on a Celestron equatorial mount, to follow the Sun throughout the day. The output of the fiber was re-imaged on the entrance of the 50 μm fiber (with a large loss of light), feeding the EMILIE spectrometer. Observations were performed during 5–7 hours per day, on four consecutive days. The 80 s measurement cycle included one 20 s solar exposure, one 20 s Fabry–Pérot exposure, 20 s of computing and display, plus another 20 s overhead. The data reduction was restricted to 16 orders of the cross-dispersed spectra, of 25 Å each. The spectral dispersion was such that one 24 μm corresponded to 1416 m s^{-1}. But the spectral resolution was only 50 000 because the prism was with plane faces at this time (no aspheric face to compensate for spherical aberration of the spherical mirror of the spectrometer, which was included later).

In Figure 2 are displayed the v_r measurements of the Sun obtained during 8 hours (during another day than the sequence of four days analyzed further below). The main effect is the rotation and orbital motion of the Earth, which is computed (descending regular line) and subtracted from the measurements

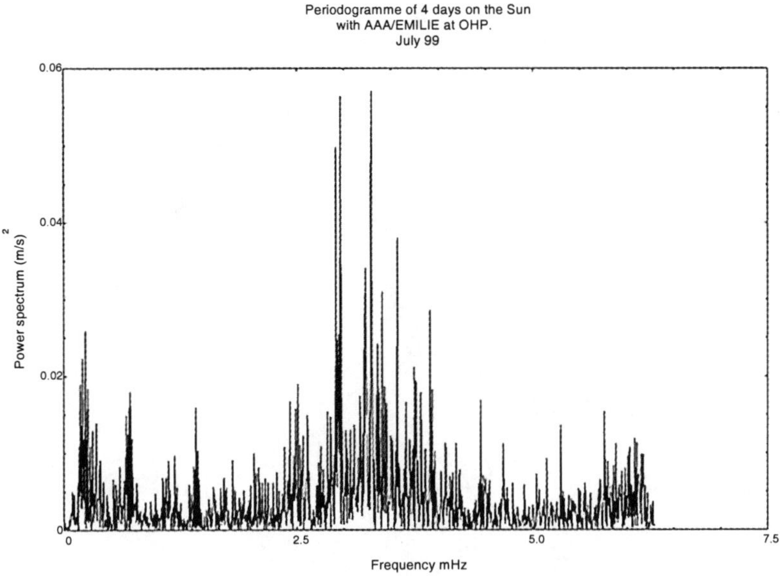

Figure 3. Periodogram of solar measurements obtained on four consecutive days. Peaks are conspicuous around the 5 min period.

(wiggly descending line following the model) to yield the v_r motion of the solar surface, plotted at the bottom of the figure and enlarged in vertical scale by a factor of ten. The 5 min solar oscillations are readily visible in the solar-surface time series of v_r. It is noted that there is a residual long term variation of solar surface v_r, which can be described as a slow increase from 8 to 12 hours UT, followed in the afternoon by a much slower increase. Possible causes may be ascribed to a combination of two effects: the difference of atmospheric transmission between the preceding and the receding solar limb (velocity 2 km s^{-1}), which contains a well known systematic trend throughout the day (from morning to afternoon, with a leveling-off around noon), and another effect depending on the Sun's elevation, with a maximum around noon. This effect could be due to telluric extinction, but it has not been studied in great detail up to now. When these slow variations are eliminated, the rms noise of the v_r measurements is computed to be 1.4 m s^{-1}, while the photon noise is estimated to be 0.7 m s^{-1}, considering the quality factor of the solar spectrum (Connes 1984). A substantial part of the difference may be ascribed to solar oscillations.

In Figure 3 is displayed a power-spectrum analysis of the series of four consecutive days of solar surface v_r measurements. The noise around 1 mHz is only 2.5 cm s^{-1}, while in the region where oscillations are present, their amplitude amounts to about 20 cm s^{-1}. While these results are significant as a proof of the quality of the instrumental combination of EMILIE spectrometer associated with the AAA calibration system, they also open the door to the possibility

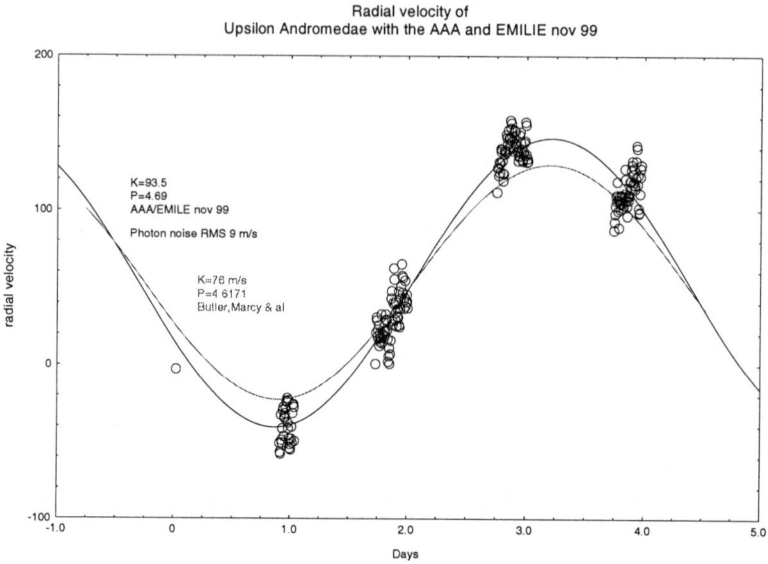

Figure 4. v_r measurements obtained on the star Ups And (HR 458) on five consecutive days (one point only for the first day, serving as a reference for the following points). The cloud of points is fitted by an orbital solution with an amplitude of 93.5 ms^{-1} and a period of 4.69 days (*thicker line*). The solution published by Butler et al. (1997) has a period of 4.6171 days, and a smaller amplitude of 76 m s^{-1} (*thinner line*; see text for comments).

of possibly greatly improving solar-oscillation studies from the ground with the new technique of "full" spectrum at high resolution and stability. Clearly, this type of investigation would also benefit from the use of a coordinated ground network of such instruments distributed in longitude, in order to avoid the day–night interruptions, as done already for the IRIS and GONG networks.

3. Observations of the Upsilon Andromedae Planetary System

As a bench test of end-to-end performance of the system for the search of extrasolar planets, we observed the star Upsilon Andromedae, known to have one planetary companion orbiting at 0.054 AU, with a period of 4.62 days and a mass $m \sin i = 0.61\ M_{\mathrm{Jup}}$ (Butler et al. 1997; Cummings et al. 1999). There are longer-term variations of the v_r curve that were interpreted as the signature of two other planets, forming the first exoplanetary system ever identified. But with only five days of observations in 1999 November, we could only try to check the signature of the planet closest to the star. We had only one measurement (taken as the reference spectrum) on the first night (November 22), and several tenths of short exposurers (5 min each) on each of the following days. All measurements are plotted in Figure 4 as a function of time. They were fitted by a forward model with orbital parameters as unknowns. The photon noise was estimated to correspond to 9 m s^{-1}.

Though the number of nights was very limited, in comparison with the measurements of the US group, the best-fit solution gave an excellent match with the determined phase and period. Our best-fit period was 4.69 days, as compared to 4.62 days established previously. In our data fit the semi-amplitude of the v_r curve was found to be $K = 93.5$ m s^{-1}, larger than the previously determined amplitude of 71.9 m s^{-1}. The mean square root difference of data to the fit is 12 m s^{-1}, showing a performance not much above the estimated photon noise (9 m s^{-1}).

The discrepancy between the two amplitudes is considerably larger than the one-sigma error in this parameter as derived from our fitting process (about 3 m s^{-1}). The reason of the discrepancy is not clear, and we cannot exclude some kind of systematic error unnoticed in our measurements or in our data-reduction procedure. However, the US individual v_r measurements were kindly provided to us by G. Marcy. We noted that, when fitting their data (one point per night) by series of four measurements as in our case, we found that the derived amplitude was also quite variable, from 63 to 95 m s^{-1} over the few weeks prior to our own measurements. It is unlikely that this is caused by the two other planets, whose orbits are too far away. Still, the high value found in our limited data is also found sometimes in their data. One possibility is that motions of the stellar surface or stellar spots contribute spuriously to the v_r signal. Clearly more observations should be carried out on this star.

4. Conclusion

Though the AAA system has not yet been proven to be able to reach a long-term accuracy of 1 m s^{-1} in the measurements of stellar v_r, the first results presented here on the Sun and on a known planetary system (Ups And) are very encouraging. We will continue to explore this technique further, which is quite different from the two other techniques that have proven successful up to now: absolute calibration with a thorium lamp spectrum recorded simultaneously on a second fiber (the first one to discover an extrasolar planet orbiting 51 Pegasi from Observatoire de Haute Provence, Mayor & Queloz 1995; Noyes et al. 1997), and absolute calibration by imprinting absorption features of an iodine cell through which is channeled the stellar spectrum before entering the spectrometer (Butler & Marcy 1996).

Acknowledgments. We wish to thank Jean-Pierre Sivan and his staff for his support at Observatoire de Haute Provence for the implementation of this new EMILIE spectrometer and its AAA system. We also wish to thank Bob Noyes and Tim Brown for useful discussions, as well as Geoffrey Marcy for the communication of his v_r measurements of Upsilon Andromedae and Jean-Claude Lebrun for support in data reduction and editing.

References

Butler, R. P., & Marcy, G. W. 1996, ApJ, 464, L153

Butler, R .P., Marcy, G. W., Williams, E., Hauser, H., & Shirts, P. 1997, ApJ, 474, L115

Connes, P. 1984, ApSS, 212, 357

Cummings, A., Marcy, G. W., & Butler, R. P. 1999, ApJ, 526, 890

Mayor, M., & Queloz, D. 1995, Nat, 378, 355

Noyes, R. W., Jha, S., Korzennick, S. G., Krockenberger, M., & Nisenson, P. 1997, ApJ, 483, 319

Disks, Planetesimals, and Planets
ASP Conference Series, Vol. 219, 2000
F. Garzón, C. Eiroa, D. de Winter, and T. J. Mahoney, eds.

Spectra, Lightcurves, and Polarization of the Close-In Extrasolar Giant Planets

S. Seager

Institute for Advanced Study, 1 Einstein Drive, Princeton, NJ, 08540, USA

Abstract. The discovery of 51 Peg b in 1995, only 0.05 AU from its parent star, heralded an unexpected, new class of planets. The recent transit detection of HD 209458b confirms that this class of EGPs—now with several members—are gas giants. Five times closer to their parent stars than Mercury is to our Sun, the CEGPs are highly irradiated by their parent stars. At effective temperatures of 1100 to 1600 K their atmospheres are closer to brown-dwarf and cool L-dwarf atmospheres than to those of our own Solar System planets. Because the CEGPs are hot and potentially bright planets, many observational programs are under way. Here I discuss the theoretical spectra, photometric lightcurves, and polarization curves of the CEGPs.

1. Introduction

To date, more than 30 extrasolar giant planets have been detected by the radial-velocity method. Several of these planets are $\lesssim$ 0.05 AU from their parent stars (CEGPs). The recent transit detection of HD 209458b in 1999 November (Charbonneau et al. 2000; Henry et al. 2000a) confirms that the CEGPs are gas giants and provides a planet radius $R_{\rm p} = 1.40 \pm 0.17\ R_{\rm Jup}$, mass $M_{\rm p} = 0.69 \pm 0.05\ M_{\rm Jup}$, and average density $\overline{\rho} = 0.31 \pm 0.07$ g cm^{-3} (Mazeh et al. 2000). For the first time we have unambiguous physical parameters of an extrasolar planet.

A CEGP and parent star are not spatially separable due to the small orbital distances, so EGP atmosphere detection and characterization requires extracting the planet signature from the combined star–planet light. The close-in EGPs are potentially the easiest to observe of all the known EGPs, because their proximity to the central star means they could have a high amount of optically reflected light, and they are hot thermally. In reflected light the planet could be as high as four orders of magnitude fainter than the parent star. With effective temperatures ($T_{\rm eff}$) $\sim$ 1100 to $\sim$ 1600 K, the CEGPs are only a few orders of magnitude fainter than the parent star in the infrared (see Figure 1). See Table 1 in Charbonneau & Noyes (this volume, p. 462) for the CEGP properties.

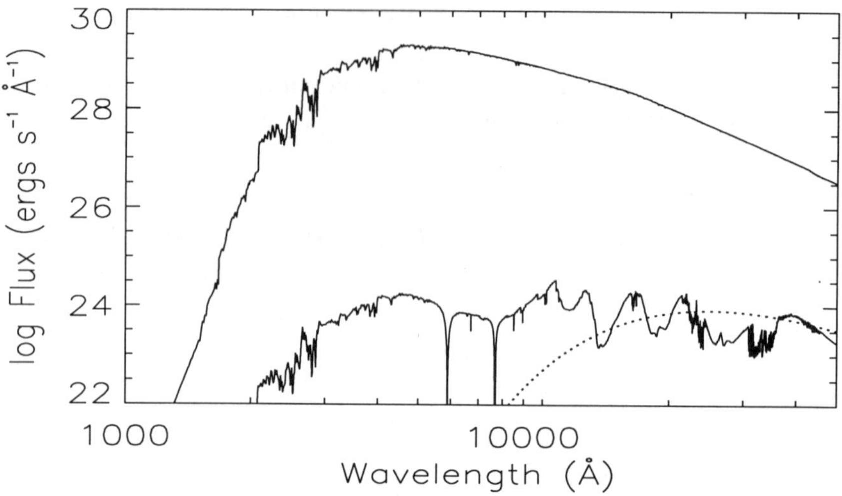

Figure 1. Flux of 51 Peg A (*upper solid curve*), 51 Peg b (*lower solid curve*), and a blackbody of 1170 K (*dotted curve*).

2. Spectra

Figure 1 shows the flux of 51 Peg A (*upper solid curve*) at the surface of the star and the flux of 51 Peg b at "opposition" (*lower solid curve*) at the surface of the planet. We compute the CEGP spectra and temperature–pressure profiles simultaneously with a 1D, plane–parallel model atmosphere code which solves the equations of radiative transfer, hydrostatic equilibrium (i.e., ignoring winds), and radiative and convective equilibrium. Chemical equilibrium (involving a few hundred species) is also computed, and gas opacities and condensate indices of refraction are taken from a variety of sources. The external radiation incident on the planet is the stellar flux, which is computed using the Kurucz model atmospheres (Kurucz 1992). The CEGPs are assumed to be tidally locked gas giants with rapid heat redistribution. Irradiation is important because when the planet is heated from the outside a shallower temperature–pressure profile and shallower and broader spectral features result as compared to an isolated planet of the same $T_{\rm eff}$ (Seager & Sasselov 1998). See Seager (1999) and Seager et al. (2000) for more details on the atmosphere model.

The spectrum of 51 Peg b in Figure 1 is only one example out of a large range of parameter space. This particular model has a homogeneous $MgSiO_3$ cloud with a Gaussian particle size distribution with mean radius ($\overline{r}$) 0.01 μm. The blackbody curve in Figure 1 has $T_{\rm eff}$ 1170 K, the same as the planet, illustrating that the optical flux is reflected, not thermally emitted, radiation. In 51 Peg b, the features in the blue and UV are reflected stellar features, the absorption features between 5000 Å and 1 μm are alkali-metal lines from the planetary atmosphere, and the absorption features at $\lambda > 1\mu$m are H_2O and CH_4.

The flux and temperature–pressure structure of the CEGP depends entirely on the type and size distribution of condensates in the planetary atmosphere because the specific absorption properties of the grains are crucial for determining the frequency and depth dependence of absorbed radiation. Different condensate assumptions can cause a flux variation in the optical of a few orders of magnitude, and can produce a difference in the shape and depth of the infrared molecular absorption bands. If enough stellar radiation is absorbed, the temperature will be too warm for CH_4, whose absorption bands will not be present in the spectra.

The wavelength region around 5000 Å of τ Boo has been probed via spectral separation (Charbonneau et al. 1999; Cameron et al. 1999). A probable detection by Cameron et al. suggests τ Boo b is very bright in scattered light. If confirmed this narrows the viable models considerably to those with highly reflective clouds, such as $MgSiO_3$, that are located high in the planetary atmosphere. See Charbonneau & Noyes (this volume, p. 461) for details.

2.1. Transmission Spectra

During an EGP transit, the planet passes in front of the star and occults the stellar flux in the amount equal to the ratio of the planet-to-star areas. During the transit, some of the stellar flux will pass through the optically thin part of the planet atmosphere. Some of the planet absorption features will be superimposed on the stellar flux at the 10^{-3} to the 10^{-4} level. Observations of these features will constrain the line-of-sight column density, temperature and pressure, which in turn will constrain cloud depth (see Figure 2). Observations of the alkali-metal lines will confirm the very basic postulate that the CEGP atmospheres are similar to those of brown dwarfs and cool L dwarfs.

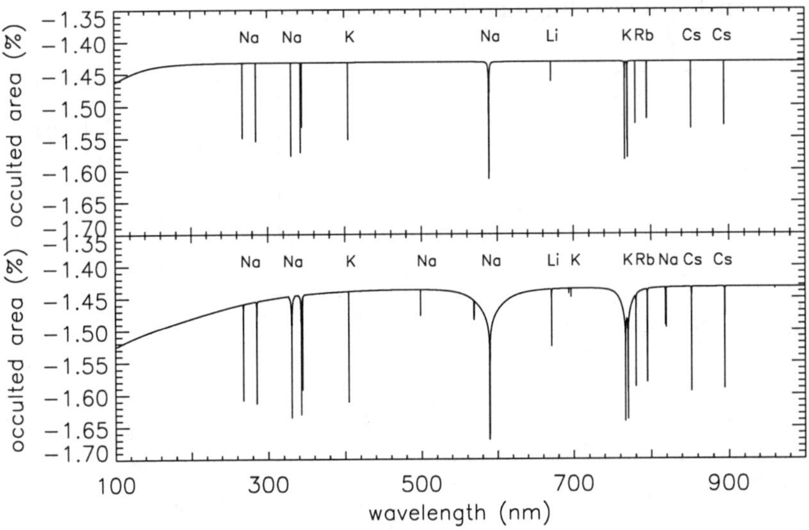

Figure 2. Theoretical transmission spectra of HD 209458b, shown as the percentage occulted area of the star. The *upper plot* is the case of a cloud base at 2.4 $\times 10^{-3}$ bar and the *lower plot* with a cloud base at 0.1 bar. See text for details.

Figure 2 shows the transmission spectra as the occulted area of the star at different wavelengths. In transmitted light the planet is different sizes at different wavelengths. The zero point (at $\sim 1.43\%$) is set by the atmosphere depth at which the planet is optically thick at all wavelengths. The planet appears largest in the line cores, where photons passing very high in the planet atmosphere are still absorbed along the line of sight due to the strong absorption by the Na I and K I resonance lines. The upper plot is for a model with a cloud at 2.4×10^{-3} bar. The lower plot is for a model with a cloud at 0.1 bar, where the stellar flux passes through higher pressures, densities, and temperatures of the planet atmosphere compared to the model in the upper plot. In addition, a larger transparent atmosphere makes the planet-to-star flux ratio larger. Observations should be able to distinguish between these models. Other lines (not shown), such as the He 10830 nm line and infrared molecular lines, will also be useful diagnostics. See Seager & Sasselov (2000) for more details.

3. Photometric Lightcurves

As the CEGP orbits its parent star, the planet changes phase as seen from Earth. The planet and star are too close together to be spatially resolved, but the optical flux from the combined system will vary with phase. With effective temperatures > 1000 K, the CEGPs have no optically emitted flux; the optical light is only scattered stellar light and will show a definite phase effect. Ground-based automatic photometric telescopes may reach a precision of 50 μmag (Henry et al. 2000b). Future space satellites such as *MOST* ($\sim$ 2002), *COROT* ($\sim$ 2004), and *MONS* ($\sim$ 2003), initially intended for asteroseismology, will be capable of detecting signals of several μmag.

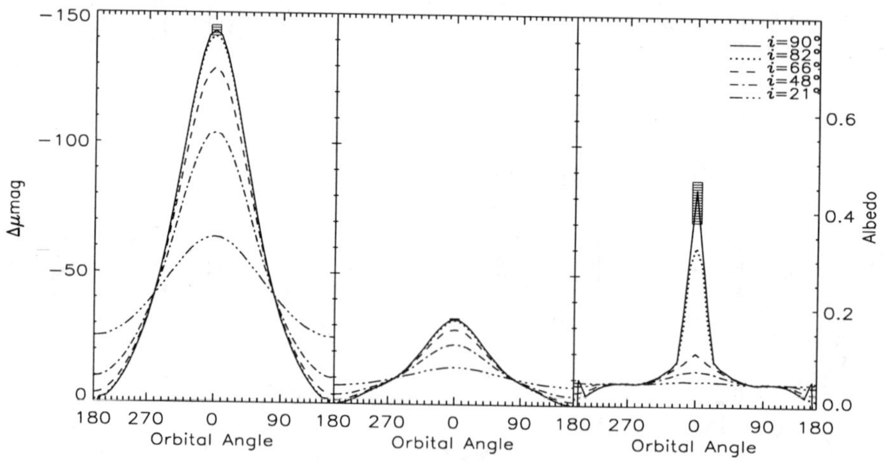

Figure 3. Photometric lightcurves of 51 Peg b at 5500 Å. The different figures represent different cloud particle types and sizes. Observations should be able to distinguish between these extreme cases.

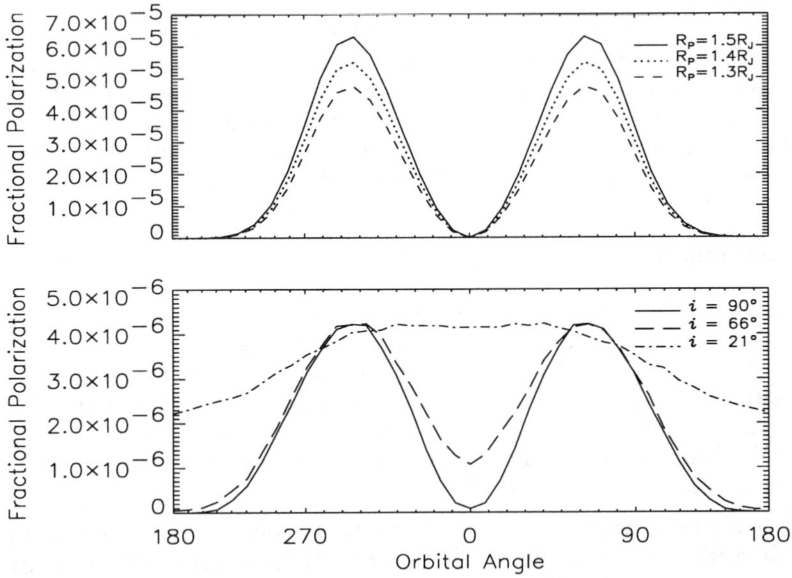

Figure 4. *Top:* Rayleigh scattering analytical estimate at 5500 Å of the fractional polarization of a CEGP at $D = 0.05$ AU. *Bottom:* Simulation of fractional polarization. See text for details.

Analytical estimates for a Lambert sphere, assuming that all stellar photons are singly isotropically scattered by the planet atmosphere, give an amplitude variation from 100 to 200 μmag (for an edge-on system), depending on the planet radius and orbital distance (D). In reality the amount of scattered light could be far lower when considering that grains absorb as well as scatter light, that grains scatter anisotropically, and that multiple scattering reduces the amount of emergent scattered light.

The simulations shown in Figure 3 are computed with a 3D Monte Carlo code (see Seager et al. 2000, and references therein) that uses the theoretical density profile of the planet, gas and grain opacities, and most importantly the scattering matrix elements of the condensates (i.e., the directional probability of scattering and polarization). The three different models in Figure 3 involve different assumptions of a homogeneous distribution of grain size and type. From left to right they are $MgSiO_3$ particles with $\overline{r} = 0.1$ μm, a mix of $MgSiO_3$–Fe–Al_2O_3 particles with $\overline{r} = 0.1$ μm, and the same mix of particles with $\overline{r} = 10$ μm. In all cases a Gaussian size distribution of particles was used with $\sigma = 0.1 \times \overline{r}$. While observations should be able to distinguish between these three extreme cases, extraction of specific particle type and size will be difficult. See Seager et al. (2000) for more details.

4. Polarization

Figure 4a shows a Rayleigh scattering analytical estimate of the fractional polarization of the total white light the CEGP system for a planet with $D = 0.05$ AU for different $R_{\rm p}$. Even the strongest signal in this figure is too small to be detected with current techniques which can reach a few $\times 10^{-4}$. In reality polarization should be smaller than this best-case estimate because multiple scattering, absorption, and interference effects weaken the signal. This is apparent in the simulation shown in Figure 4b, for a 51 Peg b model with a homogeneous $MgSiO_3$–Fe–Al_2O_3 cloud with particles with $\overline{r} = 0.1\ \mu$m.

5. Summary

In summary the CEGPs are the brightest in reflected light and thermally emitted flux of the known EGPs, and offer promise for a variety of atmosphere direct detections and study. The spectra, transmission spectra, and photometric lightcurves, all of which are detectable in the next few years are extremely varied depending on the type and size of condensates present in the planetary atmosphere. This variation suggests that observations should be able to distinguish among them.

Acknowledgments. Thanks are due to my collaborators Barbara Whitney and Dimitar Sasselov for useful discussions and for a variety of contributions.

References

Cameron, A. C., Horne, K., Penny, A., & James, D. 1999, Nat, 402, 751

Charbonneau, D., Brown, T., Latham, D., & Mayor, M. 2000, ApJ, 529, L45

Charbonneau, D., Noyes, R. W., Korzennik, S. G., Nisenson, P., Jha, S., Vogt, S. S., & Kibrick, R. I. 1999, ApJ, 522, L145

Henry, G. W., Marcy, G. W., Butler, R. P., & Vogt, S. S. 2000a, ApJ, 529, L41

Henry, G. W., Baliunas, S. L., Donahue, R. A., Fekel, F. C., & Soon, W. H. 2000b, ApJ, 531, 415

Kurucz, R. L. 1992, in IAU Symp. 149, The Stellar Populations of Galaxies, ed. B. Barbuy & A. Renzini (Dordrecht: Kluwer), p. 225

Mazeh, T., et al. 2000, ApJ, 532, L55

Seager, S. 1999, PhD Thesis, Harvard University

Seager, S., & Sasselov, D. D. 1998, ApJ, 502, L157

Seager, S., & Sasselov, D. D. 2000, ApJ, 537, 916

Seager, S., Whitney, B. A., & Sasselov, D. D. 2000, ApJ, 540, 504

Disks, Planetesimals, and Planets
ASP Conference Series, Vol. 219, 2000
F. Garzón, C. Eiroa, D. de Winter, and T. J. Mahoney, eds.

Shadow of the Planet Orbiting HD 209458

R. A. Street, T. A. Lister, A. C. Cameron, and K. Horne

School of Physics and Astronomy, University of St. Andrews, North Haugh, St. Andrews, Fife, KY16 9SS, Scotland

Abstract. We report observations of a planetary transit of HD 209458 on 1999 November 18, made in response to the IAU Circular by Henry et al. (1999). The observations were made from 17:02 UT to 22:59 UT with the 0.94 m James Gregory Telescope, St. Andrews, Scotland, using a narrow band filter with a bandpass width of 11.3 nm FWHM centered at 640 nm. Despite patchy cloud, the complete transit was observed; ingress and egress were estimated at 18:15 ± 30 min UT and 21:40 ± 45 min UT respectively, while the transit depth was determined to be 0.015 ± 0.002 mag. The fit to the data yielded a planetary radius of 1.30 ± 0.04 $R_{\rm Jup}$ and an orbital inclination of $90^\circ.0 \pm 1^\circ.7$. Further observations of this target have been confounded by bad weather.

1. Introduction

Most of the currently known extrasolar planetary systems were originally discovered by radial-velocity measurements. Consequently, the majority of these planets are of the "hot jupiter" or "51 Peg" type, with orbital inclinations close to 90° to the sky plane. This pre-selection of likely orbital inclinations, coupled with high planet-to-star mass ratio and small orbital radii makes these systems good potential targets for transit observations.

The planet orbiting HD 209458 was first discovered by radial-velocity observations conducted at Keck I with HIRES and ELODIE at the Observatoire de Haute Provence. The characteristics of the star and planet system are summarized in Table 1. It was subsequently selected as a target for two independent follow-up transit surveys. In 1999 August–September, Charbonneau et al. (2000) observed two transit events, followed by Henry's detection (Henry et al. 2000) of a transit ingress, announced in early November (Henry et al. 1999) along with predictions of the times for subsequent events.

In response to this circular, we made observations of HD 209458 at the time of the next predicted transit available to our longitude (Henry et al. 1999): 1999 November 18, 18:53 ± 1 hr UT.

2. Observations at St. Andrews

We observed HD 209458 using the 0.94 m James Gregory Telescope (hereafter JGT) at St. Andrews. The Wright Instruments CCD camera gives us a $17' \times 11'$

Table 1. Characteristics of HD 209458. Sources: The orbital period was determined by Robichon & Arenou (2000); all other parameters are from Mazeh et al. (2000)

Parameter	Determined value
Names	HD 209458, HIP 108859, SAO 107623
Stellar type	G0 V, no known stellar companions
Magnitude	$V = 7.645 \pm 0.002$, $(B-V) = 0.574 \pm 0.014$
Absolute mag. (M_V)	4.28 ± 0.10
Stellar mass	$1.1 \pm 0.1\ M_\odot$
Stellar radius	$1.2 \pm 0.1\ R_\odot$
$T_{\rm eff}$	6000 ± 50 K
Metallicity	0.00 ± 0.02 [Fe/H]
$\log g$	4.25 ± 0.2
Planetary mass	$0.69 \pm 0.05\ M_{\rm Jup}$
Orbital period	3.524739 ± 0.000014 days
Eccentricity	$0.016 \pm 0.018\ R_{\rm Jup}$
Planetary radius	$1.4 \pm 0.17\ R_{\rm Jup}$
Orbital inclination	$86°.1 \pm 1°.6$
Density	0.31 ± 0.07 g cm^{-3}
Surface gravity	870 ± 160 cm s^{-2}
Escape velocity	42 ± 4 km s^{-1}

field of view. This field is shown annotated in Figure 1. Given the magnitude of the target we used 15–25 s exposures with a narrow-band filter, centered on 640 nm with a bandpass width of 11.3 nm, to prevent CCD saturation. Readout times resulted in gaps of approximately 20 s between frames. Due to an uncertainty of ± 1 hr in the predicted timing of the transit, observations were carried out from 17:02–22:59 UT on 1999 November 18.

3. Transit-Event Observation and Data Reduction

Automated aperture photometry was performed on some five hundred frames taken during the transit event; details of our reduction procedure for JGT data can be found in Bell, Hilditch, & Edwin (1993). The curves in Figure 2 show these data binned up to approximately 120 s.

The lower curve, displaced in magnitude for the sake of clarity, is the comparison-check lightcurve which remains flat for the duration of the transit dip, illustrated by the upper curve. The mean differential magnitudes both before ingress and between ingress and egress have been calculated for the target-comparison and comparison-check curves. In the latter case, there was little difference between the two, whereas the transit is highlighted in the former.

Unfortunately there were gaps in the data during ingress and egress due to cloud, so only approximate timing of the event was possible. Ingress was estimated to occur at 18:15 ± 30 min UT and egress at 21:40 ± 45 min UT.

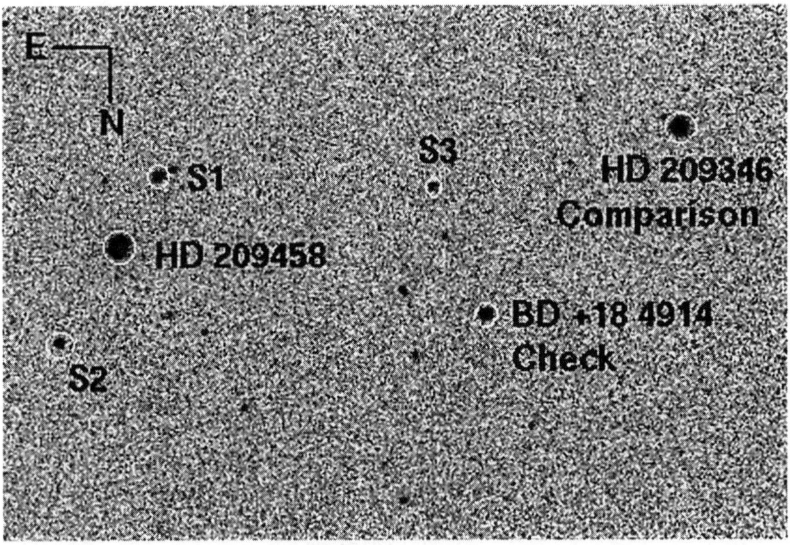

Figure 1. The JGT field of view around HD 209458, highlighting the other stars used in the photometry.

The depth of the transit was found to be 0.015 ± 0.002 mag, consistent with the value of 0.017 ± 0.002 mag measured by Henry et al. (2000).

3.1. Improving the Photometric Analysis

Our initial photometry of the transit data showed a scatter in the differential magnitude that was greater than that expected from photon statistics alone. Given that widespread cirrus cloud was present at the time, we suspected that patchy cloud was causing differential extinction across the $\sim 14'$ separation between the target and comparison stars.

To address this problem, we measured the extinction relative to the average magnitudes of the five brightest stars in the field of view, excepting HD 209458, and then fitted a differential extinction model, assuming a linear gradient. The stars used consisted of the comparison, the check and three other stars in the frame, labeled "S1–3" in Figure 1.

Frames which did not give a satisfactory fit, or those with a target-comparison uncertainty greater than 0.12 mag were rejected.

3.2. Theoretical Fitting

The full dataset was phased according to the period derived by Robichon & Arenou (2000), 3.524739 ± 0.000014 days and to a mid-transit epoch of HJD 2451501.3375 ± 0.02 determined from the times of ingress and egress. A lightcurve-fitting code was written to find the minimum χ^2 solution of a grid search in the parameter space of planetary radius and orbital inclination. All other parameters were fixed to the best available published values. Stellar radius (R_*) was

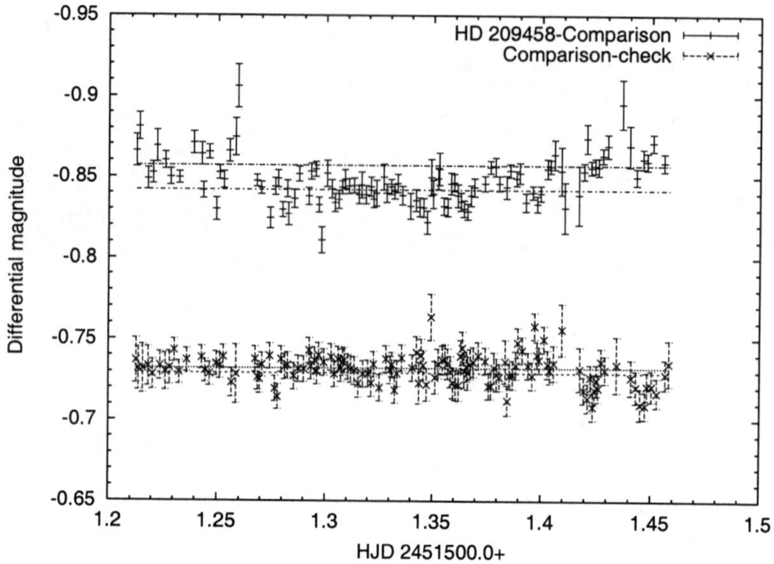

Figure 2. The binned lightcurve of the transit, showing both target-comparison and comparison-check curves. The *dashed lines* indicate the mean differential magnitude both prior to transit ingress, and during the transit. For clarity, the lower curve has been displaced by −3.1 mag.

set to 1.2 solar radii (Mazeh et al. 2000) and the orbital radius in units of R_* was determined to be 7.749 R_* from the period above, assuming a circular orbit. Claret, Díaz-Cordovés, & Giménez (1995) quote linear limb-darkening parameters for the Cousins R band for *logg* values of 4.0 and 4.5 for a range of effective stellar temperatures ($T_{\rm eff}$). As $\log g$ for HD 209458 was given to be 4.25, with a $T_{\rm eff}$ of 6000 K (Mazeh et al. 2000), we adopted a mean value of 0.553 for the linear limb-darkening parameter. This resulted in a minimum χ^2 solution at a planetary radius of 1.47 ± 0.04 $R_{\rm Jup}$ and an orbital inclination of $90^\circ.0 \pm 1^\circ.6$. The fit is shown in Figure 3.

Unfortunately, the lack of ingress and egress data in particular hinders the reliability of the fit. Nevertheless, our values for the planetary radius and orbital inclination complement those determined by other authors.

4. Conclusions

We observed a transit of HD 209458 on 1999 November 18, consistent with the prediction of Henry et al. (1999), and of comparable depth to their observations. From the duration of the transit we measure an orbital inclination of $90^\circ.0 \pm 1^\circ.6$, and a planetary radius of 1.47 ± 0.04 $R_{\rm Jup}$.

Unfortunately, we have been unable to obtain further transit data since that time, due to the RA of the target and inopportune cloud.

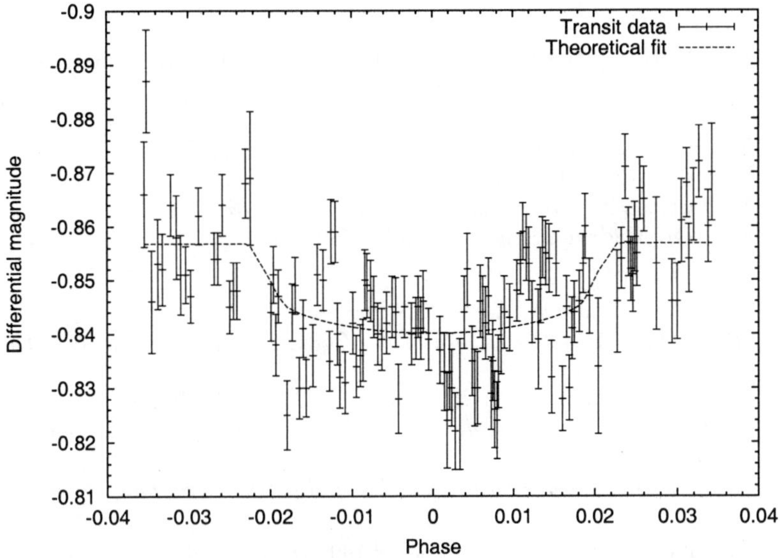

Figure 3. The phase diagram of the transit event showing the theoretical lowest χ^2 fit.

References

Bell, S.,Hilditch, R. W., & Edwin, R. P. 1993, MNRAS, 260, 478

Charbonneau, D., et al. 2000, ApJ, 529, L45

Claret, A., Díaz-Cordovés, J., & Giménez, A. 1995, A&AS, 114, 247

Henry, G. W., et al. 1999, IAUC, 7307

Henry, G. W., et al. 2000, ApJ, 529, L41

Mazeh, T., et al. 2000, ApJ, 532, L55

Robichon, N., & Arenou, F. 2000, A&A, 355, 295

Disks, Planetesimals, and Planets
ASP Conference Series, Vol. 219, 2000
F. Garzón, C. Eiroa, D. de Winter, and T. J. Mahoney, eds.

EXPORT Observations of 1998 Microlensing Events

Yiannis Tsapras[1], Rachel A. Street[1], Keith Horne[1], Alan Penny[2], Fraser Clarke[3], Hans Deeg[4,5], Francisco Garzon[4], Simon Kemp[4], Maria Rosa Zapatero Osorio[4], Alejandro Oscoz Abad[4], Santiago Madruga Sanchez[4], and EXPORT

[1]*School of Physics and Astronomy, University of St Andrews, North Haugh, St. Andrews, KY16 9SS, Scotland*
[2]*Rutherford Appleton Laboratories, Chilton, Didcot, Oxon, OX11 0QX, England*
[3]*Institute of Astronomy, Madingley Road, Cambridge, CB3 0HA, UK*
[4]*Instituto de Astrofísica de Canarias, E-38205 La Laguna, Tenerife, Spain*
[5]*Instituto de Astrofísica de Andalucía, E-18080 Granada, Spain*

Abstract. During the summer of 1998 the EXPORT collaboration monitored microlensing event lightcurves using CCD cameras on the 0.8 m IAC-80 Telescope to evaluate the prospect of hunting for Jupiter-mass planets orbiting stars that are lensing Galactic bulge sources. The results obtained are presented here. We argue that daily monitoring of microlensing events in the Galactic bulge using small telescopes from northern sites is a feasible method for detecting jupiters.

1. Introduction

Microlensing (Paczyński 1996) involves the gravitational deflection of light from a background star (source) as a massive stellar object (lens) passes in front of it. This results in two images of the background source on either side of the lens positions. For sources in the Galactic bulge, the image separation is of the order of milliarcseconds and thus unresolvable. What is actually observed in microlensing events is a variation of the brightness of the source as the lens moves in front of it.

The range of influence of a point lens is given by its Einstein-ring radius

$$R_{\rm E} = \sqrt{\frac{4GM_{\rm L}D_{\rm L}D_{\rm LS}}{c^2 D_{\rm S}}}. \tag{1}$$

where $D_{\rm LS}$, $D_{\rm S}$, and $D_{\rm L}$ are the lens–source, observer–source, and observer–lens distances, respectively. The single-point lens equation of gravitational lensing has two solutions corresponding to the positions of the images. If the source and the lens are perfectly aligned then the images merge into a ring of radius $R_{\rm E}$ around the lens. For all other positions of the source, one image lies inside and the other outside the ring (see Figures 1 and 2).The total amplification is

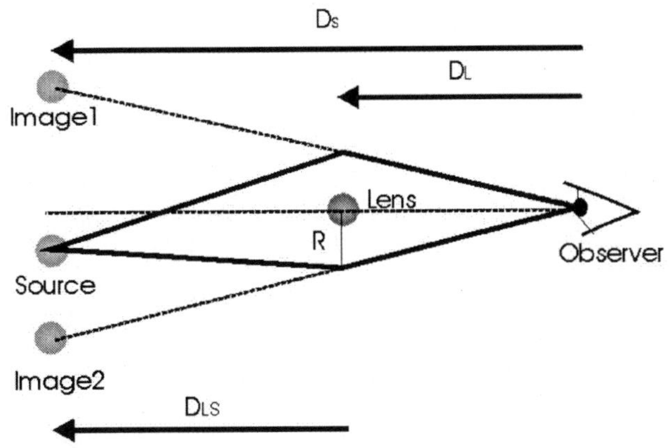

Figure 1. Schematic diagram of microlensing geometry. $D_{\rm L}$, $D_{\rm S}$, and $D_{\rm LS}$ are the observer–lens, observer–source and lens–source distances, respectively. R denotes the Einstein-ring radius.

given by:

$$A = \frac{u^2+2}{u(u^2+4)^{1/2}}, \tag{2}$$

where $u = R_{\rm S}/R_{\rm E}$, $R_{\rm S}$ is the separation on the lens plane between the lens and the source. $R_{\rm E}$ is the Einstein ring radius of the lens.

2. Microlensing Lightcurves

Sources that lie inside the Einstein ring will have peak magnifications, A, in excess of 1.34. The resulting lightcurve is axially symmetric and A increases as the source impact parameter, $u_{\rm min}$, decreases. Galactic bulge lensing events have typical timescales $t_{\rm E}$=10–100 days (Bennet & Rhie 1996).

If two or more lensing masses are involved in the lensing event, the magnification pattern is changed so that at certain positions on the sky their individual lensing strengths combine. These positions are called caustics, and when the source crosses them the event lightcurve becomes distorted. By correctly assessing such lightcurve anomalies, the presence (or absence) of planetary companions can be deduced.

3. Monitoring Microlensing Events

To quantify planetary anomalies in a lightcurve, events in progress must be imaged very frequently. The planetary effects last for 1–3 days when a Jupiter-mass companion is involved (Gould & Loeb 1992). To correctly estimate the duration and structure of the anomalous peak, and thus place restrictions on the planet-to-star mass ratio and the planetary position relative to the lens, we

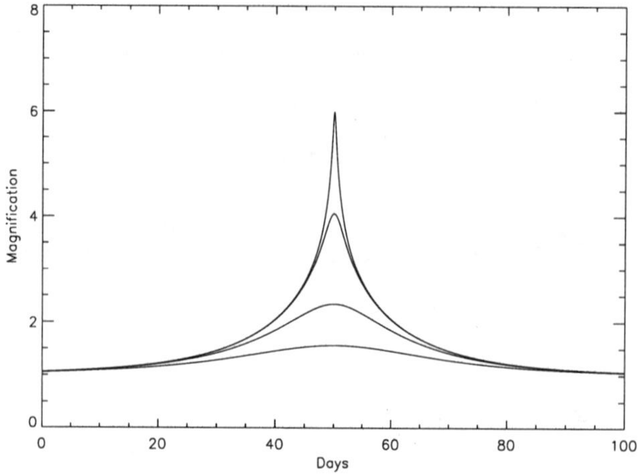

Figure 2. Lightcurves produced by a source star passing behind a lens for four different impact parameters. The impact parameters for the curves from *top* to *bottom* are: 0.01, 0.06, 0.3, and 0.7

require the fitting of several points. Ideally, a search for jupiters would employ hourly imaging, which also increases the possibility of detecting deviations caused by Earth-mass planetary companions, whose planetary deviations last for a few hours only. However, daily sampling from a northern site should suffice to detect jupiters, with one or two data points deviating from the unperturbed lightcurve assuming the event was followed closely.

In the summer of 1998 the EXPORT team decided to test this assumption by following up on microlensing events in the Galactic bulge. The 0.8 m IAC-80 Telescope on Tenerife was used for one hour per night for a period of four-and-a-half months (May 15–September 30, see Table 1).

Several ongoing microlensing events were monitored. On each observing night, three 600 s R-band images were taken. The CCD size was 1024×1024 pixels, covering a sky area of $7.3' \times 7.3'$ and the typical airmass was between 1.5 and 2. The microlensing events were recorded often with accuracies at best $\sim$ 1–2% for a star of 16th magnitude and more typically $\sim$ 3–4% at 18th magnitude. No planetary deviations from the event lightcurves were found. This was not unexpected since the gaps in temporal sampling were of appreciable size.

4. Crowded-Field Photometry at High Airmass

Our crowded-field photometry of the CCD data was performed using the STARMAN stellar photometry package (Penny 1995) in a semi-automated data-reduction pipeline. All frames were debiased and flat-fielded in the usual manner and the target was identified from finder charts. Several stars were selected manually on which photometry was to be performed. Apart from the target, these included i) several bright, unsaturated stars that were used for the PSF generation and

Table 1. 1998 IAC-80 Observations

Event	Nights	Filter	Dates
98-BLG-07	3	R	May 31, June 04, 20
98-BLG-18	6	R	May 15, 18, June 03, 08, 12
98-BLG-26	5	R	Jun 18, 21, 27, 29, Aug 16
98-BLG-35	12	R	Jul 04, 05, 06, 10, 11, 27, 29, 30, 31
			Aug 01, 07, 10
98-BUL-26	4	R	Aug 13, 15, 18, 19
98-BLG-42	15	R	Aug 24, 25, 26, 27, 28, 29, 30
			Sept 02, 03, 04, 05
			Sep 06, 15, 19, 20
98-BUL-34	7	R	Sep 21, 22, 23, 24, 26, 29, 30

ii) a selection of "error stars" of constant brightness, comparable to that of the target at each stage of the lensing event, which were used to estimate the rms uncertainty on the measured target magnitude at each stage of its brightness variation. All images were aligned and crowded field PSF-fitting photometry was performed on the stars on all frames. Stars with poor fits were excluded from further processing. The measured star images were expressed as differential magnitudes relative to a set of 25 reference stars.

The theoretical errors were calculated using the standard formula

$$\sigma^2 = \sigma_0^2 A_{\rm psf} + \frac{f_* t}{G} + \frac{f_{\rm s} t A_{\rm psf}}{G}, \tag{3}$$

where σ_0 is the readout noise (in ADU), t the exposure time, $A_{\rm psf}$ is the effective area of the point spread function, G is the gain (e^-/ADU) and f_*, $f_{\rm s}$ the flux from the star and flux per pixel from the sky, respectively.

5. Results for 98-BLG-35 and 98-BLG-42

Our best sampled lightcurves were those for the events MACHO 98-BLG-35 (Figure 3) and 98-BLG-42 (Figure 4). The observations of these events started approximately at the time of maximum amplification. Unfortunately, all of the events that we followed during the observing period were sampled only in the decline, with the exception of 98-BLG-42, where we had one point before maximum amplification. For this reason our fits to the data, although in agreement with the fits by other groups, do not yield useful constraints on the event parameters.

Figure 5 shows the accuracy achieved for 390 stars in all the frames and the theoretical error curves. Due to poorer seeing and brighter sky at airmass 1.5–2, the dominant source of noise in our data was sky noise. Photon noise from the star is important only for stars brighter than 15th magnitude. The measured rms scatter of repeat measurements is generally close to theoretical estimates. However, some of the data points scattered by more than the expected value.

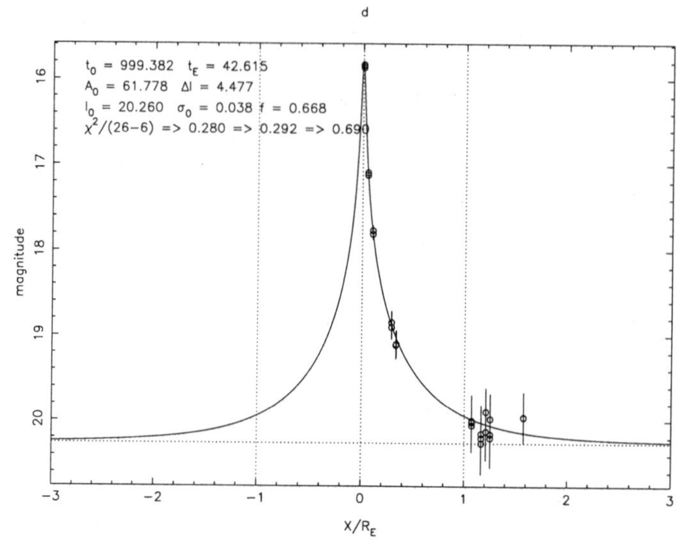

Figure 3. Fitted lightcurve for the event 98BLG35. *R* magnitude is plotted versus time in units of t_{E}. The estimated event parameters are shown on the *top left* of the plot.

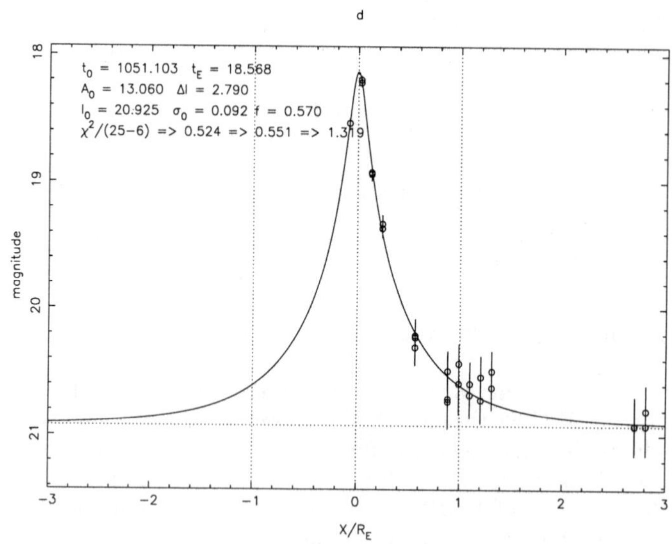

Figure 4. Fitted lightcurve for the event 98-BLG-42. *R* magnitude is plotted versus time in units of t_{E}. The slight increase in brightness in the region $x/R_{\mathrm{E}} \sim 1$ of the plot is probably a blending effect from a star that lies almost on top of the target. As the target becomes very faint the program has difficulty distinguishing between the two stars. The estimated event parameters are shown on the *top left* of the plot.

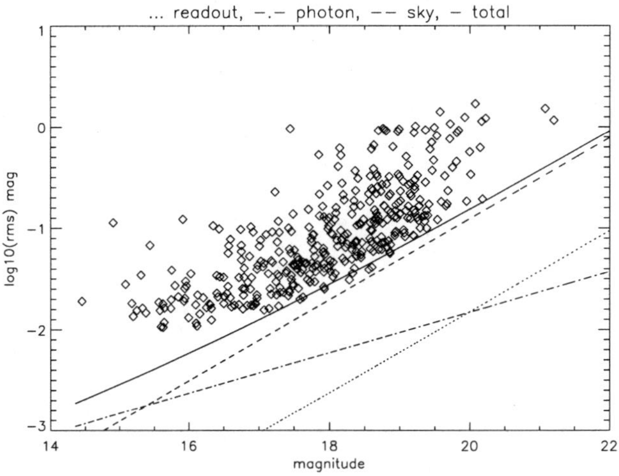

Figure 5. *R* magnitude values versus the corresponding rms values for 390 stars. The plot looks noisier than expected which is due to the overcrowding of some stars.

These stars were identified later in the frames and were seen to be suffering from overcrowding.

6. Significance of the Results

We have achieved photometric measurement errors with a lowest value of $\sim$ 1–2%. This is also comparable to the photometric accuracy reported by other microlensing teams that operate at low airmass from the advantageous (for Galactic bulge observations) environment of southern observatories. We believe that our results show that microlensing observations in the Galactic bulge looking for jupiters can also be successfully performed from northern sites and with small telescopes.

References

Paczyński, B. 1996, ARA&A, 34, 419

Bennet, D., & Rhie, S. 1996, ApJ, 472, 660

Gould, A., & Loeb, A. 1992, ApJ, 396, 104

Penny, A. 1995, Starlink Documentation

Part 5

Future Projects

Disks, Planetesimals, and Planets
ASP Conference Series, Vol. 219, 2000
F. Garzón, C. Eiroa, D. de Winter, and T. J. Mahoney, eds.

Planet Detection from the Ground: Can the VLTI Help?

Vincent Coudé du Foresto

Observatoire de Paris (DESPA), 92195 Meudon Cedex, France

Abstract. A brief overview of interferometry and the VLTI is presented in the context of exoplanet detection. Ground-based observations are limited mostly by atmospheric turbulence, yet exoplanet observations are possible with the VLTI. Spectrophotometry of hot jupiters can be achieved with high dynamic range visibility or phase measurements, while astrometric techniques will permit the detection of classical giant planets around most nearby stars.

1. Introduction

The VLTI is the interferometric mode of ESO's Very Large Telescope, being built on the Mount Paranal in northern Chile. It is a facility which can combine several telescopes into a coherent array, ultimately providing an imaging capability with milliarcsecond resolution, or relative astrometry with a precision of a few tens of μas.

Although exoplanet detection did not appear explicitly in the scientific plan of the VLTI when the observatory was formally proposed in 1987, it was gradually realized afterwards that this tool could provide unique contributions to the field.

1.1. Basic Principles

In Michelson interferometry, aperture synthesis is achieved by collecting light at two separated apertures and transferring it to a laboratory where coherent recombination occurs and creates a fringe pattern detected by the instruments. From the contrast and the phase of the fringe pattern can be measured the complex degree of coherence between the two beams which is nothing else, for an ideal instrument, than the Fourier component of the source's intensity distribution for the spatial frequency (u, v) defined by the baseline projected onto the plane of the sky. When n telescopes are recombined, they generate $n(n-1)/2$ baselines and as many Fourier-component measurements.

Therefore an interferometer acts as a sampler of the (u, v) plane, and, in principle, imaging can be performed by direct Fourier inversion. There are, in practice, several limitations:

- In optical interferometry, because of the small number of apertures, the sampling of the (u, v) plane is very scarce so that a blind inversion is usually not efficient, or even possible. Therefore, it is more appropriate to derive information from the morphology of the source by fitting a model to the visibility measurements.

- The raw phase is actually not a direct observable, however it is possible to measure the *differential* phase (i.e., the differential delay) between for example the target source and a nearby object, or between two fringe patterns generated by the source at different wavelengths. If three or more telescopes are recombined simultaneously, phase closures can be measured which also provide phase information (Haniff 1988).

In astrometric applications of interferometry, one is only interested in the differential phase, $\Delta\phi$, between two targets, which is related to their angular separation, $\delta\theta$, by the relationship:

$$\Delta\phi = 2\pi B.\delta\theta/\lambda,$$

where B is the baseline vector projected onto the plane of the sky.

1.2. Ground-Based Limitations

Limitations for ground-based interferometry occur because of the presence of atmospheric turbulence. Turbulence includes the spatial modes responsible for the seeing and which degrade the coherence over each individual subaperture. It also includes a piston mode (variation of the average phase over each subpupil), which does not affect image quality at each subpupil. However, the differential piston between the two telescopes generates a random differential phase delay (fringe jitter). The consequences of atmospheric turbulence for ground-based interferometers are twofold:

- It reduces the sensitivity of the interferometer. The effective collecting area of each subpupil is limited to πr_0^2, where r_0 is Fried's parameter, and the integration time is limited to exposures shorter than τ_0 (the coherence time of the atmosphere) to freeze the turbulence.

- It introduces a seeing noise into the fringe visibility or phase measurements, therefore limiting the astrometric accuracy or the quality of the imaging data.

Since r_0 and τ_0 scale as $\lambda^{6/5}$, these consequences are less severe for longer wavelengths.

Atmospheric turbulence can be counteracted by active optical systems (i.e., adaptive optics for the spatial modes, and fringe tracking for the differential piston mode). In addition, the seeing modes can be eliminated using spatial filters (e.g., single-mode fibers or pinholes).

1.3. VLTI Overview

The VLTI (Glindemann et al. 2000) is a modular infrastructure whose purpose is to achieve the coherent recombination of separated telescopes on Paranal. Depending on the needs, it involves different subsystems which will be progressively installed and commissioned on the mountain:

Figure 1. Aerial view of the VLTI on Paranal showing the four Unit Telescopes and the Auxiliary Telescope stations. The delay-line tunnel and the recombination laboratory building appear in the middle.

Light Collectors

- For maximum sensitivity, the VLT 8.2 m Unit Telescopes (UTs) are used, which form the VLTI Main Array (VIMA). The first coherent recombination of two UTs is expected in 2001. The VLTI will compete for UT time with other VLT programs. Therefore, it is expected that access to VIMA will be possible only for some fraction of the nights.

- Alternatively, when the maximum angular resolution or a more complete coverage of the (u, v) plane is needed, the VLTI Sub Array (VISA) will be used. The VISA light collectors are three dedicated 1.8 m Auxiliary Telescopes (ATs), which can be relocated on any of the AT stations on the mountain, arranged so that the potential (u, v) coverage is as dense as possible. The maximum baseline also will be larger than for VIMA, 200 m vs. 120 m. The Auxiliary Telescopes will be commissioned on Paranal starting mid-2002.

- During an initial period and for most nights until the implementation of VISA, a pair of two 0.4 m siderostats will be used as a test array. Their light-gathering power is limited, but sufficient nevertheless to permit observation of the nearest main-sequence stars.

Turbulence Compensation Both the UTs and the ATs will be equipped from the start with a fast angle tracking unit to compensate for the seeing tip-tilt modes. Under median Paranal seeing conditions, this level of correction is sufficient to permit interferometric observations at $\lambda \geq 2\,\mu$m with the ATs and $\lambda \geq 8\,\mu$m with the UTs. In order to use the full UT apertures at shorter wavelengths, a adaptive optics system with 60 actuators and curvature sensing will be implemented in 2003. In addition to these active systems, the residual atmospheric seeing modes and instrumental phase errors on each subpupil are filtered out within the instruments.

The full fringe tracking capability will be implemented with PRIMA (see below) starting 2003, using either an on-axis or off-axis reference source. A temporary fringe tracking unit (with an on-axis only reference and a limited sensitivity) will be installed by the end of 2001.

Beam Transport and Delay The light is transferred underground as an afocal beam from the oudé foci of the telescopes to the recombination laboratory through a series of relay mirrors. In addition, a set of delay lines maintains an equal optical path to the recombination point through all telescopes, despite the diurnal motion of the source.

Recombination and Detection Once in the common laboratory, the beams are redirected via a switchyard to a beam-combining instrument. First-generation VLTI instruments are:

- MIDI (Leinert et al. 2000), a two-way, thermal infrared (N band: $\lambda =$ 8–12 μm) beam combiner with medium spectral resolution ($\mathcal{R} = 300$), will be implemented towards the end of 2001. In a second phase, MIDI might be extended towards the longer wavelengths (Q band).

- AMBER (Petrov et al. 2000) is designed for the K band ($\lambda = 2$–$2.4\ \mu$m), with extensions foreseen at shorter wavelengths. AMBER will provide medium- to high-spectral resolution ($\mathcal{R} = 10^2$–10^4). Although it will start operating in 2002 with two telescopes only, AMBER will eventually be able to accommodate the simultaneous recombination of three telescopes and perform phase-closure measurements. One other interesting property of AMBER is that it will be able to make relative phase measurements between different wavelengths.

- VINCI (Kervella et al. 2000) is a K-band, two-way beam combiner installed early in 2001 to commission the VLTI. It will not provide spectral resolution, but nevertheless will produce accurate fringe-visibility amplitude measurements.

The angular resolution of the VLTI is set by the ratio λ/B, and reaches typically 2 milliarcsec in K and 10 milliarcsec at 10 μm for the longest VISA baseline, and respectively 3.5 and 17 milliarcsec for the largest separation between two UTs. It is possible to access spatial information at a lower scale, however, but then the SNR degrades quickly.

2. Exoplanet Detection

The VLTI can be used to detect exoplanets through either direct or indirect methods.

2.1. Direct Methods

With direct techniques, the imaging capability of the interferometer is used to observe the binary system formed by the star–planet pair. Direct techniques offer a unique way not only to fully determine the planet's orbital parameters, but also to derive the relative photometry with the parent star. If these observations are performed simultaneously at several wavelengths, this enables a spectrophotometric characterization of the planet and interferometry makes it possible to constrain the theoretical models of the planetary atmospheres.

Because the star–planet system presents a very simple morphology (it can be well modeled by a central disk and an orbiting point source), it can be well constrained even though the measurements are limited to a few number of points in the (u, v) plane. The challenge is rather set by the enormous contrast ratio between the two sources, and a direct detection is conceivable only in the most favorable case, i.e., a main-sequence star with a Jupiter-sized companion whose temperature is greater than 1000 K. Even so, the required dynamic range is considerable, of the order of 10^3 to 10^5 (Figure 2).

Generally speaking, the brightness ratio decreases as the wavelength increases, so that observations should in principle be easier at the longest available wavelengths. However, other factors have to be taken into account. Hot jupiters are also close to their parent star and the system is fully resolved by the interferometer only at shorter wavelengths. For example, if one considers a planet with a 0.05 AU orbit at 10 pc, the wavelength needs to be shorter than 4.8 μm with the longest VISA baseline (200 m), otherwise the planetary signature is weakened by the unsufficient spatial resolution. The atmospheric window with the best compromise may very well be the L band (3.4 $\mu\text{m} \leq \lambda \leq 4.1$ μm).

Direct-detection methods attempt to isolate the planetary signature either in the amplitude of the visibility function, its phase, or both. A third technique, nulling, may also eventually be implemented on the VLTI.

Visibility Amplitude Demodulation The proposed direct-detection technique with visibility amplitudes (Coudé du Foresto 1999) combines high-accuracy visibility measurements and radial-velocity information to reduce the noise by synchronous detection of the planetary signal.

If the planet ephemeris is known from radial velocities, the brightness ratio, α, the orbit orientation, Ω, and its inclination i (which leads to mass determination since $M \sin i$ is known from spectroscopy), are the only three parameters to retrieve from the interferometric observations.

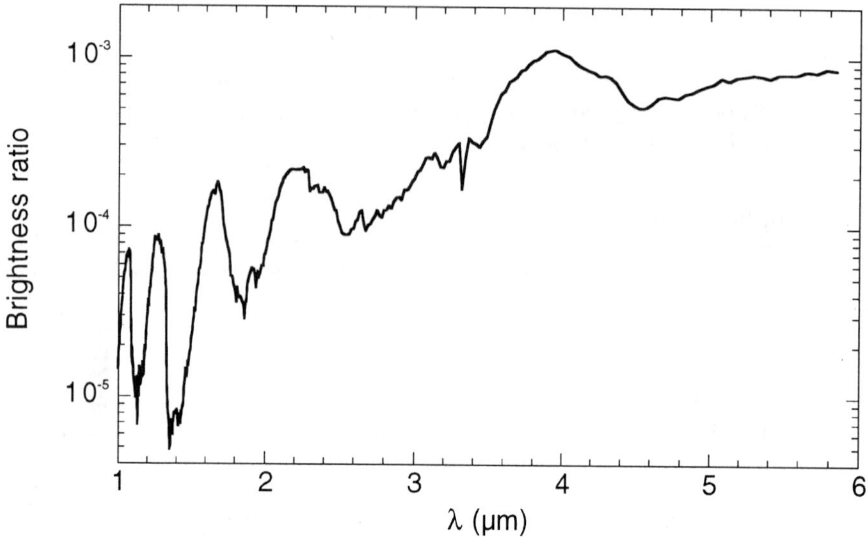

Figure 2. Predicted contrast spectrum between 51 Peg A and B (adapted from Goukenleuque et al. 2000).

The visibility amplitude of the binary system (Figure 3) shows, on top of the traditional stellar Airy function, a fringe pattern whose peak-to-peak amplitude is 2α, and whose periodicity and orientation in the (u, v) plane depend on the instantaneous position of the planet, i.e., the known phase, ϕ, of the radial-velocity oscillations. Thus any visibility point can be located into a (u, v, ϕ) data cube (or, rather, a data tore), which can be filled in progressively as the source is observed during several (not necessarily consecutive) nights and with different baselines. Once enough data have been accumulated, they can be solved for α, i, and Ω.

To obtain a high dynamic range, the visibilities need to be measured with a high degree of accuracy, though if a lot of data points are collected the error on each individual point can be substantially larger than the brightness ratio, α. In the bright source regime (which will be the case, for nearby main-sequence stars, with the ATs in K and L and for typical $\mathcal{R} = 100$ spectral resolutions), the visibility errors are dominated by the seeing and the accuracy depends on the level of correction of the turbulence modes:

- With no correction, the typical accuracy is 5% for 5 min of integration;

- When the seeing modes are removed (with full correction adaptive optics and/or spatial filtering), only the piston noise remains and the errors go down to 0.1–0.3%, i.e., the dynamic range can reach 10^3 in 5 min;

- It is not known currently what the precision would be if both the spatial and piston modes of the turbulence are eliminated, since no instrument has ever worked in that regime.

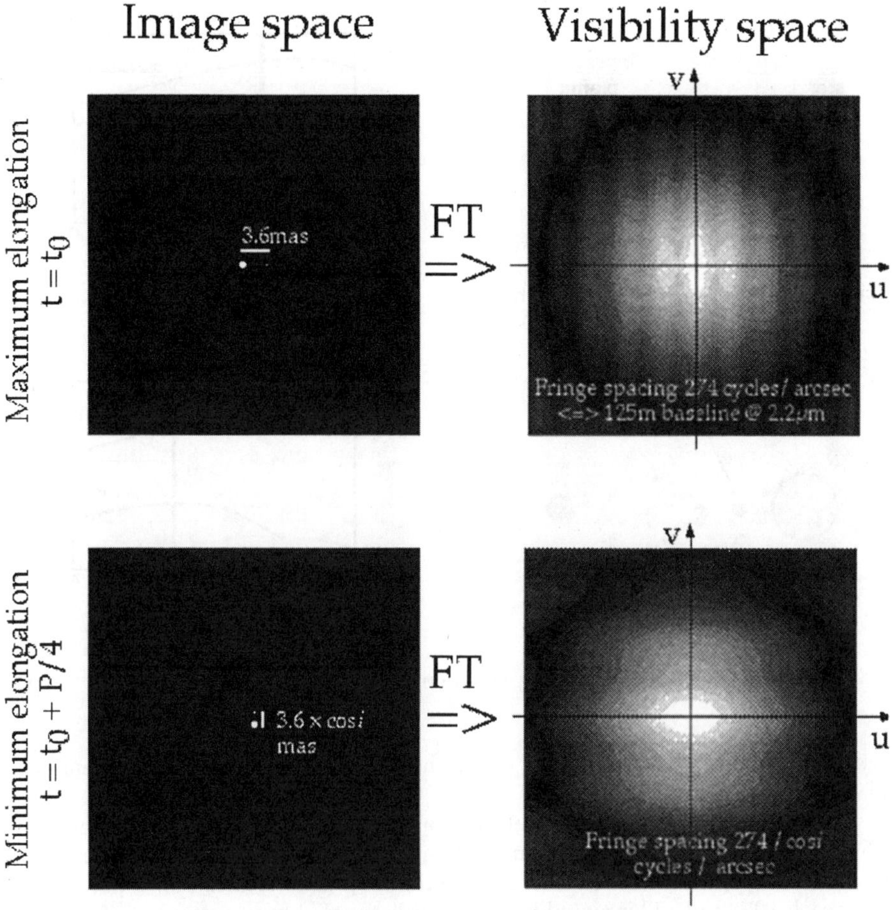

Figure 3. Modeling of the 51 Peg binary system in the image and visibility-amplitude spaces, at two different epochs. For clarity, the brightness ratio in this example has been lowered to the unrealistic value of 20.

Simulations show that, even with piston noise, hot jupiters around nearby stars can be marginally detected with VISA. Integration times and signal-to-noise ratios are drastically improved when the dynamic range is enhanced by the presence of a fringe tracker.

Differential Interferometry Differential interferometry (Lopez & Petrov 1999) exploits the color dependence of the star–planet binary system. Because of the differential effective temperature, the average star/planet brightness ratio decreases from the J, H, and K to the N band. Even within an atmospheric window, the expected absorption bands in the planet's atmosphere by water va-

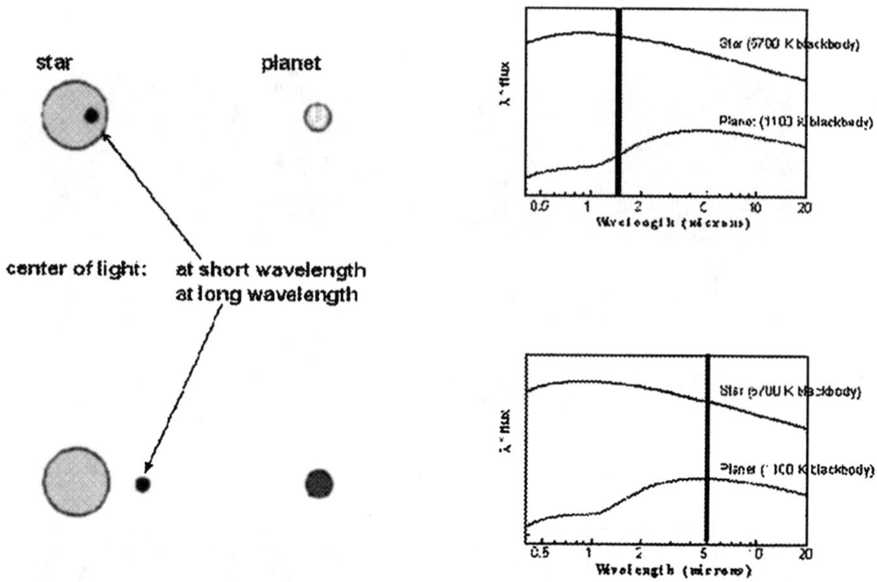

Figure 4. Photocenter displacement with wavelength.

por and methane (Burrows et al. 1997) should create a strong color dependence of the contrast.

At wavelengths where the relative brightness of the planet is highest, the photocenter of the binary system is slightly displaced towards the companion and the fringe pattern is shifted. The maximum phase variation for a hot jupiter can reach 10^{-3} rad. This signal has to be disentangled from a number of parasite atmospheric and instrumental dispersion effects. Instrumental effects are taken care of by a device which inverses the field and therefore the signal without affecting the instrumental contribution. Random external dispersion fluctuations seem to be dominated by variations in the water-vapor content of the atmosphere (Akeson, Swain, & Colavita 2000). They are expected to have zero mean so that a time integration can average them out.

Careful modeling of the atmospheric and instrumental effects still needs to be done, but simulations based on fundamental noises are very encouraging. They predict with VIMA/AMBER on 51 Peg a 10 σ detection for 30 min of integration in each of the 0.1 μm channel of the K band.

Nulling Interferometric coronagraphy or nulling consists in introducing a phase shift in one of the arms of a stabilized interferometer so that the system blocks the on-axis stellar light (dark central fringe), while the transmission is maximized for the off-axis position of the planet. This technique is considered for the space-based detection of telluric planets (ESA's *DARWIN* and NASA's *TPF* projects), and a ground-based nulling demonstrator is planned as a second generation VLTI instrument around 2005. The atmosphere will limit its rejection rate to levels that do not permit exoplanet detection from the ground: its main scientific driver will be the characterization of exozodiacal light.

2.2. Indirect Methods

Indirect methods use the astrometric capability of the array to infer the presence of a companion from the transverse wobble of the star with respect to a background reference. Because they are mostly sensitive to massive planets with large orbits, they provide a coverage of the mass/orbital period parameter space which is complementary to that of radial-velocity techniques.

Narrow-angle astrometry will be feasible with the VLTI, using the dual beam facility (PRIMA: Phase Referenced Imaging and Micro-arcsecond Astrometry), to be implemented around 2003–2004.

The VLTI Dual-Beam Facility The normal field of view of the VLTI is too small (2 arcsec) to accept both the star and a reference. The objective of PRIMA is to enable simultaneous interferometric observations of two objects that are separated by up to 1 arcmin. The brighter source is used for stabilization of the fringe motion induced by the atmospheric turbulence, while a comprehensive laser metrology system inside the interferometer makes it possible to measure and/or to control the differential delay between the two targets.

- If the differential delay is pre-set and maintained constant it becomes possible to integrate on stabilized fringes and perform observations on much fainter sources in the second field of view (PRI: Phase Referenced Imaging);
- Conversely, precisely measuring the differential delay makes it possible to determine with great accuracy the angular separation between the two stars (MA: Microarcsecond Astrometry).

The fundamental limit on astrometric measurements is set by the anisoplanicity of the atmosphere which generates a residual differential fringe jitter. For small angles, $\delta\theta$, on the sky, the atmospheric error for a baseline, B, scales as $\delta\theta B^{-2/3}$ (Shao & Colavita 1992). Under typical seeing conditions for Mount Paranal, the error for the longest VLTI baselines approaches 10 μas in 30 minutes of integration for a 10 arcsec distance between the target star and its reference (von der Lühe, Quirrenbach, & Koehler 1995). In order to achieve this precision, however, several technical challenges have to be met; for example, the differential instrumental errors have to be controlled to a precision of about 5 nm, and the baseline vector has to be known within 50 μm.

Even if VLTI/PRIMA does not quite reach its atmospheric performance limit, the potential for planet detection is considerable. At 10 pc, the astrometric signature of a jupiter at 5 AU is 220 μas (for an A5 main-sequence star) to 2400 μas (for an M5 dwarf). The wobble introduced by a uranus at 20 AU for the same spectral types has an amplitude of 40 and 440 μas, respectively. It should be possible to perform a statistically significant census of planetary companions (down to Uranus-mass) along the main sequence, which would provide hindsight into the planet formation as a function of stellar mass.

3. Conclusion

Despite the fundamental limitations imposed by atmospheric turbulence, the VLTI appears as a unique tool for the discovery and characterization of exoplanetary systems. It is complementary from space missions in several respects: for astrometry, its sensitivity is lower but it is not limited by a mission duration. Its main contribution will probably be to survey nearby stars for long-period planets. Also, the VLTI may provide within the next five years the first spectra of exoplanetary atmospheres (for hot jupiters) and it appears as a good testbench to demonstrate techniques (such as nulling) that will eventually be used in space for the spectroscopy of telluric planets.

References

Akeson, R., Swain, M. R., & Colavita, M. 2000, SPIE 4006, in press

Burrows, A., et al. 1997, ApJ491, 856

Coudé du Foresto, V. 1999, in VLT Opening Symposium, From Extrasolar Planets to Cosmology, ed. J. Bergeron, & A. Renzini (Antofagasta: Springer), p. 560

Glindemann, A., et al. 2000, SPIE 4006, in press

Goukenleuque, C., Bézard, B., Joguet, B., & Lellouch, E. 2000, Icarus, 143, 308

Haniff, C. A. 1988, in NATO ASI, Vol. 274, Diffraction-Limited Imaging with Very Large Telescopes, ed. D. M. Alloin & J.-M. Mariotti (Dordrecht: Kluwer), p. 171

Leinert, C., et al. 2000, SPIE, 4006, in press

Lopez, B., & Petrov, R. 1999, in VLT Opening Symposium, From Extrasolar Planets to Cosmology, ed. J. Bergeron & A. Renzini (Antofagasta: Springer), p. 565

Petrov, R., et al. 2000, SPIE, 4006, in press

Shao, M., & Colavita, M. M. 1992, A&A, 262, 353

von der Lühe, O., Quirrenbach, A., & Koehler, B. 1995, in Science with the VLT, ed. J. R. Walsh & I. J. Danziger (Garching: Springer), p. 445

Disks, Planetesimals, and Planets
ASP Conference Series, Vol. 219, 2000
F. Garzón, C. Eiroa, D. de Winter, and T. J. Mahoney, eds.

The ALMA Project

Stéphane Guilloteau[1,2]

Institut de Radio Astronomie Millimétrique, 300 rue de la piscine F-38406 Saint-Martin-D'Hères, France

Abstract. The Atacama Large Millimeter/submillimeter Array (ALMA)[3] will be the first instrument allowing very high angular resolution (down to 0.01″) with sufficient sensitivity to image thermal emission of protoplanetary disks at wavelengths where these disks are optically thin. I present here the projected characteristics of ALMA, in particular the expected sensitivities and frequency coverage, and illustrate by simple simulations possible use of ALMA for planetary-system formation studies.

1. Introduction

With the spectacular images of dusty disks around young stars provided by the *Hubble Space Telescope*, and the many recent discoveries of extrasolar planets by induced Doppler shift on the stars around which they orbit, the astrophysicist interested in planetary-system formation may wonder why should we bother with mm and submm waves.

The classic answers to this question apply to the whole Universe. First, the universe is on average rather dark and currently quite cold (only a few K, except in very optically thin regions). Second, by an appropriate coincidence, most of the abundant lighter molecules have their lower rotational lines in the mm window. Third, the interstellar dust, though only present with tiny densities, becomes easily optically thick even in the near IR.

These considerations apply even more strongly to protoplanetary disks, in which the temperature is about 30 K at 100 AU from a T Tauri star. Such disks are quite optically thick in the visible, but at mm wavelengths, the opacity becomes low enough to be able to see throughout the disk (except perhaps in the innermost regions). Alas, protoplanetary disks are also very small: planetary-system formation is expected to occur on scales of a few AU or even smaller. Large optical telescopes and interferometers will provide such angular resolution, but will not penetrate the hidden, optically thick regions. For such studies, a

[1] ALMA European Project Scientist.

[2] Also at European Southern Observatory, Karl Schwarzschild Str-2, Garching, Germany.

[3] References on ALMA can be found in Wootten (2000). The project description and up-to-date information are available on the WEB at the ESO Web site, http://eso.org/project/alma or at the NRAO Web site, http://www.alma.nrao.edu.

large mm/submm interferometer is required: this is one of the major science drivers for ALMA (Atacama Large Millimeter/submillimeter Array).

2. The ALMA Genesis

Existing mm arrays (BIMA, NRO, OVRO, and IRAM-PdB) suffer from severe limitations in sensitivity (IRAM-PdB being the most sensitive by a factor of about three), angular resolution (BIMA providing 2 km baselines, while only $\simeq$ 500 m baselines are available elsewhere), imaging speed (with only five to nine antennas), and frequency coverage (up 270 GHz at OVRO). The SMA will open the submm window, but with a very limited collecting area. Considering these limitations, the best angular resolution that can reasonably be reached is $0.5''$. At the distance of the nearest star-forming regions, this is about 80 AU, i.e., almost twice the size of our Solar System. Accordingly, only a significant improvement in angular resolution would enable planetary-system formation to be studied at the required scale.

Unfortunately, in an interferometer, an improvement in angular resolution by a factor F requires an improvement in sensitivity by a factor F^2. ALMA is intended to provide a factor ten better angular resolution than current arrays and hence will require a hundred times better sensitivity. The noise performance of current mm receivers is close to the quantum limit, and at these wavelengths, the atmosphere dominates the noise. Hence, such a gain in sensitivity must come from the only possible areas we can improve in mm interferometers: the detection bandwidth (for continuum, but not for spectral lines where it is limited by the Doppler width) and the collecting area.

The ALMA project is the result of studies for several original projects:

- The US-MMA with a collecting area of 2000 m^2 (forty 8 m antennas), emphasis on wide field imaging, and submm capabilities
- The European LSA, with a collecting area of 10000 m^2 (sixty 15 m antennas), emphasis on high angular resolution, in the mm domain only
- The Japanese LMSA, with a collecting area of 4000 m^2 (fifty 10 m antennas), emphasis on submm frequencies

These preliminary studies have led to an ALMA project consisting of

- 64 antennas of 12 meter diameter, providing 7000 m^2 (seven times larger than the IRAM-PdB)
- 20 μm surface accuracy for the antennas to provide high efficiency at the shortest wavelengths (near 350 μm)
- $0.6''$ pointing accuracy, to allow high-quality imaging at all frequencies
- A receiver concept allowing observations at all frequencies between 30 GHz and 900 GHz
- A system design allowing full polarization measurements

- Array configurations ranging from a dense, compact array of $\simeq$ 160 m diameter to an extended configuration of more than 12 km diameter. Combined with the observing frequencies, this provides angular resolution from $12''$ to $0.006''$
- Wide-field imaging capabilities, by the inclusion of mosaicking techniques and total-power measurements
- High-detection bandwidth (16 GHz instantaneous)

Compared to current instruments, ALMA will represent a major step forward. It will be 50 to 100 times more sensitive than the IRAM-PdB for continuum observations (15 to 25 times more sensitive for spectral lines). In the submm domain, ALMA will be 40 to 100 times more sensitive than the SMA. ALMA will be the first *imaging* array at mm and submm wavelengths: rather than having to rely on Earth rotation for aperture synthesis, its 2048 baseline will provide snapshot images of high quality. ALMA will provide routinely angular resolutions of order $0.1''$, and allow $0.01''$ to be reached. Finally, thanks to the combination of instantaneous imaging capability and high sensitivity, ALMA will allow complete surveys to be carried out.

3. ALMA Challenges

In order for ALMA to stand up its (great) promises, a number of challenges have to be overcome.

- *International Collaboration:* With the VLT, the Keck Observatory, Gemini, and the LBT, the optical community has access to a large number of competing facilities. On the contrary, there will be a single ALMA. The United States (NRAO) and a consortium or organizations from many European countries (ESO member states, the UK, Spain) have joined the ALMA project. Canada and Japan are participating in the studies and expected to join ALMA in the relatively near future. Although this allows a more ambitious project than any partner could afford alone, the management of such a large international consortium is a real challenge.
- *An Outstanding Site:* Since atmosphere is the main contribution to noise, it is primordial to select the best possible site to minimize its impact. The anticipated site is Llano de Chajnantor in the high Atacama Desert in Chile, is located at 5000 m altitude. Construction and operation at such an elevation are clear concerns. However, despite its remoteness, the site is only 10 km from a major, paved, international road and is crossed by a gas pipeline.
- *Low-Cost but High-Quality Antennas:* Antennas will certainly be the largest industrial contribution to the project. The performance specifications are quite tight (an ALMA antenna will be better than the best submm telescope currently in operation, yet it must operate without any radome or astrodome). However, for the ALMA project to become financially possible, the antenna cost must be kept to a minimum: the projected production cost is less than \$ 3M per antenna. To match these contradictory

requirements, the ALMA project has ordered two prototypes from two manufacturers. This will minimize the technological risk by allowing different technologies to be explored and also allow a better price control by avoiding a captive-market situation.

- *Large Series of (Excellent) Receivers:* ALMA will be able to accommodate ten different frequency bands, each in dual polarization. The total number of receivers far exceeds the receiver production of the whole world in the last ten years. Yet, each ALMA receiver should be comparable to (or even better than) the best existing prototype today.

- *High-Frequency Capabilities:* Although fringes have been obtained up to 450 GHz with the CSO-JCMT interferometer, no imaging has yet been performed at such frequencies. The signal between telescopes will need to be transported using optical fibers, with reconnection each time an antenna is moved, preserving the phase information. Such a technology has not been proven yet.

- *"Blind" Adaptive Optics:* Even on the excellent site, the atmospheric seeing at mm and submm wavelengths is of order 0.2 to $0.6''$. Obtaining the ultimate angular resolution will thus require some analogy to adaptive optics. However, it is almost impossible to find in the field of view of a single-antenna a source bright enough to serve as a phase reference. Correction of the pathlength variations due to the atmosphere will have to be done in a completely blind mode. This is possible because the pathlength fluctuations are dominated by fluctuations of the water-vapor content of the atmosphere, which also dominates (in clear sky conditions) the emissivity variations. Hence, by monitoring the sky brightness, preferably within a spectral line of water (e.g., near 183 GHz), it is possible to predict pathlength fluctuations.

- *Huge Data Rate:* ALMA will have a large number of baselines and spectral channels. The minimum dump-time is set up by Earth rotation, and is of only a few seconds. Hence the typical data rate for ALMA is of order 100 GBytes/hour and can be even larger for specific experiments such as on-the-fly mosaicking.

- *An "All-Purpose" Instrument:* ALMA will cover very different areas of astronomy. Hence, it must be designed to be easy to use for any astronomer. The design goal is to provide an instrument delivering high-quality images just as easily as a simple camera. While the simple applications (mm frequencies, narrow field of view) can already be automatically treated by current softwares, the more complex images (high frequencies, mosaics) will require advances in image processing. The simplicity of the user interface will also be an important issue.

4. ALMA and Stellar-System Formation

Star and planetary-system formation will be one of the major topics covered with ALMA. Our current understanding is heavily biased by instrumental limitations.

Detailed studies of protoplanetary disks (including chemical composition, high-angular-resolution images, multiwavelength analysis) are available for only a handful of sources located in the Taurus molecular cloud. This site of isolated star formation may not be representative of the whole process. Thanks to its fast imaging capability and high sensitivity, ALMA will be the first instrument to provide large-scale surveys of pre-stellar cores (in dust emission and spectral lines), allowing studies of the initial mass function. The combination of the wide-field imaging and high angular resolution will be essential in studying infall motions in protostellar envelopes. Proper-motion studies in molecular outflows will become possible. Because ALMA will be able to detect objects ten times further than current mm arrays, exhaustive studies of protostellar clusters of different characteristics will become possible.

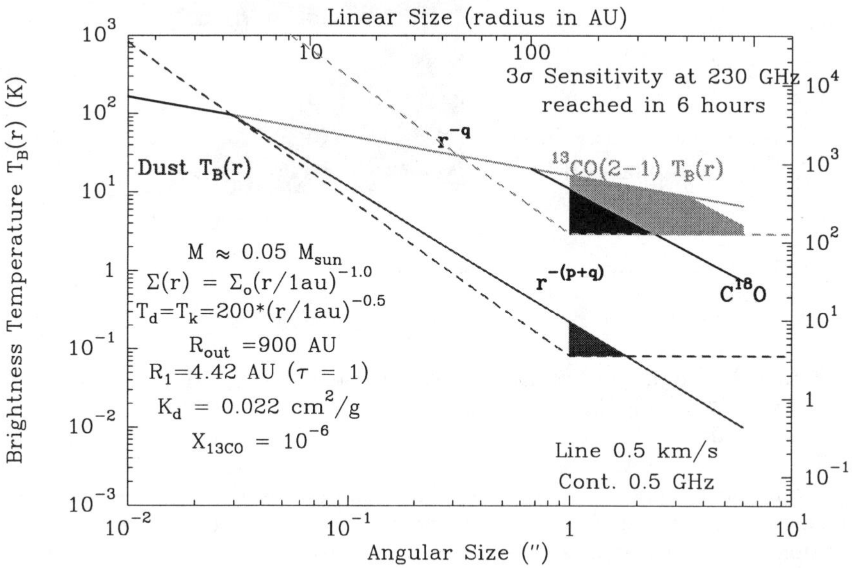

Figure 1. Brightness distribution of dust emission at 1.3 mm (*dark gray*), ^{13}CO $J = 2 \rightarrow 1$ (*gray*) and C^{18}O $J = 2 \rightarrow 1$ (*black*) lines in a circumstellar disk. The sensitivity curves at 3 σ for the continuum (*dashed line in dark gray*) and the lines (*dashed line in gray*) are given for the IRAM PdB array, for and angular resolution of $1''$. The angular size ($''$) is scaled to the Taurus distance (150 pc) with radius in AU. The interferometer is only sensitive to disk emission above the sensitivity curves. *Filled areas* correspond to the regions of the disk where the emissions are detectable and resolved. See also Dutrey (2000).

To illustrate the potential gain offered by ALMA, Figures 1–4. present the accessible parameter space for the IRAM-PdB and for ALMA, for a “representative” circumstellar disk at a distance of 150 pc. One can see that the IRAM-PdBI allows comparison between the dust emission and C^{18}O line only in a very restricted region (between 150 and 200 AU) and is insensitive to dust emission from the inner 5 AU. ALMA will be able to compare molecular emission to dust emission from 30 AU almost up to the outer edge of the disk (500 AU). In continuum, ALMA will also be sensitive to the presence of dust in the inner regions, down to 0.25 AU.

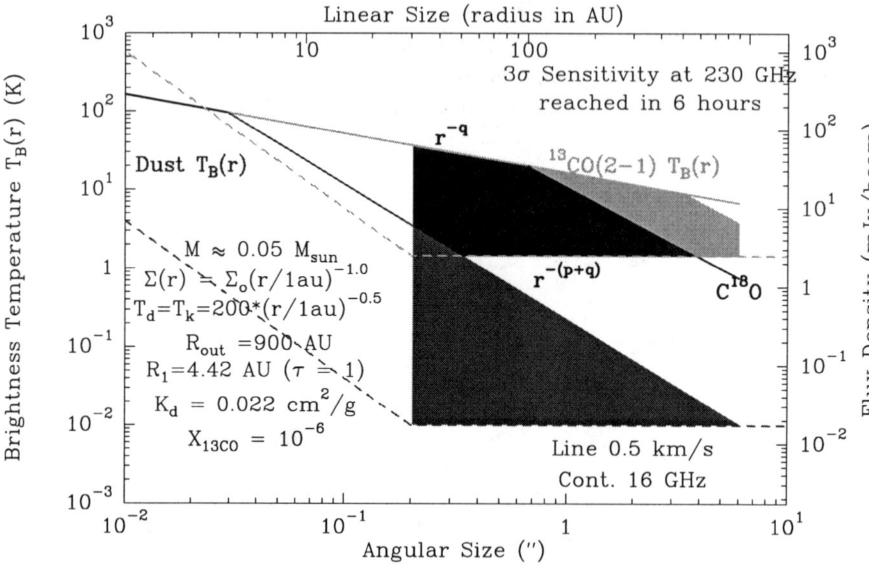

Figure 2. Same as Figure 1, but with sensitivities for the ALMA array and 0.2″ angular resolution.

The expected system temperature as function of observing frequency are given in Figures 4. and 4., with reasonable assumptions about the receiver performances (not too optimistic, but not too conservative either) and atmospheric conditions. The expected sensitivity in flux density can be derived using the classic interferometer noise formula,

$$\Delta S_\nu = \frac{\mathcal{J}}{\eta} \frac{T_{\rm sys}}{\sqrt{n(n-1)\Delta\nu}}, \tag{1}$$

where $\Delta\nu$ is the observing bandwidth, t the integration time, n the number of antennas, and $\mathcal{J}$ is the antenna gain (Jy per Kelvin) given by

$$\mathcal{J} = \frac{2k}{\eta_a A} \approx 33 \text{ Jy K}^{-1}, \tag{2}$$

η_a being the aperture efficiency and A the area of one antenna. In Equation 1, η is an efficiency factor including atmospheric decorrelation, quantization efficiency, instrumental phase noise, etc. The *brightness sensitivity* can be derived from the flux density and synthesized beam width, θ, by

$$\Delta S = \frac{2k\pi\theta^2}{\lambda^2}\Delta S. \tag{3}$$

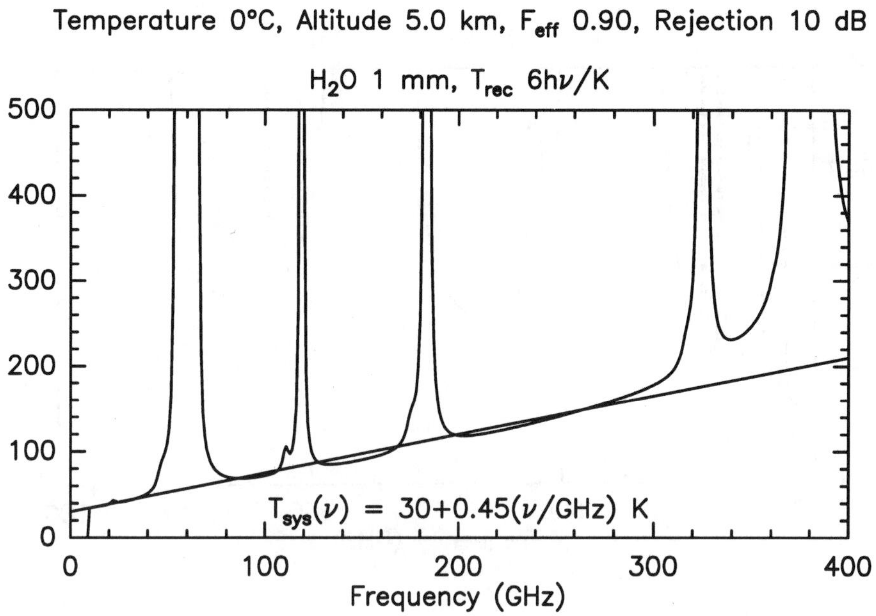

Figure 3. Typical system temperatures for ALMA in the millimeter range.

Plugging into Equation 1, the numbers from Figures 4. and 4. give a sensitivity of 10 μJy in 1 hour at 230 GHz, corresponding to 0.23 mK in brightness for 1″ resolution, and 0.5 K for the highest possible resolution (0.025″). At 900 GHz, the sensitivity in 6 hours is $\simeq$ 40 μJy, or 1.2 K at 0.007″. These numbers have to be compared to the typical surface brightness for dust emission, 0.1 to 10–100 K, as in Figure 4.. A useful way of expressing the sensitivity is in terms of surface density of material. Using the typical dust emissivity for such environment, $K_\nu = 0.1(\nu/1000\text{ GHz})\text{ cm}^2/\text{g}$ (including the mass of gas), and appropriate temperature range, e.g., 30 to 300 K, ALMA could detect surface densities of order 0.15 g/cm^2 at its higher resolution, well below the values required for giant-planet formation (a few g/cm^2).

Figures 4. and 4. give a simple simulation of how ALMA would see the inner part of a circumstellar disk in which a tidal gap has been cleared by a (proto)planet. The gap width is 1 AU, its radius 30 AU, and the gap–disk contrast 90%. The source is assumed to be located at 150 pc. Figure 4. shows that, although the angular resolution is only 4 AU, the gap is visible at 230 GHz. No noise has been added to the simulation. The expected noise level is 0.01 mJy/beam, or 0.5 K, well below the measured gap intensity, which is a few K. At higher frequencies (Figure 4.), the increased angular resolution allows the gap to be seen more clearly. The noise level, around 1.2 K, is again much smaller than the expected difference between the gap and the surrounding disk. Note that this simulation is intentionally very primitive: it includes a single

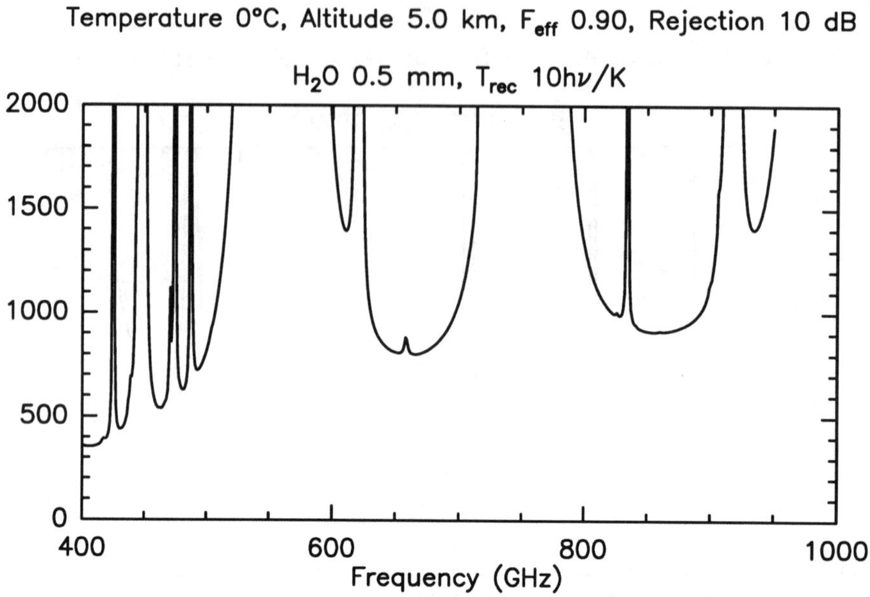

Figure 4. System temperatures for ALMA in the submillimeter range, for a precipitable water-vapor content of 0.5 mm. Such conditions are expected about 20% of the time.

configuration of ALMA, so that the "large"-scale structures (those larger than 0.2″) are not properly recovered. Inclusion of data from the more compact arrays (the 3 km ring) would avoid the resulting artifacts. Despite this simplification, the gap is nevertheless well detected.

The previous examples can be scaled for closer objects and protoplanets. The nearest star-forming region is located about 60 pc from the Sun. Gaps by a proto-jupiter orbiting a few AU from the central star in the disk of TW Hya would be detectable. Protoplanets may be difficult to detect, however. To fix orders of magnitude, let us assume a radius of 100 000 km for a proto-jupiter. At 60 pc, this is an angular size of $2.2 \times 10^{-5}\,''$. With an expected temperature of, for example, 2000 K, it would have a flux density of 0.04 μJy at 230 GHz, and 0.7 μJy at 900 GHz. Detection at such sensitivity levels would require years of observations. However, protoplanets may be surrounded by a much larger envelope/halo or even disk. Such an envelope or disk could occupy a substantial fraction of the Hill radius of the protoplanet, i.e., as large as 0.1–1 AU, and reach a luminosity of order $10^{-4}\,L_\odot$. The expected flux density would then be well above the ALMA sensitivity.

A domain in which ALMA will undoubtedly make a major contribution is the study of close binary systems and proto-binaries. While current instruments have revealed circumbinary disks around binaries of 0.3–0.8″ separation, ALMA will be able to explore a much wider domain of binary separation, at the smaller end because of its increased angular resolution, and at the larger end because of its high surface-brightness sensitivity. Since 60 to 90% of all stars form in

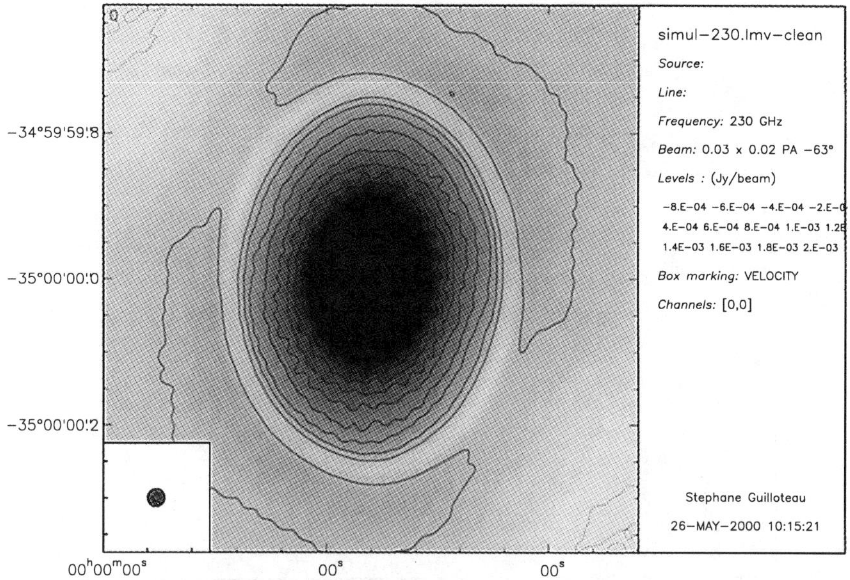

Figure 5. Simulation of a gap in a protoplanetary disk observed with ALMA at 230 GHz. Contour spacing is 0.2 mJy/beam or 10 K, about 20 times larger than the expected noise level.

binary systems, this is a key in the star and planet formation process. ALMA will be able to provide statistics on the (proto-)binary fraction in star formation regions with very different properties: from isolated stars as in the Taurus region to dense clusters as in Orion.

For the study of protoplanetary disks, another useful number is the sensitivity for spectral-line observations. For a velocity resolution of 0.15 km s^{-1}, the expected brightness sensitivity is 0.07 K in 1 hour of integration at $1''$ resolution near 230 GHz. For $0.2''$ resolution, this would be 0.7 K in 6 hours. Since the kinetic temperature in protoplanetary disks is 30–100 K or more, ALMA will be able to provide resolved images of even optically thin lines. Also, due to its wide frequency coverage, many spectral lines from different molecules, sampling widely different excitation conditions, will be observable with ALMA. Radiative transfer and excitation models can then be used to infer the gas density. These capabilities will be essential in studying the gas dissipation phase, and trying to find out whether dust and gas disappear on similar time scales or not.

5. ALMA Development and Calendar

ALMA is currently in a design and development phase until end of 2001. Funding for the construction is expected to start in 2002. One of the major activities in phase I is the construction of two prototype antennas, which should be delivered to the Very Large Array site for testing at the end of 2001. A second big activity has been the detailed costing of the project and the adjustment of the scope of

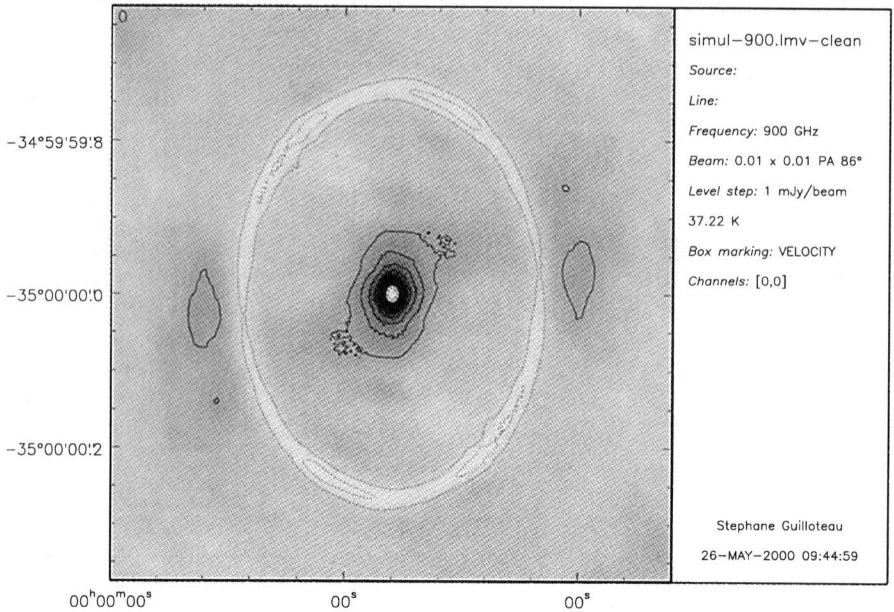

Figure 6. Same as Figure 4., but at 900 GHz. Contour spacing is 1 mJy/beam or 37 K.

the project according to the cost evaluation. ALMA is currently expected to cost ~ $550M including contingency. For such a total cost, ALMA will be equipped to cover 4 receiver bands, covering the 85–115 GHz frequency, 210–275 GHz, 275–370 GHz, and 600–720 GHz. However, the dewar will be build in order to accommodate 10 receivers, for bands from 30 GHz up to 900 GHz; the available slots may be equipped at a later time, using funds left over from the contingency. Note that the scope and cost of the project are still subjects to change. In particular, if Japan joins as a major partner, more ambitious goals would be set for ALMA. The "extended" ALMA project would certainly cover all frequency bands from 30 to 900 GHz, but also perhaps be complemented by a "compact" array to enhance its wide-field imaging capability. The compact array might even be capable of observations at even higher frequencies (above 1 THz).

In all cases, the first antennas are expected to be delivered on site in 2005. Construction will then proceed at a rate of eight to twelve antennas per year, depending mostly on financial aspects. With this scenario, ALMA first science operation would start near 2006, with an array that would already surpass in capacity all other instruments, and progressively reach its full completion near 2010.

Acknowledgments. ALMA is an international partnership between the United States (National Science Foundation) and Europe. European participants include the European Southern Observatory, the Centre National de la Recherche Scientifique (France), the Max-Planck Gesellschaft (Germany), the

Netherlands Foundation for Research in Astronomy, the United Kingdom Particle Physics and Astronomy Research Council, the Oficina de Ciencia y Tecnología/Instituto Geográfico Nacional (Spain), and the Swedish Natural Science Research Council.

References

Dutrey, A. 2000, in ASP Conf. Ser., Science with the Atacama Large Millimeter Array, ed. A. Wootten (San Francisco: ASP), in press

Wootten, A. (ed.) 2000, ASP Conf. Ser., Science with the Atacama Large Millimeter Array (San Francisco: ASP), in press

Discussion

Unidentified: What are the perspectives for constraining both the gas and dust surface density of disks inside 10–20 AU?

Stéphane Guilloteau: The limiting angular resolution to get sufficient sensitivity for spectral lines will be about $0.1''$. This is 15 AU at the typical distance of nearby star-forming regions, but $\sim$ 5 AU only for closer objects like TW Hya.

Disks, Planetesimals, and Planets
ASP Conference Series, Vol. 219, 2000
F. Garzón, C. Eiroa, D. de Winter, and T. J. Mahoney, eds.

COROT: A Space Mission for the Search for Extrasolar Planets

M. Deleuil, P. Barge, and C. Defay,

Laboratoire d'Astronomie de Marseille, BP 8, 13376, Marseille cedex 12, France

A. Léger

Institut d'Astrophysique Spatiale, Bt 121, 91405 Orsay cedex, France

D. Rouan

Observatoire de Paris, DESPA, 92195 Meudon cedex, France

J. Schneider

Observatoire de Paris, DARC, 92195 Meudon cedex, France

Abstract. *COROT* is a space mission, mainly funded by CNES, which is devoted to the search for exosolar planets and to stellar seismology. During its lifetime, about 50 000 dwarf stars will be monitored continuously, by time periods of 150 days, in order to detect possible occultations by planets. In one hour's exposure time the relative accuracy of the flux measurements is expected to reach some 10^{-4}. This high photometric precision will enable the detection of planets with radii as small as 1.5 Earth radii. Besides, one or two hundred giant 51 Peg-like planets should also be discovered. The dispersive system implemented on the exoplanet channel will help to discriminate between single- or double-transit events and false alarms; it will give access to planets much cooler than those responsible for periodic transit events and, probably, to planets in the habitable zone.

1. The *COROT* Space Mission

Planetary companions to the stars were first detected using radial-velocity measurements (Mayor & Queloz 1995); this method, however, is limited to the most massive objects, similar to the gas-giant planets of our Solar System. Planets can be also detected by taking advantage of the small decreases in stellar fluxes produced during periodic eclipses (Brown, this volume, p. 557). When performed in space, this so called "transit metho", allows high-precision photometry and gives access to the much smaller terrestrial planets. Such is the challenge of the *COROT* space mission, mainly funded by the French space agency CNES as part of the *Petites Missions* program (Baglin 1998). It is now recommended by

ESA and is opened to new European partners who contribute about 15% of the total cost. Launch is planned in 2004.

Basically, the scientific equipment of *COROT* is composed of an optical telescope with CCDs in its focal plane. The complete payload will be mounted on PROTEUS, a new-generation of platforms developed by CNES under contract with Aerospatiale, which provides a pointing accuracy of the order of $0.15''$. The satellite will be put in a quasi-polar orbit around the Earth, at an altitude of about 900 km. An important constraint imposed by this peculiar orbit, however, is that the line of sight must form an angle of more than 90° with the Sun direction and will have to be changed after a six-month period. The planned lifetime of the mission (2.5 years) will provide five periods, 150 days long each, of continuous observations.

2. The Instrument and its Performance

The instrument features a 27 cm telescope without obscuration. It consists of a two off-axis parabola plus a dioptric objective equipped with a long baffle in order to reduce very efficiently the scattered light from the Earth.

The focal plane is paved with for 2048×2048 pixel CCDs, each cooled to $-40°$. Two CCDs are dedicated to the exoplanet program located at the best focus plane. The two others are slightly out of focus for the needs of the seismology program. On the exoplanet beam, close to the plane of the CCDs, in order not to affect the asteroseismology program, is implemented a small bi-prism (see Sections 3 and 4) with very low ($R \simeq 4$) spectral resolution and zero deviation. As a result, the PSF of each star in the field of view looks like a small spectrum.

Estimates of the performance of the instrument show that the main limitations come from:

- The pointing accuracy, $0.15''$, provided by the Proteus platform which servo-controls the instrument attitude using the scientific data of a few bright stars in the asteroseismology field;

- Scattered light from the Earth which is reduced to 1.5 e^-/s/pix due to the afocal design of the telescope and the high-performance baffle;

- The zodiacal light, whose flux is expected to range from 6 to 10 e^-/s/pix, to be stable on a transit timescale and to vary in accordance with the motion of the Earth around the Sun;

- The readout noise of the CCDs, which will be less than 15 e^- and should be reduced to 10 e^-.

The consequences of the thermo-elastic deformations of the telescope, during a satellite orbit, to the form of the PSF and, finally, to the photometry should be small and will eventually be taken into account through appropriate corrections.

The main types of noise are plotted in Figure 1 as a function of the magnitude of the target star. The photon noise limit is reached for a magnitude on

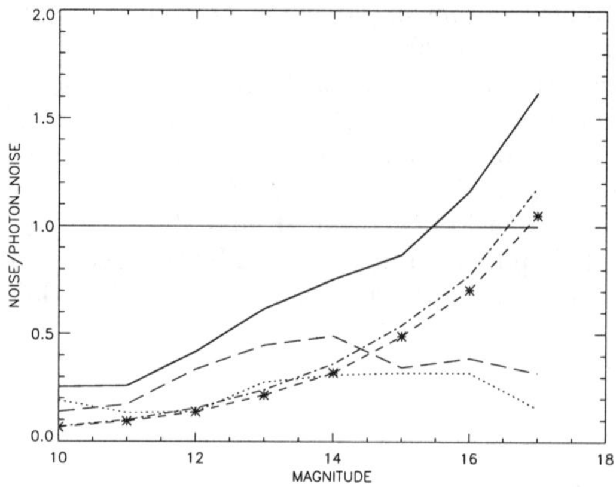

Figure 1. Noise estimates for the *COROT* instrument, given in photon noise units, as a function of the magnitude for a Sun-like star: readout noise of the CCD (*dashed-dotted line*); background noise, including scattered light from the Earth and the zodiacal light (*dashed line* with *asterisks*); defocus (*long dashes*), and pointing noise (*dotted line*). The *thick solid line* gives the total noise and the *thin solid line* is the photon noise limit.

the order of 15.5. Relatives fluxes as small as 7×10^{-4} will be measured in one hour of integration time for stars whose magnitudes range from 11 to 15.

3. The Exoplanet Program

The total field of view available for the exoplanet program is 3.5 deg^2. During the lifetime of the mission, five such stellar fields will be monitored, each one during a continuous time period of 150 days. Field selection was made in accordance with two main requirements: i) to observe regions in the vicinity of the Galactic plane, due to the constraints imposed by the satellite orbit, and ii) to satisfy simultaneously the constraints on both scientific programs: the search for exoplanets and the asteroseismology.

In the selected fields the number of stars per square degree can vary from 1500 to 4000. Such values of the stellar density provide a sufficient number of targets for the exoplanet objectives, while being compatible with a weak crowding effect among the PSFs of nearby stars.

Before starting an observational period a maximum of 12 000 targets will be selected among field stars whose magnitudes range from 11.0 to 16.0 (due to saturation and photon-noise limitation, respectively) and, as much as possible, among dwarf-type stars. *COROT* will thus monitor up to 60 000 targets during the whole lifetime of the mission. Photometry is performed on board for each target star through an aperture whose shape depends on the star's magnitude and spectral type. The elementary exposure time is 32 s, a trade-off between saturation and photon-noise limitation, but the data are co-added on board over a 16 min time period before being downloaded to the ground. For a large fraction

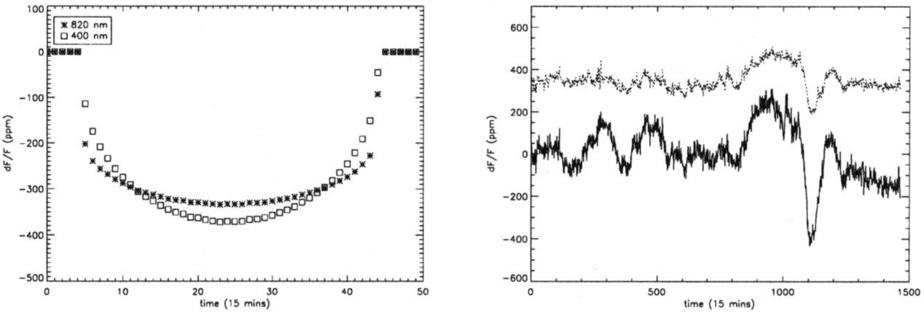

Figure 2. *Left:* Model of planetary transit in ppm at 400 nm (*rectangles*) and 820 nm (*stars*), accounting for limb-darkening effect. *Right:* The Sun's relative flux variations in ppm observed by VIRGO (Fröhlich et al. 1997) at 862 nm (*dotted line*) and 402 nm (*solid line*). The elative flux at 862 nm is shifted by 350 ppm.

of these targets, the aperture photometry will be not only realized over the total star image but over different areas, (Rouan et al. 1999) properly defined so as to provide colored fluxes. Lightcurves will be obtained in this way in different spectral bands for the largest number of target stars compatible with the actual performances of the DPU within a field of view.

4. Chromatic Information

The study of lightcurves to identify planetary transits can be undertaken using various signal-analysis methods (e.g., Defay et al. 2000a). Identification is based on three major characteristics of a transit: i) its specific shape, ii) its weak chromaticity, which is markedly different from that of the stellar activity (see Figure 2), and iii) its periodicity, which is related to the orbital radius of the planet. Each of these can be used as a criterion for the detection of transits. The main difficulty in identifying a planetary transit among other probable events in the lightcurves lies in the possible confusion between the expected signal and features of the stellar activity, for example dark spots on a rotating star's surface. In fact two cases can be distinguished depending on whether the number of transit-like features is greater or less than three. In the first case (an event with at least three transit-like features) each of the above criteria can be used simultaneously, and identification with planetary transits have a high level of confidence. With an observation period of 150 days, *COROT* is well suited to detecting unambiguously planets with orbital periods of less than 50 days. In the second case (one- or two-transit events), the periodicity criterion becomes unavailable, and the confusion between a probable planetary transit and stellar-activity features can be removed thanks to the chromatic information. For example the VIRGO data from the *SOHO* satellite show that the brightness variations of the Sun are stronger in the blue than in the red, by a factor of the order of 2.4, whereas they would be nearly the same for transit observations. As a consequence, one- or two-transit events will be identified with a good level of confidence, thus giving access to smaller orbital periods or to cooler planets.

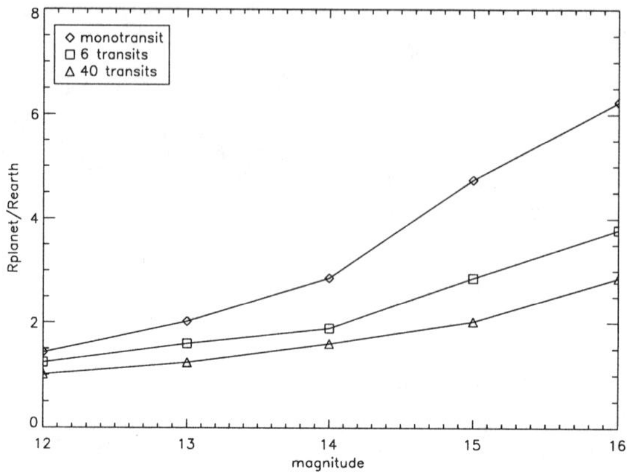

Figure 3. Minimum planetary radius (given in Earth units) detectable with an 85% confidence level as a function of the magnitude of a Sun-like parent star. The curves correspond to n-transit detection ($n = 1, 6, 40$).

Simulations (Rouan et al. 1999) show that planets in the habitable zone could be detected by *COROT* thanks to the chromatic information.

Colors will also help to discriminate in certain confusing situations such as, for example, in the case of planetary transits detected in unresolved binary systems. Indeed, colors enable the star effectively responsible for the photometric decrease to be identified and, therefore the actual radius of the planet to be determined (Rouan et al. 1999).

5. Detection Efficiency

The performance of the various methods used to identify planetary transits was tested on simulated lightcurves, such as those that will be delivered by *COROT*. These simple simulations account for:

- The readout noise
- A mean background component that includes the various contributions presented in Section 2
- The stellar variability modeled using *SOHO* data from the VIRGO mission
- The photon noise
- A planetary-transit model accounting for limb-darkening effects (Rosenblatt 1971)

The ability to detect planetary transits has been tested in a large number of cases: planets with different diameters and orbit sizes, parent stars with a wide range of apparent magnitudes and various spectral types (Defay et al. 2000b). Figure 3 shows the minimum planetary radius (given in Earth units) detectable with an 85% confidence level as a function of the magnitude of the star. In the case of single-feature events, the chromatic information allows planets with radii as small as $2R_\oplus$ to be detected for the brightest stars.

Detection is favored for small orbital radii and large transit numbers: in this case, Earth-sized planets can be detected.

6. Scientific Impact

Due to its ability to detect transit events over a wide range of planet sizes and orbital radii, *COROT* will have a broad impact on our knowledge of extrasolar planet systems and on the field of planetary formation.

It will probably be the first space mission to bring results on the challenging question of the existence of big terrestrial exoplanets, with radii of the order of a few Earth radii. If detected, these planets will most probably be the first terrestrial planets to be discovered; if not, they will provide us with a lower value on their formation probability. In any case, these results will help to put additional constraints on the way terrestrial planets form. Even if the detection of events with one or two transit features have a lower confidence level than multi-transit cases and, thus will need further confirmations, *COROT* will provide the first information on the existence of planets in the habitable zones of solar-type stars.

Hot gas giants will be very easy to detect with *COROT* and will provide good statistics concerning their frequency and radial distribution, as well as their occurrence as a function of the parent-star spectral type. It will also give access to statistical information on the radii of hot gas giants such as 51 Peg B. Of course, further studies will be possible with complementary ground-based observations. In particular, radial-velocity measurements with HARPS will give access, in a number of cases, to some important physical parameters such as the mass, and therefore the density, which are key inputs to the modeling of the internal structure of these planets.

For a small number of targets *COROT* will also be able to increase the cadence of data downloading, especially in the case of hot gas giants. For such planets, the transit profile will be sampled every two or three minutes so that a profile analysis will become possible, enabling a number of deeper studies to be made, such as the detection of possible rings or satellites (Sartoretti & Schneider 1999) and the determination of the planetary albedo.

References

Baglin, A. 1998, in IAU Symp. 185, New Eyes to See Inside the Sun and Stars: Pushing the Limits of Helio- and Asteroseismology with New Observations from the Ground and from Space, ed. F.-L. Deubner, J. Christensen-Dalsgaard, & D. Kurtz (Dordrecht: Kluwer), p. 301

Defay, C., Barge, P., Deleuil, M., & Zamkotsian, F. 2000a, in preparation

Defay, C., Deleuil, M., & Barge, P., 2000b, in preparation

Fröhlich, C., et al. 1997, Solar Phys., 170, 1

Mayor, M., & Queloz, D. 1995, Nat, 378, 355

Rosenblatt, F. 1971, Icarus, 14, 71

Rouan, D., Baglin, A., Barge, P., Copet, E., Deleuil, M., Leger, A., Schneider, J., & Toublanc, D. 1999, Phys. Chem. Earth, 24, 5, 567

Sartoretti, P., & Schneider, J. 1999, A&AS, 134, 553

Disks, Planetesimals, and Planets
ASP Conference Series, Vol. 219, 2000
F. Garzón, C. Eiroa, D. de Winter, and T. J. Mahoney, eds.

Infrared Space Interferometry—The *DARWIN* Mission

C. V. M. Fridlund

Astrophysics Division, Space Science Department, European Space Agency, ESTEC, PO Box 299, NL-2200AG Noordwijk ZH, The Netherlands

Abstract. *DARWIN* is a proposed cornerstone mission of the European Space Agency, with the express purpose of achieving unprecedented spatial resolution, in the infrared wavelength region, leading to new astrophysics and the carrying out of the first *direct* search for terrestrial exoplanets. The detection and study of the latter promises to usher in a new era in science and will affect a broad spectrum of science and technology. Further, the time line for implementation of such an instrument is now likely to be of the close order of 10 years, leading to possible answers to one of mankind's most fundamental questions in the second decade of the 21st century.

1. Introduction

The European Space Agency have already identified *interferometry in space* as an important topic in the original Horizon 2000 plan. Consequently, a number of conferences/workshops were held during the 1980's, in order to define the scientific case. Even then did the search for terrestrial exoplanets figure prominently. When an external survey committee made its recommendations for the extension of the Horizon 2000 program (H2000+), in response to technology developments worldwide, they identified interferometry from space as a potential cornerstone candidate for the new program. Within the interferometric context, three topics were identified for further study:

1. Astrometry
2. The search for terrestrial exoplanets, including the characterization of their properties and atmospheres and the possible detection of biospheres through remote sensing
3. Imaging at a spatial resolution 2–3 orders of magnitude higher than that foreseen with the *Next Generation Space Telescope* (*NGST*).

Of these the first was considered to be (relatively) "simpler" in implementation and resulted in the *GAIA* proposal and study carried out by the European Space Agency. As a result of this study, *GAIA* is, however, no longer an interferometric mission.

The *DARWIN* study concerns itself with the two latter concepts. Already during the definition phase of the *DARWIN* mission, the possibility of addressing both topics (exoplanets and imaging) in one mission was given great weight. In this paper I nevertheless focus on the search for exoplanets, and refer to the *DARWIN* report (ESA 2000) for the imaging part.

2. The Problem

To properly address the mandate of the H2000+ survey committee, it appears necessary to:

- Directly detect exoplanets
- Define and observe the fundamental requirements for life *as we know it*
- Define and observe signposts of the existence of life *as we know it*

This paper describes what may be called the "model mission". A system-level scientific and industrial study that describes a spacecraft/instrument complex that can fulfill the requirements has been carried out. It has to be realized that the version of *DARWIN* that finally flies *could* depart significantly from this model mission, particularly taking into account that possible collaborations (see below) have not been allowed to influence the current design. The background and present status of the baseline model mission is presented in this paper. I also outline the intended technology development schedule over the next few years and precursor programs, and finally provide a framework for collaboration with other space agencies.

3. Ground-Based Observations of Exoplanets–the State of the Art in the Context of *DARWIN*

The last five years have seen the detection of planets beyond our Solar System finally become a *fait accompli*. Since the first detection of a planet around 51 Peg by Queloz & Mayor (1995) several groups have together put about 30 planets onto the map of our immediate neighborhood in the Galaxy. The technique utilized is *indirect*, and we get very little information about the planet itself (essentially its minimum mass, since we do not know a priori the planetary orbit's inclination to our line of sight). The reflex motion in the parent star, with respect to the common center of mass of the star–planet system caused by the gravity of the planet is observed as a periodic Doppler shift in the stellar spectrum over a relatively long period of time. The precision of the method is therefore dependent on the accuracy, with which we can measure the spectrum of the star. This is usually done by introducing a stable calibrating source into the light path, and by utilizing regions of the spectrum where the star itself shows a suitable complement of spectral lines. This also limits the number of types of star where one can carry out this kind of search. The method is currently limited to a precision of a few m s^{-1} allowing the detection of a jupiter (12 m s^{-1}) or a saturn (3 m s^{-1}). Since Earth itself causes a reflex motion of order 0.1 m s^{-1} in the Solar spectrum, and since the acoustic oscillations of the atmospheres of solar-type stars have an amplitude of about 0.5 m s^{-1} (based on solar data), it appears unlikely that this method could be used to infer the presence of Earth-type planets. The lower limit eventually expected to be obtainable is more plausibly about the same as that of the planet Uranus (but at significantly smaller distances – the revolutionary period of Uranus is about 85 years). Since the mass determined in this fashion is dependent of the planetary orbit's inclination to the line of sight, we can usually only determine a lower limit. In conclusion, the planets found with this method are, then, objects more akin to the planet Jupiter in our system or even an intermediary step between "proper" stars (i.e., bodies generating energy through nuclear processes) and bona fide planets—-the so called brown dwarfs.

Nevertheless, important results have been, and are being reached with this method. These results include the detection of the first planetary system—υ Andromedae which contains three planets with minimum masses of 0.77 $M_{\rm Jup}$, 2.11 $M_{\rm Jup}$ and 4.61 $M_{\rm Jup}$. The planets so far discovered are also found mainly to have short orbital periods—something that has been clearly demonstrated as a selection effect caused by the period of time during which observations were carried out being relatively short. Consequently, as time passes, longer and longer periods are picked up, and there are now a number of confirmed planets in orbits with periods of a few years. Theorists, however, have problems explaining the formation of the so-called "hot jupiters"—the massive planets orbiting very near their parent stars—as well as the high value of eccentricity possessed by the majority of confirmed planets.

Another *indirect* method, is to obtain astrometric data and thus track a star's path across the sky, measuring the wobble introduced by the rotation around the common center of mass of the star–planet system. Reports of the detection of planetary companions to some nearby stars have been legion during the last century. Both Barnard's Star and 61 Cyg have several times become the central objects of planetary systems. None of these observations has ever been confirmed. In contrast, *Hipparcos* data have recently determined *upper* mass limits for a number of planets, including the first detected, 51 Peg (Perryman et al. 1996). A first attempt at an absolute mass determination for the outermost planet of υ And (10.1 $\pm$ 4.7 $M_{\rm Jup}$ (Mazeh et al. 1999) has also been carried out with *Hipparcos* data. The *GAIA* mission and the *Space Interferometer Mission* (*SIM*) promises statistical surveys of massive planets over large distances. The micro-arcsecond-order wobble introduced in the solar proper motion across the sky, as viewed over distances of $\approx$ 10 pc are, however, too small to be detected by this mission. Further, if a system consists of more than one planet, there is difficulty in characterizing the planets (orbital period, mass) uniquely and unambiguously.

4. The Scientific Case for Studying Terrestrial Exoplanets

The scientific case for *DARWIN* is easy to state but complex to describe. The phrase *To detect and study Earth-type planets and characterize them as possible abodes of life* summarizes the case nicely, but nevertheless does not project the complete picture. The simple fact is that *DARWIN* is not only an astronomy mission, but also contains elements from geophysics (including atmospheric physics), biophysics, organic chemistry, philosophy, and the humanities. Being thus totally cross-disciplinary, the preparation to answer the question put in simple terms as 'Are we alone in the Universe' necessarily involves a host of disciplines. The data from a successful *DARWIN* mission are expected to impact on all of these and more. It is no doubt one of mankind's longest-standing questions. Although for centuries it has been the topic of vigorous philosophical and religious debate, we have finally arrived at a time when technology has advanced to a state that allows this question to be properly addressed.

In order to answer question such as, How unique is the Earth as a planet? and, How unique is life in the Universe? we need to basically observe other stars and *directly* determine the existence and characteristics of any accompanying bodies. This has hitherto been impossible because of the influence of the star on attempts to observe any planets circling it.

The detection problem is essentially a matter of contrast and dynamic range. Here, picture a simple old fashioned wax candle placed very near a modern lighthouse. Now place this "model olar system" on the coast of the island of Mallorca and try to observe the wax candle from Madrid (this simile adapted very freely from M. Ollivier's (1999) thesis. The planet need to be characterized with regard to mass, orbital parameters, atmospheric composition, and temperature. All of this requires the repeated detection of the planet with a good *signal-to-noise ratio*.

The definition of an unambiguous signpost of life is another matter completely. First we need to define what life is. Then we need to determine how life affects its environment. Finally we need to define observables that can be obtained with the level of technology foreseen for the mission. In the context of *DARWIN*, we have so far avoided the first of these questions by specifying instead the requirements of the mission based on life as we have observed it on Earth. We thus disregard speculations about forms of life based on a chemistry different from that found on our planet. We then try to imagine the difference that would occur if life were nonexistent on Earth. Following Owen (1980) and Angel, Cheng, & Wolf (1986), we can then use information on how life disturbs the equilibrium in the Earth's atmosphere as our criterion.

To fully answer the questions raised above it appears to be necessary to:

- Detect planets within the "life zone" (a term coined by Frank Drake), i.e., the orbital radii where water is found in a liquid state surrounding other stars. In the *DARWIN* study, the life zone is defined in terms of a blackbody temperature and a range, and does not a priori take into account atmospheric pressure, etc. This of course assumes that life is based on the existence of liquid water.

- Determine the planet's orbital characteristics. This means we need to repeat the observations several times.

- Observe the spectrum of the planet. Does it have an atmosphere? Determine the effective temperature and total flux, as well as the diameter of planet (emitting area $\times$ albedo)

- Determine the composition of the atmosphere, viz., the presence of water and ozone/oxygen in an Earth-type planet, mainly inert gases in a Mars-/Venus-type planet and hydrogen/methane atmospheres in Jupiter-type planets (ee Figure 1*a*).

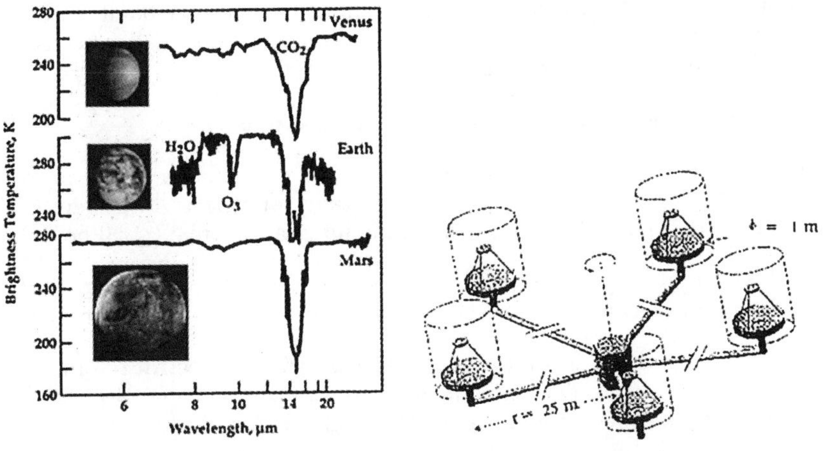

Figure 1. Representative spectra (*left*) of the planets Venus, Earth, and Mars at the relevant wavelengths. A graphical description (*right*) of the original, structured *DARWIN* concept, proposed to ESA in 1993, and taken from the paper of Legér et al. (1996).

5. The *DARWIN* Science Advisory Group and the Industrial Study

ESA received the *DARWIN* proposal in response to a call for ideas in 1993, and as mentioned above, it was among the mission concepts selected for a system-level industrial study. The *DARWIN* proposal that reached the agency has been described by Legér et al. (1996), but can briefly be described here as a nulling interferometer comprising five one-meter-class telescopes, flying at a distance of about five astronomical units from the Sun (in order to diminish the background radiation from the zodiacal dust). It was planned as a structured concept and was specifically designed for the detection and study of terrestrial exoplanets (See Figure 1*b*). Parenthetically, we can mention that also the NASA equivalent to *DARWIN* (the *Terrestrial Planet Finder—TPF*) started out at a great distance from the Sun. A re-evaluation of the dust emission in the inner Solar System led in both cases to the currently favored orbit at about 1 AU. At this point in time the community prepared for the coming study activity by having the first (European) workshop in Toledo, Spain, in March, 1996.

In order to advise ESA on *DARWIN*-related matters, a temporary external advisory body—the *DARWIN* Science Advisory Group (SAG) was formed in early 1997. The first task of this group was to aid in the preparation of the scientific specification for an Invitation To Tender (ITT) for industry. This work was carried out during the rest of 1997 and eventually, the SAG arrived at the following specification:

- *Major Goal I:* To detect Earth-like planets orbiting nearby stars and to set constraints on the possibility of the existence of *life as we know it* on those planets

- *Major Goal II:* To provide imaging in the 5μ to 28μ band, at a spatial resolution that is an order of magnitude better than that expected of the *Next Generation Space Telescope*

More specifically, these targets were quantified as follows:

1. Direct detection of an Earth-like planet at a blackbody temperature close to 300 K, circling a G0 V star at a distance of at least 10 pc (desirable 20 pc) with $S/N \geq 5$–10 in a reasonable integration time ($\leq$ 30 hours).

2. To characterize the detected planet physically through a determination of its orbital elements. This then requires observations at several epochs.

3. Obtaining the planet's thermal spectrum with a spectral resolution large enough ($\geq$ 20) to determine its atmospheric composition. See Figure 4 for a simulated spectrum of the Earth at interstellar distances.

The only item here that requires more comment is the selection of the thermal spectrum. In principle, one could also chose to utilize the reflected spectrum of the planet and detect signatures of life within the visual or ultraviolet spectral band. As detailed below, observations in these wavelength bands would, however, severely constrain the mission in a number of ways.

This specification was provided in the ITT offered to European industry in 1997 September. After a tender evaluation of industrial proposals to ESA, Alcatel Space Division (at the time Aerospatiale Space Division) was selected, thus concluding the evaluation phase. Their study began in early 1998 and ended in mid-April of 2000 (ESA 2000).

The feasibility study was initiated with the specific goals of:

- Defining a model mission that would be able to achieve the scientific goals

- Identifying a technology development program

- Identifying the necessary precursor missions (ground- and space-based)

The current baseline for the model mission—which was at the core of the final report of Alcatel in April 2000 (ESA 2000)—consists of six free-flying 1.5 meter telescopes, flying in a hexagonal configuration. In the center of the array a beam-combining satellite equipped with optical benches for both the "nulling" and the "imaging" parts of the mission, is located (see Figure 2*a*). All components found in the optical path, from the main mirrors to the output of the beam-combination station need to be passively cooled to $\leq$ 40 K, something which the thermal modeling performed by Alcatel indicates is possible if the mission is flown in an orbit far away from sources of heat input such as the Earth. The simulations then indicate that the detection of the Earth out to a distance of 20 pc and the observation of its spectrum with a resolution $\leq$ 20 (see Figure 4), is possible. In order to minimize the influence of the zodiacal dust emission at 10μm, the observing zone is a 40 degree cone around the anti-Sun axis. This configuration goes under the name of the Robin Laurance configuration in order to honor the first study manager of *DARWIN* at ESTEC, who tragically passed away in 1999. It was developed by A. Karlsson at ESTEC and is illustrated in

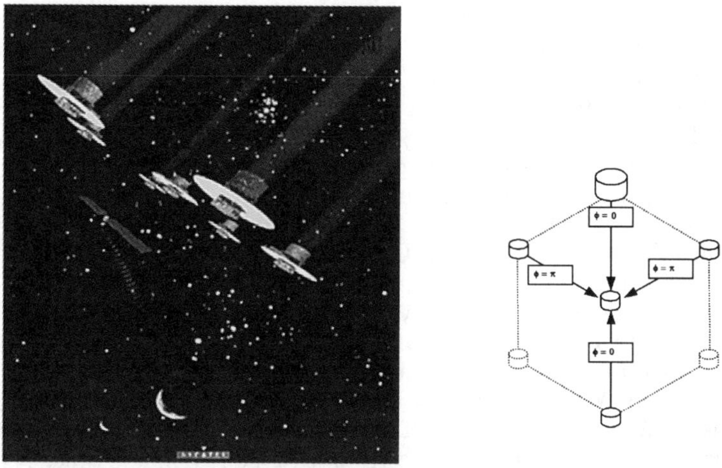

Figure 2. *Left:* An artist's impression of the current *DARWIN* configuration, seen orbiting at the Sun–Earth *L*2 point (picture courtesy of Alcatel). *Right:* Graph depicting the arrangement of the Robin Laurance configuration described in the text as the current baseline of *DARWIN*. It is a planar configuration and thus has a good thermal situation for the passive cooling.

Figure 2*b*. This planar configuration (good thermal environment) combines a manageable size of the transfer optics with a good rejection ratio of the central null, and can use both internal modulation and simple translational movement for discriminating, e.g., exozodiacal light from the signature of an Earth-type planet. There are no long delay lines and the optical benches of the hub satellite can be made relatively more simple. Finally, it allows, in principle, for both an imaging and a nulling mission to be flown on the same space craft.

The six telescopes, the hub beam-combiner satellite and a separate power and communications spacecraft are all foreseen to be launched by a single Ariane 5 launch vehicle into a direct transfer orbit to the *L*2 Lagrangian point in the Sun–Earth system.

6. The Technique of Nulling Interferometry and *DARWIN*

As mentioned before, directly detecting exoplanets is essentially a matter of contrast and dynamical range. The star around which the planet revolves is going to be $\geq 10^9$ times brighter than the planet if we choose visual light to observe the planet in. The planet is also going to be very close. An Earth-type planet in an orbit in the "life zone", i.e., 1 AU from its sun, is going to be 1 arcsec away at a distance of 1 parsec. Unfortunately, the closest potential target we have is α Centauri at 1.1 pc. In order to get a suitable number of targets, we have to reach at least 10 pc out and preferably 25 pc, thus making the life zone viewable at a distance of 0.1 and .04 arc seconds respectively. And this is for the solar-type targets. For K-type and M-type stars, which will dominate our sample, the separation distance will be much smaller, because of the demand for observing planets within the zone where we expect water to be liquid. It is clear that, using conventional telescopes, our detectors would have to handle an impossible dynamical range in order to separate out the planetary light.

The problem is alleviated by going to the infrared. As can be seen from Figure 3*a*, the relative contrast between primary and planet drops several orders of magnitude by going to the thermal infrared. This was first pointed out by Bracewell (1978) and Angel et al. (1986). These authors also pointed out the advantage of utilizing lines of the H_2O, O_3, and CO_2 molecules as tracers of life as we know it. Kasting, Holland, & Pinto (1985) and Legér, Pirre, & Marceau (1993) have demonstrated that O_3 is a very nice tracer for O_2 since the former has a logarithmic dependence on the concentration of the latter. We then have suitable tracers for Earth-type atmospheres in a wavelength band between 6 μm and about 20μm. We can still allow only 10^{-6} of the stellar light to remain in the input feed to our spectrograph if we would have any hope of detecting the planetary light and registering its spectrum within a reasonable time. It can be shown that coronagraphic methods on monolithic telescopes do not suffice to accomplish such an extinguishing of light at the relevant spatial scales. To overcome this hurdle, the technique of "nulling interferometry", as suggested by, for example, Bracewell & McPhie (1979) and Angel (1989) is baselined for *DARWIN*. A nulling (Bracewell) interferometer can best be described by considering two telescopes. By restricting the beams coming from each of the telescopes to the diameter of the point spread function and after parallelizing the beams. We then introduce a phase shift of $\phi = \pi$ in one of the ray paths, we will achieve destructive interference on the optical axis of the system (in the combined beam). At the same time, we will have constructive interference a small angle θ away. This angle θ depends on the distance between the two input telescopes. This is demonstrated in Figure 3*b*. The output of this system is a set of interference fringes (a "map"), with a sharp "null" in the center of the "map". By placing the central star under this null, and the zone where, for example, H_2O will be liquid under a bright fringe, one can in principle search for planets in the 'life zone". Now, by using more telescopes we can achieve a symmetric pattern around the star, with a deep central null on top of the star. The actual shape and transmission properties of the pattern around the central "null" depends on the configuration, and the distance between the telescopes. Essentially, if we had an ideal case with a star and a single planet and no disturbing sources, such as exosolar zodiacal dust, i.e., dust left over from collisions between asteroids, comets and suchlike in the target system, the detection of a positive flux would imply that a planet is present, if the star is well and truly "nulled out". This means that the detector could consist of a single element. We wish, however, to also carry out spectroscopy with as high a resolution as possible, and a linear array is thus indicated. In the real world, there is of course a significant amount of background radiation coming from dust. The zodiacal dust in our own Solar System is strong enough to be seen (in visible light) as a bright band, from the ground in dark locations. At a wavelength of 10 μm, this radiation is dominant (the zodiacal dust temperature within the "life zone" will of course also be close to 300 K and thus the peak of the emission is radiated around 10 μm), and a significant background signal is thus present in the inner Solar System. When observing the Solar System at interstellar distances, the zodiacal dust is

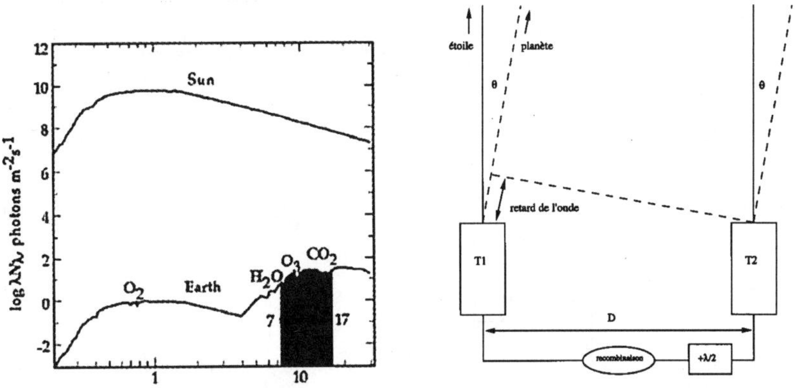

Figure 3. *Left:* The contrast between the solar flux and the Earth as a function of wavelength in microns. The *DARWIN* spectral band is marked. Adapted from Angel, Cheng, & Woolf (1986). *Right:* A two-telescope nulling interferometer. The angle θ depends on the distance D. Drawing courtesy IAS, Orsay, France.

actually about 400 times brighter than the Earth at these wavelengths (Legér et al. 1996). In order to separate out signal of the planet from this background one needs to modulate the signal from the planet and from the zodiacal dust at different frequencies. This is done either by switching between different combination schemes (certain geometrical arrangements of the telescope array), or by moving the individual telescopes around, or both (see Legér et al. 1996 for a thorough discussion of this topic).

7. Technology Development and Precursor Missions

In the specifications for the study carried out by Alcatel, one prominent item was to identify a technology development program leading to a *DARWIN* mission as soon as possible. This is also the topic of an ongoing effort within ESA at ESTEC. As a consequence, an ambitious technology development plan has been initiated—including space qualification and verification—within the context of the European Space Agency's SMART program. Activities planned for the next three years include development and construction of the following:

- High-stability optical benches.
- Optics active control.
- Fiber-optic wavefront filtering, Single-mode fibers—operating in the 10 μm region of the spectrum. An investigation into the phasing capabilities of fibers.
- Integrated optics, optical components for nulling and imaging interferometry.

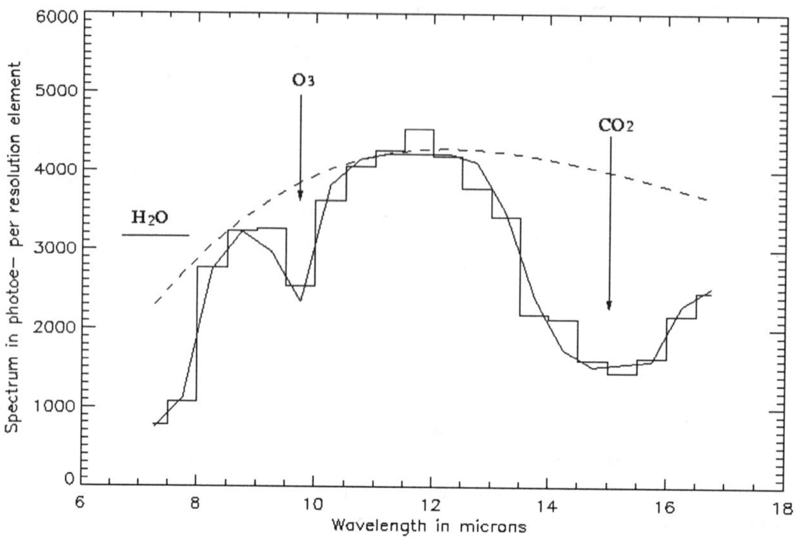

Figure 4. The Earth's spectrum as it would appear when observed with *DARWIN* from a distance of 10 pc. The spectral range has here been specified as 7–17 μm. As can easily be seen, ideally, one would like to observe a larger part of the spectrum in order to detect the thermal continuum on both sides of the H_2O and CO_2 lines.

- Achromatic phase shifters for nulling interferometry.
- Detectors and cooling systems.
- Satellite formation flying: Deployment and control. A local GPS system for 1 cm positional accuracy.
- Ultra-high precision (laser-)metrology.
- Field electric emission propulsion (FEEP) technology

One of the key issues is to develop and verify nulling interferometry. It is currently not deemed necessary to demonstrate this in space, but a representative breadboard with an associated simulator providing the necessary input signals have to be designed and built within the next few years. An associated precursor activity is the observation of the exozodiacal light from target systems. This can be carried out from the ground, using the breadboard in conjunction with a large enough interferometer. A possible instrument would be the Very Large Telescope Interferometer (VLTI). By using the breadboard together with the

VLTI, its qualification would then also provide needed, and in itself interesting, scientific information.

One of the most challenging technological aspects of the *DARWIN* mission, is that it requires to fly several spacecraft in the close vicinity of each other and with centimeter precision. The control of the space craft are foreseen—during an observation—to be performed with μnewton FEEP thrusters (the largest disturbing force in an $L2$ orbit is the differential solar photon pressure on the individual flyers). Deployment and acquisition of sources are carried out by using mN FEEP thrusters. A local GPS system keeps the spacecraft to within one centimeter of their intended position, while a laser metrology system measures the actual positions to within 5 nm. A separate channel in the interferometric system will observe the fringes from the central star in the target system—which is nulled out in the "science channel"—and since the observed systems are all relatively nearby, there is no lack of photons for tracking the fringes. This information is fed into the control loop of the interferometers two nulling and imaging circuits, as well as into the attitude and control system of the individual space craft. It is indicated that this technology requires a precursor mission to actually test:

- Deployment, acquisition of observing positions and control of a flotilla of space craft
- ˇhe metrology components—fringe tracker, laser system
- Software and hardware of the control system
- milli- and μnewton thrusters

Thus we are currently introducing this technology into the SMART precursor mission program. It should be noted in this context that most of the new technology required for *DARWIN* also has other valuable uses. As obvious examples we can mention flotilla flying (for communication-satellite purposes), interferometry (Earth observations of many kinds) and the ultra-precision laser metrology with many different uses. This then provides an additional source of support for *DARWIN*, since many technology programs will be funded (sometimes partly) outside science. It is also an added factor in ensuring that the programs will be carried out relatively fast.

8. International Collaboration

The implementation of space interferometry is also pursued in places other than Europe. Disregarding the radio VLBI-type experiments (such as the Japanese *HALCA*, which is already flying), which does not have the capability of searching for terrestrial exoplanets, the two most obvious examples are the *SIM* (*Space Interferometry Mission*) and *TPF*. Of these, SIM is an optical interferometer which will use relative astrometry to attempt the detection of "super-earths", i.e., planets of 5–10 Earth masses—thus through an indirect method. Another of *SIM*'s goals is to carry out nulling interferometry to one part in ten thousand—at visual wavelengths. Because of these targets, *SIM* can be considered a precursor mission to *DARWIN* and *TPF*. The NASA counterpart of *DARWIN* is *TPF*.

TPF has the same objectives and utilizes the same technology as that foreseen for *DARWIN*. Given that both projects—*DARWIN* and *TPF*—are extremely complex, will require very ambitious technology development programs, and are likely to be very expensive, it would make sense if NASA and ESA were to cooperate. In this context, discussions have begun in order to see if a reasonable joint mission could be designed. If these discussions are successful, a joint mission could be launched sometime around 2012.

Acknowledgments

This paper reflects the work of the complete SAG of *DARWIN* over several years, including a number of experts who have participated in the development of the concept. We all pay tribute to our colleagues Jean Marie Mariotti and Robin Laurance who passed away during the study but who will be with us in spirit at the completion of the task set before us.

References

Angel, J. R., Cheng, A. Y., & Woolf, N. J. 1986, Nat, 322, 341

Angel, R. 1989, in The Next Generation Space Telescope, ed. P. Bely, C. Burrows & G. Illingworth (Baltimore: STScI), p. 81

Bracewell, R. N. 1978, Nat, 274, 780

Bracewell, R. N., & McPhie, R. H. 1979, Icarus, 38, 136

ESA 2000, DARWIN The InfraRed Space Interferometer—Concept and Feasibility Study Report, ESA-SCI

Kasting, J. F., Holland, H. D., & Pinto, J. P. 1985, J. Geophys. Res., 90, 10, 497

Léger, A., Pirre, M., & Marceau, F. J. 1993, A&A, 277, 309

Léger, A., Mariotti, J. M., Mennesson, B., Ollivier, M., Puget, J. L., Rouan, D., & Schneider, J. 1996, Icarus, 123, 249

Mayor, M., & Queloz, D. 1995, Nat, 274, 521

Mazeh, T., Zucker, S., Dalla Torre, A., & van Leeuwen, F. 1999, ApJ, 522, L149

Ollivier, M. 1999, PhD Thesis, Paris XI

Owen, T. 1980, in Strategies for the Search of Life in the Universe, ed. M. D. Papagiannis (Dordrecht: Reidel), p. 177

Perryman, M. A. C., et al. 1996, A&A, 310, L21

Discussion

Jean-Loup Bertaux: Who is managing the joint ESA–NASA study of *DARWIN*/TPF?

Carlos Eiroa: As far as I know, ESA and NASA people are directly involved in both agencies.

In Conclusion . . .

Disks, Planetesimals, and Planets
ASP Conference Series, Vol. 219, 2000
F. Garzón, C. Eiroa, D. de Winter, and T. J. Mahoney, eds.

A Mini-History of Disks and Their Progeny

Antonella Natta

Osservatorio di Arcetri, Largo Fermi 5, 50125 Firenze, Italy

Abstract. In this talk, I will provide a very brief and personal account of the recent history of the research on disks, disk evolution, and planet formation.

1. Introduction

This conference has been extremely interesting, and rich in new ideas and results. The task of providing a summary of it is correspondingly difficult, and I have decided that I will not even try to do it. These proceedings collect all the invited talks and posters presented at the conference, and any interested reader can get a feeling for the wealth of work in the various areas presented during the meeting.

What I would like to do instead is to provide a brief history of the main areas we have discussed, describing briefly how scientific communities that have progressed almost independently in the last decades are now beginning to interact with one another.

Since I am not a historian, but am relying on my own recollection of 'how things happened' during my own working life, my summary will be highly incomplete and necessarily biased toward those areas closest to my own research, which has centered on circumstellar disks and early stellar evolution. In what follows, there is no effort to assess priorities. I just thought that it could be interesting, and not just for the youngest in the audience, to recollect those results that have been more influential in establishing some of the ideas we have discussed in the past few days. With very few exceptions, I will quote observations rather than theories. I apologize in advance to all of you who will feel that I am misrepresenting facts and people.

2. Disks around Pre-Main-Sequence Stars

The idea of circumstellar disks has been around for at least two centuries, from the Solar Nebula of Laplace onwards. However, I think that from the point of view of modern research, we can start in 1974 when Lynden-Bell and Pringle applied the already-existing models of accretion disks to the formation of stars. Nobody objected to the fact that the collapse of a rotating cloud would naturally result in the formation of a disk. However, it took about ten years before accretion disks could find observational support. The first evidence was indirect. Among the various clues, I would like to quote the bipolarity of molecular outflows (Snell, Loren, & Plambeck 1980) and jets (see, for a detailed list of references, the review by Mundt 1985), the asymmetry of the forbidden emission lines in T Tauri stars (Appenzeller 1983), and the low extinction toward T Tauri stars with strong near-infrared excess (Rydgren & Cohen 1985). In 1983 *IRAS* was launched and detected infrared emission from a large number of T Tauri stars. To my knowledge, the first paper that summarized the *IRAS* discoveries and proposed a disk origin for the infrared emission was by Rucinski (1985). I do not think that anyone in the community had at that point much doubt that accretion disks exist and play an important role in the formation of stars, at least as far as low-mass stars are concerned. However, we had to wait several years before we could actually "see" disks, in the first beautiful *HST* images of the Orion Nebula (O'Dell et al. 1992). Several of them, in fact, had already been detected through their radio emission and identified with evaporating protostellar accretion disks by Churchwell et al. (1987). However, their "black and white" science did not have the same impact on the public!

Today, we can "see" disks at all wavelengths from visual to several millimeters and we can study not only their morphology but also their dynamical properties using molecular lines resolved by millimeter interferometers or water-maser proper motions. In some cases, *ISO* has given us information on the mineralogy of the dust in the disks, revealing the close connection between pre-main–sequence disks and the Solar System comets.

3. Secondary Disks

IRAS detected infrared excess emission not only around many T Tauri stars but also, much more unexpectedly, around several nearby main-sequence stars. The paper reporting the discovery of such an excess around Vega was published in 1984 March (Aumann et al. 1984). A few months later, Smith & Terrile (1984) published the first image in optical light of the disk around β Pic, which had also been detected by *IRAS* in the infrared. This image has been for many years the only image of a circumstellar disk, and, as such, loved by all of us!

It was clear from the beginning that the β Pic disk (and all the others associated with main-sequence stars) had to be very different from the disks around T Tauri stars. Because they lack gas and are optically thin, the dust we observe has to be continuously replenished, probably by fragmentation of larger solid bodies. These disks, which I have called "secondary", are more closely related to planetesimals and planets than to disks around T Tauri stars. Their geometry, structure, and dust properties are likely to reflect the processes occurring in very young planetary systems. In discussing disk evolution, we should remember that the formation of planetesimals and the evaporation of the gas introduce a discontinuity in the disk history.

A further indication of the close relationship of secondary disks to planetary systems comes from the discovery by Jewitt and Luu of the first Kuiper Belt object (Jewitt, Luu, & Marsden 1992). The fact that Kuiper Belt dust can be the analog of the dusty disks seen around some main-sequence stars is gaining growing support from recent observations, such as, the detection of a secondary disk in 55 Cnc (Dominik et al. 1998; Trilling & Brown 1998), a system which hosts also a jovian planet (Butler et al. 1997).

At this conference, we have seen how the study of secondary disks has progressed. From having only one star to study, β Pic, we have now several. We have seen images in scattered light and in the submillimeter emission of a few objects, which are beginning to give us hints on the evolution of such systems from the young, rather massive secondary disk of β Pic to the Solar System.

4. Exosolar Planets

The announcement by Michel Mayor of the discovery of a jovian planet in 51 Peg was an emotional moment for many of us. It happened in Florence, during the 9th Cool Stars and Stellar Systems conference in 1995 October. This announcement was the culmination of a search conducted for several years by various groups, with no apparent success. Just few months earlier, the group that had been leading the way in this search announced a discouraging negative result (Walker et al. 1995). As it turned out, success was determined by the very large number of stars monitored by the Geneva group and opened the way to a sequence of discoveries, that has not yet ended. The recent detection of a transit of the planet in HD 209458 (Charbonneau et al. 2000), which has allowed us to measure the planet density, has added confidence to our belief that we are indeed detecting many Jupiter-like planets of unexpected properties.

Here, we have heard of the many techniques that are currently used to discover new planets and to explore other areas of parameter space. I am sure that in a few years our ideas about planetary systems will have changed greatly from our viewpoint a few years ago, when the only known planetary system was the solar one.

5. The EXPORT Project

This conference has been organized within the framework of the EXPORT project. EXPORT is a consortium of scientists from different countries who prepared a proposal for the La Palma 1998 International Time Programme to investigate disk evolution, planet formation, and exosolar planets using the La Palma telescopes. The program was successful in obtaining time (the first non-extragalactic program to be approved, as we have been told) and we have heard here from several of the EXPORT participants a description of the data they have obtained. The success of this conference proves that the EXPORT approach was timely. The analysis of the data has just started, and we will hear more about EXPORT in the coming months.

6. Conclusions

At the end of a conference, we generally ask ourselves what new results we have heard; if the meeting was good and productive; if we have had fun or not. My impression, which I think is shared by most of us, is that this was a delightful conference, one that we all enjoyed. Why? Not only because new, very interesting results have been presented, but, more importantly, because we feel that we have been witnesses to a time when connections between different areas of astrophysical research have become real. For many years I have known that I should pay more attention to the work of my colleagues working on the Solar System, without ever really doing so. I am sure that we all have had the same experience, of sitting through talks of very gifted scientists, trying to find out what was relevant to our own research in what they were saying (we knew that there should have been something) and finding nothing!

I have the feeling that this has changed utterly. The story of disks has now a better-defined beginning and many ends, as new planetary systems with unexpected properties are discovered. I have been impressed by the theoretical talks we have heard at this conference and by the need they all showed to address disk evolution and planet formation using all the observational information we have, from the early stages of accretion disks to the Kuiper Belt properties.

Let me finish by saying that I find the discovery of exosolar planets at a crucial moment in this area of astrophysics not only because we, human beings, wish to know about other earths. After all, not one of us would like to live on any of the systems found so far! More important to me is the fact that the discovery of exosolar planets and their unexpected properties have acted as a catalyst to various lines of research that had proceeded up till now independently of one another. The crossings and mergings that we have discussed in the past week make this an exciting and happy time.

References

Appenzeller, I. 1983, Rev .Mex. Asron. Astrofis., 7, 151

Aumann, H. H., et al. 1984, ApJ, 278, L23

Butler, R. P., Marcy, G. W., Williams, E., Hauser, H., & Shirts, P. 1997, ApJ, 474, L115

Charbonneau, D., Brown, T. M., Latham, D. W., & Mayor, M. 2000, ApJ, 529, L45

Churchwell, E., Felli, M., Wood, D. O. S., & Massi, M. 1987, ApJ, 321, 516

Dominik, C., Laureijs, R. J., Jourdain de Muizon, M., & Habing, H.J. 1998, A&A, 329, L53

Jewitt, D., Luu, J., & Marsden, B.G. 1992, IAUC, 5611

Lynden-Bell, D., & Pringle, J. 1974, MNRAS, 168, 603

Mayor, M., & Queloz, D. 1996, in ASP Conf. Ser., Vol. 109, Cool Stars, Stellar Systems and the Sun, ed. R. Pallavicini & A. K.Dupree (San Francisco: ASP), p. 35

Mundt, R. 1985 in Protostars and Planets II, ed. D. C. Black & M. S. Matthew (Tucson: University of Arizona Press), p. 414

O'Dell, C. R., Wen, Z., Hu, X., & Hester, J. J. 1992, AAS, 181, n.17.01

Rucinski, S. M. 1985, AJ, 90, 2321

Rydgren, A. E., & Cohen, M. 1985 in Protostars and Planets II, ed. D. C. Black & M. S. Matthew (Tucson: University of Arizona Press), p. 371

Smith, B. A., & Terrile, R. J. 1984, Sci, 226, 1421

Snell, R. L., Loren R. B., & Plambeck, R. L. 1980, ApJ, 239, L17

Trilling, D. E., & Brown, R. H. 1998, Nat, 395, 775

Walker, G. A. H., Walker, A. R., Irwin, A. W., Larson, A. M., Yang, S. L. S., & Richardson, D. C. 1995, Icarus, 116, 359

Author Index

ASP CONFERENCE SERIES VOLUMES

and

INTERNATIONAL ASTRONOMICAL UNION (IAU) VOLUMES

Published by

Astronomical Society of the Pacific

ASP CONFERENCE SERIES VOLUMES

Published by the Astronomical Society of the Pacific

PUBLISHED: 1988 (* asterisk means OUT OF STOCK)

Vol. CS -1 PROGRESS AND OPPORTUNITIES IN SOUTHERN HEMISPHERE OPTICAL ASTRONOMY: CTIO 25TH Anniversary Symposium
ed. V. M. Blanco and M. M. Phillips
ISBN 0-937707-18-X

Vol. CS-2 PROCEEDINGS OF A WORKSHOP ON OPTICAL SURVEYS FOR QUASARS
ed. Patrick S. Osmer, Alain C. Porter, Richard F. Green, and Craig B. Foltz
ISBN 0-937707-19-8

Vol. CS-3 FIBER OPTICS IN ASTRONOMY
ed. Samuel C. Barden
ISBN 0-937707-20-1

Vol. CS-4 THE EXTRAGALACTIC DISTANCE SCALE:
Proceedings of the ASP 100th Anniversary Symposium
ed. Sidney van den Bergh and Christopher J. Pritchet
ISBN 0-937707-21-X

Vol. CS-5 THE MINNESOTA LECTURES ON
CLUSTERS OF GALAXIES AND LARGE-SCALE STRUCTURE
ed. John M. Dickey
ISBN 0-937707-22-8

PUBLISHED: 1989

Vol. CS-6 SYNTHESIS IMAGING IN RADIO ASTRONOMY:
A Collection of Lectures from the Third NRAO Synthesis Imaging Summer School
ed. Richard A. Perley, Frederic R. Schwab, and Alan H. Bridle
ISBN 0-937707-23-6

Vol. CS-7 PROPERTIES OF HOT LUMINOUS STARS: Boulder-Munich Workshop
ed. Catharine D. Garmany
ISBN 0-937707-24-4

PUBLISHED: 1990

Vol. CS-8* CCDs IN ASTRONOMY
ed. George H. Jacoby
ISBN 0-937707-25-2

Vol. CS-9 COOL STARS, STELLAR SYSTEMS, AND THE SUN: Sixth Cambridge Workshop
ed. George Wallerstein
ISBN 0-937707-27-9

Vol. CS-10* EVOLUTION OF THE UNIVERSE OF GALAXIES: Edwin Hubble Centennial Symposium
ed. Richard G. Kron
ISBN 0-937707-28-7

Vol. CS-11 CONFRONTATION BETWEEN STELLAR PULSATION AND EVOLUTION
ed. Carla Cacciari and Gisella Clementini
ISBN 0-937707-30-9

Vol. CS-12 THE EVOLUTION OF THE INTERSTELLAR MEDIUM
ed. Leo Blitz
ISBN 0-937707-31-7

PUBLISHED: 1991

Vol. CS-13 THE FORMATION AND EVOLUTION OF STAR CLUSTERS
ed. Kenneth Janes
ISBN 0-937707-32-5

ASP CONFERENCE SERIES VOLUMES

Published by the Astronomical Society of the Pacific

PUBLISHED: 1991 (* asterisk means OUT OF STOCK)

Vol. CS-14 ASTROPHYSICS WITH INFRARED ARRAYS
ed. Richard Elston
ISBN 0-937707-33-3

Vol. CS-15 LARGE-SCALE STRUCTURES AND PECULIAR MOTIONS IN THE UNIVERSE
ed. David W. Latham and L. A. Nicolaci da Costa
ISBN 0-937707-34-1

Vol. CS-16 Proceedings of the 3rd Haystack Observatory Conference on
ATOMS, IONS, AND MOLECULES: NEW RESULTS IN SPECTRAL LINE ASTROPHYSICS
ed. Aubrey D. Haschick and Paul T. P. Ho
ISBN 0-937707-35-X

Vol. CS-17 LIGHT POLLUTION, RADIO INTERFERENCE, AND SPACE DEBRIS
ed. David L. Crawford
ISBN 0-937707-36-8

Vol. CS-18 THE INTERPRETATION OF MODERN SYNTHESIS OBSERVATIONS OF SPIRAL GALAXIES
ed. Nebojsa Duric and Patrick C. Crane
ISBN 0-937707-37-6

Vol. CS-19 RADIO INTERFEROMETRY: THEORY, TECHNIQUES, AND APPLICATIONS,
IAU Colloquium 131
ed. T. J. Cornwell and R. A. Perley
ISBN 0-937707-38-4

Vol. CS-20 FRONTIERS OF STELLAR EVOLUTION: 50th Anniversary McDonald Observatory (1939-1989)
ed. David L. Lambert
ISBN 0-937707-39-2

Vol. CS-21 THE SPACE DISTRIBUTION OF QUASARS
ed . David Crampton
ISBN 0-937707-40-6

PUBLISHED: 1992

Vol. CS-22 NONISOTROPIC AND VARIABLE OUTFLOWS FROM STARS
ed. Laurent Drissen, Claus Leitherer, and Antonella Nota
ISBN 0-937707-41-4

Vol CS-23 ASTRONOMICAL CCD OBSERVING AND REDUCTION TECHNIQUES
ed. Steve B. Howell
ISBN 0-937707-42-4

Vol. CS-24 COSMOLOGY AND LARGE-SCALE STRUCTURE IN THE UNIVERSE
ed. Reinaldo R. de Carvalho
ISBN 0-937707-43-0

Vol. CS-25 ASTRONOMICAL DATA ANALYSIS, SOFTWARE AND SYSTEMS I - (ADASS I)
ed. Diana M. Worrall, Chris Biemesderfer, and Jeannette Barnes
ISBN 0-937707-44-9

Vol. CS-26 COOL STARS, STELLAR SYSTEMS, AND THE SUN: Seventh Cambridge Workshop
ed. Mark S. Giampapa and Jay A. Bookbinder
ISBN 0-937707-45-7

Vol. CS-27 THE SOLAR CYCLE:
Proceedings of the National Solar Observatory/Sacramento Peak 12th Summer Workshop
ed. Karen L. Harvey
ISBN 0-937707-46-5

ASP CONFERENCE SERIES VOLUMES

Published by the Astronomical Society of the Pacific

PUBLISHED: 1992 (asterisk means OUT OF STOCK)

Vol. CS-28 AUTOMATED TELESCOPES FOR PHOTOMETRY AND IMAGING
ed. Saul J. Adelman, Robert J. Dukes, Jr., and Carol J. Adelman
ISBN 0-937707-47-3

Vol. CS-29 Viña del Mar Workshop on CATACLYSMIC VARIABLE STARS
ed. Nikolaus Vogt
ISBN 0-937707-48-1

Vol. CS-30 VARIABLE STARS AND GALAXIES
ed. Brian Warner
ISBN 0-937707-49-X

Vol. CS-31 RELATIONSHIPS BETWEEN ACTIVE GALACTIC NUCLEI AND STARBURST GALAXIES
ed. Alexei V. Filippenko
ISBN 0-937707-50-3

Vol. CS-32 COMPLEMENTARY APPROACHES TO DOUBLE AND MULTIPLE STAR RESEARCH, IAU Colloquium 135
ed. Harold A. McAlister and William I. Hartkopf
ISBN 0-937707-51-1

Vol. CS-33 RESEARCH AMATEUR ASTRONOMY
ed. Stephen J. Edberg
ISBN 0-937707-52-X

Vol. CS-34 ROBOTIC TELESCOPES IN THE 1990s
ed. Alexei V. Filippenko
ISBN 0-937707-53-8

PUBLISHED: 1993

Vol. CS-35* MASSIVE STARS: THEIR LIVES IN THE INTERSTELLAR MEDIUM
ed. Joseph P. Cassinelli and Edward B. Churchwell
ISBN 0-937707-54-6

Vol. CS-36 PLANETS AROUND PULSARS
ed. J. A. Phillips, S. E. Thorsett, and S. R. Kulkarni
ISBN 0-937707-55-4

Vol. CS-37 FIBER OPTICS IN ASTRONOMY II
ed. Peter M. Gray
ISBN 0-937707-56-2

Vol. CS-38 NEW FRONTIERS IN BINARY STAR RESEARCH: Pacific Rim Colloquium
ed. K. C. Leung and I.-S. Nha
ISBN 0-937707-57-0

Vol. CS-39 THE MINNESOTA LECTURES ON THE STRUCTURE AND DYNAMICS OF THE MILKY WAY
ed. Roberta M. Humphreys
ISBN 0-937707-58-9

Vol. CS-40 INSIDE THE STARS, IAU Colloquium 137
ed. Werner W. Weiss and Annie Baglin
ISBN 0-937707-59-7

Vol. CS-41 ASTRONOMICAL INFRARED SPECTROSCOPY: FUTURE OBSERVATIONAL DIRECTIONS
ed. Sun Kwok
ISBN 0-937707-60-0

Vol. CS-42 GONG 1992: SEISMIC INVESTIGATION OF THE SUN AND STARS
ed. Timothy M. Brown
ISBN 0-937707-61-9

ASP CONFERENCE SERIES VOLUMES

Published by the Astronomical Society of the Pacific

PUBLISHED: 1993 (* asterisk means OUT OF STOCK)

Vol. CS-43 SKY SURVEYS: PROTOSTARS TO PROTOGALAXIES
ed. B. T. Soifer
ISBN 0-937707-62-7

Vol. CS-44 PECULIAR VERSUS NORMAL PHENOMENA IN A-TYPE AND RELATED STARS, IAU Colloquium 138
ed. M. M. Dworetsky, F. Castelli, and R. Faraggiana
ISBN 0-937707-63-5

Vol. CS-45 LUMINOUS HIGH-LATITUDE STARS
ed. Dimitar D. Sasselov
ISBN 0-937707-64-3

Vol. CS-46 THE MAGNETIC AND VELOCITY FIELDS OF SOLAR ACTIVE REGIONS, IAU Colloquium 141
ed. Harold Zirin, Guoxiang Ai, and Haimin Wang
ISBN 0-937707-65-1

Vol. CS-47 THIRD DECENNIAL US-USSR CONFERENCE ON SETI -- Santa Cruz, California, USA
ed. G. Seth Shostak
ISBN 0-937707-66-X

Vol. CS-48 THE GLOBULAR CLUSTER-GALAXY CONNECTION
ed. Graeme H. Smith and Jean P. Brodie
ISBN 0-937707-67-8

Vol. CS-49 GALAXY EVOLUTION: THE MILKY WAY PERSPECTIVE
ed. Steven R. Majewski
ISBN 0-937707-68-6

Vol. CS-50 STRUCTURE AND DYNAMICS OF GLOBULAR CLUSTERS
ed. S. G. Djorgovski and G. Meylan
ISBN 0-937707-69-4

Vol. CS-51 OBSERVATIONAL COSMOLOGY
ed. Guido Chincarini, Angela Iovino, Tommaso Maccacaro, and Dario Maccagni
ISBN 0-937707-70-8

Vol. CS-52 ASTRONOMICAL DATA ANALYSIS SOFTWARE AND SYSTEMS II - (ADASS II)
ed. R. J. Hanisch, R. J. V. Brissenden, and Jeannette Barnes
ISBN 0-937707-71-6

Vol. CS-53 BLUE STRAGGLERS
ed. Rex A. Saffer
ISBN 0-937707-72-4

PUBLISHED: 1994

Vol. CS-54* THE FIRST STROMLO SYMPOSIUM: THE PHYSICS OF ACTIVE GALAXIES
ed. Geoffrey V. Bicknell, Michael A. Dopita, and Peter J. Quinn
ISBN 0-937707-73-2

Vol. CS-55 OPTICAL ASTRONOMY FROM THE EARTH AND MOON
ed. Diane M. Pyper and Ronald J. Angione
ISBN 0-937707-74-0

Vol. CS-56 INTERACTING BINARY STARS
ed. Allen W. Shafter
ISBN 0-937707-75-9

Vol. CS-57 STELLAR AND CIRCUMSTELLAR ASTROPHYSICS
ed. George Wallerstein and Alberto Noriega-Crespo
ISBN 0-937707-76-7

ASP CONFERENCE SERIES VOLUMES

Published by the Astronomical Society of the Pacific

PUBLISHED: 1994 (* asterisk means OUT OF STOCK)

Vol. CS-58* THE FIRST SYMPOSIUM ON THE INFRARED CIRRUS AND DIFFUSE INTERSTELLAR CLOUDS
ed. Roc M. Cutri and William B. Latter
ISBN 0-937707-77-5

Vol. CS-59 ASTRONOMY WITH MILLIMETER AND SUBMILLIMETER WAVE INTERFEROMETRY, IAU Colloquium 140
ed. M. Ishiguro and Wm. J. Welch
ISBN 0-937707-78-3

Vol. CS-60 THE MK PROCESS AT 50 YEARS: A POWERFUL TOOL FOR ASTROPHYSICAL INSIGHT, A Workshop of the Vatican Observatory --Tucson, Arizona, USA
ed. C. J. Corbally, R. O. Gray, and R. F. Garrison
ISBN 0-937707-79-1

Vol. CS-61 ASTRONOMICAL DATA ANALYSIS SOFTWARE AND SYSTEMS III - (ADASS III)
ed. Dennis R. Crabtree, R. J. Hanisch, and Jeannette Barnes
ISBN 0-937707-80-5

Vol. CS-62 THE NATURE AND EVOLUTIONARY STATUS OF HERBIG Ae/Be STARS
ed. Pik Sin Thé, Mario R. Pérez, and Ed P. J. van den Heuvel
ISBN 0-9837707-81-3

Vol. CS-63 SEVENTY-FIVE YEARS OF HIRAYAMA ASTEROID FAMILIES: THE ROLE OF COLLISIONS IN THE SOLAR SYSTEM HISTORY
ed. Yoshihide Kozai, Richard P. Binzel, and Tomohiro Hirayama
ISBN 0-937707-82-1

Vol. CS-64* COOL STARS, STELLAR SYSTEMS, AND THE SUN: Eighth Cambridge Workshop
ed. Jean-Pierre Caillault
ISBN 0-937707-83-X

Vol. CS-65* CLOUDS, CORES, AND LOW MASS STARS: The Fourth Haystack Observatory Conference
ed. Dan P. Clemens and Richard Barvainis
ISBN 0-937707-84-8

Vol. CS-66* PHYSICS OF THE GASEOUS AND STELLAR DISKS OF THE GALAXY
ed. Ivan R. King
ISBN 0-937707-85-6

Vol. CS-67 UNVEILING LARGE-SCALE STRUCTURES BEHIND THE MILKY WAY
ed. C. Balkowski and R. C. Kraan-Korteweg
ISBN 0-937707-86-4

Vol. CS-68* SOLAR ACTIVE REGION EVOLUTION: COMPARING MODELS WITH OBSERVATIONS
ed. K. S. Balasubramaniam and George W. Simon
ISBN 0-937707-87-2

Vol. CS-69 REVERBERATION MAPPING OF THE BROAD-LINE REGION IN ACTIVE GALACTIC NUCLEI
ed. P. M. Gondhalekar, K. Horne, and B. M. Peterson
ISBN 0-937707-88-0

PUBLISHED: 1995

Vol. CS-70* GROUPS OF GALAXIES
ed. Otto-G. Richter and Kirk Borne
ISBN 0-937707-89-9

Vol. CS-71 TRIDIMENSIONAL OPTICAL SPECTROSCOPIC METHODS IN ASTROPHYSICS, IAU Colloquium 149
ed. Georges Comte and Michel Marcelin
ISBN 0-937707-90-2

ASP CONFERENCE SERIES VOLUMES

Published by the Astronomical Society of the Pacific

PUBLISHED: 1995 (* asterisk means OUT OF STOCK)

Vol. CS-72 MILLISECOND PULSARS: A DECADE OF SURPRISE
ed. A. S Fruchter, M. Tavani, and D. C. Backer
ISBN 0-937707-91-0

Vol. CS-73 AIRBORNE ASTRONOMY SYMPOSIUM ON THE GALACTIC ECOSYSTEM: FROM GAS TO STARS TO DUST
ed. Michael R. Haas, Jacqueline A. Davidson, and Edwin F. Erickson
ISBN 0-937707-92-9

Vol. CS-74 PROGRESS IN THE SEARCH FOR EXTRATERRESTRIAL LIFE:1993 Bioastronomy Symposium
ed. G. Seth Shostak
ISBN 0-937707-93-7

Vol. CS-75 MULTI-FEED SYSTEMS FOR RADIO TELESCOPES
ed. Darrel T. Emerson and John M. Payne
ISBN 0-937707-94-5

Vol. CS-76 GONG '94: HELIO- AND ASTERO-SEISMOLOGY FROM THE EARTH AND SPACE
ed. Roger K. Ulrich, Edward J. Rhodes, Jr., and Werner Däppen
ISBN 0-937707-95-3

Vol. CS-77 ASTRONOMICAL DATA ANALYSIS SOFTWARE AND SYSTEMS IV – (ADASS IV)
ed. R. A. Shaw, H. E. Payne, and J. J. E. Hayes
ISBN 0-937707-96-1

Vol. CS-78 ASTROPHYSICAL APPLICATIONS OF POWERFUL NEW DATABASES: Joint Discussion No. 16 of the 22nd General Assembly of the IAU
ed. S. J. Adelman and W. L. Wiese
ISBN 0-937707-97-X

Vol. CS-79* ROBOTIC TELESCOPES: CURRENT CAPABILITIES, PRESENT DEVELOPMENTS, AND FUTURE PROSPECTS FOR AUTOMATED ASTRONOMY
ed. Gregory W. Henry and Joel A. Eaton
ISBN 0-937707-98-8

Vol. CS-80* THE PHYSICS OF THE INTERSTELLAR MEDIUM AND INTERGALACTIC MEDIUM
ed. A. Ferrara, C. F. McKee, C. Heiles, and P. R. Shapiro
ISBN 0-937707-99-6

Vol. CS-81 LABORATORY AND ASTRONOMICAL HIGH RESOLUTION SPECTRA
ed. A. J. Sauval, R. Blomme, and N. Grevesse
ISBN 1-886733-01-5

Vol. CS-82* VERY LONG BASELINE INTERFEROMETRY AND THE VLBA
ed. J. A. Zensus, P. J. Diamond, and P. J. Napier
ISBN 1-886733-02-3

Vol. CS-83* ASTROPHYSICAL APPLICATIONS OF STELLAR PULSATION, IAU Colloquium 155
ed. R. S. Stobie and P. A. Whitelock
ISBN 1-886733-03-1

ATLAS INFRARED ATLAS OF THE ARCTURUS SPECTRUM, 0.9 - 5.3 μm
ed. Kenneth Hinkle, Lloyd Wallace, and William Livingston
ISBN: 1-886733-04-X

Vol. CS-84 THE FUTURE UTILIZATION OF SCHMIDT TELESCOPES, IAU Colloquium 148
ed. Jessica Chapman, Russell Cannon, Sandra Harrison, and Bambang Hidayat
ISBN 1-886733-05-8

Vol. CS-85* CAPE WORKSHOP ON MAGNETIC CATACLYSMIC VARIABLES
ed. D. A. H. Buckley and B. Warner
ISBN 1-886733-06-6

ASP CONFERENCE SERIES VOLUMES

Published by the Astronomical Society of the Pacific

PUBLISHED: 1995 (* asterisk means OUT OF STOCK)

Vol. CS-86 FRESH VIEWS OF ELLIPTICAL GALAXIES
ed. Alberto Buzzoni, Alvio Renzini, and Alfonso Serrano
ISBN 1-886733-07-4

PUBLISHED: 1996

Vol. CS-87 NEW OBSERVING MODES FOR THE NEXT CENTURY
ed. Todd Boroson, John Davies, and Ian Robson
ISBN 1-886733-08-2

Vol. CS-88* CLUSTERS, LENSING, AND THE FUTURE OF THE UNIVERSE
ed. Virginia Trimble and Andreas Reisenegger
ISBN 1-886733-09-0

Vol. CS-89 ASTRONOMY EDUCATION: CURRENT DEVELOPMENTS, FUTURE COORDINATION
ed. John R. Percy
ISBN 1-886733-10-4

Vol. CS-90 THE ORIGINS, EVOLUTION, AND DESTINIES OF BINARY STARS IN CLUSTERS
ed. E. F. Milone and J. -C. Mermilliod
ISBN 1-886733-11-2

Vol. CS-91 BARRED GALAXIES, IAU Colloquium 157
ed. R. Buta, D. A. Crocker, and B. G. Elmegreen
ISBN 1-886733-12-0

Vol. CS-92 FORMATION OF THE GALACTIC HALO .... INSIDE AND OUT
ed. Heather L. Morrison and Ata Sarajedini
ISBN 1-886733-13-9

Vol. CS-93 RADIO EMISSION FROM THE STARS AND THE SUN
ed. A. R. Taylor and J. M. Paredes
ISBN 1-886733-14-7

Vol. CS-94 MAPPING, MEASURING, AND MODELING THE UNIVERSE
ed. Peter Coles, Vicent J. Martinez, and Maria-Jesus Pons-Borderia
ISBN 1-886733-15-5

Vol. CS-95 SOLAR DRIVERS OF INTERPLANETARY AND TERRESTRIAL DISTURBANCES:
Proceedings of 16th International Workshop National Solar Observatory/Sacramento Peak
ed. K. S. Balasubramaniam, Stephen L. Keil, and Raymond N. Smartt
ISBN 1-886733-16-3

Vol. CS-96 HYDROGEN-DEFICIENT STARS
ed. C. S. Jeffery and U. Heber
ISBN 1-886733-17-1

Vol. CS-97 POLARIMETRY OF THE INTERSTELLAR MEDIUM
ed. W. G. Roberge and D. C. B. Whittet
ISBN 1-886733-18-X

Vol. CS-98 FROM STARS TO GALAXIES: THE IMPACT OF STELLAR PHYSICS ON GALAXY EVOLUTION
ed. Claus Leitherer, Uta Fritze-von Alvensleben, and John Huchra
ISBN 1-886733-19-8

Vol. CS-99 COSMIC ABUNDANCES: Proceedings of the 6th Annual October Astrophysics Conference
ed. Stephen S. Holt and George Sonneborn
ISBN 1-886733-20-1

ASP CONFERENCE SERIES VOLUMES

Published by the Astronomical Society of the Pacific

PUBLISHED: 1996 (* asterisk means OUT OF STOCK)

Vol. CS-100 ENERGY TRANSPORT IN RADIO GALAXIES AND QUASARS
ed. P. E. Hardee, A. H. Bridle, and J. A. Zensus
ISBN 1-886733-21-X

Vol. CS-101 ASTRONOMICAL DATA ANALYSIS SOFTWARE AND SYSTEMS V – (ADASS V)
ed. George H. Jacoby and Jeannette Barnes
ISBN 1080-7926

Vol. CS-102 THE GALACTIC CENTER, 4th ESO/CTIO Workshop
ed. Roland Gredel
ISBN 1-886733-22-8

Vol. CS-103 THE PHYSICS OF LINERS IN VIEW OF RECENT OBSERVATIONS
ed. M. Eracleous, A. Koratkar, C. Leitherer, and L. Ho
ISBN 1-886733-23-6

Vol. CS-104 PHYSICS, CHEMISTRY, AND DYNAMICS OF INTERPLANETARY DUST, IAU Colloquium 150
ed. Bo Å. S. Gustafson and Martha S. Hanner
ISBN 1-886733-24-4

Vol. CS-105 PULSARS: PROBLEMS AND PROGRESS, IAU Colloquium 160
ed. S. Johnston, M. A. Walker, and M. Bailes
ISBN 1-886733-25-2

Vol. CS-106 THE MINNESOTA LECTURES ON EXTRAGALACTIC NEUTRAL HYDROGEN
ed. Evan D. Skillman
ISBN 1-886733-26-0

Vol. CS-107 COMPLETING THE INVENTORY OF THE SOLAR SYSTEM:
A Symposium held in conjunction with the 106th Annual Meeting of the ASP
ed. Terrence W. Rettig and Joseph M. Hahn
ISBN 1-886733-27-9

Vol. CS-108 M.A.S.S. -- MODEL ATMOSPHERES AND SPECTRUM SYNTHESIS: 5th Vienna - Workshop
ed. Saul J. Adelman, Friedrich Kupka, and Werner W. Weiss
ISBN 1-886733-28-7

Vol. CS-109 COOL STARS, STELLAR SYSTEMS, AND THE SUN: Ninth Cambridge Workshop
ed. Roberto Pallavicini and Andrea K. Dupree
ISBN 1-886733-29-5

Vol. CS-110 BLAZAR CONTINUUM VARIABILITY
ed. H. R. Miller, J. R. Webb, and J. C. Noble
ISBN 1-886733-30-9

Vol. CS-111 MAGNETIC RECONNECTION IN THE SOLAR ATMOSPHERE:
Proceedings of a Yohkoh Conference
ed. R. D. Bentley and J. T. Mariska
ISBN 1-886733-31-7

Vol. CS-112 THE HISTORY OF THE MILKY WAY AND ITS SATELLITE SYSTEM
ed. Andreas Burkert, Dieter H. Hartmann, and Steven R. Majewski
ISBN 1-886733-32-5

PUBLISHED: 1997

Vol. CS-113 EMISSION LINES IN ACTIVE GALAXIES: NEW METHODS AND TECHNIQUES,
IAU Colloquium 159
ed. B. M. Peterson, F.-Z. Cheng, and A. S. Wilson
ISBN 1-886733-33-3

ASP CONFERENCE SERIES VOLUMES

Published by the Astronomical Society of the Pacific

PUBLISHED: 1997 (* asterisk means OUT OF STOCK)

Vol. CS-114 YOUNG GALAXIES AND QSO ABSORPTION-LINE SYSTEMS
ed. Sueli M. Viegas, Ruth Gruenwald, and Reinaldo R. de Carvalho
ISBN 1-886733-34-1

Vol. CS-115 GALACTIC CLUSTER COOLING FLOWS
ed. Noam Soker
ISBN 1-886733-35-X

Vol. CS-116 THE SECOND STROMLO SYMPOSIUM: THE NATURE OF ELLIPTICAL GALAXIES
ed. M. Arnaboldi, G. S. Da Costa, and P. Saha
ISBN 1-886733-36-8

Vol. CS-117 DARK AND VISIBLE MATTER IN GALAXIES
ed. Massimo Persic and Paolo Salucci
ISBN-1-886733-37-6

Vol. CS-118 FIRST ADVANCES IN SOLAR PHYSICS EUROCONFERENCE: ADVANCES IN THE PHYSICS OF SUNSPOTS
ed. B. Schmieder. J. C. del Toro Iniesta, and M. Vázquez
ISBN 1-886733-38-4

Vol. CS-119 PLANETS BEYOND THE SOLAR SYSTEM AND THE NEXT GENERATION OF SPACE MISSIONS
ed. David R. Soderblom
ISBN 1-886733-39-2

Vol. CS-120 LUMINOUS BLUE VARIABLES: MASSIVE STARS IN TRANSITION
ed. Antonella Nota and Henny J. G. L. M. Lamers
ISBN 1-886733-40-6

Vol. CS-121 ACCRETION PHENOMENA AND RELATED OUTFLOWS, IAU Colloquium 163
ed. D. T. Wickramasinghe, G. V. Bicknell, and L. Ferrario
ISBN 1-886733-41-4

Vol. CS-122 FROM STARDUST TO PLANETESIMALS:
Symposium held as part of the 108th Annual Meeting of the ASP
ed. Yvonne J. Pendleton and A. G. G. M. Tielens
ISBN 1-886733-42-2

Vol. CS-123 THE 12th 'KINGSTON MEETING': COMPUTATIONAL ASTROPHYSICS
ed. David A. Clarke and Michael J. West
ISBN 1-886733-43-0

Vol. CS-124 DIFFUSE INFRARED RADIATION AND THE IRTS
ed. Haruyuki Okuda, Toshio Matsumoto, and Thomas Roellig
ISBN 1-886733-44-9

Vol. CS-125 ASTRONOMICAL DATA ANALYSIS SOFTWARE AND SYSTEMS, VI – (ADASS VI)
ed. Gareth Hunt and H. E. Payne
ISBN 1-886733-45-7

Vol. CS-126 FROM QUANTUM FLUCTUATIONS TO COSMOLOGICAL STRUCTURES
ed. David Valls-Gabaud, Martin A. Hendry, Paolo Molaro, and Khalil Chamcham
ISBN 1-886733-46-5

Vol. CS-127 PROPER MOTIONS AND GALACTIC ASTRONOMY
ed. Roberta M. Humphreys
ISBN 1-886733-47-3

Vol. CS-128 MASS EJECTION FROM AGN (Active Galactic Nuclei)
ed. N. Arav, I. Shlosman, and R. J. Weymann
ISBN 1-886733-48-1

ASP CONFERENCE SERIES VOLUMES

Published by the Astronomical Society of the Pacific

PUBLISHED: 1997 (* asterisk means OUT OF STOCK)

Vol. CS-129 THE GEORGE GAMOW SYMPOSIUM
ed. E. Harper, W. C. Parke, and G. D. Anderson
ISBN 1-886733-49-X

Vol. CS-130 THE THIRD PACIFIC RIM CONFERENCE ON RECENT DEVELOPMENT ON BINARY STAR RESEARCH
ed. Kam-Ching Leung
ISBN 1-886733-50-3

PUBLISHED: 1998

Vol. CS-131 BOULDER-MUNICH II: PROPERTIES OF HOT, LUMINOUS STARS
ed. Ian D. Howarth
ISBN 1-886733-51-1

Vol. CS-132 STAR FORMATION WITH THE INFRARED SPACE OBSERVATORY (ISO)
ed. João L. Yun and René Liseau
ISBN 1-886733-52-X

Vol. CS-133 SCIENCE WITH THE NGST (Next Generation Space Telescope)
ed. Eric P. Smith and Anuradha Koratkar
ISBN 1-886733-53-8

Vol. CS-134 BROWN DWARFS AND EXTRASOLAR PLANETS
ed. Rafael Rebolo, Eduardo L. Martin and Maria Rosa Zapatero Osorio
ISBN 1-886733-54-6

Vol. CS-135 A HALF CENTURY OF STELLAR PULSATION INTERPRETATIONS: A TRIBUTE TO ARTHUR N. COX
ed. P. A. Bradley and J. A. Guzik
ISBN 1-886733-55-4

Vol. CS-136 GALACTIC HALOS: A UC SANTA CRUZ WORKSHOP
ed. Dennis Zaritsky
ISBN 1-886733-56-2

Vol. CS-137 WILD STARS IN THE OLD WEST: PROCEEDINGS OF THE 13th NORTH AMERICAN WORKSHOP ON CATACLYSMIC VARIABLES AND RELATED OBJECTS
ed. S. Howell, E. Kuulkers, and C. Woodward
ISBN 1-886733-57-0

Vol. CS-138 1997 PACIFIC RIM CONFERENCE ON STELLAR ASTROPHYSICS
ed. Kwing Lam Chan, K. S. Cheng, and H. P. Singh
ISBN 1-886733-58-9

Vol. CS-139 PRESERVING THE ASTRONOMICAL WINDOWS:
Proceedings of Joint Discussion No. 5 of the 23rd General Assembly of the IAU
ed. Syuzo Isobe and Tomohiro Hirayama
ISBN 1-886733-59-7

Vol. CS-140 SYNOPTIC SOLAR PHYSICS --18th NSO/Sacramento Peak Summer Workshop
ed. K. S. Balasubramaniam, J. W. Harvey, and D. M. Rabin
ISBN 1-886733-60-0

Vol. CS-141 ASTROPHYSICS FROM ANTARCTICA:
A Symposium held as a part of the 109th Annual Meeting of the ASP
ed. Giles Novak and Randall H. Landsberg
ISBN 1-886733-61-9

Vol. CS-142 THE STELLAR INITIAL MASS FUNCTION: 38th Herstmonceux Conference
ed. Gerry Gilmore and Debbie Howell
ISBN 1-886733-62-7

ASP CONFERENCE SERIES VOLUMES

Published by the Astronomical Society of the Pacific

PUBLISHED: 1998 (* asterisk means OUT OF STOCK)

Vol. CS-143 THE SCIENTIFIC IMPACT OF THE GODDARD HIGH RESOLUTION SPECTROGRAPH (GHRS)
ed. John C. Brandt, Thomas B. Ake III, and Carolyn Collins Petersen
ISBN 1-886733-63-5

Vol. CS-144 RADIO EMISSION FROM GALACTIC AND EXTRAGALACTIC COMPACT SOURCES, IAU Colloquium 164
ed. J. Anton Zensus, G. B. Taylor, and J. M. Wrobel
ISBN 1-886733-64-3

Vol. CS-145 ASTRONOMICAL DATA ANALYSIS SOFTWARE AND SYSTEMS VII – (ADASS VII)
ed. Rudolf Albrecht, Richard N. Hook, and Howard A. Bushouse
ISBN 1-886733-65-1

Vol. CS-146 THE YOUNG UNIVERSE GALAXY FORMATION AND EVOLUTION AT INTERMEDIATE AND HIGH REDSHIFT
ed. S. D'Odorico, A. Fontana, and E. Giallongo
ISBN 1-886733-66-X

Vol. CS-147 ABUNDANCE PROFILES: DIAGNOSTIC TOOLS FOR GALAXY HISTORY
ed. Daniel Friedli, Mike Edmunds, Carmelle Robert, and Laurent Drissen
ISBN 1-886733-67-8

Vol. CS-148 ORIGINS
ed. Charles E. Woodward, J. Michael Shull, and Harley A. Thronson, Jr.
ISBN 1-886733-68-6

Vol. CS-149 SOLAR SYSTEM FORMATION AND EVOLUTION
D. Lazzaro, R. Vieira Martins, S. Ferraz-Mello, J. Fernández, and C. Beaugé
ISBN 1-886733-69-4

Vol. CS-150 NEW PERSPECTIVES ON SOLAR PROMINENCES, IAU Colloquium 167
ed. David Webb, David Rust, and Brigitte Schmieder
ISBN 1-886733-70-8

Vol. CS-151 COSMIC MICROWAVE BACKGROUND AND LARGE SCALE STRUCTURES OF THE UNIVERSE
ed. Yong-Ik Byun and Kin-Wang Ng
ISBN 1-886733-71-6

Vol. CS-152 FIBER OPTICS IN ASTRONOMY III
ed. S. Arribas, E. Mediavilla, and F. Watson
ISBN 1-886733-72-4

Vol. CS-153 LIBRARY AND INFORMATION SERVICES IN ASTRONOMY III -- (LISA III)
ed. Uta Grothkopf, Heinz Andernach, Sarah Stevens-Rayburn, and Monique Gomez
ISBN 1-886733-73-2

Vol. CS-154 COOL STARS, STELLAR SYSTEMS AND THE SUN: Tenth Cambridge Workshop
ed. Robert A. Donahue and Jay A. Bookbinder
ISBN 1-886733-74-0

Vol. CS-155 SECOND ADVANCES IN SOLAR PHYSICS EUROCONFERENCE: THREE-DIMENSIONAL STRUCTURE OF SOLAR ACTIVE REGIONS
ed. Costas E. Alissandrakis and Brigitte Schmieder
ISBN 1-886733-75-9

PUBLISHED: 1999

Vol. CS-156 HIGHLY REDSHIFTED RADIO LINES
ed. C. L. Carilli, S. J. E. Radford, K. M. Menten, and G. I. Langston
ISBN 1-886733-76-7

ASP CONFERENCE SERIES VOLUMES

Published by the Astronomical Society of the Pacific

PUBLISHED: 1999 (* asterisk means OUT OF STOCK)

Vol. CS-157 ANNAPOLIS WORKSHOP ON MAGNETIC CATACLYSMIC VARIABLES
ed. Coel Hellier and Koji Mukai
ISBN 1-886733-77-5

Vol. CS-158 SOLAR AND STELLAR ACTIVITY: SIMILARITIES AND DIFFERENCES
ed. C. J. Butler and J. G. Doyle
ISBN 1-886733-78-3

Vol. CS-159 BL LAC PHENOMENON
ed. Leo O. Takalo and Aimo Sillanpää
ISBN 1-886733-79-1

Vol. CS-160 ASTROPHYSICAL DISCS: An EC Summer School
ed. J. A. Sellwood and Jeremy Goodman
ISBN 1-886733-80-5

Vol. CS-161 HIGH ENERGY PROCESSES IN ACCRETING BLACK HOLES
ed. Juri Poutanen and Roland Svensson
ISBN 1-886733-81-3

Vol. CS-162 QUASARS AND COSMOLOGY
ed. Gary Ferland and Jack Baldwin
ISBN 1-886733-83-X

Vol. CS-163 STAR FORMATION IN EARLY-TYPE GALAXIES
ed. Jordi Cepa and Patricia Carral
ISBN 1-886733-84-8

Vol. CS-164 ULTRAVIOLET–OPTICAL SPACE ASTRONOMY BEYOND HST
ed. Jon A. Morse, J. Michael Shull, and Anne L. Kinney
ISBN 1-886733-85-6

Vol. CS-165 THE THIRD STROMLO SYMPOSIUM: THE GALACTIC HALO
ed. Brad K. Gibson, Tim S. Axelrod, and Mary E. Putman
ISBN 1-886733-86-4

Vol. CS-166 STROMLO WORKSHOP ON HIGH-VELOCITY CLOUDS
ed. Brad K. Gibson and Mary E. Putman
ISBN 1-886733-87-2

Vol. CS-167 HARMONIZING COSMIC DISTANCE SCALES IN A POST-HIPPARCOS ERA
ed. Daniel Egret and André Heck
ISBN 1-886733-88-0

Vol. CS-168 NEW PERSPECTIVES ON THE INTERSTELLAR MEDIUM
ed. A. R. Taylor, T. L. Landecker, and G. Joncas
ISBN 1-886733-89-9

Vol. CS-169- 11th EUROPEAN WORKSHOP ON WHITE DWARFS
ed. J.-E. Solheim and E. G. Meištas
ISBN 1-886733-91-0

Vol. CS-170 THE LOW SURFACE BRIGHTNESS UNIVERSE, IAU Colloquium 171
ed. J. I. Davies, C. Impey, and S. Phillipps
ISBN 1-886733-92-9

Vol. CS-171 LiBeB, COSMIC RAYS, AND RELATED X- AND GAMMA-RAYS
ed. Reuven Ramaty, Elisabeth Vangioni-Flam, Michel Cassé, and Keith Olive
ISBN 1-886733-93-7

ASP CONFERENCE SERIES VOLUMES

Published by the Astronomical Society of the Pacific

PUBLISHED: 1999 (* asterisk means OUT OF STOCK)

Vol. CS-172 ASTRONOMICAL DATA ANALYSIS SOFTWARE AND SYSTEMS VIII – (ADASS VIII)
ed. David M. Mehringer, Raymond L. Plante, and Douglas A. Roberts
ISBN 1-886733-94-5

Vol. CS-173 THEORY AND TESTS OF CONVECTION IN STELLAR STRUCTURE: First Granada Workshop
ed. Álvaro Giménez, Edward F. Guinan, and Benjamín Montesinos
ISBN 1-886733-95-3

Vol. CS-174 CATCHING THE PERFECT WAVE:
ADAPTIVE OPTICS AND INTERFEROMETRY IN THE 21st CENTURY,
A Symposium held as a part of the 110th Annual Meeting of the ASP
ed. Sergio R. Restaino, William Junor, and Nebojsa Duric
ISBN 1-886733-96-1

Vol. CS-175 STRUCTURE AND KINEMATICS OF QUASAR BROAD LINE REGIONS
ed. C. M. Gaskell, W. N. Brandt, M. Dietrich, D. Dultzin-Hacyan, and M. Eracleous
ISBN 1-886733-97-X

Vol. CS-176 OBSERVATIONAL COSMOLOGY: THE DEVELOPMENT OF GALAXY SYSTEMS
ed. Giuliano Giuricin, Marino Mezzetti, and Paolo Salucci
ISBN 1-58381-000-5

Vol. CS-177 ASTROPHYSICS WITH INFRARED SURVEYS: A Prelude to SIRTF
ed. Michael D. Bicay, Chas A. Beichman, Roc M. Cutri, and Barry F. Madore
ISBN 1-58381-001-3

Vol. CS-178 STELLAR DYNAMOS: NONLINEARITY AND CHAOTIC FLOWS
ed. Manuel Núñez and Antonio Ferriz-Mas
ISBN 1-58381-002-1

Vol. CS-179 ETA CARINAE AT THE MILLENNIUM
ed. Jon A. Morse, Roberta M. Humphreys, and Augusto Damineli
ISBN 1-58381-003-X

Vol. CS-180 SYNTHESIS IMAGING IN RADIO ASTRONOMY II
ed. G. B. Taylor, C. L. Carilli, and R. A. Perley
ISBN 1-58381-005-6

Vol. CS-181 MICROWAVE FOREGROUNDS
ed. Angelica de Oliveira-Costa and Max Tegmark
ISBN 1-58381-006-4

Vol. CS-182 GALAXY DYNAMICS: A Rutgers Symposium
ed. David Merritt, J. A. Sellwood, and Monica Valluri
ISBN 1-58381-007-2

Vol. CS-183 HIGH RESOLUTION SOLAR PHYSICS: THEORY, OBSERVATIONS, AND TECHNIQUES
ed. T. R. Rimmele, K. S. Balasubramaniam, and R. R. Radick
ISBN 1-58381-009-9

Vol. CS-184 THIRD ADVANCES IN SOLAR PHYSICS EUROCONFERENCE:
MAGNETIC FIELDS AND OSCILLATIONS
ed. B. Schmieder, A. Hofmann, and J. Staude
ISBN 1-58381-010-2

Vol. CS-185 PRECISE STELLAR RADIAL VELOCITIES, IAU Colloquium 170
ed. J. B. Hearnshaw and C. D. Scarfe
ISBN 1-58381-011-0

Vol. CS-186 THE CENTRAL PARSECS OF THE GALAXY
ed. Heino Falcke, Angela Cotera, Wolfgang J. Duschl, Fulvio Melia, and Marcia J. Rieke
ISBN 1-58381-012-9

ASP CONFERENCE SERIES VOLUMES

Published by the Astronomical Society of the Pacific

PUBLISHED: 1999 (* asterisk means OUT OF STOCK)

Vol. CS-187 THE EVOLUTION OF GALAXIES ON COSMOLOGICAL TIMESCALES
ed. J. E. Beckman and T. J. Mahoney
ISBN 1-58381-013-7

Vol. CS-188 OPTICAL AND INFRARED SPECTROSCOPY OF CIRCUMSTELLAR MATTER
ed. Eike W. Guenther, Bringfried Stecklum, and Sylvio Klose
ISBN 1-58381-014-5

Vol. CS-189 CCD PRECISION PHOTOMETRY WORKSHOP
ed. Eric R. Craine, Roy A. Tucker, and Jeannette Barnes
ISBN 1-58381-015-3

Vol. CS-190 GAMMA-RAY BURSTS: THE FIRST THREE MINUTES
ed. Juri Poutanen and Roland Svensson
ISBN 1-58381-016-1

Vol. CS-191 PHOTOMETRIC REDSHIFTS AND HIGH REDSHIFT GALAXIES
ed. Ray J. Weymann, Lisa J. Storrie-Lombardi, Marcin Sawicki, and Robert J. Brunner
ISBN 1-58381-017-X

Vol. CS-192 SPECTROPHOTOMETRIC DATING OF STARS AND GALAXIES
ed. I. Hubeny, S. R. Heap, and R. H. Cornett
ISBN 1-58381-018-8

Vol. CS-193 THE HY-REDSHIFT UNIVERSE: GALAXY FORMATION AND EVOLUTION AT HIGH REDSHIFT
ed. Andrew J. Bunker and Wil J. M. van Breugel
ISBN 1-58381-019-6

Vol. CS-194 WORKING ON THE FRINGE:
OPTICAL AND IR INTERFEROMETRY FROM GROUND AND SPACE
ed. Stephen Unwin and Robert Stachnik
ISBN 1-58381-020-X

PUBLISHED: 2000

Vol. CS-195 IMAGING THE UNIVERSE IN THREE DIMENSIONS:
Astrophysics with Advanced Multi-Wavelength Imaging Devices
ed. W. van Breugel and J. Bland-Hawthorn
ISBN 1-58381-022-6

Vol. CS-196 THERMAL EMISSION SPECTROSCOPY AND ANALYSIS OF DUST, DISKS, AND REGOLITHS
ed. Michael L. Sitko, Ann L. Sprague, and David K. Lynch
ISBN: 1-58381-023-4

Vol. CS-197 XVth IAP MEETING DYNAMICS OF GALAXIES:
FROM THE EARLY UNIVERSE TO THE PRESENT
ed. F. Combes, G. A. Mamon, and V. Charmandaris
ISBN: 1-58381-24-2

Vol. CS-198 EUROCONFERENCE ON "STELLAR CLUSTERS AND ASSOCIATIONS:
CONVECTION, ROTATION, AND DYNAMOS"
ed. R. Pallavicini, G. Micela, and S. Sciortino
ISBN: 1-58381-25-0

Vol. CS-199 ASYMMETRICAL PLANETARY NEBULAE II: FROM ORIGINS TO MICROSTRUCTURES
ed. J. H. Kastner, N. Soker, and S. Rappaport
ISBN: 1-58381-026-9

Vol. CS-200 CLUSTERING AT HIGH REDSHIFT
ed. A. Mazure, O. Le Fèvre, and V. Le Brun
ISBN: 1-58381-027-7

ASP CONFERENCE SERIES VOLUMES

Published by the Astronomical Society of the Pacific

PUBLISHED: 2000 (* asterisk means OUT OF STOCK)

Vol. CS-201 COSMIC FLOWS 1999:
TOWARDS AN UNDERSTANDING OF LARGE-SCALE STRUCTURES
ed. Stéphane Courteau, Michael A. Strauss, and Jeffrey A. Willick
ISBN: 1-58381-028-5

Vol. CS-202 PULSAR ASTRONOMY – 2000 AND BEYOND, IAU Colloquium 177
ed. M. Kramer, N. Wex, and R. Wielebinski
ISBN: 1-58381-029-3

Vol. CS-203 THE IMPACT OF LARGE-SCALE SURVEYS ON PULSATING STAR RESEARCH, IAU Colloquium 176
ed. L. Szabados and D. W. Kurtz
ISBN: 1-58381-030-7

Vol. CS-204 THERMAL AND IONIZATION ASPECTS OF FLOWS FROM HOT STARS: OBSERVATIONS AND THEORY
ed. Henny J. G. L. M. Lamers and Arved Sapar
ISBN: 1-58381-031-5

Vol. CS-205 THE LAST TOTAL SOLAR ECLIPSE OF THE MILLENNIUM IN TURKEY
ed. W. C. Livingston and A. Özgüç
ISBN: 1-58381-032-3

Vol. CS-206 HIGH ENERGY SOLAR PHYSICS – *ANTICIPATING HESSI*
ed. Reuven Ramaty and Natalie Mandzhavidze
ISBN: 1-58381-033-1

Vol. CS-207 NGST SCIENCE AND TECHNOLOGY EXPOSITION
ed. Eric P. Smith and Knox S. Long
ISBN: 1-58381-036-6

Vol. CS-208 POLAR MOTION: HISTORICAL AND SCIENTIFIC PROBLEMS
IAU Colloquium 178
ed. Steven Dick, Dennis McCarthy, Brian Luzum
ISBN: 1-58381-039-0

Vol. CS-209 SMALL GALAXY GROUPS
IAU Colloquium 174
ed. Mauri J. Valtonen and Chris Flynn
ISBN: 1-58381-040-4

Vol. CS-210 DELTA SCUTI AND RELATED STARS:
Reference Handbook and Proceedings of the 6th Vienna Workshop in Astrophysics
ed. Michel Breger and Michael Houston Montgomery
ISBN: 1-58381-043-9

Vol. CS-211 MASSIVE STELLAR CLUSTERS
ed. Ariane Lançon and Christian M. Boily
ISBN: 1-58381-042-0

Vol. CS-212 FROM GIANT PLANETS TO COOL STARS
ed. Caitlin A. Griffith and Mark S. Marley
ISBN: 1-58381-041-2

Vol. CS-213 BIOASTRONOMY `99: A NEW ERA IN BIOASTRONOMY
ed. Guillermo A. Lemarchand and Karen J. Meech
ISBN: 1-58381-044-7

Vol. CS-214 THE Be PHENOMENON IN EARLY-TYPE STARS
IAU Colloquium 175
ed. Myron A. Smith, Huib F. Henrichs and Juan Fabregat
ISBN: 1-58381-045-5

ASP CONFERENCE SERIES VOLUMES

Published by the Astronomical Society of the Pacific

PUBLISHED: 2000 (* asterisk means OUT OF STOCK)

Vol. CS-215 COSMIC EVOLUTION AND GALAXY FORMATION: Structure, Interactions and Feedback
ed. José Franco, Elena Terlevich, Omar López-Cruz, and Itziar Aretxaga
ISBN: 1-58381-046-3

Vol. CS-216 ASTRONOMICAL DATA ANALYSIS SOFTWARE AND SYSTEMS IX
ed. Nadine Manset, Christian Veillet, and Dennis Crabtree
ISBN: 1-58381-047-1 ISSN: 1080-7926

Vol. CS-217 IMAGING AT RADIO THROUGH SUBMILLIMETER WAVELENGTHS
ed. Jeffrey G. Mangum and Simon J. E. Radford
ISBN: 1-58381-049-8

Vol. CS-218 MAPPING THE HIDDEN UNIVERSE: THE UNIVERSE BEHIND THE MILKYWAY THE UNIVERSE IN HI
ed. Renée C. Kraan-Korteweg, Patricia A. Henning, and Heinz Andernach
ISBN: 1-58381-050-1

Vol. CS-219 DISKS, PLANETESIMALS, AND PLANETS
ed. F. Garzón, C. Eiroa, D. de Winter, and T. J. Mahoney
ISBN: 1-58381-051-X

Vol. CS-220 AMATEUR - PROFESSIONAL PARTNERSHIPS IN ASTRONOMY
ed. John R. Percy and Joseph B. Wilson
ISBN: 1-58381-052-8

Vol. CS-221 STARS, GAS AND DUST IN GALAXIES: EXPLORING THE LINKS
ed. Danielle Alloin, Knut Olsen, and Gaspar Galaz
ISBN: 1-58381-053-6

All other book orders or inquiries concerning volumes listed should be directed to the:

Astronomical Society of the Pacific Conference Series
390 Ashton Avenue
San Francisco CA 94112-1722 USA

Phone: 415-337-2126
Fax: 415-337-5205
E-mail: catalog@aspsky.org
Web Site: http://www.aspsky.org

INTERNATIONAL ASTRONOMICAL UNION (IAU)
SYMPOSIA

Published by the Astronomical Society of the Pacific

PUBLISHED: 1999

Vol. No. 190 — NEW VIEWS OF THE MAGELLANIC CLOUDS
eds. You-Hua Chu, Nicholas B. Suntzeff, James E. Hesser, and David A. Bohlender
ISBN: 1-58381-021-8

Vol. No. 191 — ASYMPTOTIC GIANT BRANCH STARS
eds. T. Le Bertre, A. Lèbre, and C. Waelkens
ISBN: 1-886733-90-2

Vol. No. 192 — THE STELLAR CONTENT OF LOCAL GROUP GALAXIES
eds. Patricia Whitelock and Russell Cannon
ISBN: 1-886733-82-1

Vol. No. 193 — WOLF-RAYET PHENOMENA IN MASSIVE STARS AND STARBURST GALAXIES
eds. Karel A. van der Hucht, Gloria Koenigsberger, and Philippe R. J. Eenens
ISBN: 1-58381-004-8

Vol. No. 194 — ACTIVITY IN GALAXIES AND RELATED PHENOMENA
eds. Yervant Terzian, Daniel Weedman, and Edward Khachikian
ISBN: 1-58381-008-0

PUBLISHED: 2000 (*) Asterisk indicates editors are still working on the volume as of September 2000

Vol. No. 195 — HIGHLY ENERGETIC PHYSICAL PROCESSES AND MECHANISMS FOR EMISSION FROM ASTROPHYSICAL PLASMAS
eds. P. C. H. Martens, S. Tsuruta, and M. A. Weber
ISBN: 1-58381-038-2

Vol. No. 196* — PRESERVING THE ASTRONOMICAL SKY
eds. R. J. Cohen and W. T. Sullivan

Vol. No. 197 — ASTROCHEMISTRY: FROM MOLECULAR CLOUDS TO PLANETARY SYSTEMS
eds. Y. C. Minh and E. F. van Dishoeck
ISBN: 1-58381-034-X

Vol. No. 198 — THE LIGHT ELEMENTS AND THEIR EVOLUTION
eds. L. da Silva, R. de Medeiros, and M. Spite
ISBN: 1-58381-048-X

Vol. No. 199* — THE UNIVERSE AT LOW RADIO FREQUENCIES
eds. A. Pramesh Rao

****

Vol. XXIVA — TRANSACTIONS OF THE INTERNATIONAL ASTRONOMICAL UNION REPORTS ON ASTRONOMY 1996-1999
ed. Johannes Andersen
ISBN: 1-58381-035-8

****

Complete lists of proceedings of past IAU Meetings are maintained at the IAU Web site at the URL: http://www.iau.org/publicat.html

Volumes 32 - 189 in the IAU Symposia Series may be ordered from
Kluwer Academic Publishers
P. O. Box 117
NL 3300 AA Dordrecht
The Netherlands